Figures available in three downloadable sizes (resolutions)

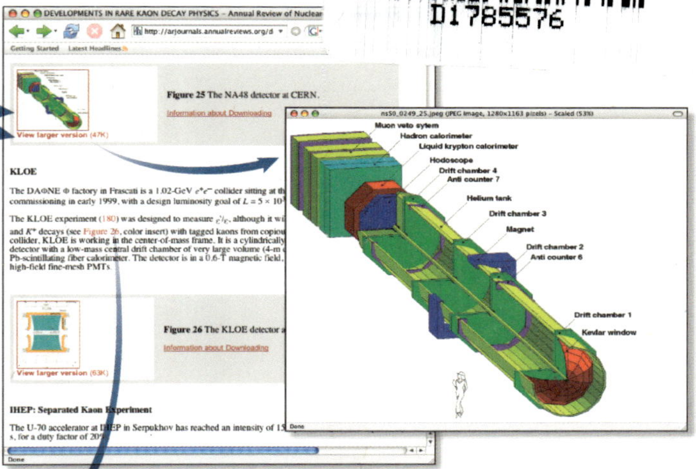

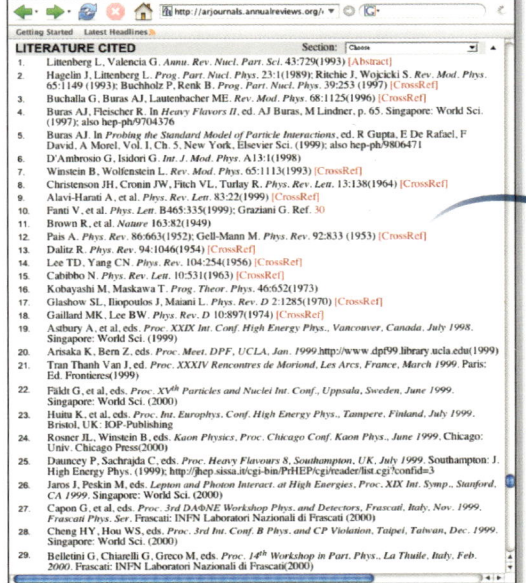

Citations in text link to references in bibliography

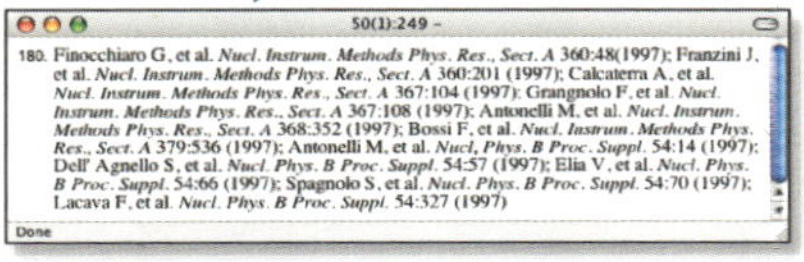

References in Annual Reviews chapter bibliography link out to sources of cited articles online

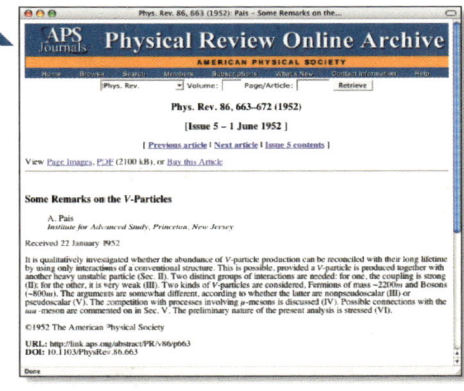

ANNUAL REVIEW OF NUCLEAR AND PARTICLE SCIENCE

ANNUAL REVIEW OF NUCLEAR AND PARTICLE SCIENCE

VOLUME 55, 2005

BORIS KAYSER, *Editor*
Fermi National Accelerator Laboratory

BARRY R. HOLSTEIN, *Associate Editor*
University of Massachusetts

ABOLHASSAN JAWAHERY, *Associate Editor*
University of Maryland

www.annualreviews.org   science@annualreviews.org   650-493-4400

ANNUAL REVIEWS
4139 El Camino Way • P.O. Box 10139 • Palo Alto, California 94303-0139

ANNUAL REVIEWS
Palo Alto, California, USA

International Standard Serial Number: 0163-8998
International Standard Book Number: 0-8243-1555-3
Library of Congress Catalog Card Number: 53-995

TYPESET BY TECHBOOKS, FAIRFAX, VA
PRINTED AND BOUND BY MALLOY INCORPORATED, ANN ARBOR, MI

PREFACE

In the present volume, Don Perkins recounts, with his signature charm, numerous adventures in a distinguished career in experimental elementary particle physics. Reviews on ultra-peripheral nuclear collisions, direct photon production and femtoscopy in relativistic heavy-ion collisions, the physics that could be explored with a future electron-ion collider, and the present and future of small-x hard QCD physics deal with timely topics in nuclear physics and, especially, the blossoming interface of nuclear and particle physics. A review on neutrons describes the current and future use of these particles to test our understanding of the weak interactions and other facets of physics. A pedagogic review of little Higgs theories explains a class of alternatives to the standard picture of the Higgs particle. Discussions of blind analysis and of tools for the Monte Carlo simulation of hadronic collisions cover techniques that play central roles in our field. A review at the interface between astrophysics and cosmology and another between nuclear and particle physics treat the efforts to understand supernova explosions and leptogenesis—an intriguing hypothesis that explains the matter-antimatter asymmetry of the universe. A detailed overview of intersecting D-brane models provides an initiation into this active area of string-theoretic phenomenology.

This volume was assembled by the work of previous and present Editors and by three Production Editors. On behalf of all these people and the Editorial Committee, I would like to thank the authors of these reviews for sharing their insight, understanding, and wisdom with the readers of the *Annual Review of Nuclear and Particle Science*.

Boris Kayser
Editor

Annual Review of Nuclear and Particle Science
Volume 55, 2005

CONTENTS

INDEXES

ERRATA
An online log of corrections to *Annual Review of
Nuclear and Particle Science* chapters may be found
at http://nucl.annualreviews.org/errata.shtml

RELATED ARTICLES

From the ***Annual Review of Astronomy and Astrophysics***, Volume 43 (2005)

An Education in Astronomy, Riccardo Giacconi

Astrobiology: The Study of the Living Universe, Christopher F. Chyba and Kevin P. Hand

The Classification of Galaxies: Early History and Ongoing Developments, Allan Sandage

From the ***Annual Review of Fluid Mechanics***, Volume 38 (2006)

Experimental Fluid Mechanics of Pulsatile Artificial Blood Pumps, Steven Deutsch, John M. Tarbell, Keefe B. Manning, Gerson Rosenberg, and Arnold A. Fontaine

Fluid Mechanics and Homeland Security, Gary S. Settles

From the ***Annual Review of Materials Research***, Volume 35 (2005)

Spatial Order and Diffraction in Quasicrystals and Beyond, Denis Gratias, Lionel Bresson, and Marianne Quiquandon

Analytical Transmission Electron Microscopy, Wilfried Sigle

Electron Microscopy in the Catalysis of Alkane Oxidation, Environmental Control, and Alternative Energy Sources, Pratibha L. Gai and Jose J. Calvino

From the ***Annual Review of Physical Chemistry***, Volume 56 (2005)

Femtosecond Laser Photoelectron Spectroscopy on Atoms and Small Molecules: Prototype Studies in Quantum Control, M. Wollenhaupt, V. Engel, and T. Baumert

Ab Initio Quantum Chemical and Mixed Quantum Mechanics/Molecular Mechanics (QM/MM) Methods for Studying Enzymatic Catalysis, Richard A. Friesner and Victor Guallar

Fourier Transform Infrared Vibrational Spectroscopic Imaging: Integrating Microscopy and Molecular Recognition, Ira W. Levin and Rohit Bhargava

ANNUAL REVIEWS is a nonprofit scientific publisher established to promote the advancement of the sciences. Beginning in 1932 with the *Annual Review of Biochemistry*, the Company has pursued as its principal function the publication of high-quality, reasonably priced *Annual Review* volumes. The volumes are organized by Editors and Editorial Committees who invite qualified authors to contribute critical articles reviewing significant developments within each major discipline. The Editor-in-Chief invites those interested in serving as future Editorial Committee members to communicate directly with him. Annual Reviews is administered by a Board of Directors, whose members serve without compensation.

D. H. Perkins

Annu. Rev. Nucl. Part. Sci. 2005. 55:1–26
doi: 10.1146/annurev.nucl.55.102703.130016

FROM PIONS TO PROTON DECAY: Tales of the Unexpected

D.H. Perkins

Sub-Department of Particle Physics, University of Oxford, Oxford OX1 3RH, England;
email: d.perkins1@physics.ox.ac.uk

Key Words   cosmic rays, nuclear emulsions, pions, muons, neutrino interactions, proton decay, atmospheric neutrinos

■ **Abstract**   This account recalls early observations of elementary particles from cosmic ray experiments, using the nuclear emulsion technique. Discoveries in this field in the 1940s and 50s led to the development of high energy particle accelerators and associated detectors, resulting eventually in the observation of the quark and lepton constituents of matter and of the fundamental interactions between them, as described in the Standard Model. The concept of unification of the fundamental interactions led to the prediction of proton decay, and although this has not been observed, the unwanted background due to atmospheric neutrino interactions led to the discovery of neutrino oscillations and neutrino mass, and the first indications of new physics beyond that of the Standard Model. In all this research, unexpected developments have often played an important role.

CONTENTS

BEGINNING IN HIGH ENERGY PHYSICS

This account recalls some of my early experiences in research on cosmic rays, and later in experiments on neutrino physics using accelerators at CERN. Research in experimental particle physics really commenced in earnest after World War II, receiving a big boost from the discoveries of pions and muons and strange particles in cosmic rays in the late 1940s and early 1950s. That led to 50 years of accelerator development, providing the intense, controlled, and ever higher energy beams, plus the associated detectors, which were necessary to put the subject of particle physics on a sound quantitative basis. I was fortunate to have entered the field as all this was beginning, when progress was easy. In fact, after over half a century in the field of particle physics, what has impressed me, almost as much as the logical development of the subject, is the extent to which chance (both long shots that came off and near misses that did not) and serendipity have often played a role. If past experience is anything to go by, we can all continue to look forward to the unexpected.

My own entry into the field was indeed smoothed by chance events. Medicine, rather than physics, was my first choice as a career subject. I had no physics teacher at high school (he had gone off to war), so I was taught by an old chemistry teacher whose physics notes went back to the time of the Michelson-Morley experiment. I learned about relativity and quantum theory only from books. To launch into a medical career, I would first have had to qualify in mathematics and physical sciences. I was offered a scholarship to Trinity College, Cambridge, where I hoped to switch to medicine after taking a science degree. However, Trinity insisted that I should first pass a Latin examination. My foreign language qualifications were in French and German; I had thought Latin a waste of time. Imperial College in London also offered me a scholarship to read physics. I found that I would not need Latin, and the required examination in scientific German was likely to be straightforward. So that was how I ended up reading physics as a direct result of shortcomings in my classical education. Ironically, a few decades later, Cambridge dropped the Latin requirement, and Imperial College started degree courses in medicine! I know of several well-known physicists who originally never intended a career in the subject. For example, the late Fred Reines (awarded the 1995 Nobel Prize for his first detection of the neutrino) told me that he had always wanted to be an engineer.

To graduate in physics at Imperial College, it was necessary to qualify in a preliminary mathematics examination, one year before the final physics examination. Again I was fortunate. This was taking place during World War II, and the examination was in July 1944, a few weeks after D-day. There were four mathematics papers which had to be passed, and I knew that I might fail the last one, which was on analysis. But on just that morning of the examination, the Germans started sending over V1 flying bombs, not in small numbers, as previously, but in hundreds. The V1—an early version of the cruise missile—flew at heights up to 5000 ft, and its journey was terminated by a time switch, which cut out the ramjet engine and sent it into a dive to Earth. At the typical heights at which the V1 operated, there followed up to 14 seconds of complete silence before it hit the ground. Thus one heard the noise of the ramjet (audible for several miles), then a sudden silence, and after a delay, a loud bang. This was happening every minute or so, and one can imagine that under these conditions—and especially for those V1s passing almost overhead—the atmosphere was hardly conducive to solving cubic equations. After more than half an hour of this racket, the invigilators suddenly announced that they had decided to abandon the examination, and that we would be assigned the average mark from the other three papers. What an escape!

Chance also played a role in my choice in 1945 of cosmic ray studies as a research topic for a PhD degree at Imperial College. I had just happened to read a statement by Karl K. Darrow, onetime president of the American Physical Society, which described the study of cosmic rays as remarkable "for the minuteness of the phenomena, the delicacy of the apparatus, the adventurous excursions of the experimenters and the grandeur of the inferences." I thought that sounded quite promising. My research supervisor was G.P. Thomson (son of J.J.), a somewhat forbidding person, known to everyone as "G.P." After graduating, I had expected to be directed into war work (although the war was just over) and had already been provisionally allocated to research in the steel industry in Sheffield. A week before I was due to go, I heard from G.P. that I had obtained a first class degree and qualified for a postgraduate research scholarship. I immediately went down from Yorkshire to London. As I was the first student he saw, he was able to offer me a choice of research topics, ranging from infrared spectroscopy, plasma physics (aiming toward fusion), crystal growth, nuclear structure, to the study of nuclear disintegrations by cosmic rays using the photographic method. I told him that my interests were in cosmic rays, so he sent me off to read the 1937 papers in Zeitschrift für Physik by two Austrian physicists, Marietta Blau and Hertha Wambacher. They had recorded approximately 30 nuclear disintegrations in Ilford plates exposed to cosmic rays on mountains in the Bavarian Alps. Although there was no indication of any momentous discoveries, it appeared to be a fairly straightforward technique, and I am sure the prospect of some Alpine skiing also played a part in my choice!

Several physicists worldwide had been experimenting with the nuclear emulsion technique during the 1920s and 30s (the method went back to the recording of alpha-particles from radioactive sources by Kinoshita in 1910, and even before that the effect of radiation on photographic plates had led Becquerel to the discovery of radioactivity in 1896). In addition to Blau and Wambacher in Austria, other scientists were studying nuclear emulsion, including Schopper and Schopper in Germany, Heitler and Powell in England, Zhdanov in Russia, and Wilkins, Rumbaugh, and Locher in the USA. All had succeeded in recording the tracks of low energy (few MeV) protons, using standard emulsions (Ilford Half-Tone and Agfa K) specially sensitized with dyes such as pinakryptol yellow. I first experimented with Ilford Half-Tone emulsions, without much success, but fortunately I was invited to join a panel set up by Patrick Blackett (1948 Nobel Physics Prize) under the auspices of the then Ministry of Supply and under the chairmanship of Joseph Rotblat (1995 Nobel Peace Prize). Its purpose was to promote development of special photographic emulsions to record nuclear particles. The members of the panel included Cecil Powell from Bristol, who was to receive the 1950 Nobel Prize for discovery of the pion and his work on the nuclear emulsion technique, Otto Frisch from Cambridge, George Rochester from Manchester, myself, and a few others, including, most importantly, two industrial chemists, Mr. Waller and Dr. Berriman from the photographic firms of Ilford Ltd. and Kodak Ltd., respectively.

By mid-1946, Waller at Ilford had succeeded in producing thick (50 μm) layers of these so-called nuclear emulsions on glass backing, with four times the normal halide to gelatin ratio and with sensitivities to charged particles with ionization above about six times the minimum value. They were labeled A, B, C. . . in order of increasing size of the grains (silver halide microcrystals) and 1, 2, 3. . . in order of increasing sensitivity. These were the first emulsions able to record the tracks of mesons. Although it was obvious that these new emulsions would be greatly superior to what had been available before, no one had the slightest idea of what would be found when they were exposed to the cosmic radiation. Considerable uncertainty and guesswork was involved in the technique; no one knew the level of background that could be tolerated, the amount of fading of the latent image with time, and so on. So it was just a question of trying things out to see what would happen.

Exposing emulsions to cosmic rays at high altitudes was something of a hit-and-miss affair. Mountains, balloons, and aircraft were the obvious choices. I quickly abandoned hydrogen-filled rubber balloons, as used by the Meteorological Office. They often went up to 60,000 ft or so, blew up, and came straight down. I had dreams of a trip to the Andes, but G.P. decided that was far too expensive; he told me bluntly not to waste his time on trivialities over which mountain to choose: simply buy an atlas of Europe and find an alp! In due course I would end up at the HFS (Hochalpine Forschungsstation) at the Jungfraujoch in Switzerland. I chose this because it was the only European mountain site that I could find above 10,000 ft with a railway all the way up. After all, why walk if you can ride? Other people were more energetic. Guiseppe ("Beppo") Occhialini at Bristol had been a mountain guide in Brazil during World War II, and he toiled with boxes of C2 emulsion up to the Pic du Midi in the Pyrenees.

However, to arrange my use of the Jungfraujoch laboratory with the secretariat in Bern and to prepare a series of experiments there (for example with various types of absorber) would take some time. Meanwhile, I was fortunate to have a supervisor who was not only a knight of the realm and a Nobel Prize winner (with Davisson and Germer, for the discovery of electron diffraction), but also, far more importantly, had been chairman of the famous and crucial Maud Committee. Formed in 1940, the Maud Committee had reported in mid-1941 on why and how construction of an atomic bomb, based on uranium 235—and possibly plutonium—would be feasible. Their report was sent to the United States, and as a result American scientists, originally unconvinced, began to take the matter seriously and persuaded their government to set up what was to become the Manhattan Project. So G.P. had considerable influence in the corridors of power, and he contacted the Air Ministry to ask if they could arrange exposures of my emulsions on aircraft. Eventually I was put into contact with the RAF Photographic Reconnaissance Unit (PRU) stationed at Benson, near Oxford. During the war they had flown unarmed aircraft at high altitude deep over enemy territory (suffering appalling losses in the process) and had, for example, produced the first crucial photographs in 1943 of the V2 rocket development site at Peenemünde in the Baltic. In the course of their peacetime duties, they very obligingly carried my emulsions to 35,000–40,000 ft on their photographic sorties over the British Isles that had been requested, so they told me, by the Ordnance Survey people. I believe this actually led to finding some of the first aerial evidence for previously unknown sites of Iron Age settlements. However, for cosmic ray work, the exposures were much too short, just an hour or so per day. That was always the problem with military aircraft, and I was to encounter it again, many years later, with U2 flights in the United States.

FIRST RESULTS WITH EMULSIONS

In November 1946 I got back some 50 μm thick B1 emulsions from the PRU. After developing and scanning them with an ancient monocular microscope, I found several cosmic ray "stars"—disintegrations of nuclei in the emulsion, with emission of protons and α-particles. In one of these, the disintegration was clearly produced by an incoming negative meson (later to be identified as the pion), undergoing nuclear capture (1), and leading to disintegration of a light nucleus (as deduced from the ranges of the secondary protons produced). My observation was confirmed just two weeks later by Occhialini and Powell in Bristol, who found six such events in the much improved C2 emulsions exposed on the Pic du Midi (2). I had heard, third hand, but not seen, the results from the famous 1946 experiment by Conversi, Pancini, and Piccioni in Rome at sea level (3). In those days, it took months for copies of the Physical Review to cross the Atlantic—they were sent by sea—and the prospect of contacting Rome University direct by telephone from the UK would have been quite hopeless. Marcello Conversi and his colleagues, using a Rossi-type array of Geiger counters and magnetic lenses built from bars of magnetized iron, were able to select positive or negative mesons stopping in an absorber under the magnet. Their counting rate was abysmally low, less than one event per hour. Those were the days of heroic experiments! With negative mesons stopping in an iron absorber, they observed no secondaries from decays or other interactions. They asked the theoretical physicist Bruno Ferretti what negative mesons were expected to produce (apart from low energy protons and α-particles, which would stay in the absorber) and were told that probably there should be gamma rays from nuclear excitation. Because gamma rays would, however, also be absorbed in the iron, they added a carbon absorber of graphite blocks, and were amazed to find that essentially all the negative mesons stopping in the carbon then decayed to electrons, with the same lifetime (2.2 μsec) as the positives. At that time, I thought they had been influenced by a 1941 paper by Auger, Maze, and Chaminade, who reported erroneously that negative mesons stopping in aluminum all underwent decay. However, I learned later from Oreste Piccioni that it was the quest for nuclear gamma rays that had caused the change to carbon (4).

Clearly, there was a big difference between my negative meson stopping in a light element at 35,000 ft, with the mass energy of the meson disrupting the nucleus, and the negative mesons of the Italian group all decaying to electrons when stopping in a light element at sea level. Obviously I was aware of Yukawa's

Figure 1   Cecil Powell (on the left) and Guiseppe ("Beppo") Occhialini in discussion at Bristol in 1947. The two women are operating a projection microscope used at that time for photomicrography and to study events at high magnification (courtesy of R.R. Hillier, University of Bristol).

1935 paper on the strong nuclear quantum and in my paper denoted the negative meson by a "Y" in his honor. But at that time I had no clear idea of what it all meant, or how the two experiments could be reconciled. I did not even know then of the existence of the Sakata and Inoue paper of November 1946 (11), proposing the two-meson hypothesis, that a parent, strongly interacting meson would decay into a daughter, weakly interacting meson (what was later termed the pi-mu decay). This paper did not reach England until months later. What I did learn from all this was that, compared with the expertise and organization of the large Bristol group working with emulsions, I was just a one-man band.

During this period I met two well-known physicists from abroad. The first was the late Louis Leprince-Ringuet, visiting from École Polytechnique, who came to see G.P. Thomson and took the opportunity to scrutinize my negative meson event. Later in that year, he exposed some B2 emulsions for me at the Vallot refuge near Chamonix high in the French Alps, and in those I observed my first pi-mu decay, in July 1947. My second visitor was Viki Weisskopf. He had seen my first results on the energy spectra of protons and α-particles from nuclear disintegrations induced by high-energy cosmic rays and how well they fitted a Maxwellian distribution, as predicted by his model likening the excited nucleus to an evaporating liquid drop. At this time I also exchanged correspondence with Yoshio Yamaguchi (5), who noted that, in the disintegrations I had observed of heavy nuclei (silver and bromine, as indicated from the total charge of the emitted protons and α-particles), a small proportion of the secondary protons had energies well below the Coulomb barrier height. This he ascribed to radioactive decay via proton emission, in analogy with α-decay. Some 30 or 40 years later, I would meet both Leprince-Ringuet and Yamaguchi on a regular basis as fellow members of the CERN Scientific Policy Committee. Yamaguchi spent some time in the CERN Theory Division and wrote one of the first (1959) papers predicting proton decay (28), and he also became the first president of ICFA (International Committee for Future Accelerators). Later on, I met Weisskopf again when he became CERN Director-General, at the time of our first neutrino experiment.

For some reason, my thesis advisor G.P. Thomson did not display very much interest in mesons, and he thought I ought to focus on studying the high energy nuclear disintegration "stars" produced by cosmic ray protons. The traditions of nuclear physics, which had started with Rutherford 30 years before, indeed died hard. So nuclear disintegrations were the main subject of my thesis. Later, I continued their study for a time as a joint effort with two other graduate students, Sam Lattimore and Brian Harding (6). The energy spectra and angular distributions of protons and α-particles emitted from the heavy (silver and bromine) nuclei in the emulsion, typically involving excitation energies of 200–700 MeV, were studied in considerable detail. They were found to be in excellent agreement with nuclear evaporation models, taking account of cooling as the nucleus de-excited. However, these were rather pedestrian results that did not lead to anything very fundamental. In retrospect, it certainly turned out to be a wrong track.

The big event in May 1947 had been the publication by the Bristol physicists of two events in C2 emulsions (7) in which a positive particle, the pion, came to rest in the emulsion and decayed into a muon (and neutrino). Although in the second event, the muon just passed out of the emulsion surface, it clearly had only a small residual range and it was obvious that, within the expected small (4%) variations due to straggling, the full range of about 600 μm was the same in the two cases, indicating a simple two-body decay. As I have said, I found the third such event shortly afterward, but never published it (except in my thesis). At the time, I did not even know that my event was only the third pion decay ever to be observed, but I do distinctly remember being totally convinced by those two Bristol events, so that I thought that published confirmation—even from a different laboratory, with different emulsions and at a different altitude—was quite unnecessary. Today, when formal confirmation for new phenomena from independent sources is usually required, that attitude would never do!

The nature of the two-body decay of the pion was put beyond doubt in October 1947, when the Bristol group (8) published a total of ten events, all with the same range of the muon secondary. Again, this involved a chance train of circumstances. A box of C2 emulsions was taken out to Mount Chacaltaya in Bolivia by Giulio Lattes, who had been brought over to Bristol from Brazil by Occhialini. Arthur Tyndall, the director of the physics department at Bristol, wanted Lattes to fly out on a British plane, but Lattes preferred to travel by Varig on one of the new

Super-Constellations. The British plane that he would have taken crashed in bad weather at Dakar, killing all aboard. If Lattes had taken Tyndall's advice, not only would he have died, but those 10 crucial pi-mu decay events would not have been found.

In between the publications of the first two pi-mu decays in May and the next ten in October, the Shelter Island conference took place in June 1947. There Marshak and Bethe (9) as well as Weisskopf (10) proposed the two-meson hypothesis, unaware not only that it had already been proposed by Sakata and Inoue in 1946, but also that it had been discovered experimentally by the Bristol physicists. Fifty years ago, communications were not very satisfactory! Obviously, the experimental verification of the two-meson hypothesis—that strongly interacting pions, produced in the high atmosphere, decayed to weakly interacting muons penetrating to sea-level and producing the Rome events—was an important step. It clarified all the confusion and mystery of the observations of mesons in the 1930s, when mesons, assumed to be identified with the strong Yukawa quanta, were never observed to interact as they traversed metal plates in cloud chambers.

Perhaps I should comment here on the meson nomenclature and how it arose. Val Fitch and Jon Rosner, in their excellent review in "Twentieth Century Physics," state that Powell named the two particles he found as pi (π) and mu (μ) simply because these were the only two Greek symbols on his typewriter. A nice story, but unfortunately quite untrue. In fact, in the beginning there were several Greek symbols in use, to describe the different but not obviously related phenomena observed in the early emulsions. Mesons which came to rest in the emulsion and apparently did nothing were termed rho (ρ): they were mostly positive or negative muons (with the odd case of nuclear capture of a negative pion with emission of neutrons only). Mesons giving nuclear disintegrations at the end of the range were termed sigma (σ) mesons, identified later as negative pions. The nomenclature π (standing for pi-meson, now called pion) and μ (standing for mu-meson, now called muon) was reserved exclusively for the particles in the pi-mu decay chain. Only later, with electron-sensitive emulsions and after measurements of the sign of the particle charges, could all the phenomena be confidently ascribed to either pions or muons.

At the time when they were first observed, it was not immediately clear to everyone that the pi-mu events actually represented decay processes at all. Charles Frank (12) at Bristol considered the possibility that the events could represent the capture of a negative meson in a Bohr-type orbit, which then catalyzed a fusion reaction of the nuclei in the molecule, with enough energy release to eject the meson again with a few MeV of energy. He argued that this could not happen in the emulsion—and was in any case disproved when the pion and muon masses were shown to be different—but Andrei Sakharov concluded that such a process would be possible in hydrogen isotopes ^{2}H and ^{3}H. Ten years later, precisely such a process was discovered quite independently by Luis Alvarez in observations of muons in a hydrogen bubble chamber at Berkeley. When the muon comes to rest in the liquid, it displaces an electron and forms a mesic μH$_2$ molecule, which diffuses

Figure 2   Marcello Conversi, flanked on the left by Niels Bohr and on the right by Carlo Rubbia, at a conference in 1961 (courtesy CERN Information Services).

around until (because of the reduced mass effect) it can switch over and bind itself in a mesic μHD molecule. The ensuing reaction is $\mu^- + p + d \rightarrow {}^3He + \mu^- + 5.5$ MeV, called muon-catalyzed fusion. Because the muon is (usually) ejected, it can repeat the process some 200 times for suitable concentrations of deuterium. As was explained by Dave Jackson however, this process as a source of (very clean) nuclear power was unfortunately impossible because the 0.5% probability that the muon sticks to the helium 3 (and finally decays) is about a factor 10 too high. Another near miss!

The next advance in the emulsion technique was in producing thick electron-sensitive emulsions, which would record minimum-ionizing charged particles. This was achieved first in 1948 by Kodak with the NT4 emulsion, with Ilford following shortly afterward in 1949 with the G5. An early success for these emulsions was the recording of the first Kπ3 decay [in those days called the τ-decay (13)]. Interestingly, Kodak (London) sent the exact recipe for making the NT4 emulsions to their parent company, Eastman Kodak in Rochester, New York, but they were never able to reproduce them. Perhaps there was a little "black magic" in the photographic technology. Waller at Ilford once told me that he was worried about running out of gelatin, which was a prewar stock and had come from the hooves of Argentinian cattle (gelatin from English cattle didn't work so well). The gelatin not only suspended the microcrystals of silver halide as an emulsion, but it also supplied crucial trace elements, such as phosphorus and sulphur, which were occluded onto the microcrystals and formed the traps (so-called f-centers) necessary for capture by the microcrystals of the electrons liberated by the ionizing particle, thus forming the latent image. In any case, while the Ilford G5 emulsions were very reproducible, Kodak had problems in that respect and in the end, they abandoned the project as not commercially viable.

These early emulsions were made in 50 or 100 μm thicknesses, but it was clear that in order to record more details of the events, much thicker emulsions would be required. The processing of thick emulsions (up to 400 μm) needed a special technique, namely the temperature cycle method invented by Dilworth, Occhialini, and Payne (14). In this technique, the emulsion was first immersed in developer solution, which was kept cold and chemically inactive until it had penetrated throughout the emulsion; then it was heated so that development proceeded uniformly with depth. I can remember Beppo Occhialini once explaining all this to me, with dramatic gestures to emphasize the shock when the heating process started. Everything that Beppo dealt with was nearly always described in dramatic and unusual terms. He frequently spoke in parables, some of them hard to follow. Discussing the road to success in physics, he was fond of saying "Paul Revere was a very successful man, but the horse didn't like it." He was also fascinated by the English Civil War, and classified all the physicists he knew as either Cavaliers or Roundheads.

Another development was of so-called stripped emulsions. This technique had actually been used over 30 years before by Kinoshita, who didn't know how to develop thick emulsions and so settled for stacks of thin ones. The emulsion was poured and dried on a glass backing in the usual way, then stripped off the glass. In this way, large sensitive volumes could be obtained by stacking a pile of stripped emulsion sheets like a pack of cards, registration of adjacent sheets being made by using a narrow X-ray beam marking the edges.

RESEARCH AT JUNGFRAUJOCH

In the early days, people did some unusual experiments. We understood little of the underlying physics, and much of the research was just trial and error. Figure 3 shows a picture of a pipe being lowered into the Aletsch glacier near the Jungfraujoch research station. The idea was to measure the absorption of the star-producing radiation in ice, using emulsions lowered down the pipe, and compare it with that in air. Both substances had roughly equal nuclear absorption lengths (measured in gm cm^{-2}) but very different densities, so if the initiating radiation had any significant component particles with a short lifetime, there could be a difference (of course there was not, because as we now know, in the GeV energy region there are no strongly interacting unstable particles with a decay length comparable with the scale height of the atmosphere). Another long-shot experiment on the glacier, by Ugo Camerini and others at Bristol, consisted of tying cocoa tins containing

Figure 3   Norman Barford (left) and the author carry out one of the Imperial College cosmic ray experiments, lowering emulsions down a 100 ft pipe which had been sunk into the Aletschgletscher near the Jungfraujoch in the Swiss Alps.

emulsions to a long vertical pole stuck into the ice, the idea being to measure the pion lifetime by recording the relative numbers of upward-traveling pions and muons as a function of height above the ice. I recall that their answer of 6 ± 3 ns was wrong, but so also were the competing measurements by Richardson, Panofsky, and others at the Berkeley synchrocyclotron. Indeed, the first four measurements of the charged pion lifetime turned out later to be all wrong by six times the stated errors!

In the late 1940s and throughout the 1950s, we thought of the pion as a fundamental particle. This inspired physicists to push ahead with a worldwide accelerator building program to study pions and their interactions, which was to lead eventually to the discovery of the elementary quark and lepton structure of matter. That was one of the two main legacies of the emulsion technique. The other is that it set a pattern of collaboration on an international scale, which is universal in "big science" today. The technique was simple and tailor-made for international collaboration. Small research groups, with the meager resources available in a Europe emerging from the catastrophe of World War II, were provided with an instrument allowing them to easily contribute at the forefront of physics research. The impact for European collaboration on a major scale, in the formation of CERN in 1953, hardly needs to be emphasized. The 1953 balloon flights from Sardinia, mentioned below, involved 22 laboratories from 12 countries, ranging from Dublin in the west to Warsaw in the east, and from Trondheim in the north to Catania in the south. These early collaborative efforts were, to my mind, the greatest achievement of the emulsion technique.

AN UNWELCOME DIVERSION:
MEASURING NEUTRON FLUXES

Every researcher has the experience of going up blind alleys, and I was no exception. My supervisor, G.P. Thomson, told me that he felt the study of cosmic rays by the photographic technique was all very well, but it was not really testing my practical or innovative abilities. While I am sure he was right in principle, he proposed that I should remedy this by performing measurements of neutron fluxes from a suitable radioactive source, for reasons that totally escaped me. For him, I believe it was a throwback to the years preceding World War II, when he and many other physicists had been carrying out experiments with slow neutrons. I had heard lurid tales of physicists going out in rowing boats on the Serpentine lake in nearby Hyde Park, trailing radioactive neutron sources and thermal neutron detectors in the water! In the set-up which I used, a phial containing a radon-beryllium neutron source was placed in a large tank of water to thermalize the neutrons. The tank was located in the yard just outside the main workshop, and I was constantly having to assure the workshop foreman that his technicians would be adequately shielded from gamma radiation. I had to integrate the neutron flux as a function of distance from the source, using an array of indium foils from which the induced beta activity could be measured, using a thin window Geiger counter to record the betas. I was expected to build the experiment, including all the electronics, from scratch—and those were the pretransistor days of vacuum tubes—as well as the thin-window counter. After spending weeks on this, I still had not made a satisfactory counter, so I asked G.P. if I could buy a counter off the shelf—total cost £17, or less than $500 in today's money. He eventually agreed, but made it a condition that I should give some help to a fellow graduate student, Alec Hester, who was actually building a small Van de Graaff accelerator in order to study (p, α) reactions in light nuclei. With only a little help from me, Hester finally obtained his PhD degree and a few years later became the CERN librarian—an absolutely vital post. So, in a way I can claim that I was contributing a little to the success of CERN even before it was formed!

Eventually I managed to get the neutron source calibrated. Despite G.P.'s evident conviction that making research students build all their apparatus was good for them, I was no electronics wizard like Piccioni in Rome, and had really learned very little and regarded the exercise as a waste of time, as I was not able to apply it to anything useful, and the final result could not be of the remotest interest to anyone. All my hand-built power supplies, scalers, etc. were just thrown away afterward. A lot of the time was also lost every week in traveling with my phials of powdered beryllium up to Amersham, north of London. There a Dr. Ross, housed in a cave in the chalk hills, filled them with radon pumped from radium needles sent from the London hospitals. Eventually I asked him if he could not send them with someone at a designated day and time, whom I could meet, for example at Euston station in northern London. He said that although that was not possible, he could put the sources on a train if I could arrange to meet it. To my concern that it was illegal to send radioactive substances by rail (and these had strengths of around 0.5 curies), Ross told me, "That's no problem. I'll put the source in a large wooden box and label it 'Medical Supplies.' "

Of course, the inevitable happened. One day there was a major hold-up on the Underground (subway) and I arrived at Euston more than an hour late. No train. No box. Then, in the distance, I saw a railway porter actually sitting on it. Only about 5 mm of lead and less than a foot of packing material separated my source from his genitals! I remember making some attempt to discover if he had any children (without raising his suspicions and without success), collected my box, got back to Imperial College and immediately telephoned Ross. Having established that the unfortunate railway employee could not have been in contact with the source for more than an hour, he told me blithely, "Quit worrying; he will very probably survive."

In those days, despite the lessons of Hiroshima and Nagasaki, people were much too relaxed about the dangers of radiation. On top of all that, after I had been regularly smashing beryllium nuts to powder for my sources, I learned of the death in Berkeley of Eugene Gardner from beryllium poisoning, following his work on the Manhattan Project. Incidentally, the service provided by Dr. Ross and his small staff was later to grow into Amersham International, now a major (and safety conscious) global supplier of radio-isotopes.

THE STUDY OF KAONS AND ANTIPROTONS

The year 1947 was memorable not only for the discovery of the pion, but also for the publication by George Rochester and Clifford Butler of two V-events found in a small cloud chamber triggered on air showers at sea-level in Manchester (they were most likely examples of $K^0 \to \pi^+ + \pi^-$ and $K^+ \to \mu^+ + \nu_\mu$). The events were published (15) more than six months after they were found, and it was only years later that Clifford Butler told me why. Patrick Blackett was the professor and head of department, and like many professors in those days, required that he should see and approve all papers before they were sent to the publisher. Blackett insisted on checking all the calculations on the V-events himself, and the paper apparently had to be sent back no less than fourteen times before he was satisfied with it! Anyone reading that paper today would be impressed by the quality of the presentation, both of the physics and the English. After those two V-events, no more were to be found anywhere for another two years. Eventually the results were confirmed and extended in cloud-chamber studies at Cal Tech, Indiana, MIT, Princeton and École Polytechnique, as well as by the Manchester group, who took their chamber and magnet to the Pic du Midi. The Jungfrau laboratory would have been easier, but the Swiss were not willing to risk taking the massive magnet sections up in the lift from the railway tunnel to the Sphinx ridge, where the cloud chamber would have been located.

By the early 1950s, the study of K-mesons was getting into full swing. I had moved to Bristol, which at that time was the most prominent emulsion group. As

Figure 4   Moments before the launch of a hydrogen-filled plastic balloon at Cardington, Bedfordshire. At sea level, the hydrogen occupied only about 1% of the total volume. Flying balloons in England eventually became too difficult because of the need to avoid the air lanes.

Louis Leprince-Ringuet said at the 1953 Rochester Conference, "En Europe, il y a pour les emulsions Bristol, le grand soleil, et puis un tout petit nombre de petite satellites, tres inferieure." There we began a program of construction of hydrogen-filled plastic balloons, in order to float large stacks of emulsion for many hours in the stratosphere (see Figure 4). Initially they were launched from an airship hangar at Cardington near Bedford. But in the increasingly crowded air lanes over southern England, getting permission to fly balloons became increasingly difficult, and eventually the decision was made to relocate to the Mediterranean area, with launching from Cagliari in Sardinia (with recovery at sea) and from Nove Ligure near Genoa. In her 2002 contribution to this series, Milla Baldo-Ceolin has already described those developments, and I will limit myself to a few remarks.

As I have mentioned above, a providential consequence of relocating and internationalizing the cosmic ray balloon program was the initiation of large collaborative experiments, with many university groups, even those with modest means, able to take a full part in the joint effort. This European collaboration on flying large plastic balloons (with the actual fabrication in Bristol and Padua) was formally launched as a result of a meeting in Rome hosted by Eduoardo Amaldi in 1952. At a reception at the university I ran into Bruno Touschek. When the reception finished, he offered me a lift back to my hotel on his motor bike. At that time he was busy designing what was in fact the world's first electron-positron

Figure 5 A few seconds after launch of one of the Bristol balloons in Italy. The emulsion stacks carried on these balloons provided some of the early information on decay modes of the strange particles (kaons and hyperons).

collider (Anello Di Accumulazione, or ADA), and on our journey through Rome he described his ideas, gesticulating with one hand. He was also very proud of the fact that his bike was Austrian, like him, and could easily beat any bike made in Italy. He proceeded to demonstrate this by passing every motor cycle or scooter he caught sight of, giving them a rude sign with his other hand as he did so. So for a lot of the journey, Bruno's bike more or less steered itself. For me, that was indeed a lost opportunity. I am afraid I felt compelled to concentrate more on the traffic than on his exposition of the principles of electron-positron colliders. Touschek was a brilliant physicist, full of optimism and charm, who unfortunately died young (16).

The study of strange particles in cosmic rays was the province of the nuclear emulsion and the cloud chamber, and this was beautifully illustrated in the 1953 conference at Bagnères de Bigorre in the Pyrenees. Everyone who was there has testified that this was the very best conference they had ever attended, before or since. It had everything going for it. First, there was a big fight. Gregory and Peyrou from École Polytechnique had made mass measurements, from magnetic curvature (momentum) and range in a multiplate cloud chamber, of mesons decaying in what is now called the $K\mu 2$ decay mode, with a mean value of $920 \pm 40\ m_e$, while Menon and O'Ceallaigh from Bristol had mass values from scattering measurements in emulsion for the $K\pi 2$ and other decay modes of $1075 \pm 100\ m_e$. The average of the two sets would have been close to the well-measured mass for the $K\pi 3$ of $965\ m_e$, as Bruno Rossi suggested in a masterly summary of the meeting. This conference also saw the first presentation of the Dalitz Plot in the analysis of $K\pi 3$ decay. Undoubtedly, however, the prize for the best experimental contribution went to Bob Thompson of Indiana (17). He had developed cloud chamber technology to a fine art, practically eliminating gas distortions, and produced the first clean separation between the decays of lambda hyperons and neutral kaons, and the first precision measurements of the decay $K^\circ{}_s \rightarrow \pi^+ + \pi^-$.

A unique feature of the conference venue at Bagnères was that it had a dance hall on one side and a casino on the other, so there were plenty of distractions if one got tired of the physics. And the hospitality extended to us by our hosts, the University of Toulouse, was out of this world. Milla Baldo-Ceolin has already described how the work on kaons and hyperons in cosmic rays developed at Bagnères and at subsequent meetings in Italy. The study of the charged kaon mass(es) and decay modes was to continue for some time, in my case ending in 1955 working with the Richman Group with an accelerator kaon beam from the Bevatron at LBL, using (for the first time) quadrupole focusing magnets and again using large stacks of nuclear emulsion to record the events. Just two years before parity violation in weak decays was accepted, people went to extraordinary lengths to get around the problem. Our experiment at LBL showed that the mass of the particle responsible for the $K\pi 2$ decay mode differed by less than $1\ \mathrm{MeV}/c^2$ from that for $K\pi 3$, although the final states concerned were of even and odd parity, respectively. Undaunted, Luis Alvarez (as well as many others) suggested that two different kaon states of opposite parity were involved, and that one underwent radiative decay to the other: $K \rightarrow K' + \gamma$, with $K \rightarrow 2\pi$ and $K' \rightarrow 3\pi$. We couldn't actually disprove that because it would have been hard to detect Compton scattering or conversion of the gamma ray. As happened in many laboratories at that time, at the end of such discussions, people would suggest that perhaps parity was not conserved after all, but they never stated it with any real belief, and it was left to Lee and Yang later on to make the convincing argument.

An experiment to detect antiprotons, again with emulsions, was made by the same group in 1956. It would parasite on the experiment of Owen Chamberlain, Emilio Segre, Tom Ypsilantis, and others, who had designed the beam at the Bevatron. The Bevatron energy of over 6 GeV per proton was just above threshold for production of a proton-antiproton pair in the collision of a beam proton with a stationary proton. The laboratory momentum of the antiproton would then be about 1 GeV/c. Their experiment used Cerenkov and time of flight counters to discriminate the antiprotons from the much more abundant negative pions, and was tuned to this secondary momentum. For the emulsion experiment, using the same beam, it was necessary therefore to use a degrader to slow down the antiprotons, so that by ionization they could be distinguished from the relativistic negative pion background. Unfortunately, because of the high cross-section for annihilation in flight, the degrader employed led to the loss of nearly all of the antiprotons before they ever got to the emulsion stack and only one antiproton event was found. Later researchers realized that the calculations of thresholds and secondary beam momenta had neglected the full effects of Fermi motion in the internal copper target of the Bevatron, which would not only reduce the threshold energy but also meant that the spectrum of the antiprotons produced would peak at lower momentum. So, running at a reduced secondary beam momentum of 600 MeV/c, there was no need for any degrader at all and many annihilation events were observed (18). But that experience showed me once again how very easy it was to get things completely wrong through simple mistakes.

STARTING ON CERN NEUTRINO EXPERIMENTS

My entry into the field of accelerator neutrino physics was indirect, by way of my experience in cosmic rays. The CERN laboratory was inaugurated in 1953, and by late 1959, the construction of the proton synchrotron (PS) was completed and CERN was beginning to consider embarking on their next project, a proton-proton collider called the ISR (Intersecting Storage Rings), which had been specified as one of the goals of CERN in the 1953 Convention. What sort of new physics might the ISR be expected to uncover? For practical guidance, the only indication could come from cosmic rays, and I was invited to a meeting in CERN to explain what amazing new things—if any—could be expected in this new energy region (which corresponded to the interactions of cosmic ray protons of a few TeV incident energy). My research at Bristol had been into the study of meson production at such energies, my principal colleague in this work being the late Peter Fowler, son of R.H. Fowler and grandson of Ernest Rutherford. We used stacks of emulsions interleaved with thin sheets of heavy element (tungsten alloy). Electromagnetic showers were generated from the decay of neutral pions produced in the nuclear interactions of primary protons. The showers developed rapidly in the sheets of heavy elements and at these energies could be easily detected (without a microscope). The exposures of these quite massive (up to 0.25 ton) stacks were on balloon flights in Texas and on Comet and VC10 commercial jets on proving flights to Beirut and Sydney. The object of this research was to relate the high energy gamma ray flux in the TeV region to the sea-level muon spectrum (measured with spectrometers and from the range spectrum deep underground) via the production spectrum of the parent pions and kaons (19).

All this worked out very well, but unfortunately, my general message at the CERN meeting had to be that there was no evidence from the available data that anything dramatically new was occurring at some magic energy threshold. All the measurable quantities—cross-sections, transverse secondary momenta, kaon/pion ratios, meson multiplicity, etc.—were either constant or varying very slowly and smoothly with energy. Of course, with only a few hundred events, these experiments were quite incapable of finding evidence for quark substructure by observing the very rare wide-angle jets. In any case, it was fairly clear that a decision to embark on the ISR had almost been made, largely, I believe, because the accelerator physicists thought it would be a great challenge to build it. But I guess that some compelling physics reasons would have been useful for the CERN Council. For me, one positive result of this meeting was that I got to know Colin Ramm and the people in his group. Colin played a major role in constructing the magnet system of the PS and was then head of the NPA (nuclear physics apparatus) division at CERN. A year or so later, this division would be the center for neutrino experiments.

Almost a quarter of a century afterward, by one of those ironies of fate, I was chairman of their Scientific Policy Committee, and the CERN Council asked our advice about closing the ISR. Although a magnificent technical achievement, the ISR had produced very little in the way of new physics, and reluctantly we had to recommend shutting it down, in order to release vital funds and manpower for the LEP collider project. Other quite fruitful activities, ranging from bubble chamber operations to those of the radiation biology group, also had to be terminated for the same reason. So the scope of operations at CERN became ever narrower. Today it is much the same story: essentially all the effort at CERN has to go into building the LHC and associated detectors.

Shortly after the CERN meeting, Sula Goldhaber invited me to attend a summer study in 1961 in Berkeley, in connection with the proposal to build a 200 GeV proton accelerator near Sacramento, which by 1973 was to mature as a 400 GeV machine at Fermilab. Again I was there as a cosmic ray expert, and my task was to calculate the characteristics of the secondary beams to be expected. On this I worked with Guiseppe Cocconi and Lou Koester. We produced some simple analytical formulae for secondary particle fluxes, which worked reasonably well until much better calculations by Hagedorn and others came along. As an extension of this work I calculated neutrino fluxes in a narrow-band beam (as far as I know, the first time such a beam was ever proposed). This, together with the paper on the proposed neutrino experiment at Brookhaven written by Mel Schwartz, sparked my interest in accelerator neutrino physics. A year later, when Colin Ramm suggested participating in the bubble chamber part of the forthcoming CERN experiment, I was eager to do so.

This 1963 neutrino experiment was actually the second one at CERN. The first, in 1961, with an internal proton target, was characterized by zero events and zero flux, and the less said about that the better. It was abandoned, and the way left open for the Brookhaven experiment (20) to discover neutrino flavor in 1962. The second CERN experiment had an extracted proton beam from the PS, incident on an external target placed inside Van der Meer's magnetic horn, and with muon shielding borrowed from the Swiss national steel reserve. The neutrino beam went into a small heavy liquid bubble chamber holding about a ton of heavy freon (CF_3Br), and thereafter into a spark chamber detector. Technically the beam, the monitoring, and both detectors worked perfectly. Colin Ramm had his own novel method for ensuring high beam intensities. If, in a particular shift, the PS machine operator had performed very conscientiously, he would nip round to the control room, and when no one was looking, slip a bottle of champagne behind one of the electronic racks, then go back to his office, telephone the control room and suggest a search. In time, like Pavlov's dogs, the operators learned that if they did their job well, they would be rewarded.

The confirmation of the BNL result, with the identification of two distinct neutrino flavors, ν_e and ν_μ, took only a couple of days' running. Unfortunately, however, the physics interpretation of the other results was fairly disastrous. After missing out on the discovery of neutrino flavor, CERN was desperate for a discovery, and hoped to find the intermediate W-boson, possibly with a mass as low as 1 or 2 GeV. The spark chamber results on possible W-decay events were presented at the Siena Conference in July 1963. Near the end of the meeting, Luis Alvarez got up and asked: "When I get back to Berkeley, how do I reply to the first question when I step off the plane, which will be: 'Has CERN found the W-boson?' " Gilberto Bernardini, CERN research director, thereupon made a five-point declaration, to the effect that CERN had dilepton events that might or might not be due to decay of the W-boson; but, if it did exist with a mass of 1 or 2 GeV, CERN had discovered it! I've often wondered if that wasn't the low point of neutrino physics at CERN.

Measurements were made in the bubble chamber events of weak nucleon form factors from the elastic events, on weak pion production, and in early attempts to check the CVC and PCAC hypotheses. Although thought to be important at the time, these are now long forgotten. Apart from the W search, there were plenty of other things that went wrong with both the spark chamber and bubble chamber experiments. The failure to find neutral currents in the spark chamber was ensured because a secondary charged lepton was one of the trigger requirements (and a proposed run without a lepton trigger had been voted out). The bubble chamber was just too small to discriminate between neutral current events and neutron background, and the group could only give limits, but we didn't even get those right (21). Our limit (5% of the charged current cross-section) for the elastic neutral current process $\nu + p \rightarrow \nu + p$ was incorrect, due to a book-keeping error, eventually discovered by a research student from Strasbourg (22). Even seven years later, when I was visiting UCLA, J.J. Sakurai was to castigate me over this. "Perkins," he said, "you set back particle physics by 20 years!" I think he may have been joking, but with J.J. one was never quite sure.

Another major misfortune, for which I have to take the main responsibility, was the failure to interpret—or at least to ponder and think seriously about—the rapid increase of the total neutrino cross-section with energy in the bubble chamber data. In retrospect, we now know that this was the first indication of pointlike structure inside the nucleon. However, I could hardly classify that as a near miss. The results did provide clues which might have been followed up, and were not; but one has to remember that, even years after the invention of the quark model by Gell-Mann and Zweig in 1964, high energy physicists worldwide thought of quarks—assuming they were real things rather than mathematical fictions—as extremely massive and strongly bound objects. So the discovery of scaling and pointlike behavior by Jerome Friedman, Henry Kendall, and Richard Taylor in inelastic electron-nucleon scattering at SLAC in 1968 came as a revelation. The idea that in high energy collisions, quarks could behave as light, weakly bound particles required a stretch of the imagination of which, in 1963 or 1964, we were just not capable. It needed a Bjorken and a Feynman for that. Even today, it is a mystery why the pointlike scattering from quasi-free constituents, as indicated by the linear rise of the cross-sections with energy, applies at neutrino energies that are far below the asymptotic regions where perturbative QCD could be expected to apply. This precocious scaling was indeed a piece of good fortune, if only we had had the sense to see it!

THE GARGAMELLE EXPERIMENTS

One positive outcome from the Siena meeting was that André Lagarrigue (whom I knew from my cosmic ray days) was sufficiently impressed by these early bubble chamber results that he embarked on a project to build a much larger heavy liquid chamber (Gargamelle) in which many of the limitations of the small CERN chamber could be overcome. Gargamelle's great contribution was to be the discovery of neutral currents, the search for which—way down at number eight (out of ten!) in the collaboration's priority list in 1970—suddenly became a top priority following the 1971 't Hooft paper (23) proving the renormalizability of the electroweak theory. Gargamelle was a complicated device; the optical system involved transporting the images through the magnet yoke via a meter-long lens train to the eight cameras, through all of which the film had to be threaded. However, the geometry of the chamber, a cylinder 5 m long by 2 m diameter, was ideal for the study of neutral currents.

The 1971–73 neutrino experiments in Gargamelle benefited not only from the much larger chamber, but also from the higher intensity with the PS fast cycling booster and from the much improved two-component magnetic lens, invented by Fred Ašner, which replaced Simon Van der Meer's single horn lens. This meant that one was able to select events with high energy transfers, greatly simplifying the calculation of neutron background. The other advantage was that the experimental analysis became a joint effort by several groups (Aachen, Bruxelles, CERN, Ecole Polytechnique, Milano, Orsay, and University College London). The result was that several independent analyses could be run in parallel and the results compared, and this was vitally important in producing convincing evidence for neutral currents. The announcement of their discovery was made by Gerald Myatt at the Bonn international conference in July 1973, where he also included the (at that time) positive but as yet unpublished results from the Harvard, Pennsylvania, Wisconsin, Fermilab counter experiment at Fermilab. Later, this group reconfigured their detector and unfortunately succeeded in wiping out the signal, with the result that the claim of the Gargamelle collaboration (24) was not generally accepted—and certainly not by most people in CERN!—until it had been confirmed by other experiments at Argonne, Brookhaven, and Fermilab, almost a year later.

Confidence in the electroweak theory prior to the Gargamelle discovery was not very high, even among its strongest protagonists. I recall presenting the very early Gargamelle data at the Chicago/Fermilab conference in summer 1972. The emphasis was on a comparison of the neutrino and antineutrino charged current results with those from deep inelastic electron scattering at SLAC, which provided a unique test of the quark-parton model, and in particular measured the mean square valence quark charge (5/18) in the nucleon and also revealed the presence in it of the quark-antiquark "sea." At that time, the analysis of hadronic neutral current events was still at a very early stage. The safest prediction regarding the electroweak theory was on muon antineutrino-electron scattering, for which the 't Hooft paper gave the cross-section in terms of the mixing angle, and for which the background could be accurately calculated and was expected to be very small. However, at that time we had no events, so I could only give limits on the weak mixing angle, which I referred to as the Weinberg angle. At the coffee break, Abdus Salam—who was to share the Nobel prize with Glashow and Weinberg for the invention of the electroweak theory—rushed up to me in great agitation. He pushed under my nose a reprint of his 1964 paper with John Ward, in which they had also introduced a mixing angle. He asked me. "Why do you keep referring to it as the Weinberg angle?" I apologized and assured him that, in the written version of my talk, I would correct this, including all the other names. But I also told him that I did not know why he was so upset. We had at that time absolutely no definitive evidence in support of the electroweak model, and I thought it might very well be complete rubbish. "You really think so?" replied Salam. "In that case, better keep my name out of it!"

There were several amazing coincidences involved in the neutral current story. In the spring of 1973, before we were quite ready to publicly claim an effect, I was asked to give some lectures on neutrino physics at a high energy physics school run by the Paul Scherrer Institut laboratory. The venue was a private high school at Zuoz, in the Engadine valley. The lecture theater was actually in the basement, and artificially illuminated. I eventually got to talking about neutral weak currents, and I said that definitive results were expected soon. Exactly on my second mention of the words "neutral currents," all the lights went out! It turned out that during that weekend, reservists in the Swiss army were on manuevers nearby, and some idiot had managed to cut through a power cable. It took more than an hour to restore power. I believe the incident made a deep impression on the audience—but whether that was in favor of or against neutral currents, I could not say. One of them, Norbert Straumann (a onetime postdoc of Pauli's), reminded me of the occasion 30 years later, when he gave a lecture in Oxford on a more modern topic, the dark energy in the universe.

In a second incident, I recall presenting the Gargamelle results at a summer institute in Hawaii, in August 1973. Richard Feynman was at this meeting, and initially he did not like either neutral currents or the electroweak theory, although in the end he came to accept them. He certainly gave me a very hard time when I presented the experimental data. I remember one amusing (and amazing) coincidence. Because the signal was now apparently so clear, Feynman wanted to know why we had not found neutral currents in the previous neutrino experiments. I explained that it had been hard to tell genuine neutral current events from neutron background because the neutron absorption length was comparable with the diameter of the smaller bubble chamber. In 1963 we had indeed observed more of these (allegedly) neutron events than expected, and we looked first for other sources of neutron background, in particular skyshine neutrons leaking through the shielding. I set Enoch Young, a Chinese graduate student from Hong Kong, to make an estimate using a Monte Carlo simulation. His conclusion was that the calculated shield leakage was some three times less than the observed rate, but because of uncertainties in parameterizing the nuclear cascade in the shield, the difference could not be considered significant. I had got to the point of explaining all this, when who should walk into the back of the auditorium but Enoch Young himself! This left me somewhat speechless, and Feynman wondered why, but I was quite unable to tell him! Apparently Enoch was on his way from Hong Kong to a cosmic ray meeting in Denver, and had stopped in Honolulu to change planes, and just looked in at the meeting by pure chance. That was indeed a very long shot, with the Hawaii theorist San Fu Tuan calculating the odds against such a happening as about 100,000 to one. Another tale of the unexpected!

QUARKS AND QUANTUM CHROMODYNAMICS

The last neutrino experiments in bubble chambers at CERN took place when the SPS started operating in 1976. One experiment used the BEBC bubble chamber filled with a mixture of liquid neon and hydrogen, with a narrowband beam from the SPS, and was a collaboration of Aachen, Bonn, CERN, Imperial College London, Oxford, and Saclay. One of the main aims of the experiment, and that of the CDHS (CERN, Dortmund, Heidelberg, Saclay) counter experiment located directly behind BEBC, was to measure neutrino and antineutrino cross-sections on nucleons up to the highest possible energies (around 250 GeV). These results confirmed the earlier SLAC data on deep inelastic electron-nucleon scattering, identifying the parton constituents of the nucleon with the long-sought quarks. We attempted to go further and measure the deviations from exact scaling predicted by perturbative quantum chromodynamics (QCD), notably for the q^2 dependence of the moments of the nonsinglet (valence quark) distributions of quarks in the nucleon. The momentum transfers involved were in the range of $q^2 \sim 2$–100 GeV^2, so hardly in the perturbative region. Nevertheless, when one included the effect of those wonderful Nachtmann mass corrections, the results (25) were in astonishingly good agreement with perturbative QCD and the anomalous dimensions of color SU(3).

PROTON DECAY AND ATMOSPHERIC NEUTRINOS

At about the same time, some experiments were conducted with wideband beams in BEBC filled with hydrogen and deuterium, the idea being to measure the neutral current couplings of the up and down quarks separately by comparing the results with those using a neon filling. The values of $\sin^2\theta_w$ measured in these experiments came out quite low—probably as a result of an unlikely fluctuation—and the result was that by 1980, the world average value (26) of $\sin^2\theta_w$ was only 0.21 (compared with the presently accepted value of 0.23). This was to have far-reaching consequences which no one could have foreseen. In my opinion, it was the most important wrong result ever obtained at the CERN laboratory.

At this time, in the late 1970s and early 1980s, and following the success of the electroweak theory, the unification of the fundamental interactions—or at least the strong and the electroweak interactions—became an important goal of particle physics. Proton instability followed as a consequence of these grand unification (GUT) schemes, such as that proposed by Pati and Salam, and the SU(5) scheme of Georgi and Glashow (27). Of course, strong limits (above 10^{26} years) on the proton lifetime already existed, from experiments going back to the early 60s, including a 1960 experiment by a CERN group using Cerenkov counters in the Lötschberg railway tunnel, inspired by the speculations of Yamaguchi mentioned above (28). Theoretical estimates of lifetime ranged widely, the record being that by Sakharov of 10^{50} years (assuming the decay to be mediated by particles of the Planck mass). However, with the anomalously low value of the weak mixing angle as found in 1980, it seemed that the SU(5) model was the most serious contender, because the three running couplings of the strong, weak, and electromagnetic interactions appeared to meet at a unification energy of around 3×10^{14} GeV. This predicted a rather definite proton lifetime of 10^{30} years. When a number of people (including myself) realized that this would provide a unique test of grand unification, and that the expected rate was of the order of one proton decay per day in a kiloton of material, several experiments were started with kiloton-size detectors situated deep underground (to reduce the cosmic ray muon background). This rush for the mines was indeed a very long shot, based on an upward extrapolation from known energies by over ten orders of magnitude, assuming that no new physics occurred in the famous "desert" between the electroweak scale and the GUT scale.

Unfortunately, the observed lower limit on the lifetime soon turned out to be more than two orders of magnitude larger than the prediction. Worse, it soon became clear that interactions of atmospheric neutrinos would pose a serious background, because their rate was of order 0.5 per kiloton per day, not much less than that for the original proton decay prediction (29) Some people thought of quite desperate measures; Salam and Pati even proposed doing the experiments on the Moon which, with no atmosphere, should have very much smaller background.

Atmospheric neutrino interactions had first been observed back in 1963 and reported at the IUPAP Cosmic Ray Conference in Jaipur at that time. The atmospheric muon flux had been measured with Conversi tube arrays in deep (6000 ft) gold mines, at the Witwatersrand mine in South Africa, and the Kolar mine in India. A few multiprong interactions from the mine walls, due presumably to atmospheric neutrino interactions, were also observed. There was interest then in comparing their rate with that expected using the cross-sections measured in our 1963 CERN PS experiment. By 1965, the observed rate of atmospheric neutrino interactions with muon secondaries appeared to be in fair agreement with expectations, although there was some indication (30) that the number was somewhat smaller than that expected—perhaps a harbinger of future results! Atmospheric neutrinos were thereafter quietly laid to rest, at least until 1982.

At the International Conference on High Energy Physics in Paris in 1982, I was given the task of reviewing the situation and concluded that, despite a claim from the Kolar group, there was no clear signal from the various experiments. However, I did mention that one ought at least to study the background carefully, to check that it was really understood. If one could not even understand that, one would not be able to lay claim to interpreting any proton decay events. In any case, the background of neutrino interactions were all that we had. At the time, I had absolutely no idea how prescient my remarks were to prove! But by 1993, five experiments from three continents (IMB and Soudan 2 in the USA, NUSEX and Frejus in Europe, and Kamiokande in Japan) were presenting ratios of numbers of events with muons (due to ν_μ) and electrons (due to ν_e), and there was clear evidence for a discrepancy, particularly in the Kamiokande, IMB, and the Soudan 2 experiment (a collaboration of ANL, Minnesota, Oxford, RAL, and Tufts). The ratio of muon to electron events in the GeV energy region was found to be approximately 0.6 of the value which was expected just by counting the numbers of muon and electron neutrinos in the pi-mu-e decay chain (31).

However, not everyone was convinced that this was a real effect or that the effect was due to neutrino oscillations. At the 1992 Dallas Conference on High Energy Physics, Hamish Robertson gave a review which concluded that the interpretation of the low value of the ratio of muon to electron events might be uncertain because the cross-sections in water (for the Cerenkov experiments) of ν_e and ν_μ might be different. At the energies involved, the relative magnitudes of ν_e and ν_μ cross-sections had already been checked at the few percent level in freon and propane in the 1960s CERN PS bubble chamber experiments, as one test of electron-muon universality. So a factor two discrepancy in water appeared to be quite impossible. The relative fluxes of atmospheric electron and muon neutrinos were reasonably well known, and they could be deduced directly from Conversi's 1950 measurements of high altitude muon fluxes as a function of altitude and geomagnetic latitude (32). Conversi had exclusive use of a B29 aircraft that carried him and his Geiger counter array back and forth between Alaska and Bolivia, measuring the latitude effect. From these muon fluxes, assuming them due to pion decay, it was possible to estimate directly the sea-level ν_e and ν_μ fluxes at different latitudes, at least up to energies of about 1 GeV. These results were later reinforced and greatly extended by calculations using the primary proton spectrum and a Monte Carlo of the nuclear cascade in the atmosphere. The absolute fluxes might be uncertain by as much as 30%—the typical variation in the absolute normalization of the measured primary proton spectrum—but the ν_e/ν_μ flux ratios were not, and in fact had been calculated as a function of energy by many people, including Osborne et al., Volkova, Tam, and Young (33), and more recently by Gaisser et al. and Honda et al. (34). All were in substantial agreement at the few percent level. Of course, because particle physicists had been looking for evidence of neutrino oscillations at accelerators and reactors for almost 25 years without success, it was difficult for some to accept that evidence for such oscillations had finally been found, using the rather feeble atmospheric fluxes and massive but quite crude detectors originally intended for a quite different purpose.

The discovery of oscillations of atmospheric neutrinos was indeed another tale of the unexpected. They would not have been found at all if people had not been searching for proton decay in massive detectors deep underground. In the 1960s (after the Jaipur conference mentioned above), and even after the original suggestion of oscillations by Pontecorvo and Maki et al. (35), nobody had requested (and probably would not have obtained) funding to put massive and expensive kiloton detectors deep underground on the off chance that they were right. Accelerator neutrino beams were much more intense, and they were the "obvious" way to check for such phenomena. Even with the later proton decay detectors, oscillations would probably not have been found, had the geomagnetic field been three or four times larger or the Earth's diameter two or three times smaller. Then both primary proton and secondary neutrino energies would have been higher and the fluxes very much lower, and the typical ratio of L/E, the ratio of neutrino path length to energy, might have been too small for the effect of oscillations to show. It is just a happy accident that the relevant neutrino mass difference is well matched to the Earth's magnetic field and diameter and that the relevant mixing angle is large (in fact, maximal). So the search for proton decay, although it failed in its original purpose, was a long shot which actually paid off in the end.

The measured values of the neutrino mass differences, from both atmospheric and solar neutrino observations, indicate tiny neutrino masses, of millivolts or less. In the Standard Model, neutrinos are left-handed and massless, and it is proposed that small but finite masses may be due to mixing with very massive right-handed Majorana neutrinos at the GUT energy scale, according to the "seesaw" mechanism. Thus the atmospheric results give indications of physics beyond that of the Standard Model.

There was even an unexpected bonus to the proton decay investigation. The IMB and Kamiokande detectors recorded the 1987A supernova, the first detection of a neutrino source outside the solar system. This observation provided confirmation of the correctness of our description of the final stages of evolution of very massive stars. So sometimes experiments in high energy physics turn out quite differently, and perhaps even better, than what had been originally expected.

The *Annual Review of Nuclear and Particle Science* is online at
http://nucl.annualreviews.org

LITERATURE CITED

1. Perkins DH. *Nature* 159:126 (1947)
2. Occhialini GPS, Powell CF. *Nature* 159:186 (1947)
3. Conversi M, Pancini E, Piccioni O. *Phys. Rev.* 71:209 (1947)
4. Piccioni O. In *The Birth of Particle Physics*, ed. LM Brown and L Hoddeson, p. 222. New York: CUP (1983)
5. Fujimoto Y, Yamaguchi Y. *Phys. Rev.* 75:1776(L) (1949)
6. Harding JB, Lattimore S, Perkins DH. *Proc. Roy. Soc.* A196:325 (1949)
7. Lattes CMG, Muirhead H, Occhialini GPS, Powell CF. *Nature* 159:694 (1947)
8. Lattes CMG, Occhialini GPS, Powell CF. *Nature* 160:486 (1947)
9. Marshak RE, Bethe HA. *Phys. Rev.* 72:506 (1947)
10. Weisskopf VF. *Phys. Rev.* 72:510 (1947)
11. Sakata S, Inoue T. *Prog. Theor. Phys.* 1:143 (1946)
12. Frank FC. *Nature* 160:525 (1947)
13. Brown R, et al. *Nature* 163:82 (1949)
14. Dilworth CC, Occhialini GPS, Payne RM. *Nature* 162:102 (1948)
15. Rochester GD, Butler CC. *Nature* 160:855 (1947)
16. Amaldi E. The Bruno Touschek Legacy. *CERN Report* 81–19 (1981)
17. Kim Y, Burwell J, Huggett R, Thompson R. *Phys. Rev.* 96:229 (1954)
18. Barkas WH, et al. *Phys. Rev.* 105:1037 (1957)
19. Duthie J, et al. *Il. Nuov. Cim.* 24:122 (1962)
20. Danby GT, et al. *Phys. Rev. Lett.* 9:36 (1962)
21. Block MM, et al. *Phys. Lett.* 12:281 (1964)
22. Paty M. *CERN Report* 65–11 (1965)
23. 't Hooft G. *Nucl. Phys.* B33:173 (1971)
24. Hasert FJ, et al. *Phys. Lett.* 46B:121 (1973); 138; *Phys. Lett.* B73:1 (1974)
25. Bosetti PC, et al. *Nucl. Phys.* B142:1 (1978); B203:362 (1982)
26. Dimopoulos A. *Proc. 28th Intl. Conf. High Energy Phys., Glasgow* (1994)
27. Pati JC, Salam A. *Phys. Rev. Lett.* 31:661 (1973); Georgi H, Glashow S. *Phys. Rev. Lett.* 32:438 (1974)
28. Yamaguchi Y. *Prog. Theor. Phys.* 22:373 (1959); Backenstoss GK, et al. *Nuovo. Cim.* 16:749 (1960)
29. Perkins DH. *Ann. Rev. Nucl. Part. Sci.* 34:1 (1984)
30. Miyake S, et al. *Phys. Lett.* 18:196 (1965); Reines F, et al. *Phys. Rev. Lett.* 15:429 (1965)
31. Olbert S. *Phys. Rev.* 96:1400 (1954); Perkins DH. *Nucl. Phys.* B399:3 (1993)
32. Conversi M. *Phys. Rev.* 79:749 (1950)
33. Osborne JL, et al. *Proc. Phys. Soc.* 86:93 (1965); Volkova LV. *Sov. J. Nucl. Phys.* 31:784 (1980); Tam AC, Young ECM. *Acta Phys. Acad. Sci. Hung.* 29:S4,307 (1970)
34. Barr G, Gaisser TK, Stanev T. *Phys. Rev.* D39:3532 (1989); Honda M, et al. *Phys. Rev.* D52:4985 (1995)
35. Pontecorvo B. *JETP* 26:984 (1968); Maki Z, Nakagawa M, Sakata S. *Prog. Theor. Phys.* 28:870 (1962)

Annu. Rev. Nucl. Part. Sci. 2005. 55:27–69
doi: 10.1146/annurev.nucl.55.090704.151611

FUNDAMENTAL NEUTRON PHYSICS

Jeffrey S. Nico[1] and W. Michael Snow[2]

[1]*National Institute of Standards and Technology, Gaithersburg, MD 20899-8461;*
email: jnico@nist.gov
[2]*Indiana University and Indiana University Cyclotron Facility, Bloomington, IN 47408;*
email: snow@iucf.indiana.edu

Key Words   cold neutrons, CKM matrix, discrete symmetries, electroweak
interactions, parity violation, ultracold neutrons

■ **Abstract**   Experiments using slow neutrons address a growing range of scientific
issues spanning nuclear physics, particle physics, astrophysics, and cosmology. The
field of fundamental physics using neutrons has experienced a significant increase
in activity over the last two decades. This review summarizes some of the recent
developments in the field and outlines some of the prospects for future research.

CONTENTS

1. INTRODUCTION

1.1. Overview

The field of neutron physics has become an integral part of investigations into an array of important issues that span fields as diverse as nuclear and particle physics, fundamental symmetries, astrophysics and cosmology, fundamental constants, gravitation, and the interpretation of quantum mechanics. The experiments employ a diversity of measurement strategies and techniques, including condensed matter and low temperature physics, optics, and atomic physics, as well as nuclear and particle physics, and they address a wide range of issues. Nevertheless, the field possesses a coherence that derives from the unique properties of the neutron as an electrically neutral, strongly interacting, long-lived unstable particle that can be used either as the probe or as an object of study. This review covers some of the important new contributions that neutrons have made in these diverse areas of science. By "fundamental" neutron physics, we mean that class of experiments using slow neutrons which primarily address issues associated with the Standard Model (SM) of the strong, weak, electromagnetic, and gravitational interactions and their connection with issues in astrophysics and cosmology.

Neutrons experience all known forces in strengths that make them accessible to experimentation. It is an amusing fact that the magnitude of the average neutron interaction energy in matter, in laboratory magnetic fields, and near the surface of the Earth is the same order of magnitude for all forces except the weak interaction. This coincidence leads to unique and occasionally bizarre experimental strategies for measurements and a unique opportunity to search for gravitational effects on an elementary particle. The experiments include measurement of neutron-decay parameters, the use of parity violation to isolate the weak interaction between nucleons, and searches for a source of time reversal violation beyond the SM. These experiments provide information that is complementary to that available from existing accelerator-based nuclear physics facilities and high-energy accelerators. Neutron physics measurements also address questions in astrophysics and cosmology. The theory of Big Bang Nucleosynthesis needs the neutron lifetime and the vector and axial vector weak couplings as input, and neutron cross sections on unstable nuclei are necessary for a quantitative understanding of element creation in the universe.

TABLE 1 Common terminology and spectrum of neutron energies

Term	Energy	Velocity (m/s)	Wavelength (nm)	Temperature (K)
ultracold	$<0.2\ \mu\mathrm{eV}$	<6	>64	<0.002
very cold	$0.2\ \mu\mathrm{eV} \leq E < 50\ \mu\mathrm{eV}$	$6 \leq v < 100$	$4 < \lambda \leq 64$	$0.002 \leq T < 0.6$
cold	$0.05\ \mathrm{meV} < E \leq 25\ \mathrm{meV}$	$100 < v \leq 2200$	$0.18 \leq \lambda < 4$	$0.6 < T \leq 300$
thermal	$25\ \mathrm{meV}$	2200	0.18	300
epithermal	$25\ \mathrm{meV} < E \leq 500\ \mathrm{keV}$	$2200 < v \leq 1 \times 10^7$		
fast	$>500\ \mathrm{keV}$	$>1 \times 10^7$		

Free neutrons are unstable with a 15 minute lifetime but are prevented from decaying while bound in nuclei through the combined effects of energy conservation and Fermi statistics. They must be liberated from nuclei using nuclear reactions with MeV-scale energies in order to be used and studied. We define "slow" neutrons to be neutrons whose energy has been lowered well below this scale. The available dynamic range of neutron energies for use in laboratory research is quite remarkable, as shown in Table 1. Thermodynamic language is used to describe different regimes; a neutron in thermal equilibrium at 300 K has a kinetic energy of only 25 meV. Because its de Broglie wavelength (0.18 nm) is comparable to interatomic distances, this energy also represents the boundary below which coherent interactions of neutrons with matter become important. The most intense sources of neutrons for experiments at thermal energies are nuclear reactors, although accelerators can also produce higher energy neutrons.

Neutron decay is an important process for the investigation of the Standard Model of electroweak interactions. As the prototypical beta decay, it is sensitive to certain SM extensions in the charged-current electroweak sector. Neutron decay can determine the Cabibbo-Kobayashi-Maskawa (CKM) matrix element $|V_{ud}|$ through increasingly precise measurements of the neutron lifetime and the decay correlation coefficients.

Searches for violations of time-reversal symmetry and/or CP symmetry address issues which lie at the heart of cosmology and particle physics. Among the important issues that can be addressed by neutron experiments is the question of what mechanisms might have led to the observed baryon asymmetry of the universe. Big Bang cosmology and the observed baryon asymmetry of the universe appear to require significantly more T-violation among quarks in the first generation than is predicted by the SM. The next generation of neutron electric dipole moment (EDM) searches, which plan to achieve sensitivities of $10^{-27}\ e \cdot \mathrm{cm}$ to $10^{-28}\ e \cdot \mathrm{cm}$, is the most important of a class of experiments aiming to search for new physics in the T-violating sector.

The last decade has also seen qualitative advances in both the quantitative understanding of nuclei, especially few-body systems, and in the connection between nuclear physics and quantum chromodynamics (QCD). Low energy properties of nucleons and nuclei, such as weak interactions in n-A systems, low energy n-A scattering lengths, and the internal electromagnetic structure of the neutron (its electric polarizability and charge radius) are becoming calculable. These theoretical developments are motivating renewed experimental activity to measure undetermined low energy properties, such as the weak interaction amplitudes between nucleons, and to improve the precision of other low energy neutron measurements. The ultimate goal is to illuminate the strongly interacting ground state of QCD, the most poorly understood sector of the SM.

This review presents and discusses the status of the experimental efforts to confront these physics questions using slow neutrons. The improvements in precision required to address these questions are technically feasible and have spurred both new experimental efforts and the development of new neutron sources. We also discuss some of the new proposed facilities under construction. It is not possible to cover the large volume of work in a review of this scope, so we refer the reader to a number of more specialized reviews wherever appropriate. Instead, we emphasize recent experiments and those planned for the near future. Neutron experiments form part of a larger subclass of low energy precision measurements which test the SM (1). There are texts that cover a broader survey of topics and provide historical context (2–4). Tests of quantum mechanics using neutron interferometry are not discussed but are covered in detail in a recent comprehensive text (5).

1.2. Neutron Sources

Most fundamental neutron physics experiments are conducted with slow neutrons for two main reasons. First, slower neutrons spend more time in an apparatus. Second, slower neutrons can be more effectively manipulated through coherent interactions with matter and external fields. Free neutrons are usually created through either fission reactions in a nuclear reactor or through spallation in high Z targets struck by GeV proton beams. We briefly examine these neutron sources and the process by which cold and ultracold neutrons (UCN) are produced starting from neutrons with energies several orders of magnitude greater.

Neutrons are produced from fission in a research reactor at an average energy of approximately 2 MeV. They are slowed to thermal energy in a moderator such as heavy or light water, graphite, or beryllium, surrounding the fuel. The peak core fluence rate of research reactors is typically in the range 10^{14} cm^{-2} s^{-1} to 10^{15} cm^{-2} s^{-1}. To maximize the neutron density, it is necessary to increase the fission rate per unit volume, but the power density is ultimately limited by heat transfer and material properties. In the spallation process, protons typically are accelerated to energies in the GeV range and strike a high Z target, producing approximately 20 neutrons with energies in the fast and epithermal region (6). This is an order of magnitude with more neutrons per nuclear reaction than from fission. Present spallation sources yield neutron rates of 10^{16} s^{-1} and 10^{17} s^{-1}. Although the time-averaged fluence from spallation neutron sources is about an order of magnitude lower than for fission reactors, there is potentially more room for technical improvements in the near-term future.

The main feature that differentiates spallation sources from reactors is the possibility for operation in pulsed mode. At reactors one obtains continuous beams with a thermalized Maxwellian energy spectrum. In a spallation source, neutrons arrive at the experiment while the production source is off, and the frequency of the pulsed source can be chosen so that slow neutron energies can be determined by time-of-flight methods. The lower radiation background and convenient neutron energy information can be advantageous for certain experiments.

Fast neutrons reach the thermal regime most efficiently through a cascade of roughly 20 collisions with matter rich in hydrogen or deuterium. Cold neutrons are produced by a cryogenic neutron moderator adjacent to the reactor core or spallation target held at a temperature of $\approx$20 K. One generally wants the moderator as cold as possible to increase the phase space density of the neutrons. As the neutron wavelengths become large compared to the atomic spacings, the total scattering cross sections in matter are dominated by elastic or quasielastic processes, and it becomes more difficult for the neutrons to thermalize.

It is not practical to describe specific neutron facilities in any detail in this review, but we note a few where the bulk of research efforts have been carried out. For 30 years the most active facility for fundamental neutron research has been the Institut Max von Laue—Paul Langevin (ILL) in Grenoble, France (7). Its 58 MW reactor is the focal point of neutron beta-decay and UCN physics in the world. The new FRM-II reactor has come online in Munich with a predicted cold neutron fluence comparable to the ILL; its beamline for fundamental neutron physics is under construction (8). The most active institutions in the United States are the National Institute of Standards and Technology (NIST) (9) and Los Alamos National Laboratory (LANL) (10). In Russia, there are significant efforts at the Petersburg Nuclear Physics Institute (PNPI), Joint Institute for Nuclear Research (JINR) in Dubna, and the Kurchatov Institute and in Japan at the High Energy Accelerator Research organization (KEK). Many smaller sources play an essential role in the development of experimental ideas and techniques.

In addition to the existing sources, the last decade has seen tremendous growth in the construction of facilities and beamlines devoted to fundamental neutron physics. Many of these new facilities are at spallation sources. The Paul Scherrer Institut (PSI), which operates a continuous spallation source, has constructed a cold neutron beamline dedicated to fundamental physics (11). In the United States, the 2 MW Spallation Neutron Source (SNS) is under construction, and the fundamental physics beamline (FNPB) should be operational some time in 2008 (12). The Japanese Spallation Neutron Source (JSNS) is in the construction phase and is also anticipated to become operational in 2008. Tables 2 and 3 give a few of the measured (or projected) cold-neutron beam properties for some of the facilities with active fundamental physics programs.

In the neutron energy spectrum from a cold moderator, there is a very small fraction whose energies lie below the $\approx$100 neV neutron optical potential of matter.

TABLE 2 Some operating parameters for major cold neutron reactor-based user facilities with active (and proposed) fundamental physics programs. Fluence rates are given as neutron capture fluence

Parameter	ILL PF1	ILL PF2	NIST NG-6	(FRM-II) (Mephisto)
Power (MW)	58	58	20	(20)
Guide length (m)	60	74	68	(30)
Guide radius (m)	4000	4000	∞	(460)
Guide type ($m=$)	1.2	2	1.2	(3)
Cross section (cm^2)	6×12	6×20	6×15	(5×11.6)
Fluence rate ($\times 10^9$ cm^{-2} s^{-1})	4	14	2	(20)

Such neutrons are called ultracold neutrons, and they can be trapped by total external reflection from material media. The existence of such neutrons was established experimentally in the late 1960s (13, 14). The UCN facility at the ILL employs a turbine to mechanically convert higher energy neutrons to UCN (15). Although the density of neutrons is bounded by the original phase space in the source (Liouville's theorem), this technique produces enough UCN to conduct a number of unique and fundamental experiments described in part below. During the last decade new types of UCN converters have been developed that can increase the phase space density through the use of "superthermal" techniques (16). They involve energy dissipation in the moderating medium (through phonon or magnon creation) and therefore are not limited by Liouville's theorem. Superfluid helium (17) and solid deuterium (18) have been used successfully as superthermal UCN sources, and solid oxygen is also being studied (19). The lack of neutron absorption in ^{4}He along with its other unique properties makes possible experiments in

TABLE 3 Some operating parameters for major cold neutron spallation-source user facilities with active (and proposed) fundamental physics programs. Fluence rates are given as neutron capture fluence

Parameter	SINQ FunSpin	LANSCE FP12	(SNS) (FNPB)	(JSNS)
Time-averaged current (mA)	1.2	0.1	(1.4)	(0.3)
Source rep. rate (Hz)	dc	20	(60)	(25)
Guide length (m)	7	8	(15)	(10 to 20)
Guide radius (m)	∞	∞	(117)	(∞)
Guide type ($m=$)	3	3	(3.5)	(3)
Cross section (cm^2)	4×15	9.5×9.5	(10×12)	(10×10)
Fluence rate ($\times 10^8$ cm^{-2} s^{-1})	8	1	(10)	(5)

which the measurement is conducted within the moderating medium. These developments have led to proposals for new UCN facilities at LANL, PSI, FRM-II (Forschungreaktor München II), KEK, North Carolina State, Mainz, and other sources. Extensive treatments of UCN physics are found in References (20, 21).

2. NEUTRON DECAY AND STANDARD MODEL TESTS

Several reviews discuss weak interaction physics using slow neutrons in greater detail or provide additional information (22–24). A recent publication addresses the issue of CKM unitarity (25). A comprehensive review of measurements in neutron and nuclear beta decay to test the SM in the semileptonic sector and its possible extensions along with a comparison with other probes of similar physics will appear in the near future (26).

2.1. Theoretical Framework

The neutron is composed of two down quarks and an up quark, and it is stable under the strong and electromagnetic interactions, which conserve quark flavor. The weak interaction can convert a down quark into an up quark through the emission of the W gauge boson. The mass difference of the neutron and proton is so small that the only possible decay products of the W are an electron and antineutrino with the release of energy distributed among all the decay products: $n \rightarrow p + e^- + \bar{v}_e + 0.783$ MeV. Neither of the other available decay modes, radiative neutron decay with a photon in the final state or decay to a hydrogen atom and an antineutrino, have been seen yet, although the first searches for radiative decay are in progress (28, 29). Experiments test the assumptions of the SM by performing precision measurements on the proton and electron energies and momenta and the neutron spin.

To leading order, free neutron decay in the SM is described by a mixed vector/axial-vector current characterized by two coupling strengths, g_V and g_A, the vector and axial-vector coupling coefficients. Because the momentum transfers involved in neutron beta decay are small compared to the W and Z masses, one can write an effective Lagrangian that describes neutron decay in the SM as a four-fermion interaction

$$\mathcal{L}_{int} = \frac{G_F V_{ud}}{2\sqrt{2}}(V_\mu - \lambda A_\mu)(v^\mu - a^\mu),\qquad\qquad 1.$$

where $V_\mu = \overline{\psi_p}\gamma_\mu\psi_n$, $v^\mu = \overline{\psi_e}\gamma^\mu\psi_v$, $A_\mu = \overline{\psi_p}\gamma_\mu\gamma_5\psi_n$, and $a^\mu = \overline{\psi_e}\gamma^\mu\gamma_5\psi_v$ are the hadronic and leptonic vector and axial vector currents constructed from the neutron, proton, electron, and neutrino fermion fields, G_F is the Fermi decay constant, V_{ud} is a CKM matrix element, and λ is the ratio of the axial vector and vector couplings.

The V-A structure for the weak currents is incorporated directly into the standard electroweak theory by restricting the weak interaction to operate only on the left-handed components of the quark and lepton fields. A more fundamental understanding of the reason for this parity-odd structure of the weak interaction is still lacking. There is also no understanding for the values of the CKM mixing matrix elements between the quark mass eigenstates and their weak interaction eigenstates. The fact that the matrix is unitary is ultimately a consequence of the universality of the weak interaction gauge theory. Extensions to the SM which either introduce non V-A weak currents or generate violations of universality can therefore be tested through precision measurements in beta decay. A recent reanalysis of the constraints on non V-A charged currents showed that improved neutron decay measurements have set new direct limits on such couplings, which are typically constrained at the 5% level (26). Complementary constraints on non V-A charged currents in neutron beta decay from neutrino mass limits have recently appeared (27).

The probability distribution for beta decay in terms of the neutron spin and the energies and momenta of the decay products (30) can be written

$$dW \propto \left(g_V^2 + 3g_A^2\right) F(E_e)$$

$$\left[1 + a\frac{\vec{p}_e \cdot \vec{p}_\nu}{E_e E_\nu} + b\frac{m_e}{E_e} + \vec{\sigma}_n \cdot \left(A\frac{\vec{p}_e}{E_e} + B\frac{\vec{p}_\nu}{E_\nu} + D\frac{\vec{p}_e \times \vec{p}_\nu}{E_e E_\nu}\right)\right], \qquad 2.$$

where one defines

$$\tau_n = \frac{2\pi^3\hbar^7}{m_e^5 c^4} \frac{1}{f(1 + \delta_R)\left(g_V^2 + 3g_A^2\right)} \qquad = (885.7 \pm 0.8)\,\text{s} \qquad \text{neutron lifetime}$$

$$\lambda = \left|\frac{g_A}{g_V}\right| e^{i\phi} \qquad\qquad\qquad = -1.2695 \pm 0.0029 \qquad \text{coupling constant ratio}$$

$$a = \frac{1 - |\lambda|^2}{1 + 3|\lambda|^2} \qquad\qquad\qquad = -0.103 \pm 0.004 \qquad \text{electron-antineutrino asymmetry}$$

$$b = 0 \qquad\qquad\qquad\qquad\qquad = 0 \qquad\qquad\qquad \text{Fierz interference}$$

$$A = -2\frac{|\lambda|^2 + |\lambda|\cos\phi}{1 + 3|\lambda|^2} \qquad\quad = -0.1173 \pm 0.0013 \qquad \text{spin-electron asymmetry}$$

$$B = 2\frac{|\lambda|^2 - |\lambda|\cos\phi}{1 + 3|\lambda|^2} \qquad\quad = 0.983 \pm 0.004 \qquad \text{spin-antineutrino asymmetry}$$

$$D = 2\frac{|\lambda|\sin\phi}{1 + 3|\lambda|^2} \qquad\qquad = (-0.6 \pm 1.0) \times 10^{-3} \qquad \text{T-odd triple-product.}$$

In these equations, which neglect small corrections such as weak magnetism, $F(E_e)$ is the electron energy spectrum, $\vec{p}_e$, $\vec{p}_\nu$, E_e, and E_ν are the momenta and kinetic energies of the decay electron and antineutrino, $\vec{\sigma}_n$ is the initial spin of the decaying neutron, ϕ is the phase angle between the weak coupling constants g_A and g_V, and $f(1 + \delta_R) = 1.71489 \pm 0.00002$ is a theoretically calculated phase space factor (31). The spin-proton asymmetry correlation coefficient C is proportional to the quantity $A + B$. The values represent the world averages as compiled by the Particle Data Group (PDG) (32). The parameter λ can be extracted from measurement of either a, A, or B. If the neutron lifetime τ_n is also measured, g_V and g_A can be determined uniquely under the assumption that $D = 0$. Figure 1 (see color insert) shows the recent history of measured values of the lifetime and correlation coefficients as used by the PDG (32).

One strong motivation for more accurate measurements of neutron decay parameters is to measure $|V_{ud}|$. The most precise number comes from the $\mathcal{F}t$ values of superallowed $0^+ \to 0^+ \beta$ transitions between iscbaric analog states (33). This gives $|V_{ud}| = 0.0738 \pm 0.0004$ with the uncertainty dominated by theoretical corrections. Using values of $|V_{us}|$ and $|V_{ub}|$ taken from the current recommendations of the PDG, $|V_{ud}|^2 + |V_{us}|^2 + |V_{ub}|^2 = 0.9966 \pm 0.0014$ value differs from unitarity by 2.1 standard deviations.

Neutron beta decay offers a theoretically cleaner environment for extracting g_V due to the absence of other nucleons (although some radiative corrections are common to both systems). Using the PDG values of τ_n and λ, the same unitarity test gives $\sum_i |V_{ui}|^2 = 0.9971 \pm 0.0039$, consistent with unity but less precise. This result agrees with both the nuclear result and unitarity. The present situation regarding unitarity is summarized in Figure 2 (see color insert). Pion beta decay is theoretically the cleanest system in which to measure $|V_{ud}|$, but the small branching ratio has so far precluded a measurement with enough sensitivity to compete with superallowed beta decay and neutron decay. The latest measurement from pion beta decay gives $|V_{ud}| = 0.9728 \pm 0.0030$ (34).

The possible deviation from unitarity has motivated a number of new precise measurements of semileptonic kaon decay rates which promise to determine $|V_{us}|$ more precisely (35). If one were use the value of $|V_{us}|$ from some recent evaluations (36), the discrepancy with unitarity disappears. There are also renewed theoretical investigations to extract $|V_{us}|$ from hyperon decay (37). A precision determination of $|V_{ud}|$ should be seen in the context of the overall effort to determine with high precision all the parameters of the CKM matrix. The CLEO-c collaboration of the Cornell Electron Storage Ring should measure the CKM matrix element $|V_{cd}|$ to 1% accuracy if lattice gauge theory calculations of the required form factors can match the expected precision of the data (36, 38). This would make possible another independent check of CKM unitarity using the first column, $|V_{ud}|^2 + |V_{cd}|^2 + |V_{td}|^2 = 1$.

2.2. Neutron Lifetime Experiments

Seven experiments (39–45) contribute to a neutron lifetime world average of $\tau_n = (885.7 \pm 0.8)$ s (32). The experiments employ one of two distinct experimental strategies for measuring the neutron lifetime. The four more precise measurements use ultracold neutrons that are confined using a combination of material walls and gravity. One fills the trap and measures the number of neutrons remaining as a function of time to extract τ. An advantage of this technique is that one avoids the necessity of knowing the absolute neutron density and detector efficiency. The measured value of τ is $(1/\tau_n + 1/\tau_{loss})^{-1}$ and includes losses from the trap as well as neutron decay. To isolate τ_{loss}, which is typically dominated by nonspecular processes in the neutron interaction with the trap walls, one measures τ in bottles with different surface-to-volume ratios and performs an extrapolation to an infinite volume. These losses depend on the UCN energy spectrum, which can change during the storage interval, so much work has been done to understand the spectrum evolution and loss mechanisms and to find surface materials with lower loss probabilities.

To address losses experimentally, Arzumanov et al. simultaneously measured the UCN storage time and the inelastically scattered neutrons (44), thus monitoring the primary loss process. An experiment by Serebrov et al. (46) achieved a significant reduction in wall losses by using low temperature fomblin oil. This coating produced very long storage times and permitted much shorter extrapolations in collision frequency. The result is very different (6.5σ) from the PDG average, as illustrated in Figure 1 (see color insert). The group intends to make additional measurements with a variable-volume trap to change the collision frequency while maintaining the same trap surface and vacuum conditions.

The second method measured simultaneously both the rate of neutron decays dN/dt and the average number of neutrons N in a well-defined volume of a neutron beam. The neutron lifetime was determined from the differential form of the radioactive decay function, $dN/dt = -N/\tau_n$. Such a measurement requires accurate absolute counting of neutrons and neutron decay products (protons) from a cold neutron beam and must overcome its own set of technical challenges. The two more precise experiments used a segmented proton trap (47) and a neutron detector with an efficiency that was proportional to $1/v$ (48, 49). Both experiments produced a value in good agreement with the PDG average.

Accurate measurements using each of these completely independent methods are important for establishing the reliability of the results for τ_n. The latest measurement is in dramatic disagreement with existing values, and the situation must be resolved by new experiments.

2.2.1. FUTURE PROSPECTS IN NEUTRON LIFETIME MEASUREMENTS   A third approach to measuring τ_n avoids many of these problems. The most natural way to measure exponential decay is to acquire an ensemble of radioactive species and register the decay products. One can then simply fit the time spectrum for the slope, or decay rate, of the exponential function. Such a measurement using neutrons has only become feasible after the demonstration of magnetically trapping UCNs in superfluid ^{4}He (50). The UCNs fill a magnetic trap through the inelastic scattering of 0.89 nm neutrons in superfluid ^{4}He (the superthermal process). As the trapped neutrons beta decay, the energetic electrons are registered via scintillations in the helium, thus allowing one to fit directly for the exponential decay.

The experiment initially observed a short lifetime and attributed it to the presence of neutrons with energies higher than the magnetic potential of the trap. When the magnetic field was ramped to eliminate these neutrons, the result is in agreement with the currently accepted value of the free neutron lifetime, but the statistical uncertainty is large (60 s) (51). Upgrades to the apparatus are in progress to increase the number of trapped neutrons. The collaboration anticipates that a statistical precision of a few seconds will be possible in the near future.

There are two new bottle-type UCN experiments in the developmental stage. The first uses a low temperature fomblin oil to reduce the collisional losses.

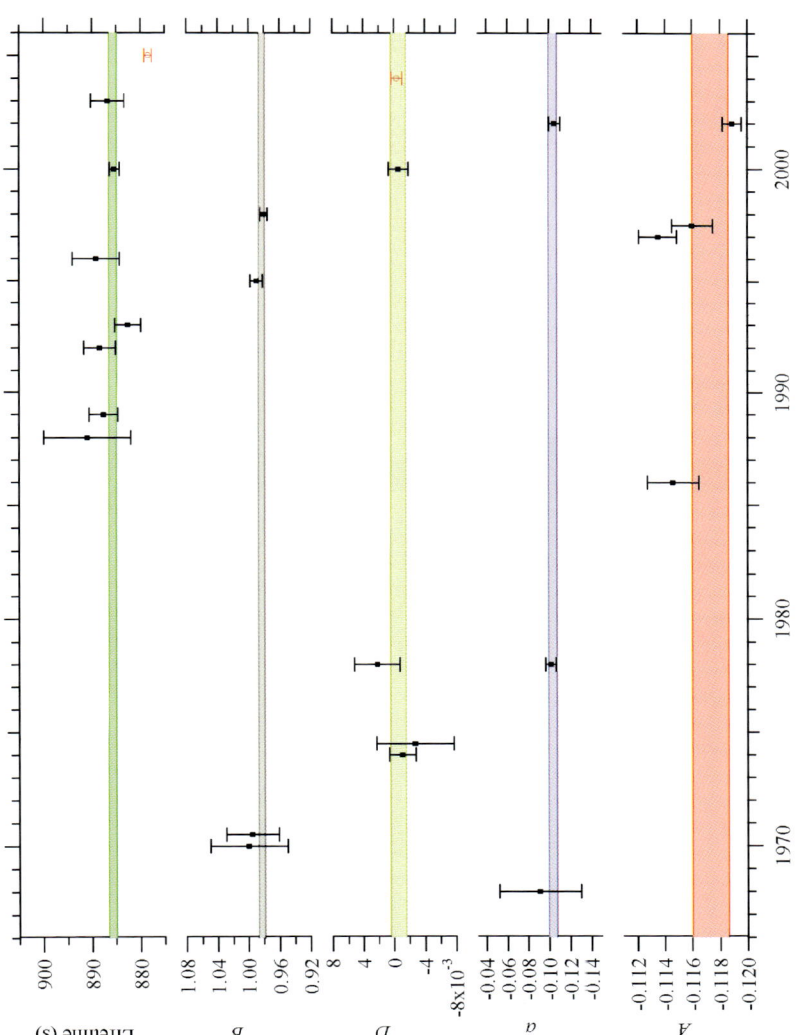

Figure 1 A summary plot of the measurements of the neutron lifetime and correlation coefficients that are used in the 2004 compilation of the Particle Data Group (PDG). Data points with open circles have not yet been included in the evaluation. The shaded bands represent the ±1σ region.

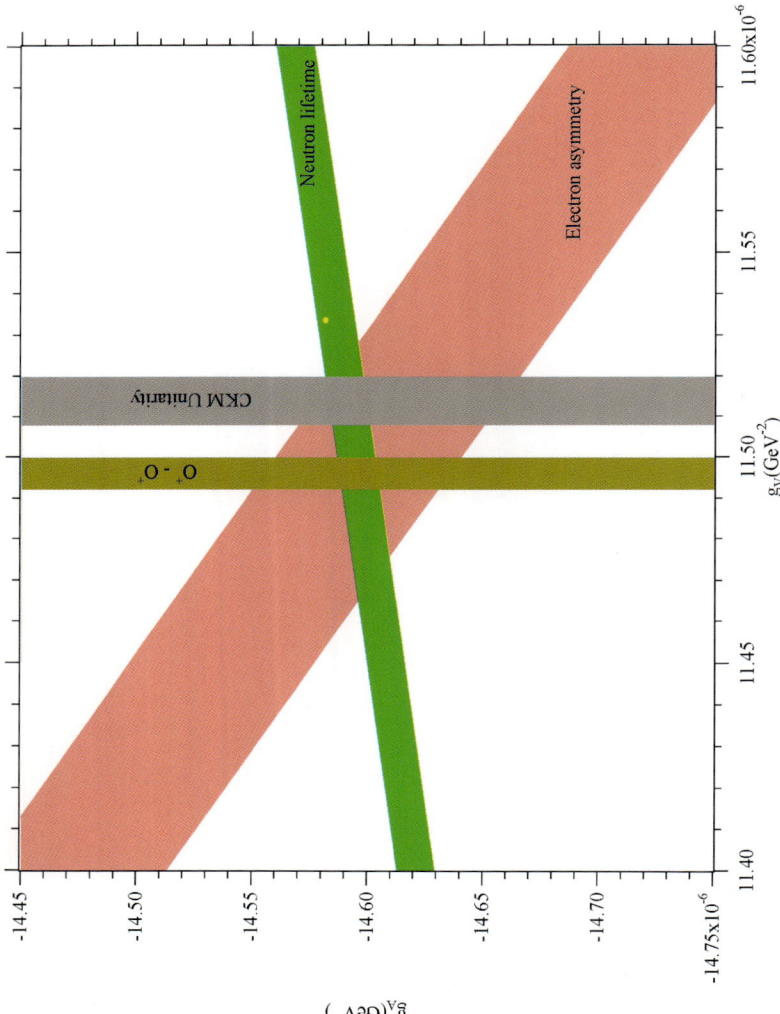

Figure 2 The weak coupling constants g_A and g_V determined from neutron decay parameters, $\mathcal{F}t$ values of superallowed $0^+ \to 0^+$ transitions, and CKM unitarity. For the neutron decay parameters, the width of the one-sigma band is dominated by experimental uncertainties. For the superallowed $0^+ \to 0^+$ transitions the uncertainty is dominated by radiative corrections.

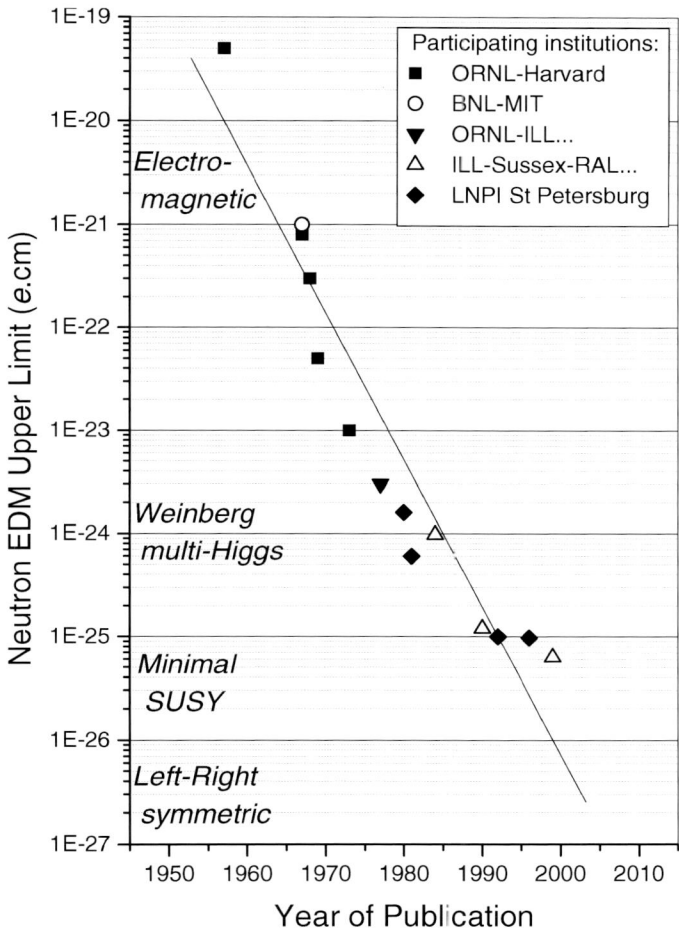

Figure 3 History of the neutron electric dipole moment limit and some of the ranges for different theoretical predictions. (Plot courtesy of P. Harris, University of Sussex.)

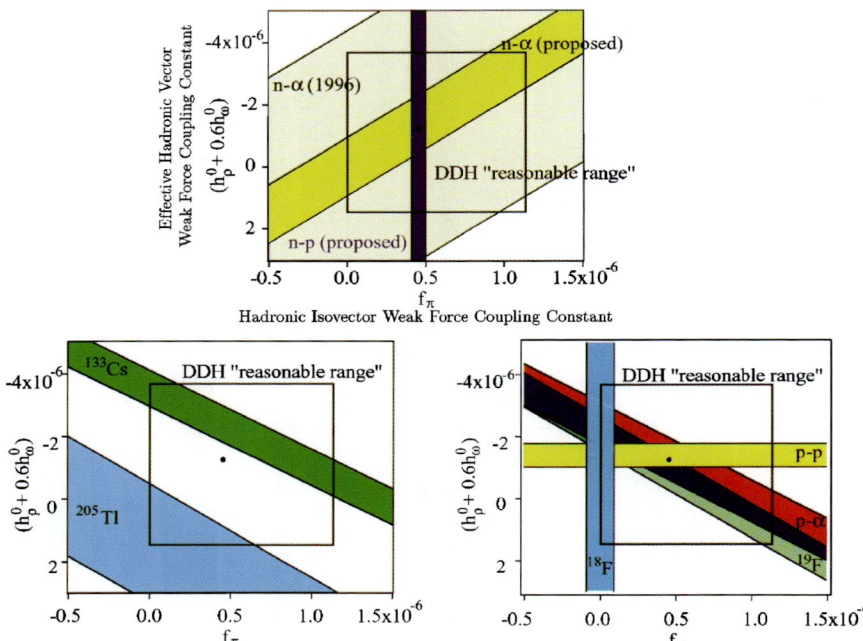

Figure 4   Constraints on linear combinations of isoscalar and isovector nucleon-nucleon weak meson couplings (178). The bottom graphs show constraints from measurements of anapole moments in ^{133}Cs and ^{205}Tl and from measurements in p-p and p-^{4}He scattering and from ^{18}F and ^{19}F gamma decay. The top graph shows the anticipated constraints from proposed measurements of the *P*-odd asymmetry p($\vec{\text{n}}$, γ)d to 5×10^{-9} accuracy and the *P*-odd neutron spin rotation in ^{4}He to 2×10^{-7} rad/m accuracy. In each plot, the box indicates the reasonable range obtained by Desplanques et al. for the couplings.

Recent data demonstrate a UCN reflection loss coefficient of 5×10^{-6} when the vessel temperature is in the range of 105 K to 150 K. They intend to use this surface coating in an "accordion-like" storage vessel, thus allowing one to vary the trap volume while keeping the surface area and characteristics constant (52). The collaboration expects to achieve a precision of 1 s. The second experiment will store UCN magnetically in vacuum using an arrangement of permanent magnets and superconducting solenoids and extract the protons electrostatically. The lifetime is measured by real-time detection of the decay protons and counting the integral number of neutrons using different storage times (53). The collaboration anticipates that a measurement with 0.1 s uncertainty is possible.

2.3. Angular Correlation Experiments

2.3.1. SPIN-ELECTRON ASYMMETRY A With the neutron lifetime and one of the correlation coefficients a, A, or B, one can determine values for g_A and g_V. Because it has the greatest sensitivity to λ and is more accessible experimentally, the spin-electron asymmetry A has been measured more frequently and with greater precision. Four independent measurements used in the PDG evaluation are not in good agreement with each other, so the PDG uses a weighted average for the central value and increases the overall uncertainty by a scale factor of 2.3 (32). We discuss the two more recent measurements, one using a time projection chamber and one using an electron spectrometer, and the prospects for future improvement.

In the experiment of Schreckenbach et al. (54), a beam of polarized cold neutrons was surrounded by a time projection chamber (TPC). Decay electrons passed through the drift chamber and were incident on plastic scintillators. The drift chamber recorded the ionization tracks in three dimensions while the scintillator gave the electron energy and start signal for the drift chamber. The TPC provided good event identification and reduced gamma ray backgrounds. The result was $A = -0.1160 \pm 0.0015$ (55). The contributions to the overall uncertainty were roughly split between statistical and systematic uncertainties with the largest systematic contribution coming from the background subtraction.

The PERKEO II experiment also used a beam of cold polarized neutrons, but the decay electrons were extracted using a superconducting magnet in a split pair configuration. The field was transverse to the beam, so neutrons passed through the spectrometer but electrons were guided by the field to one of two scintillator detectors on each end. This arrangement had the advantage of achieving a 4π acceptance of electrons. An asymmetry is formed from the electron spectra in the two detectors as a function of the electron energy; the difference in those quantities for the two detectors is directly related to the electron asymmetry. Their run produced $A = -0.1178 \pm 0.0007$, where the main contributions to the uncertainty were in the neutron polarimetry, background subtraction, and electron detector response (56).

The next version of PERKEO II will use the new ballistic supermirror guide at the ILL with four times the fluence rate (57). The collaboration intends to use a new configuration of crossed supermirror polarizers to make the neutron polarization more uniform in phase space (58). The beam polarization can also be measured with a completely different method using an opaque ^{3}He spin filter. PERKEO II anticipates reducing the main correction and uncertainty in the polarization analysis from 1.1% to less than 0.25% with an uncertainty of 0.1% in that value.

There are several other efforts underway to perform independent measurements of the electron asymmetry. The UCNA collaboration has made progress toward measuring A using a superconducting solenoidal spectrometer (59). UCNs are produced in a solid deuterium moderator at the Los Alamos Neutron Science Center (LANSCE) and transported to the spectrometer using diamond-coated guides. In the spectrometer, one produces highly polarized (>99.9%) neutrons by passing them through a 6 T magnetic field and into an open ended cylinder which increases the dwell time of the polarized UCN in the decay spectrometer. The decay electrons will be transported along the field lines to detectors at each end of the spectrometer. The detectors consist of multiwire proportional counters backed by plastic scintillator. The collaboration believes that a 0.2% measurement is possible with three weeks of running.

Two other groups propose measuring the A coefficient at the 10^{-3} level. The detector designs allow the possibility of measuring other decay correlation coefficients with the same apparatus. A group at PNPI is working on a magnetic spectrometer to be used with a highly collimated cold neutron beam. The field guides decay particles to an electron detector at one end and a proton detector at the other. Their neutron polarimeter agrees with ^{3}He-based spin filter methods at the $\approx 2 \times 10^{-3}$ level (60). The spectrometer should also be able to measure the coefficients A and B simultaneously, thus eliminating the need for precision polarimetry.

The abBA collaboration proposes to use an electromagnetic spectrometer that guides both decay electrons and protons to detectors at each end of the spectrometer (61). The detector would be able to measure a, A, B, and the Fierz interference term b, which is zero in the SM. The detectors would be large-area segmented silicon detectors with thin entrance windows that allow the detection of both the proton and electron. The ability to detect coincidences greatly suppresses backgrounds and allows the measurement of residual backgrounds. The magnetic field guides the decay products to conjugate points on the segmented Si detectors and provides 4π detection of both electrons and protons and suppression of backgrounds by use of coincidences. The apparatus is being designed for use at a pulsed spallation source to exploit background reduction and perform neutron polarimetry. The neutrons can be polarized by transmission through polarized ^{3}He, whose spin-dependent absorption cross section possesses an accurately known neutron energy dependence that can be exploited for accurate neutron polarization measurement (62).

2.3.2. SPIN-ANTINEUTRINO ASYMMETRY B The electron asymmetry and antineutrino asymmetry provide complementary information. g_A is equal to -1 in the SM Lagrangian at the quark level but is renormalized in hadrons by the strong interaction. Because g_A is nearly -1, A is close to zero, and B is near unity. Thus B

is not particularly sensitive to λ but provides more attractive ground for searching for non-SM signatures, such as extended left-right symmetric models.

Left-right symmetric models, which are motivated in part by the desire to restore parity conservation at high energy scales, add a new right-handed charged gauge-boson W_2 with mass M_2 and four new parameters to be constrained by experiments: a mixing angle ζ, $\delta = (M_1/M_2)^2$, $r_g = g_R/g_L$ the ratio of the right- and left-handed gauge coupling strengths g_R and g_L, and $R_K = V_{ud}^R / V_{ud}^L$ where R and L designate the right and left sectors. In the manifest left-right symmetric model (MLRM), $r_g = r_K = 1$, and in the SM $\delta = 0$.

The mass limit on a right-handed vector boson comes from muon decay and is 406 GeV/c^2 (63). In the MLRM where there are only two parameters (ζ and δ), constraints from other systems are better than the neutron constraints. For the extended left-right model, however, neutron-derived constraints are complementary to the other searches. Another area in which to search for right-handed currents is the decay of the neutron into a hydrogen atom and antineutrino, because one of the hyperfine levels of hydrogen cannot be populated unless right-handed currents are present (64). The small branching ratio has precluded a search so far.

In the last three decades, there have been only two new measurements of the antineutrino asymmetry. Because the antineutrino cannot be conveniently detected, its momentum was deduced from electron-proton coincidence measurements. Electrons from the decay of polarized neutrons were detected by plastic scintillators, and protons were detected by an assembly of two microchannel plates. From the electron energy and proton time-of-flight, one can reconstruct the antineutrino momentum. The first measurement was carried out at PNPI and produced a result of $B = 0.9894 \pm 0.0083$ (65). A second run at the ILL used largely the same apparatus and measured $B = 0.9821 \pm 0.0040$ (66), where the largest reduction in the overall uncertainty came from improved statistics.

A recent measurement of B was performed using the PERKEO II apparatus. Typically, one detects electron-proton coincidences using one detector for each particle. The PERKEO II measurement uses two detectors, one in each hemisphere of the detector, that can detect both electrons and protons. Electrons are detected using plastic scintillator, while the protons are accelerated on a thin carbon foil placed in front of the scintillator. The resulting secondary electrons are guided onto the electron detectors. This technique reduces systematics and increases the sensitivity to B.

2.3.3. ELECTRON-ANTINEUTRINO ASYMMETRY a Although the electron-antineutrino asymmetry a has approximately the same sensitivity to λ as A, it is only known to 4%. Since 1978 (67), there has been only one new measurement. The experimental difficulty lies with the energy measurement of the recoil protons, whose spectral shape is slightly distorted for nonzero a. Unlike A, it does not require neutron polarimetry. A precision measurement of a would produce an independent measurement of λ, an improved test of CKM unitarity, and model-independent tests of new physics. The values of a, A, and B can be related to the strength of hypothetical right-handed weak forces and scalar and tensor forces (68, 69), and it was recently shown that a precise comparison of a and A can place stringent limits on possible conserved-vector-current (CVC) violation and second class currents in neutron decay (70).

The most recent determination of a comes from measurements of the integrated energy spectrum of recoil protons stored in an ion trap (71). A collimated beam of cold neutrons passed though a proton trap consisting of annual electrodes coaxial with a magnetic field whose strength varied from 0.6 T to 4.3 T over the length of the trap. Protons created inside the volume were trapped, and those created in a high field region were adiabatically focused onto a mirror in the low field region. The trap was periodically emptied and the protons counted as a function of the mirror potential. The result of $a = -0.1054 \pm 0.0055$ is in good agreement with the previous measurement and of comparable precision.

The precision of a measurements must be improved to the level of A experiments to constrain λ. There are two major efforts underway to improve the precision of a, aSPECT (72) and aCORN (73). aCORN relies on the measurement of an asymmetry in the coincidence detection of electrons and recoil proton that is proportional to a (74). The asymmetry is formed by carefully restricting the phase space for the decay in a magnetic spectrometer so that decay events with parallel and antiparallel electron and antineutrino momenta are separated in the coincidence timing spectrum. a is directly proportional to the relative number of events, and there is no need for precise spectroscopy of the low energy protons. The experiment will be built and tested at the Low Energy Neutron Source (LENS) (75) and then run at NIST where a measurement of approximately 1% accuracy is feasible.

In the aSPECT experiment, one again measures a proton energy spectrum as a function of a potential, similar to the idea used for the proton trap experiment. One increases the statistical power by completely separating the source part and the spectroscopy part of the apparatus. A cold neutron beam will pass through a region of strong, homogeneous magnetic field transverse to the beam. The decay protons with initial momentum component along the field direction will be directed toward a detector. Near the detector is a region of weaker magnetic field and electrostatic retardation potentials, and only those protons with sufficient energy to overcome the barrier continue on to the detector. Registering the protons as a function of the retardation potential gives the recoil proton spectrum, which one fits to extract a. The collaboration believes that a statistical uncertainty of approximately 0.25% is achievable.

3. SEARCHES FOR NONSTANDARD *T* AND *B* VIOLATION

The physical origins of the observed *CP* violation in nature, first seen in the neutral kaon system (76), remain obscure. *CP* violation implies *T* violation (and vice versa) through the *CPT* theorem. Recent experiments have reported measuring

CP violation in the $K_L^o \rightarrow 2\pi$ amplitudes (77, 78) and in the decays of the neutral B-mesons (79, 80). The SM can accommodate the possibility of CP violation through a complex phase δ_{KM} in the CKM quark mixing matrix. To date there is no firm evidence against the possibility that the observed CP-violation effects are due to this phase (81), but the question remains whether or not there exist sources of CP-violation other than δ_{KM}. There is indirect evidence for this possibility from cosmology; it appears that δ_{KM} is not sufficient to generate the baryon asymmetry of the universe in the Big Bang model. One area to probe for the existence of new CP-violating interactions is systems involving first-generation quarks and leptons for which the contribution from δ_{KM} is typically suppressed. Examples of observables of this kind are electric dipole moments of the neutron, leptons, as well as atoms and T-odd correlations in leptonic and semileptonic decays.

3.1. EDM Theoretical Framework

The search for the neutron electric dipole moment addresses issues which lie at the heart of modern cosmology and particle physics. The current limit on the permanent EDM of the neutron represents one of the most sensitive null measurements in all of physics and has eliminated many theories and extensions to the SM (Figure 3, see color insert). The reader is directed to a comprehensive review of EDM experiments by Ramsey (82) and more recently in References (24, 83).

The energy of a neutral spin-1/2 particle with an EDM d_n in an electric field $\vec{E}$ is $E_n = -d_n \vec{\sigma} \cdot \vec{E}$, where $\vec{\sigma}$ is the Pauli spin matrix. This expression is odd under T and P. The current experimental bound on the neutron EDM is $d_n < 0.63 \times 10^{-25}$ $e \cdot$ cm (90% CL) (32, 84). In the SM, there are two sources of CP violation. One source is the complex phase δ_{KM} in the CKM matrix. The other source is a possible term in the QCD Lagrangian itself, the so-called θ-term

$$\mathcal{L}_{QCD} = \mathcal{L}_{QCD,\theta=0} + \frac{\theta g_s^2}{32\pi^2} G_{\mu\nu} \tilde{G}^{\mu\nu}, \qquad\qquad 3.$$

which explicitly violates CP symmetry because of the appearance of the product of the gluonic field operator G and its dual $\tilde{G}$. Because G couples to quarks but does not induce flavor change, d_n is much more sensitive to θ than it is to δ_{KM}. Thus, measurement of d_n determines an important parameter of the SM. Calculations have shown that $d_n \sim O(10^{-16}\theta)$ $e \cdot$ cm (85, 86).

Although θ is unknown, the observed limit on d_n allows one to conclude that $\theta < 10^{-(9\pm1)}$ (87). Because the natural scale is $\theta \sim O(1)$, the very small value for θ (known as the strong CP problem) requires an explanation. One attempt augments the SM by a global U(1) symmetry (referred to as the Peccei-Quinn symmetry), whose spontaneous breakdown gives rise to Goldstone bosons called axions (88). The θ-term is then essentially eliminated by the vacuum expectation value of the axion. No axions have yet been observed.

Since CP violation through the phase in the CKM matrix involves flavor mixing of higher generation quarks, d_n is very small in the SM; calculations predict it to be 10^{-32} e · cm to 10^{-31} e · cm (89, 90), several orders of magnitude beyond the reach of any experiment being considered at present. Models of new physics, including left-right symmetric models, non-minimal models in the Higgs sector, and supersymmetric (SUSY) models, allow for *CP* violating mechanisms not found in the SM, including terms that do not change flavor. Searches for electric dipole moments in the neutron, leptons, and atoms, which are particularly insensitive to flavor-changing parameters, can strongly constrain such models.

3.1.1. BARYON ASYMMETRY   Antimatter appears to be rare in the universe, and there exists a substantial asymmetry between the number of baryons and antibaryons. Although the SM possesses a nonperturbative mechanism to violate the baryon number B, no experiments have seen B violation, and it is natural to speculate on the origin of the baryon asymmetry of the universe. There are two outstanding facts: baryons make up only 5% of the total energy density of the universe and the ratio of baryons to photons is very small. The ratio $n_B/n_\gamma = (6.1 \pm 0.3) \times 10^{-10}$ is known independently both from Big Bang Nucleosynthesis and fluctuations in the microwave background (91). Sakharov first raised the possibility of calculating the baryon asymmetry from basic principles (92). He identified three criteria that, if satisfied simultaneously, will lead to a baryon asymmetry from an initial $B = 0$ state: baryon number violation, *CP* violation, and departure from thermal equilibrium. One way to explain the asymmetry assumes that all three of these conditions were met at some very early time in the universe and that this physics will remain inaccessible to us, with the B asymmetry effectively an initial condition. However, the existence of inflation in the early universe—a scenario that generates a flat universe, solves various cosmological problems, and generates a spectrum of primordial density fluctuations consistent with observation—would dilute any such early B asymmetry to a negligible level. In this case the B asymmetry must be regenerated through later processes, and there is hope that it is calculable from first principles (93).

Although the SM contains processes that satisfy the first two conditions and the Big Bang satisfies the third, it fails by many orders of magnitude in its estimate of the size of the baryon asymmetry. Grand unified theory (GUT) baryogenesis at $T \sim 10^{29}$ K corresponding to a mass scale on the order of 10^{16} GeV is disfavored by inflation. Electroweak baryogenesis (95, 96), which relies on a nonperturbative $B - L$-violating mechanism (where L is lepton number) present in the SM due to nonperturbative electroweak fields (97) combined with *CP* violation and a departure from equilibrium at the electroweak phase transition, is now very close to being ruled out (98). Leptogenesis (99, 100) combined with $B - L$ conserving processes to get the B asymmetry and the Affleck-Dine mechanism are the most favored speculations at the moment (101).

It appears that some physics beyond the SM, including new sources of *CP* violation that may lead to a measurable value for d_n, must exist if the observed baryon asymmetry is to be understood. The minimal supersymmetric extension of the SM (MSSM) (102) can possess small values of the *CP*-violating phases (consistent with constraints from d_n) that generate the baryon asymmetry. Within the broad framework of non-minimal SUSY models, including GUTs, there are numerous new sources of *CP* violation in complex Yukawa couplings and other Higgs parameters that may have observable effects on the neutron EDM (103–105). The present limit on the neutron EDM already severely constrains many SUSY models.

3.2. Electric Dipole Moment Experiments

EDM experiments employ polarized neutrons with the Ramsey interferometric technique of separated oscillatory fields. Static electric E_0 and magnetic B_0 fields are applied to the polarized neutrons. The neutron spin state is then governed by the Hamiltonian $H = -\vec{\mu} \cdot \vec{B}_0 \pm \vec{d}_n \cdot \vec{E}_0$. A radio frequency (RF) magnetic field of frequency ω_a is applied to tilt the neutron polarization normal to E_0 and B_0 and it starts to precess with a frequency ω_R. After a free precession time T a second tilt pulse in phase with the first is applied and the neutron polarization direction is proportional to $(\omega_R - \omega_a)T$. The Larmor precession frequency of the neutron depends on the direction of the applied field E_0 relative to $\vec{\mu}$. An EDM would appear as a change in ω_R as the electric field is reversed.

The two most stringent limits on d_n come from Altarev et al. at PNPI (106) and Harris et al. at ILL (84). Both experiments used stored UCN. The PNPI apparatus contained two UCN storage chambers with oppositely directed electric fields. A nonzero EDM would cause frequency shifts of opposite sign in each of the chambers, and some sources of magnetic field noise are suppressed with simultaneous measurements with both fields. Nevertheless, slowing varying magnetic fields remained a significant source of systematic uncertainty. In the experiment of Harris et al., a polarized ^{199}Hg comagnetometer occupying approximately the same volume as the neutrons was introduced into the storage volume to continuously monitor the magnetic field. The comagnetometer was essential in eliminating stray magnetic fields as a major source of systematic uncertainty. The experimental accuracy was limited by neutron counting statistics.

There are ambitious efforts underway to improve the current neutron EDM limit by one to two orders of magnitude. All of the experiments attempt to increase the number of UCN, the observation time, and the size of the applied electric field. The CryoEDM collaboration intends to produce UCNs through the superthermal process and transport them to a separate measurement chamber containing superfluid helium. Liquid helium should allow electric field values that are several times larger than used in past experiments, and the cryogenically pure environment should permit longer UCN storage times. The collaboration proposes to use a multichamber spectrometer for compensation of field fluctuations by means of SQUID (superconducting quantum interference device) magnetometers.

The nEDM collaboration proposes to search for the neutron EDM with a double chamber storage cell to suppress magnetic field fluctuations, thus allowing one to extract d_n from the simultaneous measurement in chambers with opposite electric field values. The magnetic field would be inferred from a set of laser optically pumped Cs magnetometers placed outside the storage cells. The UCN would be produced in the solid deuterium UCN source under construction at PSI.

A LANSCE-based EDM experiment under development (83) also proposes to increase the UCN density using downscattering of UCN in superfluid ^{4}He and exploit the large electric fields achievable in helium. The experiment will use polarized ^{3}He atoms as the comagnetometer in a bath of superfluid helium at a temperature of approximately 300 mK. The strong spin dependence of the ^{3}He neutron absorption cross section allows the relative orientation of the neutron and ^{3}He spins to be continuously monitored through the intensity of the scintillation light in the helium. One obtains d_n by measuring the difference in the neutron and ^{3}He precession frequencies for the different orientation of electric field.

There are also preparations underway to search for the neutron EDM using dynamical diffraction from noncentrosymmetric perfect crystals. In dynamical diffraction the incident neutron plane wave state $|k\rangle$ is split as it enters the crystal into two coherent branches $|k_+\rangle$ and $|k_-\rangle$ with slightly different momenta and energies. The probability density of these two states is concentrated along and in between the lattice planes, respectively. In noncentrosymmetric crystals the position of the electric field maxima can be displaced with respect to the nuclei. Therefore, one of the branches can experience interplanar electric field (10^9 V/cm), which are orders of magnitude larger than can be achieved through application of external fields (107, 108). The presence of a neutron EDM would produce an extra relative phase shift between the two interfering branches. Such experiments must contend with potentially large systematic effects such as those from neutron spin-orbit scattering from the atoms (109). A number of experiments which investigate neutron optical issues relevant for an eventual EDM measurement of this type have been performed recently.

3.3. T-Violation in Neutron Beta Decay

With its small SM values of time-reversal violating observables, neutron beta decay also provides an excellent laboratory in which to search for T violation. Leptoquark, left-right symmetric models, and exotic fermion models can all lead to violations of time-reversal symmetry at potentially measurable levels (110). One possible T-odd correlation in polarized neutron decay is $D\vec{\sigma}_n \cdot (\vec{p}_e \times \vec{p}_p)$, where $\vec{p}_p$ is the momentum of the recoil proton. The D coefficient is sensitive only to T-odd interactions with vector and axial vector currents. In a theory with such currents, the coefficients of the correlations depend on the magnitude and phase of $\lambda = |\lambda| e^{-i\phi}$.

D has T-even contributions from phase shifts caused by pure Coulomb and weak magnetism scattering. The Coulomb term vanishes in lowest order in V-A theory (30), but scalar and tensor interactions could contribute. Fierz interference coefficient measurements (111, 112) can be used to limit this possible contribution to $|D^{EM}| < (2.8 \times 10^{-5}) \frac{m_e}{p_e}$. Interference between Coulomb scattering amplitudes and the weak magnetism amplitudes produce a final state effect of order $E_e^2/p_e m_n$. This weak magnetism effect is predicted to be $|D^{WM}| = 1.1 \times 10^{-5}$ (113). Reference (110) summarizes the current constraints on D from analyses of data on other T-odd observables for the SM and extensions.

The EDM violates both T and P symmetries, whereas a D coefficient violates T but conserves P. This makes the two classes of experiments sensitive to different SM extensions. Although constraints on T-violating, P-conserving interactions can be derived from EDM measurements, these constraints may be model dependent (114), and EDM and neutron decay searches for T violation are complementary in some aspects.

3.4. D- and R-Coefficient Measurements

In the last decade, there have been two major experimental efforts, EMIT and Trine, to improve the limit on the D coefficient in neutron decay. Each requires an intense, longitudinally polarized beam of cold neutrons around which one places alternating proton and electron detectors. Coincidence data are collected in electron-proton pairs as a function of the neutron spin state to search for the triple correlation.

In the EMIT experiment, the detector consisted of four electron detectors and four proton detectors arranged octagonally around the neutron beam (115). The octagonal geometry maximized the experiment's sensitivity to D by balancing the sine dependence of the cross product $\sigma_n \cdot \vec{p}_e \times \vec{p}_\nu$ with the large angles between the proton and electron momenta that are favored by kinematics. The decay protons drifted in a field free region before being focused by a 30 kV to 37 kV potential into an array of PIN (positive-intrinsic-negative) diode detectors. With its maximum recoil energy of 750 eV, most of the protons arrived approximately 1 μs after the electrons. Detector pairs were grouped in the analysis to reduce potential systematic effects from neutron transverse polarization. The result from the first run of EMIT yielded an improved limit of $D = [-0.6 \pm 1.2(\text{stat}) \pm 0.5(\text{sys})] \times 10^{-3}$ (115).

Currently, the best constraint on D comes from the Trine collaboration, which reports $D = [-2.8 \pm 6.4(\text{stat}) \pm 3.0(\text{sys})] \times 10^{-4}$ (116). They used two proton detectors and two electron detectors in a rectangular geometry. The proton detectors were comprised of arrays of thin-window, low-noise PIN diodes. The detectors were held at ground while the neutron beam was set to a potential of 25 kV by surrounding it with a high voltage electrode; the field was shaped to focus the decay protons onto the PIN arrays. The electrons were detected by plastic scintillators in coincidence with multi-wire proportional chambers. This coincidence provides reduction in the gamma-ray background rates and positional information on the decay, thus minimizing some sources of systematic uncertainty.

With the current PDG limit and the Trine result, one obtains a new value for the neutron D coefficient of $(-3.9 \pm 5.9) \times 10^{-4}$, which constrains the phase of g_A/g_V to $180.05° \pm 0.08°$. Neither experiment produced a statistically limited result, and both collaborations upgraded their detectors and performed second runs (117, 118). In the near future it is reasonable to anticipate new results that will put a limit on D very near 10^{-4}. Although there have been discussions and ideas for experiments using UCNs, there are currently no concrete proposals to further improve the limit on D.

Another T-odd correlation that may be present in neutron decay is the R correlation, $R\vec{\sigma} \cdot (\vec{\sigma}_e \times \vec{p}_e)$, where σ is the neutron spin and $\vec{\sigma}_e$ is the spin of the decay electron. A nonzero R requires the presence of scalar or tensor couplings and is sensitive to different SM extensions than D. An effort is underway at PSI to measure R in neutron decay by measuring the neutron polarization and the momentum and transverse polarization of the decay electron at the level of 5×10^{-3} (119).

Neutron decay is a mixed Fermi and Gamov-Teller decay, so a measurement of R would produce a limit on both scalar and tensor T-odd couplings. The limit on R achieved in ^{8}Li Gamov-teller decay of $R = (0.9 \pm 2.2) \times 10^{-3}$ now sets the most stringent limits for time-reversal violating tensor couplings in semileptonic weak decays, $-0.022 < Im(C_T + C'_T)/C_A < 0.017$ (120).

3.5. *T*-Violation in Neutron Optics

T violation can lead to terms in the forward scattering amplitude for polarized neutrons in polarized or aligned targets of the form $\vec{s} \cdot (\vec{k} \times \vec{I})$ and the fivefold correlation $\vec{s} \cdot (\vec{k} \times \vec{I})(\vec{k} \cdot \vec{I})$ (121), where $\vec{s}$ is the neutron spin and $\vec{I}$ is the nuclear polarization. Because the enhancement mechanisms for parity violation in compound resonances of heavy nuclei are also applicable to T-odd interactions (122) (see Section 4), it is possible for such searches to reach interesting levels of sensitivity. Although in principle T-odd observables in forward scattering are motion-reversal invariant and therefore not subject to final state effects, in practice the large spin dependence of the neutron-nucleus strong interaction in a polarized target can induce large potential sources of systematic errors which require careful study.

These systematic effects are smaller in aligned targets, and a search for the fivefold correlation in MeV polarized neutron transmission in an aligned holmium target has set the best direct limit on such interactions (123). This P-even, T-odd correlation is especially interesting, because there exist no renormalizable gauge theories with P-even T-odd tree-level gauge boson couplings between quarks (124). Although EDM limits can also be used to constrain P-even T-odd interactions in many models, in general only direct measurements can set model-independent bounds (125). Searches for the threefold P-odd T-odd correlation require a polarized target. Nuclei have been identified (^{139}La, ^{131}Xe) that are polarizable in macroscopic quantities and possess large parity-odd asymmetries at low energy p-wave resonances (126–128). The first steps toward such an experiment are in progress at KEK (129). A JINR-ITEP (Institute for Theoretical and Experimental Physics) collaboration is also preparing to perform a search for the P-even T-odd fivefold correlation with low energy neutrons on p-wave resonances using microwave-induced dynamical nuclear alignment to order the nuclei (130, 131).

3.6. Neutron-Antineutron Oscillations

An observation of neutron-antineutron oscillations would constitute a discovery of fundamental importance (132). The existence of such an effect requires a change of baryon number by 2 units and no change in lepton number and therefore must be mediated by an interaction outside the SM of particle physics. Among neutral mesons and leptons with distinct particle and antiparticle species and sufficiently long lifetimes (neutrinos, kaons, B mesons), oscillations are no longer a surprising phenomenon. The observation of oscillations in these systems has yielded information on aspects of physics (lepton number violation, T violation, neutrino mass) that are not accessible using less sensitive techniques. It is reasonable to hope that a search for oscillations in the neutron, the only neutral baryon which is sufficiently long-lived to conduct a practical experiment, may uncover new processes in nature.

In the SM there are no renormalizable interactions one can write down which violate B, and any nonrenormalizable operator that can induce B violation must be suppressed by some heavy mass scale. The effective operator for neutron-antineutron oscillations involves a dimension 9 operator to change the 3 quarks in the neutron into 3 antiquarks and is suppressed by some mass scale to the 5th power. Some SM extensions lead to B violation by 2 units and not 1 unit. Examples include left-right symmetric models (133) with a local B-L symmetry needed to generate small Majorana neutrino masses by the seesaw mechanism, SUSY models with spontaneously broken $B - L$ symmetry (134), and theories with compactified extra dimensions which attempt to solve hierarchy problems by introducing a much lower scale (TeV) for the onset of quantum gravity (135). In these cases proton decay is unobservably small but neutron-antineutron oscillations can occur close to the present limit. A general analysis of all operators with scalar bilinears that couple to two SM fermion fields uncovers operators that can only lead to neutron-antineutron oscillations and not to proton decay (136).

The last experiment in the free neutron system at the ILL set an upper limit of 8.6×10^7 s (90% confidence level) on the oscillation time (137). Translated into a mass scale, this limit excludes mass scales for the effective operator that induces oscillations below ~ 100 TeV. A similar indirect limit is set by the absence of evidence for spontaneous neutron-antineutron oscillations in nuclei in large underground detectors built for proton decay and neutrino oscillation studies (138). Although there has been some discussion of possible strategies to improve on the bounds from direct searches using cold and ultracold neutrons (139, 140), there are no new free neutron-antineutron oscillation searches underway.

4. NEUTRON-NUCLEON WEAK INTERACTIONS

The most obvious consequence of the weak interaction for neutrons is that it makes neutrons unstable. In addition to the coupling of quarks to leptons that allows neutrons to decay, electroweak theory also predicts (and experiments confirm) that there are weak interactions between the quarks in the neutron with couplings comparable in size to those involved in neutron decay. The weak nucleon-nucleon (NN) interaction is a unique probe of strongly interacting systems. This section presents an overview of the importance of NN interactions for QCD and the status of the experimental efforts. Reviews of aspects of the field can be found in References (141–143).

4.1. Overview

The dynamics of the quarks in the nucleon are dominated by momentum transfers that are less than that set by the QCD scale of 1 GeV/c. In this regime QCD becomes so strong that quarks are permanently confined, and therefore the quark-quark weak interactions appear through the NN weak interactions that they induce. At these energies quark-quark weak amplitudes are of order 10^{-7} of strong amplitudes primarily because of the short range of the quark weak interactions through W and Z exchange.

Assuming that it is correctly described by the electroweak theory at low energy scales, the quark-quark weak interaction can be viewed as an internal probe of strongly interacting systems. Collider measurements have verified the SM predictions for quark-quark weak couplings for large momentum transfers at the 10% level (144). The interaction is too weak to significantly affect the strong dynamics or to excite the system, and therefore it probes quark-quark correlations in the QCD ground state. The effects of the quark-quark weak interaction can be isolated from the strong interaction using parity violation. The short range of the quark-quark weak interaction and its ability to violate parity make it visible and sensitive to interesting aspects of strongly interacting systems, as seen in four cases.

1. The ground state of the strongly interacting limit of QCD is a problem of fundamental importance. Although the dynamics that lead to the spontaneous breakdown of chiral symmetry in QCD are not yet understood, one of the leading models assumes the importance of fluctuating nonperturbative gluon field configurations called instantons (145). They induce four-quark vertices that flip the quark helicity and localize the quark wave function through a mechanism similar to Anderson localization of electrons in disordered metals (146). Some aspects of QCD spectroscopy and the high density limit can be understood by assuming that quark-diquark configurations in the nucleon are important (147). The mechanism for the phenomenon of color superconductivity in the high density limit of QCD consists of a BCS-like condensation of diquarks (148). The quark-quark weak interaction in the nucleon in the low energy limit induces four-quark operators with a known spin and flavor dependence whose relative sizes are in principle sensitive to these and other correlation phenomena in the ground state of QCD.

2. With experimental information on the low energy parity-violating (PV) partial waves in the NN system, there is a chance to understand quantitatively for the first time the extensive observations performed in many systems of

PV phenomena in nuclei (149). Nuclear parity violation is linearly sensitive to small components of the nuclear wavefunction because successive shell model levels alternate in parity, and parity-odd operators directly connect adjacent shells (150). Ideas from quantum chaos (151) and nuclear statistical spectroscopy have been used to analyze parity violation in neutron reactions in heavy nuclei in terms of the effective isovector and isoscalar weak NN interaction, and knowledge of PV in the NN system would allow a quantitative test of the predictive power of these ideas (152, 153). The matrix elements for weak NN interactions in nuclei also bear many similarities to the types of matrix elements that must be calculated to interpret limits on neutrino masses from double beta decay searches (154).

3. In atoms, the effect of NN parity violation was seen for the first time in ^{133}Cs (155) through its contribution to the anapole moment of the nucleus, which is an axial vector coupling of the photon to the nucleus induced mainly by the PV NN interaction (156, 157). Anapole moment measurements in other atoms are possible, and experiments are in progress (158). In heavy nuclei for which the anapole moment is a well-defined observable, the main contribution comes from PV admixtures in the nuclear ground state wave function (159). In electron scattering from nucleons, PV effects are sensitive to both Z exchange between the electron and the quarks in the nucleon as well as the coupling of the virtual photon to the axial current from PV interactions among the quarks in the nucleon. As PV effects in electromagnetic processes are used to learn about the nucleon (160), it will be important to know enough about the weak NN interaction to extract the information of interest.

4. The NN weak interaction is also the only practical way to study quark-quark neutral currents at low energy. The neutral weak current conserves quark flavor to high accuracy in the standard electroweak model (due to the GIM mechanism) and is not seen at all in the well-studied strangeness-changing nonleptonic weak decays. We know nothing experimentally about how QCD modifies weak neutral currents.

There are theoretical difficulties in trying to relate the underlying electroweak currents to low-energy observables in the strongly interacting regime of QCD. One expects the strong repulsion in the NN interaction to keep the nucleons sufficiently separated for a direct exchange of W and Z bosons between quarks to represent an accurate dynamical mechanism. If one knew weak NN couplings from experiment, they could be used to interpret parity violation effects in nuclei. The current approach is to split the problem into two parts. The first step maps QCD to an effective theory expressed in terms of the important degrees of freedom of low energy QCD, mesons and nucleons. In this process, the effects of quark-quark weak currents appear as PV meson-nucleon couplings (161). The second step uses this effective theory to calculate electroweak effects in the NN interaction in terms of weak couplings. The couplings themselves also become challenging targets for calculation from the SM.

4.2. Theoretical Description

In the work of Desplanques, Donoghue, and Holstein (DDH) (161), the authors used a valence quark model in combination with SU(6) symmetry relations and data on hyperon decays to produce a range of predictions for effective PV meson-nucleon couplings from the SM. At low energy, the weak interaction between nucleons in this approach is parameterized by the weak pion coupling constant f_π, and six other meson coupling denoted as h_ρ^0, h_ρ^1, $h_\rho^{\prime 1}$, h_ρ^2, h_ω^0, and h_ω^1, where the subscript denotes the exchanged meson and the superscript indicates the isospin change. Due to uncertainties in the effects of strong QCD, the range of predictions for the size of these weak couplings is rather broad. For the weak pion coupling, neutral currents should play a dominant role. Another strategy is to perform a systematic analysis of the weak NN interaction using an effective field theory (EFT) approach to classify the interaction in a manner that is consistent with the symmetries of QCD and does not assume any specific dynamical mechanism. Such an EFT approach has recently appeared (162). Preparations have also been made for an eventual calculation of the weak NN interaction vertices using lattice gauge theory in the partially quenched approximation (163).

At the low energies accessible with cold neutrons with $k_n R_{strong} \ll 1$, parity-odd effects in the two-nucleon system can be parameterized in terms of the five independent amplitudes for $S - P$ transitions involving the following nucleons and isospin exchanges: $^1S_0 \rightarrow {}^3P_0$(p-p, p-n, n-n, $\Delta I = 0, 1, 2$), $^3S_1 \rightarrow {}^1P_1$(n-p, $\Delta I = 0$), and $^3S_1 \rightarrow {}^3P_1$(n-p, $\Delta I = 1$). Thus, from the point of view of a phenomenological description of the weak NN interaction, at least five independent experiments are required. The PV longitudinal analyzing power in p-p scattering, which determines a linear combination of the $^1S_0 \rightarrow {}^3P_0$ amplitudes, has been measured at 15 MeV and 45 MeV in several experiments with consistent results (164–167) and remains the only nonzero observation of parity violation in the pure NN system dominated by p-waves.

Parity violation in few-nucleon systems should be cleanly interpretable in terms of the NN weak interaction due to recent theoretical and computational advances for the strong interaction in few nucleon systems (168). Weak effects can be included as a perturbation. These calculations have recently been done for n-p and p-p parity violation (169–171) and can be done in principle for all light nuclei. It is also possible that these microscopic calculations can be applied to systems with somewhat larger A, such as ^{10}B and ^{6}Li, where measurements of P-odd observables with low energy neutrons have reached interesting levels of sensitivity (172, 173).

The longest-range part of the interaction is dominated by the weak pion-nucleon coupling constant f_π. f_π has been calculated using QCD sum rules (174) and in a SU(3) Skyrme model (175). Measurements of the circular polarization of photons in the decay of ^{18}F (176, 177) provide a value for f_π that is considerably smaller than the DDH best value though still within the reasonable range. A precision atomic physics measurement of the ^{133}Cs hyperfine structure (anapole moment) has been analyzed to give a constraint on f_π and the combination $(h_\rho^0 + 0.6 h_\omega^0)$. This result would seem to favor a value for f_π that is larger than the ^{18}F result. Figure 4 (see color insert) presents an exclusion plot that summarizes the current situation.

4.3. Parity-Odd Neutron Spin Rotation and Capture Gamma Asymmetries

There are a few general statements that apply to the low energy weak interactions of neutrons with low A nuclei. In the absence of resonances, the PV helicity dependence of the total cross section vanishes if only elastic scattering is present, and both the PV neutron spin rotation and the PV helicity dependence of the total cross section with inelastic channels are constant in the limit of zero neutron energy (121). These results depend only on the requirement for parity violation in an $S \rightarrow P$ transition amplitude involving two-body channels. The two practical classes of neutron experiments are PV neutron spin rotation and PV gamma asymmetries.

The NPDGamma experiment will measure the parity-violating directional gamma ray asymmetry A_γ in the capture of polarized neutrons on protons (179, 180). The unique feature of this observable is that it is sensitive to the weak pion coupling f_π, $A_\gamma = -0.11 f_\pi$ (181–183). The recently commissioned beamline at LANSCE delivers pulsed cold neutrons to the apparatus, where they are polarized by transmission through a large-volume polarized ^{3}He spin filter and are transported to a liquid parahydrogen target. A resonant RF spin flipper reverses the direction of the neutron spin on successive beam pulses using a sequence that minimizes susceptibility to some systematics. The 2.2 MeV gamma rays from the capture reaction are detected in an array of CsI(Tl) scintillators read out by vacuum photodiodes operated in current mode and coupled via low-noise I–V preamplifiers to transient digitizers. The current-mode CsI array possesses an intrinsic noise two orders of magnitude smaller than the shot noise from the gamma signal and has been shown in offline tests to possess no false instrumental asymmetries at the 5×10^{-9} level (184). The pulsed beam enables the neutron energy to be determined by time-of-flight, which is an important advantage for diagnosing and reducing many types of systematic uncertainty. This apparatus has been used to conduct measurements of parity violation in several medium and heavy nuclei (185).

Another experiment in preparation is a search for parity violation in neutron spin rotation in liquid ^{4}He (186). A transverse rotation of the neutron spin vector about its momentum manifestly violates parity (187) and can be viewed from a neutron optical point of view as due to a helicity-dependent neutron index of refraction. For ^{4}He, the calculated PV neutron spin rotation in terms of weak couplings is (188)

$$\phi = \left(0.97 f_\pi + 0.32 h_\rho^0 - 0.11 h_\rho^1 + 0.22 h_\omega^0 - 0.22 h_\omega^1\right) \times 10^{-6} \, \text{rad/m}. \qquad 4.$$

To measure the small parity-odd rotation, a neutron polarizer-analyzer pair with axes at right angles transmitted only the component of a polarized beam that rotated as it traversed the target. The challenge was to distinguish small PV rotations from rotations that arise from residual magnetic fields. The first measurement achieved a sensitivity of 14×10^{-7} rad/m at NIST (189), and no systematic effects were seen at the 2×10^{-7} rad/m level.

Four plausible experiments employ beams of cold neutrons and involve targets with $A < 5$: measurement of the PV gamma asymmetries in p($\vec{\text{n}}$, γ)d and in d($\vec{\text{n}}$, γ)t and of the PNC neutron spin rotations in ^{4}He and H. Successful measurements in all of these systems, in combination with existing measurements, would have a strong impact on the knowledge of the NN weak interaction (190).

4.4. Test of Statistical Theories for Heavy Nuclei Matrix Elements

One might assume that a quantitative treatment of NN parity violation in neutron reactions with heavy nuclei would not be feasible. A low energy compound nuclear resonance expressed in terms of a Fock space basis in a shell model might possess a million components with essentially unknown coefficients, and the calculation of a parity-odd effect would involve a matrix element between such a state and another equally complicated state. However, one can imagine a theoretical approach which exploits the large number of essentially unknown coefficients in such complicated states. If we assume that it is possible to treat these components as random variables, one can devise statistical techniques to calculate, not the value of a particular parity-odd observable, but the width of the distribution of expected values. A similar strategy has been used to understand properties of complicated compound nuclear states. The distribution of energy spacings and neutron resonance widths obeys a Porter-Thomas distribution (191) in agreement with the predictions of random matrix theory, and statistical approaches have been used to understand isospin violation in heavy nuclei (192). Statistical analyses have been applied to an extensive series of measurements of parity violation in heavy nuclei performed mainly at JINR, KEK, and LANSCE (193, 194).

The TRIPLE collaboration at LANSCE measured 75 statistically significant PV asymmetries in several compound nuclear resonances in heavy nuclei. In the case of parity violation in compound resonances in neutron-nucleus reactions there are amplification mechanisms which can enhance parity-odd observables by factors as large as 10^5. These amplification mechanisms, which are interesting phenomena in themselves, depend in an essential way on the complexity of the states involved. Part of the amplification comes from the decrease in the spacing between levels as the number of nucleons increases, which brings opposite parity states closer together and increases their weak mixing amplitudes (195, 196); for low energy neutron reactions in heavy nuclei, it leads to a generic amplification of order 10^2 in parity-odd amplitudes. In addition, for low energy neutron-nucleus interactions the resonances are mainly $l = 0$ and $l = 1$, with the scattering amplitudes in s-wave resonances larger than for p-wave resonances by a factor of order 10^2 to 10^3. At an energy close to a p-wave resonance, the weak interaction mixes in an s-wave component that is typically much larger, and this factor amplifies the asymmetry. These amplification mechanisms were predicted theoretically (197) before they were measured (198, 199).

A basic tenet of the statistical approach is that there should be, on average, equal numbers of negative and positive PV asymmetries in a given nucleus. This condition appears to be satisfied provided that one removes the ^{232}Th data, wherein all ten measured parity-odd asymmetries have the same sign. At present, this result is ascribed to some poorly understood nuclear structure effect. Its observation illustrates the potential for the use of NN parity violation to discover unsuspected coherent effects in complicated many-body systems.

The values of the weak matrix elements determined by the statistical analysis varied in the range 0.5 meV to 3.0 meV, in rough agreement with theory. The accuracy of this analysis was improved through new measurements of the required spectroscopic information on compound nuclear levels (200). Theoretical calculations that use this data to extract the weak isoscalar and isovector couplings in ^{238}U obtain results in qualitative agreement with DDH expectations (153). If the weak NN couplings were known, we would be able to see if there is any evidence for nuclear medium effects.

4.4.1. PARITY-ODD AND TIME REVERSAL-ODD CORRELATIONS IN NEUTRON-INDUCED TERNARY FISSION   An example where symmetry violation in neutron-induced reactions has led to progress in the understanding of many-body nuclear dynamics is fission. *P*-odd effects in binary fission induced by polarized neutron capture have been observed for a long time (201, 202). Although one might expect that a treatment of parity violation in nuclear fission would be even more difficult than for compound nuclear resonances in heavy nuclei, there is a compelling understanding of parity-odd asymmetries observed in fission after capture by polarized neutrons (201–203) based on interference of amplitudes of opposite parity from parity doublets in the cold pear-shaped transition states of open channels. Since this interference occurs among the small number of fission channels in the initial state near the saddle point, it can survive the inevitable averaging over the enormous number of final states later produced in the rupture.

In the case of ternary fission, wherein a third light-charged particle (usually an alpha) is emitted in addition to the two main fragments, recent parity violation measurements have given support to a specific mechanism for the emission of the ternary particle (204). Consider two generic mechanisms for the emission of the ternary particle: the simultaneous emission of the three particles (three-body compound nucleus decay) and "double neck rupture" in which the ternary particle is emitted after the first rupture of the neck from its remnants. In the first case, because all three objects originate from the same system where the dominant parity violation comes from the mixing of opposite parity compound nuclear states, one expects all of the PV asymmetries in various channels to be about the same size. In the second case, however, the mechanism for the emission of the ternary particle does not possess the same intermediate states that are known to exist in binary fission, and upon averaging over the large number of fragment states one would expect the parity-odd correlations that involve the ternary particle to be much smaller. This is what was observed experimentally in ^{233}U (205–209). Furthermore, the parity-odd asymmetries of the two large fragments were seen to be independent of the energy of the ternary particle. Since different ternary particle energies are presumably coming from different Coulomb repulsion effects from different shapes of the neck, this independence would also seem to indicate that the parity-odd asymmetry is established before the scission process.

In ternary fission one can also look for a T-odd triple correlation between the momenta of the light fragment and ternary particle and the neutron spin. This measurement has recently been done (210, 211) and a large nonzero effect of order 10^{-3} was seen in both ^{233}U and ^{235}U for both alphas and tritons as the ternary products. It is believed that in the ternary fission system, this correlation is due to a final state effect and not to a fundamental source of T violation. The fact that the size of the observed triple correlation depends on the ternary particle energy also suggests that a final state effect is responsible. One model (212, 213) can reproduce the order of magnitude of the effect if one assumes that the projection of the orbital angular momentum of the recoiling ternary particle changes the spin projections and therefore the level densities of the larger fragments. If the emission probabilities of the ternary particle are proportional to these level densities and the angular momentum of the initial system is correlated with the neutron polarization, this mechanism can generate a nonzero triple correlation. Semiclassically this can be viewed in terms of the Coriolis interaction of the ternary particle with the rotating compound nucleus (214). Future work will attempt to confirm this mechanism in plutonium.

5. LOW ENERGY QCD TESTS

One of the long-term goals of strong interaction physics is to see how the properties of nucleons and nuclei follow from QCD. For nuclei the first step in this process is to see if one can start from QCD and calculate the well-measured NN strong interaction scattering amplitudes and the properties of the deuteron. During the last decade a number of theoretical developments have started to show the outlines of how this connection between QCD and nuclear physics can be made. In this section, we discuss some of the theoretical developments in few nucleon systems along with several precision scattering length experiments. The status of two fundamental properties of the neutron, its polarizability and the neutron-electron scattering length, are also discussed.

5.1. Theoretical Developments in Few Nucleon Systems and the Connection to QCD

Based on a suggestion by Weinberg (215), one strategy to develop an effective field theory (216) for QCD that is valid in the low energy limit relevant for nuclei and incorporates the most important low energy symmetry of QCD is through chiral symmetry. This alone is not enough because some of the important energy scales of nuclear physics, such as the deuteron binding energy and its correspondingly large low energy scattering lengths, seem to be the result of a delicate cancellation between competing effects which will need more than chiral symmetry alone to understand. Recent efforts to understand the emergence of smaller energy scales in nuclear physics not set by chiral dynamics have led to interesting suggestions that the low energy limit of QCD is not described by the usual renormalization group fixed point but rather is close to a limit cycle which can be reached by a fine tuning of the values of the current quark masses in the Lagrangian (217, 218).

Recently, significant insight into certain features of few nucleon systems has come from the EFT approach based on the chiral symmetry of QCD (219–221). The value of the EFT approach is that it is a well-defined field theoretical procedure for the systematic construction of a low energy Lagrangian consistent with the symmetries of QCD. To a given level of accuracy the Lagrangian contains all possible terms accompanied by symmetries with a number of arbitrary coefficients which, once fixed by experiment, can be used to calculate other observables. EFTs based on the chiral symmetry of QCD have been used to develop an understanding for the relative sizes of many quantities in nuclear physics, such as that of nuclear N-body forces (222) and in particular the nuclear three-body force (3N), which is the subject of much activity. Although it is well understood that 3N forces must exist with a weaker strength and shorter range than the NN force, little else about them is known.

EFT has been used to solve the two- and three-nucleon problems with short-range interactions in a systematic expansion of the small momentum region set by $kb \leq 1$, where k is the momentum transfer and b is the bound scattering length (163, 221). For the two-body system, EFT is equivalent to effective range theory and reproduces its well-known results for NN forces (223–225). The chiral EFT expansion does not require the introduction of an operator corresponding to a 3N force until next-to-next-to leading order in the expansion, and at this order it requires only two low energy constants (226, 227). Significant advances have been made in other approaches to the computation of the properties of few-body nuclei with modern potentials (228) such as the AV18 potential (168, 229), which includes electromagnetic terms and terms to account for charge-independence breaking and charge symmetry breaking.

All of these developments show that precision measurements of low energy strong interaction properties, such as the zero energy scattering lengths and electromagnetic properties of small A nuclei, are becoming more important for strong interaction physics both as precise data that can be used to fix parameters in the EFT expansion and also as new targets for theoretical prediction. It is possible now to envision the accurate calculation of low energy neutron scattering lengths for systems with $A > 2$.

5.2. Precision Scattering Length Measurements Using Interferometric Methods

In parallel with these theoretical developments, two interferometric methods have been perfected to allow high-precision measurements of n-A scattering lengths. One is neutron interferometry using diffraction from perfect silicon crystals (5), which measures the coherent scattering length. The other is pseudomagnetic precession of polarized neutrons in a polarized nuclear target, which measures the incoherent scattering length. Together these two measurements can be used to determine the scattering lengths in both channels.

Neutron interferometry can be used to measure the phase shift caused by neutron propagation in the optical potential of a medium. For a target placed in one arm of an interferometer, the phase shift is given by the expression $\phi = bND\lambda$, where N is the target density, D is the thickness of the sample, λ is the neutron de Broglie wavelength, and b is the bound coherent scattering length. High absolute accuracy in the determination of N, D, and λ are required but possible at the 10^{-4} level. Recent results from measurements at the NIST Neutron Interferometry and Optics Facility yielded the coherent scattering lengths $b_{np} = (-3.738 \pm 0.002)$ fm, $b_{nd} = (6.665 \pm 0.004)$ fm (230, 231), and $b_{n^3He} = (5.857 \pm 0.007)$ fm (232). These experiments showed that almost all existing theoretical calculations of the n-d and n-^{3}He coherent scattering lengths are in disagreement with experiment and that the accuracy of present measurements is sensitive to such effects as nuclear three-body forces and charge symmetry-breaking (233, 234).

If the neutron-nucleus interaction is spin dependent, a polarized neutron moving through a polarized medium possesses a contribution to the forward scattering amplitude proportional to $(b_+ - b_-)\vec{\sigma}_n \cdot \vec{I}$ where b_+ and b_- are the scattering lengths in the two channels, $\vec{\sigma}_n$ is the neutron spin, and $\vec{I}$ is the nuclear polarization. The angle of the polarization of a neutron polarized normal to the target polarization precesses as it moves through the medium. This phenomenon is referred to as nuclear pseudomagnetic precession (235) and has used to measure scattering length differences in many nuclei. Recently, a high-precision measurement of this precession angle was performed in polarized ^{3}He using a neutron spin-echo spectrometer at the ILL (236). A new experiment to determine the spin-dependence of the n-d scattering length is in preparation at PSI (237).

In combination with the n-D coherent scattering length determined by neutron interferometry, this experiment should determine both n-D scattering lengths to 10^{-3} accuracy. The $^2S_{1/2}$ scattering length in the n-d system is especially interesting. The quartet s-wave scattering length ($^4S_{3/2}$) can be unambiguously determined from current theory. Because the three nucleons in this channel exist in a spin-symmetric state, and hence have an antisymmetric space-isospin wavefunction, the scattering in this state is completely determined by the long range part of the triplet s-wave NN interaction in the n-p channel, i.e., by n-p scattering and the properties of the deuteron. By contrast the Pauli principle does not deter the doublet channel from exploring the shorter-range components of the NN interaction, where 3N forces should appear.

Perhaps the single most interesting scattering length to measure is the neutron-neutron scattering length a_{nn}. No direct measurements exist. An experiment to determine a_{nn} by viewing a high-density neutron gas near the core of a reactor and measuring a quadratic dependence of the neutron fluence on source power is currently being designed (239). An experiment to let the neutrons in an extracted beam scatter from each other has been considered (238). The motivation for such a measurement might increase if low energy effective field theories of QCD are able to predict a_{nn} accurately. An EFT analysis to extract a_{nn} from the $\pi^- d \rightarrow nn\gamma$ reaction has recently appeared (240).

5.3. Neutron-Electron Interaction

Although the neutron has a net zero electric charge, it is composed of charged quarks which possess a nontrivial radial charge distribution. This distribution produces a nonzero value of the neutron mean-square charge radius $\langle r_n^2 \rangle$. To the first order in the electromagnetic coupling α, the neutron-electron scattering length

$$a_{ne} = \frac{2\alpha m_n c}{\hbar} \frac{dG_{eN}}{dq^2} \qquad 5.$$

is proportional to the slope of the electric form factor of the neutron, G_{eN}, in the $q^2 \rightarrow 0$ limit, where q is the momentum transfer. For the proton (neutron), this limiting value for the electric form factor is normalized to one (zero). This leads to $G_{eN}(-q^2) \rightarrow \frac{1}{6}\langle r_n^2 \rangle q^2$, where $-q^2 = \vec{q}^2$ is the four-momentum transfer. Although defined for arbitrary q^2, in the Breit frame G_{eN} has an interpretation as the spatial Fourier transform of the charge distribution of the neutron (241).

The sign of this slope, or equivalently the sign of the charge radius, has physical significance. For the neutron one expects a negative charge radius from its virtual pion cloud (242). From the QCD point of view, the neutron charge radius is especially interesting because it is more sensitive to sea quark contributions than the proton charge radius, which has a large valence quark contribution. With the advent of improved lattice gauge theory calculations of nucleon properties (243) and the ability to use chiral extrapolation procedures to ensure proper treatment of the nonanalytic chiral corrections (244), it is possible that the neutron-electron scattering length may be calculable in the near future directly from QCD.

There are two clusters of values in a_{ne} measurements. One set comes from measurements of the asymmetric angular distribution of neutron scattering in noble gases, $a_{ne} = (-1.33 \pm 0.03) \times 10^{-3}$ fm (245), and the total cross section of lead, $a_{ne} = (-1.33 \pm 0.03) \times 10^{-3}$ fm (246). The other set comes from measurements of the total cross section of bismuth, $a_{ne} = (-1.55 \pm 0.11) \times 10^{-3}$ fm (247), and neutron diffraction from a tungsten single crystal, $a_{ne} = (-1.60 \pm 0.05) \times 10^{-3}$ fm (248). Two new experiments are in preparation; one exploits dynamical diffraction in a perfect silicon crystal (F.E. Wietfeldt, personal communication), and a second attempts to improve on the technique of scattering in noble gases (249). An experiment using Bragg reflections in perfect silicon crystals to determine a_{ne} has also been proposed (250).

The precision of the charge radii of the proton and the deuteron has greatly improved during the last decade from theoretical and experimental advances in electron scattering and atomic physics. The charge radius of the proton is well-determined from both electron scattering data, $\sqrt{r_p^2} = (0.895 \pm 0.018)$ fm (251), and from high precision atomic spectroscopy in hydrogen, $\sqrt{r_p^2} = (0.890 \pm 0.014)$ fm (252). The charge radius of the deuteron is also well-determined from electron scattering data $\sqrt{r_d^2} = (2.128 \pm 0.011)$ fm (253), a value consistent with theoretical calculations of deuteron structure from the deuteron wave function and the triplet n-p scattering length ($\sqrt{r_d^2} = 2.131$ fm) (254) and from high-precision atomic spectroscopy measurements of the 2P-2S transition in deuterium ($\sqrt{r_p^2} = (2.133 \pm 0.007)$ fm) (255). Atomic physics measurements of the H-D isotope shift of the 1S-2S two-photon resonance were used to derive an accurate value for the difference between the mean-square charge radii of the deuteron and proton of $r_d^2 - r_p^2 = (3.8212 \pm 0.0015)$ fm^2 (256). Because the neutron charge radius is simply related to the proton and deuteron charge radii, it is very timely for a theoretical analysis that uses these precise values as input and predicts the neutron mean-square charge radius in an EFT analysis.

5.4. Neutron Polarizability

The electric and magnetic polarizabilities of the neutron are fundamental properties which characterize how easily the neutron deforms under external electromagnetic fields. The quarks in the neutron are confined by the strong interaction with a tension equivalent to about one ton over their distance of separation of one fermi, so the polarizabilities are very small. To measure neutron polarizability, one may exploit the electric fields accessible on the surface of heavy nuclei which polarize the neutron to give a calculable contribution to the neutron-nucleus scattering length with a linear dependence on the neutron momentum k. A measurement using ^{208}Pb observed a term whose size and neutron energy dependence was consistent with a nonzero polarizability of $\alpha_n = [12.0 \pm 1.5(\text{stat}) \pm 2.0(\text{sys})] \times 10^{-4}$ fm^3 (257). Subsequent analyses assert that the data analysis is not definitive (258–260). Another approach using deuteron Compton scattering gave $\alpha_n = [8.8 \pm 2.4(\text{stat} + \text{sys}) \pm 3.0(\text{theo})] \times 10^{-4}$ fm^3 (261) whereas quasi-free Compton scattering from the deuteron gave $\alpha_n = [12.5 \pm 1.8(\text{stat})^{+1.1}_{-0.6}(\text{sys}) \pm 1.1(\text{theo})] \times 10^{-4}$ fm^3 (262). The theoretical uncertainties come from different treatments of strong interaction effects.

QCD effective field theory is developing into a quantitative theory for the calculation of low energy nucleon properties. The lowest-order prediction of chiral perturbation theory for the neutron polarizability is $\alpha_n = 12.2 \times 10^{-4}$ fm^3 (263), and analysis of the extensive new Compton scattering data on the proton and deuteron using an EFT analysis is in progress (264). The result should be a sharp prediction for the neutron electric polarizability from QCD. Lattice gauge theory calculations of the polarizability are improving (265). This work should motivate further efforts to improve the neutron polarizability measurement.

6. NEUTRONS IN ASTROPHYSICS AND GRAVITY

This section discusses some of the ways in which neutrons and nuclear reactions involving neutrons play vital roles in several astrophysical processes. Neutrons play a decisive role in determining the element distribution in the universe. The decay rate of the neutron determines the amount of primordial ^{4}He in Big Bang theory, and neutron reactions in stars form most heavy nuclei beyond iron. In addition, one can use the fact that neutrons in the gravitational field of the Earth are sensitive to potential differences comparable to those from the strong and electromagnetic interactions as an opportunity to search for gravitational effects on an elementary particle.

6.1. Big Bang Nucleosynthesis

Neutron decay influences the dynamics of Big Bang Nucleosynthesis (BBN) through both the size of the weak interaction couplings g_A and g_V and the lifetime. The couplings determine when weak interaction rates fall sufficiently below the Hubble expansion rate to cause neutrons and protons to fall out of chemical equilibrium, which occurs on the scale of a few seconds, and thus the n/p ratio decreases as the neutrons decay. The lifetime determines the fraction of neutrons available as the universe cools, most of which end up in ^{4}He (266), and occurs on the scale of a few minutes. The neutron lifetime remains the most uncertain nuclear parameter in cosmological models that predict the cosmic ^{4}He abundance (267, 268). With the recent high-precision determination of the cosmic baryon density reported by the Wilkinson Microwave Anisotropy Probe (WMAP) measurement of the microwave background (269), there is a growing tension between the BBN prediction for the ^{4}He abundance, which is quite sensitive to the neutrino sector of the SM, and that inferred from observation (270). BBN calculations predicted that the number of light neutrinos that couple to the Z was about 3 before the Large Electron-Positron (LEP) storage ring measurements ended all doubt. The quantitative success of BBN is now routinely used to constrain various aspects of physics beyond the SM. The small size of the baryon density relative to the density required to close the universe is one of the observational cornerstones of the dark matter problem in astrophysics.

The main concern in deviations between BBN theory and experiment remains the astronomical determinations of the ^{4}He abundance, whose systematic errors are perhaps not yet fully understood. As the neutron lifetime measurements improve, other neutron-induced reactions in the early universe, such as the $n + p \rightarrow d + \gamma$ cross section, will become more important to measure precisely. Some reactions that are difficult to measure would benefit from the application of EFT methods for calculation in the relevant energy regime, and again low energy neutron measurements will be useful to fix the EFT parameters. It is therefore likely that neutron measurements will continue to be relevant for BBN.

6.2. Stellar Astrophysics

An examination of the observed elemental abundances in the solar system, together with rudimentary nuclear physics considerations, reveals that neutron capture reactions are essential for the origin of the elements heavier than iron (271). Almost all these elements are thought to have been synthesized inside stars, supernovae, or other more exotic environments through sequences of neutron capture reactions and beta decays during the so-called slow neutron capture ("s") (272) and rapid neutron capture ("r") (273) processes. The s and r processes are each responsible for roughly half of the observed heavy element abundances. The remaining neutron-deficient isotopes that cannot be reached via neutron capture pathways are thought to have been formed in massive stars or during supernova explosions through the photodissociation ("p") process.

In some cases, further progress in these areas is hampered by the lack of accurate rates for nuclear reactions governing stellar nucleosynthesis. Many of these astrophysical reaction rates can be determined by measuring neutron-induced cross sections in the energy range between approximately 1 eV and 300 keV. For about 20 radionuclides along the s-process path, the neutron-capture and β-decay time scales are roughly equal. The competition between neutron capture and β decay occurring at these isotopes causes branches in the s-process reaction path that, if measured, could be used to directly constrain dynamical parameters of s-process models. There is very little data on the (n,γ) reaction rates for such radioactive branching points.

Measurements of cross sections in an energy range relevant to astrophysics constitute an important program for a number of neutron beamlines at facilities such as the Oak Ridge Electron Linear Accelerator (ORELA) (274), the Geel Electron Linear Accelerator (GELINA) (275), nTOF, and the Detector for Advanced Neutron Capture Experiments (DANCE) (276), along with many others. The intensities now suffice to conduct cross section measurements on small quantities of unstable isotopes. As the understanding of the origins and processes which lead to element formation improves, we can more effectively exploit astrophysical observations to constrain the understanding of what phenomena may lie beyond the SM.

6.3. Gravitationally Induced Phase Shift

The contribution of precision neutron measurements to gravitation are few in number but notable in conceptual impact. The equivalence principle for free neutrons has been verified at the 10^{-4} level (277), and although one might do better with ultracold neutrons (278), experiments with bulk matter are several orders of magnitude more precise. Another connection between neutron physics and gravity is the observation of the gravitational phase shift by neutron interferometry, which was the first verification that the principles of quantum mechanics seem to apply to the gravitational potential as well as the potentials produced by other interactions (279, 280). This measurement has been performed with an accuracy at the 1% level, and there is a slight disagreement between theory and experiment (281).

The source of the difference is believed to lie in gravitationally induced distortions in the perfect crystal interferometer as it is rotated to change the relative height of the two paths in the interferometer. There are two plans underway, one to conduct the experiment with the interferometer suspended in a neutrally buoyant fluid to eliminate possible distortions of the interferometer crystal (H. Kaiser, personal communication) and another to use a recently developed Mach-Zehnder interferometer with cold neutrons (282).

6.4. UCN Gravitational Bound States

Experimental tests of gravity and searches for new long-range forces have attracted more interest in recent decades. A reanalysis (283) of an experiment on the principle of equivalence by Eötvös motivated a series of precise tests of the principle of equivalence culminating in the torsion balance experiments of the Eöt-Wash group, which set stringent new limits to violations of the equivalence principle (284). Recently, speculations involving the propagation of the gravitational field into extra dimensions produce as a natural consequence a modification of the inverse square law for gravity on submillimeter scales (285). Several experiments are searching for such modifications (286) and have already set useful limits. Experiments to probe extra-dimensional gravity theories using the angular and neutron energy dependence of neutron scattering from spin zero nuclei have been discussed (287).

At first glance, experiments with neutrons would not seem to offer a productive technique to conduct sensitive searches for new short-range forces of gravitational strength. Although the very small polarizability of the neutron is an advantage relative to atoms, whose van der Waals attraction poses a background issue for such searches, neutron beam densities are very small compared to bulk matter. The recent measurements in search of gravitational bound states of ultracold neutrons are of some interest for constraints on new forces at smaller distance scales (288).

For neutron kinetic energies smaller than the neutron optical potential of a plane surface, the potential seen by a neutron moving above the plane in the gravitational field of the Earth should possess neutron bound states with energies given to good approximation by the zeroes of Airy functions, which are solutions to the Schrödinger equation in a linear potential. The lowest bound state, of energy 1.4 peV, hovers above the medium at a distance of order 10 μm, and any nonstandard attractive interaction of the neutron with the matter on this length scale could create another bound state.

The experiments were designed to populate these bound states by forcing the UCN to pass through a narrow gap above a planar medium and to detect their presence by measuring the transmission of the UCNs through the gap as the separation is varied. Intensity oscillations in the transmission are observed which can be fit to a model which includes the effect of the spatial extent of the bound states. The agreement of this data with theory was used to set an interesting bound on gravitational-strength forces in the nanometer range (289).

7. SUMMARY

Slow neutron experiments address fundamental scientific issues in a surprisingly large range of physics subfields. The development of new types of ultracold neutron sources, pulsed and continuous spallation sources, and the continued increase in the fluence of reactor beams are making possible new types of experiments and opening new scientific areas. We anticipate significant progress in the areas of neutron decay measurements and neutron EDM searches and also in the field of weak NN interactions and the measurement of other low energy neutron properties. We also expect more use of neutron measurements for applications in astrophysics. As is typical, many of the new opportunities made possible by technical developments and the emerging scientific issues were not clearly foreseen a decade ago. The steady progress during the last decade in quantitative theoretical understanding of the strong interaction is an underappreciated development with exciting applications to neutron physics.

Although still a relatively small field compared to other areas in nuclear and particle physics, the expanding scientific opportunities in fundamental neutron physics are attracting a growing number of young researchers. This growth is driving the increasing number and variety of new facilities. The diverse applications and the location of neutron physics at facilities whose main purpose is generally neutron scattering have combined to obscure somewhat its accomplishments. We hope that the reader has gained an appreciation for the breadth of activity in the field.

ACKNOWLEDGMENTS

We would like to thank Torsten Soldner of the ILL and Yasuhiro Masuda of KEK for their assistance in supplying some of the parameters for neutron facilities, and Scott Dewey for his careful reading of the manuscript. W.M. Snow gratefully acknowledges support from the National Science Foundation, the Department of Energy, and the Indiana 21st Century Fund. J.S. Nico acknowledges the support of the NIST Physics Laboratory and Center for Neutron Research.

**The *Annual Review of Nuclear and Particle Science* is online at
http://nucl.annualreviews.org**

LITERATURE CITED

1. Erler J, Ramsey-Musolf MJ. *Prog. Nucl. Part. Phys.* 54:351 (2005)
2. Krupchitsky PA. *Fundamental Research with Polarized Slow Neutrons*. Berlin: Springer-Verlag (1987)
3. Alexandrov Yu A. In *Fundamental Properties of the Neutron*, ed. SW Lovesey,

EWJ Mitchell. Oxford: Clarendon Press (1992)
4. Byrne J. In *Neutrons, Nuclei, and Matter*. Bristol, UK: Institute of Physics Publishing (1994)
5. Rauch H, Werner S. In *Neutron Interferometry: Lessons in Experimental*

Quantum Mechanics, ed. SW Lovesey, EWJ Mitchell. Oxford: Oxford Univ. Press (2000)

6. Windsor CG. *Pulsed Neutron Scattering.* London: Taylor & Francis (1981)
7. Soldner T, Nesvizhevsky V. http://www.ill.fr/pages/science/IGroups/yb.pdf
8. www.frm2.tum.de/mephisto/index_en.shtml
9. Nico JS, et al. *J. Res. Natl. Inst. Stand. Technol.* 110:137 (2005)
10. Seo PN, et al. *Nucl. Instrum. Meth.* A517:285 (2004)
11. Zejma J, et al. *Nucl. Inst. Meth.* A539:622 (2005)
12. Greene GL, et al. *J. Res. Natl. Inst. Stand. Technol.* 110:149 (2005)
13. Luschikov VI, Pokotolovsky YN, Strelkov AV, Shapiro FL. *Sov. Phys. JETP Lett.* 9:23 (1969)
14. Steyerl A. *Phys. Lett.* B29:33 (1969)
15. Steyerl A, et al. *Phys. Lett.* A116:347 (1986)
16. Golub R, Pendlebury JM. *Phys. Lett.* A62:337 (1977)
17. Ageron P, Mampe W, Golub R, Pendelbury JM. *Phys. Lett.* A66:469 (1978)
18. Saunders A, et al. *Phys. Lett.* B593:55 (2004)
19. Liu CY, Young AR. *Phys. Rev. B.* In press (2005)
20. Ignatovich VK. In *The Physics of Ultracold Neutrons*, ed. SW Lovesey, EWJ Mitchell. Oxford: Clarendon Press (1990)
21. Golub R, Richardson D, Lamoreaux SK. In *Ultra-Cold Neutrons*. Bristol, UK: Adam Hilger (1991)
22. Byrne J. *Rep. Prog. Phys.* 45:115 (1982)
23. Dubbers D. *Prog. Part. Nucl. Phys.* 26:173 (1991)
24. Pendlebury JM. *Ann. Rev. Nucl. Part. Sci.* 43:687 (1993)
25. Mund D, Abele H, eds. *Quark Mixing, CKM Unitarity.* Heidelberg: Mattes Verlag (2003)
26. Severijns N, Beck M, Naviliat-Cuncic O. *Rev. Mod. Phys.* In press (2005)
27. Ito TM, Prezeau G. *Phys. Rev. Lett.* 94:161802 (2005)
28. Beck M, et al. *JETP. Lett.* 76:332 (2002)
29. Fisher BM, et al. *J. Res. Natl. Inst. Stand. Technol.* In press (2005)
30. Jackson JD, Treiman SB, Wyld HW. *Phys. Rev.* 106:517 (1957)
31. Towner IS, Hardy JC. In *Symmetries and Fundamental Interactions in Nuclei*, eds. WC Haxton, EM Henley, pp. 183–249. Singapore: World Sci. (1995)
32. Eidelman A, et al. *Phys. Lett.* B592:1 (2004)
33. Hardy JC, Towner IS. *Phys. Rev. Lett.* 94:092502 (2005)
34. Pocanic D, et al. *Phys. Rev. Lett.* 93:181803 (2004)
35. Ball P, Flynn J, Kluit P, Stocchi A. In *Proceedings of the Second Workshop on the CKM Unitarity Triangle*. Durham, NC: IPPP, hep-ph/0304132 (2003)
36. Franzini P. hep-ex/0203033
37. Cabbibo N, Swallow E, Winston R. hep-ph/0307214
38. Shipsey I. In *Quark Mixing, CKM Unitarity*, eds. D Mund, H Abele, pp. 175–88. Heidelberg: Mattes Verlag (2003)
39. Spivak PE. *Zh. Eksp. Fiz.* 94:1 (1988)
40. Mampe W, Ageron P, Bates C, Pendlebury JM, Steyerl A. *Phys. Rev. Lett.* 63:593 (1989)
41. Nezvizhevskii VV, et al. *JETP* 75:405 (1992)
42. Mampe W, et al. *JETP Lett.* 57:82 (1993)
43. Byrne J, et al. *Europhys. Lett.* 33:187 (1996)
44. Arzumanov S, et al. *Phys. Lett.* B483:15 (2000)
45. Dewey MS, et al. *Phys. Rev. Lett.* 91:152302 (2003)
46. Serebrov A, et al. *Phys. Lett.* B605:72 (2005)
47. Byrne J. *Nucl. Instrum. Meth.* 284:116 (1989)
48. Williams AP. *The Determination of the Neutron Lifetime by Trapping Decay Protons.* Ph.D. thesis. Univ. Sussex (1989)

49. Nico JS, et al. *Phys. Rev.* C71:055502 (2005)
50. Huffman PR, et al. *Nature* 403:62 (2001)
51. Dzhosyuk SN, et al. *J. Res. Natl. Inst. Stand. Technol.* In press (2005)
52. Yerozolimsky B, et al. *J. Res. Natl. Inst. Stand. Technol.* In press (2005)
53. Picker R, et al. *J. Res. Natl. Inst. Stand. Technol.* In press (2005)
54. Liaud P, et al. *Nucl. Phys.* A612:53 (1997)
55. Schreckenbach K, et al. *Phys. Lett.* B349:427 (1995)
56. Abele H, et al. *Phys. Rev. Lett.* 88:211801 (2002)
57. Häse H, et al. *Nucl. Instrum. Meth.* A485: 453 (2002)
58. Petoukhov A, et al. In *Quark Mixing, CKM Unitarity*, eds. D Mund, H Abele, pp. 123–27. Heidelberg: Mattes Verlag (2003)
59. Young AR, et al. In *Fundamental Physics with Pulsed Neutron Beams*, ed. C Gould, et al., pp. 164–80. Singapore: World Sci. (2001)
60. Serebrov AP, et al. *Nucl. Instrum. Meth.* A357:503 (1995); Zimmer O, et al. *Nucl. Instrum. Meth.* A440:764 (2000)
61. Wilburn WS, et al. In *Fundamental Physics with Pulsed Neutron Beams*, ed. C. Gould, et al. pp. 214–23. Singapore: World Sci. (2001)
62. Rich DR, et al. *Nucl. Instrum. Meth.* A481:431 (2002)
63. Jodido A, et al. *Phys. Rev. D* 34:1967 (1986)
64. Byrne J. *Europhys. Lett.* 56:633 (2001)
65. Kuznetzov IA, et al. *Phys. Rev. Lett.* 75: 794 (1995)
66. Serebrov AP, et al. *JETP* 86:1074 (1998)
67. Stratowa Chr, Dobrozemsky R, Weinzierl, P. *Phys. Rev. D* 18:3970 (1978)
68. Dubbers D. *Nucl. Phys.* A527:239c (1991)
69. Yerozolimsky B, Mostovoy Y. *Sov. J. Nucl. Phys.* 53:260 (1991)
70. Gardner S, Zhang C. *Phys. Rev. Lett.* 86: 5666 (2001)
71. Byrne J, et al. *J. Phys.* G28:1325 (2002)
72. Zimmer O, et al. *Nucl. Instrum. Meth.* 440:548 (2000)
73. Wietfeldt FE, et al. *Nucl. Instrum. Meth.* A545:181 (2005)
74. Yerozolimsky B. nucl-ex/0401014 (2004)
75. Baxter DB, et al. *AIP Conference Proceedings* 680:265 (2003)
76. Christenson JH, Cronin JW, Fitch VL, Turlay R. *Phys. Rev. Lett* 13:138 (1964)
77. Alavi-Harati A, et al. *Phys. Rev. Lett.* 83: 22 (1999)
78. Fanti V, et al. *Phys. Lett.* B465:335 (1999)
79. Aubert B, et al. *Phys. Rev. Lett.* 89:201802 (2002)
80. Abe K, et al. *Phys. Rev. D* 66:071102(R) (2002)
81. Herczeg P. *J. Res. Natl. Inst. Stand. Technol.* In press (2005)
82. Ramsey NF. *Ann. Rev. Nucl. Part. Sci.* 32:211 (1982)
83. Golub R, Lamoreaux SK. *Phys. Rep.* 237:1 (1994)
84. Harris PG, et al. *Phys. Rev. Lett.* 85:904 (1999)
85. Baluni V. *Phys. Rev. D* 19:2227 (1979)
86. Crewther RJ, Di Vecchia P, Veneziano G, Witten E. *Phys. Lett.* B88:123 (1979); errata 91:487 (1980)
87. Bigi II, Sanda AI. In *CP Violation*. Cambridge: Cambridge Univ. Press (2000)
88. Peccei R, Quinn H. *Phys. Rev. Lett.* 38:1440 (1977); *Phys. Rev. D* 16:1791 (1977)
89. Khriplovich IB, Zhitnitsky AR. *Phys. Lett.* B109:490 (1982)
90. Gavela MB. *Phys. Lett.* B109:215 (1982)
91. Bennett CL, et al. (WMAP Collaboration). astro-ph/0302207 (2003)
92. Sakharov AD. *JETP Lett.* 5:24 (1967)
93. Riotto A, Trodden M. *Ann. Rev. Nucl. Part. Sci.* 49:35
94. Weinberg S. In *The First Three Minutes*. New York: Harper-Collins (1977)
95. Kuzmin VA, Rubakov VA, Shaposhnikov ME. *Phys. Lett.* B155:36 (1985)
96. Shaposhnikov ME. *Nucl. Phys.* B287:757 (1987)
97. 't Hooft G. *Phys. Rev. Lett.* 37:8 (1976)

98. Cline JM. *Pramana—J. Phys.* 55:1 (2000)
99. Fukugita M, Yanagida T. *Phys. Lett.* B174:45 (1986)
100. Büchmuller W, Peccei R, Yanagida T. *Ann. Rev. Nucl. Part. Sci.* 55: (2005)
101. Dine M, Kusenko A. *Rev. Mod. Phys.* 76:1 (2004)
102. Riotto A. *Phys. Rev. D* 58:0950099 (1998)
103. Barbieri R, Romanino A, Strumia A. *Phys. Lett.* B369:283 (1996)
104. Dimopoulos S, Hall LJ. *Phys. Lett.* B344: 185 (1995)
105. Khriplovich IB, Zyablyuk KN. *Phys. Lett.* B383:429 (1996)
106. Altarev IS, et al. *Phys. At. Nucl.* 59:1152 (1996)
107. Voronin VV, Lapin EG, Semenikhin S Yu, Federov VV. *JETP Lett.* 71:76 (2000)
108. Zeyen CME, Otake Y. *Nucl. Instrum. Meth.* A440:489 (2000)
109. Golub R, Lamoreaux SK. nucl-ex/991-0001 (1999)
110. Herczeg P. *Prog. Part. Nucl. Phys.* 46:413 (2001)
111. Wenninger H, Steiwe J, Leutz H. *Nucl. Phys.* A109:561 (1968)
112. Carnoy AS, Deutsch J, Quin P. *Nucl. Phys.* A568:265 (1994)
113. Callan CG, Treiman SB. *Phys. Rev.* 162: 1494 (1967)
114. Ramsey-Musolf MJ. In *Fundamental Physics with Pulsed Neutron Beams* eds. C Gould, et al., pp. 97–107. Singapore: World Sci. (2001)
115. Lising LJ, et al. *Phys. Rev. C* 62:055501 (2000)
116. Soldner T, et al. *Phys. Lett.* B581:49 (2004)
117. Mumm HP, et al. *Rev. Sci. Instrum.* 75: 5343 (2004)
118. Plonka C. *Untersuchung der Zeitumkehrinvarianz am D-Koeffizienten des freien Neutronenzerfalls mit TRINE.* Ph.D. thesis. Technical Univ. Munich (2004)
119. Bodek K, et al. *J. Res. Natl. Inst. Stand. Technol.* In press (2005)
120. Huber R, et al. *Phys. Rev. Lett.* 90:202301 (2003)
121. Stodolsky L. *Nucl. Phys.* B197:213 (1982)
122. Bunakov VE, Gudkov VP. *Z. Phys. A* 308:363 (1982)
123. Huffman PR, et al. *Phys. Rev. C* 55:2684 (1997)
124. Herczeg P. *Hyperfine Interactions* 75:127 (1992)
125. Ramsay-Musolf MJ. *Phys. Rev. Lett.* 83: 3997 (1999)
126. Alfimenkov VP, et al. *JETP Lett.* 34:308 (1982)
127. Szymanski JJ, et al. *Phys. Rev. C* 53: R2576 (1996)
128. Skoy VR, et al. *Phys. Rev. C* 53:R2573 (1996)
129. Masuda Y. *Nucl. Inst. Meth.* A440:632 (2000)
130. Atsarkin VA, Barabanov AL, Beda AG, Novitsky VV. *Nucl. Instrum. Meth.* A440:626 (2000)
131. Barabanov AL, Beda AG, Volkov AF. *Czech. J. Phys.* B53: 371 (2003)
132. Kuzmin VA. *JETP Lett.* 12:228 (1970)
133. Mohapathra R, Marshak R. *Phys. Rev. Lett.* 44:1316 (1980)
134. Babu KS, Mohapathra RN. *Phys. Lett.* B518:269 (2001)
135. Dvali GR, Gabadadze G. *Phys. Lett.* B460:47 (1999)
136. Klapdor-Kleingrothaus HV, Ma E, Sarkar U. *Mod. Phys. Lett.* A17:2221 (2002)
137. Baldo-Ceolin M, et al. *Z. Phys. C* 63:409 (1994)
138. Kamyshkov Y, Kolbe E. *Phys. Rev. D* 67:076007 (2003)
139. Kamyshkov Y. hep-ex/0211006 (2002)
140. *International Workshop on N-Nbar Transition Search with Ultracold Neutrons.* Bloomington, IN: Indiana Univ. http://www.iucf.indiana.edu/Seminars/NNBAR/workshop.shtml (2002)
141. Haeberli W, Holstein B. In *Symmetries and Fundamental Interactions in Nuclei*, eds. WC Haxton, EM Henley, pp. 17–66. Singapore: World Sci. (1995)

142. Desplanques B. *Phys. Rep.* 297:1 (1998)
143. Dmitriev VF, Khriplovich IB. *Phys. Rep.* 391:243 (2004)
144. Arnison G, et al. *Phys. Lett.* B166:484 (1986)
145. Belavin A, Polyakov A, Schwartz A, Tyupkin Yu. *Phys. Lett.* 59:85 (1975)
146. Schafer T, Shuryak E. *Rev. Mod. Phys.* 70:323 (1998)
147. Anselmino M, et al. *Rev. Mod. Phys.* 65:1199 (1993)
148. Alford M, Rajagopal K, Wilczek F. *Nucl. Phys.* B537:433 (1999)
149. Adelberger EG, Haxton WC. *Ann. Rev. Nucl. Part. Sci.* 35:501 (1985)
150. Adelberger EG. *J. Phys. Soc. Jpn.* 54:6 (1985)
151. Zelevinsky V. *Ann. Rev. Nucl. Part. Sci.* 46:238 (1996)
152. Bowman JD, Garvey GT, Johnson MB. *Ann. Rev. Nucl. Part. Sci.* 43:829 (1993)
153. Tomsovic S, Johnson MB, Hayes A, Bowman JD. *Phys. Rev. C* 62:054607 (2000)
154. Prezeau G, Ramsay-Musolf M, Vogel P. *Phys. Rev. D* 68:034016 (2003)
155. Wood CS, et al. *Science* 275:1759 (1997)
156. Zeldovich YB. *Sov. Phys. JETP* 6:1184 (1957)
157. Flambaum VV, Khriplovich IB. *Sov. Phys. JETP* 52:835 (1980)
158. Aubin S, et al. In *Proceedings of the XV International Conference on Laser Spectroscopy*, eds. S Chu, V Vuletic, AJ Kemand, C Chin, pp. 305–9. Singapore: World Sci. (2002)
159. Haxton W, Weiman CE. *Ann. Rev. Nucl. Part. Sci.* 51:261 (2001)
160. Beck DH, Holstein B. *Int. J. Mod. Phys. E* 10:1 (2001)
161. Desplanques B, Donoghue J, Holstein B. *Ann. Phys.* 124:449 (1980)
162. Zhu SL, et al. nucl-th/0407087, *Nucl. Phys.* A748:435 (2005)
163. Beane SR, Savage MJ. *Nucl. Phys.* B636:291 (2002)
164. Potter JM, et al. *Phys. Rev. Lett.* 33:1594 (1974)
165. Balzer R, et al. *Phys. Rev. Lett.* 44:699 (1980)
166. Kistryn S, et al. *Phys. Rev. Lett.* 58:1616 (1987)
167. Eversheim PD, et al. *Phys. Lett.* B256:11 (1991)
168. Pieper SC, Wiringa RB. *Ann. Rev. Nucl. Part. Sci.* 51:53 (2001)
169. Carlson J, Schiavalla R, Brown VR, Gibson BF. *Phys. Rev. C* 65035502 (2002)
170. Schiavilla R, Carlson J, Paris M. *Phys. Rev. C* 67:032501 (2003)
171. Schiavilla R, Carlson J, Paris M. *Phys. Rev. C* 70:044007 (2004)
172. Vesna VA, et. al. *Phys. At. Nucl.* 62:565 (1999)
173. Vesna VA, et al. *Izves. AN Ser. Phys.* 67:118 (2003)
174. Henley EM, Hwang WY, Kisslinger L. *Phys. Lett.* B271:403 (1998)
175. Meissner U-G, Weigel H. *Phys. Lett* B447:1 (1999)
176. Page SA, et al. *Phys. Rev. C* 35:1119 (1987)
177. Bini M, Fazzini TF, Poggi G, Taccetti N. *Phys. Rev. Lett* 55:795 (1985)
178. Haxton WC, Liu C-P, Ramsey-Musolf MJ. *Phys. Rev. Lett.* 86:5247 (2001)
179. Snow WM, et al. *Nucl. Instrum. Meth.* A440:729 (2000)
180. Snow WM, et al. *Nucl. Instrum. Meth.* A515:563 (2003)
181. Desplanques B. *Nucl. Phys.* A242:423 (1975)
182. Desplanques B. *Nucl. Phys.* A335:147 (1980)
183. Desplanques B. *Phys. Lett.* B512:305 (2001)
184. Gericke M, et al. *Nucl. Instrum Meth.* A540:328 (2005)
185. Mitchell GS, et al. *Nucl. Instrum. Meth.* A521:468 (2004)
186. Bass CD, et al. *J. Res. Natl. Inst. Stand. Technol.* 110:205 (2005)
187. Michel FC. *Phys. Rev.* 133:B329 (1964)
188. Dmitriev VF, Flambaum VV, Sushkov OP, Telitsin VB. *Phys. Lett.* B125:1 (1983)

189. Markoff DM. *Measurement of the Parity Nonconserving Spin-Rotation of Transmitted Cold Neutrons Through a Liquid Helium Target.* Ph.D. thesis. Univ. Washington (1997)

190. Snow WM. *J. Res. Natl. Inst. Stand. Technol.* 110:189 (2005)

191. Porter CE. *Statistical Theories of Spectra: Fluctuations.* New York: Academic Press (1965)

192. Harney HJL, Richter A, Weidenmuller HA. *Rev. Mod. Phys.* 58:607 (1986)

193. Mitchell GE, Bowman JD, Weidenmuller HA. *Rev. Mod. Phys.* 71:445 (1999)

194. Mitchell GE, Bowman JD, Penttila SI, Sharapov EI. *Phys. Rep.* 354:157 (2001)

195. French JB, Kota VKB, Pandey A, Tomsovic A. *Ann. Phys.* 181:198 (1988)

196. French JB, Kota VKB, Pandey A, Tomsovic A. *Ann. Phys.* 181:235 (1988)

197. Sushkov OP, Flambaum VV. *JETP Lett.* 32:352 (1980)

198. Alfimenkov VP, et al. *JETP Lett.* 35:51 (1982)

199. Alfimenkov VP, et al. *Nucl. Phys.* A398:93 (1983)

200. Corvi F, Przytula M. *Phys. Part. Nuclei* 35:767 (2004)

201. Danilyan GV, et al. *JETP Lett.* 26:186 (1977)

202. Vodennikov BD, et al. *JETP Lett.* 27:62 (1978)

203. Shuskov OP, Flambaum VV. *Sov. Phys. Usp.* 25:1 (1982)

204. Bunakov VE, Goennenwein F. *Phys. At. Nucl.* 65:2036 (2002)

205. Petrov G, et al. *Nucl. Phys.* A502:297 (1989)

206. Belozerov AV, et al. *JETP Lett.* 54:132 (1991)

207. Goennenwein F, et al. *Nucl. Phys.* A567:303 (1994)

208. Jesinger P, et al. *Phys. At. Nucl.* 65:630 (2002)

209. Koetzle A, et al. *Nucl. Instrum. Meth.* A440:618 (2000)

210. Jesinger P, et al. *Phys. At. Nucl.* 62:1608 (1999)

211. Jesinger P, et al. *Nucl. Instrum. Meth.* A440:618 (2000)

212. Bunakov VE. *Phys. At. Nucl.* 65:616 (2002); 65:648 (2002)

213. Kadmensky SG. *Phys. At. Nucl.* 65:1785 (2002)

214. Bunakov VE, Kadmensky SG. *Phys. At. Nucl.* 66:1846 (2003)

215. Weinberg S. *Physica A* 96:327 (1979)

216. Georgi H. *Ann. Rev. Nucl. Part. Sci.* 43:209 (1993)

217. Braaten E, Hammer HW. *Phys. Rev. Lett.* 91:102002 (2003)

218. Braaten E, Hammer HW. cond-mat/0410417 (2004)

219. Bedaque PF, van Kolck U. *Phys. Lett.* B428:221 (1998)

220. Hammer HW. nucl-th/9905036 (1999)

221. Bedaque PF, van Kolck U. *Ann. Rev. Nucl. Part. Sci* 52:339 (2002)

222. Friar J. In *Nuclear Physics with EFT*, eds. R Seki, U van Kolck, MJ Savage, pp. 145–161. Singapore: World Sci. (2001)

223. van Kolck U. *Nucl. Phys.* A645:273 (1999)

224. Kaplan DB, Savage MJ, Wise MB. *Phys. Lett.* B424:390 (1998)

225. Gegelia J. nucl-th/9802038 (1998)

226. Epelbaum E, et al. *Phys. Rev. Lett.* 86:4787 (2001)

227. Epelbaum E, et al. *Phys. Rev. C* 66:064001 (2002)

228. Carlson J, Schiavilla R. *Rev. Mod. Phys.* 70:743 (1998)

229. Wiringa RB, Stoks VGJ, Schiavilla R. *Phys. Rev. C* 51:38 (1995)

230. Black T, et al. *Phys. Rev. Lett.* 90:192502 (2003)

231. Schoen K, et al. *Phys. Rev. C* 67:044005 (2003)

232. Huffman PR, et al. *Phys. Rev. C* 70:014004 (2004)

233. Hofmann HM, Hale GM. *Phys. Rev. C* 68:021002(R) (2003)

234. Witala H, et al. nucl/th 0305028 (2003)

235. Abragam A, et al. *Phys. Rev. Lett.* 31:776 (1973)

236. Zimmer O, et al. *EJPdirect* A1:1 (2002)
237. van der Brandt B, et al. *Nucl. Instrum. Meth.* A526:91 (2004)
238. Pokotolovskii YuN. *Phys. At. Nucl.* 56: 524 (1993)
239. Furman WI, et al. *J. Phys. G Nucl. Part. Phys.* 28:2627 (2002)
240. Gardestig A, Phillips DR. nucl-th/0501049 (2005)
241. Isgur N. *Phys. Rev. Lett.* 83:272 (1999)
242. Hand L, Miller DG, Wilson R. *Rev. Mod. Phys.* 35:335 (1963)
243. Tang A, Wilcox W, Lewis R. *Phys. Rev. D* 68:094503 (2003)
244. Leinweber DB, Thomas AW, Young RD. *Phys. Rev. Lett.* 86:5011 (2001)
245. Krohn VE, Ringo GR. *Phys. Rev.* 148:1303 (1966); *Phys. Rev. D* 8:1305 (1973)
246. Kopecky S, et al. *Phys. Rev. C* 56: 2229 (1997)
247. Alexandrov YA, et al. *Yad. Fiz.* 44:1384 (1986)
248. Alexandrov YA, et al. *Zh. Eksp. Teor. Fiz.* 89:34 (1985)
249. Mitsyna LV, et al. In *JINR*, Dubna preprint E3-2003-183 (2003)
250. Sparenberg J-M, Leeb H. *Phys. Rev. C* 66:055210 (2002)
251. Sick I. *Phys. Lett.* B576:62 (2003)
252. Udem T, et al. *Phys. Rev. Lett.* 79:2646 (1997)
253. Sick I, Trautmann D. *Phys. Lett.* B375:16 (1996)
254. Klarsfeld S, et al. *Nucl. Phys.* A456:373 (1986)
255. Pachucki K, Weitz M, Hänsch TW. *Phys. Rev. A* 49:2255 (1994)
256. Huber A, et al. *Phys. Rev. Lett.* 80:468 (1998)
257. Schmiedmayer J, Riehs P, Harvey JA, Hill NW. *Phys. Rev. Lett.* 66:1015 (1991)
258. Koester L, et al. *Phys. Rev. C* 51:3363 (1995)
259. Aleksejeva A, et al. *Phys. Scripta* 56:20 (1997)
260. Wissmann F, Levchuk MI, Schumacher M. *Euro. Phys. J.* A1:193 (1998)
261. Lundin M, et al. *Phys. Rev. Lett.* 90: 192501 (2003)
262. Kossert K, et al. *Phys. Rev. Lett.* 88: 162301 (2002)
263. Bernard V, Kaiser N, Meissner UG. *Phys. Rev. Lett.* 67:1515 (1991)
264. Beane SR, et al. nucl-th/0403088 (2004)
265. Christensen J, Wilcox W, Lee FX, Zhou L. hep-lat/0408024 (2004)
266. Schramm DN, Wagoner RV. *Ann. Rev. Nucl. Part. Sci.* 27:37 (1977)
267. Lopez RE, Turner MS. *Phys. Rev. D* 59:103502–1 (1999)
268. Burles S, Nollett KM, Truran JW, Turner MS. *Phys. Rev. Lett.* 82:4176 (1999)
269. Spergel DN, et al. (WMAP collaboration). *Astrophys. J. Suppl.* 148:175 (2003)
270. Cyburt RH, Fields BD, Olive KA. *Phys. Lett.* B567:227 (2003)
271. Käppeler F, Thielemann F-K, Wiescher M. *Ann. Rev. Nucl. Part. Sci.* 48:175 (1998)
272. Kappeler F. *Prog. Part. Nucl. Phys.* 43:419 (1999)
273. Qian YZ. *Prog. Part. Nucl. Phys.* 50:153 (2003)
274. Koehler PE. *Nucl. Instrum. Meth.* A460: 352 (2001)
275. Flaska M, et al. *Nucl. Instrum. Meth.* A531:392 (2004)
276. Reifarth R, et al. *Nucl. Instrum. Meth.* A531:528 (2004)
277. Schmiedmayer J. *Nucl. Instrum. Meth.* A284:59 (1989)
278. Pokotilovskii YN. *Phys. At. Nucl.* 57:390 (1994)
279. Colella R, Overhauser AW, Werner SA. *Phys. Rev. Lett.* 34:1472 (1975)
280. Rauch H, Treimer W, Bonse U. *Phys. Lett.* A47:369 (1974)
281. Littrell KC, Allman BE, Werner SA. *Phys. Rev. A* 56:1787 (1997)
282. Funahashi H. *Nucl. Instrum. Meth.* A529: 172 (2004)
283. Fischbach E, et al. *Phys. Rev. Lett* 56:3 (1986)

284. Gundlach JH, et al. *Phys. Rev. Lett.* 78: 2523 (1997)

285. Sundrum R. *J. High Energy Phys.* 07:001 (1999)

286. Adelberger EG, Heckel BR, Nelson AE. *Ann. Rev. Nuc. Part. Sci.* 53:77 (2003)

287. Frank A, van Isacker C, Gomez-Camacho C. *Phys. Lett.* B582:15 (2004)

288. Nesvizhevsky VV, et al. *Nature* 415:297 (2002)

289. Nesvizhevsky VV, Protasov KV. *Class. Quant. Gravity* 21:4557 (2004)

Annu. Rev. Nucl. Part. Sci. 2005. 55:71–139
doi: 10.1146/annurev.nucl.55.090704.151541

TOWARD REALISTIC INTERSECTING D-BRANE MODELS

Ralph Blumenhagen,[1] Mirjam Cvetič,[2]
Paul Langacker,[2] and Gary Shiu[3,4]
[1]Max-Planck-Institut für Physik, D-80805 München, Germany;
email: blumenha@mppmu.mpg.de
[2]Department of Physics and Astronomy, University of Pennsylvania, Philadelphia,
Pennsylvania 19104; email: cvetic@cvetic.hep.upenn.edu,
pgl@electroweak.hep.upenn.edu
[3]Department of Physics, University of Wisconsin, Madison, Wisconsin 53706;
email: shiu@physics.wisc.edu
[4]Perimeter Institute for Theoretical Physics, Waterloo, Ontario N2L 2Y5, Canada

Key Words string, orbifold, orientifold, flux

■ **Abstract** We provide a pedagogical introduction to a recently studied class of phenomenologically interesting string models known as Intersecting D-Brane Models. The gauge fields of the Standard Model are localized on D-branes wrapping certain compact cycles on an underlying geometry, whose intersections can give rise to chiral fermions. We address the basic issues and also provide an overview of the recent activity in this field. This article is intended to serve non-experts with explanations of the fundamental aspects of string phenomenology and also to provide some orientation for both experts and non-experts in this active field.

CONTENTS

0163-8998/05/1208-0071$20.00

71

1. INTRODUCTION

By now we have ample evidence that the Standard Model of particle physics extended by right-handed neutrinos or other mechanisms of neutrino masses describes nature with very high accuracy up to the energy range of the weak scale $E_W = 10^2$ GeV. The only missing ingredient is the Higgs particle itself, which is expected to be detected at the Tevatron or the Large Hadron Collider (LHC). However, from a more formal point of view, the Standard Model is not completely satisfactory for two reasons. First, it contains 26 free parameters (not counting arbitrary electric charges) like the masses and couplings of fermions and bosons, which have to be measured and among which no relation is apparent. Second, the Standard Model is formulated as a local four-dimensional quantum field theory and as such it does not include gravity. In fact, the Einstein theory of general relativity cannot simply be quantized according to the rules of local quantum field theory. Therefore, the physics we know of cannot describe our universe at very high energies where quantum effects of gravity become important.

Given these two formal shortcomings, in an ideal world we might hope that both problems actually have the same solution. Maybe there exists a fundamental quantum theory, which combines the Standard Model and general relativity into a unified framework and at the same time substantially reduces the number of independent parameters in the Standard Model. As this unified theory may be geometric in nature, one might envision that some of the structure of the Standard Model turns out to have a geometric origin as well.

Currently we do not what this final theory is, but at least we have a good candidate for it, which still has to reveal many of its secrets. This candidate theory of quantum gravity is called superstring theory and has been studied intensively during the last three decades. Since superstring theory is anomaly-free only in ten space-time dimensions, to make contact with the universe surrounding us we have to explain what happened to the other six dimensions without contradicting experiments. Compactifying à la Kaluza-Klein string theory on a compact six-dimensional space of very tiny dimensions, our visible world would be interpreted as an effective four-dimensional theory, where one only keeps the states of lowest mass. The question immediately arising is whether the formal string equations of motion allow for six-dimensional spaces such that the low-energy four-dimensional world resembles the Standard Model of particle physics. As a first approach, it would be too ambitious to require that all the couplings come out correctly. Instead, to begin with, one has to think about stringy mechanisms for generating gauge theories with chiral matter organized in replicated families.

The subfield of string theory concerned with such questions is called string phenomenology and has been pursued since the mid-1980s. By that time it was clear that there exist two different types of ten-dimensional superstring theories containing gauge fields on the perturbative level. The heterotic string theories contain only closed-oriented strings and can support $SO(32)$ or $E_8 \times E_8$ gauge groups, whereas in the non-oriented Type I string theory, the gauge degrees of freedom arise from open strings, which can only support the gauge group $SO(32)$. In Type I string theory, the two possible orientations of the string are identified; in other words, one is gauging the word-sheet parity transformation.

From the mid-1980s to the mid-1990s, string theorists were mostly studying $E_8 \times E_8$ heterotic string compactifications, as it seemed to be more natural to embed the Standard Model gauge group $SU(3)_C \times SU(2)_W \times U(1)_Y$ into one of the E_8 factors and consider the second E_8 as a hidden gauge group, which might provide the infrared physics for supersymmetry breaking. In fact, it turned out that six-dimensional manifolds with $SU(3)$ holonomy, the so-called Calabi-Yau manifolds, also give rise to chiral fermions, which in the most simple scenario come in identical families where the multiplicity is given by one-half of the Euler number of the Calabi-Yau manifold. Many examples of such Calabi-Yau manifolds were constructed, including for instance toroidal orbifolds, hypersurfaces in weighted projective space, or toric varieties.

In the mid-1990s, string theory encountered an intellectual phase transition triggered by the realization that not only supersymmetric gauge theories but also string theories can be related by various dualities, some of which exchange weak and strong coupling. While previously scientists had been merely studying perturbative aspects of string theory, now it was possible to move beyond the perturbative framework and to catch a glimpse of the non-perturbative physics of string theory. The conjectured web of string dualities relied on a speculative theory in eleven space-time dimensions, which was called M-theory. String theorists believe that this M-theory is actually the fundamental theory, of which the various string theories arise in certain perturbative limits.

In the process of establishing these dualities, it became clear that at the non-perturbative level, string theory is not only a theory of strings but also contains even higher dimensional objects called p-branes, which have p space-like and one time-like dimension. Surprisingly, the fluctuations of a certain subset of these p-branes, so-called D-branes, are again described by a string theory, which in this case is an open string theory with endpoints on the brane. Because at the massless level these D-branes support gauge fields, they are natural candidates for string phenomenology. The question is whether one can construct consistent string compactifications with D-branes in the background. The simplest example is the aforementioned Type I string itself, which contains space-time filling D9-branes placed on top of the topological defect introduced by the gauging of the world-sheet parity. This already indicates that for getting models with D-branes, one should consider generalizations of the Type I string. Such models, now called orientifolds, have been studied in the conformal field theory framework before [see (1) and references therein] and were reinvented during the mid-1990s from a space-time point of view. The aforementioned defects were called orientifold planes.

Since their discovery, many orientifold models have been constructed, and there exists an extensive literature on this subject including some review articles, e.g., (1, 2). This article is not intended to be an additional review on general orientifold models, but instead focuses on a phenomenologically interesting class of orientifold models, which comes with its own intertwined history.

The class of models covered here has its origin in the observation that two generically intersecting D-branes can support chiral fermions on the intersection locus (3). Therefore, one is led to models that not only contain D-branes on top of or parallel to orientifold planes, but also allow these D-branes to be placed such that there exist chiral intersections, as long as they do not violate the stringy consistency conditions. Historically, the first models of this kind were discussed in a T-dual formulation with magnetic fluxes in Type I string theory by Bachas (4). Providing the complete stringy picture of this early idea and showing its dual formulation in terms of intersecting D-branes, the first really intersecting D-brane models were constructed in (5, 6). Independently, supersymmetric compactifications of the Type I string to six-dimensions with magnetic fluxes were discussed in (7, 8). Intersecting branes and magnetized branes are equivalent descriptions (3). Therefore, without loss of generality, in this review article we stick to the more intuitive picture of geometrically intersecting D-branes. Non-chiral orientifold models with D-branes intersecting at angles had been considered even before the chiral ones (9–14).

Intersecting branes provide a stringy mechanism for generating not only gauge symmetries but also chiral fermions, where family replication is achieved by multiple topological intersection numbers of various D-branes. Therefore, these models provide a beautiful geometric picture of some of the fundamental ingredients of the Standard Model.

After the introduction of these kinds of models, some generalizations and additional profound issues were discussed in (15–17). In the original models of (5), the closed string background was simply a flat torus, for which it could be shown that flat non-trivially intersecting D-branes always break supersymmetry explicitly at the string scale. Therefore, chiral models were necessarily non-supersymmetric.

For a field theorist this is not a problem, as the Standard Model as we know is non-supersymmetric anyway. However, from the stringy point of view, supersymmetry is generally the mechanism which guarantees that string compactifications are stable. In order for a string vacuum to have a lifetime longer than the Planck (or better string) time, it seems desirable to start with a supersymmetric vacuum and then break supersymmetry softly in a controlled way. Though many papers in the literature deal with non-supersymmetric intersecting D-brane models, the reader should keep in mind that for these models, even though the open string sectors look amazingly similar to the Standard Model (18), one generically encounters stability problems in the closed string sector (19). Due to their popularity and some issues that carry over to the supersymmetric models, we will also cover the non-supersymmetric models in this review, but our main focus will be on chiral supersymmetric models, first constructed in References 20, 21. For them, Standard-like Models are much harder to construct and one has to consider more general than purely toroidal backgrounds, e.g., orbifolds.

In the original setting, one considered orientifolds of Type IIA string theory, which contain only orientifold six-planes, whose charges are canceled by introducing intersecting D6-branes. Such models can be defined on general six-dimensional manifolds, where the requirement of supersymmetry however implies this to be a Calabi-Yau manifold. Various generalizations with D-branes of other dimensions have been contemplated, but we think that the original models are the most natural class of intersecting D-brane models, for instance, they are related to M-theory compactifications on G_2 manifolds. Therefore, throughout this article we will mainly work in this framework and only mention the possible generalizations.

Different aspects of these intersecting D-brane models have been discussed during the last four years in a large number of papers, which can be mainly categorized into three classes (we will provide the references in the appropriate sections of the main text). First, there are the stringy model building aspects, which in particular include the derivation of the stringy consistency conditions (R-R tadpole cancellation conditions) and the computation of the massless spectrum. Second, tools have been developed to compute for a given string model the four-dimensional low-energy effective action, which includes tree-level expressions for Yukawa couplings, higher point correlation functions, gauge couplings, Fayet-Iliopoulos terms, and Kähler potentials. Moreover, for the gauge couplings, one-loop corrections have been computed. This program of determining the low-energy effective action is not complete yet and has mainly been applied to purely toroidal (orbifold) string backgrounds. Finally, using the results about the effective action, people discussed the phenomenological low-energy implications of intersecting D-brane models; some of them turn out to be rather model-independent, whereas others are not and might be used to discard certain models for phenomenological reasons. These are the three main aspects, but of course there exist relations of intersecting D-brane models to other branches of recent research such as M-theory compactification on G_2 manifolds or compactifications with non-trivial background fluxes. These latter developments will also be covered in this article, although we do not provide a general introduction to G_2 manifolds or flux compact-ifications, as this would fill another review article. Another possible connection is to the phenomenological brane world ideas associated with possible large extra dimensions (22–24) that have been popular in recent years. Although most intersecting D-brane constructions involve only small extra dimensions (within a few orders of magnitude of the inverse Planck scale), it is possible (and probably necessary for non-supersymmetric constructions) to consider internal spaces with large dimensions, providing a stringy realization of those ideas.

The aim of this article is twofold. First, it is intended to give a pedagogical introduction to the subject and to provide the main technical tools for the construction of intersecting D-brane models. It should allow non-experts to understand the main aspects of the subject and enable students to get started in this field. Second, we attempt to give as broad an overview as possible of developments in the field and to point out open questions. Of course, to be as complete as possible, we had to neglect many details, and we are aware that the topics we put special emphasis on reflect in some way our own preferences. We apologize to all those authors who feel that their work has not been covered to a degree they believe it deserves. Several articles of review type with slightly different emphases have appeared during the last few years (25–31).

2. ORIENTIFOLDS WITH INTERSECTING D-BRANES

Throughout this technical introduction into intersecting D-brane models, we assume that the reader is familiar at least at a textbook level (32–37) with the basic notions of string theory including the concept of D-branes.

String compactifications from ten to four space-time dimensions have been studied throughout the history of string theory, but in the mid-1990s the second string theory revolution provided new insights into the constructions of four-dimensional vacua from M-theory. As with all the progress made during this exciting epoch, this had to do with the realization that string theory is not only a theory of either closed or open strings but also contains in its non-perturbative sector extended objects of higher dimensions, so called D-branes (see (36, 38, 39) for reviews on D-branes). These D-branes are charged under some of the massless fields appearing in the Ramond-Ramond (R-R) sector of the ten-dimensional Type IIA/B string theories. More concretely, a p-brane is an extended object with p space-like directions and one time-like direction and it couples to a $(p + 1)$ form potential A_{p+1} as follows:

$$\mathcal{S}_p = Q_p \int_{D_p} A_{p+1}, \qquad\qquad 1.$$

where the integral is over the $(p + 1)$-dimensional world-volume of the D-brane and Q_p denotes its R-R charge. For supersymmetric, so-called BPS, D-branes in Type IIA string theory, p is an even number and in Type IIB, an odd one. Polchinski was the first to realize that the fluctuations of such D-branes can by themselves be described by a string theory (40), which in this case are open strings attached to the D-brane, i.e., with Dirichlet boundary conditions transversal to the D-brane and Neumann boundary conditions along the D-brane

$$\mu = 0, \ldots, p \qquad \partial_\sigma X^\mu|_{\sigma=0,\pi} = 0,$$
$$\mu = p+1, \ldots, 9 \qquad \partial_\tau X^\mu|_{\sigma=0,\pi} = 0, \qquad\qquad 2.$$

where (σ, τ) denote the world-sheet space and time coordinates and X^μ the space-time coordinates. Their world-sheet superpartners are denoted as ψ^μ in the following. Upon quantization of an open string, the massless excitations $\psi^\mu_{-\frac{1}{2}}|0\rangle$ give rise to a $U(1)$ gauge field, which can only have momentum along the D-brane and is therefore confined to it. It is precisely the occurrence of these gauge fields which makes D-branes interesting objects for string model building. If one can construct string models with D-branes in the background, then one has a natural source of gauge fields, which are of fundamental importance in the Standard Model of particle physics. Placing N D-branes on top of each other the gauge fields on the branes transform in the adjoint representation of the gauge group $U(N)$.

In this section, we will discuss the general rules for constructing intersecting D-brane models. In Subsection 2.1, employing the effective gauge and gravitational couplings, we discuss how the string scale of the intersecting D-brane models depends on the closed string moduli. Then in Subsection 2.2, we discuss how chiral fermions arise at the intersection of D-branes. The fact that D-branes can intersect more than once in a compact space gives rise to the interesting feature of family replication, which will be discussed in Subsection 2.3. In addition to R-R charges, D-branes also couple gravitationally which means that they have tension. To cancel the positive contribution to the vacuum energy from the tension of D-branes, we need to introduce negative tension objects known as orientifold planes. The notion of orientifolds will be discussed in Subsection 2.4. The total R-R charge carried by the D-branes and orientifold planes has to vanish for consistency. Such tadpole cancellation conditions are derived in Subsection 2.5. With the configuration of D-branes and orientifold planes that satisfy the tadpole conditions, one can derive the spectrum of massless open strings ending on the D-branes. The chiral part of the spectrum is summarized in Subsection 2.6. In general, there are anomalous $U(1)$s in intersecting D-brane models whose anomalies are canceled by the generalized Green-Schwarz mechanism as explained in Subsection 2.7. In Subsection 2.8, we discuss the conditions for the configuration of D6-branes to be supersymmetric. It turns out that they have to wrap around three-cycles known as special Lagrangian (sLag) cycles. Interestingly, the intersecting D6-brane models that preserve $\mathcal{N} = 1$ supersymmetry in four dimensions can be lifted to eleven-dimensional M-theory as compactifications on singular G_2 manifolds. The lift and the connection to how chiral fermions arise in the G_2 context are discussed in 2.9. As two warmup examples for later use, intersecting D-branes on T^5 and the $T^6/\mathbb{Z}_2 \times \mathbb{Z}_2$ orbifold are presented in Subsection 2.10.

2.1. The String Scale

The localization of gauge fields on D-branes provides a concrete stringy realization of the brane world scenario in which the Standard Model fields are confined on the branes, while gravity propagates in the bulk. As a result, the four-dimensional gauge couplings are determined by the volume of the cycles that the D-branes wrap around, whereas the gravitational coupling depends on the total internal volume. This opens up the possibility of lowering the string scale. More specifically, by dimensional reduction to four dimensions[1]:

$$\frac{1}{g_{YM}^2} = \frac{M_s^{p-3} V_{p-3}}{(2\pi)^{p-2} g_s}$$

$$M_P^2 = \frac{M_s^8 V_6}{(2\pi)^7 g_s^2},$$ 3.

where V_{p-3} is the volume of the $p-3$ cycle wrapped by a Dp-brane (which is in general different for different branes) and V_6 is the total internal volume. In this article, we will focus on models with intersecting D6-branes so that

$$g_{YM}^2 M_P = \sqrt{2\pi} M_s \frac{\sqrt{V_6}}{V_3}.$$ 4.

The experimental bounds on the masses of Kaluza-Klein replicas of the Standard Model gauge bosons imply that the volume of three-cycles cannot be larger than the inverse TeV scale generically. For a general internal space (such as a Calabi-Yau manifold), the volumes of the three-cycles are not directly constrained by the scale of the total internal volume, and can be much smaller than $\sqrt{V_6}$. In this case, a large Planck mass can be generated from a large total internal volume. This is precisely the idea of the large extra dimension scenario.

However, for intersecting D6-brane models in toroidal backgrounds, V_3 is of the same order as $\sqrt{V_6}$ (since for chiral models, there is no dimension transverse to all the branes) so the string scale is of the order of the Planck scale M_P. There is, however, more freedom than in theories with only closed strings (e.g., the heterotic string), and this could be used to lower the string scale to, e.g., 10^{16} GeV, a certainly desirable choice for Grand Unified Models.

2.2. Chirality

One of the main features of the Standard Model is that the light fermionic matter fields appear in chiral representations of the $SU(3)_C \times SU(2)_W \times U(1)_Y$ gauge symmetry such that all gauge anomalies are canceled. Considering just parallel D-branes in flat space one does not get chiral matter on the branes, so that one has to invoke an additional mechanism to realize this phenomenologically very important feature. Essentially, two ways have been proposed to realize chirality

---

[1]The factors of 2π were carefully worked out in (41).

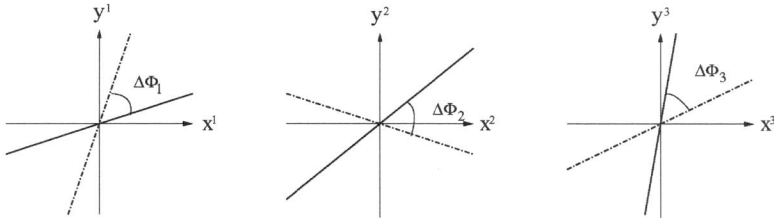

Figure 1   Intersecting D6-branes.

for the D-brane matter spectrum. The first one is to place the D-brane not in flat space, but on so-called orbifold (or conifold) singularities; the second is to let the D-branes intersect at non-trivial angles (3). We will discuss here only the second mechanism in more detail and refer the interested reader to the existing literature for discussion of the first mechanism [see for instance (1, 42) and references therein].

To be more precise, consider two D6-branes sharing the four-dimensional Minkowskian space-time. This means that in the six-dimensional transversal space the branes are three-dimensional and wrap a three-dimensional cycle. In a general position, two such branes do intersect in a point in the internal space. Consider the simple case of a flat six-dimensional internal space. Choosing light cone gauge, let us introduce complex coordinates $z^i = x^i + iy^i$ with $i = 0, \ldots, 3$. Then two D6-branes cover the z^0 plane and intersect in the other directions as shown in Figure 1.

Placing for convenience one D-brane along the x^i axes, an open string stretched between two intersecting D-branes has the following boundary conditions

$$\sigma = 0: \quad \partial_\sigma X^i = \partial_\tau Y^i = 0$$

$$\sigma = \pi: \quad \partial_\sigma X^i + \tan(\Delta\Phi_i)\partial_\sigma Y^i = 0$$

$$-\tan(\Delta\Phi_i)\partial_\tau X^i + \partial_\tau Y^i = 0 \qquad 5.$$

which in complex coordinates read

$$\sigma = 0: \quad \partial_\sigma(Z^i + \bar{Z}^i) = \partial_\tau(Z^i - \bar{Z}^i) = 0$$

$$\sigma = \pi: \quad \partial_\sigma Z^i + e^{2i\Delta\Phi_i}\partial_\sigma \bar{Z}^i = 0$$

$$\partial_\tau Z^i - e^{2i\Delta\Phi_i}\partial_\tau \bar{Z}^i = 0. \qquad 6.$$

Now, implementing these boundary conditions in the mode expansion of the fields Z^i and $\bar{Z}^i$, one finds (3)

$$Z^i(\sigma, \tau) = \sum_{n\in\mathbb{Z}} \frac{1}{(n-\epsilon_i)}\alpha^i_{n-\epsilon_i}\, e^{-i(n-\epsilon_i)(\tau+\sigma)} + \sum_{n\in\mathbb{Z}} \frac{1}{(n+\epsilon_i)}\tilde{\alpha}^i_{n+\epsilon_i}\, e^{-i(n+\epsilon_i)(\tau-\sigma)}$$

$$7.$$

with $\epsilon_i = \Delta\Phi_i/\pi$ for $i \in \{1, 2, 3\}$. Therefore the bosonic oscillator modes of the fields $Z^1, \ldots, Z^3$ are given by

$$\alpha^i_{n-\epsilon_i}, \quad \tilde{\alpha}^i_{n+\epsilon_i} \qquad\qquad 8.$$

Similarly, for the world-sheet fermions the modes are $\psi^i_{n-\epsilon_i}$ and $\tilde{\psi}^i_{n+\epsilon_i}$ in the R-R sector and with an additional $1/2$-shift in the Neveu-Schwarz Neveu-Schwarz (NS-NS) sector. Therefore, in analogy to the closed string sector, an open string between two intersecting D-branes can be considered as a twisted open string. As a consequence, for all ϵ_i non-vanishing, there are only two zero modes in the R-R sector, $\psi^1_0, \tilde{\psi}^1_0$, which give rise to a twofold degenerate R-R ground state. The GSO projection eliminates one half of these states, so that one is left with only one fermionic degree of freedom. Taking into account also the open string with the opposite orientation between the two D6-branes, one finally gets two fermionic degrees of freedom corresponding to one chiral Weyl-fermion from the four-dimensional space-time point of view. To summarize, we have found that two generically intersecting D6-branes give rise to one chiral fermion at the intersection point. If we now consider the intersection between a stack of M D6-branes with another stack of N D6-branes, it is clear that, for example, the left-handed chiral fermion transforms in the bi-fundamental representation of the $U(M) \times U(N)$ gauge symmetry. We choose the convention that this is the $(\bar{M}, N)$ representation of the gauge group. As such this result is not invariant under the exchange of the roles of M and N. This can be remedied by giving an orientation to the branes and by assigning a sign to the intersection on each plane as shown in Figure 2. A negative intersection simply means that one gets a left-handed chiral fermion transforming in the conjugate representation of the gauge group. The intersection defined this way is anti-symmetric under exchange of the two branes.

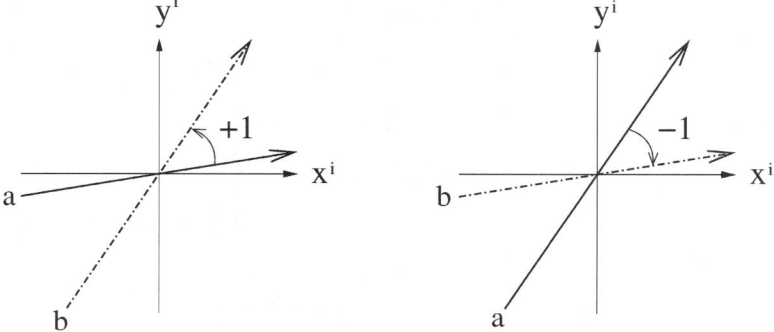

Figure 2   Oriented intersection.

2.3. Family Replication

In the last section we have seen that intersecting D-branes can be a source for chiral fermions, which makes them very interesting candidates for model building. However, chiral fermions in the Standard Model come in three families differing only by their mass scale. Therefore, it is important to search for a mechanism for family replication. As we will see by considering intersecting branes on compact backgrounds, such a mechanism automatically arises.

In the non-compact flat background depicted in Figure 1, it is clear that the intersection number can only be ± 1. However, in the compact case such as a torus, it can be easily seen that the intersection number can be larger than one. Assuming for simplicity that the background is a six-dimensional torus with complex structure chosen such that it can be written as $T^6 = T^2 \times T^2 \times T^2$, a large class of D6-branes cover only a one-dimensional cycle on each factor T^2. Such D6-branes have been called factorizable in the literature and are described by three pairs of wrapping numbers (n^i, m^i) along the fundamental 1-cycles of three T^2s. In Figure 3, we have shown two such wrapped D6-branes with wrapping numbers $(1, 0)(1, 1)(2, 1)$ for the first D-brane and $(0, 1)\,(1, -1)\,(1, -1)$ for the second one.

Apparently, the intersection number between the two D6-branes is $I_{ab} = 6$, which is just one simple example of the general expression for the intersection number

$$I_{ab} = \prod_{i=1}^{3} \left(n_a^i m_b^i - m_a^i n_b^i\right). \qquad 9.$$

By deforming the D6-branes, one can easily generate additional intersections, but they always come in pairs with a positive and negative sign, so that the net number of chiral fermions remains constant. Therefore, what really counts the net number of chiral fermions is the topological intersection number, which only depends on the homology classes of the two branes. In our case, the homological three-cycles are simply products of three one-cycles, where the homological one-cycles on each T^2 are characterized just by the wrapping numbers (n_a, m_a).

Generalizing the set-up we have introduced so far, we consider compactifications of Type IIA string theory on a six-dimensional manifold $\mathcal{M}$. To preserve

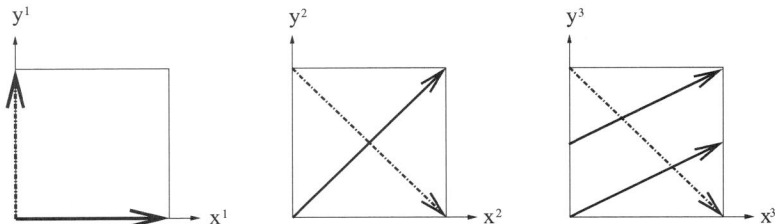

Figure 3 Intersecting D6-branes on a torus.

$\mathcal{N} = 1$ supersymmetry, the manifold $\mathcal{M}$ is a Calabi-Yau manifold. As a topological space, $\mathcal{M}$ has homological three-cycles π_a, $a \in \{1, \dots, K\}$, on which we can wrap N_a D6-branes. From the effective four-dimensional point of view, we obtain gauge fields of $\Pi_{a=1}^K U(N_a)$ localized on the seven-dimensional world-volume of the D6-branes. Additionally, one gets chiral fermions localized on the four-dimensional intersection locus of two branes which come with multiplicity given by the topological intersection number $\pi_a \circ \pi_b$ and transform in the $(\bar{N}_a, N_b)$ representation of the gauge group.

As we will discuss in the following sections, tadpole cancellation and supersymmetry impose certain constraints on the three-cycles that the D6-branes are wrapped around.

2.4. Orientifolds

As has been pointed out in (19), non-supersymmetric models, though easy to handle, are unstable in the sense that their perturbative scalar potential, which is due to the so-called NS-NS tadpoles, gives rise to runaway behavior for many of the closed string moduli fields including the dilaton. In addition, for constructions based on toroidal-type compactifications, the volume of three-cycles is of the same order of magnitude as the square root of the volume of the internal space. Thus, for the case of intersecting D6-branes there is no direction in the internal space that can be taken to be large compared to the Planck radius while keeping the correct values of gauge couplings and the Planck scale in four-dimensions (see Equation 3 in Section 2.1). Therefore, in this case the string scale cannot be much below the Planck scale and thus both the NS-NS tadpole contributions to the potential as well as radiative corrections in the effective theory are large, i.e., of the order of the Planck scale. There is a chance that NS-NS tadpoles might be stabilized by non-perturbative effects or by turning on fluxes, but nevertheless, in order to stay on firm ground from the string theory perspective, we prefer to mainly consider supersymmetric models.

The set-up introduced so far with intersecting D6-branes in Type IIA compactifications always breaks supersymmetry. This can be seen as follows. For a globally supersymmetric background the vacuum energy has to vanish. However, all D6-branes have a positive contribution to the vacuum energy, as their tension is always positive and therefore they break supersymmetry. The only way to finally find non-trivial supersymmetric models is by introducing objects of negative tension into the theory. It is well known that such objects exist in string theory and that they naturally occur in so-called orientifold models.

An orientifold is the quotient of Type II string theory by a discrete symmetry group G including the world-sheet parity transformation Ω: $(\sigma, \tau) \rightarrow (-\sigma, \tau)$. As a consequence, the resulting string models contain non-oriented strings and their perturbative expansion also involves non-oriented surfaces like the Klein-bottle. Dividing out by such a symmetry, new objects called orientifold planes arise, whose presence can be detected for instance by computing the Klein-bottle amplitude

$$K = \int_0^\infty \frac{dt}{t} \mathrm{Tr} \left(\frac{\Omega}{2} e^{-2\pi t(L_c + \bar{L}_0)} \right).$$ 10.

These objects, though non-dynamical, do couple to the closed string modes and in particular they carry tension and charge under some of the R-R fields. In other words, there exist non-vanishing tadpoles of the closed modes on the orientifold planes which, as it turns out, can have opposite sign than the corresponding terms for D-branes. Since the overall charge one puts on a compact space has to vanish by Gauss's law, the contributions from the orientifold planes and the D-branes have to cancel. We would like to emphasize that for orientifolds, the presence of D-branes in the background is in most cases not an option but a necessity.

Which are the appropriate orientifolds to consider so that intersecting D6-branes might cancel the tadpoles? Clearly we need O6-planes, meaning that the world-sheet parity has to be dressed with an involution, locally reflecting three out of the six internal coordinates, and should be a symmetry of the internal space. Let us assume that $\mathcal{M}$ admits a complex structure so that we locally can introduce complex coordinates z^i. Now, we consider Type IIA string theory divided out by $\Omega\bar{\sigma}(-1)^{F_L}$, where F_L denotes the left-moving space-time fermion number[2] and $\bar{\sigma}$ an isometric anti-holomorphic involution of $\mathcal{M}$. This acts on the Kähler class J and the holomorphic covariantly constant three-form Ω_3 as

$$\bar{\sigma} J = -J, \quad \bar{\sigma}\Omega_3 = e^{2i\varphi}\bar{\Omega}_3$$ 11.

with $\varphi \in \mathbb{R}$. For $\varphi = 0$ in local coordinates this can be thought of as complex conjugation. As a result we get an orientifold O6-plane localized at the fixed point locus of $\bar{\sigma}$, which topologically is a three-cycle π_{O6} in $H_3(\mathcal{M}, \mathbb{Z})$. To cancel the resulting massless tadpoles, we introduce appropriate configurations of intersecting D6-branes wrapping homological three-cycles π_a. For $\bar{\sigma}$ to be a symmetry of the brane configuration, one also needs to wrap D6-branes on the $\bar{\sigma}$ image three-cycles π'_a. As a new feature, in orientifold models it is also possible to get orthogonal and symplectic gauge symmetries. The rule is very simple. If a three-cycle is invariant under the anti-holomorphic involution one gets either $SO(2N_a)$ or $SP(2N_a)$ gauge symmetry; if the cycle is not invariant, one gets $U(N_a)$. Figure 4 (see color insert) depicts in a simplified way the set-up discussed in this section.

2.5. R-R Tadpole Cancellation

As we have mentioned already, tadpole cancellation provides some constraints on the positions of the O6-planes and D6-branes, which we now summarize. Historically, for deriving the tadpole cancellation conditions, one used an indirect method by first computing, using conformal field theory techniques, the one-loop Klein-bottle, annulus and Möbius strip diagrams, and extracting from

---

[2]Note that the $(-1)^{F_L}$ factor was not explicitly written down in many of the papers on intersecting D-brane models.

the corresponding tree-channel amplitudes the infrared divergences due to massless tadpoles. Employing a direct method using the Dirac Born Infeld action, here we essentially follow (43, 44), where more details of the derivation can be found.

Consider the part of the supergravity Lagrangian where the R-R field C_7 appears

$$
S = -\frac{1}{4\kappa^2} \int_{\mathbb{R}^{3,1} \times \mathcal{M}} dC_7 \wedge \star dC_7 + \mu_6 \sum_a N_a \int_{\mathbb{R}^{3,1} \times \pi_a} C_7
$$

$$
+ \mu_6 \sum_a N_a \int_{\mathbb{R}^{3,1} \times \pi_a'} C_7 - 4\mu_6 \int_{\mathbb{R}^{3,1} \times \pi_{O6}} C_7, \qquad 12.
$$

where the ten-dimensional gravitational coupling is $\kappa^2 = \frac{1}{2}(2\pi)^7(\alpha')^4$ and the R-R charge of a D6-brane reads $\mu_6 = (\alpha')^{-\frac{7}{2}}/(2\pi)$.[6] Note that here we have assumed that the orientifold planes are of type $O^{(-,-)}$, i.e., they carry negative tension and R-R charge. Recall that D-branes in this convention carry positive tension and R-R charge. Such models have also been called orientifolds without vector structure. The resulting equation of motion for the R-R field strength $G_8 = dC_7$ is

$$
\frac{1}{\kappa^2} d \star G_8 = \mu_6 \sum_a N_a \delta(\pi_a) + \mu_6 \sum_a N_a \delta(\pi_a') - 4\mu_6 \delta(\pi_{O6}), \qquad 13.
$$

where $\delta(\pi_a)$ denotes the Poincaré dual three-form of π_a. Since the left-handed side in Equation 13 is exact, the R-R tadpole cancellation condition boils down to just a simple condition on the homology classes

$$
\sum_a N_a(\pi_a + \pi_a') - 4\pi_{O6} = 0. \qquad 14.
$$

This condition implies that the homological sum of the cycles that all the D-branes and the orientifold planes wrap is trivial. This is a restrictive condition but it is moderate enough to admit non-trivial solutions with branes are not simply placed right on top of the orientifold plane. Note that so far we have not assumed supersymmetry and that therefore Equation 14 does not automatically guarantee the NS-NS tadpoles to be canceled as well.

However, it is important to note that the above method, using the Dirac Born Infeld action together with supergravity, does not take into account all of the R-R charges carried by the D-branes. The reason is that D-brane charges are classifed by K-theory groups rather than homology groups, and the R-R fields in general are not simply p-forms (see, e.g., (45) for a more detailed discussion). Indeed, as pointed out in (46), the cancellation of homological R-R charges [i.e., the conditions (15) above] are not sufficient to ensure that all the R-R tadpoles vanish and hence the consistency of the models. The inconsistencies due to uncanceled K-theory charges would show up as discrete global anomalies (47) either in the low-energy spectrum or on the world-volume of a probe D-brane (46). One way to

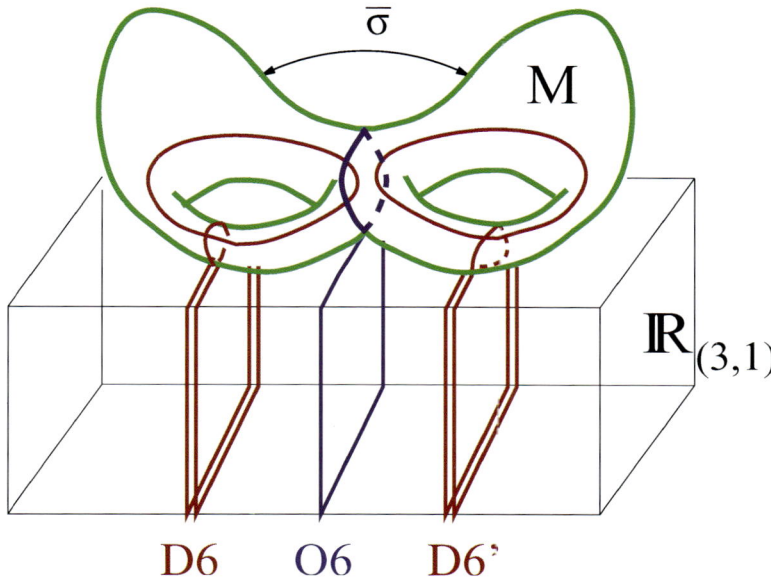

Figure 4 Schematic image of an $\Omega\bar{\sigma}(-1)^{F_L}$ orientifold with O6-planes and intersecting D6-branes. In reality the O6-plane and the $D6$-branes would cover the entire flat Minkowski space.

heuristically derive these constraints is to introduce probe D-branes with an $Sp(2n)$ gauge group, and require the total number of fundamental representations in their world-volume theory to be even. These K-theory constraints are widely unnoticed in the model building literature because for simple models, they are automatically satisfied. However, these consistency constraints are far from trivial. For example, such K-theory constraints for the $\mathbb{Z}_2 \times \mathbb{Z}_2$ orientifold were derived in (48, 49) and have been shown to play an important role in the construction of more realistic models. We will discuss such constraints in more detail in Subsection 2.10.

2.6. The Massless Spectrum

For model building purposes, it is very important to have control over the massless spectrum arising from any kind of string compactification. For the orientifold models with intersecting D6-branes, the chiral spectrum arising from the various open string sectors can be determined just from the intersection numbers of the three-cycles that the D6-branes are wrapped around. For simplicity let us assume that all D6-branes wrap three-cycles not invariant under the anti-holomorphic involution, so that the gauge symmetry is $\Pi_a U(N_a)$. For this case the general rule for determining the massless left-handed chiral spectrum is presented in Table 1. Open strings stretched between a D-brane and its $\bar{\sigma}$ image are the only ones left invariant under the combined operation $\Omega\bar{\sigma}(-1)^{F_L}$. Therefore, they transform in the antisymmetric or symmetric representation of the gauge group, indicating that the price that we have to pay by considering intersecting D-branes in an orientifold background is that more general representations are possible for the chiral fermions. Sometimes this is an advantage, like for constructing $SU(5)$ Grand Unified Models, but sometimes the absence of such fermions imposes new conditions on the possible D-brane set-ups.

The rule for the chiral spectrum in Table 1 is completely general and, as was demonstrated in (43), the chiral massless spectra from many orientifold models discussed using conformal field theory methods in the existing literature can be understood in this framework.

Moreover, one can easily check that the R-R tadpole cancellation condition (15) together with the chiral spectrum listed in Table 1 guarantees the absence of non-Abelian gauge anomalies. Naively, there exist Abelian and mixed Abelian-non-Abelian anomalies, as well as gravitational anomalies. However, we shall see

TABLE 1 Chiral spectrum for intersecting D6-branes

Representation	Multiplicity
$\boxminus_a$	$\frac{1}{2}(\pi'_a \circ \pi_a + \pi_{O6} \circ \pi_a)$
$\boxplus_a$	$\frac{1}{2}(\pi'_a \circ \pi_a - \pi_{O6} \circ \pi_a)$
$(\bar{\Box}_a, \Box_b)$	$\pi_a \circ \pi_b$
$(\Box_a, \Box_b)$	$\pi'_a \circ \pi_b$

in the subsequent section that all of these are canceled by a generalized Green-Schwarz mechanism.

To apply Table 1 to concrete models, one has to compute the intersection numbers of three-cycles, which by itself is not an easy task. However, there exist backgrounds for which generic rules can be presented. In addition to the simplest case of just a torus T^6, toroidal orbifolds, such as $T^6/\mathbb{Z}_N$ or $T^6/\mathbb{Z}_N \times \mathbb{Z}_M$, are natural candidates for string backgrounds. Therfore, let us discuss the application of Table 1 to such orbifolds in some more detail.

The spectrum in Table 1 is meant to be computed using the intersection numbers on the resolved orbifold and not on the ambient torus. There are some three-cycles π_a on the orbifold space that are inherited from the torus. In the Kaluza-Klein reduction on the orbifold, they correspond to massless modes in the untwisted closed string sector of the theory. In general, three-cycles π_a^t on the torus are arranged in orbits of length N under a $\mathbb{Z}_N$ orbifold group, i.e.,

$$\pi_a^o = \sum_{j=0}^{N-1} \Theta^j \, \pi_a^t, \qquad\qquad 15.$$

where Θ denotes the generator of $\mathbb{Z}_N$. Such an orbit can then be considered as a three-cycle of the orbifold, where the intersection number is given by

$$\pi_a^o \circ \pi_b^o = \frac{1}{N} \left(\sum_{j=0}^{N-1} \Theta^j \pi_a^t \right) \circ \left(\sum_{k=0}^{N-1} \Theta^k \pi_b^t \right). \qquad\qquad 16.$$

In addition to these untwisted three-cycles, certain twisted sectors of the orbifold action can give rise to so-called twisted three-cycles, which correspond to massless fields in the twisted sectors of the orbifold. Since these twisted three-cycles are not explicitly needed in this article, we refer the reader to the existing literature (43, 50–52) to see how these twisted cycles can be appropriately dealt with.

Table 1 only gives the chiral spectrum of an intersecting D6-brane model. To compute the generally moduli-dependent non-chiral spectrum, one has to employ the usual techniques of conformal field theory. Therefore, the Higgs sector of a given model is under less analytic control than the chiral matter sector.

2.7. Generalized Green-Schwarz Mechanism

Given the chiral spectrum of Table 1, we have stated that the non-abelian gauge anomalies of all $SU(N_a)$ factors in the gauge group vanish. On the other hand, the Abelian, the mixed Abelian-non-Abelian, and the mixed Abelian-gravitational anomalies naively do not. However, as string theory is a consistent theory, it provides another mechanism to cancel these anomalies. This is the so-called Green-Schwarz mechanism (53), which can be generalized to the intersecting D-brane case (16). Here let us discuss in some more detail the mixed Abelian-non-Abelian anomalies.

Computing the $U(1)_a - SU(N_b)^2$ anomalies in the effective four-dimensional gauge theory one finds

$$A_{ab} = \frac{N_a}{2}(-\pi_a + \pi_a') \circ \pi_b. \qquad 17.$$

for each pair of stacks of D-branes. On each stack of D6-branes there exist Chern-Simons couplings of the form

$$\int_{\mathbb{R}^{1,3} \times \pi_a} C_3 \wedge \mathrm{Tr}(F_a \wedge F_a), \quad \int_{\mathbb{R}^{1,3} \times \pi_a} C_5 \wedge \mathrm{Tr}\,(F_a) \qquad 18.$$

where F_a denotes the gauge field on the D6$_a$-brane. Now we expand every three-cycle π_a and π_a' into an integral basis (α^I, β_J) of $H_3(M, \mathbb{Z})$ with $I, J = 0, \ldots,$ h_{21}.

$$\pi_a = e_I^a \alpha^I + m_a^J \beta_J, \quad \pi_a' = \left(e_I^a\right)' \alpha^I + \left(m_a^J\right)' \beta_J. \qquad 19.$$

This allows us to define the four-dimensional axions Φ_I and 2-forms B^I as

$$\Phi_I = \int_{\alpha^I} C_3, \quad \Phi^{I+h^{(2,1)}+1} = \int_{\beta_I} C_3,$$

$$B^I = \int_{\beta_I} C_5, \quad B_{I+h^{(2,1)}+1} = \int_{\alpha^I} C_5. \qquad 20.$$

In four dimensions $(d\Phi_I, dB^I)$ and $(d\Phi^{I+h^{(2,1)}+1}, dB_{I+h^{(2,1)}+1})$ are Hodge dual to each other. The general couplings (19) can now be dimensionally reduced to four dimensions and yield axionic couplings of the form

$$\int_{\mathbb{R}^{1,3}} \Phi_I \wedge \mathrm{Tr}\,(F_a \wedge F_a), \qquad \int_{\mathbb{R}^{1,3}} B^I \wedge F_a. \qquad 21.$$

The tree-level contribution to the mixed gauge anomaly described by these couplings takes the form depicted in Figure 5, and, adding up all these terms taking the R-R tadpole conditions into account, one can show that the result has precisely the form (18) and cancels the field theoretic anomaly (43). By the same mechanism, the Abelian and mixed gravitational-gauge anomalies are canceled, where the latter arise from the $U(1)_a - G - G$ triangle diagram and are given by

$$A_a^{(G)} = 3 N_a \pi_{O6} \circ \pi_a. \qquad 22.$$

This anomaly is canceled by the Chern-Simons coupling

$$\int_{\mathbb{R}^{1,3} \times \pi_a} C_3 \wedge \mathrm{Tr}\,(R \wedge R). \qquad 23.$$

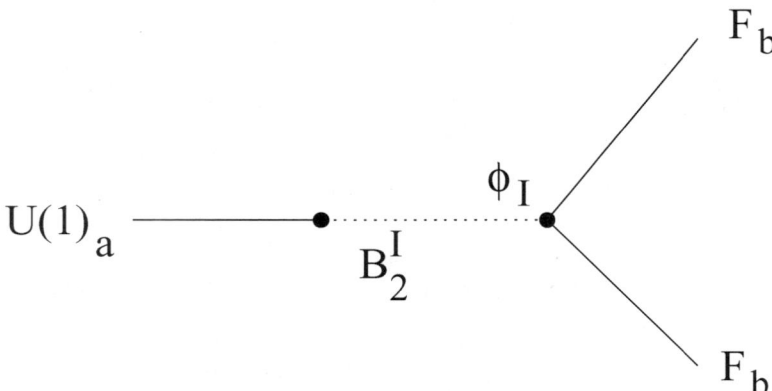

Figure 5   Green-Schwarz mechanism.

A second important effect of these couplings is that some of the $U(1)$ gauge fields pair up with the axions to become massive. The axionic couplings have the detailed form

$$\int_{\mathbb{R}^{1,3}} N_a(\pi_{a,I} - \pi'_{a,I}) \, B^I \wedge F_a \qquad 24.$$

with $\pi_{a,I} \in \{e_I^a, m_a^I\}$ depending on the index I. It has been pointed out in (18, 54) that in general, not only the anomalous $U(1)$s receive a mass, but also some of the anomaly-free ones, which are given by the kernel of the matrix

$$M_a^I = N_a(\pi_{a,I} - \pi'_{a,I}). \qquad 25.$$

Therefore, to determine the low-energy spectrum, one has to carefully analyze these quadratic couplings. The massive $U(1)$s still give rise to perturbative global $U(1)$ symmetries of the low-energy effective action (55), which severely constrain the allowed couplings.

2.8. Supersymmetric D-Branes

So far we were not assuming anything more about the D6-branes than that they are wrapping some homological three-cycles in the background geometry. If one is interested in supersymmetric models, further constraints on the bulk geometry and the cycles on which the D6-branes wrap have to be imposed. Throughout this section we assume that $\mathcal{M}$ is a Calabi-Yau manifold so that the closed string bulk sector of the Type IIA orientifold preserves $\mathcal{N} = 1$ supersymmetry. First of all, one has to require that each D-brane by itself preserves supersymmetry, i.e., it has to be a BPS brane. As was shown in (56), this implies that the three-cycles the

D6-branes are allowed to wrap have to be special Lagrangian (sLag) cycles, which are defined as follows.

On a Calabi-Yau manifold, there exist a covariantly constant holomorphic three-form, Ω_3, and a Kähler 2-form J. A three-cycle π_c is called Lagrangian if the restriction of the Kähler form on the cycle vanishes

$$J|\pi_a = 0. \qquad 26.$$

If the three-cycle in addition is volume minimizing, which can be expressed as the property that the imaginary part of the three-form Ω_3 vanishes when restricted to the cycle,

$$\Im\left(e^{i\varphi_a}\Omega_3\right)|_{\pi_a} = 0, \qquad 27.$$

then the three-cycle is called a sLag cycle. The parameter φ_a determines which $\mathcal{N} = 1$ supersymmetry is preserved by the brane. Thus, different branes with different values for φ_a preserve different $\mathcal{N} = 1$ supersymmetries. One can show that (28) implies that the volume of the three-cycle is given by

$$\mathrm{Vol}(\pi_a) = \left|\int_{\pi_a} \Re(e^{i\varphi_a}\Omega_3)\right|. \qquad 28.$$

A shift of $\varphi_a \rightarrow \varphi_a + \pi$ corresponds to exchanging a D-brane by its anti-D-brane, where the D-brane really satisfies (29) without taking the absolute value. Therefore a supersymmetric cycle π_a is calibrated with respect to $\Re(e^{i\varphi_a}\Omega_3)$.

Let us define locally the holomorphic 3-form Ω_3 and the Kähler form J by

$$\Omega_3 = dz_1 \wedge dz_2 \wedge dz_3, \quad J = i\sum_{i=1}^{3} dz_i \wedge d\bar{z}_i. \qquad 29.$$

Let us choose $\bar{\sigma}$ to be just complex conjugation in local coordinates. Then from $\bar{\sigma}(\Omega_3) = \bar{\Omega}_3$ and $\bar{\sigma}(J) = -J$ it follows that the fixed three-cycle of the anti-holomorphic involution is a sLag cycle with $\varphi_a = 0$. Therefore, to finally obtain a globally $\mathcal{N} = 1$ supersymmetric intersecting D-brane model, all D6-branes have to wrap sLag three-cycles, which are calibrated with respect to the same three-form $\Re(\Omega_3)$. It has been checked in (43, 57) that for such globally supersymmetric configurations, the NS-NS tadpoles cancel precisely if the R-R tadpoles are canceled. One can show that if two branes are relatively supersymmetric, one of the four complex world-sheet bosons becomes massless and extends the massless chiral fermion at the intersection point to a complete $\mathcal{N} = 1$ chiral supermultiplet.

In section 2.10.1 we shall see that for the toroidal (orbifold) compactifications, this condition becomes a simple geometric condition on the intersection angles of each D-brane with respect to the orientifold plane.

2.9. Lift to G_2 Compactifications of M-Theory

In this section we briefly discuss the relation between intersecting D6-brane models and M-theory compactifications on G_2 manifolds. This needs some more advanced mathematical notions and is not really relevant for understanding the rest of the review. For completeness, however, we summarize some key ideas here.

Globally $\mathcal{N} = 1$ supersymmetric intersecting D6-brane models have also shed light on how chiral fermions arise in G_2 compactifications of M theory. D6-branes and O6-planes are special because they correspond to pure geometry at strong coupling (unlike other branes which carry additional sources, i.e., M-branes or G-fluxes). Therefore, from the number of supercharges that the background preserves, the globally $\mathcal{N} = 1$ supersymmetric intersecting D6-brane models are expected to lift up to eleven-dimensional M-theory compactification on singular G_2 manifolds (21, 58–61). In the Type IIA picture, chiral fermions are localized at the intersection of D6-branes. Away from the intersections of IIA D6-branes and/or O6-planes, the IIA configuration corresponds to D6-branes and O6-planes wrapped on (disjoint) smooth supersymmetric three-cycles, which we denote generically by Q. The corresponding G_2 holonomy space hence corresponds to fibering a suitable Hyperkähler four-manifold over each component of Q. That is an A-type ALE space singularity for N overlapping D6-branes, and a D-type ALE space for D6-branes on top of O6-planes [with the Atiyah-Hitchin manifold for no D6-brane, and its double covering for two D6-branes etc., as follows from (62, 63)]. Intersections of objects in type IIA therefore lift to co-dimension 7-singularities, which are isolated up to orbifold singularities. It is evident from the IIA picture that the chiral fermions are localized at these singularities.

The structure of these singularities has been studied directly in the G_2 context in (58). One starts by considering the (possibly partial) smoothing of a Hyperkähler ADE singularity to a milder singular space, parameterized by a triplet of resolution parameters (D-terms or moment maps in the Hyperkähler construction of the space). The kind of 7-dimensional singularities of interest are obtained by considering a three-dimensional base parameterizing the resolution parameters, on which one fibers the corresponding resolved Hyperkähler space. The geometry is said to be the unfolding of the higher singularity into the lower one. This construction guarantees that the total geometry admits a G_2 holonomy metric. To determine the matter content arising from the singularity, one decomposes the adjoint representation of the ADE group associated with the higher singularity with respect to that of the lower. One obtains chiral fermions with quantum numbers in the corresponding coset, and multiplicity is given by an index which for an isolated singularity is one. This construction arises in the M-theory lift of the intersecting D6-brane models. For example, at points where two stacks of N D6-branes and M D6-branes intersect, the M-theory lift corresponds to a singularity of the G_2 holonomy space that represents the unfolding of an A_{M+N-1} singularity into a 4-manifold with an A_{M-1} and an A_{N-1} singularity. By the decomposition of the adjoint representation of A_{M+N-1}, we expect the charged matter to be in the bi-fundamental representation of the $SU(N) \times SU(M)$ gauge group, in agreement with the IIA picture. A different kind of intersection arises when N D6-branes intersect with an O6-plane, and consequently with the N D6-brane images. The M-theory lift corresponds to the unfolding of a D_N type singularity into an A_{N-1} singularity. The decomposition of the adjoint representation predicts the appearance of chiral fermions in the antisymmetric representation of $SU(N)$, in agreement with the IIA picture.

2.10. Examples

So far we have presented the main conceptual ingredients for constructing intersecting D6-brane models in a fairly general way. In order to see how this formalism works, let us work out two simple examples in more detail.

2.10.1. INTERSECTING D6-BRANES ON THE TORUS   As in section 2.2, we assume that the six-dimensional torus factorizes as $T^6 = T^2 \times T^2 \times T^2$. Introducing complex coordinates $z^i = x^i + iy^i$ on the three T^2 factors, the anti-holomorphic involution $\bar{\sigma}$ is chosen to be just complex conjugation $z^i \to \bar{z}^i$. Then as shown in Figure 6, on each T^2 there exist two different choices of the complex structure, which are consistent with the anti-holomorphic involution.

Next we introduce factorizable D6-branes, which are specified by wrapping numbers (n^i, m^i) along the fundamental cycles $[a^i]$ and $[b^i]$ respectively $[a'^i]$ and $[b^i]$ on each T^2. It is useful to express also the branes for the tilted tori in terms of the untilted 1-cycles $[a^i]$ and $[b^i]$ by writing $[a'^i] = [a^i] + \frac{1}{2}[b^i]$. Then a three-cycle can be written as a product of three 1-cycles

$$\pi_a = \prod_{i=1}^{3} \left(n_a^i [a^i] + \widetilde{m}_a^i [b^i] \right). \qquad\qquad 30.$$

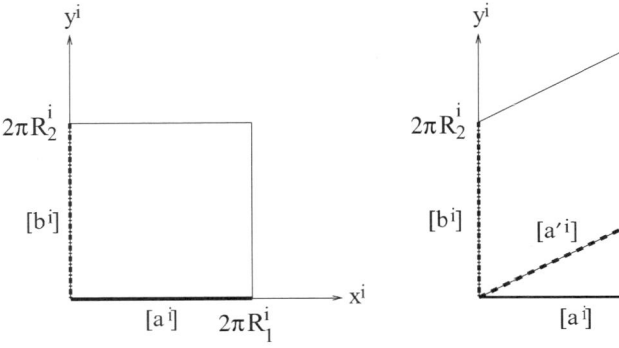

Figure 6   Choices of T^2s.

with $\tilde{m}_a^i = m_a^i$ for untilted tori and $\tilde{m}_a^i = m_a^i + \frac{1}{2}n_a^i$ for tilted ones. Using the fundamental intersection number $[a^i] \circ [b^i] = -1$ with all the remaining ones vanishing, the intersection number between two three-cycles can be computed as

$$I_{ab} = \prod_{i=1}^{3} \left(n_a^i \tilde{m}_b^i - \tilde{m}_a^i n_b^i \right) = \prod_{i=1}^{3} \left(n_a^i m_b^i - m_a^i n_b^i \right). \tag{31.}$$

To work out the tadpole cancellation conditions, one has to determine the three-cycle of the O6-plane and the action of the anti-holomorphic involution on the D6-branes.

Independent of the tilt on each T^2, the O6-plane is wrapping the cycle $2[a^i]$, so that the entire three-cycle reads $\pi_{O6} = 8\,\Pi_i[a^i]$. The action of $\bar{\sigma}$ on a general three-cycle is simply $(n^i, \tilde{m}^i) \to (n^i, -\tilde{m}^i)$. Expanding the general tadpole cancellation condition for the homological R-R charges (15), one obtains the four independent equations

$$[a^1][a^2][a^3]: \sum_{a=1}^{K} N_a \prod_i n_a^i = 16$$

$$[a^i][b^j][b^k]: \sum_{a=1}^{K} N_a n_a^i \tilde{m}_a^j \tilde{m}_a^k = 0, \quad \text{with} \quad i \neq j \neq k \neq i. \tag{32.}$$

These formulas were derived initially in (5) using a conformal field theory approach. As discussed in Subsection 2.5, these tadpole conditions should be supplemented with some additional K-theory constraints. The K-theory constraints for the toroidal orientifold were derived in (26), and together with the tadpole conditions (33) above provide the main constraints for building semi-realistic Standard-like Models in toroidal orientifolds.

In evaluating the supersymmetry conditions we first consider the non-compact situation and a factorizable D-brane which intersects the x^i-axes on each T^2 at an angle φ_a^i. With J and Ω_3 chosen as in (30), we notice that a factorizable D-brane always satisfies $J|_{\pi_a} = 0$. Expanding the second condition $\Im(\Omega_3)|_{\pi_a} = 0$ leads to

$$0 = (dy^1\,dy^2\,dy^3 - dy^1\,dx^2\,dx^3 - dx^1\,dy^2\,dx^3 - dx^1\,dx^2\,dy^3)|_{\pi_a}. \tag{33.}$$

Using $\frac{dy^i}{dx^i}\big|_{\pi_a} = \frac{\tilde{m}_a^i}{n_a^i} u^i$ with $u^i = \frac{R_2^i}{R_1^i}$, this can be brought to the form

$$\prod_{i=1}^{3} \tilde{m}_a^i - \sum_{i \neq j \neq k \neq i} \tilde{m}_a^i\, n_a^j n_a^k (u^j u^k)^{-1} = 0. \tag{34.}$$

A further constraint arises from the condition $\Re(\Omega_3)|_{\pi_a} > 0$, which takes the form

$$\prod_{i=1}^{3} n_a^i - \sum_{i \neq j \neq k \neq i} n_a^i\, \tilde{m}_a^j \tilde{m}_a^k (u^j u^k) > 0. \tag{35.}$$

These two conditions are equivalent to the maybe more familiar supersymmetry condition

$$\phi_a^1 + \phi_a^2 + \phi_a^3 = 0 \text{ mod } 2\pi. \tag{36.}$$

We conclude that for a given D-brane with definite wrapping numbers, the supersymmetry condition (37) puts a constraint on the complex structure moduli $u^i = \frac{R_2^i}{R_1^i}$. If all $\phi_a^i \neq 0$ then the D6-branes preserve $\mathcal{N} = 1$ supersymmetry, and if some angles are vanishing either $\mathcal{N} = 2$ or the maximal $\mathcal{N} = 4$ supersymmetry is preserved.

Using factorizable branes on T^6, no non-trivial, globally supersymmetric, tadpole-cancelling configuration of intersecting D6-branes exists. The physical reason for this is that the moment the D6-branes do not lie entirely on the x-axes, the tension of the branes in the perpendicular y-directions cannot be compensated, as there are no orientifold planes with negative tension along these directions. In the T-dual picture with magnetic fluxes, more general non-factorizable configurations of branes were investigated (64, 65).

2.10.2. INTERSECTING D6-BRANES ON THE $\mathbb{Z}_2 \times \mathbb{Z}_2$ ORBIFOLD   To obtain nontrivial supersymmetric models, one needs more orientifold planes extending also along y-directions. The easiest way to obtain these is by considering not just tori but toroidal orbifolds, of which the $\mathbb{Z}_2 \times \mathbb{Z}_2$ orbifold is the simplest (13, 20, 21). The orbifold action of the two $\mathbb{Z}_2$ symmetries is defined as

$$\Theta: \begin{cases} z_1 \rightarrow -z_1 \\ z_2 \rightarrow -z_2 \\ z_3 \rightarrow z_3 \end{cases} \quad \Theta': \begin{cases} z_1 \rightarrow z_1 \\ z_2 \rightarrow -z_2 \\ z_3 \rightarrow -z_3. \end{cases} \tag{37.}$$

As it stands, this model is not completely defined; there are two possible choices for the signs of the action of Θ' in the Θ twisted sector and vice versa. This freedom is called discrete torsion, and here we consider the model in which one keeps the $(1, 1)$ forms in the twisted sectors and kills the $(2, 1)$ forms. Therefore, this model has the Hodge numbers $(h_{21}, h_{11}) = (3, 51)$, which means that there are precisely eight three-cycles in the untwisted sector. These are

$$[a^1][a^2][a^3]^t, \ [a^i][a^j][b^k]^t, \ [a^i][b^j][b^k]^t, \ [b^1][b^2][b^3]^t \tag{38.}$$

with $i \neq j \neq k \neq i$, and where the upper index indicates that so far these three-cycles are defined on the ambient T^6.

Given the rules from section 2.6 about how to deal with three-cycles in the orbifold case, we have to carefully distinguish between three-cycles in the ambient T^6 and three-cycles on the orbifold space. Under the action of $\mathbb{Z}_2 \times \mathbb{Z}_2$, a three-cycle on T^6 has 3 images, which homologically are identical to the original cycles. Therefore a three-cycle in the bulk of the orbifold space can be identified with $\pi^B = 4\pi^t$. Applying the rule for the intersection number we get $\pi_a^B \circ \pi_b^B = 4\pi_a^t \circ \pi_b^t$.

Therefore, the cycles π_a^B do not span the integral homology lattice $H_3(M, \mathbb{Z})$, which suggests that there exist smaller three-cycles in the orbifold space. This is indeed the case. By choosing the three-cycles to run through the origin, we obtain three-cycles which are given by $\pi_a^o = \frac{1}{2}\pi_a^B$, which have intersections on the orbifold $\pi_a^o \circ \pi_b^o = \pi_a^t \circ \pi_b^t$. Therefore, the untwisted three-cycles on the orbifold space have the same form as in (31) and the same intersection form as in (32), with the only difference that the basis of three-cycles is now defined on the orbifold space instead of the torus.

Working out the fixed point locus of the four orientifold projections $\Omega\bar{\sigma}(-1)^{F_L}$, $\Omega\bar{\sigma}\Theta(-1)^{F_L}$, $\Omega\bar{\sigma}\Theta'(-1)^{F_L}$, and $\Omega\bar{\sigma}\Theta\Theta'(-1)^{F_L}$ and expressing everything in terms of three-cycles in the orbifold, we obtain

$$\pi_{O6} = 4\prod_i [a^i]^o - \sum_{i \neq j \neq k \neq i} 4^{1-\beta_j-\beta_k}[a^i][b^j][b^k]^o, \qquad 39.$$

where $\beta_j = 0$ for an untilted T^2 factor and $\beta_j = 1/2$ for a tilted one. Therefore, the four tadpole cancellation conditions for the homological R-R charges now read[3]

$$[a^1][a^2][a^3]^o: \sum_{a=1}^K N_a \prod_i n_a^i = 8,$$

$$[a^i][b^j][b^k]^o: \sum_{a=1}^K N_a n_a^i \tilde{m}_a^j \tilde{m}_a^k = -2^{3-2\beta_j-2\beta_k}, \quad \text{with} \quad i \neq j \neq k \neq i. \qquad 40.$$

Note the changes on the right-hand side of (41) as compared to the purely toroidal case (33). To ensure consistency of the models, these tadpole conditions should be supplemented with additional K-theory constraints. For untilted tori, the K-theory constraints for this $\mathbb{Z}_2 \times \mathbb{Z}_2$ orientifold (49) read:

$$\sum_{a=1}^K N_a \prod_i m_a^i \in 2\mathbb{Z},$$

$$\sum_{a=1}^K N_a n_a^i n_a^j m_a^k \in 2\mathbb{Z}, \quad \text{with} \quad i \neq j \neq k \neq i. \qquad 41.$$

It is straightforward to generalize these conditions to cases where some or all of the tori are tilted.

Finally, for the intersection number between a D-brane and the orientifold plane one obtains

---

[3]In (61) a different convention was used such that there appeared an overall factor of two in all four R-R tadpole cancellation in conditions (41). This is consistent with the rank of the gauge group, as the rule in (61) was that a stack of N_a branes carries a gauge group $U(N_a/2)$.

$$\pi_{O6} \circ \pi_a^o = 4 \prod_i \widetilde{m}_a^i - \sum_{i \neq j \neq k \neq i} 4^{1-\beta_:-\beta_k} \widetilde{m}_a^i n_a^j n_a^k. \qquad 42.$$

The supersymmetry conditions are the same as for the toroidal case.

The equations developed in the last two subsections provide the main tools for constructing quasi-realistic intersecting D-brane models in these two most simple backgrounds.

3. SEMI-REALISTIC INTERSECTING D-BRANE MODELS

In this section we give an overview of the different intersecting D-brane world models explicitly constructed so far. Essentially, there are two philosophical attitudes toward approaching this problem, which differ in their assumptions about the size of the string scale, i.e., the energy scale where stringy effects become relevant. In particular, because of stability and phenomenological considerations it is usually assumed that M_s is low (e.g., $1 - 100$ TeV) for non-supersymmetric constructions, although supersymmetric studies usually assumed that M_s is much closer to the Planck scale.

3.1. Non-Supersymmetric Standard-Like Models

Here we review different approaches to construct semi-realistic non-supersymmetric Standard-like Models and highlight some fairly general phenomenological features of such models. The first explicit chiral intersecting D-brane models were constructed in (5, 7, 15–18), where, except in (7, 15, 16), the background space was simply chosen to be an $\Omega \bar{\sigma} (-1)^{F_L}$ orientifold of a factorizable torus T^6. These articles triggered a lot of subsequent work using essentially the same framework and ideas but generalizing the D-brane set-ups in certain ways. Before we list all these different constructions, we would like to present a prototype model, which shows that the particle content of intersecting D-brane models can come quite close to that of the Standard Model.

3.1.1. A SIMPLE SEMI-REALISTIC MODEL   In (18) the authors were considering the simple toroidal orientifold set-up mentioned above. Using a bottom-up approach, they introduced four stacks of D6-branes with the wrapping numbers chosen as shown in Table 2. The intersection numbers between these four stacks of D6-branes give rise to the chiral fermions listed in Table 3, which transform in the various bi-fundamental representations of the (naive) gauge group $SU(3)_C \times SU(2)_W \times U(1)_a \times U(1)_b \times U(1)_c \times U(1)_d$. The hypercharge is given by the linear combination $Q_Y = \frac{1}{6} Q_a - \frac{1}{2} Q_c + \frac{1}{2} Q_d$. For more details and the phenomenological implications, we refer the reader to (18). Here we would like to simply list some features typical for such intersecting D-brane models:

TABLE 2 Wrapping numbers for a semi-realistic non-supersymmetric model. The parameters are defined as $\bar{\beta}^{1,2} = 1 - \beta^{1,2}$, $\beta^3 = 1/2$, $\rho = 1, 1/3$, $\epsilon = \pm 1$ and $n_a^2, n_b^1, n_c^1, n_d^2 \in \mathbb{Z}$

N_a	$(n^1, \widetilde{m}^1)$	$(n^2, \widetilde{m}^2)$	$(n^3, \widetilde{m}^3)$
$N_a = 3$	$(1/\bar{\beta}^1, 0)$	$\left(n_a^2, -\epsilon\bar{\beta}^2\right)$	$(1/\rho, -1/2)$
$N_b = 2$	$\left(n_b^1, \epsilon\bar{\beta}^1\right)$	$(1/\bar{\beta}^2, 0)$	$(1, -3\rho/2)$
$N_c = 1$	$\left(n_c^1, -3\rho\epsilon\bar{\beta}^1\right)$	$(1/\bar{\beta}^2, 0)$	$(0, -1)$
$N_d = 1$	$(1/\bar{\beta}^1, 0)$	$\left(n_d^2, \bar{\beta}^2\epsilon/\rho\right)$	$(1, -3\rho/2)$

- There are many (infinite) non-supersymmetric intersecting D-brane constructions with the Standard Model particle spectrum, where in most cases one needs additional "hidden" branes to satisfy tadpole cancellation.

- Models with only bi-fundamental matter necessarily contain right-handed neutrinos.

- One obtains additional $U(1)$ factors, which partly receive a mass via the generalized Green-Schwarz mechanism (see section 2.7); the condition that $U(1)_Y$ remains massless imposes further conditions on the parameters in Table 2.

- All the massive former $U(1)$ gauge symmetries survive as perturbative global symmetries and can be identified with baryon number Q_a and lepton number Q_d, stabilizing the proton and preventing Majorana neutrino masses.

- One can show that in the (bc) sector there are additional non-chiral (tachyonic) fields, which might have an interpretation as Higgs particles; condensation of these fields corresponds to D-brane recombination in string theory (see for instance (16, 66–70)).

TABLE 3 Chiral massless spectrum of the semi-realistic four stack model. $(.)^c$ denotes the charge conjugated field

Intersection	Matter	Rep.	Y
(a, b)	Q_L	$(3, 2)_{(1,-1,0,0)}$	$1/6$
(a', b)	q_L	$2 \times (3, 2)_{(1,1,0,0)}$	$1/6$
(a, c)	$(U_R)^c$	$3 \times (\bar{3}, 1)_{(-1,0,1,0)}$	$-2/3$
(a', c)	$(D_R)^c$	$3 \times (\bar{3}, 1)_{(-1,0,-1,0)}$	$1/3$
(b', d)	L_L	$3 \times (1, 2)_{(0,-1,0,-1)}$	$-1/2$
(c, d)	$(E_R)^c$	$3 \times (1, 1)_{(0,0,-1,1)}$	1
(c', d)	$(N_R)^c$	$3 \times (1, 1)_{(0,0,1,1)}$	0

Because the models are not supersymmetric, there are typically uncanceled NS-NS tadpoles, contributing to the (dilaton-dependent) cosmological constant which is of the order of M_s^4. In addition, in the effective theory below the string scale, there are large radiative corrections of the order of M_s. Therefore, typically these models require M_s of the order of the TeV scale. However, as emphasized in Section 2.1, for the toroidal constructions with intersecting D6-branes, the internal space cannot be much larger than the Planck volume, and the string scale M_s is restricted to be of the order of the Planck scale.[4]

3.1.2. GENERALIZATIONS   Many generalizations of the above construction have been considered in the literature. Here we only list, in non-chronological order, the ones for which only non-supersymmetric models are possible or have been considered. For more details on the various constructions we refer the reader to the original literature.

The straightforward generalization of the above set-up is to introduce more than four stacks of D6-branes (72–74) to realize directly the Standard Model gauge group. Similarly, one can try to find Grand Unified–like Models in this toroidal set-up (75, 76).

If one is giving up supersymmetry, then of course there is no need to introduce orientifold planes in the first place, and one can simply start with intersecting D6-branes in Type IIA (15, 16).

Another approach is not to work with D6-branes but instead with D4-branes, or with D5-branes in the Type IIB string theory, where in order to achieve chirality one has to perform an additional orbifold in the transverse space (15, 16). Therefore, the models constructed in (15, 16, 71, 77–85) can be regarded as a hybrid of the two ways to obtain chiral fermions, as intersecting branes at singularities.

Giving up supersymmetry one can also start with orientifolds of Type O string theory (86).

A peculiarity about intersecting D-branes has been pointed out in (57, 67, 87), namely that one can build models in which at each intersection between two branes an $\mathcal{N} = 1$ supersymmetry is preserved, even though it is not preserved globally. In such models the absence of one-loop corrections to the Higgs mass weakens the gauge hierarchy problem and allows one to enhance the string scale up to 10 TeV. Such so-called quasi-supersymmetric models have also been studied in (88, 89) from a field theory perspective, and additional models have been constructed in (90).

In (19, 91) Type IIA orientifolds on the $\mathbb{Z}_3$ orbifold were considered and a non-supersymmetric three-generation flipped $SU(5)$ Grand Unified Model (92) was constructed explicitly.

---

[4]This is not, however, a fundamental problem of the intersecting brane world scenario because the chiral spectrum of Table 3 can be achieved in models where the string scale can be lowered to a TeV (71).

TABLE 4   D6-brane configuration for the three-family $\mathbb{Z}_2 \times \mathbb{Z}_2$ orientifold model

Type	N_a	$(n_a^1, m_a^1) \times (n_a^2, m_a^2) \times (n_a^3, \widetilde{m}_a^3)$	Gauge group
A_1	4	$(0, 1) \times (0, -1) \times (2, \widetilde{0})$	$Q_8, Q_{8'}$
A_2	1	$(1, 0) \times (1, 0) \times (2, \widetilde{0})$	$Sp(2)_A$
B_1	2	$(1, 0) \times (1, -1) \times (1, \widetilde{3/2})$	$SU(2), Q_2$
B_2	1	$(1, 0) \times (0, 1) \times (0, \widetilde{-1})$	$Sp(2)_B$
C_1	$3 + 1$	$(1, -1) \times (1, 0) \times (1, \widetilde{1/2})$	$SU(3), Q_3, Q_1$
C_2	2	$(0, 1) \times (1, 0) \times (0, \widetilde{-1})$	$Sp(4)$

3.2. Supersymmetric Models on the $\mathbb{Z}_2 \times \mathbb{Z}_2$ Orientifold

In order to show how semi-realistic supersymmetric intersecting D-brane models can arise and what their salient features are, we now describe in more detail the four-dimensional chiral $\mathcal{N} = 1$ supersymmetric intersecting D-brane models constructed in (20, 21). Unlike the non-supersymmetric models discussed so far, these supersymmetric intersecting D-brane models are stable because both the NS-NS and the R-R tadpoles are canceled. As discussed in section 2.9, these models also have the additional interesting feature that when lifted to M-theory, they correspond to chiral G_2 compactifications (21, 61).

The background geometry of this class of models is the $T^6/\mathbb{Z}_2 \times \mathbb{Z}_2$ orbifold as described in Subsection 2.10.2. As explained, there exist two choices of complex structure of T^2 that are compatible with the orientifold symmetry: rectangular or tilted (see Figure 6). If the Standard Model sector D-branes are not on top of the orientifold planes and the T^2 are rectangular, as in the toroidal models discussed in (5), the number of chiral families is even.[5] Hence, we consider models with one tilted T^2. This slightly modifies the closed string sector but has an important impact on the open string sector because the number of chiral families can now be odd. Owing to the smaller number of O6-planes in tilted configurations, the R-R tadpole conditions (41) are very stringent for more than one tilted T^2, so we focus on models with only one tilted T^2.

To simplify the supersymmetry conditions within our search for realistic models, we consider a particular Ansatz for the intersection angles of the branes with the x-axes: $(\phi_1, \phi_2, 0)$, $(\phi_1, 0, \phi_3)$ or $(0, \phi_2, \phi_3)$ with $\sum_i \phi_i = 0$ for each brane. Focusing on tilting just the third torus, the search for theories with $U(3)$ and $U(2)$ gauge factors carried by branes at angles and three left-handed quarks turns out to be very constraining, at least within our Ansatz. A D6-brane configuration with wrapping numbers $(n_a^i, \widetilde{m}_a^i)$ which gives rise to a three-family supersymmetric Standard-like Model is presented in Table 4.

---

[5]The weak sector can come from D-branes on top of an orientifold plane. Because $Sp(2) \simeq SU(2)$, in which case an odd number of families can be obtained without tilted tori (48, 49).

The four D6-branes labeled C_1 are split into two parallel but not overlapping stacks of three and one branes leading to an adjoint breaking of $U(4)$ into $U(3) \times U(1)$. Consequently, a linear combination of the two $U(1)$s is actually a generator within the non-abelian $SU(4)$ arising for coincident branes. This ensures that this $U(1)$ is automatically non-anomalous and massless (free of linear couplings to untwisted moduli) (15, 16, 18), which turns out to be crucial for the appearance of the Standard Model hypercharge.

For convenience we consider the four D6-branes labelled A_1 to be away from the O6-planes in all three complex planes. This leads to two D6-branes that can move independently, giving rise to a gauge group $U(1)^2$, plus their Θ, Θ' and $\Omega\bar{\sigma}(-1)^{F_L}$ images. These $U(1)$s are also automatically non-anomalous and massless. In the effective theory, this corresponds to Higgsing of $USp(8)$ down to $U(1)^2$.

The surviving non-Abelian gauge group is $SU(3)_C \times SU(2)_W \times Sp(2) \times Sp(2) \times Sp(4)$. The $SU(3)_C \times SU(2)_W$ corresponds to the minimal supersymmetric Standard Model (MSSM), whereas the last three factors form a quasi-hidden sector, i.e., most states are charged under one sector or the other, but there are a few which couple to both. In addition, there are three non-anomalous $U(1)$ factors and two anomalous ones. The generators Q_3, Q_1 and Q_2 refer to the $U(1)$ factor within the corresponding $U(N)$, while Q_8, Q_8' are the $U(1)$s arising from the higgsed $USp(8)$. $Q_3/3$ and Q_1 are essentially baryon (B) and lepton (L) number, respectively, while $(Q_8 + Q_8')/2$ is analogous to the generator T_{3R} occurring in left-right symmetric extensions of the Standard Model. The hypercharge is defined as:

$$Q_Y = \frac{1}{6}Q_3 - \frac{1}{2}Q_1 + \frac{1}{2}(Q_8 + Q_8').$$ 43.

From the above comments, Q_Y is non-anomalous guaranteeing that $U(1)_Y$ remains massless. There are two additional surviving non-anomalous $U(1)$s, i.e., $B - L = Q_3/3 - Q_1$ and $Q_8 - Q_8'$. The gauge bosons corresponding to the anomalous $U(1)$ generators $B+L$ and Q_2 acquire string-scale masses, so those generators act like perturbative global symmetries on the effective four-dimensional theory.

The spectrum of chiral multiplets in the open string sector is tabulated in Table 5. There are also vector-like multiplets in the model, but they are generically massive so we do not tabulate them here [they can be found in (93)]. The theory contains three Standard Model families, multiple Higgs candidates, a number of exotic chiral (but anomaly-free) fields, and multiplets which transform in the adjoint or singlet representation of the Standard Model gauge group.

For more details and phenomenological features, please consult the original literature (93, 94). Here, we would like to highlight some of the special features of this supersymmetric model:

- The model involves an extended gauge structure, including two additional $U(1)$ factors, one of which has family non-universal and therefore

TABLE 5   The chiral spectrum of the open string sector in the supersymmetric three-family model. To be complete, we also list in the bottom part of the table, below the double horizontal line, the non-chiral massless states from the aa sectors, which are not localized at the intersections and correspond to deformation and Wilson line moduli of the $D6$ branes

Sector	$SU(3)_C \times SU(2)_Y \times$ $Sp(2)_B \times Sp(2)_A \times Sp(4)$	$(Q_3, Q_1, Q_2, Q_8, Q_8')$	Q_Y	$Q_8 - Q_8'$	Field
$A_1 B_1$	$3 \times 2 \times (1, \bar{2}, 1, 1, 1)$	$(0, 0, -1, \pm 1, 0)$	$\pm \frac{1}{2}$	± 1	H_U, H_D
	$3 \times 2 \times (1, \bar{2}, 1, 1, 1)$	$(0, 0, -1, 0, \pm 1)$	$\pm \frac{1}{2}$	∓ 1	H_U, H_D
$A_1 C_1$	$2 \times (\bar{3}, 1, 1, 1, 1)$	$(-1, 0, 0, \pm 1, 0)$	$\frac{1}{3}, -\frac{2}{3}$	$1, -1$	$\bar{D}, \bar{U}$
	$2 \times (\bar{3}, 1, 1, 1, 1)$	$(-1, 0, 0, 0, \pm 1)$	$\frac{1}{3}, -\frac{2}{3}$	$-1, 1$	$\bar{D}, \bar{U}$
	$2 \times (1, 1, 1, 1, 1)$	$(0, -1, 0, \pm 1, 0)$	$1, 0$	$1, -1$	$\bar{E}, \bar{N}$
	$2 \times (1, 1, 1, 1, 1)$	$(0, -1, 0, 0, \pm 1)$	$1, 0$	$-1, 1$	$\bar{E}, \bar{N}$
$B_1 C_1$	$(3, \bar{2}, 1, 1, 1)$	$(1, 0, -1, 0, 0)$	$\frac{1}{6}$	0	Q_L
	$(1, \bar{2}, 1, 1, 1)$	$(0, 1, -1, 0, 0)$	$-\frac{1}{2}$	0	L
$B_1 C_2$	$(1, 2, 1, 1, 4)$	$(0, 0, 1, 0, 0)$	0	0	
$B_2 C_1$	$(3, 1, 2, 1, 1)$	$(1, 0, 0, 0, 0)$	$\frac{1}{6}$	0	
	$(1, 1, 2, 1, 1)$	$(0, 1, 0, 0, 0)$	$-\frac{1}{2}$	0	
$B_1 C_1'$	$2 \times (3, 2, 1, 1, 1)$	$(1, 0, 1, 0, 0)$	$\frac{1}{6}$	0	Q_L
	$2 \times (1, 2, 1, 1, 1)$	$(0, 1, 1, 0, 0)$	$-\frac{1}{2}$	0	L
$B_1 B_1'$	$2 \times (1, 1, 1, 1, 1)$	$(0, 0, -2, 0, 0)$	0	0	
	$2 \times (1, 3, 1, 1, 1)$	$(0, 0, 2, 0, 0)$	0	0	
$A_1 A_1$	$3 \times 8 \times (1, 1, 1, 1, 1)$	$(0, 0, 0, 0, 0)$	0	0	
	$3 \times 4 \times (1, 1, 1, 1, 1)$	$(0, 0, 0, \pm 1, \pm 1)$	± 1	0	
	$3 \times 4 \times (1, 1, 1, 1, 1)$	$(0, 0, 0, \pm 1, \mp 1)$	0	± 2	
	$3 \times (1, 1, 1, 1, 1)$	$(0, 0, 0, \pm 2, 0)$	± 1	± 2	
	$3 \times (1, 1, 1, 1, 1)$	$(0, 0, 0, 0, \pm 2)$	± 1	∓ 2	
$A_2 A_2$	$3 \times (1, 1, 1, 1, 1)$	$(0, 0, 0, 0, 0)$	0	0	
$B_1 B_1$	$3 \times (1, 3, 1, 1, 1)$	$(0, 0, 0, 0, 0)$	0	0	
	$3 \times (1, 1, 1, 1, 1)$	$(0, 0, 0, 0, 0)$	0	0	
$B_2 B_2$	$3 \times (1, 1, 1, 1, 1)$	$(0, 0, 0, 0, 0)$	0	0	
$C_1 C_1$	$3 \times (8, 1, 1, 1, 1)$	$(0, 0, 0, 0, 0)$	0	0	
	$3 \times (1, 1, 1, 1, 1)$	$(0, 0, 0, 0, 0)$	0	0	
$C_2 C_2$	$3 \times (1, 1, 1, 1, 5 + 1)$	$(0, 0, 0, 0, 0)$	0	0	

flavor-changing couplings. Extended gauge structure is quite generic among string models, and more so for intersecting D-brane models.

- There are additional Higgs doublets, suggesting such effects as a rich spectrum of Higgs particles, neutralinos, and charginos, perhaps with nonstandard couplings due to mixing and flavor-changing effects.

- In addition to the three chiral families of the Standard Model, there are chiral exotic states, i.e., chiral states with unconventional Standard Model quantum numbers. It was argued in (93) that these states may decouple from the low-energy spectrum due to hidden sector charge confinement.

- There exists a quasi-hidden non-Abelian sector, which becomes strongly coupled above the electroweak scale. The dynamics of the strongly coupled hidden sector leads to dynamical supersymmetry breaking with dilaton and untwisted complex structure moduli stabilization, as studied in detail in (95). Charge confinement modifies the low-energy spectrum by causing some exotics to disappear, while anomaly considerations imply that new composite states may emerge (93) (see also (96, 97)).

Like the non-supersymmetric constructions of Standard-like Models, the model does not have the conventional form of gauge unification, as each gauge factor is associated with a different set of branes. However, the string-scale couplings are predicted in terms of the ratio of the Planck and string scales and a geometric factor as discussed in section 2.1. The explicit dependence of the tree-level holomorphic gauge kinetic function on the dilaton and the complex structure moduli will be discussed in section 4.2. For all intersecting D6-brane constructions, the Yukawa couplings among chiral matter are due to world-sheet instantons associated with the string world-sheet stretching among the intersections where the corresponding chiral matter fields are localized (15). The details of the Yukawa coupling interactions will be discussed in Subsection 4.1.1.

3.2.1. SUPERSYMMETRIC GRAND UNIFIED MODELS   The set-up with intersecting D6-branes on orientifolds also allows for the construction of Grand Unified Models, based on the Georgi-Glashow $SU(5)$ gauge group (98). Such non-supersymmetric Grand Unified Models were constructed in (19, 92, 99, 100) and supersymmetric ones in (21, 101, 102). (For additional work on non-supersymmetric Grand Unified Models see also (76, 103).)

The supersymmetric constructions of such models have a lift on a circle to M-theory and provide examples of Grand Unified Models of strongly-coupled M-theory compactified on singular seven-dimensional manifolds with G_2 holonomy (21, 58–61). (See also section 2.9.)

The key point in these constructions is the appearance of anti-symmetric representations, i.e., **10** of $SU(5)$, which can emerge at the intersection of the D-brane with its orientifold image (see Table 1). Thus, **10**-plets, along with the bi-fundamental representations $(\bar{\mathbf{5}}, \mathbf{N_b})$ at the intersections of $U(5)$ branes with $U(N_b)$ branes, form the chiral particle content of the quark and lepton families. It turns out that the gauge boson for the diagonal $U(1)$ factor of $U(5)$ is massive, and the anomalies associated with this $U(1)$ are canceled via the generalized Green-Schwarz mechanism, as explained in section 2.7.

For toroidal and $\mathbb{Z}_2 \times \mathbb{Z}_2$ orbifold compactifications, there are three copies of the adjoint representations on the world-volume of the branes. They are moduli associated with the splitting and Wilson lines of the branes that wrap the same three-cycles which in these two cases are not rigid. Turning on appropriate vacuum expectation values (VEVs) of these adjoint representations can spontaneously break $SU(5)$ down to the Standard Model gauge group. Such VEVs have a geometric interpretation in terms of the appropriate parallel splitting of the $U(5)$ branes.

Because all the Standard Model gauge group factors arise from branes wrapped on parallel, but otherwise identical, cycles, this construction provides a natural framework for gauge coupling unification and thus a natural embedding of the traditional grand unification (98) into intersecting D-brane models.

The explicit supersymmetric constructions of Grand Unified Models were given for the $\mathbb{Z}_2 \times \mathbb{Z}_2$ orbifold models. The first such example (21) was a four-family model. Further systematic analysis (101) revealed that within $\mathbb{Z}_2 \times \mathbb{Z}_2$ orbifold models with factorizable three-cycles, all the three-family models necessarily also contain three copies of **15**-plets, the symmetric representations of $SU(5)$. (Analogous observations have been reported in (104) for supersymmetric $SU(5)$ models in the $\mathbb{Z}_4 \times \mathbb{Z}_2$ orbifold background.) There are approximately twenty such models (101), which are not fully realistic:

- The additional **15**-plets decompose under $SU(3)_C \times SU(2)_Y \times U(1)_Y$ as $(\mathbf{6}, \mathbf{1})(-\frac{2}{3}) + (\mathbf{1}, \mathbf{3})(+1) + (\mathbf{3}, \mathbf{2})(+\frac{1}{6})$ and thus contain additional exotic Standard Model particles.

- Because the chiral states are charged under the $U(1)$ factor of $U(5)$ and the only candidates for the Higgs fields are in the adjoint **24** and fundamental **5** representations, the fermion masses can arise only from the Yukawa couplings of the type: $\bar{\mathbf{5}}\,\mathbf{10}\,\bar{\mathbf{5}}_H$ (subscript H refers to the Higgs fields), while the couplings of the type $\mathbf{10}\,\mathbf{10}\,\mathbf{5}_H$ are absent due to the $U(1)$ charge conservation (19). The absence of perturbative Yukawa couplings to the up-quark families is generic for these constructions (supersymmetric or not).

- Within this framework one can address the long-standing problem of doublet-triplet splitting, i.e., ensuring that after the breaking of $SU(5)$, the doublet of $\mathbf{5}_H$, responsible for the electroweak symmetry breaking, remains light, while the triplet becomes heavy. The mechanism, suggested within M-theory on G_2 holonomy manifolds (105), allows for the $SU(5)$ breaking via Wilson lines with different discrete quantum numbers for the doublet and the triplet, which in turn forbids the mass term for the doublet. However, the Wilson lines in the present context are continuous rather than discrete, owing to the generic non-rigid nature of three-cycles on orbifold compactifications. Although the current constructions of the models are not fully realistic, generalizations to examples with rigid three-cycles may provide an avenue to address the appearance of genuinely discrete Wilson lines.

3.2.2. SYSTEMATIC SEARCH FOR SUPERSYMMETRIC STANDARD-LIKE MODELS   In the previous two subsections, we have seen that supersymmetric Standard-like and Grand Unified Models can be constructed within the intersecting D6-brane framework. Subsequently, systematic searches for supersymmetric three-family Standard-like Models have been carried out within $\mathbb{Z}_2 \times \mathbb{Z}_2$ orbifold constructions with factorizable three-cycles.

Within this framework, the systematic search for three-family $SU(5)$ Grand Unified Models (101) produced three-family models which contain three copies of **15**-plets. However, this feature is specific to this specific orbifold and it remains to be seen whether it persists for more general models.

As for the Standard-like Model constructions with gauge group factors arising from different intersecting D6-branes, sets of models with fewer Higgs doublets (106) were obtained. However, all these three-family models still possess additional exotics. Subsequently, a systematic search for supersymmetric Pati-Salam models based on the left-right symmetric gauge symmetry $SU(4)_C \times SU(2)_L \times SU(2)_R$ was presented in (107). The gauge symmetry can be broken down to the Standard Model one via D6-brane splitting and a further D- and F-flatness preserving Higgs mechanism from massless open string states in an $\mathcal{N} = 2$ subsector. Among the models that also possess at least two confining hidden gauge sectors, where gaugino condensation can in turn trigger supersymmetry breaking and (some) moduli stabilization, the search revealed eleven models. Two models realize gauge coupling unification of $SU(2)_L$ and $SU(2)_R$ at the string scale. However, all these models still possess additional exotic matter.

In another related work (108), the study of splitting of D6-branes parallel to orientifold planes, within $\mathbb{Z}_2 \times \mathbb{Z}_2$ orientifolds, led to the examples of four-family Standard-like orientifold models without chiral exotics. The starting point is a one-family $U(4) \times Sp(2f)_L \times Sp(2f)_R$, $(f = 4)$ model, which is broken down to a four-family $U(4) \times U(2)_L \times U(2)_R$ model by parallel splitting of the D-branes, originally positioned on the O-planes. The chirality of the model is changed because the original branes were positioned on top of an orientifold singularity. Both the string theory and field theory aspects of these specific D-brane splittings are discussed in detail in (108).

These systematic searches for realistic models seem to suggest that an extended Higgs sector is ubiquitous in intersecting D-brane models. It is therefore quite remarkable that a simple D-brane configuration, introduced in (109), yields just the MSSM chiral spectrum and its minimal Higgs content. As the authors of (109) pointed out, in toroidal compactifications this model must be seen as a local construction, where extra R-R sources such as hidden sector branes and/or background fluxes should be added. Several attempts have been made to embed this local construction into a global model. First, it was shown in (110) that this local model can be embedded into an abstract conformal field theory construction known as Gepner orientifold (see Subsection 3.3). These Gepner constructions are located at special points in the Calabi-Yau moduli space where the geometric intuition is lost,[6] and so it is therefore desirable to find an embedding into a geometrical construction. However, it proves difficult to do so for a toroidal (orbifold) background without introducing anti-branes because the cancellation of R-R tadpoles requires some R-R charges of a D-brane to have the same sign as that of an O6-plane. Peculiar as it might seem, D-branes with this property do exist (20, 21). Armed with this observation, two independent attempts, (108) and (48, 49), were made to embed the local model of (109) into a $\mathbb{Z}_2 \times \mathbb{Z}_2$ orientifold, and indeed a consistent global realization of (109) was found in (48, 49). Unfortunately, in the original version of (108), only the homological R-R charges are canceled, but the K-theory constraints (49) are not satisfied, resulting in the massless spectrum with discrete global anomalies (47). In the revised version of (108), employing the K-theory constraints derived in (49), a consistent model was obtained with minor modifications. The Higgs sector of the model is no longer minimal, unlike the construction in (48, 49). It should be noted that the hidden sector D-branes introduced in these global models (48, 49, 108) have non-trivial intersections with the Standard Model sector D-branes and so there are chiral exotics.

3.3. Supersymmetric Models on More General Backgrounds

In recent years, other supersymmetric intersecting D-brane models have been constructed with the aim of finding realizations of the MSSM. Essentially, two different classes of string backgrounds were considered. First, using the methods reviewed in section 2, more complicated orbifold backgrounds like a $\mathbb{Z}_4$, $\mathbb{Z}_4 \times \mathbb{Z}_2$ or $\mathbb{Z}_6$ orbifold have been studied. Second, to move beyond toroidal orbifolds and to consider intersecting branes on more general Calabi-Yau spaces, methods to treat Gepner model orientifolds were developed. Let us briefly review these activities in the following two sections.

3.3.1. OTHER TOROIDAL ORBIFOLDS   One way to generalize the $\mathbb{Z}_2 \times \mathbb{Z}_2$ orientifolds studied above is to include additional shift symmetries in the $\mathbb{Z}_2$ actions (111–113). These have the effect of eliminating some of the orientifold planes present in the original models, which would make it much harder to find interesting supersymmetric models. On the other hand, it also gives rise to twisted sector three-cycles, which allows for more general fractional D6-branes. Some of these models were constructed in (111–113).

Employing the topological methods introduced in section 2 (43), chiral supersymmetric intersecting D-brane models have been studied so far on the $\mathbb{Z}_4$ (50), $\mathbb{Z}_4 \times \mathbb{Z}_2$ (51, 104, 114) and $\mathbb{Z}_6$ (52) toroidal orbifolds. In the first two cases, semi-realistic MSSM-like models could only be achieved after certain

---

[6]For instance, it is not straightforward how to introduce background fluxes.

D-brane recombination processes were taken into account (see the original papers for more details). For the $\mathbb{Z}_6$ model (52), the authors were performing an exhaustive search for MSSM-like models and found a class of interesting D-brane configurations, which gave rise to the MSSM spectrum without the complication of brane recombinations.

Moreover, there are both four- and six-dimensional toroidal backgrounds, where so far only the non-chiral solutions to the tadpole cancellation condition, with D6-branes placed parallel to the orientifold planes, have been considered (9–14, 115, 116).

3.3.2. GEPNER MODEL ORIENTIFOLDS   One of the unattractive phenomenological features of all the toroidal orbifold models discussed above is that they give rise to too many adjoint scalars. Geometrically this means that the three-cycles π_a one is considering have too many deformations, which are counted by $b^1(\pi_a)$. For this reason, among others, it is desirable to have many more backgrounds available. However, for more general algebraic Calabi-Yau spaces, not very much is known about sLag three-cycles, which prevents a direct geometric approach to the problem as pursued for instance in (43, 44, 117, 118).

One solution is to use Gepner models, which are exactly solvable conformal field theories known to describe certain symmetric points in the moduli space of distinguished Calabi-Yau manifolds. The description of D-branes and orientifold planes in this context is a subject of its own, but here we mention only that after some first attempts (119, 120), methods have recently been developed to treat Gepner model orientifolds very efficiently (110, 121–126), allowing a systematic computer search for MSSM like models. Specifically, the impressive results of (127) provide large classes of three-family Standard-like Models with no chiral exotics. It remains to be seen whether some of these models also satisfy the more refined Standard Model constraints. Note, however, that these exact conformal field theory models are located at very special points in the Kähler and complex structure moduli space, where the geometric intuition is lost. Because one expects that all radii are of string-scale size, couplings, such as Yukawa couplings, are not expected to possess hierarchies associated with the size of the internal spaces, as in the case of the toroidal orbifolds with D-branes. In addition, the introduction of supergravity fluxes is not straightforward to perform (see section 5).

4. LOW-ENERGY EFFECTIVE ACTIONS

The models presented in the previous section 3 provide a starting point for the study of couplings in the effective low-energy theory whose massless spectrum was determined by techniques presented in section 2. For the orientifold models with intersecting D6-branes compactified on orbifolds, the calculation of such couplings can be done by employing the conformal field theory techniques on orbifolds (128). The tree-level calculations can in principle be performed both for the supersymmetric and the non-supersymmetric constructions; for the one-loop calculations, supersymmetry is a necessary ingredient to obtain an unambiguous finite answer. The summary of the explicit results will therefore focus primarily on the supersymmetric constructions. (Alternatively, part of the low-energy effective action can also be determined by a dimensional reduction of the ten-dimensional supergravity theory (129–131).)

The calculation of couplings for chiral superfields at the intersection of D6-branes and, in particular, the Yukawa couplings for such states are clearly of phenomenological interest. As discussed in section 2.2, the states at such intersections correspond to the open string excitations stretched between the two intersecting D6-branes. As a consequence, the bosonic string oscillator modes are like those given in Equation 8. Therefore, physical string excitations at the two D6-brane intersections are associated with the twisted open-string sectors, which are analogous to the closed string twisted sectors on orbifolds (128).

The string amplitudes for these excitations can in turn be calculated employing conformal field theory techniques (128). The correlation functions of the fermionic string excitations can be obtained in a straightforward manner by employing a world-sheet bosonization procedure. On the other hand, the correlation functions for the bosonic excitations involve the calculation of the correlation functions for the so-called bosonic twist fields, i.e., $\sigma_{\epsilon_i}(x)$, evaluated at the world-sheet location x on the disc. The bosonic twist field ensures that the bosonic open string fields $X^i(z)$ (in the i-th toroidal direction) have the correct twisted boundary conditions. Here z is the world-sheet coordinate. These boundary conditions are encoded in the following operator product expansion (128, 132):

$$\partial X^i(z)\sigma_{\epsilon_i}(x) \sim (z-x)^{\epsilon_i-1}\tau_{\epsilon_i}(x) + \cdots$$

$$\partial \tilde{X}^i(z)\sigma_{\epsilon_i}(x) \sim (z-x)^{-\epsilon_i}\tau'_{\epsilon_i}(x) + \cdots$$

$$\bar{\partial} X^i(\bar{z})\sigma_{\epsilon_i}(x) \sim -(\bar{z}-x)^{-\epsilon_i}\tau'_{\epsilon_i}(x) + \cdots$$

$$\bar{\partial} \tilde{X}^i(\bar{z})\sigma_{\epsilon_i}(x) \sim -(\bar{z}-x)^{\epsilon_i-1}\tau_{\epsilon_i}(x) + \cdots \qquad 44.$$

and similarly for $\sigma_{-\epsilon_i}(x)$. Here τ_{ϵ_i} and τ'_{ϵ_i} correspond to the excited bosonic twist fields. Employing the so-called stress-energy method (128), which allows one to determine the correlation functions of bosonic twist fields by employing the properties of the operator product expansion of the conformal field theory stress-energy tensor with the twist fields, along with the above operator product expansions (45), enables one (132) to determine the bosonic twisted sector string amplitudes. The application of these calculations to the four-point couplings and Yukawa couplings will be discussed in Subsection 4.1.1.

Another set of couplings involves the calculation of the Kähler potential for the states at the intersection, in particular the explicit dependence of the leading term which is bi-linear in powers of chiral superfields at the intersection. The corresponding string amplitudes involve the correlation functions containing both states at the intersection (open string states) and toroidal moduli fields (closed string states) (133). Such couplings will be discussed in Subsection 4.1.2.

Another important topic is the calculation of the gauge couplings. In particular, determination of the holomorphic gauge kinetic function in terms of the dilaton and the toroidal moduli both at the tree-level and one-loop-level is an important task and will be discussed in section 4.2.

4.1. Correlation Functions for States at D-Brane Intersections

In this section we summarize the results for the tree-level calculations for the chiral matter appearing at the D6-brane intersections (see section 2.2). In the supersymmetric constructions, the states at intersections correspond to the full massless chiral supermultiplet. The couplings of most interest are the tri-linear superpotential couplings, such as the coupling of quarks and leptons to the Higgs fields. On the other hand, the four-point couplings are also of interest, because they indicate the appearance of higher order terms in the effective Lagrangian; for example, certain four-fermion couplings could contribute to the flavor-changing neutral currents in the Standard-like Model constructions (see (134, 135)) and in the Grand Unified Models, triggering proton decay (see (136)).

4.1.1. THE FOUR-POINT AND THREE-POINT FUNCTIONS: YUKAWA COUPLINGS The explicit calculations of the three-level four-point and three-point correlation functions for the states appearing at the D-brane intersections were done in (132, 134, 136, 137). Generalizations to n-point functions were addressed in (138).

As discussed in the introduction of this section, the non-trivial part in the calculation involves the evaluation of the correlation functions of four (three) bosonic twist fields, which signify the fact that the states at the intersection arise from the sector with twisted boundary conditions on the bosonic and fermionic string fields. The conformal field theory techniques employed are related to the study of bosonic twist fields of the closed string theory on orbifolds (128). For technical details of the specific calculation of the four- and three-point functions, employing conformal field theory techniques, we refer the reader to references (132, 133), and for a detailed calculation of the classical part of Yukawa couplings to reference (137).

The calculations have been done in the case of intersecting D6-branes wrapping factorizable three-cycles of a six-torus $T^6 = T^2 \times T^2 \times T^2$. Thus, in each T^2, the D6-branes wrap one-cycles, and the problem reduces to a calculation of correlation functions of bosonic twist fields associated with the twisted sectors at intersections of D6-branes wrapping the one-cycles of a T^2. The final answer is therefore a product of contributions from correlation functions on each of the three T^2 (137).

In particular, the following four-point correlation functions of bosonic twist fields are of interest:

$$\langle \sigma_\nu(x_1)\sigma_{-\nu}(x_2)\sigma_\nu(x_3)\sigma_{-\nu}(x_4)\rangle \tag{45.}$$

and

$$\langle \sigma_\nu(x_1)\sigma_{-\nu}(x_2)\sigma_{-\lambda}(x_3)\sigma_\lambda(x_4)\rangle. \tag{46.}$$

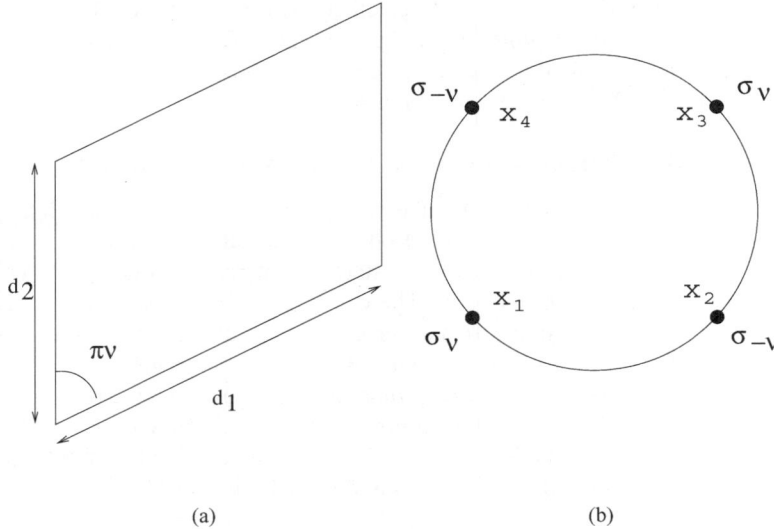

(a)   (b)

Figure 7   Target space: the intersection of two parallel branes separated by respective distances d_1 and d_2 and intersecting at angles $\pi \nu$ (a). World-sheet: a disk diagram of the four twist fields located at $x_{1,2,3,4}$ (b). The calculation involves a map from the world-sheet to target space.

The first one corresponds to the bosonic twist field correlation function of states appearing at the intersection of two pairwise parallel branes with intersection angle $\pi \nu$ (see Figure 7). This correlation function is a key ingredient in the calculation of the four-fermion couplings, an ingredient that contributes to the flavor-changing neutral currents in the Standard-like models (135) and to the proton decay amplitudes in the Grand Unified Models (136).

The second amplitude (47) corresponds to the bosonic twist field correlation function of states appearing at the intersection of two branes intersecting at respective angles $\pi \nu$ and $\pi \lambda$ with the third set of parallel branes (See Figure 8). This correlation function is specifically suited for taking the limit of the world-sheet coordinate $x_2 \rightarrow x_3$ which factorizes to a three-point function associated with the intersection of three branes. This latter result is particularly interesting because it provides a key element in the calculation of the Yukawa coupling.

By employing the stress-energy conformal field theory techniques and the properties of the operator product expansions of the bosonic twist fields (45) one can determine (132) both the classical part and the quantum part of such amplitudes and thus the exact tree-level answer for the corresponding couplings. The calculation of the quantum part depends only on the intersection angles and is thus insensitive to the scales of the internal space and relative position of the branes. On the other hand the classical part carries information on the actual separation among the

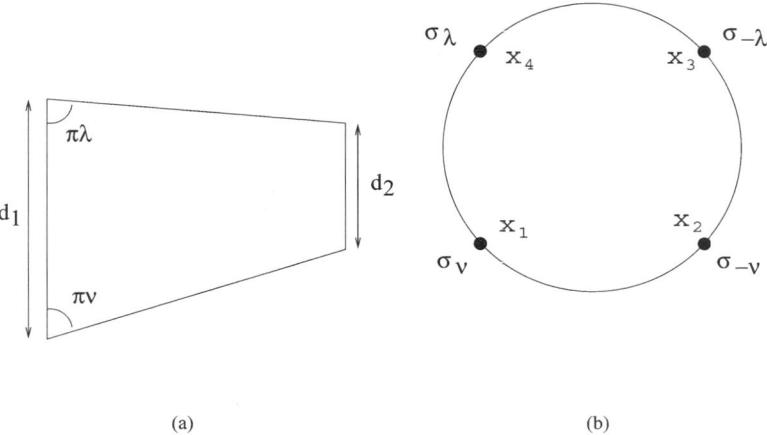

(a)   (b)

Figure 8   Target space: the intersection of two branes with the two parallel branes at angles $\pi\nu$ and $\pi\lambda$, respectively (*a*). World-sheet: a disk diagram of the four twist fields located at $x_{1,2,3,4}$ (*b*). The calculation involves a map from the world-sheet to target space, allowing for a factorization to a three-point function.

branes and the overall volume of T^2 as well. An important part in the calculation is the determination of an overall normalization of the four-point amplitude, which can be done by factorizing the amplitude in the limits $x \to 0$ or $x \to 1$, where it reduces to a product of the two three-point amplitudes. Namely, in these limits, the four-point amplitude contains a dominant contribution from the exchanges of the intermediate open string winding states around the compact directions. The dominant contribution can be interpreted as the s (or t)-channel exchange of the massless gauge bosons living on the world-volume of the D6-branes. Thus, in this limit the amplitude is completely determined in terms of the gauge-coupling which in turn determines the normalization of the full four-point amplitude. For details see (132, 136).

We shall skip the technical details and in the following only quote the result for the exact (string) tree-level expression for the Yukawa coupling for two massless fermionic states and one massless bosonic state, appearing at the intersections of three D6-branes (132):

$$Y = \sqrt{2}g_0 2\pi \prod_{j=1}^{3} \left[\frac{16\pi^2 \Gamma(1-\nu_j)\Gamma(1-\lambda_j)\Gamma(\nu_j+\lambda_j)}{\Gamma(\nu_j)\Gamma(\lambda_j)\Gamma(1-\nu_j-\lambda_j)} \right]^{\frac{1}{4}}$$

$$\times \sum_{m\in\{\text{ws}-\text{inst.}\}} \exp\left(-\frac{A_j(m)}{2\pi\alpha'} \right), \qquad\qquad 47.$$

where $A_j(m)$ is the area of the m-th triangle (world-sheet instanton) formed by the three intersecting branes on the j-th two-torus and $g_0 = e^{\Phi/2}$, with Φ

corresponding to the Type IIA dilaton. In order to derive Equation (48) it was assumed that all three fields have canonically normalized kinetic energies. For a discussion of phenomenological implications of the above results, see Subsection 6.4.

In earlier works (94, 109) the leading classical contribution to such Yukawa couplings was calculated. Also previously, a comprehensive analysis and computation of the full classical contribution (which contains all the open string moduli dependence) was performed in (137). Intriguingly, in (139) the three-point function was calculated in the mirror-dual Type IIB theory. Here a purely classical, leading order in α', computation gave already the full world-sheet instanton corrected Type IIA superpotential contribution to the Yukawa couplings [see Equation (48)]. This can be considered a nice confirmation of mirror symmetry. As expected, the Kähler potential contributions to the Yukawa couplings only agreed to leading order in α'.

The prefactor in (48) corresponds to the quantum part of the bosonic twist correlator; it has a suggestive factorizable form associated with the angles of states appearing at each intersection, resulting in the following complex structure dependence of the Kähler potential for chiral superfields at D6-brane intersections:

$$K = \frac{1}{4\pi} \sum_{\nu} \prod_{j=1}^{3} \sqrt{\frac{\Gamma(\nu_j)}{\Gamma(1 - \nu_j)}} \Phi_{\nu} \Phi_{\nu}^*. \qquad 48.$$

(For simplicity in the above expression the Planck scale was set to $M_P = 1$.)

On the other hand, the classical part of the Yukawa coupling is proportional to the area of the intersection triangles and thus depends on the toroidal Kähler structure moduli. It includes a contribution from the part of the Kähler potential that depends on toroidal Kähler moduli (139). Thus the full Kähler potential takes the form displayed in the next Subsection [see Equation (49)]. After the inclusion of toroidal two-form field potentials and the asscoiated Wilson lines (137), the remaining part of the Yukawa coupling describes the superpotential tri-linear coupling as a holomorphic function of toroidal Kähler moduli; this coupling typically takes a form of modular theta functions [for further details, see (137, 139)]. This splitting of the three-point function (48) into the leading Kähler potential contribution and the superpotential contribution has been confirmed as a part of the calculations described in the following subsection.

Higher tree-level n-point correlation functions for chiral superfields at D-brane intersections have been studied in (134, 138). The one-loop calculation of the three-point functions for such states was done in (140) and it leads to new results for the one-loop corrections to the Kähler potential for the corresponding chiral superfields at D-brane intersections.

4.1.2. THE CLOSED-OPEN STRING AMPLITUDES—KÄHLER POTENTIAL   A direct calculation of the tree-level leading order Kähler potential for chiral superfields at D-brane intersections and their dependence on the closed string sector moduli involves the determination of the string amplitudes for the two open string sector vertices and an arbitrary number of closed sector moduli vertices. For the toroidal (orbifold) backgrounds, these calculations have been carried out explicitly for any number of toroidal complex and Kähler structure moduli in (133). First, explicit results for the four-point string amplitudes were derived and then, by further employing the symmetry structure of higher n-point functions, sets of differential equations were obtained for the n-point functions with an arbitrary number of vertices for the toroidal moduli. These could be explicitly solved, thus resulting in the explicit string amplitudes with any number of toroidal closed sector moduli. As a consequence, the leading order tree-level Kähler potential for chiral superfields at D-brane intersections, and its explicit dependence on both the toroidal Kähler and complex structure moduli, could be derived. Specifically, the Kähler potential for the open string sector chiral superfields $\Phi_{\nu_{ab}}$, appearing at the intersection of the stack a and stack b of D6-branes, takes the form (133):

$$
K = \frac{1}{4\pi} \left[\prod_{i=1}^{3} (T_i + T_i^*)^{-\nu_{ab}^i} \sqrt{\frac{\Gamma(\nu_{ab}^i)}{\Gamma(1 - \nu_{ab}^i)}} \right] \Phi_{\nu_{ab}} \Phi_{\nu_{ab}}^*,
\qquad 49.
$$

where again $\pi \nu_{ab}^i$ denotes the angle of the a- and b-D6-brane intersection in the i-th two-torus and T_i is the Kähler modulus of the i-th two-torus. (For simplicity, again, the Planck scale is set to $M_{pl} = 1$.) Note that the dependence of the above Kähler potential on the angles and thus implicitly on the toroidal complex structure moduli is the same as the one obtained from the Yukawa coupling calculation (48). In addition, (50) also contains information on the toroidal Kähler moduli.

4.2. Gauge Couplings

The last function that specifies the effective four-dimensional $\mathcal{N} = 1$ supersymmetric theory is the gauge kinetic function. The tree-level gauge kinetic function for each stack of D6-branes can be determined in a straightforward way by reducing the D6-brane world-volume kinetic energy action along the three-cycle wrapped by the stack of D6-branes in the internal space. For a supersymmetric three-cycle, π_a, the tree-level gauge kinetic function is a holomorphic function of complex structure moduli fields and it is of the form (43, 57, 141)

$$
f_a = \frac{M_s^3}{(2\pi)^4} \left[e^{-\varphi} \int_{\pi_a} \Re(\Omega_3) + 2i \int_{\pi_a} C_3 \right],
\qquad 50.
$$

where C_3 denotes the R-R three-form. For supersymmetric three-cycles on toroidal orbifolds, this holomorphic gauge kinetic function takes an explicit form in terms of the toroidal complex structure moduli U^i and the dilaton field S

$$
S = \frac{M_s}{2\pi} e^{-\varphi} \prod_i R_1^i + \frac{i}{4\pi} C^0,
$$

$$
U^i = \frac{M_s}{2\pi} e^{-\varphi} R_1^i R_2^j R_2^k + \frac{i}{4\pi} C^i,
\qquad 51.
$$

with $i \neq j \neq k \neq i$. For example in the case of the $\mathbb{Z}_2 \times \mathbb{Z}_2$ orientifold, the gauge coupling function takes the form [see, e.g., (57, 95)]:

$$f_a(U^i, S) = \tfrac{1}{4}\left[n_a^1 n_a^2 n_a^3 S - n_a^1 \widetilde{m}_a^2 \widetilde{m}_a^3 U^1 - \widetilde{m}_a^1 n_a^2 \widetilde{m}_a^3 U^2 - \widetilde{m}_a^1 \widetilde{m}_a^2 n_a^3 U^3 \right], \quad 52.$$

where as usual n_a^i and $\widetilde{m}_a^i$ are the wrapping numbers of the three-cycle π_a and the pre-factor $\tfrac{1}{4}$ is the dimension of the orbifold/orientifold group.

Because the gauge coupling for each gauge group factor depends on the volume of the corresponding three-cycle π_a, in general the intersecting D-brane constructions do not have gauge coupling unification in the sense of Grand Unified Models. However, for each gauge group factor the tree-level gauge coupling is calculable in terms of the toroidal complex structure moduli, the dilaton, and the wrapping numbers of the three-cycle π_a. The results for the tree-level gauge kinetic functions were employed as the starting point to address the renormalization group running of gauge couplings from the string scale to the electroweak regime for the semi-realistic constructions (93, 141, 142), as well as in the study of the moduli stabilization and supersymmetry breaking due to gaugino condensation in the hidden sector of supersymmetric semi-realistic constructions (95) (for the respective phenomenological implications see Subsections 6.2 and 6.6.1).

The tree-level couplings receive corrections at the one-loop level due to the so-called threshold corrections of the heavy string modes. These explicit calculations are involved because the complete massive string spectrum is needed. For the perturbative heterotic string theory on orbifolds, these threshold corrections were first calculated in (143). The calculation of the gauge coupling threshold corrections for the intersecting D6-branes on a toroidal orbifold backgrounds amounts to similar complexity and the explicit results have been computed in (144). While the one-loop corrections for open string sectors preserving $\mathcal{N} = 4$ supersymmetry vanish, the $\mathcal{N} = 2$ sectors depend on both the toroidal complex and Kähler moduli. For explicit expressions, please consult (144). These corrections bear similarities with the heterotic orbifold corrections (143). For details and specific explicit calculation for the $\mathbb{Z}_2 \times \mathbb{Z}_2$ orientifold, see again (144). In addition, there are $\mathcal{N} = 1$ sector corrections; they can be cast in a compact expression which for the $SU(N_a)$ gauge couplings takes the form:

$$\Delta_{ab} = -b_{ab} \ln \frac{\Gamma\!\left(1 - v_{ba}^1\right)\Gamma\!\left(1 - v_{ba}^2\right)\Gamma\!\left(1 + v_{ba}^1 + v_{ba}^2\right)}{\Gamma\!\left(1 + v_{ba}^1\right)\Gamma\!\left(1 + v_{ba}^2\right)\Gamma\!\left(1 - v_{ba}^1 - v_{ba}^2\right)}, \qquad 53.$$

where $b_{ab} = N_b I_{ab} \mathrm{Tr}(Q_a^2)$ and πv_{ba}^i denote the intersection angles of a- and b-D6-branes in the i-th two-torus. These angles can be expressed in terms of the wrapping numbers of the π_a and π_b three-cycles and the toroidal complex structure moduli U^i. [For the explicit formula, see, e.g., (132).] The total correction from the $\mathcal{N} = 1$ sector is obtained as a summation over all bs, i.e., stacks of all the other D6-branes wrapping the three-cycles π_b.

These threshold corrections could play an important role in the study of the renormalization group running of gauge couplings; in particular, they can modify the effective string scale. In addition, because the threshold corrections depend on both the Kähler and complex structure moduli, they could play an important role in the strong infrared "hidden sector" dynamics and the possibility of stabilizing all toroidal moduli.

5. FLUX VACUA WITH MAGNETIZED D-BRANES

As we have seen in the previous sections, intersecting D-brane worlds provide a simple geometric framework within which many semi-realistic particle physics models can be constructed. However, just like other supersymmetric string constructions, these intersecting D-brane models suffer from the usual moduli problem. Typically, these models contain a lot of moduli (from both closed and open string sectors) which remain massless before supersymmetry is broken and hence, if not stabilized, would lead to serious phenomenological problems as well as loss of predictivity. Deeply related to the moduli problem is the question of how supersymmetry is broken. In studying the phenomenological consequences of string theory, one traditionally starts with an $\mathcal{N} = 1$ supersymmetric string vacuum whose low-energy spectrum contains the Standard Model. The hope is that the same mechanism that breaks supersymmetry and gives masses to the superpartners of the Standard Model could also lift all the moduli. Strong D-brane gauge dynamics, for example, can result in gaugino and matter condensations, generating a non-perturbative Veneziano-Yankielowicz-type potential (145, 146) that can in principle stabilize certain closed string moduli. However, such non-perturbative dynamics can be analyzed only at the level of an effective super Yang-Mills theory with the leading instanton contribution. Moreover, without fine-tuning, the vacua stabilized by non-perturbative effects typically have a large non-vanishing cosmological constant, thus rendering these models unrealistic for further phenomenological studies (see section 6.6). Therefore, in practice, one often treats the supersymmetry breaking sector as a black box and simply parameterizes our ignorance of supersymmetry breaking with the VEVs of the auxiliary fields of some moduli without specifying how they acquire a VEV [see, e.g., (147)].

Recently there have been some interesting attempts to understand simultaneously these two central problems in string phenomenology—moduli stabilization and supersymmetry breaking—by considering compactification with background flux (148–153). The idea, which is most conveniently expressed in the framework of Type IIB string theory, is that a superpotential can be generated by the NS-NS and R-R three-form flux background (148, 150). The superpotential thus generated depends on the dilaton and the complex structure moduli, and so these moduli are generically lifted. Interestingly, depending on how the gauge and chiral sectors are embedded, supersymmetry can be softly broken by the flux. More importantly, the resulting soft SUSY breaking terms can be calculated in a systematic way, perturbatively (154–158). We shall see toward the end of this section how such soft terms are generated from the local D-brane physics point of view. From the effective four-dimensional supergravity perspective, the effect of the background flux is to introduce a microscopic source for the auxiliary fields of the moduli, which in turn signals supersymmetry breaking.

Thus, flux compactification provides a rather attractive framework for string phenomenology. However, in order to explore quantitatively its phenomenological features, it is important to construct some concrete examples in which realistic features of the Standard Model, such as chirality, can be incorporated. The general techniques of constructing chiral flux vacua have been developed in (159–161), although no chiral models which are free of tadpole instability have been found. More recently, it was recognized by (48, 49) that a crucial step in constructing stable chiral flux vacua is to introduce additional pairs of $D9 - \overline{D9}$-branes, which nonetheless are BPS because of the magnetic flux on their world-volumes. Chiral flux vacua (both supersymmetric and non-supersymmetric) that are free of tadpoles have been constructed in (48, 49). Furthermore, the low-energy spectrum of these models is remarkably close to that of the MSSM, and hence they provide a proof of concept that realistic particle physics features can be embedded in flux compactification.[7] Interestingly, in cases where supersymmetry is broken softly by the background flux, the vacuum remains Minkowski after supersymmetry breaking (at least to lowest order) because the NS-NS tadpoles are absent. Subsequently, there have been further interesting attempts to construct realistic models within this framework, and more examples of three- and four-family chiral flux vacua, including supersymmetic ones, have been found in (163, 164).

Although this review has so far focused on Type IIA string theory with intersecting D6-branes, the intersecting D-brane models discussed here are related (in the absence of flux) by a simple duality to Type IIB orientifolds with magnetized D9-branes [see, e.g., (160) for the details of such map]. Hence, there is an alternative, albeit less geometrical, description of the same models in Type IIB string theory. In fact, the techniques of building intersecting D-brane models that we have discussed can be readily adapted to construct chiral flux vacua, which we will review below.

5.1. Three-Form Fluxes in Type IIB String Theory

Various aspects of flux compactifications have been discussed in the literature. Instead of providing a comprehensive overview of flux compactifications (which is not the main purpose of this review), we will only sketch here some of the basic results (149, 153, 165, 166) relevant to string model building. Consider Type IIB string theory in the presence of a non-trivial three-form background flux $G_3 = F_3 - \tau H_3$. Here F_3 denotes the R-R and H_3 the NS-NS three-form flux, and τ is the complex dilaton-axion field. The background fluxes must obey the Bianchi identity and be properly quantized, i.e., they take values in $H^3(\mathcal{M}, \mathbb{Z})$. In toroidal

---

[7]Local models of flux compactification with realistic particle physics features have been considered in (162).

(orbifold) backgrounds, such quantization conditions are particularly simple:

$$\frac{1}{(2\pi)^2\alpha'} \int_\Sigma H_3 \in N_{min} \times \mathbb{Z}, \qquad \frac{1}{(2\pi)^2\alpha'} \int_\Sigma F_3 \in N_{min} \times \mathbb{Z}, \qquad 54.$$

where Σ is a three-cycle in the *covering space*, N_{min} is a positive integer which reflects the fact that in an orientifold (orbifold), there can exist three-cycles which are smaller that in the covering space. For example, the $T^6/\mathbb{Z}_2 \times \mathbb{Z}_2$ orientifold, taking into account also the orientifold projection, $N_{min} = 8$, and N_{flux}, defined in (57), is a multiple of 64.

Turning on F_3 and H_3 fluxes has two important effects. First, the Chern-Simons terms in the Type IIB effective supergravity action

$$S_{CS} = \frac{1}{2\kappa_{10}^2} \int d^{10}x \frac{C_4 \wedge G_3 \wedge \bar{G}_3}{4i\,\mathrm{Im}\tau} \qquad 55.$$

when integrated over the six-dimensional manifold $\mathcal{M}$ induces a tadpole for the R-R four-form gauge potential C_4. In particular, the D3 charge contribution to the tadpoles is of the form

$$N_{flux} = \frac{1}{(4\pi^2\alpha')^2} \int_\mathcal{M} H_3 \wedge F_3 . \qquad 56.$$

The second effect is that the kinetic term for G_3 (suppressing the warp factor)

$$V = \frac{1}{4\kappa_{10}^2\mathrm{Im}\tau} \int_\mathcal{M} d^6y\, G_3 \wedge \star_6 \bar{G}_3, \qquad 57.$$

induces a scalar potential, which can be written as

$$V = \frac{1}{2\kappa_{10}^2\mathrm{Im}\tau} \int_\mathcal{M} d^6y G_3^- \wedge \star_6 \bar{G}_3^- - \frac{i}{4\kappa_{10}^2\mathrm{Im}\tau} \int_\mathcal{M} d^6y G_3 \wedge \bar{G}_3, \qquad 58.$$

where again we suppress the warp factor. Here, $G_3^\pm$ is the imaginary self-dual/anti-self-dual (ISD/IASD) part of G_3, i.e., it satisfies $\star_6 G_3^\pm = \pm i G_3^\pm$. The second term in (59) is a topological term equal in magnitude to N_{flux} and gives rise to the NS-NS tadpole of the flux. Contrarily, the first term in (59) is a positive semi-definite F-term potential (148, 150), which precisely vanishes if the flux is imaginary self-dual, i.e., $G_3^- = 0$.

It has been shown in (150) that this F-term potential V_F can be derived from the Gukov-Vafa-Witten superpotential (148)

$$W = \int_\mathcal{M} \Omega_3 \wedge G_3, \qquad 59.$$

which apparently depends only on the complex structure moduli and the dilaton and vanishes if these moduli are chosen such that G_3 is imaginary self-dual. Self-duality implies that the three-form flux has $(2, 1)$ and $(0, 3)$ components with respect to the complex structure of the underlying Calabi-Yau, i.e., $G_3 = G_3^{(2,1)} + G_3^{(0,3)}$. For supersymmetric minima, one gets the additional conditions that (i) $G_3^{(0,3)} = 0$, and (ii) primitivity of G_3, i.e., $G_3 \wedge J = 0$; the latter condition is automatically satisfied on a Calabi-Yau manifold. Taking the back-reaction of the fluxes on the geometry into account, one finds it is quite moderate in the (topological) sense that one gets a warped Calabi-Yau metric [although the metric can be strongly warped as in (165, 167)]. To summarize, the flux-induced scalar potential allows one to freeze the complex structure moduli and the dilaton at its minima.

Although the discussion here is in the context of Type IIB string theory, there should be an alternative description in Type IIA string theory where intersecting D6-branes (the subject of this review) can be introduced. However, under duality, the three-form fluxes that we consider here become metric fluxes on the Type IIA side and the underlying geometry becomes non-Kähler (166, 168, 169). The types of three-cycles that the D6-branes can wrap around in such non-Kähler geometries are not well understood. Alternatively, one can study directly Type IIA orientifolds with background fluxes [see (170–172) and references therein]. For recent efforts in obtaining examples of flux compactifications in massive Type IIA supergravity with intersecting D6-branes, see (173–175).

5.2. Semi-Realistic Flux Vacua

In the following, we summarize the techniques for constructing consistent Type IIB orientifolds with magnetized D9-branes in toroidal (orbifold) backgrounds, developed in (159, 160).

A stack of N_a magnetized D9-branes on toroidal orbifolds is characterized by three pairs of integers (n_a^i, m_a^i) which satisfy

$$\frac{m_a^i}{2\pi} \int_{T_i^2} F_a^i = n_a^i, \qquad 60.$$

where the m_a^i denote the wrapping numbers of the D9-brane around the i-th two-torus T_i^2, n_a^i is the magnetic flux, and the F_a^i is the corresponding $U(1)$ magnetic field-strength on the D9-brane. The orientifold projection on these quantum numbers acts as: $\Omega\bar{\sigma}(-1)^{F_L} : (n_a^i, m_a^i) \rightarrow (n_a^i, -m_a^i)$. In the T-dual intersecting D6-branes picture, these quantum numbers correspond to the wrapping numbers (n_a^i, m_a^i) of the homology one-cycles $([a_i], [b_i])$ of the i-th two-torus T_i^2 that the D6-branes wrap around. [For a detailed dictionary between these two T-dual descriptions, see, e.g., (160). More general aspects of the T-dual picture were discussed in (176).]

Because of the orientifold projection, the magnetized D9-branes set-up as a whole does not carry any net D5- and D9-brane charges. However, there are additional discrete K-theory charges that need to be taken into account (49). Other than this subtlety, the tadpole cancellation conditions for the magnetized D9-brane sector simply amount to the cancellation of the D3- and three types of D7-brane charges. Such conditions can be deduced from the conditions (41) in the T-dual intersecting D6-brane picture in section 2. Note, however, that the background fluxes introduce an additional contribution to the D3 tadpole (57). The corresponding tadpole conditions, here specifically written for the $\mathbb{Z}_2 \times \mathbb{Z}_2$ orientifold, read (159, 160):

$$\text{D3} - \text{charge} \sum_a N_a n_a^1 n_a^2 n_a^3 = 8 - \frac{N_{flux}}{4},$$

$$\text{D7}_\text{i} - \text{charge} \sum_a N_a n_a^i m_a^j m_a^k = -8 \quad \text{for} \quad i \neq j \neq k \neq i. \qquad 61.$$

[For consistency conditions as applied to other orbifolds, see (177).] The large positive D3-charge contribution from N_{flux} makes it hard to satisfy these tadpole conditions without introducing anti-D3 branes which give rise to instabilities (159, 160).

Fortunately, it was recognized by (48, 49) that the negative D3-brane charge needed for the cancellation of R-R tadpoles can be accounted for by introducing additional pairs of $D9 - \overline{D9}$-branes which are nonetheless BPS because of the magnetic flux supported on their world-volumes. This observation led to the first example (48, 49) of a three-family supersymmetric Standard Model in flux compactification. In this construction, the Standard Model sector is based on the local MSSM module introduced in (109). Therefore, as in (109), there is only a pair of Higgs doublets in the low-energy spectrum and thus precisely the minimal Higgs content of the MSSM. However, the $D9 - \overline{D9}$ pairs which carry the needed negative D3-charge have a non-vanishing "intersection product" with the Standard Model building block. Hence in addition to the gauge and matter content of the MSSM, there are also some additional chiral exotics.

Subsequently, additional semi-realistic flux vacua have been constructed in (163, 164). In particular, by considering such BPS $D9 - \overline{D9}$ pairs as part of the observable sector, a broader class of Standard Model-like vacua with three and four families of chiral matter, and larger units of flux, including supersymmetric ones, were constructed in (164). These models typically have chiral exotics and more than one-pair of Higgs doublets. It is fair to say that a fully realistic model of flux compactification has yet to be found.

Let us now briefly comment on two related issues pertinent to these chiral flux vacua: generation of soft supersymmetry breaking terms and stabilization of open string moduli. First, we can understand heuristically how soft supersymmetry breaking terms are generated by the background flux. As discussed above, the background three-form flux carries R-R charge and tension. More precisely, ISD (IASD) flux carries the same type of R-R charge and tension as that of a D3-brane ($\overline{D3}$-brane). Therefore, a D3-brane ($\overline{D3}$-brane) will be attracted to a region where the IASD (ISD) flux is maximum. Recall that the positions of D3-branes correspond to world-volume scalars, and so the energy needed to move the D3-branes away from the maximum flux region would reflect as soft masses on the world-volume gauge theory. The analysis for the D7-brane sector is more involved, but one can again understand how soft terms are generated by studying the induced D3-brane charge (due to the background flux) carried by the D7-branes.

For the same reason that soft terms are generated, the background flux can also induce a mass to some of the open string moduli and thus provide a way to stabilize them (158, 178, 179). Finally, although the Kähler moduli do not enter the flux-induced superpotential, a linear combination of some toroidal Kähler moduli and open string moduli enters the Fayet-Iliopoulos D-term, and so we expect such linear combination of closed and open string moduli to be fixed.

Much work needs to be done before a fully realistic model of flux compactification (i.e., a model not only with a realistic low-energy spectrum and couplings, but also with all its moduli stabilized) can be found. However, the developments described here have undoubtedly pointed to an interesting direction in string phenomenology.

6. PHENOMENOLOGICAL ISSUES

No fully realistic intersecting D-brane model has been constructed yet. Furthermore, many of the phenomenological features, such as the gauge and Yukawa couplings, or the masses and other properties of the low-energy particles, are model dependent or depend on details of supersymmetry breaking. Nevertheless, it is useful to survey here some of the phenomenological features that have emerged in various constructions, with an emphasis on the difficulties, possibilities for new physics, and things to watch for in the future. Many of the technical aspects or detailed consequences of specific models were discussed in earlier sections. Here we focus on general issues.

Let us start with two general comments. The first concerns the fundamental string-scale M_s. As discussed in Subsection 2.1, most toroidal (orbifold) constructions have assumed either that M_s and the inverse sizes of the extra dimensions are very large, i.e., within a few orders of magnitude of the Planck scale, or else that M_s is much lower, e.g., in the 1–1000 TeV range. The latter can occur for either supersymmetric or non-supersymmetric constructions, when the overall volume of the extra dimensions is much larger than the volumes of the three-cycles wrapped by the D6-branes, and provides a stringy implementation of the phenomenological brane world models with large extra dimensions (23) [for recent reviews, see e.g., (180–182)].

Most non-supersymmetric constructions, including many toroidal ones, have assumed a low M_s, of the order of the TeV scale, to avoid large radiative corrections governed by M_s which aggravate the Higgs-hierarchy problem and to avoid large contributions of $\mathcal{O}(M_s^4)$ to the cosmological constant resulting from NS-NS tadpoles. However, as discussed in Subsection 2.1, for the purely toroidal constructions with intersecting D6-branes, there is typically no direction in the internal space that would be transverse to all the D6-branes, and thus the size of the internal space is constrained to be of the order of the Planck 6-volume, and

M_s is restricted to be within a few orders of magnitude to the Planck scale. The consistent implementation of the non-supersymmetric constructions with a low M_s would therefore be possible only for more general Calabi-Yau spaces (like fractional D6-branes on toroidal orbifolds). On the other hand, for the supersymmetric constructions, the NS-NS tadpoles cancel and the radiative corrections below M_s are at most logarithmic. Therefore such constructions with large M_s, as dictated by toroidal (orbifold) internal spaces, are stable, resulting in a calculable spectrum and effective Lagrangians at M_s.

We should also point out that models (57, 67) with a locally supersymmetric spectrum of the MSSM, which however do not cancel R-R tadpoles for toroidal orientifolds, are also of interest because they may provide a prototype D-brane configuration of the MSSM-sector. [R-R tadpole-free implementation of such construction was realized, at the expense of Standard Model chiral exotics, within $\mathbb{Z}_2 \times \mathbb{Z}_2$ orientifolds in (48, 108); see Subsection 3.2.2.]

Another issue to keep in mind is that there may be new physics at the TeV scale beyond the Standard Model or MSSM. Most of the explicit constructions lead, e.g., to additional Higgs or exotic matter or additional $U(1)$ gauge symmetries at the TeV scale [as do most heterotic constructions; see, e.g., (183–185)]. It is of course possible that these are defects of the models and that there is nothing beyond the MSSM at the TeV scale. On the other hand, one should keep open the possibility of a rich spectrum of "top-down"-motivated new physics, especially of the kinds that appear so commonly in constructions.

6.1. The Spectrum

Most explicit intersecting D-brane constructions that contain the MSSM spectrum also involve additional matter states. As described in section 3, it is straightforward to construct non-supersymmetric or locally supersymmetric intersecting D-brane models with only the Standard Model spectrum, but existing fully supersymmetric constructions are highly constrained and always contain additional matter. This is true of most heterotic constructions as well. It is often considered the goal of model building to come as close to the MSSM as possible, but it is also possible that additional matter really does exist at the TeV scale. Such states may either be chiral or non-chiral, although non-chiral states can typically obtain a string-scale mass after deformations of D-brane configurations. For toroidal (orbifold) models, these states involve two stacks of D-branes that wrap the same one-cycle in one two-torus; parallel splitting of the two stacks of D-branes in this two-torus renders the non-chiral matter massive (21). Chiral matter occurs in (non-Abelian) anomaly-free combinations, and is frequently necessary to cancel the anomalies associated with additional gauge factors that would not be present for the MSSM spectrum alone. There are stringent constraints from precision electroweak physics on new matter that is chiral with respect to $SU(2)_W \times U(1)_Y$ (186) that essentially exclude the possibility of a fourth ordinary or mirror family or other representations that are chiral under the Standard Model gauge group, except possibly for rather tuned ranges of masses or other compensations. However, states that are singlets or vector-pairs under the Standard Model group but chiral under additional gauge factors are allowed in those cases in which the construction allows a mechanism for generating fermion masses in the hundreds of GeV range (scalar masses may be due to soft supersymmetry breaking). Additional matter may also have important implications for gauge coupling unification, as discussed in Subsection 6.3.

6.1.1. EXTENDED HIGGS SECTOR   One ubiquitous possibility is an extended Higgs sector, involving more than one pair of Higgs doublets (often many pairs) and often Standard Model singlets whose VEVs could break additional gauge factors. Additional doublets would lead to a rich Higgs spectrum detectable at colliders and could mediate flavor-changing neutral currents. Higgs singlets S could couple to doublets with superpotential couplings $W = hSH_uH_d$, so that a TeV-scale VEV could lead to an effective μ parameter $\mu_{eff} = h\langle S \rangle$, elegantly solving the μ problem. This would be an implementation of some form of the next to minimal supersymmetric Standard Model (NMSSM) [see, for example, (187) and references therein] or its $U(1)$'s extension (188). Such models differ dramatically from the MSSM, e.g., by the expanded spectrum; possible large doublet-singlet mixing with implications for Higgs masses, production, and decays (189); a different allowed range for $\tan\beta$; expanded possibilities for electroweak baryogenesis because of a strong first order phase transition and new sources of CP violation [see (190, 191) and references therein]; and an enlarged and modified neutralino sector, extending the possibilities for cold dark matter (191–193). It should be stressed that no existing construction has all of these features or all of the couplings needed for a fully realistic model. For example, the supersymmetric construction (20) has no chiral singlet S to generate a μ_{eff}, though its role could be played by a field in the $\mathcal{N} = 2$ sector if it did not acquire a large mass (93). [Another possibility for a μ term would be a D-brane splitting in models in which the Higgs doublets are non-chiral, as in the locally supersymmetric model in (57, 67), although it is not clear why the splitting would be sufficiently small.]

6.1.2. EXOTIC QUARKS AND LEPTONS   Heavy exotic (i.e., with non-standard Standard Model representations) quarks and leptons, presumably vector-like pairs with respect to $SU(2)_W \times U(1)_Y$, are also possible, with exotic quarks especially producing distinctive effects at a hadron collider [see, e.g., (194)]. These are familiar in E_6 grand unification [see, e.g., (195)], and often emerge in string constructions as well, e.g., associated with the remnants of a would-be fourth family (20).

6.1.3. CHIRAL EXOTICS   Still more exotic (and probably unwanted) possibilities exist. These include the open string sector moduli, typically in the adjoint (anti-symmetric) representation for the $U(N)$ ($Sp(2N)$) gauge symmetry, as well as fields in the symmetric and anti-symmetric representation for the $U(N)$ gauge factors, associated with the intersection of D6-branes with its orientifold image (see Subsection 2.6). It is expected that the inclusion of fluxes would induce a back-reaction that would give a mass to some of the open string moduli (160). Another possibility is to introduce rigid three-cycles like fractional branes in toroidal orbifolds. D6-branes wrapping such three-cycles do not have massless moduli in the adjoint representation.

Many constructions also involve intersections between the ordinary and hidden sector branes, leading to states that are charged under the non-Abelian factors of both, i.e., the hidden sector is really only quasi-hidden. ($U(1)$ gauge bosons also typically couple to both sectors.) These mixed states often carry exotic electric charges (such as $1/2$). The laboratory and astrophysical constraints on fractional charges are severe (196). Fortunately, the quasi-hidden groups are often strongly coupled, so that such states may be confined, and may even lead to observable composite states with more conventional quantum numbers (93).

6.1.4. GRAND UNIFICATION EXOTICS   As discussed in Subsection 3.2.1, it is possible to construct Grand Unified gauge groups such as $SU(5)$ from intersecting branes, and both fundamental and adjoint Higgs, and fundamental and antisymmetric matter representations, appear. The supersymmetric three-family $SU(5)$ models that have been constructed always also include symmetric **15**-plet representations (101), which contain highly exotic states such as color sextets and weak triplets. Other phenomenological aspect of Grand Unification, such as doublet-triplet splitting (105) due to discrete Wilson lines, are touched on in Subsection 3.2.1.

6.2. The Gauge Group

Physics beyond the Standard Model may involve extended gauge groups. In particular, intersecting D6-brane models and D-brane models in general involve $U(N)$ ($Sp(2n)$) groups for planes not parallel (parallel) to orientifold planes. These may break to the Standard Model gauge group $G_{SM} = SU(3)_C \times SU(2)_W \times U(1)_Y$ and additional $U(1)$ and non-Abelian factors (the latter most commonly involves a quasi-hidden sector non-Abelian group, as discussed in Subsection 6.6). At intermediate stages in all explicit $\mathbb{Z}_2 \times \mathbb{Z}_2$ examples, $SU(3)_C$ is embedded into a Pati-Salam $SU(4)$ in which lepton number is the fourth color (197). In some cases, there is also an embedding of $SU(2)_W \times U(1)_Y$ into $SU(2)_W \times SU(2)_R$. Here we focus on additional $U(1)$ factors, which frequently occur in intersecting D-brane constructions, as well as other types of string constructions (198, 199) and alternative approaches to move beyond the Standard Model, such as dynamical symmetry breaking (200) and Little Higgs models (201, 202). Because of their generality, extra $U(1)$ symmetries and their associated heavy Z' bosons are probably the best motivated extension of the Standard Model after supersymmetry. Experimental limits on an extra Z' are very model dependent, depending on the Z' mass, gauge coupling, and couplings to the left- and right-handed quarks and leptons. However, typical limits from the combination of Z-pole and other precision experiments and direct searches for $p\bar{p} \rightarrow e^+ e^-, \mu^+ \mu^-$ are typically $M_{Z'} > 500\text{--}800$ GeV and the $Z - Z'$ mixing is less than a few $\times 10^{-3}$ (see, e.g., (203)).

There are several sources of extra $U(1)$ symmetries in intersecting D-brane models. One is that stacks of branes not parallel to O6-planes yield $U(N) \sim SU(N) \times U(1)$ groups, where the $U(1)$s are typically anomalous. Recall that the $U(1)$ anomalies are canceled via a generalized Green-Schwarz mechanism. As described in Subsections 2.7 and 3.1, the Z' gauge bosons associated with these extra $U(1)$ factors will typically acquire string-scale masses by Chern-Simons terms, even if they are not anomalous (18). [Field theoretic analogs have been studied recently in (204).] However, the $U(1)$s will survive as perturbative global symmetries of the theory, often restricting possible Yukawa couplings and/or leading to conserved baryon and lepton numbers, stabilizing the proton and preventing Majorana neutrino masses (18). For non-supersymmetric models with a TeV string scale, this implies new Z' gauge bosons with masses generated without a Higgs mechanism. Experimental constraints from their mixing with the Z have been examined in (205–207), where lower bounds on $M_{Z'}$ and therefore on the string scale in the TeV range were obtained, somewhat more stringent than typical bounds on E_6-motivated Z' bosons.

Non-anomalous additional $U(1)$s may arise from the breaking of $SU(N)$ factors by parallel splitting of $U(N)$ branes, such as the extra $U(1)_{B-L}$ emerging from $SU(4)$ in (20, 21) and other typical Standard-like Model constructions; or from the splitting of $Sp(2N)$ branes parallel to O6-planes, such as the $Q_8 - Q_{8'}$ in (20, 21). [See (108) for a general discussion.] As discussed in (93), such $U(1)$ factors need to be broken by the VEVs of Standard Model singlets charged under the $U(1)$. The breaking could be at the TeV scale if it is driven by the same type of terms which drive electroweak breaking (208), or at a scale intermediate between the TeV and Planck scale if it is along a D and (tree-level) F-flat direction (209), provided there are appropriate Standard Model singlet fields. In some cases, the only candidates are the bosonic partners of right-handed neutrinos (93), and the needed couplings are not always present.

Experimental implications of a TeV-scale Z' are significant. These include the effects at colliders of the Z' and associated exotics needed for anomaly cancellation (see, e.g., (203)), and the effects of the extended Higgs and neutralino sectors for colliders and cosmology, commented on in Subsection 6.1. The Z' couplings are often family-nonuniversal in both intersecting brane and heterotic constructions, implying flavor-changing neutral currents after fermion mixing is turned on [see, e.g., (210–212)]. These could be significant, e.g., for rare B, K, and μ decays.

6.3. Gauge Coupling Unification

It is well known that the observed (properly normalized) low-energy gauge couplings $\alpha_1^{-1} \equiv \frac{3}{5}\alpha_Y^{-1}$, α_g^{-1}, and α_s^{-1} associated respectively with $U(1)_Y$, $SU(2)_W$, and $SU(3)_C$ are roughly consistent with gauge unification at a scale $M_U \sim 3 \times 10^{16}$ GeV when the β functions are calculated assuming the MSSM particle content [see, e.g., (213)]. The value of $\alpha_s \sim 0.13$ predicted from α and $\sin^2 \theta_W$ is slightly larger than the observed value (~ 0.12), even accounting for uncertainties in the sparticle spectrum, but could be due to high scale threshold effects in traditional Grand Unified theories. In heterotic string constructions one expects to maintain gauge unification at the string scale, which is typically an order of magnitude larger than the apparent Grand Unified Theory (GUT) scale M_U. However, the normalization of gauge couplings at M_s is modified for higher Kač-Moody embeddings, which are common for $U(1)_Y$ but not for $SU(3)_C \times SU(2)_W$. Also, most constructions involve additional matter which can modify the β functions. These two effects can each modify the predicted α_s and $\log M_U$ by $O(1)$, compared to the 10% corrections that are needed, i.e., traditional gauge unification is lost unless these two effects are absent or somehow compensate.

As discussed in Subsection 4.2, traditional gauge unification is lost in most intersecting D-brane constructions because the gauge coupling at the string scale for each stack of D-branes depends on stack-dependent moduli, i.e., on the volume of the three-cycle wrapped by the stack. [One exception are supersymmetric Grand Unified Models (21, 101), in which the Standard Model gauge factors all are derived from a single stack. A local construction (90), based on a three-family $U(3)^3$ sector also provides a gauge coupling unification of the three gauge factors at the string scale.] Furthermore, as discussed in Subsection 6.1 the constructions (including the Grand Unified ones) typically involve exotic states that will modify the running. One must therefore hope that these effects will somehow compensate to yield the observed couplings.

One approach is to predict the low-energy couplings (including those for any additional gauge factors) for a given construction in terms of the spectrum. This was done as a function of M_s and the volume moduli in (93) for the supersymmetric model (20) described in Subsection 3.2. It was found that the predicted couplings were typically smaller than the observed ones due to the extra chiral matter. The analysis was refined in (95), in which it was shown that due to gaugino condensation, the toroidal complex structure moduli and dilaton are fixed (see also Subsection 6.6). One could then predict $\alpha_s^{-1} \sim 52.2$ and $\alpha(M_Z)^{-1} \sim 525$, much larger than the observed values ~ 8.5 and 128, due to the extra chiral matter. The weak angle, which is a ratio, came out better, with the predicted $\sin^2 \theta_W \sim 0.29$ not too far from the observed 0.23. Although not successful, this illustrates the possibility that a more realistic construction might lead to the observed couplings.

In a general D-brane construction, there is no simple relation between the three gauge couplings at M_s. However, it was observed in (141) that under certain circumstances, there is a tree-level relation

$$\alpha_1^{-1} = \frac{2}{5}\alpha_s^{-1} + \frac{3}{5}\alpha_g^{-1}, \qquad\qquad 62.$$

a special case of the canonical GUT relation in which all three are equal. This could come about in models in which the weak hypercharge satisfies the left-right symmetry relation $Q_Y = \frac{1}{2}(B - L) + Q_{3R}$, with the additional assumptions that $U(1)_{B-L}$ derives from the same stack as $SU(3)_C$ (Pati-Salam embedding) and that there is a left-right symmetry that ensures the same coupling for Q_{3R} and

$SU(2)_W$. It was shown in (141) that if (63) holds, then from the observed low-energy couplings and from the contributions to the β functions from exotic matter, one can predict the value of M_s and the volume moduli. In fact it turned out that an effective β function coefficient was always an even integer leading to a discrete set of possible values for the string scale. For example, for no exotics one finds $M_s \sim 2 \times 10^{16}$ GeV with volume radii $R_s = 2.6/M_s$ and $R_W = 3.3/M_s$ for the $SU(3)_C$ and $SU(2)_W$ branes, respectively. The addition of exotic matter can lead to very different M_s and radii. Of course, there is no guarantee that after stabilization M_s and the moduli would actually take these values. Although the first assumption (on $U(1)_{B-L}$) is satisfied by existing supersymmetric constructions because $SU(3)_C \times U(1)_{B-L} \subset SU(4)$, the second (on Q_{3R}) is not satisfied in most known constructions such as (20, 21), which are not left-right symmetric. [Examples in which it does hold were given for the locally supersymmetric construction (109), two models among supersymmetric constructions in (107) and a four-family model in (108).] In (127) the frequency of this relation was statistically investigated in the ensemble of MSSM-like Gepner model orientifolds. It was found that for approximately 10% of these models, this relation was satisfied [which could clearly be seen in the overall plot in (127)].

6.4. Yukawa Couplings

Yukawa couplings and the pattern of fermion masses and mixings are one of the least understood aspects of nature. In the context of the Standard Model or MSSM, or in simple Grand Unification extensions, it is often assumed that some sort of additional family symmetry might lead to textures (hierarchies of elements including zeroes) in the fermion Yukawa matrices to explain the observed patterns. The ratio $\tan \beta$ of the VEVs of the neutral Higgs fields from the two Higgs doublets H_u and H_d of the MSSM may also play a role. The recent observation of neutrino oscillations further complicates the situation because of the possibility of Majorana masses.

In existing string constructions (including heterotic), the possible Yukawa and other superpotential interactions are typically very much restricted by additional symmetries (e.g., the perturbative global symmetries that remain after anomalous $U(1)$s are broken by the Green-Schwarz mechanism) or by stringy selection rules such as orbifold and orientifold projections. Such restrictions may be weakened in more general constructions, but are an important feature of existing examples, and they may lead, e.g., to texture zeros. For example, one of the families of quark and lepton doublets in the supersymmetric model (20) in Subsection 3.2 has no Yukawa couplings due to the Q_2 symmetry and remains massless, or the $SU(5)$ models described in subsection 3.2.1 have no **10 10 5**$_H$ couplings owing to the $U(1)$ of $U(5)$. Many models [e.g., (18, 20)] have conserved B and L owing to global and local $U(1)$s, stabilizing the proton and preventing Majorana neutrino masses. Similarly, in existing intersecting D-brane models (unlike some heterotic constructions) H_d and lepton doublets are clearly distinguished even though they have the same Standard Model quantum numbers because two of the relevant branes are distinct.

String constructions also allow natural mechanisms for hierarchies of Yukawa couplings. For example, free fermionic models can lead to small effective couplings from higher dimensional operators. As described in Subsection 4.1.1 intersecting D-brane constructions allow for a geometrical origin of hierarchies, because allowed Yukawa couplings are caused by world-sheet instantons and are proportional to $\exp(-A)$, where A is the area of the triangle connecting the three intersecting branes.

Existing supersymmetric intersecting D-brane constructions contain more than a single $H_{u,d}$ pair, as described in subsection 6.1 (this is true for many heterotic constructions, as well), with each having different Yukawa matrices. Thus, hierarchies of their VEVs could be an additional mechanism for achieving hierarchies of masses and nontrivial mixings, and in generating otherwise vanishing masses. Of course, the actual VEVs would depend on the details of how supersymmetry is broken. In particular, in schemes of radiative electroweak breaking (in which negative Higgs mass squares are generated from positive ones at a higher scale by renormalization group running) there will be a strong tendency for only those Higgs fields with large Yukawa couplings to actually acquire VEVs. There has been relatively little phenomenological work on these sorts of extended Higgs sectors.

In specific intersecting D-brane models on toroidal (orbifold) compactifications, the Yukawa couplings often factorize in terms of the family indices for the left and right-handed fermions, e.g., the couplings $h_{i,j}^k$ between H_u^k, Q_i and $\bar{U}_j$ are proportional to products $a_i^k b_j^k$. This can occur, for example, if the non-trivial intersections for Q_i and $\bar{U}_j$ occur in different two-tori (137) or if the orientifold and orbifold projections associate each $\bar{U}_j$ with a distinct H_u^j (94). The factorization does not hold in more general examples [e.g., (107)]. Factorization could actually pose a problem for a construction with only a single pair of $H_{u,d}$ doublets, because it allows only one massive state of each fermion type (u-type, d-type, e-type). Some means must therefore be found to populate other terms in the mass matrices. Possibilities include accepting additional Higgs pairs, modifying the D-brane geometry, invoking (non-aligned) four-point interactions in non-supersymmetric models with low M_s (214), or allowing for (non-aligned) supersymmetry breaking A terms (if allowed by the supersymmetry breaking mechanism) (214). There are also potential problems with the minimal two-doublet structure if the electroweak symmetry is promoted to $SU(2)_L \times SU(2)_R$, because the $SU(2)_R$ symmetry would ensure equal Yukawa matrices for the u and d, preventing a nontrivial CKM quark mixing matrix (67, 109, 137). [This is also one reason $SO(10)$ models require more than a single Higgs multiplet coupling to fermions (215).]

A more detailed analysis of the Yukawa couplings for the supersymmetric multi-Higgs model (20) described in Subsection 3.2 was made in (94). It was shown that for appropriate values of some (unknown) volume moduli, one could obtain nontrivial masses and mixing for two families. Near the symmetric points (small splitting between stacks of branes) one obtains the GUT-like result of similar d and charged lepton mass matrices, as well as similar u and Dirac neutrino masses. The Dirac neutrino masses are problematic because the model has no non-perturbative mechanism to generate Majorana masses for a seesaw mechanism. The Yukawa structure for non-supersymmetric models with one or two pairs of Higgs fields was studied in (18, 214). A locally supersymmetric model with a single Higgs pair (whose global embedding was realized in (48, 49, 110)) was considered in (109, 137), where it was emphasized that having only one massive family is actually an excellent first approximation, because m_t, m_b, and m_τ are much larger than the other generations.

6.5. Flavor-Changing Effects and Proton Decay

In the Standard Model there are no flavor-changing neutral currents (FCNC) mediated by the Z, γ, or Higgs at tree-level, and FCNC at loop-level are suppressed (the GIM mechanism). However, there are enhanced FCNC effects in most extensions of the Standard Model, including new loop effects in supersymmetry and new interactions in dynamical symmetry breaking. Similarly, the only sources of CP violation are the phases in the quark (and lepton) mixings, possible neutrino Majorana phases, and a possible strong CP parameter θ_{QCD}. For small quark mixings, all but θ_{QCD} lead to extremely small neutron, atomic, and electric dipole moments (EDM), whereas most extensions of the Standard Model lead to enhanced effects. Therefore, experimental studies of rare decays and suppressed mixings, such as $\mu \rightarrow 3e$, $K_L - K_S$ mixing and rare B decays, as well as refined EDM experiments, are an excellent way to search for new physics.

6.5.1. FCNC   There are a number of sources of FCNC in string constructions (in addition to the standard particle loops in supersymmetric constructions). The tree-level calculation of the string four-point amplitudes (134) (see Subsection 4.1) produces flavor-changing four-point operators in the effective action. For non-supersymmetric constructions with a low M_s, the analysis of such operators was carried out in (135, 214), where it was shown that there could be significant effects from both Kaluza-Klein modes and stretched heavy string modes. For example, Kaluza-Klein excitations couple non-universally to states located at different positions, and therefore to FCNC. The authors of (135, 214) studied the constraints on these operators from experimental bounds on FCNCs, EDMs, and supernova cooling by neutrino emission induced by four-fermi operators, and they showed that the FCNC severely restrict the string scale to be higher than $\sim 10^4$ TeV. This suggests that such non-supersymmetric constructions have a severe fine-tuning problem and makes it unlikely that other effects, such as the $U(1)$ gauge bosons which acquire a string-scale mass by the Chern-Simons terms (205–207) described above, will be observable.

A number of other (field theoretic) sources of FCNC may be expected from intersecting D-brane (and other) string constructions and may be observable in future experiments. The most promising are additional TeV-scale $U(1)$s with family-nonuniversal couplings, as described in Subsection 6.2; multiple Higgs doublets, for which the neutral components can mediate FCNC; or extended non-Abelian groups that can survive down to low energies, such as the embedding of $SU(2)_W$ into $Sp(6)$ at ~ 100 TeV, leading to $K_L \to \mu^\pm e^\mp$ (108).

6.5.2. CP-VIOLATING PHASES   The CP-violating phases for supersymmetric constructions can appear in the Yukawa couplings which depend on the VEVs of the (complex) Kähler moduli [for a detailed discussion of this moduli dependence, see (137)]. Another source of the CP-violating phases can be complex soft supersymmetry breaking masses and the μ parameters; in intersecting D-brane constructions the complex soft supersymmetry breaking mass parameters are due to the complex VEVs of closed sector moduli, as discussed briefly in Subsection 6.6.

6.5.3. STRONG CP PROBLEM   In (216) a mechanism to solve the strong CP problem was proposed, which could have a realization within intersecting D6-brane models. This mechanism is reminiscent of the (chiral) anomaly inflow mechanism. Specifically, the proposal employs an additional bulk $U(1)_X$ gauge factor under which quarks are not charged, and the flux associated with the NS-NS three-form field strength H_3. The anomaly cancellation takes place owing to a Chern-Simons term of the Type IIA supergravity and terms in the expansion of the D6-brane world-volume Chern-Simons action. A specific non-supersymmetric intersecting D6-brane model that explicitly realizes this mechanism was constructed in (216). It remains an open problem to implement this mechanism for the supersymmetric intersecting D6-brane constructions with supersymmetric H_3 fluxes.

6.5.4. PROTON DECAY   Supersymmetric Grand Unified theories (186) allow proton decay by dimension 5 or dimension 6 operators (we assume that dimension 4 R parity-violating terms that could lead to unacceptable rates are absent). The dimension 5 operators (via heavy colored fermion exchange) lead to too rapid proton decay unless they are somehow forbidden, whereas the dimension 6 operators from heavy gauge boson exchange typically lead to a lifetime of the order of 10^{36} yr, too long to observe in planned experiments (the current limit of $\sim 4 \times 10^{33}$ yr for $p \to e^+ \pi^0$, which may be improved to $\sim 10^{35}$ yr).

The expectations for proton decay in supersymmetric intersecting D-brane constructions have been studied recently in (136). In many intersecting D-brane constructions, baryon number is conserved perturbatively and the proton is stable. However, the proton can decay in the Grand Unified constructions described in 3.2.1. The four-fermion contact operator for $\mathbf{10^2 \overline{10}^2}$ in intersecting D-brane $SU(5)$ models for the four states located at the same intersection (where there is no suppression from area factors) was calculated in (136) (see also Subsection 4.1). This operator has an enhancement, relative to the standard Grand Unified Models, caused by the exchange of Kaluza-Klein excitations of the color triplet gauge bosons, which leads to the decay amplitude $\propto \alpha_{GUT}^{-1/3}$. In order to further increase the decay amplitude, the string coupling was taken to be $\mathcal{O}(1)$, thus leading to the M-theory on G_2 holonomy space (see subsection 2.9), and the gauge coupling threshold corrections (217) were included. However the final result did not have additional large enhancement factors, suggesting a lifetime of around 10^{36} yr, comparable to ordinary supersymmetric Grand Unification.

6.6. Moduli Stabilization and Supersymmetry Breaking

Intersecting D-brane constructions on toroidal (orbifold) backgrounds possess a large number of closed and open string sector moduli, thus leading to a large vacuum degeneracy. In fact, the vacuum degeneracy problem is generic for supersymmetric string constructions. As mentioned before, this problem has been addressed via two mechanisms: (1) implementation of the strong D-brane gauge dynamics that can lead to gaugino and matter condensations and generates a non-perturbative superpotential for the closed string sector moduli fields; (2) introduction of supergravity fluxes whose back-reaction introduces a moduli dependent potential. It is expected that in a realistic framework, a combination of both mechanisms will play a role in obtaining string vacua with (all) moduli stabilized, broken supersymmetry and potentially realistic cosmological constant. In the following sections we shall summarize the phenomenological implications, studied for these two mechanisms.

6.6.1. STRONG D-BRANE GAUGE DYNAMICS   Explicit supersymmetric intersecting D6-brane constructions typically possess a quasi-hidden gauge sector that has a number of non-Abelian confining gauge group factors, typically with $Sp(2N)$ gauge symmetries. The non-perturbative superpotential of the Veneziano-Yankielowicz type (145, 146) is a sum of exponential factors (associated with each confining gauge factor):

$$W_a(U^i, S) = \frac{\beta_a}{32\pi^2} \frac{\Lambda^3}{e} \exp\left(\frac{8\pi^2}{\beta_a} f_a(U^i, S)\right),$$ 63.

where the dynamically generated scale Λ is roughly of the order of the string-scale M_s, β_a is the beta function of the specific gauge group factor and $f_a(U^i, S)$ denotes the corresponding gauge kinetic function, which for intersecting D6-branes depends on the complex structure moduli U^i and the dilaton field S. Equation (64) accounts only for the leading instanton contribution. One should also point out that for a specific number of "flavor" (matter) N_f and "color" (gauge) N_c degrees of freedom, there are subtleties; e.g., $Sp(2N_c)$ gauge factors can lead to the quantum lift of the moduli space ($N_f = N_c + 1$) or absence of the non-perturbatively generated global superpotential ($N_f > N_c + 2$). [For a review see, e.g., (218) and references therein; for the implementation of strong gauge dynamics in the effective actions from heterotic strings, see (219).] Classes of semi-realistic supersymmetric intersecting D6-branes constructions, e.g., (106, 107), have the property that the hidden sector gauge group factors satisfy $N_f < N_c + 1$, resulting in confining infrared dynamics and the non-perturbative superpotential of the type (64).

For toroidal (orbifold) compactifications, as discussed in Subsection 4.2, the tree-level gauge kinetic function $f_a(U^i, S)$ (53) depends on the dilaton S and three toroidal complex structure moduli U^i (some of the toroidal complex structure moduli are fixed by the supersymmetry constraints in the D6-brane sector) and the specific wrapping numbers $(n_a^i, \widetilde{m}_a^i)$ of the three-cycle π_a, wrapped by a stack of N_a D6-branes. For a specific supersymmetric semi-realistic construction (20, 21), the non-perturbative superpotential (64), associated with the confining $Sp(2) \times Sp(2) \times Sp(4)$ sector, resulted (95) in the minimum of the potential that stabilized the remaining toroidal complex structure modulus U and the dilaton S, and broke supersymmetry. It would also be interesting to implement the threshold corrections to the gauge kinetic function (144) as discussed in Subsection 4.2. For $\mathcal{N} = 2$ sectors these corrections depend also on toroidal Kähler moduli, and thus the non-perturbative superpotential (64) could in principle allow for the stabilization of the toroidal Kähler moduli as well.

When supersymmetry is broken by such a non-perturbative superpotential, the gaugino masses m_{λ_a} can be determined in terms of F-breaking terms associated with S and U^i moduli directions:

$$m_{\lambda_a} = \left(\partial_{\phi^i} f_a(\Phi^i)\right) K^{\Phi^i \bar{\Phi}^j} \bar{F}_{\bar{\Phi}^j}. \qquad 64.$$

Here $K^{\Phi^i \bar{\Phi}^j}$ is the inverse of the Kähler metric of the moduli Φ^i, and F_{Φ^j} are the F-breaking-terms for the moduli $\Phi^j = \{S, U^i\}$. Unlike the heterotic constructions and simple Grand Unified theories, the gaugino masses and gauge couplings at the string scale depend on more than one modulus, i.e., S and U_i, which in general have complex VEVs and thus lead to non-universal and complex (indicating significant CP-violating phases) gaugino masses. Unfortunately for the specific model studied in (95), these masses were too heavy, i.e., $\mathcal{O}(10^8)$ GeV.

The study of soft supersymmetry breaking parameters of the charged matter sector requires detailed information on the moduli dependence of the leading term in the Kähler potential for the charged matter; this Kähler potential was recently determined in (132, 133) and discussed in subsection 4.1.2 (specifically, see Equation 49). Unfortunately, for the specific example studied in (95), the minimum of the non-perturbative superpotential produced a large negative cosmological constant, and thus these vacua do not provide realistic backgrounds for a detailed study of the soft supersymmetry breaking parameters of the charged matter sector. However, one can assume that the non-perturbative mechanism for supersymmetry breaking does not introduce a large cosmological constant, and then one can parameterize such soft masses via F_{Φ_i}-breaking terms associated with moduli Φ_i by employing the standard supergravity techniques. Such a study was recently performed in (220). [For an earlier work see (221).] In the regime where the F_{U_i}-breaking terms that are associated with the U_i moduli are dominant, the mass parameters do not depend on the Yukawa couplings and have a pattern different from the heterotic string.

In principle, the strong gauge dynamics can also lead to composite (baryon-type) states whose constituents include states that are chiral exotics, i.e., states charged both under the Standard Model gauge factors and the hidden strong gauge sectors. This scenario could provide another mechanism to remove chiral exotics from the light spectrum (see (93)).

6.6.2. SUPERGRAVITY FLUXES   Supergravity fluxes provide another mechanism to stabilize the compactification moduli. The supersymmetric flux compactifications are better understood on the Type IIB side (for details see section 5). Semi-realistic constructions of Type IIB vacua consist of the magnetized D-brane sector (T-dual to the intersecting D6-branes) and the G_3 fluxes stabilizing the toroidal complex structure moduli (in the T-dual picture Kähler moduli) and the dilaton-axion field.

Typical semi-realistic examples have fluxes that break supersymmetry via a $(0, 3)$ component of G_3, and recent phenomenological studies focused on the implied generation of soft supersymmetry breaking terms in the low-energy effective action (154–158, 220, 222–224). These terms have been derived by employing two complementary approaches:

- The soft supersymmetry breaking mass terms due to fluxes were obtained by expanding the resulting Dirac-Born-Infeld action for the D3- and D7-branes (154, 155, 158) to the lowest order in the coordinates transverse to the D-brane world-volume.

- Employing the standard supergravity formalism, one can parameterize the soft supersymmetry breaking terms via the supersymmetry breaking VEVs of the auxiliary F and D components of the chiral and vector supermultiplets (156, 157).

Both approaches are expected to be equivalent. A third approach using the F-theory description of a certain orientifold has been pursued in (65).

We have given in section 5 a heuristic argument why such soft terms are generated. In the following we summarize the specific results. For the imaginary self-dual G_3, the soft supersymmetry breaking mass terms are absent for the matter associated with the open string states on the D3-branes. However, for the anti-D3-branes these masses are non-vanishing, and specifically they stabilize the open string modulus associated with the position of the anti-D3-brane (154, 156). This point also plays a very important role in getting de-Sitter vacua via the KKLT construction (165).

On the other hand, such mass terms for the open string states on D7-branes are due to the non-supersymmetric $(0, 3)$-components of G_3 fluxes (157, 158). Interestingly, the supersymmetric $(2, 1)$-components of G_3 fluxes can induce superpotential mass terms for the D7-brane moduli, including those associated with D7-brane "intersections" (158), thus providing a stabilization mechanism for them. Assuming a homogeneous flux, the scale of such mass terms is of the order $\frac{M_s^2}{M_{pl}}$.

For the supersymmetry breaking masses to be of the TeV scale, this implies that the string scale is in the intermediate regime. This is reminiscent of the gravity mediated supersymmetry breaking mechanism. More generally, such mass terms measure the local flux density and so M_s that appears in the above estimate should be the local string scale which can in principle be much smaller because of the non-trivial warp factor.

6.7. Cosmological Aspects

The main focus of this review has been the particle physics aspects of intersecting D-brane models. For completeness, however, let us briefly mention some cosmological aspects of this scenario as well. A comprehensive overview of string/brane cosmology is beyond the scope of this review. Here, we shall only sketch some highlights of this subject that are particularly relevant to intersecting D-brane worlds. For details and references, we refer the readers to some excellent reviews (225–227).

There has been widespread hope that string theory may provide a microscopic origin for inflation. The discovery of D-branes has opened up several new possibilities. In this review, we have focused on D-brane models that preserve $\mathcal{N} = 1$ supersymmetry, for otherwise the D-brane configurations are generically unstable. Such instability has typically too short a life-time, which would spell disaster for particle physics today. However, in the early universe, the initial configuration of D-branes is not necessarily perfectly stable. Instead the D-branes could intersect at non-supersymmetric angles, or there could be additional pairs of branes and anti-branes separated in the compact dimensions. These instabilities drive the system of D-branes to a neighboring stable configuration, so we can think of the supersymmetric models that we have discussed at length in this review as the endpoints of such dynamical processes. In fact, a natural candidate for the inflaton field in this scenario is the open string mode whose VEV describes the inter-brane separation (228). The dynamics of inflation are therefore governed by the interaction between D-branes. This idea of brane inflation (228) has been applied to construct inflationary models arising from the collision of branes and anti-branes (228–232), as well as branes intersecting at angles (233–235). In particular (232), which is by far the most detailed model of inflation from string theory, demonstrated that the brane inflation proposal can be implemented in a string model where the geometric moduli are stabilized by the background fluxes (a concrete mechanism that we discussed in section 5).

Interestingly, the cosmic string network produced at the end of D-brane inflation offers an exciting opportunity to test stringy physics from cosmological observations [see, e.g., (227, 236) for some reviews and references]. Toward the end of brane inflation, the inflaton potential becomes tachyonic. The condensation of this complex tachyon mode results in the formation of cosmic strings, rather than other cosmological defects such as monopoles or domain walls. Finally, in addition to cosmic D-strings, there could in general be stable D-branes carrying K-theory charges in the intersecting D-brane models discussed here. They could be interesting candidates for superheavy dark matter (237).

7. CONCLUSIONS AND OUTLOOK

In this review we have provided a pedagogical introduction to string theoretic intersecting D-brane models, which we hope suits the need of students to have a comprehensive though not too technical guideline for this topic. We have also tried to briefly review much of the work on intersecting D-brane models carried out so far, including an overview of model building attempts, such as recent flux compactifications, as well as the structure of the low-energy effective action. The latter of course is very important for concrete phenomenological applications of these models.

During the short history of intersecting D-brane constructions, new momentum was brought into the field from other branches of string theory research like M-theory compactifications on G_2 manifolds or flux compactifications. Clearly, all these model building schemes are intimately related. After more than four years of intense research, it has become clear that intersecting D-brane models provide a general phenomenologically appealing class of string constructions. These constructions also provide explicit string theory realizations of some of the ideas that have emerged from a more bottom-up approach.

Even though we have a nice geometric framework, we are still lacking a completely convincing model realizing the MSSM. One can find isolated mechanisms for realizing most of the features of the Standard Model, like family replication, hierarchical Yukawa couplings, absence of extra gauge symmetries and vector-like matter, etc., but all concrete models studied so far do not realize all Standard Model properties at the same time: They either have extra chiral exotic matter as is typical for supersymmetric constructions, or the models fail at the level of couplings, such as gauge and Yukawa couplings. Of course only a very few classes of models, primarily based on toroidal orbifolds, were constructed, and even fewer were studied in detail. In addition, the techniques are not yet available to study more general intersecting D-brane models on, for example, generic smooth Calabi-Yau spaces. It seems that the notorious appearance of extra vector-like matter is related to the fact that we are only considering models at very special, highly symmetric points in moduli space, like orbifold or Gepner points. Considering that the finer details of the Standard Model are far from being very natural, there is no guarantee that nature has finally stabilized in a string vacuum that is highly symmetric and treatable with the simple methods developed so far. Therefore, it would be interesting to develop the tools to study more generic intersecting D-brane models.

Alternatively, it is entirely possible that physics at the TeV scale is richer than the MSSM, and that some of the features found in existing constructions, such as extended gauge symmetries, extra chiral matter, and flavor changing neutral currents, really exist. The LHC and future experimental probes are eagerly awaited to refine the target of our theoretical investigations.

Concerning the low-energy effective field theory, considerable progress has been made in computing, for example, Yukawa couplings, the Kähler potential, or the resulting soft supersymmetry breaking terms for very simple toroidal backgrounds. However, much more work is needed to derive similar results for more general backgrounds.

Clearly, for each Calabi-Yau manifold there does exist a plethora of consistent intersecting D-brane models. In view of these, one might ask whether there does exist any chance to find the/a realistic string vacuum. This picture becomes even more severe when one also takes into account the so-called landscape of flux compactifications. It was proposed that complementary to a model-by-model search, one could study the statistical distribution of string theory vacua (238) [see (239) for a statistical analysis of intersecting D-branes] to obtain an estimate of the chances of finding a realistic model, as well as to identify a region of the parameter space in which to look.

For the moment we can only hope that continuous work on both approaches— the model-by-model search and the statistical analysis—will eventually lead us to a realistic string model from which, once the background is fixed, all features of the low-energy effective theory can be derived. However, whether such a model is in any sense unique is not guaranteed, as we will always measure the physical parameters with some finite accuracy. Having one string model which describes our world within the accuracy of our measurements would nevertheless be considered a milestone in our understanding of nature.

ACKNOWLEDGMENTS

R.B. would like to thank PPARC for financial support during the first half of this project. The research was supported in part by Department of Energy Grant DOE-EY-76-02-3071 (M.C. and P.L.), Fay R. and Eugene L. Langberg Endowed Chair (M.C.), National Science Foundation Grant INT02-03585 (M.C. and G.S.), National Science Foundation CAREER Award PHY-0348093 (G.S.), Department of Energy Grant DE-FG-02-95ER40896 (G.S.) and a Research Innovation Award from Research Corporation (G.S.). Part of this review has been written at DAMTP (University of Cambridge) and R.B., M.C. and G.S. would like to thank DAMTP for its hospitality. R.B. also thanks the University of Pennsylvania for its hospitality. G.S. thanks the Perimeter Institute for Theoretical Physics for its hospitality during the final stage of writing this review. We would like to thank G. Honecker, F. Marchesano, T. Liu, and T. Weigand for useful comments about the manuscript, and our various collaborators, including C. Angelantonj, V. Braun, J.P. Conlon, F. Gmeiner, L. Görlich, B. Greene, G. Honecker, B. Körs, T. Li, T. Liu, D. Lüst, F. Marchesano, T. Ott, I. Papadimitriou, K. Schalm, S. Stieberger, K. Suruliz, T. Taylor, H. Tye, A. Uranga, L.-T. Wang, and T. Weigand.

**The *Annual Review of Nuclear and Particle Science* is online at
http://nucl.annualreviews.org**

LITERATURE CITED

1. Angelantonj C, Sagnotti A. *Phys. Rep.* 371:1 (2002), hep-th/0204089
2. Dabholkar A. *Proc. High Energy Physics and Cosmology, Summer School 1997, Trieste, Italy,* pp. 128–191 (1998), hep-th/9804208
3. Berkooz M, Douglas MR, Leigh RG. *Nucl. Phys.* B480:265 (1996), hep-th/9606139
4. Bachas C. http://www.arXiv.org/abs/hep-th/9503030
5. Blumenhagen R, Görlich L, Körs B, Lüst D. *JHEP* 10:006 (2000), hep-th/0007024
6. Blumenhagen R, Görlich L, Körs B, Lüst D. *Fortsch. Phys.* 49:591 (2001), hep-th/0010198
7. Angelantonj C, Antoniadis I, Dudas E, Sagnotti A. *Phys. Lett.* B489:223 (2000), hep-th/0007090
8. Angelantonj C, Sagnotti A. *Proc. Int. Conf. Quantization, Gauge Theory, and Strings, Moscow* 1:21 (2000), hep-ph/0010279
9. Blumenhagen R, Görlich L, Körs B. *Nucl. Phys.* B569:209 (2000), hep-th/9908130
10. Pradisi G. *Nucl. Phys.* B575:134 (2000), hep-th/9912218
11. Angelantonj C, Blumenhagen R. *Phys. Lett.* B473:86 (2000), hep-th/9911190
12. Blumenhagen R, Görlich L, Körs B. *JHEP* 1:040 (2000), hep-th/9912204
13. Förste S, Honecker G, Schreyer R. *Nucl. Phys.* B593:127 (2001), hep-th/0008250
14. Blumenhagen R, Conlon JP, Suruliz K. *JHEP* 7:022 (2004), hep-th/0404254
15. Aldazabal G, Franco S, Ibáñez LE, Rabadán R, Uranga AM. *JHEP* 2:047 (2001), hep-ph/0011132
16. Aldazabal G, Franco S, Ibáñez LE, Rabadán R, Uranga AM. *J. Math. Phys.* 42:3103 (2001), hep-th/0011073
17. Blumenhagen R, Körs B, Lüst D. *JHEP* 2:030 (2001), hep-th/0012156
18. Ibáñez LE, Marchesano F, Rabadán R. *JHEP* 11:002 (2001), hep-th/0105155
19. Blumenhagen R, Körs B, Lüst D, Ott T. *Nucl. Phys.* B616:3 (2001), hep-th/0107138
20. Cvetič M, Shiu G, Uranga AM. *Phys. Rev. Lett.* 87:201801 (2001), hep-th/0107143
21. Cvetič M, Shiu G, Uranga AM. *Nucl. Phys.* B615:3 (2001), hep-th/0107166
22. Antoniadis I. *Phys. Lett.* B246:377 (1990)
23. Arkani-Hamed N, Dimopoulos S, Dvali GR. *Phys. Lett.* B429:263 (1998), hep-ph/9803315
24. Antoniadis I, Arkani-Hamed N, Dimopoulos S, Dvali GR. *Phys. Lett.* B436:257 (1998), hep-ph/9804398
25. Uranga AM. *Class. Quant. Gravity* 20:S373 (2003), hep-th/0301032
26. Marchesano FG. http://www.arXiv.org/abs/hep-th/0307252
27. Ott T. *Fortsch. Phys.* 52:28 (2004), hep-th/0309107
28. Kiritsis E. *Fortsch. Phys.* 52:200 (2004), hep-th/0310001
29. Görlich L. http://www.arXiv.org/abs/hep-th/0401040
30. Lüst D. *Class. Quant. Gravity* 21:S1399 (2004), hep-th/0401156
31. Blumenhagen R. *Fortsch. Phys.* 53:426 (2005), hep-th/0412025
32. Green MB, Schwarz JH, Witten E. *Superstring Theory.* Vol. 1: *Introduction.* Cambridge, UK: Cambridge Univ. Press. 469 pp. (1987)
33. Green MB, Schwarz JH, Witten E. *Superstring Theory.* Vol. 2: *Loop Amplitudes, Anomalies and Phenomenology.* Cambridge, UK: Cambridge Univ. Press. 596 pp. (1987)

34. Polchinski J. *String Theory.* Vol. 1: *An Introduction to the Bosonic String.* Cambridge, UK: Cambridge Univ. Press. 402 pp. (1998)

35. Polchinski J. *String Theory.* Vol. 2: *Superstring Theory and Beyond.* Cambridge, UK: Cambridge Univ. Press. 531 pp. (1998)

36. Johnson CV. *D-Branes.* Cambridge, UK: Cambridge Univ. Press. 548 pp. (2003)

37. Zwiebach B. *A First Course in String Theory.* Cambridge, UK: Cambridge Univ. Press. 558 pp. (2004)

38. Polchinski J, Chaudhuri S, Johnson CV. http://www.arXiv.org/abs/hep-th/9602052

39. Polchinski J. In *Fields, Strings and Duality: TASI 96*, pp. 239–356. Singapore: World Sci. (1996)

40. Polchinski J. *Phys. Rev. Lett.* 75:4724 (1995), hep-th/9510017

41. Shiu G, Tye SHH. *Phys. Rev. D* 58:106007 (1998), hep-th/9805157

42. Uranga AM. http://www.arXiv.org/abs/hep-th/0007173

43. Blumenhagen R, Braun V, Körs B, Lüst D. *JHEP* 7:026 (2002), hep-th/0206038

44. Blumenhagen R, Braun V, Körs B, Lüst D. http://www.arXiv.org/abs/hep-th/0210083

45. Witten E. *JHEP* 12:019 (1998), hep-th/9810188

46. Uranga AM. *Nucl. Phys.* B598:225 (2001), hep-th/0011048

47. Witten E. *Phys. Lett.* B117:324 (1982)

48. Marchesano F, Shiu G. *Phys. Rev. D* 71:011701 (2005), hep-th/0408059

49. Marchesano F, Shiu G. *JHEP* 11:041 (2004), hep-th/0409132

50. Blumenhagen R, Görlich L, Ott T. *JHEP* 1:021 (2003), hep-th/0211059

51. Honecker G. *Nucl. Phys.* B666:175 (2003), hep-th/0303015

52. Honecker G, Ott T. *Phys. Rev. D* 70:126010 (2004), hep-th/0404055

53. Green MB, Schwarz JH. *Phys. Lett.* B149:117 (1984)

54. Ibáñez LE. In *Proc. 9th Int. Conf. Supersymmetry and Unification of Fundamental Interactions (SUSY01)*, Dubna, Russia, June 11–17, p. 307 (2001), hep-ph/0109082

55. Ibanez LE, Quevedo F. *JHEP* 10:001 (1999), hep-ph/9908305

56. Becker K, Becker M, Strominger A. *Nucl. Phys.* B456:130 (1995), hep-th/9507158

57. Cremades D, Ibáñez LE, Marchesano F. *JHEP* 7:009 (2002), hep-th/0201205

58. Atiyah M, Witten E. *Adv. Theor. Math. Phys.* 6:1 (2003), hep-th/0107177

59. Witten E. http://www.arXiv.org/abs/hep-th/0108165

60. Acharya B, Witten E. http://www.arXiv.org/abs/hep-th/0109152

61. Cvetič M, Shiu G, Uranga AM. In *Proc. 9th Int. Conf. Supersymmetry and Unification of Fundamental Interactions (SUSY01)*, Dubna, Russia, June 11–17, p. 317 (2001), hep-th/0111179

62. Seiberg N. *Phys. Lett.* B384:81 (1996), hep-th/9606017

63. Seiberg N, Witten E. http://www.arXiv.org/abs/hep-th/9607163

64. Antoniadis I, Maillard T. *Nucl. Phys.* B716:3 (2005), hep-th/0412008

65. Lüst D. Mayr P, Reffert S, Stieberger S. http://www.arXiv.org/abs/hep-th/0501139

66. Hashimoto A, Washington TI. *Nucl. Phys.* B503:193 (1997), hep-th/9703217

67. Cremades D, Ibáñez LE, Marchesano F. *JHEP* 7:022 (2002), hep-th/0203160

68. Hashimoto K, Nagaoka S. *JHEP* (6):034 (2003), hep-th/0303204

69. Erdmenger J, Guralnik Z, Helling R, Kirsch I. *JHEP* 04:064 (2004), hep-th/0309043

70. Epple F, Lüst D. *Fortsch. Phys.* 52:367 (2004), hep-th/0311182

71. Cremades D, Ibáñez LE, Marchesano F. *Nucl. Phys.* B643:93 (2002), hep-th/0205074

72. Kokorelis C. *JHEP* 09:029 (2002), hep-th/0205147

73. Kokorelis C. *JHEP* 08:036 (2002), hep-th/0206108

74. Kokorelis C. http://www.arXiv.org/abs/hep-th/0211091
75. Kokorelis C. *JHEP* 08:018 (2002), hep-th/0203187
76. Kokorelis C. *JHEP* 11:027 (2002), hep-th/0209202
77. Förste S, Honecker G, Schreyer R. *JHEP* 06:004 (2001), hep-th/0105208
78. Bailin D, Kraniotis GV, Love A. *Phys. Lett.* B530:202 (2002), hep-th/0108131
79. Honecker G. *Fortsch. Phys.* 50:896 (2002), hep-th/0112174
80. Honecker G. *JHEP* 01:025 (2002), hep-th/0201037
81. Kokorelis C. *Nucl. Phys.* B677:115 (2004), hep-th/0207234
82. Bailin D, Kraniotis GV, Love A. *Phys. Lett.* B547:43 (2002), hep-th/0208103
83. Bailin D, Kraniotis GV, Love A. *Phys. Lett.* B553:79 (2003), hep-th/0210219
84. Bailin D. http://www.arXiv.org/abs/hep-th/0210227
85. Bailin D, Kraniotis GV, Love A. *JHEP* 02:052 (2003), hep-th/0212112
86. Blumenhagen R, Körs B, Lüst D. *Phys. Lett.* B532:141 (2002), hep-th/0202024
87. Cremades D, Ibáñez LE, Marchesano F. http://www.arXiv.org/abs/hep-ph/0212048
88. Klein M. *Phys. Rev. D* 66:055009 (2002), hep-th/0205300
89. Klein M. *Phys. Rev. D* 67:045021 (2003), hep-th/0209206
90. Li T-j, Liu T. *Phys. Lett.* B573:193 (2003), hep-th/0304258
91. Blumenhagen R, Körs B, Lüst D, Ott T. *Fortsch. Phys.* 50:843 (2002), hep-th/0112015
92. Ellis JR, Kanti P, Nanopoulos DV. *Nucl. Phys.* B647:235 (2002), hep-th/0206087
93. Cvetič M, Langacker P, Shiu G. *Phys. Rev. D* 66:066004 (2002), hep-ph/0205252
94. Cvetič M, Langacker P, Shiu G. *Nucl. Phys.* B642:139 (2002), hep-th/0206115
95. Cvetič M, Langacker P, Wang J. *Phys. Rev. D* 68:046002 (2003), hep-th/0303208
96. Kitazawa N. *Nucl. Phys.* B699:124 (2004), hep-th/0401096
97. Kitazawa N. *JHEP* 11:044 (2004), hep-th/0409146
98. Georgi H, Glashow SL. *Phys. Rev. Lett.* 32:438 (1974)
99. Kokorelis C. http://www.arXiv.org/abs/hep-th/0406258
100. Kokorelis C. http://www.arXiv.org/abs/hep-th/0412035
101. Cvetič M, Papadimitriou I, Shiu G. *Nucl. Phys.* B659:193 (2003), hep-th/0212177
102. Chen CM, Kraniotis GV, Mayes VE, Nanopoulos DV, Walker JW. *Phys. Lett.* B611:156 (2005), hep-th/0501182
103. Axenides M, Floratos E, Kokorelis C. *JHEP* 10:006 (2003), hep-th/0307255
104. Honecker G. *Mod. Phys. Lett. A* 19:1863 (2004), hep-th/0407181
105. Witten E. http://www.arXiv.org/abs/hep-ph/0201018
106. Cvetič M, Papadimitriou I. *Phys. Rev. D* 67:126006 (2003), hep-th/0303197
107. Cvetič M, Li T, Liu T. *Nucl. Phys.* B698:163 (2004), hep-th/0403061
108. Cvetič M, Langacker P, Li V, Liu T. *Nucl. Phys.* B709:241 (2005)
109. Cremades D, Ibáñez LE, Marchesano F. http://www.arXiv.org/abs/hep-ph/0212064
110. Dijkstra TPT, Huiszoon LR, Schellekens AN. *Phys. Lett.* B609:408 (2005), hep-th/0403196
111. Pradisi G. http://www.arXiv.org/abs/hep-th/0210088
112. Pradisi G. http://www.arXiv.org/abs/hep-th/0310154
113. Larosa M, Pradisi G. *Nucl. Phys.* B667:261 (2003), hep-th/0305224
114. Honecker G. http://www.arXiv.org/abs/hep-th/0309158
115. Blumenhagen R, Görlich L, Körs B. http://www.arXiv.org/abs/hep-th/0002146
116. Blumenhagen R, Görlich L, Körs B, Lüst D. *Nucl. Phys.* B582:44 (2000), hep-th/0003024

117. Uranga AM. *JHEP* 12:058 (2002), hep-th/0208014
118. Uranga AM. *Fortsch. Phys.* 51:879 (2003)
119. Angelantonj C, Bianchi M, Pradisi G, Sagnotti A, Stanev YS. *Phys. Lett.* B387:743 (1996), hep-th/9607226
120. Blumenhagen R, Wisskirchen A. *Phys. Lett.* B438:52 (1998), hep-th/9806131
121. Aldazabal G, Andres EC, Leston M, Nunez C. *JHEP* 09:067 (2003), hep-th/0307183
122. Blumenhagen R. *JHEP* 11:055 (2003), hep-th/0310244
123. Brunner I, Hori K, Hosomichi K, Walcher J. http://www.arXiv.org/abs/hep-th/0401137
124. Blumenhagen R, Weigand T. *JHEP* 02:041 (2004), hep-th/0401148
125. Aldazabal G, Andres EC, Juknevich JE. *JHEP* 05:054 (2004), hep-th/0403262
126. Blumenhagen R, Weigand T. *Phys. Lett.* B591:161 (2004), hep-th/0403299
127. Dijkstra TPT, Huiszoon LR, Schellekens AN. *Nucl. Phys.* B710:3 (2005), hep-th/0411129
128. Dixon LJ, Friedan D, Martinec EJ, Shenker SH. *Nucl. Phys.* B282:13 (1987)
129. Grimm TW, Louis J. *Nucl. Phys.* B699:387 (2004), hep-th/0403067
130. Jockers H, Louis J. *Nucl. Phys.* B705:167 (2005), hep-th/0409098
131. Grimm TW, Louis J. *Nucl. Phys.* B718:153 (2005), hep-th/0412277
132. Cvetič M, Papadimitriou I. *Phys. Rev. D* 68:046001 (2003), hep-th/0303083
133. Lüst D, Mayr P, Richter R, Stieberger S. *Nucl. Phys.* B696:205 (2004), hep-th/0404134
134. Abel SA, Owen AW. *Nucl. Phys.* B663:197 (2003), hep-th/0303124
135. Abel SA, Masip M, Santiago J. *JHEP* 04:057 (2003), hep-ph/0303087
136. Klebanov IR, Witten E. *Nucl. Phys.* B664:3 (2003), hep-th/0304079
137. Cremades D, Ibáñez LE, Marchesano F. *JHEP* 07:038 (2003), hep-th/0302105
138. Abel SA, Owen AW. *Nucl. Phys.* B682:183 (2004), hep-th/0310257
139. Cremades D, Ibáñez LE, Marchesano F. *JHEP* 05:079 (2004), hep-th/0404229
140. Abel SA. Schofield BW. *JHEP* 0506:072 (2005), hep-th/0412206
141. Blumenhagen R, Lüst D, Stieberger S. *JHEP* 07:036 (2003), hep-th/0305146
142. Blumenhagen R. http://www.arXiv.org/abs/hep-th/0309146
143. Dixon LJ, Kaplunovsky V, Louis J. *Nucl. Phys.* B355:649 (1991)
144. Lüst D, Stieberger S. http://www.arXiv.org/abs/hep-th/0302221
145. Veneziano G, Yankielowicz S. *Phys. Lett.* B113:231 (1982)
146. Taylor TR, Veneziano G, Yankielowicz S. *Nucl. Phys.* B218:493 (1983)
147. Brignole A, Ibáñez LE, Muñoz C. http://www.arXiv.org/abs/hep-ph/9707209
148. Gukov S, Vafa C, Witten E. *Nucl. Phys.* B584:69 (2000), hep-th/9906070
149. Dasgupta K, Rajesh G, Sethi S. *JHEP* 08:023 (1999), hep-th/9908088
150. Taylor TR, Vafa C. *Phys. Lett.* B474:130 (2000), hep-th/9912152
151. Greene BR, Schalm K, Shiu G. *Nucl. Phys.* B584:480 (2000), hep-th/0004103
152. Curio G, Klemm A, Lüst D, Theisen S. *Nucl. Phys.* B609:3 (2001), hep-th/0012213
153. Giddings SB, Kachru S, Polchinski J. *Phys. Rev. D* 66:106006 (2002), hep-th/0105097
154. Cámara PG, Ibáñez LE, Uranga AM. *Nucl. Phys.* B689:195 (2004), hep-th/0311241
155. Graña M, Grimm TW, Jockers H, Louis J. *Nucl. Phys.* B690:21 (2004), hep-th/0312232
156. Lüst D, Reffert S, Stieberger S. *Nucl. Phys.* B706:3 (2005), hep-th/0406092
157. Lüst D, Reffert S, Stieberger S. http://www.arXiv.org/abs/hep-th/0410074
158. Cámara PG, Ibáñez LE, Uranga AM. *Nucl. Phys.* B708:268 (2005), hep-th/0408036
159. Blumenhagen R, Lüst D, Taylor TR. *Nucl. Phys.* B663:319 (2003), hep-th/0303016

160. Cascales JFG, Uranga AM. *JHEP* 05:011 (2003), hep-th/0303024
161. Cascales JFG, Uranga AM. http://www.arXiv.org/abs/hep-th/0311250
162. Cascales JFG, Garcia del Moral MP, Quevedo F, Uranga AM. *JHEP* 02:031 (2004), hep-th/0312051
163. Cvetič M, Liu T. *Phys. Lett.* B610:122 (2005), hep-th/0409032
164. Cvetič M, Li T, Liu T. *Phys. Rev. D* 71:106008 (2005), hep-th/0501041
165. Kachru S, Kalloch R, Linde A, Trivedi SP. *Phys. Rev. D* 68:046005 (2003), hep-th/0301240
166. Kachru S, Schulz MB, Tripathy PK, Trivedi SP. *JHEP* 03:061 (2003), hep-th/0211182
167. Klebanov IR, Strassler MJ. *JHEP* 08:052 (2000), hep-th/0007191
168. Gurrieri S, Louis J, Micu A, Waldram D. *Nucl. Phys.* B654:61 (2003), hep-th/0211102
169. Cardoso GL, Curio G, Dall'Agata G, Lüst D, Manousselis P, Zoupanos G. *Nucl. Phys.* B652:5 (2003), hep-th/0211118
170. Angelantonj C, Ferrara S, Trigiante M. *JHEP* 10:015 (2003), hep-th/0306185
171. Angelantonj C, Ferrara S, Trigiante M. *Phys. Lett.* B582:263 (2004), hep-th/0310136
172. Derendinger J-P, Kounnas C, Petropoulos PM, Zwirner F. *Nucl. Phys.* B715:211 (2005), hep-th/0411276
173. Behrndt K, Cvetič M. *Nucl. Phys.* B676:149 (2004), hep-th/0308045
174. Behrndt K, Cvetič M. *Nucl. Phys.* B708:45 (2005), hep-th/0407263
175. Behrndt K, Cvetič M. *Nucl. Phys.* B708:45 (2005), hep-th/0407263
176. Rabadán R. *Nucl. Phys.* B620:152 (2002), hep-th/0107036
177. Font A. *JHEP* 11:077 (2004), hep-th/0410206
178. Gorlich L, Kachru S, Tripathy PK, Trivedi SP. *JHEP* 0412:074 (2004), hep-th/0407130
179. Cascales JFG, Uranga AM. *JHEP* 11:083 (2004), hep-th/0407132
180. Hewett J, Spiropulu M. *Annu. Rev. Nucl. Part. Sci.* 52:397 (2002), hep-ph/0205106
181. Pérez-Lorenzana A. http://www.arXiv.org/abs/hep-ph/0406279
182. Csaki C. http://www.arXiv.org/abs/hep-ph/0404096
183. Quevedo F. *ICTP Spring Sch. Superstrings Related Matters, Trieste, Italy, 18–26 Mar.* (2002)
184. Langacker P. http://www.arXiv.org/abs/hep-ph/0308033
185. Kakushadze Z, Shiu G, Tye SHH, Vtorov-Karevsky Y. *Int. J. Mod. Phys. A* 13:2551–98 (1998), hep-th/9710149
186. Eidelman S, Hayes KG, Olive KA, Aguilar-Benitez M, Amsler C, et al. (Particle Data Group Collab.) *Phys. Lett.* B592:1 (2004)
187. Ellwanger U, Gunion JF, Hugonie C, Moretti S. http://www.arXiv.org/abs/hep-ph/0305109
188. Erler J, Langacker P, Li T-J. *Phys. Rev. D* 66:015002 (2002), hep-ph/0205001
189. Han T, Langacker P, McElrath B. *Phys. Rev. D* 70:115006 (2004), hep-ph/0405244
190. Kang J, Langacker P, Li T-J, Liu T. *Phys. Rev. Lett.* 94:061801 (2005), hep-ph/0402086
191. Menon A, Morrissey DE, Wagner CEM. *Phys. Rev. D* 70:035005 (2004), hep-ph/0404184
192. de Carlos B, Espinosa JR. *Phys. Lett.* B407:12 (1997), hep-ph/9705315
193. Barger V, Kao C, Langacker P, Lee H-S. *Phys. Lett.* B600:104 (2004), hep-ph/0408120
194. Andre TC, Rosner JL. *Phys. Rev. D* 69:035009 (2004), hep-ph/0309254
195. Hewett JL, Rizzo TG. *Phys. Rep.* 183:193 (1989)
196. Chang S, Coriano C, Faraggi AE. *Nucl. Phys.* B477:65 (1996), hep-ph/9605325
197. Pati JC, Salam A. *Phys. Rev. D* 10:275 (1974)
198. Cvetič M, Langacker P. *Phys. Rev. D* 54:3570 (1996), hep-ph/9511378

199. Langacker P. http://www.arXiv.org/abs/hep-ph/0402203

200. Hill CT, Simmons EH. *Phys. Rep.* 381: 235 (2003), hep-ph/0203079

201. Arkani-Hamed N, Cohen AG, Georgi H. *Phys. Lett.* B513:232 (2001), hep-ph/0105239

202. Han T, Logan HE, McElrath B, Wang LT. *Phys. Rev. D* 67:095004 (2003), hep-ph/0301040

203. Kang J, Langacker P. *Phys. Rev. D* 71: 035014 (2005), hep-ph/0412190

204. Körs B, Nath P. *JHEP* 12:005 (2004), hep-ph/0406167

205. Ghilencea DM, Ibáñez LE, Irges N, Quevedo F. *JHEP* 08:016 (2002), hep-ph/0205083

206. Ghilencea DM. *Nucl. Phys.* B648:215 (2003), hep-ph/0208205

207. Ghilencea DM. http://www.arXiv.org/abs/hep-ph/0212120

208. Cvetič M, Demir DA, Espinosa JR, Everett LL, Langacker P. *Phys. Rev. D* 56:2861 (1997), hep-ph/9703317

209. Cleaver G, Cvetič M, Espinosa JR, Everett LL, Langacker P. *Phys. Rev. D* 57:2701 (1998), hep-ph/9705391

210. Langacker P, Plümacher M. *Phys. Rev. D* 62:013006 (2000), hep-ph/0001204

211. Leroux K, London D. *Phys. Lett.* B526:97 (2002), hep-ph/0111246

212. Barger V, Chiang C-W, Langacker P, Lee H-S. *Phys. Lett.* B580:186 (2004), hep-ph/0310073

213. Langacker P, Polonsky N. *Phys. Rev. D* 52:3081 (1995), hep-ph/9503214

214. Abel SA, Lebedev O, Santiago J. *Nucl. Phys.* B696:141 (2004), hep-ph/0312157

215. Langacker P. *Phys. Rep.* 72:185 (1981)

216. Aldazabal G, Ibáñez LE, Uranga AM. *JHEP* 03:065 (2004), hep-ph/0205250

217. Friedmann T, Witten E. *Adv. Theor. Math. Phys.* 7:577 (2003), hep-th/0211269

218. Intriligator KA, Seiberg N. *Nucl. Phys. Proc. Suppl.* 45BC:1 (1996), hep-th/9509066

219. Kaplunovsky VS, Louis J. *Phys. Lett.* B306:269 (1993), hep-th/9303040

220. Kane GL, Kumar P, Lykken JD, Wang TT. *Phys. Rev. D* 71:115017 (2005), hep-th/0411125

221. Körs B, Nath P. *Nucl. Phys.* B681:77 (2004), hep-th/0309167

222. Ibáñez LE. *Phys. Rev. D* 71:055005 (2005), hep-ph/0408064

223. Marchesano F, Shiu G, Wang L-T. *Nucl. Phys.* B712:20 (2005), hep-th/04110-80

224. Font A, Ibáñez LE. *JHEP* 0503:040 (2005), hep-th/0412150

225. Quevedo F. *Class. Quant. Gravity* 19: 5721 (2002), http://www.arXiv.org/abs/hep-th/0210292

226. Danielsson UH. *Class. Quant. Gravity* 22:S1 (2005), hep-th/0409274

227. Polchinski J. http://www.arXiv.org/abs/hep-th/0412244

228. Dvali GR, Tye SHH. *Phys. Lett.* B450:72 (1999), hep-ph/9812483

229. Burgess CP, Majumdar M, Nolte D, Quevedo F, Rajesh G, Zhang RJ. *JHEP* 07:047 (2001), hep-th/0105204

230. Dvali GR, Shafi Q, Solganik S. http://www.arXiv.org/abs/hep-th/0105203

231. Shiu G, Tye SHH. *Phys. Lett.* B516:421 (2001), hep-th/0106274

232. Kachru S, Kallosh R, Linde A, Maldacena J, McAllister L, Trivedi SP. *J. Cosmol. Astropart. Phys.* 10:013 (2003), hep-th/0308055

233. Garcia-Bellido J, Rabadán R, Zamora F. *JHEP* 01:036 (2002), hep-th/0112147

234. Blumenhagen R, Körs B, Lüst D, Ott T. *Nucl Phys.* B641:235 (2002), hep-th/0202124

235. Gomez-Reino M, Zavala I. *JHEP* 09:020 (2002), hep-th/0207278

236. Shiu G. http://www.arXiv.org/abs/hep-th/0210313

237. Shiu G, Wang L-T. *Phys. Rev. D* 69:126007 (2004), hep-ph/0311228

238. Douglas MR. *JHEP* 05:046 (2003), hep-th/0303194

239. Blumenhagen R, Gmeiner F, Honecker G, Lüst D, Weigand T. *Nucl. Phys.* B713L83 (2005), hep-th/0411173

Annu. Rev. Nucl. Part. Sci. 2005. 55:141–63
doi 10.1146/annurev.nucl.55.090704.151521
Copyright © 2005 by Annual Reviews. All rights reserved

BLIND ANALYSIS IN NUCLEAR AND PARTICLE PHYSICS

Joshua R. Klein

Department of Physics, University of Texas, Austin, Texas 78712;
email: jrk@physics.utexas.edu

Aaron Roodman

Stanford Linear Accelerator Center, Stanford University, Stanford, California 94309;
email: roodman@slac.stanford.edu

Key Words   blind analysis, experimenter's bias, experimental methods, systematic errors

■ **Abstract**   During the past decade, blind analysis has become a widely used tool in nuclear and particle physics measurements. A blind analysis avoids the possibility of experimenters biasing their result toward their own preconceptions by preventing them from knowing the answer until the analysis is complete. There is at least circumstantial evidence that such a bias has affected past measurements, and as experiments have become costlier and more difficult and hence harder to reproduce, the possibility of bias has become a more important issue than in the past. We describe here the motivations for performing a blind analysis, and give several modern examples of successful blind analysis strategies.

CONTENTS

1. INTRODUCTION

Hans von Osten, who lived at the beginning of the twentieth century, could do math. Given a pair of single-digit numbers written on a blackboard, Hans could add them together correctly nearly all of the time. What was remarkable about this

141

ability was the fact that Hans—often called "Clever Hans"—was a horse. He would demonstrate his skill by pawing the ground with his hoof until he had reached the sum of the two numbers, while those who had presented the problem looked on. Critics of Hans's ability tried to determine whether his trainer was providing him with signals, but could find none. Eventually, they asked the trainer to leave the room, but still Hans managed to add the numbers correctly more often than not. The mystery of Hans's ability was only solved in 1907 when the psychologist Oskar Pfungst proposed that a trial be done in which no one in the room with Hans knew both of the numbers presented (1). With all the observers blind to the answer, Hans was unable to produce a correct result. The conclusion was that Hans was indeed clever: He had been using subtle non-verbal cues from those in the room—cues his observers were not even aware they were providing—to decide when to stop pawing the ground.

The Clever Hans Effect[1] has left its impression on modern science, particularly on medicine. Most large-scale clinical trials of new drugs not only require that the patients be unaware of whether they are receiving a placebo or not, but also that those who administer the drug be kept blind as to which patients are in the control sample (the trials are thus doubly blind).[2]

By contrast, throughout most of their history, nuclear and particle physics have run open experiments in which an estimate of the final answer is known well before the analysis is complete. Adjustments to cuts, measurements of backgrounds and acceptances, and evaluations of systematic uncertainties are routinely made with full knowledge of the current value of the intended measurement and the effects any changes have upon it. Such an approach makes a great deal of sense in a physics experiment, as it allows us to bring to bear some of our most commonly used techniques: we check that the answer makes sense; we give particular scrutiny to results which contradict established models, previous measurements, or conventional wisdom; we are conservative in our estimates of systematic uncertainties to avoid misleading others about the significance of our result.

---

[1]The effect of experimenter expectations on the behavior of subjects is more often referred to as the Hawthorne Effect, named after a factory at which worker productivity was being studied. The productivity was a strong function of what the experimenters indicated they thought would be important (for example, light levels).

[2]Nevertheless, in some extreme cases, the results of a clinical trial have been used by the experimenters as an unacknowledged criterion for their publication: no one wants to publish the fact that their new drug does not work. In such cases, the trial may be repeated until a desired result is obtained, thus leading to a bias in the published data. The problem has so concerned the medical field that the International Committee of Medical Journal Editors (2) now formally requires all clinical trials to first be registered in a public trials registry before the trial begins. With such registration, the investigators essentially commit to publishing the results of their experiments, regardless of whether the outcome is favorable to them or their sponsors.

In his 1932 paper presenting the results of his measurement of the electron charge to mass ratio e/m (3), Frank Dunnington concludes with a warning born of his own experience applying these common sense principles:

> It is also desirable to emphasize the importance of the human equation in accurate measurements such as these. It is easier than is generally realized to unconsciously work toward a certain value. One cannot, of course, alter or change natural phenomena ... but one can, for instance, seek for those corrections and refinements which shift the results in the desired direction. Every effort has been made to avoid such tendencies in the present work.

At least part of Dunnington's effort "to avoid such tendencies" was to keep himself from actually knowing the results of his measurement until he was finished with it. To do this, he kept the value of the angle between the electron source and his detector hidden from himself, by asking his machinist to build something close— but not exactly at—the 340° that was needed (4). Without exact knowledge of this angle, he could not make the final calculations that would change his data into a measurement of e/m.

Is there any evidence over the history of nuclear and particle physics that experimentalists, in Dunnington's words, "unconsciously work toward a certain value"? It would be almost impossible to say definitively, because, of course, we do not know what any particular experimentalist was thinking (unconsciously or otherwise) during the process of a measurement. We might expect that if such a bias exists, then over time new measurements would tend to agree better with prior measurements than with their modern and much more precise value.

Perhaps the classic example of a measurement suspected of experimenter's bias is the speed of light. Measurements of the speed of light span an entire century, with improvements of five orders of magnitude in the experimental uncertainties. A summary of measurements (5), using a variety of techniques, made prior to 1960 is shown in Figure 1, along with the much more accurate value from later results using a methane absorption line frequency (6). One striking feature is the 17 km/sec shift between the series of experiments from 1930–1940 and later determinations. A fascinating post-mortem on the systematic uncertainties in these experiments (7), noting the different techniques used in the four low results, speculates about one of the sources of bias:

> the investigator searches for the source or sources of such errors, and continues to search until he gets a result close to the accepted value.
>
> *Then he stops!*

We have done a brief investigation of this issue in more modern particle physics experiments, using a selected (and hence biased!) set of historical measurements typically compiled by the Particle Data Group (PDG). The PDG's own history plots, which they have published from time to time since 1975 as part of their Reviews of Particle Physics, depict only what the PDG itself published in each year. Their numbers are typically averages over several measurements, and so the

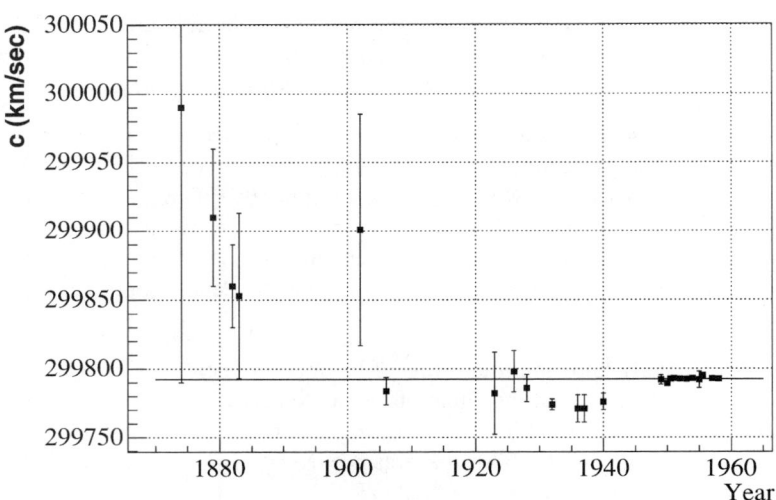

Figure 1 Summary of speed of light measurements. The line indicates the ultimate experimental value. Among other interesting features, the series of four measurements from 1930–1940 displays a 17 km/sec systematic shift from the true value (5).

published values over time are by construction correlated with one another. They therefore do not by themselves necessarily indicate that there is any bias in the data. We have looked instead at the individual measurements, compiled by the PDG or by other reviews, and compared them to the published values which existed at the time of the measurement.

Figure 2 (see color insert) shows the four measurements we have examined: the neutron lifetime, the K_S^0 lifetime, the mass of the Λ, and the value of the ratio g_A/g_V determined from neutron β decay (8–41). The measurements are shown as open circles, with the error bars depicting the uncertainties published along with the measurement. The published average which existed at the time each measurement was made (that is, not including the measurement itself) is represented by the dashed curves, where the spacing between the curves represents the published 1σ uncertainty on those averages. The horizontal dotted lines show the 2004 PDG averages (41).

Although the effect is not striking, the measurements do tend to cluster nearer the prior published averages than the final value. Grouping all the data points together, the χ^2 for the hypothesis that the measurements are normally distributed around the prior averages is 131.2 for 83 degrees of freedom, whereas the χ^2 for the hypothesis that they are normally distributed about the final average is 249.7 for $\sim$82 degrees of freedom.

Even if Figure 2 showed a strong correlation between the measurements and previously published averages, that would not necessarily mean that the experimental results were biased by experimenters' concerns about contradicting conventional wisdom. Many other explanations could account for the behavior—common techniques that were used and were later found to have systematic effects which had been neglected, for example, or physical corrections which were unknown for many years and which would have shifted all the values nearer to the modern averages. Other examples often cited as possible evidence of bias in measurements are the unusually low χ^2 values found in global fits to data sets, such as the the average B meson lifetime $\chi^2 = 4.5$ for 13 degrees of freedom (4), or global fits to solar neutrino data which have, for example, $\chi^2 = 70.2$ for 81 degrees of freedom (42).

Although we cannot say conclusively whether bias has influenced measurements in nuclear and particle physics, the way to avoid even the possibility is to follow Dunnington's and Pfungst's examples and perform measurements while staying blind to the value of our answer. Blind analysis in nuclear and particle physics experiments has its modern origins around 1990 with experiment E791 at Brookhaven National Laboratory, a search for the rare decay $K_L \rightarrow \mu e$, although the idea had been discussed at least ten years earlier (J.R. Ritchie, private communication). As a rare process experiment, E791 had good motivation to use a blind analysis: a potential discovery could easily be missed if they allowed flexibility in their final cuts to remove events suspiciously close to the edge of the signal box. We discuss this kind of hidden signal box technique in more detail in Section 3.1.

The number of experiments that analyze their data blindly has grown steadily since E791's example, and the approaches to how to successfully do a blind analysis are as varied as the experiments themselves. In the next sections we describe the fundamental philosophy behind blind analysis, and detail examples of experiments that have results obtained using blind analysis techniques. We have restricted our discussion only to techniques aimed at avoiding the kind of unintentional bias that concerned both Clever Hans's critics and Dunnington. We do not consider here intentional bias or the bias resulting from systematic effects in instrumentation or technique, none of which can be removed through blind analysis techniques.

2. EXPERIMENTER'S BIAS AND THE MOTIVATIONS FOR BLIND ANALYSES

The Oxford English Dictionary defines bias as "A systematic distortion of an expected statistical result due to a factor not allowed for in its derivation." In a nuclear or particle physics measurement there are many potential sources of bias: the trigger for the experiment, the algorithms used to reconstruct events or extract signals, or the particular instrumentation used. The accepted procedure, when such biases cannot be directly measured or eliminated, is to estimate their size and include the estimates as systematic uncertainties on the measurement. Experimenter's bias differs from these biases because its source is the human being making the measurement, who may (in Dunnington's words) "unconsciously work toward a certain value." The bias may be in the direction of previous measurements, prior theoretical expectations, or some other preconception. The crucial difference between experimenter's bias and any other bias in a measurement is that the size of an experimenter's bias cannot be estimated. Thus the only available approach is to use a methodology that prevents or suppresses it.

Experimenter's bias can creep into a measurement in several ways. The first scenario is subtle, in that the biases it may produce are of the order of the measurement's statistical uncertainty. In general, the data used for a measurement is isolated with a series of selection requirements, or cuts. Although the value of these cuts on a particular quantity may be chosen to maximize the sensitivity of the result, very often there is a wide plateau in the value of the cut, over which the quality of the result varies little. The cartoon shown in Figure 3 (see color insert) illustrates this point. In the case shown, the value of the cut may be chosen arbitrarily within the sensitivity plateau. The numerical value of the result, however, may vary as a function of the cut value, especially for cuts that significantly alter the signal efficiency or background contamination. Such variations will be statistical, with a magnitude on the order of the statistical uncertainty of the measurement. If the value of the cut is chosen with the knowledge of how that value affects the final answer, then the measurement can be biased toward the expected (or desired) result.

It does not take much of a statistical bias to produce a surprisingly large signal, even if the data set contains only background events. A typical analysis sensitive to statistical bias might be the search for a peak in an invariant mass distribution. If the cuts are chosen in a biased way, then the actual fraction of background events accepted in the region of a potential signal may be artificially high, and thus appear as a false signal peak. Consider m sequential cuts whose acceptance for any event outside the signal region is A_i but whose acceptance for any event within the signal region A_i' is biased to permit additional events to leak into the final data sample. The significance of a false peak, defined as $S = N_{signal}/\sqrt{N_{background}}$, is given by

$$S = \left(\prod_{i=1}^{m} \frac{A_i'}{A_i} - 1 \right) \times \sqrt{N \prod_{i=1}^{m} A_i}, \qquad 1.$$

where N is the number of total number of events in the signal region before any cuts are made. As a numerical example, if an analysis begins with $N = 2500$, and uses a set of 10 cuts each with acceptance $A_i = 0.9$, a geometric average bias of just 1% $((\prod_{i=1}^{m} A_i'/A_i)^{1/m} = 1.01)$ will lead to an apparent signal above background of roughly 3σ.

Even though the cuts may be asymptotically unbiased—applied to an infinite data set, $A_i' = A_i$—by having been chosen based in part on how many more apparent signal events they accept in the signal region in this data set, they can still be biased. Put another way, if an ensemble of experiments with the same number of events is analyzed using an identical set of cut values (tuned perhaps, on just one of the experimental data sets), then the mean number of observed signal events will tend toward the true value (which may be zero). But if the ensemble of experiments is analyzed and in each experiment a new value of the cuts is chosen based on the observation of events in the signal bin, the mean number of observed events in the signal bin can always be larger than the true number of signal events.

Of course in practice, finding a set of cuts whose geometric average bias is as large as 1% is not necessarily easy. For example, for the first of the $A_i = 0.9$ cuts, the variance on the number of accepted events is less than 1%, and so either one needs to work hard to tune the cuts, or else another factor is at work, such as an initial statistical fluctuation upward, or one is operating in a particularly sensitive region of the analysis where small variations in the cut position for background events lead to much larger variations in the number of accepted events in the signal region.

The next bias scenario, typically involving the search for rare processes or decays, is much less subtle. Experiments searching for small signals, at the edge of detectability in statistics or above backgrounds, are especially dependent on the exact values of the selection cuts. If the values of the cuts are chosen with the knowledge of which events are included or excluded, the results may be biased toward either observation or elimination of a signal. Such choices can be easy to make if each event is examined individually. Nearly every observed event can be found to have something unusual about it, and so can be included in a signal sample (sometimes a sample of one event) or excluded as an unexpected background. In one extreme, cuts chosen to remove individual events will yield a better upper limit than is deserved. In the other extreme, cuts chosen to retain individual events may produce a signal where none is warranted. This *selection* bias is perhaps the most dangerous.

Finally, another way in which experimenter's bias may affect a measurement is in the decision that a measurement has been completed, as the warning from the speed of light study indicates. Galison (43) notes in his historical study *How Experiments End* that ". . . there is no strictly logical termination point inherent in the experimental process," instead "the decision to end an investigation draws on the full set of skills, strategies, beliefs, and instruments at the experimentalist's disposal." If the decision to stop analyzing and publish relies on the value of the result—in particular how close it adheres to the experimentalist's preconceptions— the result may be biased toward the preconceived value. The danger of continuing the data analysis, finding mistakes or improving the analysis, until the result agrees with expectations, is well known. This *stopping* bias may affect any kind of measurement and may be a small effect or a large one. It is also probably the most common kind of bias found in nuclear or particle physics.

A *blind analysis* is a method that hides some aspect of the data or result to prevent experimenter's bias. There is no single blind analysis technique, nor is each technique appropriate for all measurements. Instead the blind analysis method must carefully match the experiment, both to prevent experimenter's bias and to allow the measurement to be made unimpeded by the method. There are several blind analysis methodologies that will be described in this review, each appropriate for a certain kind of measurement. These methods can be grouped according to exactly what is kept hidden in the measurement:

1. The signal events, when the signal occurs in a well-defined region of the experiment's phase space.

2. The result, when the numerical answer can be separated from all other aspects of the analysis.

3. The number of events in the data set, when the answer relies directly upon their count.

4. A fraction of the entire data set.

Although blind analysis techniques may not be feasible or necessary in all measurements, as a general rule the possibility of experimenter's bias should be considered in all experiments. Typical objections to the use of blind analysis techniques are that it slows the pace of data analysis, that certain aspects of the analysis become difficult, and that unexpected phenomena can be found only by full exploration of the data. The latter issue is a serious one. Consider the following anecdote from a well-known physicist:

> While looking for the decay $\pi^+ \rightarrow e^+ \nu_e$, we focused all our attention on reducing backgrounds, since a prior experiment had set a limit at the level of 10^{-6} on the branching ratio. When we heard that an experiment at CERN had seen a signal around 10^{-4} I switched from delayed to prompt. The signal was right there, and could have been seen on the first day (B. Richter, private communication).

Although some blind techniques are susceptible to this pitfall, not all are. In such cases, a method allowing a full exploration of a data subsample would not have missed such a large signal.

It is crucial that the blind analysis technique be designed as simply and narrowly as possible. A good method, appropriately used, minimizes delays or difficulties in the data analysis. In some cases, a blind analysis may delay certain aspects of an analysis until the blind procedure has been removed, or *unblinded*. For example, certain cross-checks may be possible only after the blind procedure has been removed. The trade-offs involved must be considered according to the individual merits of each case. However, by blinding only a very narrow aspect of the analysis, the methods described in this review minimize the scope of the data analysis needed after unblinding. In general, for the examples described in this review, the pace of data analysis has been slowed only to the extent that individual data analysts have worked to check their measurement more carefully before unblinding their result. We note that none of the blind techniques we describe here—and perhaps no blind technique—can be applied to an analysis in which backgrounds are cut or signals identified by event-by-event human inspection.

Blind analyses solve only one problem, the influence of experimenter's bias on the measurement. Other biases in the measurement caused by the general approach or the instrumentation are not avoided by any of the techniques we describe here. For such biases, either a correction based on a measurement or the inclusion of the bias as a systematic uncertainty still needs to be made, whether the analysis has been done blindly or not.

All sizeable collaborations have internal data analysis and publication review processes. A blind analysis, and the associated division of the data analysis into a blind and an unblind phase, gives the collaboration at large an opportunity to review the work before the transition of looking at the answer. Today large collaborations are grappling with the issue of vetting many publications for both quality and correctness. Collaborations can require that the decision to unblind a result be made as a part of the internal review process, with the consultation or approval of a wider subgroup, and not by the data analysts alone (see the *BABAR* Blind Analysis Task Force report, *BABAR* Analysis Document #91). This procedure improves the effectiveness of the internal review.

The last issue to confront in using blind analyses as a technique is what do to if the analysis strategy breaks down. For example, what should an experiment do if, after all selections cuts have been set, the events in the nominal signal region are clearly background, and additional selections to remove such background were simply omitted? It is not necessary in the blind analysis approach to insist that, because an analysis was done blindly, no additional selections may be applied. Ideally, an experiment should consider such situations in advance to prepare for such cases. One useful principle that may be adopted is that the publication simply describe the full analysis procedure, in this case to be explicit about which selections were applied after the unblinding. The blind analysis method does not require that data analysis stop after unblinding, nor does it ensure that the results of the analysis are correct. There is no reason to publish a result known to be wrong, just because the analysis was done blindly.

Multiple independent analyses are occasionally suggested as a way to prevent experimenter's bias. Although independent analyses can be a powerful tool for preventing errors in a measurement, in our opinion blind analyses prevent experimenter's bias much more directly than redundant analyses. The two methods can, however, easily be used together.

As experiments have become larger, lengthier, more expensive, and therefore harder to reproduce, the issue of whether to do a blind analysis has perhaps become more important than it was a few decades ago. In some cases, a biased result may stand for many years, possibly leading theorists and experimentalists down unproductive and expensive paths. Without the luxury of having new and important results verified quickly, the assurance that experimentalist's bias does not contribute to the many other possibilities for error is generally worth the additional time and effort.

3. BLIND ANALYSIS METHODS

3.1. Hidden Signal Box

Perhaps the most straightforward blind analysis method is the *hidden signal box*. In this technique, a subset of the data, containing the potential signal, is kept hidden until all aspects of the analysis are complete. Often the signal region is defined in terms of two experimental parameters, chosen to separate the signal from backgrounds, and this two-dimensional signal region forms a signal box. Only after the data selection requirements, the signal efficiency, and the estimated background are determined is the hidden signal box opened.

This method is very well suited to measurements searching for rare signals, as long as two criteria are met. First, the signal characteristics and location must be known. In rare decay searches, such as $K_L \rightarrow \mu^{\pm} e^{\mp}$ or $B^0 \rightarrow \mu^+ \mu^-$, the signal may be simulated, the efficiency determined, and an appropriate hidden box defined using the invariant mass and another relevant kinematic variable. Second, the experiment must be able to independently estimate the size of the background expected in the signal box. Ideally, this may be accomplished by understanding the source of background events near the signal box, and extrapolating from this sideband region into the hidden signal box. In particular, the background estimate cannot depend on the characteristics of any events that may be inside the signal box. Generally, the size and placement of the hidden box is determined after an optimization of signal efficiency and backgound rejection. With these conditions, the dependence of the signal-to-background ratio on the selection requirements is known, and the cuts may be optimized as desired, again without reference to the events in the hidden box.

The hidden box method was first used in a search for the rare decay $K_L \rightarrow \mu^{\pm} e^{\mp}$ by the E791 experiment at Brookhaven National Laboratory (BNL) (44). They formed a hidden signal box in a region of the invariant mass, $M_{\mu e}$, and the momentum transverse to the kaon beam direction, P_T^2, as shown in Figure 4. Also visible are a population of background events at lower $M_{\mu e}$, primarily from $K_L \rightarrow \pi e \nu_e$ decays where the pion decays in flight. The many cuts applied to remove backgrounds were optimized using the sideband region, $P_T^2 > 144$ MeV/c^2, and the signal box was not opened until the cuts were determined. No events were observed, so an upper limit was set on this lepton number–violating process.

The hidden signal box method is now a standard technique for searches of rare decays from known particles, and it has been used by many particle physics experiments (see, for example, Reference (45)). In most rare decay searches, the above criteria are generally satisfied, and we recommend that this blind analysis method always be used in these cases. Rare decay searches are distinguished from counting experiments where a sizable signal is present, because even if a small signal is observed, it is generally insufficient to constrain or verify the expected signal characteristics. Thus the expected signal must be characterized by simulation in any case, and a blind analysis causes few extra difficulties. The dividing line between a search for a rare process and a branching fraction or cross-section measurement is a judgment for each experiment. The important consideration is whether or not the signal events themselves must be used to ensure that the experimental efficiency and background rejection are well understood; if so, then the hidden signal box method is no longer appropriate, and one of the alternative methods described in Section 3.4 may be used instead.

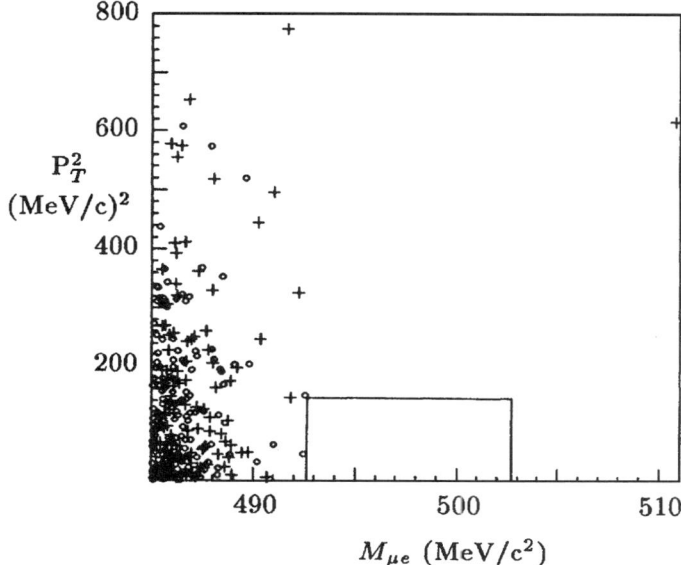

Figure 4   The P_T^2 vs. $M_{\mu e}$ distribution of events from the E791 experiment (44). Also shown is the hidden signal box used for the blind search.

There are several other considerations when using the hidden signal box technique. First, the hidden box should be chosen somewhat larger than the anticipated signal region. Because the final signal box may be made smaller than the hidden box, but not larger, the exact size of the signal box may be optimized during the course of the blind analysis to maximize the signal-to-background ratio, for instance. Next, in general some background may be in the signal box, and the number of background events should be predicted from data sideband regions in advance of opening the box.

If no events lie in the signal box, as in E791, the conclusion is straightforward. The presence of events in the box does require additional effort, however. If there is a significant signal, or if the events in the signal box are consistent with the estimated background, the results follow directly. Nevertheless, several ambiguous situations may occur. It is possible that the events in the box are clearly due to backgrounds, and remained in the signal box owing to the unfortunate omission of certain cuts. Two options are available in this circumstance: to retain the events, treating them as background, and derive a limit accordingly; or to remove them by applying the omitted cut, but also assess the statistical impact of the entire procedure on the result. Ideally, experiments will consider in advance which procedure to use. The next quandary may occur if there are more events in the signal box than expected from backgrounds, but the events are very inconsistent with the expected signal properties. The use of a blind analysis does not require that events in the signal box must absolutely be treated as signal, only that the cuts may not be adjusted to reject or include individual events. The excess events may be interpreted as background, although a posteriori estimation of the signal probability for individual events is fraught with difficulty, and must be done carefully. Some experiments use a hidden signal box in conjunction with an unbinned maximum likelihood fit to the number of signal and background events; such fits incorporate directly the signal or background probability for each event and hence avoid this issue.

3.2. Hidden Answer Methods

Measurements in which most or all of the data analysis can be separated from the numerical value of the result are most amenable to the *hidden answer* blind analysis technique. The e/m measurement of Dunnington (3) illustrates this technique. None of the data analysis, evaluations of corrections, or other features of the e/m measurement were dependent on the unknown angle in his spectrometer, only the final result. The separation between a narrowly constructed hidden feature and the bulk of the measurement permits a blind analysis with little risk or difficulty.

In general the hidden answer method works best for experiments measuring a single precise parameter, when that parameter does not depend directly on the number of observed events.

3.2.1. HIDDEN DETECTOR PARAMETERS   Most modern nuclear and particle physics experiments are too complex to keep their physics results hidden as Dunnington did. A single detector parameter is unlikely to be enough to keep the physics answer hidden, and very often these parameters need to be measured through calibration runs that cannot wait until the analysis is complete.

One area of fundamental physics in which a hidden detector parameter approach does work is the laboratory study of gravity. These experiments are often still performed in the kind of laboratory environment in which ex situ measurements of the apparatus can determine the final answer. Experimental tests of the gravitational inverse-square law (46), such as those done at the University of California at Irvine's Laboratory for Gravitation Research in 1985, employ this kind of hidden parameter approach to hide the physics answer.

The Irvine group measured the torque exerted on a torsion balance by a set of test masses. One pair of 7.3-kg masses (the "far masses") was positioned 105 cm away, while a 43-g "near mass" was positioned 5 cm away. The masses and positions were chosen so that when they were moved to opposing positions, the change in torque predicted by Newtonian gravity was nearly zero. The Irvine group aimed for a precision that would allow them to test deviations from Newtonian gravity as small as one part in 10^4.

The Irvine group's measurement relied on precise knowledge of many different detector parameters—the dimensions of the torsion balance and test masses, the positions of the test masses, and of course the masses of all test components. To prevent themselves from selecting data in a biased way, or from (in their words) "slackening of analysis effort" when their answer began to meet their expectations (what we have called a stopping bias), they kept the value of their near mass known only to 1%—the exact mass known only to someone outside their collaboration. They used the true value of the mass only when they had completed the analysis and were ready to report their initial results. Subsequent improvements to the analysis were made and later published, but they nevertheless published the measurement made before these improvements were made.

3.2.2. HIDDEN OFFSET   The *hidden offset* method inserts an unknown numerical offset into the data analysis so that the true measured value is hidden from the experimenters. This method was first used in the measurement of the direct *CP* violation parameter ϵ'/ϵ by the KTeV collaboration at Fermilab (47).

Direct CP violation is measured in neutral Kaon decays using the double ratio of decay rates for K_S and K_L into charged $\pi^+\pi^-$ and neutral $\pi^0\pi^0$ final states, according to the expression

$$\frac{\Gamma(K_L \to \pi^0\pi^0)/\Gamma(K_S \to \pi^0\pi^0)}{\Gamma(K_L \to \pi^+\pi^-)/\Gamma(K_S \to \pi^+\pi^-)} Re(\epsilon'/\epsilon) = 1 - 6. \qquad 2.$$

In practice, KTeV fit its data, in kaon energy bins, to extract a value for ϵ'/ϵ, as well as other parameters relevant to the experiment. The aim for KTeV was to determine ϵ'/ϵ with a precision of 1–2×10^{-4}. This required both a very large sample of kaon decays, including of order six million $K_L \to \pi^0\pi^0$ events, as well as exquisite control over systematic uncertainties. For instance, KTeV made an acceptance correction to the ratio of observed K_L to K_S events, derived from simulation, of roughly 10%, which had to be understood at the 1×10^{-3} level or better.

In addition, two prior experiments had measured ϵ'/ϵ to a precision of approximately $\sigma_{\epsilon'/\epsilon} \sim 7 \times 10^{-4}$, but differed by roughly 2.5σ. Theoretical estimates ranged from a few 10^{-4} to perhaps 15×10^{-4}. Therefore, KTeV used a blind analysis to prevent any experimenter's bias in what is a difficult and systematically sensitive measurement.

KTeV used a hidden offset directly in its ϵ'/ϵ fit. Instead of fitting for the value of ϵ'/ϵ, the fit used

$$\epsilon'/\epsilon(\text{Hidden}) = \begin{Bmatrix} 1 \\ -1 \end{Bmatrix} \times \epsilon'/\epsilon + C \qquad 3.$$

where C was a hidden random constant, and the choice of 1 or -1 was also hidden and random. KTeV relied on extensive comparisons of data and simulation to design event selection criteria, acceptance corrections, and background subtractions. None of these were affected or impeded by the hidden offset in the fit to ϵ'/ϵ. In addition, direct but separate comparisons were made between the distributions and number of events in data and simulation for $K \to \pi^0\pi^0$ and $K \to \pi^+\pi^-$. The one comparison that could not be made was to form the double ratio that appears in the expression for ϵ'/ϵ. Fortunately, KTeV's method for ϵ'/ϵ almost completely separated the charged $\pi^+\pi^-$ and neutral $\pi^0\pi^0$ analyses so that there was no impact from this restriction.

Both the hidden offset C and the sign choice were made by a pseudo-random number generator, with a seed chosen by the experimenters. The generator picked a value of C with a Gaussian distribution, centered at zero, with a width of approximately 60×10^{-4}. The $+1$ or -1 in the hidden value served to hide the direction ϵ'/ϵ changed as different corrections or selections were applied (48). In practice, KTeV had to remove the sign choice at an earlier stage to permit a full evaluation of systematic errors. Nevertheless, the first KTeV ϵ'/ϵ result was unblinded only one week before the result was made public.

The pseudo-random distribution used for a hidden offset must be chosen with care. A smooth Gaussian distribution with a width large enough to cover prior measurements and predictions is often a good choice. The addition of an unknown sign also hides the direction the result has moved with changes to the analysis. The hidden offset technique also permits multiple analyses within an experiment. In this case, the competing groups should begin with different seeds for the pseudo-random generation of the hidden offset C and the hidden sign. In this way the analyses are blind with respect to the final result and also with respect to each other. Because the motivation for multiple analyses is to provide a strong internal cross-check, this arrangement prevents the groups from comparing their results directly. To make comparisons, both analyses can switch to a common seed. At this stage, any problems with the internal cross-check may be addressed without unblinding the final result.

3.2.3. HIDDEN OFFSET AND HIDDEN ASYMMETRY   For many measurements, hiding the answer would be an appropriate approach to prevent a biased result, but is not sufficient for a blind analysis. In particular, the numerical result may be evident in certain experimental distributions that display the data. For example, the value of a lifetime may be inferred from the decay time distribution. A blind analysis is still possible, but more care is required in constructing it.

A good example of such a measurement is the observation of a CP-violating asymmetry in B-meson decays by the $BABAR$ experiment at the PEP-II asymmetric-energy B factory at the Stanford Linear Accelator Center. The CP-violating parameter $\sin 2\beta$ is measured by comparing the decay-time distribution for B^0 and $\bar{B}^0$ decays into CP-eigenstates, such as $J/\psi K_S^0$. The B flavor, B^0 or $\bar{B}^0$, is determined by the flavor-specific decay (or flavor tag) of the other neutral B-meson in the event. Before CP-violation had been observed, BABAR adopted a blind analysis to avoid the possibility of bias, especially with respect to the prior expectations around $\sin 2\beta \equiv 0.7$ from other weak-interaction measurements and the unitarity of the CKM matrix (4).

The value for the CP asymmetry is determined in a complex unbinned maximum likelihood fit to the decay time, Δt, along with information about the flavor tag and the kinematics of the B^0 decay (49). As such, the hidden offset method, as described above, can easily be used to hide the value of $\sin 2\beta$. The hidden offset method by itself is not enough, however: One of the distributions that must be examined during the course of the analysis is the decay time itself. For example, to ensure that the maximum likelihood fit is done correctly, it is crucial that the probability density function (PDF) used to describe the decay-time is a good match to the data. In this case, the PDFs are determined using much larger samples of other exclusively reconstructed B decays, and simulation may be used to verify that the PDFs will apply for the rarer CP eigenstates. Nevertheless, the decay time distribution for the CP sample must still be examined. The problem for a blind analysis is that the decay time distribution shown separately for B^0 and $\bar{B}^0$ flavor tags, as in Figure 5a (see color insert), uncovers the asymmetry.

To solve this problem, *BABAR* used two extra restrictions in its blind analysis. The first restriction, used in the initial $\sin 2\beta$ measurements (49), was to hide the asymmetry in the time variable used in plots. The asymmetry is evident in two ways in the time distribution: as a difference between B^0 and $\bar{B}^0$ flavor tags and as an asymmetry around $\Delta t = 0$. Both visible asymmetries can be obscured by using a hidden Δt variable, defined as:

$$\Delta t(\text{Hidden}) = \left\{ \begin{matrix} 1 \\ -1 \end{matrix} \right\} \times s_{\text{Tag}} \times \Delta t + \text{Offset}. \qquad 4.$$

The variable s_{Tag} is equal to 1 or -1 for B^0 or $\bar{B}^0$ flavor tags. Because the asymmetry is nearly equal and opposite for the different B flavors, the asymmetry is hidden by flipping one of the distributions. The asymmetry of the individual Δt distributions around zero is hidden by the unknown offset. The result is shown in Figure 5b, where the remaining difference in curves is due to the charm lifetime not CP violation. Although the asymmetric shape of the distribution is still visible in these curves, in an actual sample, limited by statistics, the asymmetry is effectively hidden by the statistical uncertainty of the mean.

Of course, in the actual fits to the data, the true Δt is used, not the hidden Δt. This technique allowed *BABAR* to look at the Δt distribution, but remain blind to any CP asymmetry. In addition, there was one extra restriction: The resulting Δt distribution from the fit could not be overlaid directly on the data, because the smooth PDF would effectively unblind the asymmetry. Instead the residuals between data and the smooth PDF were used to assess the quality of the fit.

The *BABAR* experiment has regularly updated its measurement of $\sin 2\beta$ as more data has been collected. By the third public result, a simpler method of hiding the visual asymmetry was adopted. Instead of using the hidden Δt variable, the only Δt distribution used was for the combination of B^0 and $\bar{B}^0$ flavor tags. The asymmetry completely vanishes if no distinction is made between the CP eigenstates. With the experience accumulated over the course of repeating this analysis, it was clear that the combined Δt distribution was adequate for checking the maximum likelihood fit prior to unblinding.

3.2.4. DIVIDED ANALYSES   As discussed in Section 2, the analysis of data by independent groups is a very powerful tool for uncovering errors, for encouraging creativity in the analysis process, and for instilling a healthy sense of competition to produce answers in a timely way. In this review, we do not include this approach as a method of blind analysis if each group is able to calculate a physically meaningful answer based on their own work.

One exception is the case where the independent groups cannot by themselves calculate a physics answer—only the combination of two (or more) pieces of the analysis can do so, and this combination is never made until the individual analyses have been completed.

A very nice example of this approach was the measurement of the anomalous magnetic moment of the muon, performed by the BNL $g-2$ Collaboration (50). The determination of the anomalous magnetic moment a_μ relies on two completely independent measurements: the angular frequency ω_a of the difference in the muon's spin precession frequency and the cyclotron frequency, and the free proton NMR frequency ω_p, which yields a precise measurement of the magnetic field B. Two independent groups were charged with the analysis—one that measured only ω_a and one that measured ω_p. To discuss results between the two groups, each group had its own hidden offset, which it applied to its measurement. Only when both analyses were complete were the two results combined to provide the final measurement.

3.3. Adding or Removing Events

Measurements of cross sections, branching ratios, or fluxes are typically based on counting the number of events passing all analysis cuts. As discussed in Section 3.2, a hidden answer method can be difficult to use in such cases, because there is no simple offset that can hide the number of events. Although a hidden signal box approach like that described in Section 3.1 can be used for many of these measurements, it prevents the experimenter from being able to examine the characteristics of the signal, and hence carries the kind of risk discussed in Section 2: a large and obvious signal can be missed while the experimenters examine background details that later turn out to have little impact on the measurement. In addition, a hidden signal box approach assumes that the characteristics of the backgrounds are known well enough that nothing unexpected will be discovered when the signal box is opened.

A very general approach to blind analysis which is appropriate for counting experiments is to spoil the event count itself in an unknown way. The spoiling can be done by adding an unknown set of false signal events, by removing a small unknown number of all events from the data set, or by doing both.

3.3.1. ADDING UNKNOWN NUMBERS OF EVENTS   If an unknown number of false signal events can be added to a data sample, an analysis can examine an entire data set while remaining blind to the physical measurement being made. In such

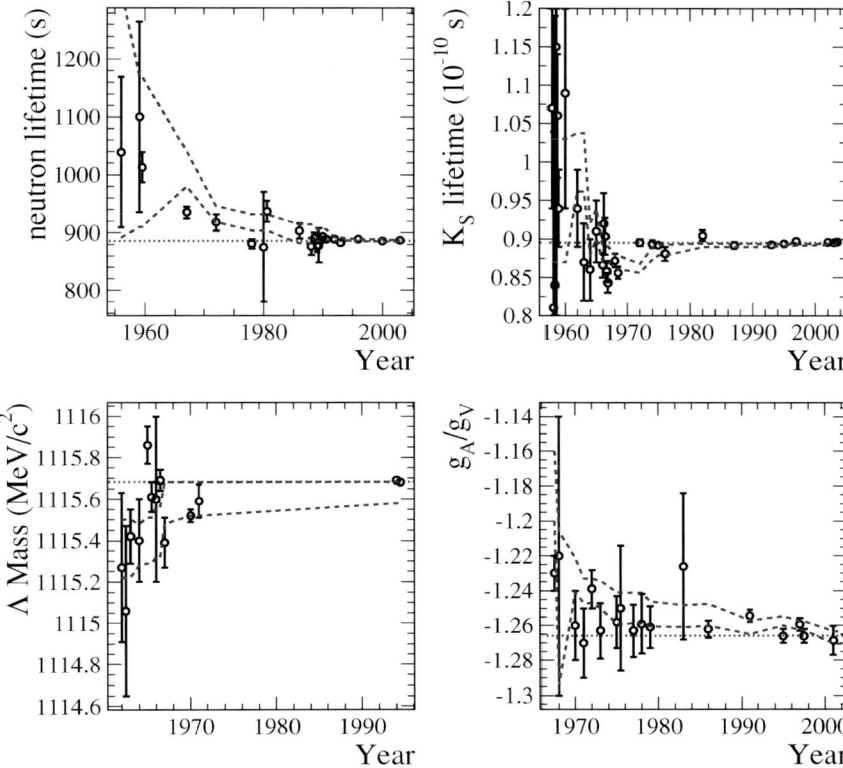

Figure 2 The history of four measurements compared to published averages before each measurement was made (dashed curves) and the currently accepted value (dotted lines). The space between the dashed curves indicates the 1σ uncertainties on the published values at the time each measurement was made.

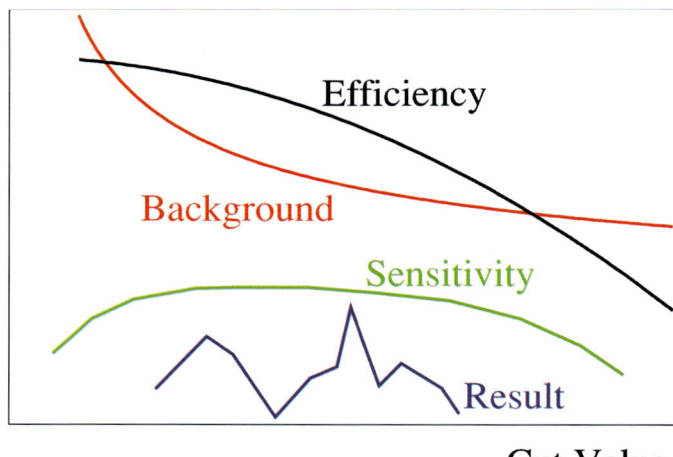

Cut Value

Figure 3 A cartoon demonstrating the variation in the central value of a measurement due to fluctuations even when the sensitivity for signal is reasonably flat.

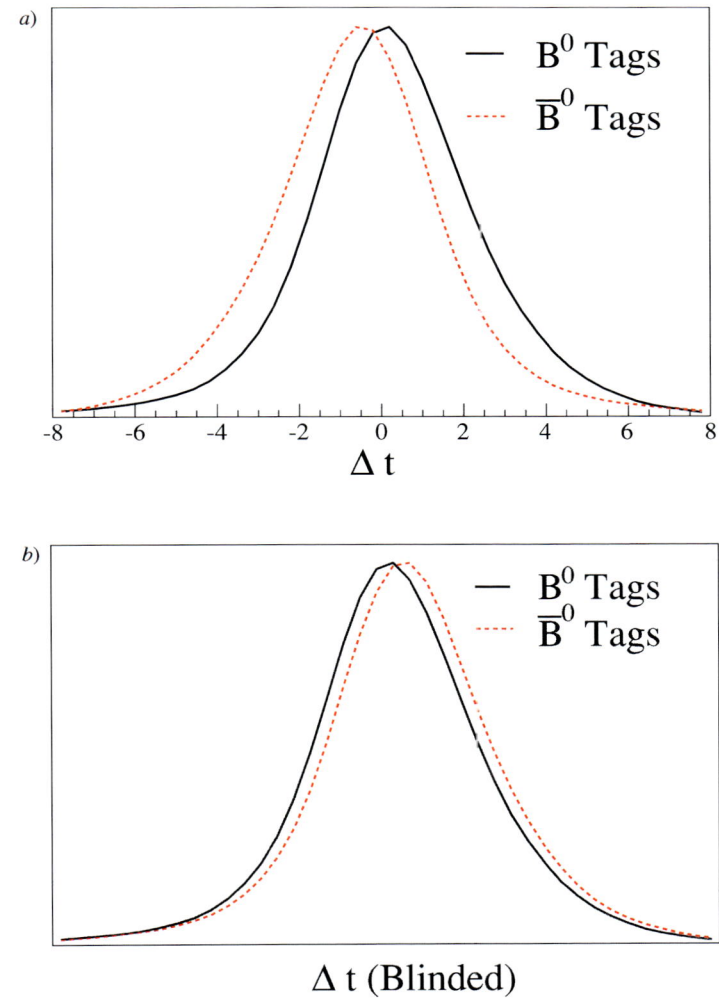

Figure 5   The Δt distributions for B decays into CP eigenstates, for $\sin 2\beta = 0.75$ and with the B^0 flavor tagging and vertex resolution which are typical for the *BABAR* experiment. a) The number of B^0 (solid line) and $\bar{B}^0$ (dashed line) decays into CP eigenstates as a function of Δt. b) The hidden Δt distributions for B^0 (solid line) and $\bar{B}^0$ (dashed line).

an approach, the experimenters tag the false events in some way and the tag is then used to remove them only when the analysis has been completed. To ensure that the number of added events remains unknown, it is critical that the false events mimic signal events as closely as possible. In some cases, a Monte Carlo simulation that can produce realistic-looking data—with all the associated noise and instrumental effects—can be used to provide a false signal event sample. The Soudan 2 experiment (51), for example, was able to produce simulated events which looked (to a human eye) indistinguishable from real detector events. They inserted the simulated events into the data set and removed them only when the analysis was complete. Very often, however, simulations are not so realistic, and a better approach may be to add true detector events from a sample that looks nearly identical to the signal.

One example of such an approach was the Sudbury Neutrino Observatory (SNO) collaboration's second direct measurement of the total active solar ^{8}B neutrino flux (42). The measurement was made in SNO's second phase of operation, in which the sensitivity to neutrons produced via neutral current (NC) neutrino interactions with heavy water ($v + d \rightarrow n + p + v$) was enhanced by the addition of $\sim$2 tons of NaCl. The goal of the blind analysis in the second phase of the experiment was to ensure that the measurement would be independent of the previously published first phase measurement.

The primary detected signal from the NC reaction is the capture of a neutron on the dissolved Cl in the heavy water. To hide the answer, SNO added an unknown number of tagged neutrons that were not the result of the NC process by solar neutrinos. Cosmic ray interactions within the SNO detector provided just such a tagged neutron sample, and except for the preceding muon, they looked nearly identical to the signal neutrons. Figure 6 illustrates the similarity of the reconstructed energy distribution of these "muon-follower" events compared to data from a deployment of a ^{252}Cf neutron source and simulated neutral current solar neutrino events (N. McCauley, SNO Collaboration, private communication).

These muon-follower events are normally removed from the final data sample by applying an offline veto to all events falling no more than 20 seconds after a muon event. With NaCl in the detector, the time between the liberation of a neutron from a deuteron in the heavy water and its subsequent capture on Cl is roughly 5 ms. To accept a small number of these neutrons into the final sample, a small window was cut out from the 20 second veto. The window was selected to be near enough to the 5 ms neutron capture time that it would add a number of neutrons that was on the order of the number expected from the neutrino neutral current interactions. Both the location of the window and its width were kept hidden from the collaboration, and so the number of added neutrons was not known until the analysis was complete and the events removed from the data set. The first phase measurement by SNO of the total active ^{8}B neutrino flux was $5.09^{+0.44}_{-0.43}$(stat.)$^{+0.46}_{-0.43}$(syst.) $\times 10^6$ cm^{-2} s^{-2}, and the result of the blind analysis, after unbinding, in the second phase yielded 4.90 ± 0.24(stat.)$^{+0.29}_{-0.27}$(syst.) $\times 10^6$ cm^{-2} s^{-2}, in excellent agreement.

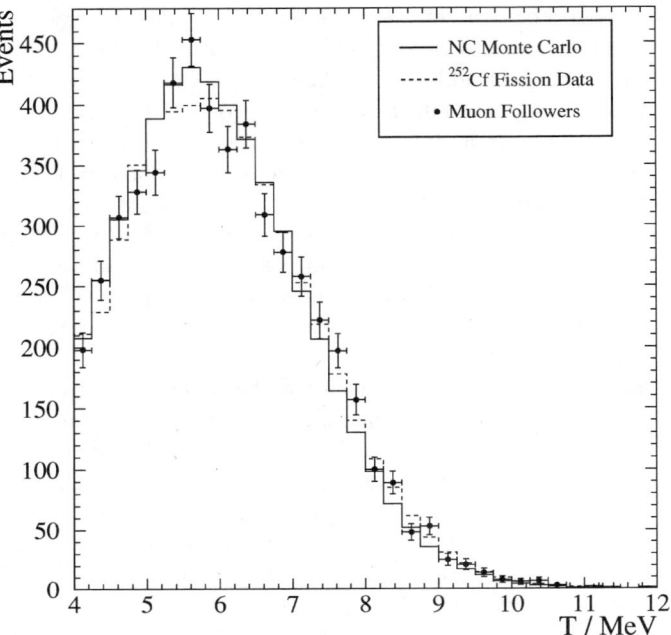

Figure 6   Comparison of the visible energy from neutrons created by cosmic ray interactions in the Sudbury Neutrino Observatory to neutron calibration source data and Monte Carlo–simulated ^{8}B solar neutrino neutral current (NC) events. (Figure courtesy of Neil McCauley, SNO Collaboration.)

3.3.2. REMOVING UNKNOWN NUMBERS OF EVENTS   When there is no tagged sample of signal-like events to add to a data set, the complementary approach— removing an unknown number of events—can be almost as effective. Unlike a "data prescaling" method in which the majority of the data set is kept blind (see Section 3.4), the fraction of removed events in this method need not be large, as long as its intentional uncertainty is large enough to prevent the experimenters from deriving a meaningful answer.

The method can be particularly powerful when used in conjunction with the approach described in the previous section. Adding an unknown number of events still allows an experimenter to have some information about the number of true signal events in the data set—they can never be larger than the number measured before the additional events are removed. There is thus a small possibility for bias; if the current count of events is too small given that events have been added to the data set by hand, then searches for problems in acceptance measurements may be given more weight than, for example, searches for additional sources of background. Removing an unknown fraction of events, however, gets around the problem: One does not know which effect is bigger and so the value of the measurement before the blind criteria are removed holds no information at all, and thus there is no possibility of bias. One of the earliest uses of an unknown prescale factor was by experiment E871 at BNL. E871 was the successor to the E791 rare K-decay search that first published a hidden signal box blind result (see Section 3.1). In measuring the branching ratio $K_L \rightarrow \mu^+\mu^-$, E871 prescaled their normalization sample, keeping a fraction of events known only to lie between 90–95% (52). Another example of this approach was the neutrino flux measurement by SNO described in the previous section: In addition to adding an unknown number of neutrons, the collaboration removed 20–40% (the exact fraction of course being unknown) of all the events from the data set. The events were then added back to the data sets, and the fits to extract the fluxes re-done, shortly before publication.

The method can also be used effectively on its own, and is simple and general enough to work in almost any experiment. The one potential problem is handling time-correlated data, which is often present in non-accelerator experiments. To deal with this, the data needs to be divided into blocks, and rather than removing individual events from the data set, a fraction of the blocks are removed.

3.4. Data Prescaling

Perhaps the most direct blind analysis is one in which the entire analysis chain— cuts, calibrations, acceptance calculations, normalizations—are developed without any reference to the physics data set at all. Such a scenario could arise if the complete analysis could be based on a Monte Carlo simulation, for example, and then applied without change to the data. A second case might be an experiment that has already analyzed a first run of data, in which the experimenters feel confident in applying the identical analysis to a new run.

Neither of these two cases is often practical—rarely is the Monte Carlo simulation trusted enough to use in the creation of an entire analysis chain, nor is a new run of data likely to be so identical to the first that one is willing to blindly (and blithely) apply an older analysis. The same benefits as this completely blind approach can be achieved, however, if the analysis is developed on a prescaled fraction of the data set and then applied to the remainder. Unlike the small removal of events described in the previous section, the prescaling fraction in this case is known, Because what is blind here is the majority of the data set.

By itself, data prescaling only avoids statistical bias—the tuning of cuts to enhance statistical fluctuations in the data. For an experiment of limited lifetime or low statistics, this can be a potentially damaging source of bias as there is no way to determine after the fact whether there has been any unintentional tuning—there is only one instance of the data set to study.

Data prescaling relies on the fact that the set of cuts applied is asymptotically unbiased—if applied to an infinite data set, they have no preference for accepting non-signal events that happen to lie in the signal region. Imagine, for example, that it is found that by cutting harder on a reconstructed track χ^2, the significance of an invariant mass peak grows slightly. The growth may simply be due to the fact that by moving the cut, the number of events in the peak fluctuates high, as depicted in

Figure 3. If the cut is then fixed and applied to a much larger data set, the cut will have almost no preference for events in the peak region if there are no real signal events.

Several assumptions are inherent in any data prescaling scheme:

1. The prescaling is done in an unbiased way.

2. Any data sample is the same as any other, or time-dependent variations have characteristic scales that can be properly sampled or contained within a subset of the data that spans those characteristic times.

3. The statistics of the prescaled sample are large enough to identify backgrounds, but small enough that they will not bias the result for the entire data set.

The first assumption almost always holds, unless one picks a prescaling scheme that is based on some criteria within the data set itself. A poor choice might be to look only at data when the trigger rates are high, for example. The safest scheme randomly decides whether an event or set of events falls within the prescale sample or not. For some experiments—particularly non-accelerator experiments—time correlations between events can be important, and thus blocks of data need to be prescaled rather than individual events.

The second assumption is not necessarily always true. It is rare that any sample of data is exactly the same as any other—data can be taken during times when detector subsystems are offline, or at different beam intensities, or even different times of day. Any external variation that can change the background levels or the detector sensitivity or acceptance can make a prescaling scheme fail if the variations are not reasonably sampled. Here "reasonably sampled" means that the sampling frequency—how often an event or block of data is selected to be in the prescaled sample—is higher than the rate of known variations in detector or beam conditions.

The third assumption, strictly speaking, is almost always false. Determining that the background in a pre-scaled sample is zero, for example, provides a very weak limit on the background levels inside the remainder of the data set, unless the prescaled sample is a large fraction of the full data set. In practice, one uses the smallest data sample possible that makes the first two assumptions true, and then determines whether the upper limit on residual backgrounds in the full data set is acceptable. If one feels that the prescale fraction needs to be large in order to measure backgrounds—in excess of 30%—then in principle the prescaled fraction should be discarded in the calculations of the final results, as otherwise any statistical bias in the analysis of the prescaled sample will have a nontrivial effect on the final answer.

Many experiments have used data prescaling either alone or in combination with other techniques. The BNL E888 experiment (53), a search for the doubly strange H dibaryon, used a hidden signal box technique in combination with a 10% prescaled sample of data within the signal box. The Neutrino Oscillation Magnetic Detector (NOMAD) experiment, which searched for $v_\mu \rightarrow v_\tau$ oscillations at CERN, had an effective prescaling scheme in which they analyzed openly a first run of data that constituted 20% of the data set, and then analyzed a much higher statistics run using a hidden signal box technique (54). The E787 experiment at BNL, which searched for the rare decay $K^+ \rightarrow \pi^+ \nu \bar{\nu}$, used a hidden signal box as their primary blind approach, but then divided the events outside the signal box into a 1/3 and 2/3 sample. E787 then developed their analysis of backgrounds on the 1/3 sample and applied that analysis to the remaining 2/3 to provide the final background measurement (45). In its first publication, SNO (55) used a prescale technique as a test of gross statistical bias—30% of the data was kept in reserve and the analysis developed on the other 70% applied to the hidden sample. When no significant differences were found between the 70% and the 30% data samples, results from the complete data set were published. The Laser Interferometer Gravitational Wave Observatory (LIGO) experiment used a 10% sample (56) for analysis optimizations in their search for gravity wave bursts. This subsample was then discarded in the final results. The MiniBooNE neutrino experiment currently uses a data prescaling scheme of just 0.5% for all data, in addition to a hidden signal box approach (R.G. Van de Water and H.A. Tanaka, private communication).

One question that can arise in treating the hidden sample as a blind data set is, how blind is blind? As the events are recorded. on-line event displays and other monitors are often viewed by collaborators. Does some blindness scheme need to be imposed on these? As discussed in Section 2, a blindness scheme is not intended to keep experimenters from looking at the data, but to keep the results of the analysis from influencing the analysis itself. For most experiments, event displays and other monitors carry no information about the physics results, and therefore do not influence the analysis itself. The MiniBooNE experiment is one example of this—low-level event and electronic channel properties can be examined throughout the data set. The possible exceptions to this guideline are open-ended rare process searches, where one is looking for unique and unusual events. Noticing a few strange events by hand-scanning is likely to influence a later search for new physics; it is hard to design a new analysis that doesn't ensure that these interesting events make it into the final sample.

Finally, in general it may be difficult to use a blind analysis method in searches for new particles. However, there are also a number of cases in which new particles, observed as bumps in invariant mass distributions, were ultimately found to be experimental artifacts (57). Often such statistical fluctuations do not have physical characteristics—the width of the bump, for example, is found to be inconsistent with the expected detector resolution. In addition, the evaluation of the statistical significance of the bump depends on the measure of the search space or the number of places a bump could have appeared (the "trials" problem). Given the vagaries of both of these issues, searches for new particles, or bump hunting, would clearly benefit from a blind approach. Unfortunately, the hidden signal box technique cannot readily be applied, because the mass of the particles is obviously not known. Of the methods described, only the data division method may be readily used. Despite the limitations described above, this method for determining experimental selections and analysis techniques may help prevent some purely statistical artifacts. However, creative new approaches may be possible, and we urge experimenters making new particle searches to consider new blind methods.

4. CONCLUSION

In his speech *Cargo Cult Science* (58), Richard Feynman warns that

> It's a thing that scientists are ashamed of—this history—because it's apparent that people did things like this: When they got a number that was too high above Millikan's, they thought something must be wrong—and they would look for and find a reason why something might be wrong. When they got a number closer to Millikan's value they didn't look so hard...

> The first principle is that you must not fool yourself—and you are the easiest person to fool.

Experimenter's bias represents one way to fool yourself, and blind analysis provides the solution. By describing the application of different blind analysis methods to a range of measurements, including both the motivations and potential problems, we hope that this review will be useful as a resource for experiments and will promote the current trend towards blind analysis in nuclear and particle physics.

ACKNOWLEDGMENTS

The work of A.R. is supported by the Department of Energy under contract DE-AC02-76-SF00515. The work of J.K. is supported by the Department of Energy under grants DE-FG02-04ER41332 and DE-FG02-93ER40757, and by the Alfred P. Sloan Foundation. The authors would like to acknowledge useful discussions, both in the development of blind analysis methods and in the preparation of this review, with Art Snyder, Pat Burchat, William Molzon, Bruce Winstein, Peter Shawhan, Jack Ritchie, Robert Cousins, Huaizhang Deng, Richard Van de Water, Hirohisa Tanaka, Brian Rebel, and Stanton Goldman.

**The *Annual Review of Nuclear and Particle Science* is online at
http://nucl.annualreviews.org**

LITERATURE CITED

1. Pfungst O. *Clever Hans (The Horse of Mr. Von Osten). A Contribution to Experimental Animal and Human Psychology.* New York: Henry Holt (1911)
2. De Angelis C, et al. *New Engl. J. Med.* 351:1250 (2004)
3. Dunnington FG. *Phys. Rev.* 43:404 (1932)
4. Roodman A. eConf C030908:TUIT001 (2003)
5. Birge RT. *Rep. Prog. Phys.* 8:90 (1941); Cohen ER, DuMond JWM. *Rev. Mod. Phys.* 37:537 (1965)

6. Cohen ER, Taylor BN. *J. Phys. Chem. Ref. Data* 2:663 (1973)

7. Birge RT. *Nuovo Cim. Suppl.* 6:39 (1957)

8. Christensen CJ, et al. *Phys. Lett.* B26:11 (1967)

9. Christensen CJ, et al. *Phys. Rev. D* 5:1628 (1972)

10. Barkas WH, Rosenfeld AH. *UCRL-8030* (1958)

11. Roos M, et al. *Rev. Mod. Phys.* 35:314 (1963)

12. Roos M, et al. *Nucl. Phys.* 52:1 (1964)

13. Rosenfeld AH, et al. *Rev. Mod. Phys.* 36:977 (1964)

14. Rosenfeld AH, et al. *Rev. Mod. Phys.* 37:633 (1965)

15. Rosenfeld AH, et al. *UCRL-8030-Rev* (1966)

16. Rosenfeld AH, et al. *13th Int. Conf. on High-energy Physics*, Berkeley, Calif. (1966)

17. Rosenfeld AH, et al. *Rev. Mod. Phys.* 39:1 (1967)

18. Rosenfeld AH, et al. *Rev. Mod. Phys.* 40:77 (1968)

19. Barash-Schmidt N, et al. *UCRL-8030-Pt-1-Rev* (1968)

20. Barash-Schmidt N, et al. *Rev. Mod. Phys.* 41:109 (1969)

21. Barbaro-Galtieri A, et al. *Rev. Mod. Phys.* 42:87 (1970)

22. Roos M, et al. *Phys. Lett.* B33:127 (1970)

23. Rittenberg A, et al. *Rev. Mod. Phys.* 43:S1 (1971)

24. Söding PS, et al. *Phys. Lett.* B39:1 (1972)

25. Lasinski TA, et al. *Rev. Mod. Phys.* 45:S1 (1973)

26. Chaloupka V, et al. *Phys. Lett.* B50:1 (1974)

27. Trippe TG, et al. *Rev. Mod. Phys.* 48:S1 (1976)

28. Bricman C, et al. *Phys. Lett.* B75:1 (1978)

29. Bricman C, et al. *Rev. Mod. Phys.* 52:S1 (1980)

30. Roos M, et al. *Phys. Lett.* B111:1 (1982)

31. Wohl CG, et al. *Rev. Mod. Phys.* 56:S1 (1984)

32. Aguilar-Benítez M, et al. *Phys. Lett.* B170:1 (1986)

33. Yost GP, et al. *Phys. Lett.* B204:1 (1988)

34. Hernández JJ, et al. *Phys. Lett.* B239:1 (1990)

35. Hikasa K, et al. *Phys. Rev. D* 45:S1 (1992)

36. Montanet L, et al. *Phys. Rev. D* 50:1173 (1994)

37. Barnett RM, et al. *Phys. Rev. D* 54:1 (1996)

38. Caso C, et al. *Eur. Phys. J.* C3:1 (1998)

39. Groom DE, et al. *Eur. Phys. J.* C15:1 (2000)

40. Hagiwara K, et al. *Phys. Rev. D* 66:010001 (2002)

41. Eidelman S, et al. *Phys. Lett.* B592:1 (2004)

42. Ahmed SN, et al. *Phys. Rev. Lett.* 92:181301 (2004)

43. Galison, P. *How Experiments End.* Chicago: Univ. Chicago Press (1987)

44. Arisaka K, et al. *Phys. Rev. Lett.* 70:1049 (1993)

45. Adler S, et al. *Phys. Rev. D* 70:37102 (2004)

46. Hoskins JK, Newman RD, Spero R, Schultz J. *Phys. Rev. D* 12:3084 (1985)

47. Alavi-Harati A, et al. (KTeV Collaboration). *Phys. Rev. Lett.* 83:22 (1999)

48. Shawhan PS. *Observation of direct CP violation in K(S,L) → pi pi decays.* PhD thesis. Univ. Chicago (1999)

49. Aubert B, et al. (BABAR Collaboration). *Phys. Rev. Lett.* 86:2515 (2001)

50. Bennett GW, et al. *Phys. Rev. Lett.* 92:161802 (2004)

51. Sanchez M, et al. *Phys. Rev. D* 68:113004 (2003)

52. Ambrose D, et al. *Phys. Rev. Lett.* 84:1389 (2000)

53. Belz J, et al. *Phys. Rev. D* 53:3487 (1996)

54. Astier P, et al. *Phys. Lett.* B453:1690186 (1999)

55. Ahmad QR, et al. *Phys. Rev. Lett.* 87:071301 (2001)

56. Abbott B, et al. (LIGO Collaboration). *Phys. Rev. D* 69:102001 (2004)

57. Stone S. In *Flavor Physics for the Millennium: TASI 2000 Proceedings*, ed. JL Rosner. Singapore: World Sci. (2001)

58. Feynman R. *Surely You're Joking, Mr. Feynman!* New York: W.W. Norton (1985)

Annu. Rev. Nucl. Part. Sci. 2005. 55:165–228
doi: 10.1146/annurev.nucl.54.070103.181218
Copyright © 2005 by Annual Reviews. All rights reserved

STUDY OF THE FUNDAMENTAL STRUCTURE OF MATTER WITH AN ELECTRON-ION COLLIDER

Abhay Deshpande,[1] Richard Milner,[2] Raju Venugopalan,[3] and Werner Vogelsang[4]

[1]Department of Physics & Astronomy, State University of New York at Stony Brook, New York 11794 and RIKEN-BNL Research Center, Brookhaven National Laboratory, Upton, New York 11973; email: abhay@bnl.gov
[2]Physics Department and Laboratory for Nuclear Science, Massachusetts Institute of Technology, Cambridge, Massachusetts 02139; email: milner@mit.edu
[3]Physics Department, Brookhaven National Laboratory, Upton, New York 11973; email: raju@quark.phy.bnl.gov
[4]Physics Department and RIKEN-BNL Research Center, Brookhaven National Laboratory, Upton, New York 11973; email: vogelsan@quark.phy.bnl.gov

This review is dedicated to the memory of Professor Vernon W. Hughes.

Key Words Quantum Chromodynamics, DIS structure functions, Polarized ep Scattering, Nucleon Spin, DIS off Nuclei, Saturation, Color Glass Condensate, EIC, eRHIC

■ **Abstract** We present an overview of the scientific opportunities that would be offered by a high-energy electron-ion collider. We discuss the relevant physics of polarized and unpolarized electron-proton collisions and of electron-nucleus collisions. We also describe the current accelerator and detector plans for a future electron-ion collider.

CONTENTS

165

1. INTRODUCTION

Understanding the fundamental structure of matter is one of the central goals of scientific research. In the closing decades of the twentieth century, physicists developed a beautiful theory, Quantum Chromodynamics (QCD), which explains all of strongly interacting matter in terms of point-like quarks interacting by the exchange of gauge bosons, known as gluons. The gluons of QCD, unlike the photons of QED, can interact with each other. The color force which governs the interaction of quarks and gluons is responsible for more than 99% of the observable mass in the physical universe and explains the structure of nucleons and their composite structures, atomic nuclei, as well as astrophysical objects such as neutron stars.

During the last 30 years, experiments have verified QCD quantitatively in collisions involving a very large momentum exchange between the participants. These collisions occur over very short distances much smaller than the size of the proton. In these experiments, the confined quarks and gluons act as if they are nearly free point-like particles and exhibit many properties that are predicted by perturbative QCD (pQCD). This experimental phenomenon was first discovered in deeply inelastic scattering (DIS) experiments of electrons off nucleons. The discovery resulted in the 1990 Nobel Prize in Physics being awarded to Friedman, Kendall, and Taylor. The phenomenon, that quarks and gluons are quasi-free at short distances, follows from a fundamental property of QCD known as *asymptotic freedom*. Gross, Politzer, and Wilczek, who first identified and understood this unique characteristic of QCD, were awarded the 2004 Nobel Prize in Physics.

When the interaction distance between the quarks and gluons becomes comparable to or larger than the typical size of hadrons, the fundamental constituents of the nucleon are no longer free. They are confined by the strong force that does not allow for the observation of any "colored" object. In this strong coupling QCD regime, where most hadronic matter exists, the symmetries of the underlying quark-gluon theory are hidden, and QCD computations in terms of the dynamical properties of quarks and gluons are difficult. A major effort is underway worldwide to carry out *ab initio* QCD calculations in the strong QCD regime using Monte-Carlo simulations on large scale computers.

The experimental underpinnings for QCD are derived from decades of work at the CERN, DESY, Fermilab, and SLAC accelerator facilities. Some highlights include the determination of the nucleon quark momentum and spin distributions and the nucleon gluon momentum distribution, the verification of the QCD prediction for the running of the strong coupling constant α_s, the discovery of jets, and the discovery that quark and gluon momentum distributions in a nucleus differ from those in a free nucleon.

However, thirty years after QCD has been established as the Standard Model of the strong force, and despite impressive progress made in the intervening decades, understanding how QCD works in detail remains one of the outstanding issues in physics. Some crucial open questions that need to be addressed are listed below.

What is the gluon momentum distribution in the atomic nucleus? QCD tells us that the nucleon is primarily made up of specks of matter (quarks) bound by tremendously powerful gluon fields. Thus atomic nuclei are primarily composed of glue. Very little is known about the gluon momentum distribution in a nucleus. Determining these gluon distributions is therefore a fundamental measurement of high priority. This quantity is also essential for an understanding of other important questions in hadronic physics. For example, the interpretation of experiments searching for a deconfined quark-gluon state in relativistic heavy ion collisions is dependent on the knowledge of the initial quark and gluon configuration in a heavy nucleus. This will be especially true for heavy ion experiments at the Large Hadron Collider (LHC) at CERN. Further, there are predictions that gluonic matter at high parton densities has novel properties that can be probed in hard scattering experiments on nuclear targets. Hints of the existence of this state may have been seen in Deuteron-Gold experiments at the Relativistic Heavy Ion Collider (RHIC) at Brookhaven.

How is the spin structure of the nucleon understood to arise from the quark and gluon constituents? High energy spin-dependent lepton scattering experiments from polarized nucleon targets have produced surprising results. The spins of the quarks account for only about 20% of the spin of the proton. The contribution of the gluons may be large. Dramatic effects are predicted for measurements beyond the capability of any existing accelerator. There are hints from other experiments that the contribution of orbital angular momentum may be large.

Testing QCD. It is imperative to continue to subject QCD to stringent tests because there is so much about the theory that remains a mystery. QCD can be tested in two ways: one is by precision measurements, and the other is by looking for novel physics which is sensitive to the confining properties of the theory. Both of these can be achieved at a high luminosity lepton-ion collider with a detector that has a wide rapidity and angular coverage. An example of precision physics is the Bjorken Sum Rule in spin-dependent lepton scattering from a polarized nucleon. This fundamental sum rule relates inclusive spin-dependent lepton scattering to the ratio of axial to vector coupling constants in neutron β-decay. Present experiments test it to about $\pm 10\%$: it would be highly desirable to push these tests to about $\pm 1\%$. Further, with lattice QCD expected to make substantial progress in the ability to make *ab initio* QCD calculations during the next decade, precise measurements of the calculable observables will be required. An example of a physics measurement sensitive to confinement is hard diffraction, where large mass final states are formed with large "color-less" gaps in rapidity separating them from the hadron or nucleus.

At the Hadron Electron Ring Accelerator (HERA) at DESY, roughly 10% of events are of this nature. The origins of these rapidity gaps, which must be intimately related to the confining properties of the theory, can be better understood with detectors that are able to provide detailed maps of the structure of events in DIS.

This article motivates and describes the next generation accelerator required by nuclear and particle physicists to study and test QCD, namely a polarized lepton-ion collider. The basic characteristics of the collider are motivated as follows:

- *Lepton beam.* The lepton probe employs the best understood interaction in nature (QED) to study hadron structure. Electrons and positrons couple directly to the quarks. The experimental conditions which maximize sensitivity to valence and sea quarks as well as probe gluons are well understood. Further, the availability of both positron and electron beams will enable experiments that are sensitive to the exchange of the parity violating Z and W-bosons.

- *Range of center-of-mass (CM) energies.* To cleanly interact with quarks, a minimum center-of-mass (CM) energy of about 10 GeV is required. To explore and utilize the powerful Q^2 evolution equations of QCD, CM energies of order 100 GeV are desirable. This consideration strongly motivates the collider geometry.

- *High luminosity.* The QED interaction between the lepton probe and the hadron target is relatively weak. Thus precise and definitive measurements demand a high collision luminosity of order 10^{33} nucleons cm^{-2} s^{-1}.

- *Polarized beams.* Polarized lepton and nucleon beams are essential to address the central question of the spin structure of the nucleon. Both polarized proton and neutron (effectively polarized ^{2}H or ^{3}He) are required for tests of the fundamental Bjorken Sum Rule. The polarization direction of at least one of the beams must be reversible on a rapid timescale to minimize systematic uncertainties.

- *Nuclear beams.* Light nuclear targets are useful for probing the spin and flavor content of parton distributions. Heavy nuclei are essential for experiments probing the behavior of quarks and gluons in the nuclear medium.

- *Detector considerations.* The collider geometry has a significant advantage over fixed-target experiments at high energy because it makes feasible the detection of complete final-states. A central collider detector with momentum and energy measurements and particle identification for both leptons and hadrons will be essential for many experiments. Special purpose detectors that provide wide angular and rapidity coverage will be essential for several specific measurements.

These considerations constrain the design parameters of the collider to be a 5 to 10 GeV energy electron (or positron) beam colliding with a nucleon beam of energy 25 GeV to 250 GeV. The collider is anticipated to deliver nuclear beams of energies ranging from 20−100 GeV/nucleon. The lepton and nucleon beams must be highly polarized and the collision luminosity must be of order 10^{33} nucleons cm^{-2} s^{-1}.

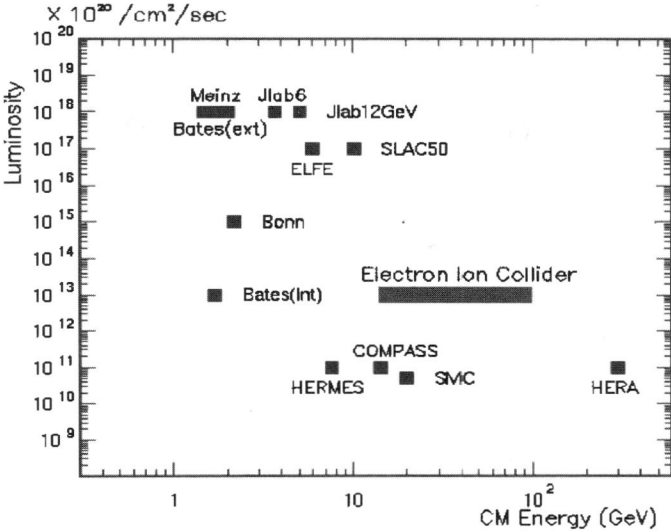

Figure 1   The center-of-mass energy vs. luminosity of the proposed Electron-Ion Collider eRHIC compared to other lepton scattering facilities.

The proposed eRHIC design (described in Section 4) realizes the required specifications in a cost effective and timely way by using the existing RHIC facility at BNL. The characteristics of eRHIC are well beyond the capability of any existing accelerator, as is clear from Figure 1.

By delivering high energies to the collision, the collider provides an increased range for investigating quarks and gluons with small momentum fraction (x) and for studying their behavior over a wide range of momentum transfers (Q^2). In deeply inelastic scattering, the accessible values of the Bjorken variable x (defined in Section 2) are limited by the available CM energy. For example, collisions between a 10 GeV lepton beam and nuclear beams of 100 GeV/nucleon provide access to values of x as small as 3×10^{-4} for $Q^2 \sim 1$ GeV^2. In a fixed-target configuration, a 2.1 TeV lepton beam would be required to produce the same CM energy. Figure 2 (see color insert) shows the x-Q^2 range possible with the proposed eRHIC machine and compares that range to the currently explored kinematic region.

In this article, the scientific case and accelerator design for a new facility to study the fundamental quark and gluon structure of strongly interacting matter are presented. Section 2 describes the current understanding of the quark and gluon structure of hadrons and nuclei. Section 3 presents highlights of the scientific opportunities available with a lepton-ion collider. Section 4 describes the accelerator design effort and Section 5 describes the interaction region and eRHIC detector design.

2. STATUS OF THE EXPLORATION OF THE PARTONIC STRUCTURE OF HADRONS AND NUCLEI

This section will summarize our current understanding of the partonic structure of hadrons and nuclei in QCD, accumulated during the past three decades from a variety of deeply inelastic and hadronic scattering experiments. We will also comment on what new information may become available from DIS as well as from RHIC and other experimental facilities around the world before a future electron-ion collider starts taking data. We will outline the status of our knowledge on i) the parton distributions in nucleons, ii) spin and flavor distributions in the nucleon, iii) nuclear modifications to the inclusive nucleon distributions such as the European Muon Collaboration (EMC) effect and quark and gluon shadowing, and iv) color coherent phenomena in nuclei that probe the space-time structure of QCD such as color transparency and opacity, partonic energy loss and the p_T broadening of partons in media. In each case, we will outline the most important remaining questions and challenges. These will be addressed further in Section 3.

2.1. Deeply-Inelastic Scattering

The cross-section for the inclusive deeply inelastic scattering (DIS) process shown in Figure 3 can be written as a product of the leptonic tensor $\mathcal{L}_{\mu\nu}$ and the hadronic tensor $\mathcal{W}^{\mu\nu}$ as

$$\frac{d^2\sigma}{dxdy} \propto \mathcal{L}_{\mu\nu}(k, q, s)\, \mathcal{W}^{\mu\nu}(P, q, S),\qquad 1.$$

where one defines the Lorentz invariant scalars, the famous Bjorken variable $x = -q^2/2Pq$, and $y = Pq/Pk$. Note that as illustrated in Figure 3, $k(k')$ is the 4-momentum of the incoming (outgoing) electron, P is the 4-momentum of the incoming hadron, and $q = k - k'$ is the 4-momentum of the virtual photon. The center of mass energy squared is $s = (P + k)^2$. From these invariants, one can deduce simply that $xy \approx Q^2/s$, where $Q^2 = -q^2 > 0$.

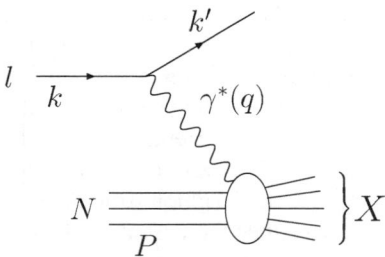

Figure 3 Deeply-inelastic lepton-nucleon scattering mediated by virtual photon exchange.

The hadronic tensor can be written in full generality as

$$
\mathcal{W}^{\mu\nu}(P, q, S) = \frac{1}{4\pi} \int d^4z \, e^{iqz} \langle P, S|[\mathcal{J}_\mu(z), \mathcal{J}_\nu(0)]|P, S\rangle = -g^{\mu\nu} F_1(x, Q^2)
$$

$$
+ \frac{P^\mu P^\nu}{Pq} F_2(x, Q^2) - i\varepsilon^{\mu\nu\rho\sigma} \frac{q_\rho P_\sigma}{2Pq} F_3(x, Q^2) + i\varepsilon^{\mu\nu\rho\sigma} q_\rho \left[\frac{S_\sigma}{Pq} g_1(x, Q^2) \right.
$$

$$
+ \left. \frac{S_\sigma(Pq) - P_\sigma(Sq)}{(Pq)^2} g_2(x, Q^2) \right] + \left[\frac{P^\mu S^\nu + S^\mu P^\nu}{2Pq} - \frac{Sq}{(Pq)^2} P^\mu P^\nu \right]
$$

$$
\times g_3(x, Q^2) + \frac{Sq}{(Pq)^2} P^\mu P^\nu g_4(x, Q^2) - \frac{Sq}{Pq} g^{\mu\nu} g_5(x, Q^2). \qquad 2.
$$

The F_i are referred to as the "unpolarized" structure functions, whereas the g_i are the "spin-dependent" ones, because their associated tensors depend on the nucleon spin vector S^μ. Note that parity-violating interactions mediated by electroweak boson exchange are required for F_3, g_3, g_4, g_5 to contribute.

Inserting Equation 2 and the straightforwardly calculated leptonic tensor into Equation 1, one obtains the DIS cross section in terms of the structure functions. If one averages over the hadronic spins and restricts oneself to parity conserving (for $Q^2 \ll M_Z^2$) electron-nucleon scattering alone, one finds the simple expression

$$
\frac{d^2\sigma}{dx \, dQ^2} = \frac{2\pi\alpha_{\rm em}^2}{Q^4} \left[\left(1 + (1 - y)^2\right) F_2(x, Q^2) - y^2 F_L(x, Q^2) \right]. \qquad 3.
$$

Here, $\alpha_{\rm em}$ is the coupling constant of Quantum Electrodynamics and F_L is the "longitudinal" structure function, defined by the relation $F_L = F_2 - 2x \, F_1$.

In the leading logarithmic approximation of QCD the measured structure function $F_2(x, Q^2)$ can be written as

$$
F_2(x, Q^2) = \sum_{q=u,d,s,c,b,t} e_q^2(xq(x, Q^2) + x\bar{q}(x, Q^2)), \qquad 4.
$$

where $q(x, Q^2)$ ($\bar{q}(x, Q^2)$) is the probability density for finding a quark (anti-quark) with momentum fraction x at a momentum resolution scale Q^2; e_q is the quark charge.

In the simple parton model one has Bjorken scaling, $F_2(x, Q^2) \to F_2(x)$. The "scaling violations" seen in the Q^2-dependence of $F_2(x, Q^2)$ arise from the fact that QCD is not a scale invariant theory and has an intrinsic scale $\Lambda_{\rm QCD} \approx 200$ MeV. They are only logarithmic in the Bjorken limit of $Q^2 \to \infty$ and $s \to \infty$ with $x \sim Q^2/s$ fixed. As one moves away from the asymptotic regime, the scaling violations become significant. They can be quantitatively computed in QCD perturbation theory using for example the machinery of the operator product expansion and the renormalization group. The result is most conveniently summarized by the Dokshitzer-Gribov-Lipatov-Altarelli-Parisi (DGLAP) evolution equations for the parton densities (1, 2):

$$\frac{d}{d \ln Q^2} \begin{pmatrix} q \\ g \end{pmatrix}(x, Q^2) = \begin{pmatrix} P_{qq}(\alpha_s, x) & P_{qg}(\alpha_s, x) \\ P_{gq}(\alpha_s, x) & P_{gg}(\alpha_s, x) \end{pmatrix} \otimes \begin{pmatrix} q \\ g \end{pmatrix}(x, Q^2), \qquad 5.$$

where $\otimes$ denotes a convolution, and the P_{ij} are known as "splitting functions" (2) and are evaluated in QCD perturbation theory. They are now known to three-loop accuracy (3). The evolution of the quark densities q, $\bar{q}$ involves the gluon density $g(x, Q^2)$. The physical picture behind evolution is the fact that the virtuality Q^2 of the probe sets a resolution scale for the partons, so that a change in Q^2 corresponds to a change in the parton state seen. The strategy is then to parameterize the parton distributions at some initial scale $Q^2 = Q_0^2$, and to determine the parameters by evolving the parton densities to (usually, larger) Q^2 and by comparing to experimental data for $F_2(x, Q^2)$.

The pioneering DIS experiments, which first measured Bjorken scaling of F_2, were performed at SLAC (4). However, because of the (relatively) small energies, these experiments were limited to the region of $x \geq 0.1$. With the intense muon beams of CERN and Fermilab, with energies in excess of 100 GeV, the DIS cross-section of the proton was measured down to and below $x \sim 10^{-3}$ (5). In the 1990's, the HERA collider at DESY extended the DIS cross-section of the proton to below $x = 10^{-4}$ (6, 7). The current experimental determination of $F_2^{\text{proton}}(x, Q^2)$ extends over 4 orders of magnitude in x and Q^2. This is shown in Figure 4 (see color insert). The left panel in Figure 4 shows next-to-leading order (NLO) QCD global fits by the ZEUS and H1 detector collaborations at HERA to F_2 as a function of Q^2 for the world DIS data. The data and the QCD fit are in excellent agreement over a wide range in x and Q^2. In the right panel of Figure 4, the x dependence of F_2 is shown for different bins in Q^2. The rapid rise in F_2 with decreasing x reflects the sizeable contribution from the sea quark distribution at small x.

In Figure 5 (see color insert) we show the valence up and down quark distributions as well as the gluon and sea quark distributions extracted by the H1 and ZEUS collaborations as functions of x for fixed $Q^2 = 10$ GeV^2. The valence parton distributions are mainly distributed at large x whereas the glue and sea quark distributions dominate hugely at small x. Indeed, the gluon and sea quark distributions are divided by a factor of 20 to ensure they can be shown on the scale of the plot. Already at $x \sim 0.1$, the gluon distribution is nearly a factor of two greater than the sum of the up and down quark valence distributions.

As follows from Equation 5, the gluon distribution in DIS may be extracted from scaling violations of F_2: $xg(x, Q^2) \propto \frac{\partial F_2(x, Q^2)}{\partial \ln Q^2}$. As one goes to low Q^2, $xg(x, Q^2)$ becomes small, and some analyses find a preference for a negative gluon distribution at low x, modulo statistical and systematic uncertainties (9,10). This is in principle not a problem in QCD beyond leading order. However, the resulting longitudinal structure function F_L also comes out close to zero or even negative for $Q^2 \sim 2$ GeV^2,[1] which is unphysical because F_L is a positive-definite quantity. A likely explanation for this finding is that contributions to F_L that are suppressed by inverse powers of Q^2 are playing a significant role at these values of Q^2 (11). These contributions are commonly referred to as higher twist effects.

It has been shown recently (12) that the HERA data on the virtual photon-proton cross-section $(\sigma^{\gamma^* p} = 4\pi^2 \alpha_{em} F_2(x, Q^2)/Q^2)$, for all $x \le 10^{-2}$ and $0.045 \le Q^2 < 450$ GeV^2, exhibit the phenomenon of "geometrical scaling" shown in Figure 6 (see color insert). The data are shown to scale as a function of $\tau = Q^2/Q_s^2$, where $Q_s^2(x) = Q_0^2(x_0/x)^{-\lambda}$ with $Q_0^2 = 1$ GeV^2, $x_0 = 3 \cdot 10^{-4}$ and $\lambda \approx 0.3$. The scale Q_s^2 is called the saturation scale. Geometrical scaling, although very general, is realized in a simple model, the Golec-Biernat-Wüsthoff model which includes all twist contributions (13). The model (and variants) provides a phenomenological description of the HERA data on diffractive cross-sections and inclusive vector meson production (14–18). The saturation scale and geometrical scaling will be discussed further in Section 3.

2.2. Spin Structure of the Nucleon

2.2.1. WHAT WE HAVE LEARNED FROM POLARIZED DIS   Spin physics has played a prominent role in QCD for several decades. The field has been driven by the successful experimental program of polarized deeply-inelastic lepton-nucleon scattering at SLAC, CERN, DESY and the Jefferson Laboratory (19). A main focus has been on measurements with longitudinally polarized lepton beam and target. For leptons with helicity λ scattering off nucleons polarized parallel or antiparallel to the lepton direction, one has (20)

$$\frac{d^2\sigma^{\lambda, \Rightarrow}}{dx dQ^2} - \frac{d^2\sigma^{\lambda, \Leftarrow}}{dx dQ^2} \propto C(G_v, G_a, \lambda) \left[\lambda x y (2 - y) g_1 + (1 - y) g_4 + x y^2 g_5 \right],$$

6.

where $C(G_v, G_a, \lambda)$ are factors depending on the vector and axial couplings of the lepton to the exchanged gauge boson. The terms involving g_4 and g_5 in Equation 6 are associated with Z and W exchange in the DIS process and violate parity. In the fixed-target regime, pure-photon exchange strongly dominates, and scattering off a longitudinally polarized target determines g_1. Figure 7 (left) (see color insert) shows a recent compilation (21) of the world data on $g_1(x, Q^2)$, for proton, deuteron, and neutron targets. Roughly speaking, g_1 is known about as well now as the unpolarized F_2 was in the mid-eighties, prior to HERA. Figure 7 (right) shows the measured Q^2-dependence of g_1; the predicted scaling violations are visible in the data.

---

[1]The leading twist expression for F_L is simply related to $\alpha_s x g(x, Q^2)$.

In leading order of QCD, g_1 can be written as

$$g_1(x, Q^2) = \frac{1}{2} \sum_q e_q^2 \left[\Delta q(x, Q^2) + \Delta \bar{q}(x, Q^2) \right],$$ 7.

where

$$\Delta q \equiv q_{\Rightarrow}^{\rightarrow} - q_{\Rightarrow}^{\leftarrow} \qquad (q = u, d, s, \ldots),$$ 8.

$q_{\Rightarrow}^{\rightarrow}$ ($q_{\Rightarrow}^{\leftarrow}$) denoting the number density of quarks of same (opposite) helicity as the nucleon. Clearly, the $\Delta q(x, Q^2)$, $\Delta \bar{q}(x, Q^2)$ contain information on the nucleon spin structure. Also in the spin-dependent case, QCD predicts Q^2-dependence of the densities. The associated evolution equations have the same form as Equation 5, but with polarized splitting functions (2, 23, 24). Also, the spin-dependent gluon density Δg, defined in analogy with Equation 8, appears.

The results of a recent QCD analysis (25) of the data for $g_1(x, Q^2)$ in terms of the polarized parton densities are shown in Figure 8 (see color insert). The shaded bands in the figure give estimates of how well we know the distributions so far. As can be seen, the valence densities are fairly well known and the sea quark densities to some lesser extent. This analysis (25) assumes flavor-SU(3) symmetry for the sea quarks; the actual uncertainties in the individual sea distributions are much larger. Finally, Figure 8 also shows that we know very little about the polarized gluon density. A tendency toward a positive Δg is seen. It is not surprising that the uncertainty in Δg is still large: at LO, Δg enters only through the Q^2-evolution of the structure function g_1. Because all polarized DIS experiments thus far have been with fixed targets, the lever arm in Q^2 has been limited. This is also seen in a comparison of Figure 7 with Figure 4.

A particular focus in the analysis of g_1 has been on the integral $\Gamma_1(Q^2) \equiv \int_0^1 g_1(x, Q^2) dx$. Ignoring QCD corrections, one has from Equation 7:

$$\Gamma_1 = \frac{1}{12} \Delta \mathcal{A}_3 + \frac{1}{36} \Delta \mathcal{A}_8 + \frac{1}{9} \Delta \Sigma,$$ 9.

where

$$\Delta \Sigma = \Delta \mathcal{U} + \Delta \bar{\mathcal{U}} + \Delta \mathcal{D} + \Delta \bar{\mathcal{D}} + \Delta \mathcal{S} + \Delta \bar{\mathcal{S}},$$

$$\Delta \mathcal{A}_3 = \Delta \mathcal{U} + \Delta \bar{\mathcal{U}} - \Delta \mathcal{D} - \Delta \bar{\mathcal{D}},$$

$$\Delta \mathcal{A}_8 = \Delta \mathcal{U} + \Delta \bar{\mathcal{U}} + \Delta \mathcal{D} + \Delta \bar{\mathcal{D}} - 2(\Delta \mathcal{S} + \Delta \bar{\mathcal{S}}),$$ 10.

with $\Delta \mathcal{Q} = \int_0^1 \Delta q(x, Q^2) \, dx$, which does not evolve with Q^2 at lowest order. The flavor non-singlet combinations $\Delta \mathcal{A}_i$ turn out to be proportional to the nucleon matrix elements of the quark non-singlet axial currents, $\langle P, S \mid \bar{q} \gamma^\mu \gamma^5 \lambda_i q \mid P, S \rangle$. Such currents typically occur in weak interactions, and by SU(3) rotations one may relate the matrix elements to the β-decay parameters F, D of the baryon octet (28, 29). One finds $\Delta \mathcal{A}_3 = F + D = g_A = 1.267$ and $\Delta \mathcal{A}_8 = 3F - D \approx 0.58$.

The first of these remarkable connections between hadronic and DIS physics corresponds to the famous Bjorken sum rule (28),

$$\Gamma_1^p - \Gamma_1^n = \frac{1}{6}\Delta\mathcal{A}_3[1 + \mathcal{O}(\alpha_s)] = \frac{1}{6}g_A[1 + \mathcal{O}(\alpha_s)], \qquad 11.$$

where the superscripts p and n denote the proton and neutron respectively. The sum rule has been verified experimentally with about 10% accuracy (19). The QCD corrections indicated in Equation 11 are known (30) through $\mathcal{O}(\alpha_s^3)$. Assuming the validity of the sum rule, it can be used for a rather precise determination of the strong coupling constant (31).

Determining Γ_1 from the polarized-DIS data, and using the information from β-decays on $\Delta\mathcal{A}_3$ and $\Delta\mathcal{A}_8$ as additional input, one may determine $\Delta\Sigma$. This quantity is of particular importance because it measures *twice the quark spin contribution to the proton spin*. The analysis reveals a small value $\Delta\Sigma \approx 0.2$. The experimental finding that the quarks carry only about 20% of the proton spin has been one of the most remarkable results in the exploration of the structure of the nucleon. Even though the identification of nucleon with parton helicity is not a prediction of QCD (perturbative or otherwise) the result came as a major surprise. It has sparked tremendous theoretical activity and has also been the motivation behind a number of dedicated experiments in QCD spin physics, aimed at further unraveling the nucleon spin.

A small value for $\Delta\Sigma$ also implies a sizable negative strange quark polarization in the nucleon, $\Delta\mathcal{S} + \Delta\bar{\mathcal{S}} \approx -0.12$. It would be desirable to have independent experimental information on this quantity, to eliminate the uncertainty in the value for $\Delta\Sigma$ due to SU(3) breaking effects in the determination of $\Delta\mathcal{A}_8$ from baryon β decays (32). More generally, considering Figure 8, more information is needed on the polarized sea quark distribution functions and their flavor decomposition. Such knowledge is also very interesting for comparisons to model calculations of nucleon structure. For example, there have been a number of predictions (33) for the $\Delta\bar{u} - \Delta\bar{d}$. Progress toward achieving a full flavor separation of the nucleon sea has been made recently, through semi-inclusive measurements in DIS (SIDIS) (34, 35). Inclusive DIS via photon exchange only gives access to the combinations $\Delta q + \Delta\bar{q}$, as is evident from Equation 7. If one detects, however, a hadron in the final state, the spin-dependent structure function becomes

$$g_1^h(x, z) = \frac{1}{2}\sum_q e_q^2 \left[\Delta q(x)\, D_q^h(z) + \Delta\bar{q}(x)\, D_{\bar{q}}^h(z)\right]. \qquad 12.$$

Here, the $D_i^h(z)$ are fragmentation functions, with $z = E^h/\nu$, where E^h is the energy of the produced hadron and ν the energy of the virtual photon in the Lab frame. Figure 9 shows the latest results on the flavor separation by the HERMES collaboration at HERA (35). Uncertainties are still fairly large; unfortunately, no further improvements in statistics are expected from HERMES. The results are not inconsistent with the large negative polarization of $\Delta\bar{u} = \Delta\bar{d} = \Delta\bar{s}$ in the sea that has been implemented in many determinations of polarized parton distributions

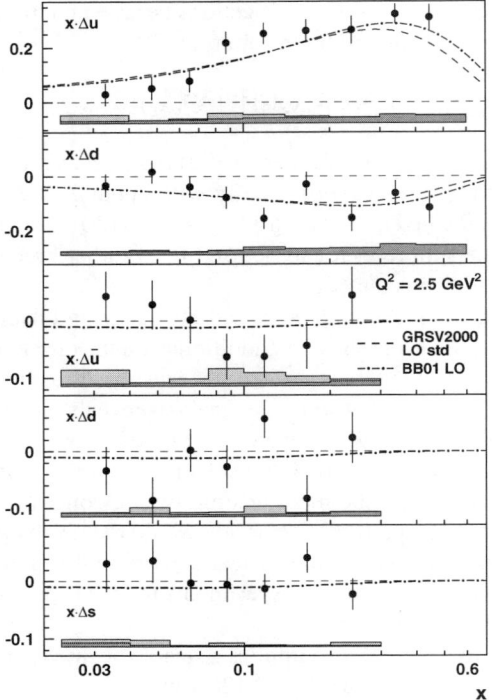

Figure 9   Recent HERMES results (35) for the quark and antiquark polarizations extracted from semi-inclusive DIS.

from inclusive DIS data (see, e.g., the curves in Figure 8). On the other hand, there is no evidence for a large negative strange quark polarization. The results have sparked much renewed theory activity on SIDIS (36). We note that at RHIC $W^{\pm}$ production will be used to determine Δu, $\Delta \bar{u}$, Δd, $\Delta \bar{d}$ with good precision, exploiting the parity-violating couplings of the W to left-handed quarks and right-handed antiquarks (37). Comparisons of such data taken at much higher scales with those from SIDIS will be extremely interesting.

A measurement of Γ_1 obviously relies on an estimate of the contribution to the integral from x outside the measured region. The extrapolation to small x constitutes one main uncertainty in the value of $\Delta\Sigma$. As can be seen from Figure 7, there is not much information on $g_1(x, Q^2)$ at $x < 0.003$. In addition, the data points at the smaller x also have Q^2 values that are below the DIS regime, making it conceivable that the "higher-twist" contributions to $g_1(x, Q^2)$ are important and contaminate the extraction of $\Delta\Sigma$. About half of the data points shown in Figure 7 are from the region $Q^2 \le 4$ GeV^2, $W^2 = Q^2(1 - x)/x \le 10$ GeV^2, which in the unpolarized case is usually excluded in analyses of parton distribution functions. Clearly, measurements of polarized DIS and SIDIS at smaller x, as well as at presently available x, but higher Q^2, will be vital for arriving at a definitive understanding of the polarized quark distributions, and of $\Delta \Sigma$ in particular.

2.2.2. CONTRIBUTORS TO THE NUCLEON SPIN   The partons in the nucleon have to provide the nucleon spin. When formulating a "proton spin sum rule" one has in mind the expectation value of the angular momentum operator (38, 39),

$$\frac{1}{2} = \langle P, 1/2 \,|\, \hat{J}_3 |\, P, 1/2 \rangle = \langle P, 1/2 \,|\, \int d^3 \left[\vec{x} \times \vec{T}\right]_3 |\, P, 1/2 \rangle, \qquad 13.$$

where $T^i \equiv T^{0i}$ with T the QCD energy-momentum tensor. Expressing the operator in terms of quark and gluon operators, one may write:

$$\frac{1}{2} = \frac{1}{2} \Delta \Sigma + \Delta G(Q^2) + L_q(Q^2) + L_g(Q^2), \qquad 14.$$

where $\Delta G(Q^2) = \int_0^1 \Delta g(x, Q^2)$ is the gluon spin contribution and the $L_{q,g}$ correspond to orbital angular momenta of quarks and gluons. Unlike $\Delta \Sigma$, ΔG and $L_{q,g}$ depend on the resolution scale Q^2 already at lowest order in evolution. The small size of the quark spin contribution implies that we must look elsewhere for the proton's spin: sizable contributions to the nucleon spin should come from ΔG and/or $L_{q,g}$.

Several current experiments are dedicated to a direct determination of $\Delta g(x, Q^2)$. High-transverse momentum jet, hadron, and photon final states in polarized pp scattering at RHIC offer the best possibilities (37). For example, direct access to Δg is provided by the spin asymmetry for the reaction $pp \to \gamma X$, owing to the presence of the QCD Compton process $qg \to \gamma q$. The Spin Muon Collaboration (SMC) and COMPASS fixed-target experiments at CERN, and the HERMES experiment at DESY, access $\Delta g(x, Q^2)$ in charm or high-p_T hadron pair final states in photon-gluon fusion $\gamma^* g \to q\bar{q}$ (40). Additional precision measurements with well established techniques will be needed to determine the integral of the polarized gluon distribution, particularly at lower x.

Orbital effects are the other candidate for contributions to the proton spin. Close analysis of the $\vec{x} \times \vec{T}$ matrix elements in Equation 13 revealed (38) that they can be measured from a wider class of parton distribution functions, the so-called generalized parton distributions (GPD) (41). These take the general form $\langle p + \Delta | \mathcal{O}_{q,g} | p \rangle$, where $\mathcal{O}_{q,g}$ are suitable quark and gluon operators and Δ is some momentum transfer. The latter is the reason that the GPDs are also referred to as "off-forward" distributions. The explicit factor $\vec{x}$ in Equation 13 forces one off the forward direction, simply because it requires a derivative with respect to momentum transfer. This is in analogy with the nucleon's Pauli form factor. In fact, matrix elements of the above form interpolate between DIS structure functions and elastic form factors.

To be more specific (42), the *total* (spin plus orbital) angular momentum contribution of a quark to the nucleon spin is given as (38)

$$J_q = \frac{1}{2} \lim_{\Delta^2 \to 0} \int dx \, x \left[H_q(x, \xi, \Delta^2) + E_q(x, \xi, \Delta^2) \right]. \qquad 15.$$

Here, $\xi = \Delta^+/P^+$, where the light-cone momentum $\Delta^+ \equiv \Delta^0 + \Delta^z$, and likewise for P^+. H_q, E_q are defined as form factors of the matrix element $\int dy \, e^{iyx}$ $\langle P' | \bar{\psi}_+(y) \, \psi_+(0) | P \rangle$. H_q reduces to the ordinary (forward) quark distribution in the limit $\Delta \to 0$, $H_q(x, 0, 0) = q(x)$, whereas the first moments (in x) of H_q and E_q give the quark's contributions to the nucleon Dirac and Pauli form factors, respectively. In addition, Fourier transforms of H_q, E_q with respect to the transverse components of the momentum transfer Δ give information on the position space distributions of partons in the nucleon (43), for example:

$$H_q\left(x, \xi = 0, -\vec{\Delta}_\perp^2\right) = \int d^2\vec{b} \, e^{-i\vec{\Delta}_\perp \cdot \vec{b}} \, q(x, b). \qquad 16.$$

$q(x, b)$ is the probability density for finding a quark with momentum fraction x at transverse distance $\vec{b}$ from the center. It thus gives a transverse profile of the nucleon. GPDs, therefore, may give us remarkably deep new insight into the nucleon.

The classic reaction for a measurement of the H_q, E_q is "deeply virtual Compton scattering (DVCS)," $\gamma^* p \to \gamma p$ (38). It is the theoretically best explored and understood reaction (44). Next-to-leading order calculations are available (45). The GPDs contribute to the reaction at amplitude level. The amplitude for DVCS interferes with that for the Bethe-Heitler process. The pure Bethe-Heitler part of the differential ep cross-section is calculable and can in principle be subtracted, provided it does not dominate too strongly. Such a subtraction has been performed in DVCS measurements at small x by H1 and ZEUS (46). A different possibility to eliminate the Bethe-Heitler contribution is to take the difference of cross sections for opposite beam or target polarization. In both cases, contributions from Compton scattering and the Compton-Bethe-Heitler interference survive. The cleanest separation of these pieces can be achieved in experiments with lepton beams of either charge. Because the Compton contribution to the ep amplitude is linear and the Bethe-Heitler contribution quadratic in the lepton charge, the interference term is projected out in the difference $d\sigma(e^+ p) - d\sigma(e^- p)$ of cross sections, whereas it is absent in their sum. Both the "beam-spin" asymmetry

$$\frac{d\sigma_+(e^- p) - d\sigma_-(e^- p)}{d\sigma_+(e^- p) + d\sigma_-(e^- p)}, \qquad 17.$$

where $\pm$ denote positive (negative) beam helicities, and the "beam-charge" asymmetry

$$\frac{d\sigma(e^+ p) - d\sigma(e^- p)}{d\sigma(e^+ p) + d\sigma(e^- p)} \qquad 18.$$

have been observed (47–49). Figure 10 shows some of the results.

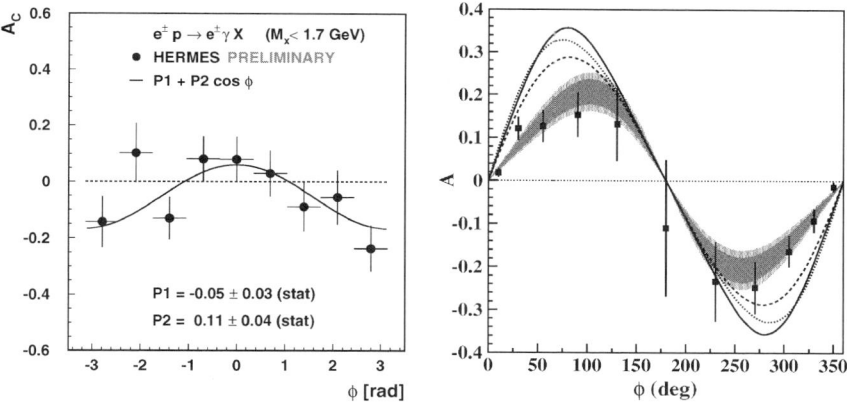

Figure 10   Data for the beam charge asymmetry in DVCS from HERMES (47) (left) and for the beam spin asymmetry from CLAS (49) (right), as functions of the azimuthal angle ϕ. For the definitions of these asymmetries, see text.

Hard exclusive meson production, $\gamma^* p \rightarrow Mp$, is another process that gives access to GPDs, and much activity has gone into this direction as well (42, 50). Both DVCS and exclusive meson production have their practical advantages and disadvantages. Real photon production is cleaner, but the price to be paid is an additional power of α_{em}. Meson production may be easier to detect; however, its amplitude is suppressed relatively by a power $1/Q$. The importance of using nucleon polarization in off-forward reactions is well established. There have also been first studies for DVCS off nuclei (51).

Practical problems are the fact that GPDs depend on three variables (plus a scale in which they evolve), and that they appear in complicated convolutions with the partonic hard-scattering kernels. We are still far from the quantitative experimental surveys of DVCS and related processes that would allow us to work backwards to new insights into off-diagonal matrix elements and angular momentum. Nevertheless, a direction for the field has been set.

2.2.3. TRANSVERSE POLARIZATION   In addition to the unpolarized and the helicity-dependent distributions, there is a third set of twist-2 parton distributions, namely transversity (52). In analogy with Equation 8 these distributions measure the net number (parallel minus antiparallel) of partons with transverse polarization in a transversely polarized nucleon:

$$\delta q(x) = q_\Uparrow^\uparrow(x) - q_\Uparrow^\downarrow(x). \qquad 19.$$

In a helicity basis (52), transversity corresponds to an interference of an amplitude in which a helicity-+ quark emerges from a helicity-+ nucleon, but is returned as a quark of negative helicity into a nucleon of negative helicity. This helicity-flip structure makes transversity a probe of chiral symmetry breaking in QCD (53). Perturbative-QCD interactions preserve chirality, and so the helicity flip must primarily come from soft non-perturbative interactions for which chiral symmetry is broken. The required helicity flip also precludes a gluon transversity distribution at leading twist (52).

Measurements of transversity are not straightforward. Again the fact that perturbative interactions in the Standard Model do not change chirality (or, for massless quarks, helicity) means that inclusive DIS is not useful. Collins, however, showed (54) that properties of fragmentation might be exploited to obtain a "transversity polarimeter": a pion produced in fragmentation will have some transverse momentum with respect to the fragmenting parent quark. There may then be a correlation of the form $i \vec{S}_T (\vec{P}_\pi \times \vec{k}_\perp)$ among the transverse spin $\vec{S}_T$ of the fragmenting quark, the pion momentum $\vec{P}_\pi$, and the transverse momentum $\vec{k}_\perp$ of the quark relative to the pion. The fragmentation function associated with this correlation is denoted as $H_1^{\perp,q}(z)$, the Collins function. If non-vanishing, the Collins function makes a *leading-power* (54–56) contribution to the single-spin asymmetry $A_\perp$ in the reaction $ep^\uparrow \rightarrow e\pi X$:

$$A_\perp \propto |\vec{S}_T| \sin(\phi + \phi_S) \sum_q e_q^2 \delta q(x) H_1^{\perp,q}(z), \qquad 20.$$

where $\phi(\phi_S)$ is the angle between the lepton plane and the $(\gamma^* \pi)$ plane (and the transverse target spin). As shown in Equation 20, this asymmetry would then allow access to transversity.

If "intrinsic" transverse momentum in the fragmentation process can play a crucial role in the asymmetry for $ep^\uparrow \rightarrow e\pi X$, a natural question is whether $k_\perp$ in the initial state can be relevant as well. Sivers suggested (57) that the $k_\perp$ distribution of a quark in a transversely polarized hadron could have an azimuthal asymmetry, $\vec{S}_T (\vec{P} \times \vec{k}_\perp)$. It was realized (58, 59) that the Wilson lines in the operators defining the Sivers function, required by gauge invariance, are crucial for the function to be non-vanishing. This intriguing discovery has been one of the most important theoretical developments in QCD spin physics in the past years. Another important aspect of the Sivers function is that it arises as an interference of wave functions with angular momenta $J_z = \pm 1/2$ and hence contains information on parton orbital angular momentum (58, 60), complementary to that obtainable from DVCS.

Model calculations and phenomenological studies of the Sivers functions $f_{1T}^{\perp,q}$ have been presented (61). It makes a contribution to $ep^\uparrow \rightarrow e\pi X$ (55),

$$A_\perp \propto |\vec{S}_T| \sin(\phi - \phi_S) \sum_q e_q^2 \, f_{1T}^{\perp,q}(x) \, D_q^\pi(z). \qquad 21.$$

This is in competition with the Collins function contribution, Equation 20; however, the azimuthal angular dependence is discernibly different. HERMES has completed a run with transverse polarization and performed an extraction of the contributions from the Sivers and Collins effects (62). There are also first results from COMPASS (63). First independent information on the Collins functions is now coming from Belle measurements in e^+e^- annihilation (64). The Collins and Sivers functions are also likely involved (65) in explanations of experimental observations of very large single-transverse spin asymmetries in pp scattering (66), where none were expected. It was pointed out (59, 67) that comparisons of DIS results and results from $p^\uparrow p$ scattering at RHIC will be particularly interesting: from the properties of the Wilson lines it follows that the Sivers functions violate universality of the distribution functions. For example, the Sivers functions relevant in DIS and in the Drell-Yan process should have opposite sign. This is a striking prediction awaiting experimental testing.

2.3. Nuclear Modifications

The nucleus is traditionally described as a collection of weakly bound nucleons confined in a potential created by their mutual interaction. It came as a surprise when the EMC experiment (68) uncovered a systematic nuclear dependence to the nuclear structure function $F_2^A(x, Q^2)$ in iron relative to that for Deuterium because the effect was as much as 20% for $x \sim 0.5$. This is significantly larger than the effect ($< 5\%$) due to the natural scale for nuclear effects given by the ratio of the binding energy per nucleon to the nucleon mass. Several dedicated fixed target experiments (69–71) confirmed the existence of the nuclear dependence observed by the EMC albeit with significant modifications of the original EMC results at small x. The upper part of Figure 11 shows an idealized version of the nuclear modification of the relative structure functions per nucleon. It is $2/A$ times the ratio of a measured nuclear structure function of nucleus A to that for Deuterium. The rise at the largest values of x is ascribed to the nucleons' Fermi momentum. The region above $x \geq 0.2$ is referred to as the EMC effect region. When $x \leq 0.05$, the nuclear ratio drops below one and the region is referred to as the nuclear shadowing region, whereas the region with the slight enhancement in between the shadowing and EMC effect regions is called the anti-shadowing region. The lower part of Figure 11 presents a sample of high precision data of ratios of structure functions over a broad range in A, x, and Q^2. We shall now discuss what is known about these regions, focusing in particular on the EMC effect and nuclear shadowing regions.

2.3.1. THE EMC EFFECT   A review of the DIS data and various interpretations of the EMC effect since its discovery in the early 1980's can be found in Ref. (73). A common interpretation of the EMC effect is based on models where inter-nucleon interactions at a wide range of inter-nucleon distances are mediated by meson exchanges. The traditional theory (74, 75) of nuclear interactions predicts a net increase in the distribution of virtual pions with increasing nuclear density relative to that of free nucleons. This is because meson interactions are attractive in nuclei. In these models, nuclear pions may carry about 5% of the total momentum to fit the EMC effect at $x \sim 0.3$. Each pion carries a light-cone fraction of about 0.2–0.3 of that for a nucleon. Sea anti-quarks belonging to these nuclear pions may scatter

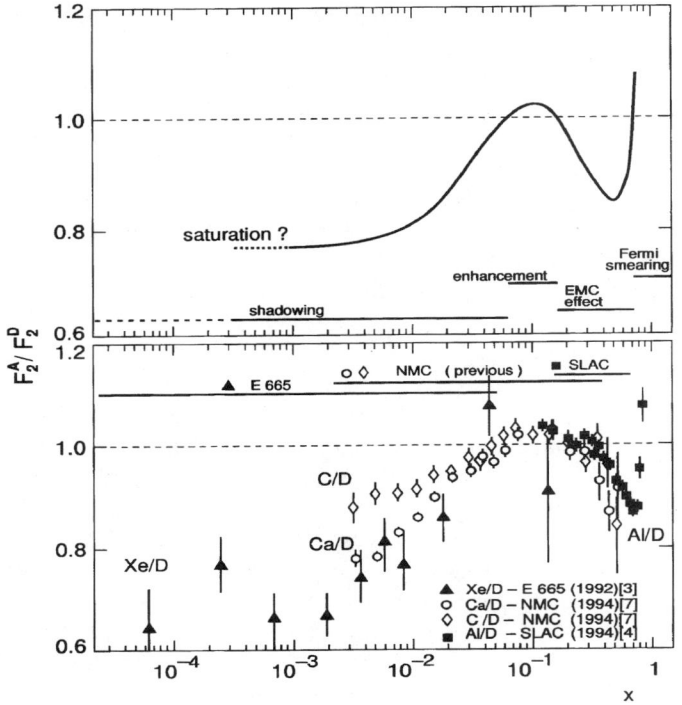

Figure 11 Upper: An idealized depiction of the ratio of the structure function of a nucleus, $F_2^A(x, Q^2)$ per nucleon to $F_2^d(x, Q^2)$ of Deuterium. Lower: Measured $F_2(x, Q^2)$ structure functions for C, Ca, and Xe relative to Deuterium. From (72).

off a hard probe. Hence the predicted enhancement of the nuclear sea of 10% to 15% for $x \sim 0.1$–0.2 and for $A \geq 40$. The conventional view of nuclear binding is challenged by the constancy with A of the anti-quark distribution extracted from the production of Drell-Yan pairs in proton-nucleus collisions at Fermilab (76). These data are shown in Figure 12. No enhancement was observed at the level of 1% accuracy in the Drell-Yan experiments. The Drell-Yan data was also compared with and showed good agreement with the DIS EMC data for the F_2 ratio of Tin to Deuterium.

Furthermore, first results (77) from the Jefferson Laboratory (TJNAF) experiment E91-003 indicate that there is no significant pion excess in the $A(e, e', B)$ reaction. (It has however been pointed out (78) that parameters of pion interactions in nuclei can be readily adjusted to reduce the pion excess to conform with the Drell-Yan data.) In addition, the energy excitation for the residual nuclear system also reduces the contribution of pions to the nuclear parton densities (79). Thus all of these observations suggest that pions may not contribute significantly to F_2^A in the EMC region.

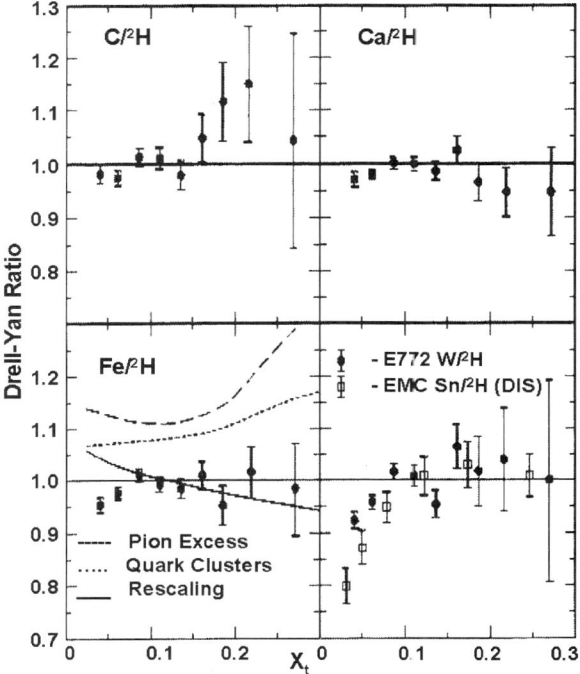

Figure 12   The ratio of the anti-quark distribution per nucleon in several nuclei relative to Deuterium. Data are shown from a Drell-Yan experiment (76) and compared to theoretical predictions. Also shown is the ratio for Tungsten to Deuterium compared to DIS data from the EMC experiment (68) for the F_2 ratio of Tin to Deuterium.

The chiral quark-soliton model (33) is a phenomenologically successful model that for instance explains the difference in the anti-quark up and down distributions as a function of x (80). Interestingly, it has been shown recently (80) to simultaneously provide a good description of both the EMC effect and the ratio of anti-quark distributions from Drell-Yan pairs. Recently, it has been argued that a key feature of the EMC effect, the factorization of the x and A dependence of the EMC ratio, can be understood in a model independent way in an effective field theory approach (81). Several joint leading order QCD analyses of the nuclear DIS and Drell-Yan data combined with the application of the baryon charge and momentum sum rules (82–85) provide further information on the nuclear effects on parton densities in this kinematic region. These analyses indicate that the valence quark distribution in nuclei is enhanced at $x \sim 0.1$–0.2. Gluons in nuclei carry practically the same fraction of the momentum (within 1%) as in a free nucleon. If one assumes that gluon shadowing is similar to that for quarks, these analyses predict a significant enhancement of the gluon distribution in nuclei at $x \sim 0.1$–0.2 (86). A recent next-to-leading order (NLO) analysis of nuclear parton distributions

(87) however finds that this gluon "anti-shadowing" is much smaller than in the LO analysis.

2.3.2. NUCLEAR SHADOWING   Nuclear shadowing is the phenomenon, shown in Figure 11, where the ratio of the nuclear electromagnetic structure function F_2^A relative to $A/2$ times the Deuteron electromagnetic structure functions F_2^D is less than unity for $x \leq 0.05$. Shadowing is greater for decreasing x and with increasing nuclear size. For moderately small x, shadowing is observed to decrease slowly with increasing Q^2. Unfortunately, because x and Q^2 are inversely correlated for fixed energies, much of the very small x data ($x \leq 10^{-3}$) is at very low values of $Q^2 \leq 1$ GeV^2. In addition, as results (72) from the fixed target E665 experiment at Fermilab and the New Muon Collaboration (NMC) experiment at CERN shown in Figure 11 suggest, good quality data exists only for $x > 4 \times 10^{-3}$. At high Q^2, the shadowing of F_2^A can be interpreted in terms of shadowing of quark and anti-quark distributions in nuclei at small x. Information on quark shadowing can also be obtained from proton-nucleus Drell-Yan experiments (88) and from neutrino-nucleus experiments—most recently from NuTeV at Fermilab (89).

The phenomenon of shadowing has different interpretations depending on the frame in which we consider the space-time evolution of the scattering. Consider for instance the rest frame of a nucleus in γ-p/A scattering. The γp cross-section is only 0.1 mb for energies in excess of 2 GeV, corresponding to a mean-free-path of well over 100 fm in nuclear matter. However, although the high-energy γA cross-section might be expected to be proportional to A, the observed increase in the cross-section is smaller than A times the γp cross-section. This is because the photon can fluctuate into a $q\bar{q}$-pair that has a cross-section typical of the strong interactions ($\sim$20 mb) and is absorbed readily (with a mean free path of $\sim$3.5 fm). If the fluctuation persists over a length greater than the inter-nucleon separation distance (2 fm), its absorption shadows it from encountering subsequent nucleons. The coherence length of the virtual photon's fluctuation is $l_{\text{coh.}} \sim 1/2m_N x$ where m_N is the nucleon mass. Therefore the onset of shadowing is expected and observed at $x \approx 0.05$. In this Gribov multiple scattering picture (90), there is a close relation between shadowing and diffraction. The so-called AGK cutting rules (91) relate the first nuclear shadowing correction to the cross-section for diffractively producing a final state in coherent scattering off a nucleon (integrated over all diffractive final states). See Figure 13(a) for an illustration of this correspondence. With these relations (and higher order re-scattering generalizations of these) and with the HERA diffractive DIS data as input, the NMC nuclear shadowing data can be reproduced as shown in the sample computation (92, 93) in Figure 13(b).

In the infinite momentum frame (IMF), shadowing arises due to gluon recombination and screening in the target. When the density of partons in the transverse plane of the nucleus becomes very large, many body recombination and screening effects compete against the growth in the cross-section, leading eventually to a saturation of the gluon density (94). In the IMF picture, one can again use

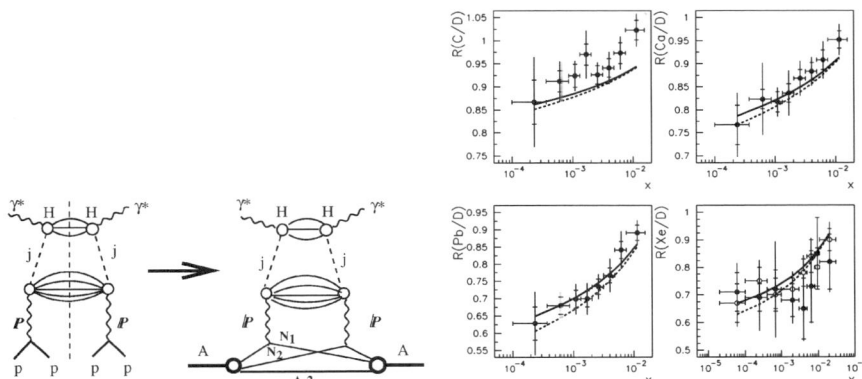

Figure 13   Left: An illustration of the AGK rules relating shadowing corrections in nuclei to diffractive scattering on nucleons. Right: A calculation (92) which uses AGK to fit nuclear F_2 data using HERA diffractive data. The two curves correspond to two different unitarization prescriptions.

the AGK rules we discussed previously to relate shadowing and diffraction (95), and the result is amenable to a partonic interpretation. The saturation regime is characterized by a scale $Q_s(x, A)$, called the saturation scale, which grows with decreasing x and increasing A. This saturation scale arises naturally in the Color Glass Condensate (CGC) framework which is discussed in section 3.

A natural consequence of saturation physics is the phenomenon of geometrical scaling. (See for instance the discussion on geometrical scaling of HERA data in section 2.1.) It has been argued that the NMC DIS data also display geometrical scaling (96)—the evidence here albeit interesting is not compelling owing to the paucity of nuclear data over a wide range of x and Q^2. It is widely believed that shadowing is a leading twist effect (97, 98), but some of the IMF discussion in the CGC saturation framework suggests higher twist effects are important for $Q^2 \leq Q_s^2$ because of the large gluon density (99). Constraints from non-linear corrections to the DGLAP framework have also been discussed recently (100). The available data on the Q^2 dependence of shadowing are inconclusive at small x.

Our empirical definition of shadowing in DIS refers to quark shadowing, likewise for quarks and anti-quarks in the Drell-Yan process in hadronic collisions. In DIS gluon distributions are inferred only indirectly because the virtual photon couples to quarks. The most precise extractions of gluon distributions thus far are from scaling violations of F_2^A. To do this properly, one needs a wide window in x and Q^2. In contrast to the highly precise data on nucleon gluon distributions from HERA, our knowledge of nuclear gluon structure functions $(g_A(x, Q^2))$ is *nearly non-existent*. This is especially so relative to our knowledge of quark distributions in nuclei. The most precise data on the modification of gluon distributions in nuclei come from two NMC high precision measurements of the ratio of the

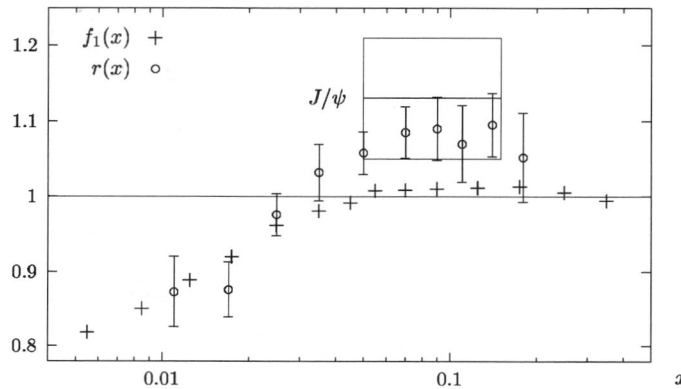

Figure 14 The ratio $r(x)$ of the gluon distributions in Sn relative to C and the ratio $f_1(x)$ of their $F_2(x)$ structure functions (101). The box represents the extraction of $r(x)$ from J/ψ electro-production in the process $\mu + A \rightarrow \mu + J/\psi + X$.

scaling violations of the structure functions of Tin (Sn) and Carbon (C). The experiments measure ratios $f_1 = F_2^{Sn}/F_2^C$ and $f_2 = \frac{\partial}{\partial \ln Q^2} f_1$. The ratio $r = g_{Sn}/g_C$ can be determined (101) from f_1 and f_2 and the scaling violations of $F_2^{Deuterium}$ (with minimal assumptions). The result for r is shown in Figure 14. At small x, gluon shadowing is observed. The trend suggests that gluon shadowing at small x is greater than that of F_2, even though the error bars are too large for a conclusive statement.

At larger x of $0.1 < x < 0.2$, one observes anti-shadowing of the gluon distributions. This result from scaling violations can be compared to the ratio of gluon distributions extracted from inclusive J/ψ production in DIS. The latter assumes the gluon fusion model of J/ψ-production. The results for r from the latter method are consistent with those from scaling violations. The large experimental uncertainties however leave the extent of anti-shadowing in doubt. Other measurements of scaling violations for the ratio of F_2^{Sn}/F_2^C showed an increase of the ratio with the increase of Q^2 consistent with predictions (82, 83).

The limited data we have may be interpreted to suggest a provocative picture of nuclear parton densities in the $x \sim 0.1$–0.2 region, which corresponds to distances of $\sim$1–1.5 fm, where medium range and short range inter-nucleon forces are expected to be important. In this region, if the gluon and valence quark fields are enhanced while the sea is somewhat suppressed, as some analyses suggest, gluon-induced interactions between nucleons, as well as valence quark interchanges between nucleons, may contribute significantly to nuclear binding (97, 101). Nuclear gluon distributions can also be further constrained by inclusive hadron distributions recently measured by the RHIC experiments (102–105) in Deuteron-Gold scattering at $\sqrt{s} = 200$ GeV/nucleon. These RHIC results will be discussed in section 3.

2.4. Space-Time Correlations in QCD

The space-time picture of DIS processes strongly depends on the value of Bjorken x. An analysis of electromagnetic current correlators in DIS reveals that one probes the target wave function at space-time points separated by longitudinal distances $l_{coh.}$ and transverse distances $\sim 1/Q$. At large $x(x > 0.2)$, the virtual photon transforms into a strongly interacting state very close to the active nucleon, typically in the middle of the nucleus. If Q^2 is large enough as well, the produced partonic state interacts weakly with the medium. At smaller $x(x < 0.05)$, the longitudinal length scale $l_{coh.}$ exceeds the nuclear size of the heaviest nuclei. At sufficiently small x ($x < 0.005$ for the heaviest nuclei) DIS processes undergo several spatially separated stages. First, the virtual photon transforms into a quark-gluon wave packet well before the nucleus. Time dilation ensures that interactions amongst partons in the wave packet are frozen over large distances. (These can be several hundred fermis at EIC energies.) The partons in the wave packet interact coherently and instantaneously with the target. At high energies, these interactions are eikonal in nature and do not affect the transverse size of the wave packet. Finally, the fast components of the wave packet transform into a hadronic final state when well past the nucleus. This interval could be as large as $2\nu/\mu^2$ where ν is the energy of the virtual photon and $\mu \leq 1$ GeV is a soft hadronic scale.

Space-time studies thus far have been limited to semi-exclusive experiments that investigate the phenomenon of color transparency, and more generic inclusive studies of quark propagation through nuclei. Both these studies involved fixed targets. They are briefly summarized below.

In pQCD, color singlet objects interact weakly with a single nucleon in the target. Additional interactions are suppressed by inverse powers of Q^2. This phenomenon is called "color transparency" because the nucleus appears transparent to the color singlet projectile (106–108). At very high energies, even the interaction of small color singlet projectiles with nuclei can be large. In this kinematic region, the phenomenon is termed "color opacity" (109, 110). The earliest study of color transparency in DIS was a study of coherent J/ψ photo-production off nuclei (111). The amplitude of the process at small t (momentum transfer squared) is approximately proportional to the nuclear atomic number A. This indicates that the pair that passes through the nucleus is weakly absorbed. For hadronic projectiles, a similar and approximately linear A-dependence of the amplitude was observed recently for coherent diffraction of 500 GeV pions into two jets (112), consistent with predictions (109).

A number of papers (113–115) predict that the onset of color transparency at sufficiently large Q^2 will give for the coherent diffractive production of vector mesons

$$\frac{d\sigma(\gamma_L^* + A \to V + A)}{dt}\Big|_{t=0} \propto A^2. \qquad 22.$$

For incoherent diffraction at sufficiently large t (> 0.1 GeV^2), they predict

$$R(Q^2) \equiv \frac{d\sigma(\gamma_L^* + A \rightarrow V + A')/dt}{A\,d\sigma(\gamma_L^* + N \rightarrow V + N)/dt} = 1. \qquad 23.$$

The first measurements of incoherent diffractive production of vector mesons were performed by the E665 collaboration at Fermilab (116). A significant increase of the nuclear transparency, as reflected in the ratio $R(Q^2)$, was observed. The limited luminosity and center-of-mass energy however do not provide a statistically convincing demonstration of color transparency. In addition, the results are complicated by large systematic effects.

Measurements of the inclusive hadron distribution for different final states as a function of the virtual photon energy ν, its transverse momentum squared Q^2, the fraction z_h of the photon energy carried by the hadron, and the nuclear size A, provide insight into the propagation of quarks and gluons in nuclear media. In addition to the time and length scales discussed previously, the "formation time" τ_h of a hadron is an additional time scale. It is in principle significantly larger than the production time $1/Q$ of a color singlet parton. If the formation time is large, the "pre-hadron" can multiple scatter in the nucleus, thereby broadening its momentum distribution, and also suffer radiative energy loss before hadronization. The QCD prediction for transverse momentum broadening resulting from multiple scattering is (for quarks) given by the expression (117)

$$\langle \Delta p_\perp^2 \rangle = \frac{\alpha_S C_F \pi^2}{2} x g(x, Q^2) \rho L \approx 0.5\,\alpha_S \left(\frac{L}{5\,\text{fm}}\right) \text{GeV}^2. \qquad 24.$$

Here, $C_F = 4/3$ is the color Casimir of the quark, ρ is the nuclear matter density and L is the length of matter traversed. The Drell-Yan data in Ref. (118) agree with this expression and with the predicted small size of the effect (empirically, $\langle \Delta p_\perp^2 \rangle \sim 0.12$ GeV^2 for heavy nuclei). One also observes a large difference in the A dependence of the transverse momentum of Drell-Yan di-muons relative to those from J/ψ and Υ production and decay (119). In the former process only the incident quark undergoes strong interactions, whereas in the latter, the produced vector mesons interact strongly as well. However, the size of the effect and the comparable broadening of the J/ψ and Υ (albeit the latter is appreciably smaller than the former) need to be better understood. The $p_\perp$ imbalance of dijets in nuclear photo-production suggests a significantly larger $p_\perp$ broadening effect than in J/ψ production (120). This suggests non-universal behavior of $p_\perp$ broadening effects but it may also occur from a contamination of the jets by soft fragments. Parton $p_\perp$ broadening due to multiple scattering may also be responsible for the anomalous behavior of inclusive hadron production in hadron-nucleus scattering at moderate $p_\perp$ of a few GeV. In this case, the ratio R_{pA} of inclusive hadron production in hadron-nucleus scattering to the same process on a nucleon is suppressed at low $p_\perp$ but exceeds unity between $1-2$ GeV. This "Cronin effect" (121) was discovered in proton-nucleus scattering experiments in the late 70's. The flavor dependence of the Cronin effect provided an early hint that scattering of projectile partons off gluons dominates over scattering off quarks (122). The Cronin effect will be discussed further in section 3 in light of the recent RHIC experiments on Deuteron-Gold scattering (102–105).

The energy loss of partons due to scattering in nuclear matter is complicated by vacuum induced energy loss in addition to the energy loss due to scattering. One computation suggests that vacuum energy loss is the dominant effect (123). For a quark jet, the medium induced energy loss increases quadratically with the length, L, and is independent of the energy for $E \to \infty$. For $L = 5$ fm, the asymptotic energy loss, ΔE, is estimated to be less than 1 GeV in a cold nuclear medium (117). This makes it difficult to empirically confirm this remarkable L-dependence of the energy loss. DIS data are qualitatively consistent with small energy loss (124–126). The data indicate that the multiplicity of the leading hadrons is moderately reduced (by 10%) for virtual photon energies of the order of 10–20 GeV for scattering off Nitrogen-14 nuclei. At higher energies, the leading multiplicities gradually become A-dependent, indicating absorption of the leading partons (125–129).

A pQCD description of partonic energy loss in terms of modified fragmentation functions is claimed to describe HERMES data (131). However, at HERMES energies, and perhaps even at EMC energies, descriptions in terms of hadronic re-scattering and absorption are at least as successful (130, 132). As previously discussed, however, the latter descriptions usually require that the color singlet "pre-hadrons" have a formation time $\tau_h \sim 0.5$ fm (123, 130, 133). Figure 15 shows the results from one such model as a function of z_h (the fraction of the parton momentum carried by a hadron) and ν (the virtual photon energy) compared to the HERMES data. At EIC energies, a pQCD approach in terms of modified fragmentation functions (134, 135) should be more applicable. The results from these

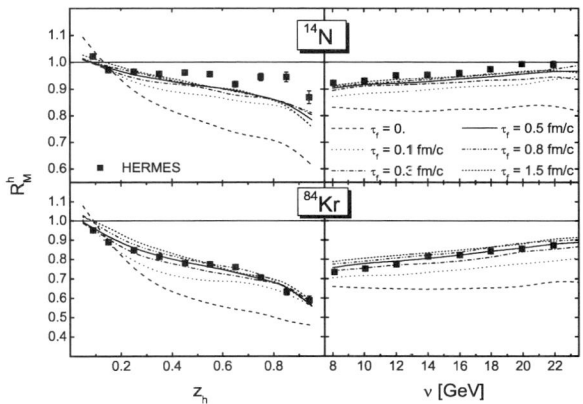

Figure 15   Data from the HERMES experiment (124) showing the ratio of the inclusive hadron cross section in a nucleus relative to that in a nucleon, plotted as a function a) of z_h, which is the fraction of the quark's momentum carried by a hadron, and b) as a function of ν, the photon energy, for two different nuclei. Curves denote results of a "pre-hadron" scattering model (130), with differing formation times.

analyses will provide an important test of jet quenching in hot matter descriptions of the RHIC data.

3. SCIENTIFIC OPPORTUNITIES WITH AN ELECTRON-ION COLLIDER

This section will discuss the exciting scientific opportunities that will be made possible by the novel features of an electron-ion collider: the high luminosity, the possibility to do scans over a wide range in energy, polarization of the electron and hadronic beams, a range of light and heavy nuclear beams and, not least, the collider geometry of the scattering. Scientific firsts for the electron-ion collider will include a) the first high energy polarized electron-polarized proton collider, and b) the first high energy electron-nucleus collider.

3.1. Unpolarized e-p Collisions at EIC

Unpolarized e-p collisions have been studied extensively most recently at the HERA collider at DESY. In the eRHIC option for an EIC, the center of mass energy in an e-p collision is anticipated to be $\sqrt{s} = 100$ GeV compared to $\sqrt{s}$ of over 300 GeV at HERA. Although the x-Q^2 reach of an EIC may not be as large as that of HERA, it has significant other advantages which we will itemize below.

- The current design luminosity is approximately 25 times the design luminosity of HERA. Inclusive observables will be measured with great precision. The additional luminosity will be particularly advantageous for studying semi-inclusive and exclusive final states.

- The EIC (particularly in the eRHIC version) will be able to vary the energies of both the electron and nucleon beams. This will enable a first measurement of F_L in the small x regime. The F_L measurement is very important in testing QCD fits of structure functions.

- Electron-Deuteron collisions, with tagging of spectator nucleons, will allow high precision studies of the flavor dependence of parton distributions.

- An eRHIC detector proposed by Caldwell et al. (136) would have a rapidity coverage nearly twice that of the ZEUS and H1 detectors at HERA. This would allow the reconstruction of the event structure of hard forward jets with and without rapidity gaps in the final state. With this detector, exclusive vector meson and DVCS measurements can be performed for a wider range of the photon-proton center of mass energy squared W^2. It also permits measurements up to high $|t|$, where t denotes the square of the difference in four-momenta of the incoming and outgoing proton. These will enable a precise mapping of the energy dependence of final states, as well as open a window into the spatial distribution of partons down to very low impact parameters.

We will briefly discuss the physics measurements that can either be done or improved upon with the above enumerated capabilities of EIC/eRHIC. For inclusive measurements, F_L is clearly a first, "gold plated" measurement. Current QCD fits predict that F_L is very small (and in some analyses negative) at small x and small $Q^2 < 2$ GeV^2. An independent measurement can settle whether this reflects poor extrapolations of data (implying leading twist interpretations of data are still adequate in this regime), or whether higher twist effects are dominant. It will also constrain extractions of the gluon distribution because F_L is very sensitive to it. Another novel measurement would be that of structure functions in the region of large $x \approx 1$. These measurements can be done with 1 fb^{-1} of data for up to $x = 0.9$ and for $Q^2 < 250$ GeV^2. This kinematic window is completely unexplored to date. These studies can test perturbative QCD predictions of the helicity distribution of the valence partons in a proton (138) as well as the detailed pattern of SU(6) symmetry breaking (139). Moments of structure functions can be compared to lattice data. These should help quantify the influence of higher twist effects. Finally ideas such as Bloom-Gilman duality can be further tested in this kinematic region (140).

At small x, very little is understood about the quark sea. For instance, the origins of the $\bar{u} - \bar{d}$ asymmetry and the suppression of the strange sea are not clear. High precision measurements of $\pi^\pm$, $K^\pm$, K_s and open charm will help separate valence and sea contributions in the small-x region. We have already discussed Generalized Parton Distributions and DVCS measurements. The high luminosity, wide coverage and measurements at high $|t|$ will quantify efforts to extract a 3-D snapshot of the distribution of partons in the proton.

3.2. Polarized *ep* Collisions at EIC

We expect the EIC to dramatically extend our understanding of the spin structure of the proton through measurements of the spin structure function g_1 over a wide range in x and Q^2, of its parity-violating counterparts g_4 and g_5, of gluon polarization ΔG, as well as through spin-dependent semi-inclusive measurements, the study of exclusive reactions, and of polarized photo-production.

3.2.1. INCLUSIVE SPIN-DEPENDENT STRUCTURE FUNCTION   We have emphasized in Section 2.2.1 the need for further measurements of $g_1(x, Q^2)$ at lower x and higher Q^2. A particularly important reason is that one would like to reduce the uncertainty in the integral $\Gamma_1(Q^2)$ and hence in $\Delta\Sigma$. However, the behavior of $g_1(x, Q^2)$ at small x is by itself of great interest in QCD.

At very high energies, Regge theory gives guidance to the expected behavior of $g_1(x)$. The prediction (141) is that $g_1(x)$ is flat or even slightly vanishing at small x, $g_1(x) \propto x^{-\alpha}$ with $-0.5 \leq \alpha \leq 0$. It is an open question how far one can increase Q^2 or decrease energy and still trust Regge theory. A behavior of the form $g_1 \sim x^{-\alpha}$ with $\alpha < 0$ is unstable under DGLAP evolution (142, 143) in the sense that evolution itself will then govern the small-x behavior at higher Q^2. Under the assumption that Regge theory expectations are realistic at some (low) scale Q_0 one then obtains "perturbative predictions" for $g_1(x, Q^2)$ at $Q \gg Q_0$. The fixed-target polarized DIS data indicate that although the non-singlet combination $g_1^p - g_1^n$ is quite singular at small x, the singlet piece does appear to be rather flat (142, 144), so that the above reasoning applies here. It turns out that the leading eigenvector of small-x evolution is such that the polarized quark singlet distribution and gluon density become of opposite sign. For a sizeable positive gluon polarization, this leads to the striking feature that the singlet part of $g_1(x, Q^2)$ is negative at small x and large Q^2 (142), driven by Δg. Figure 16 (see color insert) shows this dramatic behavior for different values of $Q^2 = 2, 10, 20, 100$ GeV^2 (145). The projected statistical uncertainties at eRHIC, corresponding to 400 pb^{-1} integrated luminosity with an almost 4π acceptance detector, are also shown. Note that 400 pb^{-1} can probably be collected within about one week of e-p running of eRHIC. Thus, at eRHIC one will be well positioned to explore the evolution of the spin structure function $g_1(x, Q^2)$ at small x. We note that at small x and toward $Q^2 \to 0$, one could also study the transition region between the Regge and pQCD regimes (144).

Predictions for the small-x behavior of g_1 have also been obtained from a perturbative resummation of double-logarithms $\alpha_s^k \ln^{2k}(1/x)$ appearing in the splitting functions (147–150) at small x in perturbative QCD. Some of these calculations indicate a very singular asymptotic behavior of $g_1(x)$. It has been shown (148), however, that subleading terms may still be very important even far below $x = 10^{-3}$.

The neutron g_1 structure function could be measured at eRHIC by colliding the electrons with polarized Deuterons or with Helium. If additionally the hadronic proton fragments are tagged, a very clean and direct measurement could be performed. As can be seen from Figure 7, information on g_1^n at small x is scarce. The small-x behavior of the isotriplet $g_1^p - g_1^n$ is particularly interesting for the Bjorken sum rule and because of the steep behavior seen in the fixed-target data (142, 144). It is estimated (151) that an accuracy of the order of 1% could be achieved for the Bjorken sum rule in a running time of about one month. One can also turn this argument around and use the accurate measurement of the non-singlet spin structure function and its evolution to determine the value of the strong coupling constant $\alpha_s(Q^2)$ (31, 146). This has been tried, and the value one gets from this exercise is comparable to the world average for the strong coupling constant. It is expected that if precision low-x data from the EIC is available and the above mentioned non-singlet structure functions are measured along with their evolutions, this may result in the most accurate value of the strong coupling constant $\alpha_s(Q^2)$.

Because of eRHIC's high energy, very large Q^2 can be reached. Here, the DIS process proceeds not only via photon exchange; also the W and Z contribute significantly. Equation 2 shows that in this case new structure functions arising from parity violation contribute to the DIS cross section. These structure functions contain very rich additional information on parton distributions (20, 152, 153). As an example, let us consider charged-current (CC) interactions. Events in the case of W exchange are characterized by a large transverse momentum imbalance caused by the inability to detect neutrinos from the event. The charge of the W boson is dictated by that of the lepton beam used in the collision. For W^- exchange one then has for the structure functions g_1 and g_5 in Equation 6:

$$g_1^{W^-}(x) = \Delta u(x) + \Delta \bar{d}(x) + \Delta \bar{s}(x), \quad g_5^{W^-}(x) = \Delta u(x) - \Delta \bar{d}(x) - \Delta \bar{s}(x). \quad 25.$$

These appear in the double-spin asymmetry as defined in Ref. (20), where the asymmetry can be expressed in terms of structure functions as

$$A^{W^-} = \frac{2bg_1^{W^-} + ag_5^{W^-}}{aF_1^{W^-} + bF_3^{W^-}}. \quad 26.$$

Here $a = 2(y^2 - 2y + 2)$, $b = y(2 - y)$ and F_3 is the unpolarized parity-violating structure function of Equation 2. Note that the typical scale in the parton densities is M_W here. Availability of polarized neutrons and positrons is particularly desirable. For example, one finds at lowest order:

$$g_1^{W^-,p} - g_1^{W^+,p} = \Delta u_v - \Delta d_v \quad 27.$$

$$g_5^{W^+,p} + g_5^{W^-,p} = \Delta u_v + \Delta d_v \quad 28.$$

$$g_5^{W^+,p} - g_5^{W^-,n} = -\left[\Delta u + \Delta \bar{u} - \Delta d - \Delta \bar{d}\,\right]. \quad 29.$$

The last of these relations gives, after integration over all x and taking into account the first-order QCD correction (152),

$$\int_0^1 dx \left[g_5^{W^+,p} - g_5^{W^-,n}\right] = -\left(1 - \frac{2\alpha_s}{3\pi}\right)g_A, \quad 30.$$

equally fundamental as the Bjorken sum rule.

A Monte Carlo study, including the detector effects, has shown that the measurement of the asymmetry in Equation 26 and the parity violating spin structure functions is feasible at eRHIC. Figure 17 shows simulations (154) for the asymmetry and the structure function g_5 for CC events with an electron beam. The luminosity was assumed to be 2 fb^{-1}. The simulated data shown are for $Q^2 > 225$ GeV^2. Similar estimates exist for W^+. Measuring this asymmetry would require a positron beam. The curves in the figure use the polarized parton distributions of (155). It was assumed that the unpolarized structure functions will have been measured well by HERA by the time this measurement would be performed at eRHIC. Standard assumptions used by the H1 collaboration about the scattered electrons for good detection were applied. The results shown could be obtained (taking into account machine and detector inefficiencies) in a little over one month with the eRHIC luminosity. It is possible that only one or both of the electron-proton and positron-proton collisions could be performed, depending on which design of the accelerator is finally chosen (see Section 4).

3.2.2. SEMI-INCLUSIVE MEASUREMENTS   As we discussed in Section 2.2.1, significant insights into the nucleon's spin and flavor structure can be gained from semi-inclusive scattering $ep \rightarrow ehX$. Knowledge of the identity of the produced

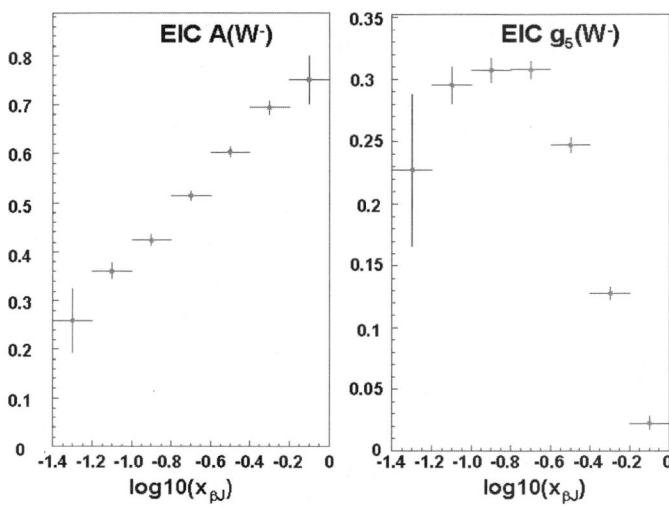

Figure 17   Simulations (154) for the spin asymmetry A^{W^-} of Equation 26 and the structure function $g_5^{W^-}$ as functions of $\log_{10}(x)$.

hadrons h allows separation of the contributions from the different quark flavors. In fixed target experiments, the so-called current hadrons are at forward angles in the laboratory frame. This region is difficult to instrument adequately, especially if the luminosity is increased to gain significant statistical accuracy. A polarized ep collider has the ideal geometry to overcome these shortfalls. The collider kinematics open up the final state into a large solid angle in the laboratory which, using an appropriately designed detector, allows complete identification of the hadronic final state both in the current and target kinematic regions of fragmentation phase space. At eRHIC energies the current and target kinematics are well separated and may be individually studied. At eRHIC higher Q^2 will be available than in the fixed-target experiments, making the observed spin asymmetries less prone to higher-twist effects, and the interpretation cleaner.

Figure 18 shows simulations (156) of the precision with which one could measure the polarized quark and antiquark distributions at the EIC. The events were produced using the DIS generator LEPTO. The plotted uncertainties are statistical only. The simulation was based on an integrated luminosity of 1 fb^{-1} for 5 GeV electrons on 50 GeV protons, with both beams polarized to 70%. Inclusive and semi-inclusive asymmetries were analyzed using the leading order "purity" method developed by the SMC (34) and HERMES (35) collaborations. Excellent precision for $\Delta q/q$ can be obtained down to $x \approx 0.001$. The measured average Q^2 values vary as usual per x bin; they are in the range $Q^2 = 1.1$ GeV^2 at the lowest x to $Q^2 \sim 40$ GeV^2 at high x. With proton beams, one has greater sensitivity to up quarks than to down quarks. Excellent precision for the down quark polarizations could be obtained by using Deuteron or Helium beams.

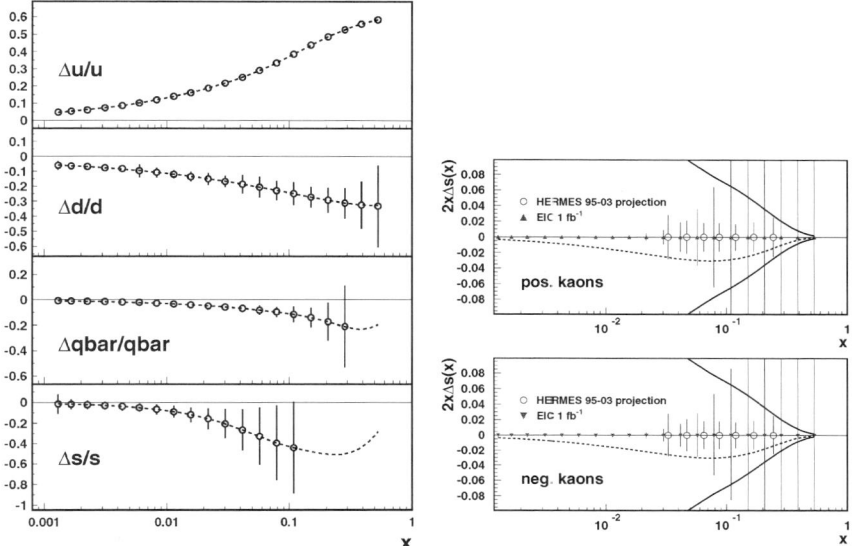

Figure 18 Left: projected precision of eRHIC measurements of the polarized quark and antiquark distributions (156). Right: expected statistical accuracy of $\Delta s(x)$ from spin asymmetries for semi-inclusive $K^{\pm}$ measurements for 1 fb^{-1} luminosity operation of eRHIC, and comparison with the statistical accuracy of the corresponding HERMES measurements.

With identified kaons, and if the up and down quark distributions are known sufficiently well, one will have a very good possibility to determine the strange quark polarization. As we discussed in Section 2.2.1, $\Delta s(x)$ is one of the most interesting quantities in nucleon spin structure. On the right-hand side of Figure 18, we show results expected for $\Delta s(x)$ as extracted from the spin asymmetries for $K^{\pm}$ production. As in the previous figure, only statistical uncertainties are indicated. The results are compared with the precision available in the HERMES experiment.

There is also much interest in QCD in more refined semi-inclusive measurements. For example, the transverse momentum of the observed hadron may be observed. Here, interesting azimuthal-angle dependences arise at leading twist (56, 157), as we discussed in Subsection 2.2.3. At small transverse momenta, resummations of large Sudakov logarithms are required (158). Measurements at eRHIC would extend previous results from HERA (159) and be a testing ground for detailed studies in perturbative QCD.

3.2.3. MEASUREMENTS OF THE POLARIZED GLUON DISTRIBUTION $\Delta g(x, Q^2)$ One may extract Δg from scaling violations of the structure function $g_1(x, Q^2)$.

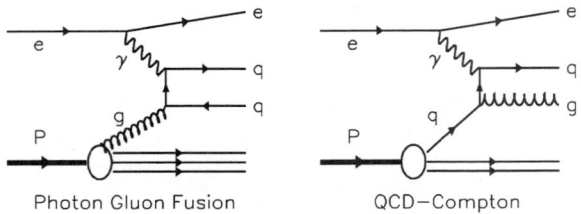

Figure 19   Feynman diagrams for the photon-gluon fusion and the QCD Compton processes.

Figure 8 shows that indeed some initial information on $\Delta g(x, Q^2)$ has been obtained in this way, albeit with very poor accuracy. The uncertainty of the integral of Δg is probably about 100% at the moment (31). Measurements at RHIC will vastly improve on this. eRHIC will offer independent and complementary information. Thanks to the large lever arm in Q^2, and to the low x that can be reached, scaling violations alone will constrain $\Delta g(x, Q^2)$ and its integral much better. Studies (160) indicate for example that the total uncertainty on the integral of ΔG could be reduced to about 5–10% by measurements at eRHIC with integrated luminosity of 12 fb^{-1} ($\sim$2–3 years of eRHIC operation).

Lepton-nucleon scattering also offers direct ways of accessing gluon polarization. Here one makes use of the photon-gluon fusion (PGF) process, for which the gluon appears at leading order. Charm production is one particularly interesting channel (161–163). It was also proposed (162–164) to use jet pairs, produced in the reaction $\gamma^*g \rightarrow q\bar{q}$, for a determination of Δg. This process competes with the QCD Compton process, $\gamma^*q \rightarrow qg$. Feynman diagrams for these processes are shown in Figure 19.

In the unpolarized case, dijet production has successfully been used at HERA to constrain the gluon density (165). Dedicated studies have been performed for dijet production in polarized collisions at eRHIC (166), using the MEPJET (167) generator. The two jets were required to have transverse momenta >3 GeV, pseudorapidities $-3.5 \leq \eta \leq 4$, and invariant mass $s_{JJ} > 100$ GeV^2. A 4π detector coverage was assumed. The results for the reconstructed $\Delta g(x)$ are shown in Figure 20, assuming luminosities of 1 fb^{-1} (left) and 200 pb^{-1} (right). The best probe would be in the region $0.02 \leq x \leq 0.1$; at higher x, the QCD Compton process becomes dominant. This region is indicated by the shaded areas in the figure. The region $0.02 \leq x \leq 0.1$ is similar to that probed at RHIC. Measurements at eRHIC would thus allow an independent determination of Δg in a complementary physics environment.

Eventually data for the scaling violations in $g_1(x, Q^2)$ and for dijet production in DIS will be analyzed jointly. Such a combined analysis would determine the gluon distribution with yet smaller uncertainties. A first preliminary study for eRHIC (168), following the lines of (169), indeed confirms this.

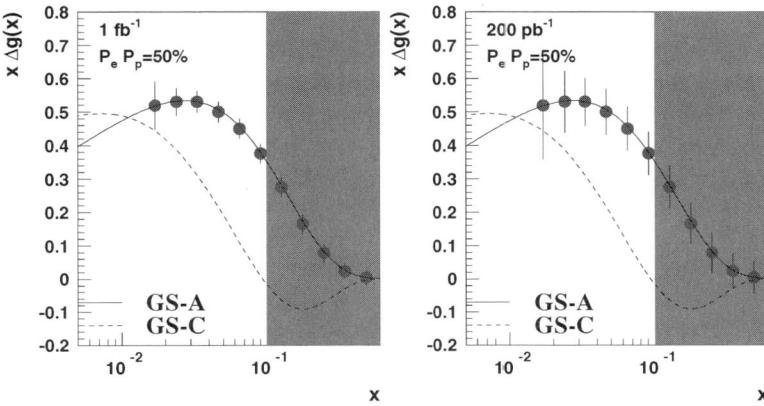

Figure 20   The statistical precision of $x \Delta g$ from dijets in LO for eRHIC, for two different luminosities, with predictions for sets A and C of the polarized parton densities of Ref. (155).

3.2.4. EXPLORING THE PARTONIC STRUCTURE OF POLARIZED PHOTONS   In the photoproduction limit, when the virtuality of the intermediate photon is small, the ep cross-section can be approximated by a product of a photon flux and an interaction cross section of the real photon with the proton. Measurements at HERA in the photoproduction limit have led to a significant improvement in our knowledge of the *hadronic structure* of the photon.

The structure of the photon manifests itself in so-called "resolved" contributions to cross sections. We show this in Figure 21 for the case of photoproduction of hadrons. On the left, the photon participates itself in the hard scattering, through "direct" contributions. On the right, the photon behaves like a hadron. This

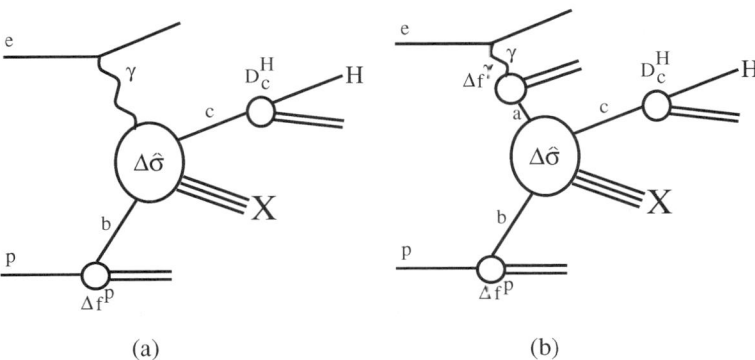

Figure 21   Generic direct (a) and resolved (b) photon contributions to the process $lp \rightarrow l'HX$.

possibility occurs because of (perturbative) short-time fluctuations of the photon into $q\bar{q}$ pairs and gluons, and because of (non-perturbative) fluctuations into vector mesons ρ, ϕ, ω with the same quantum numbers (170). The resolved contributions have been firmly established by experiments in e^+e^- annihilation and ep scattering (171).

A unique application of eRHIC would be to study the parton distributions of *polarized* quasi-real photons, defined as (172, 173)

$$\Delta f^\gamma(x) \equiv f_+^{\gamma+}(x) - f_-^{\gamma+}(x), \qquad 31.$$

where $f_+^{\gamma+}$ ($f_-^{\gamma+}$) denotes the density of a parton $f = u, d, s, \ldots, g$ with positive (negative) helicity in a photon with positive helicity. The $\Delta f^\gamma(x)$ give information on the spin structure of the photon; they are completely unmeasured so far.

Figure 22 shows samples from studies (174, 175) for observables at eRHIC that would give information on the $\Delta f^\gamma(x)$. Two models for the $\Delta f^\gamma(x)$ were considered (173), one with a strong polarization of partons in the photon ("maximal" set), the other with practically unpolarized partons ("minimal" set). On the left, we show the double-spin asymmetry for photoproduction of high-p_T pions, as a

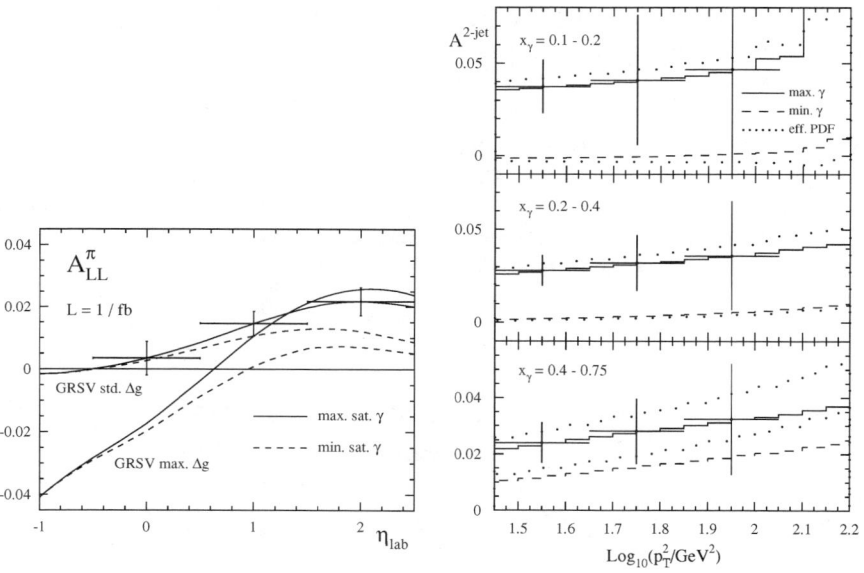

Figure 22   Left: spin asymmetry for π^0 photoproduction in NLO QCD for two sets of polarized photon densities and two different choices of spin-dependent proton distributions. The error bars indicate the statistical accuracy anticipated for eRHIC assuming an integrated luminosity of 1 fb^{-1}. Right: spin asymmetry for dijet production as a function of the jet transverse momentum, in three bins of the photon momentum fraction x_γ.

function of the pion's pseudorapidity η_{lab} in the eRHIC laboratory frame. The advantage of this observable is that for negative η_{lab}, in the proton backward region, the photon mostly interacts "directly," via the process $\gamma g \to q\bar{q}$, whereas its partonic content becomes visible at positive η_{lab}. This may be seen from the figure, for which we have also used two different sets of polarized parton distributions of the proton (26), mainly differing in $\Delta g(x)$.

The right part of Figure 22 shows predictions for the spin asymmetry in dijet photoproduction at eRHIC. If one assumes the jets to be produced by a $2 \to 2$ partonic hard scattering, the jet observables determine the momentum fractions $x_{p,\gamma}$ of the partons in the proton and the photon. Selecting events with $x_\gamma < 1$, one therefore directly extracts the "resolved"-photon contribution. At higher orders, this picture is somewhat diluted, but remains qualitatively intact. Such measurements of dijet photoproduction cross sections at HERA (176) have been particularly successful in providing information on photon structure. This makes the spin asymmetry a good candidate for learning about the Δf^γ at eRHIC. In the figure we show results for the asymmetry in three different bins of x_γ. One can see that with 1 fb^{-1} luminosity one should be able at eRHIC to establish the existence of polarized resolved-photon contributions, and distinguish between our "maximal" and "minimal" photon scenarios. For a first exploration one could also use the approach of "effective" parton densities considered in (174, 176–178).

We finally note that measurements of the polarized total photoproduction cross section at high energies would also give new valuable information on the high-energy contribution to the Drell-Hearn-Gerasimov sum rule (144). The latter relates the total cross sections with photon-proton angular momentum 3/2 and 1/2 to the anomalous magnetic moments of the nucleon (179):

$$\int_0^\infty \frac{d\nu}{\nu}[\sigma_{3/2}(\nu) - \sigma_{1/2}(\nu)] = \frac{2\pi^2\alpha}{M^2}\kappa^2 = \begin{cases} 204.5\,\mu\text{b} & p \\ 232.8\,\mu\text{b} & n \end{cases}, \qquad 32.$$

where on the right we have given the numerical values of the sum rule. Currently, the experimental result for the proton is a few percent high, and the one for the neutron about 20% low (180). There is practically no information on the contribution to the sum rule from photon energies $\nu \geq 3$ GeV; estimates based on Regge theory indicate that it is possible that a substantial part comes from this region. Measurements at eRHIC could give definitive answers here. The H1 and ZEUS detectors at DESY routinely take data using electron taggers situated in the beam pipe 6–44 meters away from the end of the detectors. They detect the scattered electrons from events at very low Q^2 and scattering angles. If electron taggers were included in eRHIC, similar measurements could be performed. The Q^2 range of such measurements at eRHIC is estimated to be $10^{-8} - 10^{-2}$ GeV^2.

3.2.5. HARD EXCLUSIVE PROCESSES   As we have discussed in Subsection 2.2.2, generalized parton distributions (GPDs) are fundamental elements of nucleon structure. They contain both the parton distributions and the nucleon form factors as limiting cases, and they provide information on the spatial distribution of partons in the transverse plane. GPDs allow the description of exclusive processes at large Q^2, among them DVCS. It is hoped that eventually these reactions will provide information on the total angular momenta carried by partons in the proton. See for example Equation 15.

The experimental requirements for a complete investigation of GPDs are formidable. Many different processes need to be investigated at very high luminosities, at large enough Q^2, with polarization, and with suitable resolution to determine reliably the hadronic final state. The main difficulty, however, for experimental measurements of exclusive reactions is detecting the scattered proton. If the proton is not detected, a "missing-mass" analysis has to be performed. In case of the DVCS reaction, there may be a significant contribution from the Bethe-Heitler process. The amplitude for the Bethe-Heitler process is known and, as we discussed in Subsection 2.2.2, one may construct beam-spin and charge asymmetries to partly eliminate the Bethe-Heitler contribution. Early detector design studies have been performed for the EIC (181). These studies indicate that the acceptance can be significantly increased by adding stations of Silicon-strip-based Roman Pot Detectors away from the central detector in a HERA-like configuration. The detector recently proposed for low-x and low-Q^2 studies at the EIC (136) (for details, see Section 5) may also be of significant use to measure the scattered proton. Further studies are underway and will proceed along with iterations of the design of the interaction region and of the beam line.

Although more detailed studies need yet to be performed, we anticipate that the EIC would provide excellent possibilities for studying GPDs. Measurements at the collider will complement those now underway at fixed target experiments and planned with the 12-GeV upgrade at the Jefferson Laboratory (137).

3.3. Exploring the Nucleus with an Electron-Ion Collider

In this section, we discuss the scientific opportunities available with the EIC in DIS off nuclei. At very high energies, the correct degrees of freedom to describe the structure of nuclei are quarks and gluons. The current understanding of partonic structure is just sufficient to suggest that their behavior is non-trivial. The situation is reminiscent of Quantum Electrodynamics. The rich science of condensed matter physics took a long time to develop even though the nature of the interaction was well understood. Very little is known about the condensed matter many-body properties of QCD, particularly at high energies. There are sound reasons based in QCD to believe that partons exhibit remarkable collective phenomena at high energies. Because the EIC will be the first electron-ion collider, we will be entering a *terra incognita* in our understanding of the properties of quarks and gluons in nuclei. The range in x and Q^2 and the luminosity will be greater than at any previous fixed target DIS experiment. Further, the collider environment is ideal for studying semi-inclusive and exclusive processes. Finally, it is expected that a wide range of particle species and beam energies will be available to study carefully the systematic variation of a wide range of observables with target size and energy.

We will begin our discussion in this section by discussing inclusive "bread and butter" observables such as the inclusive nuclear quark and gluon structure functions. As we observed previously, very little is known about nuclear structure functions at small x and $Q^2 \gg \Lambda_{QCD}^2 \sim 0.04 \text{ GeV}^2$. This is especially true of the nuclear gluon distribution. We will discuss the very significant contributions that the EIC can make in rectifying this situation. A first will be a reliable extraction of the longitudinal structure function at small x. Much progress has been made recently in defining universal diffractive structure functions (50, 182, 183). These structure functions can be measured in nuclei for the first time. Generalized parton distributions will help provide a three-dimensional snapshot of the distribution of partons in the nucleus (43).

We will discuss the properties of partons in a nuclear medium and the experimental observables that will enable us to tease out their properties. These include nuclear fragmentation functions that contain valuable information on hadronization in a nuclear environment. The momentum distributions of hadronic final states as functions of x, Q^2, and the fraction of the parton energy carried by a hadron also provide insight into dynamical effects such as parton energy loss in the nuclear medium.

A consequence of small x evolution in QCD is the phenomenon of parton saturation (94). This arises from the competition between attractive Bremsstrahlung (184) and repulsive screening and recombination (many body) effects (95), which results in a phase space density of partons of order $1/\alpha_S$. At such high parton densities, the partons in the wavefunction form a Color Glass Condensate (CGC) for reasons we will discuss later (185). The CGC is an effective theory describing the remarkable universal properties of partons at high energies. It provides an organizing principle for thinking about high energy scattering and has important ramifications for colliders. The evolution of multi-parton correlations predicted by the CGC can be studied with high precision in lepton-nucleus collisions.

Experimental observables measured at the EIC can be compared and contrasted with observables extracted in proton/Deuteron-nucleus and nucleus-nucleus scattering experiments at RHIC and LHC. The kinematic reach of the EIC will significantly overlap with these experiments. Measurements of parton structure functions and multi-parton correlations in the nuclear wave function will provide a deeper understanding of the initial conditions for the formation of a quark gluon plasma (QGP). Final state interactions in heavy ion collisions such as the energy loss of leading hadrons in hot matter (often termed "jet quenching") are considered strong indicators of the formation of the QGP. The EIC will provide benchmark results for cold nuclear matter which will help quantify energy loss in hot matter. Finally, recent results on inclusive hadron production in RHIC D-Au collisions at 200 GeV/nucleon show hints of the high parton density effects predicted by the CGC. We will discuss these and consider the similarities and differences between a p/D-A and an e-A collider.

3.3.1. NUCLEAR PARTON DISTRIBUTIONS   The range of the EIC in x and Q^2 was discussed previously (see Figure 2). It is significantly larger than for the previous fixed target experiments. The projected statistical accuracy, per inverse picobarn of data, of a measurement of the ratio $\frac{\partial R}{\partial \ln Q^2}$ versus x at the EIC relative to data from previous NMC measurements and a hypothetical future e-A collider at HERA energies is shown in Figure 23 (see color insert). Here R denotes the ratio of nuclear structure functions, $R = F_2^A/F_2^N$. As discussed previously, the logarithmic derivative with Q^2 of this ratio can be used to extract the nuclear gluon distribution. The EIC is projected to have an integrated luminosity of several hundred pb^{-1} for large nuclei, so one can anticipate high precision measurements of nuclear structure functions at small x. In particular, because the energy of the colliding beams can be varied, the nuclear longitudinal structure function can be measured for the first time at small x. At small x and large Q^2, it is directly proportional to the gluon distribution. At smaller values of Q^2, it may be more sensitive to higher twist effects than F_2 (11).

Measurements of nuclear structure functions in the low x kinematic region will test the predictions of the QCD evolution equations in this kinematic region. The results of QCD evolution with Q^2 depend on input from the structure functions at smaller values of Q^2 for a range of x values. The data on these is scarce for nuclei. These results are therefore very sensitive to models of the small x behavior of structure functions at low Q^2. A nice plot from Ref. (187) reproduced in Figure 24 clearly illustrates the problem. Figure 24 (see color insert) shows results from theoretical models for the ratio of the gluon distribution in Lead to that in a proton as a function of x. Though all the models employ the same QCD evolution equations, the range in uncertainty is rather large at small x—about a factor of 3 at $x \sim 10^{-4}$. Although one can try to construct better models, the definitive constraint can only come from experiment.

The shadowing of gluon distributions shown in Figure 24 is not understood in a fundamental way. We list here some relevant questions which can be addressed by a future electron-ion collider.

- Is shadowing a leading twist effect; namely, is it unsuppressed by a power of Q^2 ? Most models of nuclear structure functions at small x assume this is the case. (For a review, see Ref. (98).) Is there a regime of x and Q^2, where power corrections due to high parton density effects can be seen? (94, 95, 193).

- What is the relation of shadowing to parton saturation? As we will discuss, parton saturation dynamically gives rise to a semi-hard scale in nuclei. This suggests that shadowing at small x can be understood in a weak coupling analysis.

- Is there a minimum to the shadowing ratio for fixed Q^2 and A with decreasing x? If so, is it reached faster for gluons or for quarks?

- The Gribov relation between shadowing and diffraction that we discussed previously is well established at low parton densities. How is it modified at high parton densities? The EIC can test this relation directly by measuring diffractive structure functions in ep (and e-A) and shadowing in e-A collisions.

- Is shadowing universal? For instance, would gluon parton distribution functions extracted from p-A collisions at RHIC be identical to those extracted from e-A in the same kinematic regime? The naive assumption that this is the case may be false if higher twist effects are important. Later in this review, we will discuss the implications of the possible lack of universality for p-A and A-A collisions at the LHC.

We now turn to a discussion of diffractive structure functions. At HERA, hard diffractive events were observed where the proton remained intact and the virtual photon fragmented into a hard final state producing a large rapidity gap between the projectile and target. A rapidity gap is a region in rapidity essentially devoid of particles. In pQCD, the probability of a gap is exponentially suppressed as a function of the gap size. At HERA though, gaps of several units in rapidity are relatively unsuppressed; one finds that roughly 10% of the cross-section corresponds to hard diffractive events with invariant masses $M_X > 3$ GeV. The remarkable nature of this result is transparent in the proton rest frame: a 50 TeV electron slams into the proton and, 10% of the time, the proton is unaffected, even though the interaction causes the virtual photon to fragment into a hard final state.

The interesting question in diffraction is the nature of the color singlet object (the "Pomeron") within the proton that interacts with the virtual photon. This interaction probes, in a novel fashion, the nature of confining interactions within hadrons. (We will discuss later the possibility that one can study in diffractive events the interplay between strong fields produced by confining interactions and those generated by high parton densities.) In hard diffraction, because the invariant mass of the final state is large, one can reasonably ask questions about the quark and gluon content of the Pomeron. A diffractive structure function $F_{2,A}^{D(4)}$ can be defined (182, 183, 194), in a fashion analogous to F_2, as

$$\frac{d^4\sigma_{eA \to eXA}}{dx_{Bj}dQ^2 dx_p dt} = A \cdot \frac{4\pi\alpha_{em}^2}{xQ^4} \left\{ 1 - y + \frac{y^2}{2[1 + R_A^{D(4)}(\beta, Q^2, x_p, t)]} \right\}$$

$$\times F_{2,A}^{D(4)}(\beta, Q^2, x_p, t), \qquad\qquad 33.$$

where, $y = Q^2/sx_{Bj}$, and analogously to F_2, one has $R_A^{D(4)} = F_L^{D(4)}/F_T^{D(4)}$. Further, $Q^2 = -q^2 > 0$, $x_{Bj} = Q^2/2Pq$, $x_p = q(P - P')/qP$, $t = (P - P')^2$ and $\beta = x_{Bj}/x_p$. Here P is the initial nuclear momentum, and P' is the net momentum of the fragments Y in the proton fragmentation region. Similarly, M_X is the invariant mass of the fragments X in the electron fragmentation region. An illustration of the hard diffractive event is shown in Figure 25.

It is more convenient in practice to measure the structure function $F_{2,A}^{D(3)} = \int F_{2,A}^{D(4)} dt$, where $|t_{min}| < |t| < |t_{max}|$, where $|t_{min}|$ is the minimal momentum transfer to the nucleus, and $|t_{max}|$ is the maximal momentum transfer to the

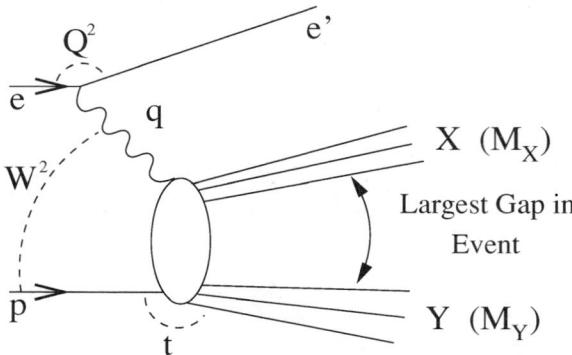

Figure 25   The diagram of a process with a rapidity gap between the systems X and Y. The projectile nucleus is denoted here as p. Figure from Ref. (195).

nucleus that still ensures that the particles in the nuclear fragmentation region Y are undetected. An interesting quantity to measure is the ratio $R_{A1,A2}(\beta, Q^2, x_p) = \frac{F_{2,A1}^{D(3)}(\beta, Q^2, x_p)}{F_{2,A2}^{D(3)}(\beta, Q^2, x_p)}$. The A-dependence of this quantity will contain very useful information about the universality of the structure of the Pomeron. In a study for e-A collisions at HERA, it was argued that this ratio could be measured with high systematic and statistical accuracy (195)—the situation for eRHIC should be at least comparable, if not better. Unlike F_2 however, F_2^D is not truly universal—it cannot be applied, for instance, to predict diffractive cross sections in p-A scattering; it can be applied only in other lepton-nucleus scattering studies (50, 182). This has been confirmed by a study where diffractive structure functions measured at HERA were used as an input in computations for hard diffraction at Fermilab. The computations vastly overpredicted the Fermilab data on hard diffraction (196). Some of the topics discussed here will be revisited in our discussion of high parton densities.

3.3.2. SPACE-TIME EVOLUTION OF PARTONS IN A NUCLEAR ENVIRONMENT   The nuclear structure functions are inclusive observables and are a measure of the properties of the nuclear wavefunction. Less inclusive observables, which measure these properties in greater detail, will be discussed in the section on the Color Glass Condensate. In addition to studying the wave function, we are interested in the properties of partons as they interact with the nuclear medium. These are often called final state interactions to distinguish them from the initial state interactions in the wavefunction. Separating which effects arise from the wavefunction is not easy because our interpretation of initial state and some final state interactions may depend on the gauge in which the computations are performed (197). Isolating the two effects in experiments is difficult. A case in point is the study of energy loss effects on final states in p-A collisions (198). These effects are not easy to

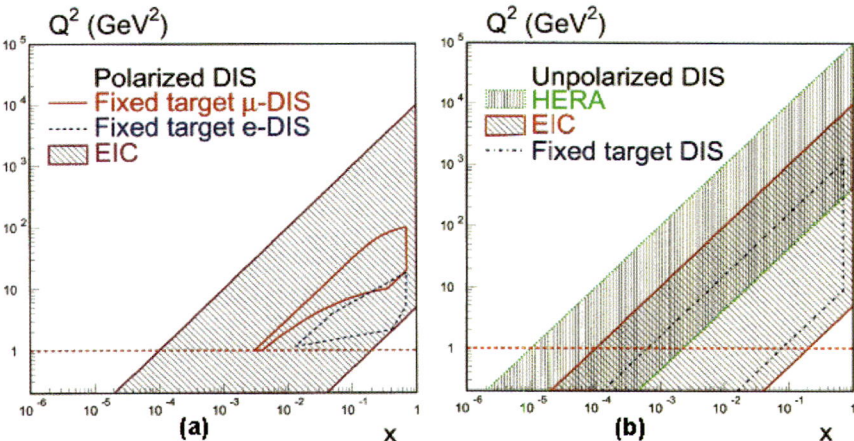

Figure 2   The x-Q^2 range of the proposed lepton-ion collider at Brookhaven National Laboratory (eRHIC) in comparison with the past and present experimental DIS facilities. The left plot is for polarized DIS experiments, and the right corresponds to the unpolarized DIS experiments.

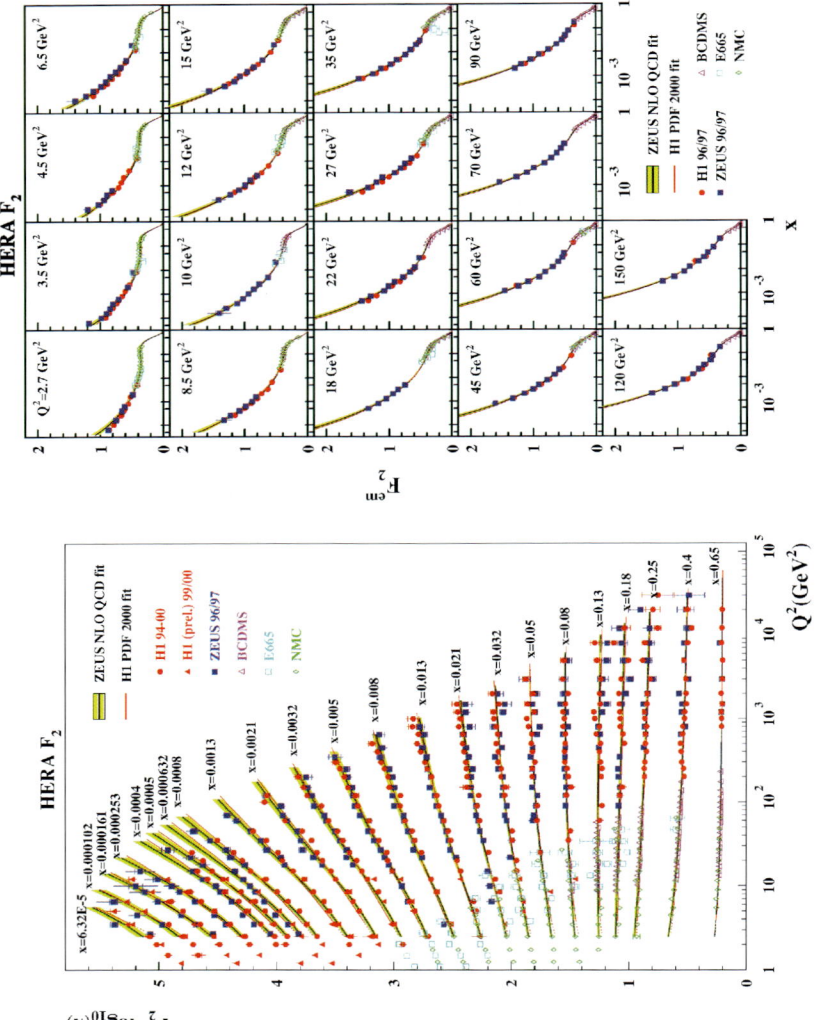

Figure 4   The plot on the left shows the world data on F_2 as a function of Q^2 for fixed values of x. On the right we show the converse: F_2 as a function of x for fixed values of Q^2. From (8).

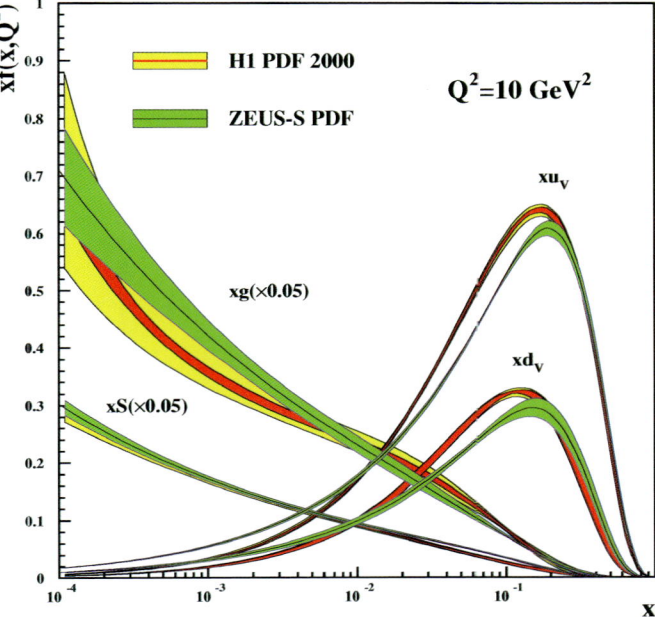

Figure 5   The valence (up and down) quark, sea quark, and gluon distributions plotted as a function of x for fixed $Q^2 = 10 \text{ GeV}^2$. Note that the sea and glue distributions are scaled down by a factor of 1/20. From (8).

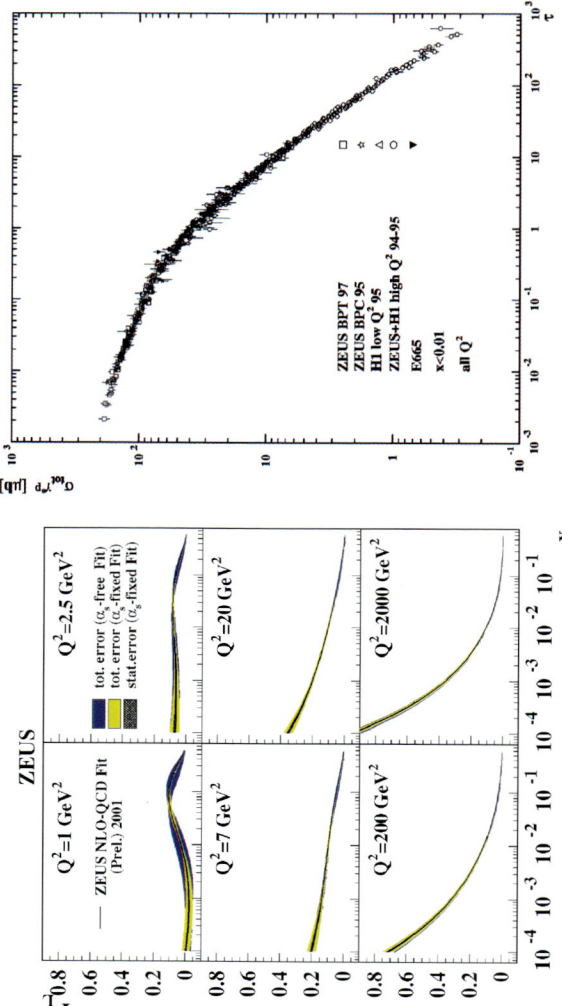

Figure 6   The plot on the left shows the longitudinal structure function F_L as a function of x for different Q^2 bins (9). On the right (from Ref. 12) is a plot of the virtual photon-proton cross-section plotted as a function of $\tau = Q^2/Q_s^2$. See text for further explanation.

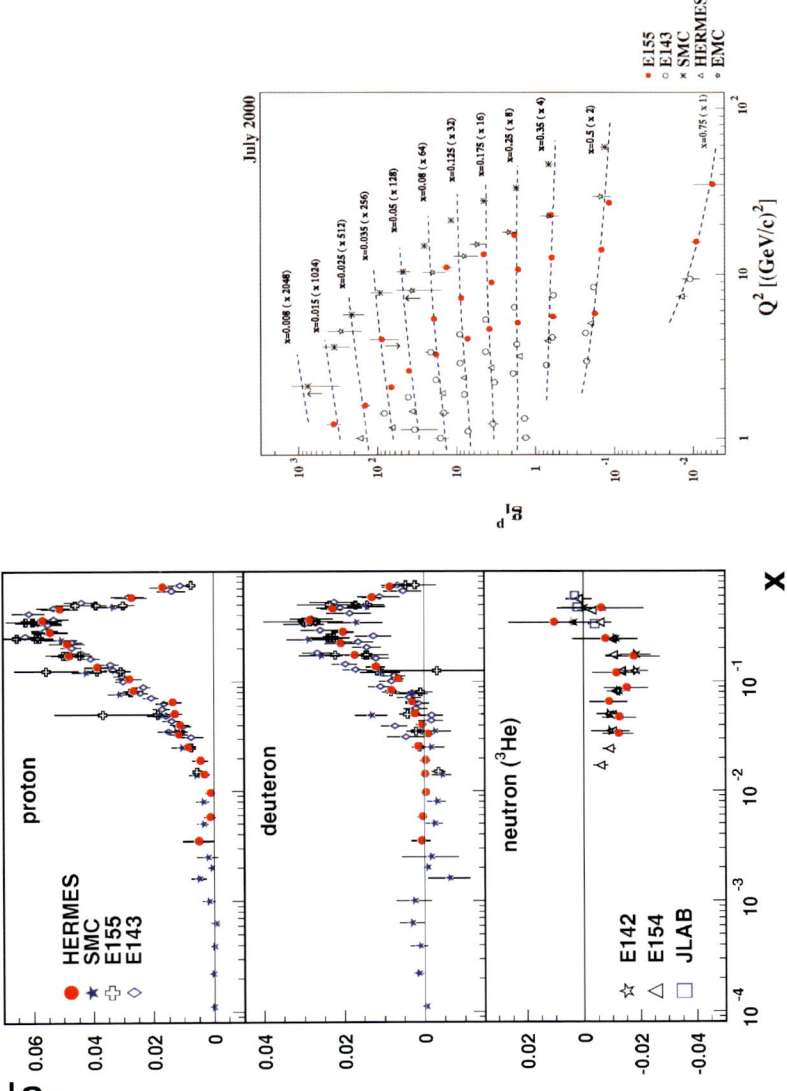

Figure 7 Left: world data on the spin structure function g_1 as compiled and shown in (21). Right: $g_1(x, Q^2)$ as a function of Q^2 for various x. The curves are from a phenomenological fit. Taken from (22).

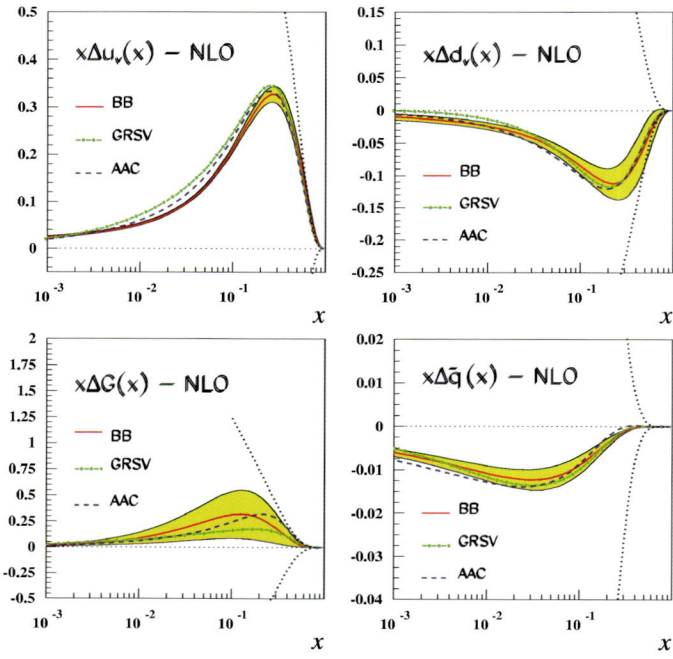

Figure 8   Recent analysis of polarized parton densities of the proton. Taken from (25) ("BB"). The additional curves represent the central fits from the analyses of (26) ("GRSV") and (27) ("AAC").

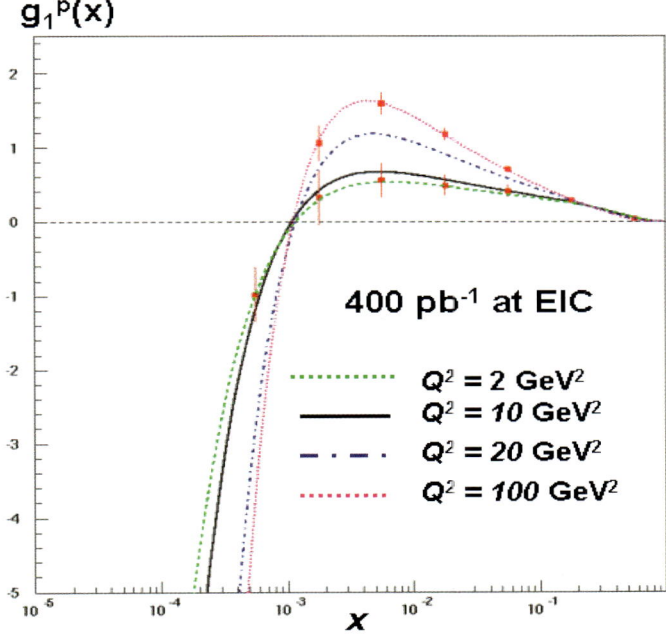

Figure 16 Possible eRHIC data (statistical accuracy) with 250×10 GeV collisions are shown for 400 pb^{-1}. Also shown is the evolution of $g_1(x, Q^2)$ at low x for different values of Q^2 for a positive gluon polarization (31, 146).

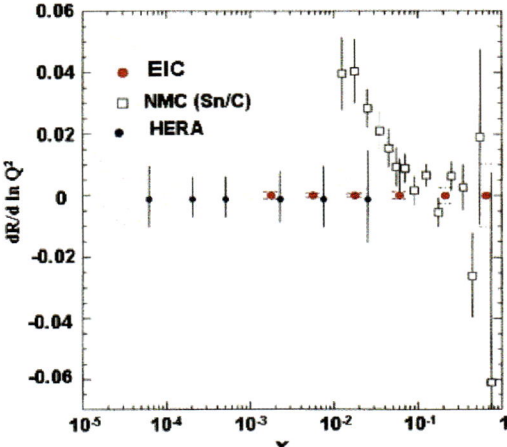

Figure 23 The projected statistical accuracy of $\frac{\partial F_2^A/F_2^N}{\partial \ln Q^2}$ as a function of x for an integrated luminosity of 1 pb^{-1} at the EIC (186). The simulated data are compared to previous data from the NMC and to data from a hypothetical e-A collider at HERA energies.

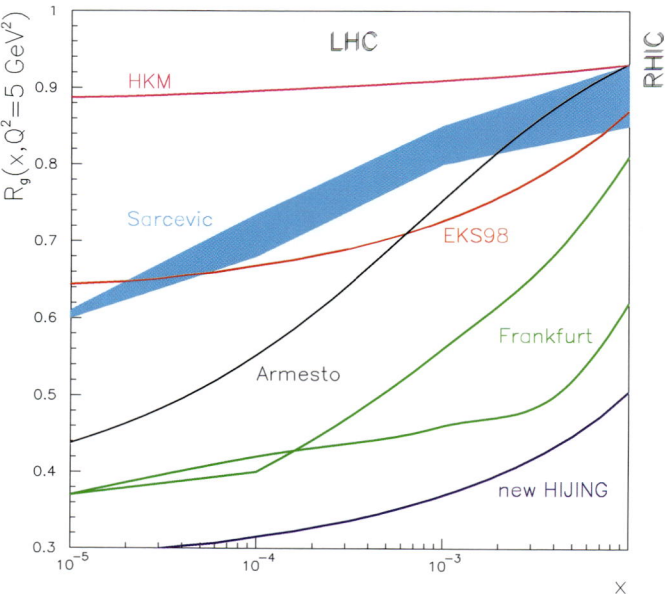

Figure 24   Ratio of the gluon distribution in Lead to that in a proton, normalized by the number of nucleons, plotted as a function of x for a fixed $Q^2 = 5$ GeV^2. From Ref. (187). Captions denote models-HKM (188), EKS98 (189), Sarcevic (190), Armesto (191), Frankfurt (93), Hijing (192). The vertical bands denote the accessible x regions at central rapidities at RHIC and LHC.

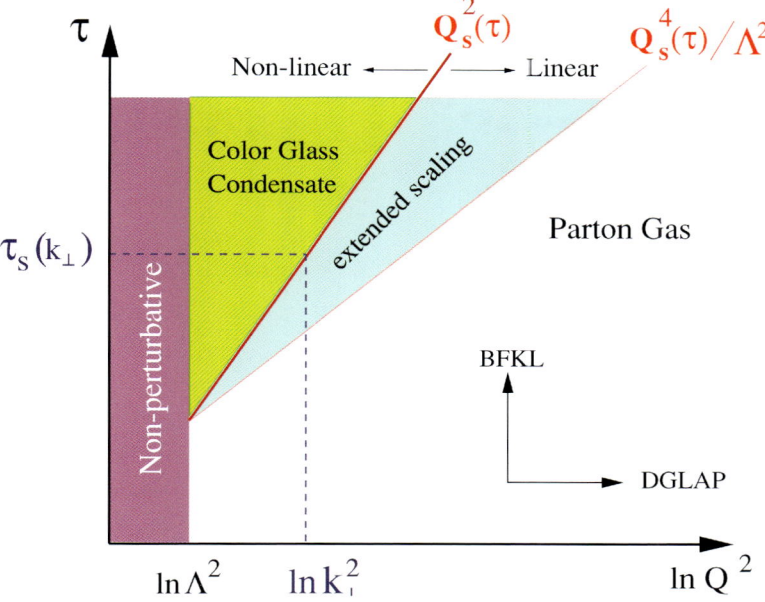

Figure 26 A schematic plot of the Color Glass Condensate and extended scaling regimes in the x-Q^2 plane. Here $\tau = \ln(1/x)$ denotes the rapidity. From Ref (185).

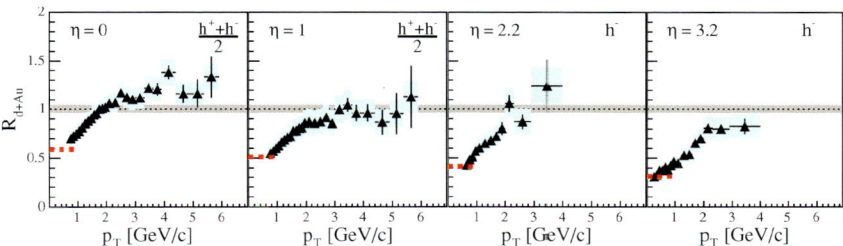

Figure 27 Depletion of the Cronin peak from $\eta = 0$ to $\eta = 3$ for minimum bias events. From Ref. (105).

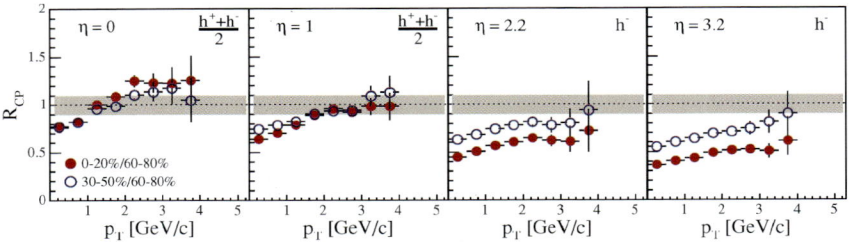

Figure 28   Centrality dependence of the Cronin ratio as a function of rapidity. From Ref. (105).

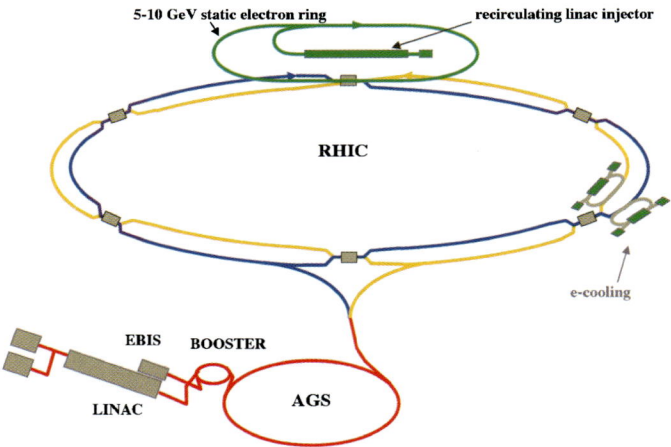

Figure 29   Schematic layout of the ring-ring eRHIC collider.

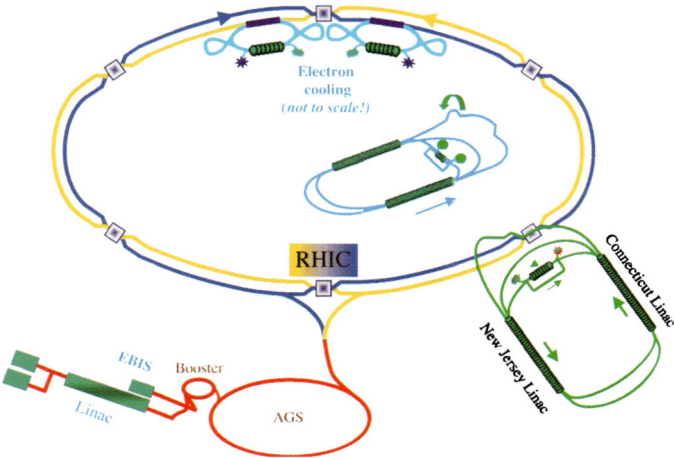

Figure 30   Schematic layout of a possible linac-ring eRHIC design.

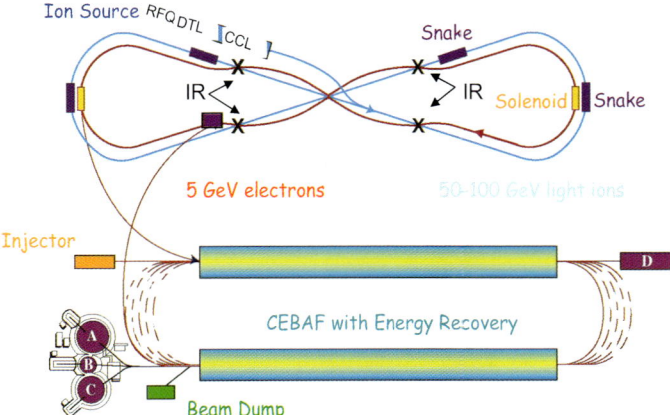

Figure 31   Schematic layout of a possible electron light ion collider at Jefferson Laboratory.

distinguish from shadowing effects in the wavefunction. Nevertheless, in the right kinematics this can be done.

In section 2, we discussed various final state in-medium QCD processes such as color transparency, parton energy loss and the medium modification of fragmentation functions. The EIC will enable qualitative progress in studies of the space-time picture of strong interactions relative to previous fixed target DIS experiments. The reasons for this are as follows.

- The high luminosity of the EIC will increase by many orders of magnitude the current data sample of final states in DIS scattering off nuclei at high energies.

- The EIC will provide a much broader range of Q^2 and x, making it possible to compare dynamics for approximately the same space-time coherence lengths as a function of Q^2. Fixing the coherence length of partons will allow one to distinguish events wherein a photon is transformed into a strongly interacting system either outside or inside the nucleus. This will help isolate initial state interactions from those in the final state.

- The collider geometry will enable measurements of final states currently impossible in fixed target kinematics. In particular, a hermetic detector would clearly isolate coherent processes as well as quasi-elastic processes in DIS off nuclei. In addition, one can study the sizes and distributions of rapidity gaps as a function of nuclear size and energy. These will provide a sensitive probe of the interplay between space-time correlations in the final state and in the nuclear wavefunction.

- The detection of nucleons produced in the nuclear fragmentation region would make it feasible to study DIS as a function of the number of the nucleons involved in the interaction. In particular, it may be possible to study impact parameter dependence of final states, which will be important to understand in detail the nuclear amplification of final state effects. In addition, the impact parameter dependence will help distinguish geometrical effects from dynamical effects in event-by-event studies of final states.

In section 2, we discussed Generalized Parton Distributions (GPDs) in the context of DIS scattering off nucleons. These GPDs can also be measured in DIS scattering off nuclei (42). The simplest system in which to study GPDs is the deuteron. The transition from $D \rightarrow p + n$ in the kinematics where the neutron absorbs the momentum transfer in the scattering is sensitive to the GPD in the neutron with the proton playing the role of a spectator (199). Some preliminary studies have been done for heavier nuclei (51). Certain higher twist correlations in nuclei which scale as $A^{4/3}$ are sensitive to nucleon GPDs (134). This leads us to a discussion of GPDs in nuclei at small x. As we will discuss, high parton density effects are enhanced in large nuclei. $k_\perp$ dependent GPDs might provide the right approach to study this novel regime (200). The study of nuclear GPDs at moderate and small x is a very promising, albeit nascent, direction for further research to uncover the detailed structure of hard space-time processes in nuclear media. These nuclear distributions can be studied for the first time with the EIC.

3.3.3. THE COLOR GLASS CONDENSATE   The Color Glass Condensate (CGC) is an effective field theory describing the properties of the dominant parton configurations in hadrons and nuclei at high energies (185). The degrees of freedom are partons, which carry color charge, hence the "Color" in CGC. The matter behaves like a glass for the following reason. The kinematics of high energy scattering dictates a natural separation between large x and small x modes (201). The large x partons at high energies behave like frozen random light cone sources over time scales that are large compared to the dynamical time scales associated with the small-x partons. One can therefore describe an effective theory where the small x partons are dynamical fields and the large-x partons are frozen sources (193). Under quantum evolution (202), this induces a stochastic coupling between the wee partons via their interaction with the sources. This stochastic behavior is very similar to that of a spin glass. Finally, the Condensate in CGC arises because each of these colored configurations is very similar to a Bose-Einstein Condensate. The occupation number of the gluons can be computed to be of order $1/\alpha_S$, and the typical momentum of the partons in the configuration is peaked about a typical momentum—the saturation momentum Q_s. These properties are further enhanced by quantum evolution in x. Because the occupation number is so large, by the correspondence principle of quantum mechanics, the small x modes can be treated as classical fields. The classical field retains its structure while the saturation scale, generated dynamically in the theory, grows with energy: $Q_s(x') > Q_s(x)$ for $x' < x$. The CGC is sometimes used interchangeably with "saturation" (203)— both refer to the same phenomenon, the behavior of partons at large occupation numbers.

The Jalilian-Marian-Iancu-McLerran-Weigert-Leonidov-Kovner (JIMWLK) renormalization group equations describe the properties of partons in the high density regime (202). They form an infinite hierarchy (analogous to the Bogoliubov-Born-Green-Kirkwood-Young (BBGKY) hierarchy in statistical mechanics) of ordinary differential equations for the gluon correlators $\langle A_1 A_2 \cdots A_n \rangle_Y$, where $Y = \ln(1/x)$ is the rapidity. Thus the evolution, with x, of multi-gluon (semi-inclusive) final states provides precise tests of these equations. The full hierarchy of equations are difficult to solve[2] though there have been major theoretical developments in that direction recently (205).

A mean field version of the JIMWLK equation, called the Balitsky-Kovchegov (BK) equation (206), describes the inclusive scattering of the quark-anti-quark dipole off the hadron in deeply inelastic scattering. In particular, the virtual photon-proton cross-section at small x can be written as (207, 208)

---

[2]For a preliminary numerical attempt, see Ref. (204).

$$\sigma_{T,L}^{\gamma^* p} = \int d^2 r_\perp \int dz |\psi_{T,L}(r_\perp, z, Q^2)|^2 \sigma_{q\bar{q}N}(r_\perp, x),$$ 34.

where $|\psi_{T,L}|^2$ is the probability for a longitudinally (L) or transversely (T) polarized virtual photon to split into a quark with momentum fraction z and an anti-quark with momentum fraction $1 - z$ of the longitudinal momentum of the virtual photon. For the quark and anti-quark located at $\vec{x}_\perp$ and $\vec{y}_\perp$ respectively from the target, their transverse size is $\vec{r}_\perp = \vec{x}_\perp - \vec{y}_\perp$, and the impact parameter of the collision is $\vec{b} = (\vec{x}_\perp + \vec{y}_\perp)/2$. The probability for this splitting is known exactly from QED and it is convoluted with the cross-section for the $q\bar{q}$-pair to scatter off the proton. This cross-section for a dipole scattering off a target can be expressed as

$$\sigma_{q\bar{q}N}(x, r_\perp) = 2 \int d^2 b \, \mathcal{N}_Y(x, r_\perp, b),$$ 35.

where $\mathcal{N}_Y$ is the imaginary part of the forward scattering amplitude. The BK equation (206) for this amplitude has the operator form

$$\frac{\partial \mathcal{N}_Y}{\partial Y} = \bar{\alpha}_S \, \mathcal{K}_{\text{BFKL}} \otimes \{\mathcal{N}_Y - \mathcal{N}_Y^2\}.$$ 36.

Here $\mathcal{K}_{\text{BFKL}}$ is the well known Balitsky-Fadin-Kuraev-Lipatov (BFKL) kernel (184). When $\mathcal{N} \ll 1$, the quadratic term is negligible and one has BFKL growth of the number of dipoles; when $\mathcal{N}$ is close to unity, the growth saturates. The approach to unity can be computed analytically (209). The BK equation is the simplest equation including both the Bremsstrahlung responsible for the rapid growth of amplitudes at small x as well as the repulsive many body effects that lead to a saturation of this growth.

Saturation models, which incorporate key features of the CGC, explain several features of the HERA data. In section 2.1, we discussed the property of geometrical scaling observed at HERA which is satisfied by the LHS of Equation 34, where it scaled as a function of the ratio of Q^2 to the saturation scale Q_s^2. We also mentioned briefly a simple saturation model, the Golec-Biernat model (13), which captured essential features of this phenomenon in both inclusive and diffractive cross sections at HERA. Geometric scaling arises naturally in the Color Glass Condensate (210, 211), and it has been studied extensively both analytically (212) and numerically (213–215) for the BK equation. The success of saturation models, as discussed in section 2.1, in explaining less inclusive features of the HERA data is also encouraging since their essential features can be understood to follow from the BK equation. Below we will discuss the implications of mean field studies with the BK equation, as well as effects beyond BK.

As mentioned previously, a very important feature of saturation is the dynamical generation of a dimensionful scale $Q_s^2 \gg \Lambda_{\text{QCD}}^2$, which controls the running of the coupling at high energies: $\alpha_S(Q_s^2) \ll 1$. From the BK equation, or more generally, from solutions of BFKL in the presence of an absorptive boundary (corresponding to a CGC-like regime of high parton densities), one can deduce that, for fixed coupling, Q_s^2 has the asymptotic form $Q_s^2 = Q_0^2 \exp(cY)$, where $c = 4.8\alpha_s$ and $Y = \ln(x_0/x)$. Here, Q_0^2 and x_0 are parameters from the initial conditions. Pre-asymptotic Y dependent corrections can also be computed and are large. The behavior of Q_s^2 changes qualitatively when running coupling effects are taken into account. The state of the art is a computation of the saturation scale to next-to-leading order in BFKL with additional resummation of collinear terms that stabilize the predictions of NLO BFKL (216). One recovers the form $Q_s^2 = Q_0^2 \exp(\lambda Y)$, now with small pre-asymptotic corrections, with $\lambda \approx 0.25$. Remarkably, this value is very close to the value extracted in the Golec-Biernat model from fits to the HERA data.

Figure 26 (see color insert) shows a schematic plot of the CGC and extended scaling regions in the x-Q^2 plane. Clearly, with the wide kinematic range of the EIC, and the large number of available measurements—to be discussed later— one has the opportunity to make this plot quantitative. One can further add an additional axis for the atomic number to see how the kinematic reach of the CGC scales with A. In principle, one can also study the impact parameter dependence of the saturation scale in addition to the A-dependence.

3.3.4. SIGNATURES OF THE CGC

Inclusive signatures. Inclusive measurements include F_2 and F_L for a wide range of nuclei, the latter measurements being done independently for the first time. The data will be precise enough to extract derivatives of these with respect to $\ln Q^2$ and $\ln x$ in a wide kinematic range in x and Q^2. Logarithmic derivatives of F_2 and F_L will enable the extraction of the coefficient λ of the saturation scale, which as discussed previously, is defined to be $Q_s^2 = Q_0^2 e^{\lambda Y}$, where $Y = \ln(x_0/x)$, and where x_0 and Q_0^2 are reference values corresponding to the initial conditions for small-x evolution. Simulations suggest that a precise extraction of this quantity may be feasible (136). Except at asymptotic energies, $\lambda \equiv \lambda(Y)$. Predictions exist for "universal" pre-asymptotic Y-dependent corrections to λ (212). Second derivatives of F_2 and F_L with respect to $\ln(x_0/x)$ will be sensitive to these corrections. The logarithmic derivatives of F_2 and F_L with Q^2, especially the latter, will be sensitive to higher twist effects for $Q^2 \approx Q_s^2(x, A)$. The saturation scale is larger for smaller x and larger A—thus deviations of predictions of CGC fits from DGLAP fits should systematically increase as a function of both. CGC fits have been shown to fit HERA data at small x (217, 218). These fits can be extended to nuclei and compared to scaling violation data relative to DGLAP fits. The A dependence of the saturation scale can also be extracted from nuclear structure functions at small x. Again, predictions exist for the pre-asymptotic scaling of the saturation scale with rapidity (or x), for different A (219), that can be tested against the data.

In the BK equation (mean field approximation of the CGC renormalization group equations), we now have a simple way to make predictions for the effects of high parton densities on both inclusive and diffractive (220) structure functions. There are now a few preliminary computations for e-A DIS in this framework (221,

222). Much more remains to be done—in particular, comparisons with DGLAP for EIC kinematics and detector cuts.

Semi-inclusive and exclusive signatures. The collider geometry of the EIC will greatly enhance the semi-inclusive final states in e-A relative to previous fixed target experiments. Inclusive hadron production at $p_\perp \sim Q_s$ should be sensitive to higher twist effects for $Q^2 \approx Q_s^2(x, A)$. For the largest nuclei, these effects should be clearly distinguishable from DGLAP based models. Important semi-inclusive observables are coherent (or diffractive) and inclusive vector meson production, which are sensitive measures of the nuclear gluon density (110, 115). Exclusive vector meson production was suggested by Mueller, Munier and Stasto (17) as a way to extract the S-matrix (and therefore the saturation scale in the Golec-Biernat–Wüsthoff parameterization) from the t-dependence of exclusive ρ-meson production. A similar analysis of J/ψ production was performed by Guzey et al. (223). These studies for e-A collisions will provide an independent measure of the energy dependence of the saturation scale in nuclei. An extensive recent theoretical review of vector meson production of HERA (relevant for EIC studies as well) can be found in Ref. (224).

In hard diffraction, for instance, one should be able to distinguish predictions based on the strong field effects of BK (or hard Pomeron based approaches in general) from the soft Pomeron physics associated with confinement. As we discussed previously, some saturation models predict that hard diffractive events will constitute 30–40% of the cross-section (225, 226). These computations can be compared with DGLAP predictions which match soft Pomeron physics with hard perturbative physics. One anticipates that the latter would result in a much smaller fraction of the cross-section and should therefore be easily distinguishable from CGC based "strong field" diffraction.

The BK renormalization group equation is not sensitive to multi-particle correlations. These are sensitive to effects such as Pomeron loops (205), although phenomenological consequences of these remain to be explored. These effects are reflected in multiplicity fluctuations and rapidity correlations over several units in rapidity (91, 227). One anticipates quantitative studies of these will be developed in the near future. A wide detector coverage able to resolve the detailed structure of events will be optimal for extracting signatures of the novel physics of high parton densities.

3.3.5. EXPLORING THE CGC IN PROTON/DEUTERON-NUCLEUS COLLISIONS   Although high parton density hot spots may be studied in pp collisions, they are notoriously hard to observe. The proton is a dilute object, except at small impact parameters, and one needs to tag on final states over a wide 4π coverage. Deuteron-nucleus experiments are more promising in this regard. They have been performed at RHIC and may be performed at LHC in the future. The Cronin effect discovered in the late 70's (121) predicts a hardening of the transverse momentum spectrum in proton-nucleus collisions, relative to proton-proton collisions at transverse momenta of order $p_\perp \sim 1 - 2$ GeV. It disappears at much larger $p_\perp$. A corresponding depletion is seen at low transverse momenta. The effect was interpreted as arising from the multiple scatterings of partons from the proton off partons from the nucleus (122).

First data from RHIC on forward D-Au scattering at $\sqrt{s} = 200$ GeV/nucleon demonstrate how the Cronin effect is modified with energy or, equivalently, with the rapidity. The x values in nuclei probed in these experiments, at $p_\perp \sim 2$ GeV, range from 10^{-2} in the central rapidity region down to 10^{-4} at very forward rapidities.[3] At central rapidities, one clearly sees a Cronin peak at $p_\perp \sim 1 - 2$ GeV. A dramatic result obtained by the BRAHMS (105) experiment at RHIC[4] is the rapid shrinking of the Cronin peak with rapidity shown in Figure 27 (see color insert). In Figure 28 (see color insert), the centrality dependence of the effect is shown. At central rapidities, the Cronin peak is enhanced in more central collisions. For forward rapidities, the trend is reversed: more central collisions at forward rapidities show a greater suppression than less central collisions!

Parton distributions in the classical theory of the CGC exhibit the Cronin effect (229–231). However, unlike this classical Glauber picture (232), quantum evolution in the CGC shows that it breaks down completely when the x_2 in the target is such that $\ln(1/x_2) \sim 1/\alpha_S$. This is precisely the trend observed in the RHIC D-Au experiments (105). The rapid depletion of the Cronin effect in the CGC picture is due to the onset of BFKL evolution, whereas the subsequent saturation of this trend reflects the onset of saturation effects (233). The inversion of the centrality dependence can be explained as arising from the onset of BFKL anomalous dimensions, that is, the nuclear Bremsstrahlung spectrum changes from $Q_s^2/p_\perp^2 \rightarrow Q_s/p_\perp$. Finally, an additional piece of evidence in support of the CGC picture is the broadening of azimuthal correlations (234) for which preliminary data now exists from the STAR collaboration (235). We note that alternative explanations have been given to explain the BRAHMS data (236). These ideas can be tested conclusively in photon and di-lepton production in D-A collisions at RHIC (237) as well as by more detailed correlation studies.

Hadronic collisions in pQCD are often interpreted within the framework of collinear factorization. At high energies, $k_\perp$ factorization may be applicable (238) where the relevant quantities are "unintegrated" $k_\perp$ dependent parton densities. Strict $k_\perp$-factorization which holds for gluon production in p-A collisions (197, 239) is broken for quark production (240, 241), for azimuthal correlations (242) and diffractive final states (243). For a review, see Ref. (244). These cross-sections can still be written in terms of $k_\perp$-dependent multi-parton correlation functions (240) and will also appear in DIS final states (220). DIS will allow us to test the universality of these correlations, that is, whether such correlations extracted from p-A collisions can be used to compute e-A final states (230, 245).

---

[3]It has been argued (228), however, that the forward D-A cross section in the BRAHMS kinematic regime receives sizable contributions also from rather large x values.

[4]The trends seen by BRAHMS are also well corroborated by the PHOBOS, PHENIX and STAR experiments at RHIC in different kinematic ranges (102–104).

3.3.6. THE COLOR GLASS CONDENSATE AND THE QUARK GLUON PLASMA   The CGC provides the initial conditions for nuclear collisions at high energies. The number and energy of gluons released in a heavy ion collision of identical nuclei can be simply expressed in terms of the saturation scale as (246–248)

$$\frac{1}{\pi R^2} \frac{dE}{d\eta} = \frac{c_E}{g^2} Q_s^3, \quad \frac{1}{\pi R^2} \frac{dN}{d\eta} = \frac{c_N}{g^2} Q_s^2, \qquad 37.$$

where $c_E \approx 0.25$ and $c_N \approx 0.3$. Here η is the space-time rapidity. These simple predictions led to correct predictions for the hadron multiplicity at central rapidities in Au-Au collisions at RHIC (246, 249) and for the centrality and rapidity dependence of hadron distributions (250). However, the failure of more detailed comparisons to the RHIC jet quenching data (251) and elliptic flow data (252) suggested that final state effects are important and significantly modify predictions based on the CGC alone. The success of hydrodynamic predictions suggests that matter may have thermalized to form a quark gluon plasma (253). Indeed, bulk features of multiplicity distributions may be described by the CGC precisely as a consequence of early thermalization—leading to entropy conservation (254). Initial-state effects will be more important in heavy ion collisions at the LHC because one is probing smaller x in the wave function. Measurements of saturation scales for nuclei at the EIC will independently corroborate equations such as Equation 37 and therefore the picture of heavy ion collisions outlined above. Further, a systematic study of energy loss in cold matter will help constrain extrapolations of pQCD (131) used to study jet quenching in hot matter.

3.3.7. PROTON/DEUTERON-NUCLEUS VERSUS ELECTRON-NUCLEUS COLLISIONS AS PROBES OF HIGH PARTON DENSITIES   Both p/D-A and e-A collisions probe the small x region at high energies. Both are important to ascertain truly universal aspects of novel physics. e-A collisions, owing to the independent "lever" arm in x and Q^2, as well as the simpler lepton-quark vertex, are better equipped for precision measurements. For example, in e-A collisions, information about gluon distributions can be extracted from scaling violations and from photon-gluon fusion processes. In both cases, high precision measurements are feasible. In p-A collisions, one can extract gluon distributions from scaling violations in Drell-Yan and gluon-gluon and quark-gluon fusion channels such as open charm and direct photon measurements respectively. However, for both scaling violations and fusion processes, one has more convolutions and kinematic constraints in p-A than in e-A. These limit both the precision and range of measurements. In Drell-Yan, in contrast to F_2, clear scaling violations in the data are very hard to see and data are limited to $M^2 > 16 \text{ GeV}^2$, above the J/ψ and ψ' thresholds.

A clear difference between p/D-A and e-A collisions is in hard diffractive final states. At HERA, these constituted approximately 10% of the total cross section. At eRHIC, these may constitute 30–40% of the cross section (225, 226). Also, factorization theorems derived for diffractive parton distributions only apply to lepton-hadron processes (182). Spectator interactions in p/D-A collisions will destroy rapidity gaps. A comparative study of p/D-A and e-A collisions thus has great potential for unravelling universal aspects of event structures in high energy QCD.

4. ELECTRON-ION COLLIDER-ACCELERATOR ISSUES

With the scientific interest in a high luminosity lepton-ion collider gathering momentum during the last several years, there has been a substantial effort in parallel to develop a preliminary technical design for such a machine. A team of physicists from BNL, MIT-Bates, DESY and the Budker Institute have developed a realistic design (255) for a machine using RHIC, which would attain an e-p collision luminosity of 0.4×10^{33} cm^{-2} s^{-1} and could with minimal R&D start construction as soon as funding becomes available. Other more ambitious lepton-ion collider concepts which would use a high intensity electron linac to attain higher luminosity are under active consideration (255, 256). This section gives an overview of the activities currently underway related to the accelerator design.

The physics program described above sets clear requirements and goals for the lepton-ion collider to be a successful and efficient tool. These goals include: a sufficiently high luminosity; a significant range of beam collision energies; and polarized beam (both lepton and nucleon) capability. On the other hand, to be realistic, the goals should be based on the present understanding of the existing RHIC machine and limitations which arise from the machine itself. Realistic machine upgrades should be considered to overcome existing limitations and to achieve advanced machine parameters, but those upgrades should be cost-effective.

The intent to minimize required upgrades in the existing RHIC rings affects the choice of parameters and the set of goals. For example, the design assumed simultaneous collisions of both ion-ion and lepton-ion beams. In the main design line, collisions in two ion-ion interaction regions, at the "6" and "8 o'clock" locations, have to be allowed in parallel with electron-ion collisions.

Taking these considerations into account, the following goals were defined for the accelerator design:

- The machine should be able to provide beams in the following energy ranges: for the electron accelerator, 5–10 GeV polarized electrons, 10 GeV polarized positrons; for the ion accelerator, 50–250 GeV polarized protons, 100 GeV/u Gold ions.
- Luminosity: in the $10^{32} - 10^{33}$ cm^{-2} s^{-1} range for e-p collisions; in the $10^{30} - 10^{31}$ cm^{-2} s^{-1} range for e-Au collisions.
- 70% polarization for both lepton and proton beams.
- Longitudinal polarization in the collision point for both lepton and proton beams.

An additional design goal was to include the possibility of accelerating polarized ions, especially polarized ^{3}He ions.

4.1. eRHIC: Ring-Ring Design

The primary eRHIC design centers on a 10 GeV lepton storage ring which intersects with one of the RHIC ion beams at one of the interaction regions (IRs), not used by any of the ion-ion collision experiments. RHIC uses superconducting dipole and quadrupole magnets to maintain ion beams circulating in two rings on a 3834 meter circumference. The ion energy range covers 10.8 to 100 GeV/u for gold ions and 25 to 250 GeV for protons. There are in total 6 intersection points where two ion rings, Blue and Yellow, cross each other. Four of these intersections points are currently in use by physics experiments.

A general layout of the ring-ring eRHIC collider is shown in Figure 29 (see color insert) with the lepton-ion collisions occurring in the "12 o'clock" interaction region. Plans have been made for a new detector, developed and optimized for electron-ion collision studies, to be constructed in that interaction region.

The electron beam in this design is produced by a polarized electron source and accelerated in a linac injector to energies of 5 to 10 GeV. To reduce the injector size and cost, the injector design includes recirculation arcs, so that the electron beam passes through the same accelerating linac sections multiple times. Two possible linac designs, superconducting and normal conducting, have been considered. The beam is accelerated by the linac to the required collision energy and injected into the storage ring. The electron storage ring is designed to be capable of electron beam storage in the energy range of 5 to 10 GeV with appropriate beam emittance values. It does not provide any additional acceleration for the beam. The electron ring should minimize depolarization effects in order to keep the electron beam polarization lifetime longer than the typical storage time of several hours.

The injector system also includes the conversion system for positron production. After production the positrons are accelerated to 10 GeV energy and injected into the storage ring similarly to the electrons. Obviously the field polarities of all ring magnets should be reversed in the positron operation mode. Unlike electrons, the positrons are produced unpolarized and have to be polarized using radiative self-polarization in the ring. Therefore, the design of the ring should allow for a sufficiently small self-polarization time. The current ring design provides a self-polarization time of about 20 min at 10 GeV. But with polarization time increasing sharply as beam energy goes down the use of a polarized positron beam in the present design is limited to 10 GeV energy.

The design of the eRHIC interaction region involves both accelerator and de-tector considerations. Figure 29 shows the electron accelerator located at the "12 o'clock" region. Another possible location for the electron accelerator and for electron-ion collisions might be the "4 o'clock" region. For collisions with elec-trons the ion beam in the RHIC Blue ring will be used, because the Blue ring can operate alone, even with the other ion ring, Yellow, being down. The inter-action region design provides for fast beam separation for electron and Blue ring ion bunches as well as for strong focusing at the collision point. In this design, the other (Yellow) ion ring makes a 3 m vertical excursion around the collision region, avoiding collisions both with electrons and the Blue ion beam. The eRHIC interaction region includes spin rotators, in both the electron and the Blue ion rings, to produce longitudinally polarized beams of leptons and protons at the collision point.

The electron cooling system in RHIC (257, 258) is one of the essential upgrades required for eRHIC. The cooling is necessary to reach the luminosity goals for lepton collisions with Gold ions and low (below 150 GeV) energy protons. Electron cooling is considered an essential upgrade of RHIC to attain higher luminosity in ion-ion collisions.

In addition, the present eRHIC design assumes a total ion beam current higher than that being used at present in RHIC operation. This is attained by operating RHIC with 360 bunches.

The eRHIC collision luminosity is limited mainly by the maximum achievable beam-beam parameters and by the interaction region magnet aperture limitations. To understand this, it is most convenient to use a luminosity expression in terms of beam-beam parameters ($\xi_e \xi_i$) and rms angular spread in the interaction point ($\sigma'_{xi}, \sigma'_{ye}$):

$$L = f_c \frac{\pi \gamma_i \gamma_e}{r_i r_e} \xi_{xi} \xi_{ye} \sigma'_{xi} \sigma'_{ye} \frac{(1+K)^2}{K}.$$

The $f_c = 28.15$ MHz is a collision frequency, assuming 360 bunches in the ion ring and 120 bunches in the electron ring. The parameter $K = \sigma_y / \sigma_x$ presents the ratio of beam sizes in the interaction point. One of the basic conditions which defines the choice of beam parameters is a requirement on equal beam sizes of ion and electron beams at the interaction point: $\sigma_{xe} = \sigma_{xi}$ and $\sigma_{ye} = \sigma_{ye}$. The requirement is based on the operational experience at the HERA collider and on the reasonable intention to minimize the amount of one beam passing through the strongly nonlinear field in the outside area of the counter-rotating beam.

According to the above expression, the luminosity reaches a limiting value at the maximum values of beam-beam parameters, or at the beam-beam parameter limits. For protons (and ions) the total beam-beam parameter limit was assumed to be 0.02, following the experience and observation from other proton machines as well as initial experience from RHIC operation. With three beam-beam interaction points, two for proton-proton and one for electron-proton collisions, the beam-beam parameter per interaction point should not exceed 0.007.

For the electron (or positron) beam a limiting value of the beam-beam parameter has been put at 0.08 for 10 GeV beam energy, following the results of beam-beam simulations, as well as from the experience at electron machines of similar energy range. Because the beam-beam limit decreases proportionally with the beam energy, the limiting value for 5 GeV is reduced to 0.04.

The available magnet apertures in the interaction region also put a limit on the achievable luminosity. The work on the interaction region design revealed considerable difficulties to provide an acceptable design for collisions of round beams. The IR has been designed to provide low beta focusing and efficient separation of elliptical beams, with beam size ratio $K = 1/2$. The main aperture

TABLE 1 Luminosities and main beam parameters for $e^{\pm}$-p collisions

High energy tune	p	e	p	e
Energy, GeV	250	10	50	5
Bunch intensity, 10^{11}	1	1	1	1
Ion normalized emittance, π mm · mrad, x/y	15/15		5/5	
rms emittance, nm, x/y	9.5/9.5	53/9.5	16.1/16.1	85/38
β^*, cm, x/y	108/27	19/27	186/46	35/20
Beam-beam parameters, x/y	0.0065/0.003	0.03/0.08	0.019/0.0095	0.036/0.04
$\kappa = \xi_y/\xi_x$	1	0.18	1	0.45
Luminosity, 1.0×10^{32} cm^{-2} s^{-1}		4.4		1.5

limitation comes from the septum magnet, which leads to the limiting values of $\sigma'_{xp} = 93$ μrad.

Another limitation which must be taken into account is a minimum acceptable value of the beta-function at the interaction point (β^*). With the proton rms bunch length of 20 cm, decreasing β^* well below this number results in a luminosity degradation due to the hour-glass effect. The limiting value $\beta^* = 19$ cm has been used for the design, which results in a luminosity reduction of only about 12%. A bunch length of 20 cm for Au ions would be achieved with electron cooling.

Tables 1 and 2 show design luminosities and beam parameters. The positron beam intensity is assumed to be identical to the electron beam intensity, hence the luminosities for collisions involving a positron beam are equivalent to electron-ion collision luminosities. To achieve the high luminosity in the low energy tune in Table 1, the electron cooling has to be used to reduce the normalized transverse emittance of the lower energy proton beam to 5 π mm · mrad. Also, in that case the

TABLE 2 Luminosities and main beam parameters for $e^{\pm}$-Au collisions

High energy tune	Au	e	Au	e
Energy, GeV	100	10	100	5
Bunch intensity, 10^{11}	0.01	1	0.0045	1
Ion normalized emittance, π mm · mrad, x/y	6/6		6/6	
rms emittance, nm, x/y	9.5/9.5	54/7.5	9.5/9.5	54/13.5
β^*, cm, x/y	108/27	19/34	108/27	19/19
Beam-beam parameters, x/y	0.0065/0.003	0.0224/0.08	0.0065/0.003	0.02/0.04
$\kappa = \xi_y/\xi_x$	1	0.14	1	0.25
Luminosity, 1.0×10^{32} cm^{-2} s^{-1}		4.4		2.0

proton beam should have collisions only with the electron beam. Proton-proton collisions in the other two interaction points have to be avoided to allow for a higher proton beam-beam parameter. The maximum luminosity achieved in the present design is 4.4×10^{32} cm^{-2} s^{-1} in the high energy collision mode (10 GeV leptons on 250 GeV protons). Possible paths to luminosities as high as 10^{33} cm^{-2} s^{-1} are being explored, with studies planned to investigate the feasibility of higher electron beam intensity operation. To achieve and maintain the Au normalized transverse beam emittances shown in Table 2, electron cooling of the Au beam will be used. For the lower energy tune of electron-Gold collisions, the intensity of the Gold beam is considerably reduced because of the reduced value of the beam-beam parameter limit for the electron beam.

4.2. eRHIC: Linac-Ring Design

A linac-ring design for eRHIC is also under active consideration. This configuration uses a fresh electron beam bunch for each collision and so the tune shift limit on the electron beam is removed. This provides the important possibility to attain significantly higher luminosity (up to 10^{34} cm^{-2} s^{-1}) than the ring-ring design. A second advantage of the linac-ring design is the ability to reverse the electron spin polarization on each bunch. A disadvantage of the linac-ring design is the inability to deliver polarized positrons. The realization of the linac beam is technically challenging and the polarized electron source requirements are well beyond present capabilities (259).

Figure 30 (see color insert) shows a schematic layout of a possible linac-ring eRHIC design. A 450 mA polarized electron beam is accelerated in an Energy Recovery Linac (ERL). After colliding with the RHIC beam in as many as four interaction points, the electron beam is decelerated to an energy of a few MeV and dumped. The energy thus recovered is used for accelerating subsequent bunches to the energy of the experiment.

4.3. Other Lepton-Ion Collider Designs: ELIC

A very ambitious electron-ion collider design seeking to attain luminosities up to 10^{35} cm^{-2} s^{-1} is underway at Jefferson Laboratory (156). This Electron Light Ion Collider (ELIC) design (see Figure 31, color insert) is based on use of polarized 5 to 7 GeV electrons in a superconducting ERL upgrade of the present CEBAF accelerator and a 30 to 150 GeV ion storage ring (polarized p, d, ^{3}He, Li and unpolarized nuclei up to Ar, all totally stripped). The ultra-high luminosity is envisioned to be achievable with short ion bunches and crab-crossing at 1.5 GHz bunch collision rate in up to four interaction regions. The ELIC design also includes a recirculating electron ring that would help to reduce the linac and polarized source requirements compared to the linac-ring eRHIC design of section 4.2.

The ELIC proposal is at an early stage of development. A number of technical challenges must be resolved, and several R&D projects have been started. These include development of a high average current polarized electron source with a high bunch charge, electron cooling of protons/ions, energy recovery at high current and high energy, and the design of an interaction region that supports the combination of high luminosity and high detector acceptance and resolution.

5. DETECTOR IDEAS FOR THE EIC

The experience gained at HERA with the H1 and ZEUS detectors (260) provides useful guidance for the conceptual design of a detector that will measure a complete event (4π-coverage) produced in collisions of energetic electrons with protons and ions, at different beam energies and polarizations. The H1 and ZEUS detectors are general-purpose magnetic detectors with nearly hermetic calorimetric coverage. The differences between them are based on their approach to calorimetry. The H1 detector collaboration emphasized the electron identification and energy resolution, whereas the ZEUS collaboration puts more emphasis on optimizing hadronic calorimetry. The differences in their physics philosophy were reflected in their overall design: H1 had a liquid Argon calorimeter inside the large diameter magnet, whereas ZEUS chose to build a Uranium scintillator sampling calorimeter with equal response to electrons and hadrons. They put their tracking detectors inside a superconducting solenoid surrounded by calorimeters and muon chambers. H1 placed their tracking chambers inside the calorimeter surrounded by their magnet. In addition, both collaborations placed their luminosity and electron detectors downstream in the direction of the proton and electron directions, respectively. Both collaborations added low angle forward proton spectrometers and neutron detectors in the proton beam direction.

Both detectors have good angular coverage (approximately, $3° < \theta < 175°$, where the angle is measured with respect to the incoming proton beam direction) for electromagnetic (EM) calorimetry, with energy resolutions of 1–3% and for electromagnetic showers $\sigma/E \sim 15\%/\sqrt{E(\text{GeV})} - 1\%$. Hadronic energy scale uncertainties of 3% were achieved for both, with some differences in the σ/E for hadronic showers, which were $\sim R/\sqrt{E(\text{GeV})} + 2\%$, where $R = 35\%$ and 50% for ZEUS and H1, respectively. With the central tracking fields ≈ 1.5 T covering a region similar to the calorimetric angular acceptance, momentum resolution $\sigma/p_T < 0.01\ p_T(\text{GeV})$ was generally achieved for almost all acceptances, except for the forward and backward directions. These directions were regarded at the beginning as being less interesting. However, the unexpected physics of low x and low Q^2 (including diffraction in e-p scattering) came from this rather poorly instrumented region. And since the luminosity upgrade program, the low β^* magnets installed close to the interaction point to enhance the luminosity of e-p collisions (HERA-II) have further deteriorated the acceptance of detectors in these specific geometric regions.

The EIC detector design ideas are already being guided by the lessons learned from the triumphs and tribulations of the HERA experience. All advantages of the HERA detectors such as the almost 4π coverage and the functionality with respect to spatial orientation will be preserved. The EIC detector will have enhanced capability in the very forward and backward directions to measure continuously the low x and low Q^2 regions that are not comprehensively accessible at HERA. The detector design directly impacts the interaction region design and hence the accelerator parameters for the two beam elements: the effective interaction luminosity and the effective polarization of the two beams at the interaction point. Close interaction between the detector design and the IR design is hence needed in the very early stage of the project, which has already started (255). It is expected that the detector design and the IR design will evolve over the next few years. The e-p and e-A collisions at EIC will produce very asymmetric event topologies, not unlike HERA events. These asymmetries, properly exploited, allow precise measurements of energy and color flow in collisions of large and small-x partons. They also allow observation of interactions of electrons with photons that are coherently emitted by the relativistic heavy ions. The detector for EIC must detect: the scattered electrons, the quark fragmentation products and the centrally produced hadrons. It will be the first collider detector to measure the fragmentation region of the proton or the nucleus, a domain not covered effectively at HERA. The detector design, in addition, should pose no difficulties for important measurements such as precision beam polarization (electron as well as hadron beam) and collision luminosity.

The EIC detector design will allow measurements of partons from hard processes in the region around 90° scattering angle with respect to the beam pipes. This central region could have a jet tracker with an EM calorimeter backed by an instrumented iron yoke. Electrons from DIS are also emitted into this region and will utilize the tracking and the EM calorimetry. Electrons from photo-production and from DIS at intermediate and low momentum transfer will have to be detected by specialized backward detectors. With these guiding ideas, one could imagine that the EIC barrel might have a time projection chamber (TPC) backed by an EM Calorimeter inside a superconducting coil. One could use Spaghetti Calorimetry (SPACAL) for endcaps and GEM-type micro-vertex detector to complement the tracking capacity of the TPC in the central as well as forward/backward (endcap) regions. This type of central and end-cap detector geometry is now fairly standard. Details of the design could be finalized in the next few years using the state of the art technology and experience from more recent detectors such as BaBar at SLAC (USA) and Belle at KEK (Japan). To accommodate tracking and particle ID requirements for the different center-of-mass energy running ($\sqrt{s} = 30$–100 GeV) resulting at different beam energies, the central spectrometer magnet will have multiple field strength operation capabilities, including radial dependence of field strengths. Possible spectrometry based on dipole and toroidal fields is also being considered at this time.

The forward and backward regions (hadron and electron beam directions) in e-p collisions were instrumented at HERA up to a pseudo-rapidity of $\eta \approx 3$. A specialized detector added later extended this range with difficulty to $\eta \approx 4$. Although acceptance enhancement in the regions beyond $\eta = 4$ is possible with conventional ideas such as forward calorimetry and tracking using beam elements and silicon strip based Roman Pot Detectors (181), it is imperative for the EIC that this region be well instrumented. A recent detector design for eRHIC developed by the experimental group at the Max-Planck Institute, Munich, accomplishes just this (136) by allowing continuous access to physics up to $\eta \approx 6$. The main difference with respect to a conventional collider detector is a dipole field, rather than a solenoid, that separates the low energy scattered electron from the beam. High precision silicon tracking stations capable of achieving $\Delta p/p \sim 2\%$, EM calorimetry with energy resolution better than $20\%/\sqrt{E}$, an excellent e/π separation over a large Q^2 range, all in the backward region (in the electron beam direction) are attainable. In the forward region, the dipole field allows excellent tracking and a combination of EM and hadronic calorimetry with $20\%/\sqrt{E(\text{GeV})}$ and $50\%/\sqrt{E(\text{GeV})}$ energy resolution, respectively. This allows access to very high $x \sim 0.9$ with excellent accuracy. This region of high x is largely unexplored both in polarized and in unpolarized DIS. A significant distance away from the EIC central detector and IR, there may be Roman Pots, high rigidity spectrometers including EM calorimetry and forward electron taggers, all placed to improve the measurement of low angle scattering at high energy.

Although significant effort will be made to avoid design conflicts, the conventional detector using a solenoid magnet and the one described above may not coexist in certain scenarios being considered for the accelerator designs of the EIC at BNL. The main design line, presently the ring-ring design, may be particularly difficult with only one IR. Options such as time sharing between two detectors at the same IR with the two detectors residing on parallel rails may be considered. In the case of the linac-ring scenario, several other options are available. Because the physics of low x and low Q^2 does not require a large luminosity, nor is presently the beam polarization a crucial requirement for the physics (136), an interaction point with sufficient beam luminosity would be possible with innovative layouts of the accelerator complex. These and other details will be worked out in the next several years. Depending on the interest shown by the experimental community, accelerator designs that incorporate up to four collision points (while still allowing two hadron-hadron collision points at RHIC) will be considered and developed.

ACKNOWLEDGMENTS

This review draws generously on the whitepaper for the Electron Ion Collider (BNL-68933-02/07-REV), and we thank all our co-authors on this publication for their efforts. We are especially grateful to Marco Stratmann and Mark Strikman for valuable advice, discussions and comments. W.V. and A.D. are grateful to RIKEN and Brookhaven National Laboratory. R.M. is supported by the Department of Energy Cooperative Agreement DE-FC02-94ER40818. W.V. and R.V. were supported by Department of Energy (contract number DE-AC02-98CH10886). R.V. 's research was also supported in part by a research award from the A. Von Humboldt Foundation.

The *Annual Review of Nuclear and Particle Science* is online at
http://nucl.annualreviews.org

LITERATURE CITED

1. Dokshitzer YL. *Sov. Phys. JETP* 46:641 (1977); Lipatov LN. *Sov. J. Nucl. Phys.* 20:95 (1975); Gribov VN, Lipatov LN. *Sov. J. Nucl. Phys.* 15:438 (1972)
2. Altarelli G, Parisi G. *Nucl. Phys.* B126: 298 (1977)
3. Moch S, Vermaseren JAM, Vogt A. *Nucl. Phys.* B688:101 (2004); *Nucl. Phys.* B691:129 (2004)
4. Whitlow LM, et al. *Phys. Lett.* B282:475 (1992); Whitlow LM. SLAC-preprint, SLAC-357 (1990)
5. Arneodo M, et al. (New Muon Collab.) *Nucl. Phys.* B483:3 (1997); Adams MR, et al. (E665 Collab.) *Phys. Rev. D* 54:3006 (1996)
6. Aid S, et al. (H1 Collab.) *Nucl. Phys.* B470:3 (1996); Adloff C, et al. (H1 collab.) *Nucl. Phys.* 497:3 (1997)
7. Derrick M, et al. (ZEUS collab.) *Z. Phys. C* 69:607 (1996); Derrick M, et al. (ZEUS collab.) *Z. Phys. C* 72:399 (1996)
8. Rizvi E. Presented at the Int. Europhys. Conf. High Energy Phys., Aachen, Germany, July (2003)
9. Breitweg J, et al. (ZEUS collab.) *Eur. Phys. J. C* 7:609 (1999); Chekanov S, et al. (ZEUS collab.) *Phys. Rev. D* 67:012007 (2003)
10. Martin AD, Roberts RG, Stirling WJ, Thorne RS. *Eur. Phys. J. C* 35:325 (2004); Huston J, Pumplin J, Stump D, Tung WK. arXiv:hep-ph/0502080
11. Bartels J, Golec-Biernat K, Peters K. *Eur. Phys. J. C* 17:121 (2000)
12. Stasto AM, Golec-Biernat K, Kwiecinski J. *Phys. Rev. Lett.* 86:596 (2001)
13. Golec-Biernat K, Wüsthoff M. *Phys. Rev. D* 59:014017 (1999)
14. Golec-Biernat K, Wüsthoff M. *Phys. Rev. D* 60:114023 (1999)
15. Bartels J, Golec-Biernat K, Kowalski H. *Phys. Rev. D* 66:014001 (2002)
16. Kowalski H, Teaney D. *Phys. Rev. D* 68: 114005 (2003)
17. Munier S, Stasto AM, Mueller AH. *Nucl. Phys.* B603:427 (2001)
18. Goncalves VP, Machado MVT. *Phys. Rev. Lett.* 91:202002 (2003)
19. Hughes EW, Voss R. *Annu. Rev. Nucl. Part. Sci.* 49:303 (1999); Filippone BW, Ji XD. *Adv. Nucl. Phys.* 26:1 (2001); Bass SD. arXiv:hep-ph/0411005
20. Anselmino M, Gambino P, Kalinowski J. *Z. Phys. C* 64:267 (1994); *Phys. Rev. D* 55:5841 (1997)
21. De Nardo L. (HERMES collab.) *Czech. J. Phys.* 52:A1 (2002)
22. Tobias WA. (E155 collab.) *Measurement of the proton and deuteron spin structure functions g_1 and g_2.* PhD thesis. Univ. Virginia, SLAC-R617 (January 2001)
23. Ahmed MA, Ross GG. *Nucl. Phys.* B111:441 (1976)
24. Mertig R, van Neerven W. *Z. Phys. C* 70:637 (1996); Vogelsang W. *Phys. Rev. D* 54:2023 (1996); *Nucl. Phys.* B475:47 (1996)
25. Blümlein J, Böttcher H. *Nucl. Phys.* B636:225 (2002)
26. Glück M, Reya E, Stratmann M, Vogelsang W. *Phys. Rev. D* 63:094005 (2001)
27. Asymmetry Analysis Collab., Hirai M, Kumano S, Saito N. *Phys. Rev. D* 69: 054021 (2004)
28. Bjorken JD. *Phys. Rev.* 148:1467 (1966); *Phys. Rev. D* 1:1376 (1970)
29. Ellis J, Jaffe RL. *Phys. Rev. D* 9:1444 (1974); *Phys. Rev. D* 10:1669 (1974)
30. Kodaira J, et al. *Nucl. Phys.* B159:99 (1979); Gorishnii SG, Larin SA. *Phys. Lett.* B172:109 (1986); *Nucl. Phys.*

B283:452 (1987); Larin SA, Vermaseren JAM. *Phys. Lett.* B259:345 (1991)

31. Ellis J, Karliner M. *Phys. Lett.* B341:397 (1995); Altarelli G, Ball RD, Forte S, Ridolfi G. *Nucl. Phys.* B496:337 (1997)

32. Ratcliffe P. arXiv:hep-ph/0402063

33. Diakonov D, Petrov V, Pobylitsa P, Polyakov MV, Weiss C. *Nucl. Phys.* B480:341 (1996); *Phys. Rev. D* 56:4069 (1997); Wakamatsu M, Kubota T. *Phys. Rev. D* 60:034020 (1999); Dressler B, Goeke K, Polyakov MV, Weiss C. *Eur. Phys. J. C* 14:147 (2000); Dressler B, et al. *Eur. Phys. J. C* 18:719 (2001); Kumano S. *Phys. Rept.* 303:183 (1998); *Phys. Lett.* B479:149 (2000); Glück M, Reya E. *Mod. Phys. Lett.* A15:883 (2000); Cao FG, Signal AI. *Eur. Phys. J. C* 21:105 (2001); Bhalerao RS. *Phys. Rev. C* 63:025208 (2001); Fries RJ, Schäfer A, Weiss C. *Eur. Phys. J. A* 17:509 (2003)

34. Adeva B, et al. (Spin Muon collab.) *Phys. Lett.* B420:180 (1998)

35. Airapetian A, et al. (HERMES collab.) *Phys. Rev. Lett.* 92:012005 (2004); *Phys. Rev. D* 71:012003 (2005)

36. Stratmann M, Vogelsang W. *Phys. Rev. D* 64:114007 (2001); Glück M, Reya E. arXiv:hep-ph/0203063; Kotzinian A. *Phys. Lett.* B552:172 (2003); arXiv:hep-ph/0410093; Navarro G, Sassot R. *Eur. Phys. J. C* 28:321 (2003); Christova E, Kretzer S, Leader E. *Eur. Phys. J. C* 22:269 (2001); Leader E, Stamenov DB. *Phys. Rev. D* 67:037503 (2003); Bass SD. *Phys. Rev. D* 67:097502 (2003); de Florian D, Navarro GA, Sassot R. *Phys. Rev. D* 71:094018 (2005)

37. Bunce G, Saito N, Soffer J, Vogelsang W. *Annu. Rev. Nucl. Part. Sci* 50:525 (2000)

38. Ji XD. *Phys. Rev. Lett.* 78:610 (1997)

39. Jaffe RL, Manohar A. *Nucl. Phys.* B337:509 (1990); Bashinsky SV, Jaffe RL. *Nucl. Phys.* B536:303 (1998)

40. Bravar A, von Harrach D, Kotzinian AM. *Phys. Lett.* B421:349 (1998); Airapetian A, et al. (HERMES collab.) *Phys. Rev. Lett* 84:4047 (2000); Hedicke S. (COMPASS collab.) Prog. Part. *Nucl. Phys.* 50:499 (2003); Mielech A. (COMPASS collab.) *Nucl. Phys.* A752:191 (2005); Adeva B. et al. (Spin Muon Collab.) *Phys. Rev. D* 70:012002 (2004); Anthony PL, et al. (E155 collab.) *Phys. Lett.* B458:536 (1999); Afanasev A, Carlson CE, Wahlquist C. *Phys. Lett.* B398:393 (1997); *Phys. Rev. D* 58:054007 (1998); *Phys. Rev. D* 61:034014 (2000); Jäger B, Stratmann M, Vogelsang W. arXiv:hep-ph/0505157

41. Dittes FM, et al. *Phys. Lett.* B209:325 (1988); Radyushkin AV. *Phys. Rev. D* 56:5524 (1997)

42. Vanderhaeghen M. *Eur. Phys. J. A* 8:455 (2000); Goeke K, Polyakov MV, Vanderhaeghen M. Prog. Part. *Nucl. Phys.* 47:401 (2001); Belitsky AV, Müller D, Kirchner A. *Nucl. Phys.* B629:323 (2002); Diehl M. *Phys. Rept.* 388:41 (2003); Ji XD. *Annu. Rev. Nucl. Part. Sci.* 54:413 (2004)

43. Burkardt M. *Phys. Rev. D* 62:071503 (2000); *Phys. Rev. D* 66:114005 (2002); *Int. J. Mod. Phys.* A18:173 (2003); Burkardt M. *Nucl. Phys.* A735:185 (2004); arXiv:hep-ph/0505189; Burkardt M, Hwang DS. arXiv:hep-ph/0309072; Ralston JP, Pire B. *Phys. Rev. D* 66:111501 (2002); Diehl M. *Eur. Phys. J. C* 25:223 (2002); Diehl M. *Eur. Phys. J. C* 31:277 (2003); Belitsky AV, Müller D. *Nucl. Phys.* A711:118 (2002); Ji XD. *Phys. Rev. Lett.* 91:062001 (2003); Belitsky AV, Ji XD, Yuan F. *Phys. Rev. D* 69:074014 (2004); Diehl M, Hägler P. arXiv:hep-ph/0504175

44. Collins JC, Freund A. *Phys. Rev. D* 59:074009 (1999); Ji XD, Osborne J. *Phys. Rev. D* 58:094018 (1998); Radyushkin AV. *Phys. Lett.* B380:417 (1996); *Phys. Rev. D* 56:5524 (1997)

45. Belitsky AV, Müller D, Niedermeier L, Schäfer A. *Phys. Lett.* B474:163 (2000); Belitsky AV, Freund A, Müller D. *Phys. Lett.* B493:341 (2000); Freund A, McDermott MF. *Phys. Rev. D* 65:091901

(2002); *Phys. Rev. D* 65:074008 (2002); *Eur. Phys. J.* C23:651 (2002)

46. Adloff C, et al. (H1 collab.) *Phys. Lett.* B517:47 (2001); Aktas A, et al. (H1 collab.) arXiv:hep-ex/0505061; Chekanov S, et al. (ZEUS collab.) *Phys. Lett.* B573:46 (2003)

47. Airapetian A, et al. (HERMES collab.) *Phys. Rev. Lett.* 87:182001 (2001)

48. Ellinghaus F, et al. (HERMES collab.) *Nucl. Phys.* A711:171 (2002)

49. Stepanyan S, et al. (CLAS collab.) *Phys. Rev. Lett.* 87:182002 (2001)

50. Collins JC, Frankfurt L, Strikman M. *Phys. Rev. D* 56:2982 (1997); Belitsky AV, Müller D. *Phys. Lett.* B513:349 (2001)

51. Cano F, Pire B. *Eur. Phys. J. A* 19:423 (2004); Strikman M, Guzey V. *Phys. Rev. C* 68:015204 (2003); Kirchner A, Müller D. *Eur. Phys. J. C* 32:347 (2003); Freund A, Strikman M. *Phys. Rev. C* 69:015203 (2004); Polyakov MV. *Phys. Lett.* B555:57 (2003)

52. Ralston JP, Soper DE. *Nucl. Phys.* B152:109 (1979); Artru X, Mekhfi M. *Z. Phys. C* 45:669 (1990); Jaffe RL, Ji X. *Phys. Rev. Lett.* 67:552 (1991); *Nucl. Phys.* B375:527 (1992)

53. Collins JC. *Nucl. Phys.* B394:169 (1993)

54. Collins JC. *Nucl. Phys.* B396:161 (1993)

55. Ji XD, Ma JP, Yuan F. *Phys. Rev. D* 71:034005 (2005); *Phys. Lett.* B597:299 (2004); *Phys. Rev. D* 70:074021 (2004)

56. Mulders PJ, Tangerman RD. *Nucl. Phys.* B461:197 (1996); Mulders PJ, Tangerman RD. *Nucl. Phys.* B484:538 (1997); Boer D, Mulders PJ. *Phys. Rev. D* 57:5780 (1998); Efremov AV, Goeke K, Schweitzer P. *Eur. Phys. J. C* 32:337 (2003)

57. Sivers DW. *Phys. Rev. D* 41:83 (1990); *Phys. Rev. D* 43:261 (1991)

58. Brodsky SJ, Hwang DS, Schmidt I. *Phys. Lett.* B530:99 (2002)

59. Collins JC. *Phys. Lett.* B536:43 (2002); Belitsky AV, Ji XD, Yuan F. *Nucl. Phys.* B656:165 (2003); Boer D, Mul-

ders PJ, Pijlman F. *Nucl. Phys.* B667:201 (2003); Bomhof C, Mulders PJ, Pijlman F. *Phys. Lett. B* 596:277 (2004); Bacchetta A, Bomhof C, Mulders PJ, Pijlman F. arXiv:hep-ph/0505268

60. Ji XD, Ma JP, Yuan F. *Nucl. Phys.* B652:383 (2003)

61. Bacchetta A, Kundu R, Metz A, Mulders PJ. *Phys. Lett.* B506:155 (2001); *Phys. Rev. D* 65:094021 (2002); Bacchetta A, Metz A, Yang JJ. *Phys. Lett.* B574:225 (2003); Gamberg LP, Goldstein GR, Oganessyan KA. *Phys. Rev. D* 67:071504 (2003); Gamberg LP, Goldstein GR, Oganessyan KA. *Phys. Rev. D* 68:051501 (2003); arXiv:hep-ph/0309137; Yuan F. *Phys. Lett.* B575:45 (2003); Bacchetta A, Schweitzer P. *Nucl. Phys.* A732:106 (2004); Efremov AV, Goeke K, Schweitzer P. *Phys. Lett.* B568:63 (2003); *Eur. Phys. J. C* 32:337 (2003); Amrath D, Bacchetta A, Metz A. arXiv:hep-ph/0504124

62. Airapetian A, et al. (HERMES collab.) *Phys. Rev. Lett.* 94:012002 (2005); Diefenthaler M. (HERMES collab.) Talk presented at the 13th International Workshop on Deep Inelastic Scattering (DIS 2005), Madison, Wisconsin, April (2005)

63. Alexakhin VY, et al. (COMPASS collab.) arXiv:hep-ex/0503002

64. Seidl R. (Belle collab.) Talk presented at the 13th International Workshop on Deep Inelastic Scattering (DIS 2005), Madison, Wisconsin, April (2005); Collins JC, Metz A. *Phys. Rev. Lett.* 93:252001 (2004); Metz A. *Phys. Part. Nucl.* 35:S85 (2004)

65. D'Alesio U, Murgia F. *Phys. Rev. D* 70:074009 (2004); arXiv:hep-ph/0412317; Anselmino M, Boglione M, D'Alesio U, Leader E, Murgia F. *Phys. Rev. D* 71:014002 (2005); Qiu JW, Sterman G. *Phys. Rev. D* 59:014004 (1999)

66. Adams DL, et al. (FNAL E704 collab.) *Phys. Lett.* B261:201 (1991); Krueger K, et al. *Phys. Lett.* B459:412 (1999); Adams J, et al. (STAR collab.) *Phys. Rev. Lett.*

92:171801 (2004); Rakness G. (STAR collab.) arXiv:nucl-ex/0501026

67. Boer D, Vogelsang W. *Phys. Rev. D* 69:094025 (2004); Efremov AV, Goeke K, Menzel S, Metz A, Schweitzer P. *Phys. Lett.* B612:233 (2005)

68. Aubert JJ, et al. (EMC collab.) *Phys. Lett.* B123:275 (1983)

69. Arnold RG, et al. *Phys. Rev. Lett.* 52:727 (1984)

70. Arneodo M, et al. (New Muon collab.) *Nucl. Phys.* B441:3, 12 (1995)

71. Adams MR, et al. (E665 collab.) *Z. Phys. C* 67:403 (1995)

72. Arneodo M. *Phys. Rept.* 240:301 (1994)

73. Geesaman DF, Saito K, Thomas AW. *Annu. Rev. Nucl. Part. Sci.* 45:337 (1995)

74. Friman BL, Pandharipande VR, Wiringa RB. *Phys. Rev. Lett.* 51:763 (1983)

75. Pandharipande VR, et al. *Phys. Rev. C* 49:789 (1994)

76. Alde DM, et al. *Phys. Rev. Lett.* 51:763 (1990)

77. Gaskell D, et al. (E91-003 collab.) *Phys. Rev. Lett.* 87:202301 (2001)

78. Brown GE, Buballa M, Li Z, Wambach B. *J. Nucl. Phys.* A593:295 (1995)

79. Kolton DS. *Phys. Rev. C* 57:1210 (1998)

80. Smith JR, Miller GA. *Phys. Rev. Lett.* 91:212301 (2003)

81. Chen JW, Detmold W. arXiv:hep-ph/0412119

82. Frankfurt LL, Strikman MI, Liuti S. *Phys. Rev. Lett.* 65:1725 (1990)

83. Eskola KJ. *Nucl. Phys.* B400:240 (1993)

84. Eskola KJ, Kolhinen VJ, Ruuskanen PV. *Nucl. Phys.* B535:351 (1998)

85. Eskola KJ, Kolhinen VJ, Salgado CA, Thews RL. *Eur. Phys. J. C* 21:613 (2001)

86. Eskola KJ, Honkanen H, Kolhinen VJ, Salgado CA. arXiv:hep-ph/0302170

87. de Florian D, Sassot R. *Phys. Rev. D* 69:074028 (2004)

88. Leitch MJ, et al. (E772 and E789 Collabs.) *Nucl. Phys.* A544:197C (1992)

89. Naples D, et al. (NuTeV collab.) arXiv:hep-ex/0307005

90. Gribov VN. arXiv:hep-ph/0006158

91. Abramovsky VA, Gribov VN, Kancheli OV. *Yad. Fiz.* 18:595 (1973); *Sov. J. Nucl. Phys.* 18:308 (1974)

92. Armesto N, Capella A, Kaidalov AB, Lopez-Albacete J, Salgado CA. *Eur. Phys. J. C* 29:531 (2003)

93. Frankfurt L, Guzey V, McDermott M, Strikman M. *JHEP* 0202:027 (2002)

94. Gribov LV, Levin EM, Ryskin MG. *Phys. Rept.* 100:1 (1983)

95. Mueller AH, Qiu JW. *Nucl. Phys.* B268:427 (1986)

96. Freund A, Rummukainen K, Weigert H, Schäfer A. *Phys. Rev. Lett.* 90:222002 (2003)

97. Frankfurt LL, Strikman MI. *Phys. Rept.* 160:235 (1988)

98. Piller G, Weise W. *Phys. Rept.* 330:1 (2000)

99. Jalilian-Marian J, Wang XN. *Phys. Rev. D* 63:096001 (2001)

100. Eskola KJ, Honkanen H, Kolhinen VJ, Qiu JW, Salgado CA. *Nucl. Phys.* B660:211 (2003)

101. Gousset T, Pirner HJ. *Phys. Lett.* B375:349 (1996)

102. Adler SS, et al. (PHENIX collab.) *Phys. Rev. Lett.* 94:082302 (2005)

103. Ablikim M, et al. (STAR collab.) arXiv:nucl-ex/0408016; Rakness G (STAR collab.) arXiv:nucl-ex/0505062

104. Back BB, et al. (PHOBOS collab.) *Phys. Rev. C* 70:061901 (2004)

105. Arsene I, et al. (BRAHMS collab.) *Phys. Rev. Lett.* 93:242303 (2004)

106. Brodsky SJ. In Proc. 13th Int. Symp. Multiparticle Dynamics, ed. Kittel W, Metzger W, Stergiou A, p. 963. Singapore: World Sci. (1982)

107. Mueller AH. In *Proc. of the 17th Rencontres de Moriond*, ed. Tran Thanh Van J., Vol. 1, p. 13. Gif-sur-Yvette, France: Editions Frontieres (1982)

108. Jain P, Pire B, Ralston JP. *Phys. Rept.* 271:67 (1996)

109. Frankfurt L, Miller GA, Strikman MI. *Phys. Lett.* B304:1 (1993)

110. Brodsky SJ, Frankfurt L, Gunion JF,

Mueller AH, Strikman MI. *Phys. Rev. D* 50:3134 (1994)

111. Sokoloff MD, et al. (Fermilab Tagged Photon Spectrometer collab.) *Phys. Rev. Lett.* 57:3003 (1986)

112. Aitala EM, et al. (E791 collab.) *Phys. Rev. Lett.* 86:4773 (2001)

113. Brodsky SJ, Mueller AH. *Phys. Lett.* B206:685 (1988)

114. Frankfurt LL, Strikman MI. *Phys. Rept.* 160:235 (1988)

115. Kopeliovich BZ, Nemchik J, Nikolaev NN, Zakharov BG. *Phys. Lett.* B309:179 (1993)

116. Adams MR, et al. (E665 collab.) *Phys. Rev. Lett.* 74:1525 (1995)

117. Baier R, Schiff D, Zakharov BG. *Annu. Rev. Nucl. Part. Sci.* 50:37 (2000); Kopeliovich BZ, Tarasov AV, Schäfer A. *Phys. Rev. C* 59:1609 (1999)

118. McGaughey PL, Moss JM, Peng JC. *Annu. Rev. Nucl. Part. Sci.* 49:217 (1999)

119. Leitch ML. (E866 collab.) *New Measurements of the Nuclear Dependence of the and Resonances in High-energy Proton-Nucleus Collisions.* Presented at the Amer. Phys. Soc. Mtg., Long Beach, Calif. May (2000)

120. Naples D, et al. (E683 collab.) *Phys. Rev. Lett* 72:2341 (1994)

121. Cronin JW, et al. *Phys. Rev. Lett* 31:1426 (1973)

122. Krzywicki A, Engels J, Petersson B, Sukhatme U. *Phys. Lett.* B85:407 (1979)

123. Kopeliovich BZ, Nemchik J, Predazzi E, Hayashigaki A. *Nucl. Phys.* A740:211 (2004)

124. Airapetian A, et al. (HERMES collab.) *Eur. Phys. J. C* 20:479 (2001)

125. Adams MR, et al. (E665 collab.) *Phys. Rev. D* 50:1836 (1994)

126. Ashman J, et al. (European Muon collab.) *Z. Phys. C* 52:1 (1991)

127. Burkot W, et al. (BEBC WA21/WA59 Collabs.) *Z. Phys. C* 70:47 (1996)

128. Strikman MI, Tverskoii MG, Zhalov MB. *Phys. Lett.* B459:37 (1999)

129. Adams MR, et al. (E665 collab.) *Phys. Rev. Lett.* 74:5198 (1995); *Phys. Rev. Lett.* 80:2020 (1995)

130. Falter T, Cassing W, Gallmeister K, Mosel U. *Phys. Lett.* B594:61 (2004)

131. Wang E, Wang XN. *Phys. Rev. Lett.* 89:162301 (2002)

132. Accardi A, Muccifora V, Pirner HJ. *Nucl. Phys.* A720:131 (2003)

133. Cassing W, Gallmeister K, Greiner CJ. *Nucl. Phys.* G30:S801 (2004)

134. Osborne J, Wang XN. *Nucl. Phys.* A710:281 (2002)

135. Majumder A, Wang E, Wang XN. arXiv:nucl-th/0412061

136. Abt I, Caldwell A, Liu X, Sutiak J. arXiv:hep-ex/0407053

137. http://www.jlab.org/12GeV/

138. Farrar GR, Jackson DR. *Phys. Rev. Lett.* 35:1416 (1975); Brodsky SJ, Burkardt M, Schmidt I. *Nucl. Phys.* B441:197 (1995)

139. Close FE. *Phys. Lett.* B43:422 (1973)

140. Melnitchouk W, Ent R, Keppel C. *Phys. Rept.* 406:127 (2005)

141. Galfi L, Kuti J, Patkos A. *Phys. Lett.* B31:465 (1970); Heimann RL. *Nucl. Phys.* B64:429 (1973); Ellis J, Karliner M. *Phys. Lett.* B213:73 (1988); Close FE, Roberts RG. *Phys. Rev. Lett.* 60:1471 (1988)

142. Ball RD, Forte S, Ridolfi G. *Nucl. Phys.* B444:287 (1995); *Nucl. Phys.* B449:680 (1995)

143. Altarelli G, Ball RD, Forte S, Ridolfi G. *Acta Phys. Polon.* B29:1145 (1998); Gehrmann T, Stirling WJ. *Phys. Lett.* B365:347 (1996)

144. Bass S, De Roeck A. *Eur. Phys. J. C* 18:531 (2001)

145. Deshpande A. *Proc. Yale-eRHIC Workshop,* BNL Report 52592 (2000)

146. Adeva B, et al. (Spin Muon Collab.) *Phys. Rev. D* 58:112002 (1998)

147. Bartels J, Ermolaev BI, Ryskin MG. *Z. Phys. C* 70:273 (1996); *Z. Phys. C* 72:627 (1996); Ermolaev B, Greco M, Troyan SI. *Phys. Lett.* B579:321 (2004)

148. Blümlein J, Vogt A. *Phys. Lett.* B370:149 (1996); *Phys. Lett.* B386:350 (1996)
149. Kiyo Y, Kodaira J, Tochimura H. *Z. Phys. C* 74:631 (1997)
150. Kwiecinski J, Ziaja B. *Phys. Rev. D* 60:054004 (1999); Badelek B, Kwiecinski J. *Phys. Lett.* B418:229 (1998); Badelek B, Kiryluk J, Kwiecinski J. *Phys. Rev. D* 61:014009 (2000)
151. Igo G. *Proc. Yale-eRHIC Workshop*, BNL Report 52592 (2000)
152. Stratmann M, Vogelsang W, Weber A. *Phys. Rev. D* 53:138 (1996)
153. Wray D. *Nuov. Cim.* 9A:463 (1972); Derman E. *Phys. Rev. D* 7:2755 (1973); Joshipura AS, Roy P. *Ann. Phys.* 104:460 (1977); Lampe B. *Phys. Lett.* B227:469 (1989); Vogelsang W, Weber A. *Nucl. Phys.* B362:3 (1991); Mathews P, Ravindran V. *Phys. Lett.* B278:175 (1992); *Int. J. Mod. Phys.* A7:6371 (1992); Ravishankar V. *Nucl. Phys.* B374:309; Blümlein J, Kochelev N. *Phys. Lett.* B381: 296 (1996)
154. Contreras J, De Roeck A, Maul M. arXiv: hep-ph/9711418
155. Gehrmann T, Stirling WJ. *Phys. Rev. D* 53:6100 (1996)
156. Kinney ER, Stösslein U. *AIP Conf. Proc.* 588:171 (2001)
157. Koike Y, Nagashima J. *Nucl. Phys.* B660: 269 (2003)
158. Meng R, Olness FI, Soper DE. *Nucl. Phys.* B371:79 (1992); *Phys. Rev. D* 54:1919 (1996); Nadolsky P, Stump DR, Yuan CP. *Phys. Rev. D* 64:114011 (2001); *Phys. Lett.* B515:175 (2001)
159. Derrick M, et al. (ZEUS Collab.) *Z. Phys. C* 70:1 (1996); Adloff C, et al. (H1 Collab.) *Eur. Phys. J. C* 12:595 (2000)
160. Lichtenstadt J. *Nucl. Phys.* B (Proc. Suppl.) 105:86 (2002)
161. Watson AD. *Z. Phys. C* 12:123 (1982); Glück M, Reya E. *Z. Phys. C* 39:569 (1988); Guillet JPh. *Z. Phys. C* 39:75 (1988); Frixione S, Ridolfi G. *Phys. Lett.* B383:227 (1996); Bojak I, Stratmann M.

Phys. Lett. B433:411 (1998); *Nucl. Phys.* B540:345 (1999)
162. Altarelli G, Stirling WJ. *Particle World* 1:40 (1989)
163. Glück M, Reya E, Vogelsang W. *Nucl. Phys.* B351:579 (1991)
164. Carlitz RD, Collins JC, Mueller AH. *Phys. Lett.* B214:229 (1988); Kunszt Z. *Phys. Lett.* B218:243 (1989); Vogelsang W. *Z. Phys. C* 50:275 (1991); Manohar A. *Phys. Lett.* B255:579 (1991)
165. Aid S, et al. (H1 Collab.) *Nucl. Phys.* B449:3 (1995); Wobisch M. (H1 collab.) *Nucl. Phys.* Proc. Suppl. 79:478 (1999); Hadig T. (ZEUS collab.) Presented at the 8th Int. Workshop on Deep Inelastic Scattering and QCD (DIS 2000), Liverpool, England, April (2000), arXiv:hep-ex/0008027
166. Rädel G. *Proc. Yale-eRHIC Workshop*, BNL Report 52592 (2000); De Roeck A, Rädel G. *Nucl. Phys.* B (Proc. Suppl.) 105:90 (2002)
167. Mirkes E, Zeppenfeld D. *Phys. Lett.* B380:205 (1996)
168. Lichtenstadt J. *Proc. Yale-eRHIC Workshop*, BNL Report 52592 (2000)
169. De Roeck A, Deshpande A, Hughes VW, Lichtenstadt J, Rädel G. *Eur. Phys. J. C* 6:121 (1999)
170. Klasen M. *Rev. Mod. Phys.* 74:1221 (2002)
171. Nisius R. *Phys. Rept.* 332:165 (2000)
172. Irving AC, Newland DB. *Z. Phys. C* 6:27 (1980); Hassan JA, Pilling DJ. *Nucl. Phys.* B187:563 (1981); Xu ZX. *Phys. Rev. D* 30:1440 (1984); Bass SD. *Int. J. Mod. Phys.* A7:6039 (1992); Narison S, Shore GM, Veneziano G. *Nucl. Phys.* B391:69 (1993); Bass SD, Brodsky SJ, Schmidt I. *Phys. Lett.* B437:417 (1998)
173. Glück M, Vogelsang W. *Z. Phys. C* 55:353 (1992); *Z. Phys. C* 57:309 (1993); Glück M, Stratmann M, Vogelsang W. *Phys. Lett.* B337:373 (1994)
174. Stratmann M, Vogelsang W. *Proc. Yale-eRHIC Workshop*, BNL Report 52592 (2000); Stratmann M, Vogelsang W. *Z.*

Phys. C 74:641 (1997); Butterworth JM, Goodman N, Stratmann M, Vogelsang W. arXiv:hep-ph/9711250

175. Jäger B, Stratmann M, Vogelsang W. *Phys. Rev. D* 68:114018 (2003)
176. Adloff C, et al. (H1 collab.) *Eur. Phys. J. C* 1:97 (1998)
177. Combridge BL, Maxwell CJ. *Nucl. Phys.* B239:429 (1984)
178. Stratmann M, Vogelsang W. arXiv:hep-ph/9907470
179. Drell SD, Hearn AC. *Phys. Rev. Lett.* 162:1520 (1966); Gerasimov SB. *Yad. Fiz.* 2:839 (1965)
180. Helbing K. (GDH collab.) *AIP Conf. Proc.* 675:33 (2003); Ahrens J. et al. (GDH Collab.) *Phys. Rev. Lett* 87:022003 (2001)
181. Deshpande A (for Sandacz A). *Acceptance studies for DVCS and DES physics using PP2PP roman pots in eRHIC hadron beam lattice.* Presented at the EIC Conf., Jefferson Lab, Newport News, Virginia, April (2004)
182. Collins JC. *Phys. Rev. D* 57:3051 (1998); *Phys. Rev. D* 61:019902 (1998)
183. Grazzini M, Trentadue L, Veneziano G. *Nucl. Phys.* B519:394 (1998)
184. Kuraev EA, Lipatov LN, Fadin VS. *Sov. Phys. JETP* 45:199 (1977); Balitsky II, Lipatov LN. *Sov. J. Nucl. Phys.* 28:822 (1978)
185. Iancu E, Venugopalan R. arXiv:hep-ph/0303204, in *Quark Gluon Plasma 3*, ed. R Hwa, XN Wang. Singapore: World Sci. (2004)
186. Sloan T. *Proc. Yale-eRHIC Workshop*, BNL Report BNL-52592 (2000)
187. Armesto N, Salgado CA. arXiv:hep-ph/0301200
188. Hirai M, Kumano K, Miyama M. *Phys. Rev. D* 64:034003 (2001)
189. Eskola KJ, Kolhinen VJ, Salgado CA. *Eur. Phys. J. C* 9:61 (1999)
190. Huang Z, Lu HJ, Sarcevic I. *Nucl. Phys.* A637:79 (1998)
191. Armesto N. *Eur. Phys. J. C* 26:35 (2002)
192. Wang XN, Li SY. *Phys. Lett. B* 527:85 (2002)
193. McLerran L, Venugopalan R. *Phys. Rev. D* 49:2233 (1994); *Phys. Rev. D* 49:3352 (1994); *Phys. Rev. D* 50:2225 (1994)
194. Berera A, Soper DE. *Phys. Rev. D* 53:6162 (1996)
195. Arneodo M, et al. In *Proc. Future Physics at HERA*, DESY, September arXiv:hep-ph/9610423 (1995)
196. Alvero L, Collins JC, Terron J, Whitmore JJ. *Phys. Rev. D* 59:074022 (1999)
197. Kovchegov YV, Mueller AH. *Nucl. Phys.* B529:451 (1998)
198. Johnson MB, et al. *Phys. Rev. C* 65:025203 (2002)
199. Cano F, Pire B. *Eur. Phys. J. A* 19:423 (2004)
200. Martin AD, Ryskin MG. *Phys. Rev. D* 64:094017 (2001)
201. Susskind L. *Phys. Rev.* 65:1535 (1968)
202. Jalilian-Marian J, Kovner A, Leonidov A, Weigert H. *Nucl. Phys.* B504:415 (1997); *Phys. Rev. D* 59:014014 (1999); Kovner A, Milhano G, Weigert H. *Phys. Rev. D* 62:114005 (2000); Jalilian-Marian J, Kovner A, McLerran LD, Weigert H. *Phys. Rev. D* 55:5414 (1997); Iancu E, Leonidov A, McLerran LD. *Nucl. Phys.* A692:583 (2001); *Phys. Lett.* B510:133 (2001); Ferreiro E, Iancu E, Leonidov A, McLerran LD. *Nucl. Phys.* A703:489 (2002)
203. Mueller AH. arXiv:hep-ph/9911289; McLerran LD. *Acta Phys. Polon.* B34:5783 (2003); Armesto N. *Acta Phys. Polon.* B35:213 (2004); Stasto AM. *Acta Phys. Polon.* B35:3069 (2004)
204. Rummukainen K, Weigert H. *Nucl. Phys.* A739:183 (2004)
205. Iancu E, Triantafyllopoulos DN. *Phys. Lett.* B610 (2005); Mueller AH, Shoshi AI, Wong SMH. *Nucl. Phys.* B715:440 (2005); Kovner A, Lublinsky M. *Phys. Rev. D* 71:085004 (2005)
206. Balitsky I. *Nucl. Phys.* B463:99 (1996); Kovchegov YV. *Phys. Rev. D* 61:074018 (2000)
207. Mueller AH. *Nucl. Phys.* B415:373 (1994)

208. Nikolaev NN, Zakharov BG. *Z. Phys. C* 49:607 (1991)
209. Levin EM, Tuchin K. *Nucl. Phys.* B573: 833 (2000)
210. Iancu E, Itakura K, McLerran LD. *Nucl. Phys.* A724:181 (2003)
211. Mueller AH, Triantafyllopoulos DN. *Nucl. Phys.* B640:331 (2002)
212. Munier S, Peschanski R. *Phys. Rev. Lett.* 91:232001 (2003); *Phys. Rev. D* 69:034008 (2004); *Phys. Rev. D* D70: 077503 (2004)
213. Braun MA. arXiv:hep-ph/0101070; Armesto N, Braun MA. *Eur. Phys. J. C* 20: 517 (2001)
214. Albacete JL, Armesto N, Milhano G, Salgado CA, Wiedemann UA. *Phys. Rev. D* 71:014003 (2005)
215. Golec-Biernat K, Motyka L, Stasto AM. *Phys. Rev. D* 65:074037 (2002)
216. Triantafyllopoulos DN. *Nucl. Phys.* B648: 293 (2003)
217. Iancu E, Itakura K, Munier S. *Phys. Lett.* B590:199 (2004)
218. Forshaw JR, Shaw G. *JHEP* 0412:052 (2004)
219. Mueller AH. *Nucl. Phys.* A724:223 (2003)
220. Kovchegov YV, Tuchin K. *Phys. Rev. D* 65:074026 (2002)
221. Albacete JL, et al. arXiv:hep-ph/0502167
222. Levin EM, Lublinsky M. *Nucl. Phys.* A730:191 (2004)
223. Rogers T, Guzey V, Strikman MI, Zu X. *Phys. Rev. D* 69:074011 (2004)
224. Ivanov IP, Nikolaev NN, Savin AA. arXiv:hep-ph/0501034
225. Levin EM, Lublinsky M. *Nucl. Phys.* A712:95 (2002)
226. Frankfurt L, Guzey V, Strikman MI. *Phys. Lett.* B586:41 (2004)
227. Kovchegov YV, Levin E, McLerran LD. *Phys. Rev. C* 63:024903 (2001)
228. Guzey V, Strikman M, Vogelsang W. *Phys. Lett.* B603:173 (2004)
229. Dumitru A, Jalilian-Marian J. *Phys. Rev. Lett.* 89:022301 (2002)
230. Gelis F, Jalilian-Marian J. *Phys. Rev.* D 66:094014 (2002); *Phys. Rev. D* 66: 014021 (2002)
231. Blaizot JP, Gelis F, Venugopalan R. *Nucl. Phys.* A743:13 (2004)
232. Accardi A. arXiv:hep-ph/0212148; Accardi A, Gyulassy M. *Phys. Lett.* B586: 244 (2004)
233. Kharzeev D, Kovchegov YV, Tuchin K. *Phys. Rev. D* 68:094013 (2003); Baier R, Kovner A, Wiedemann UA. *Phys. Rev. D* 68:054009 (2003); Albacete JL, et al. *Phys. Rev. Lett.* 92:082001 (2004); Jalilian-Marian J, Nara Y, Venugopalan R. *Phys. Lett.* B577:54 (2003); Iancu E, Itakura K, Triantafyllopoulos DN. *Nucl. Phys.* A742:182 (2004)
234. Kharzeev D, Levin EM, McLerran LD. *Nucl. Phys.* A748:627 (2005)
235. Ogawa A. (STAR collab.) arXiv:nucl-ex/0408004
236. Qiu JW, Vitev I. arXiv:hep-ph/0405068; Hwa RC, Yang CB, Fries RJ. *Phys. Rev. C* 71:024902 (2005); Kopeliovich BZ, et al. arXiv:hep-ph/0501260
237. Baier R, Mueller AH, Schiff D. *Nucl. Phys.* A741:358 (2004); Betemps MA, Gay Ducati MB. *Phys. Rev. D* 70:116005 (2004); Jalilian-Marian J. *Nucl. Phys.* A739:319 (2004)
238. Catani S, Ciafaloni M, Hautmann F. *Nucl. Phys.* B366:135 (1991); Collins JC, Ellis RK. *Nucl. Phys.* B360:3 (1991)
239. Kharzeev D, Kovchegov YV, Tuchin K. *Phys. Rev. D* 68:094013 (2003)
240. Blaizot JP, Gelis F, Venugopalan R. *Nucl. Phys.* A743:57 (2004); Gelis F, Venugopalan R. *Phys. Rev. D* 69:014019 (2004)
241. Nikolaev NN, Schäfer W. *Phys. Rev. D* 71:014023 (2005); Nikolaev NN, Schäfer W, Zakharov BG. arXiv:hep-ph/0502018
242. Jalilian-Marian J, Kovchegov YV. *Phys. Rev. D* 70:114017 (2004)
243. Kovner A, Wiedemann UA. *Phys. Rev. D* 64:114002 (2001)
244. Jalilian-Marian J, Kovchegov YV. arXiv: hep-ph/0505052

245. Kopeliovich BZ, Raufeisen J, Tarasov AV, Johnson MB. *Phys. Rev. C* 67:014903 (2003)

246. Krasnitz A, Venugopalan R. *Nucl. Phys.* B557:237 (1999); *Phys. Rev. Lett.* 84:4309 (2000); *Phys. Rev. Lett.* 86:1717 (2001)

247. Krasnitz A, Nara Y, Venugopalan R. *Phys. Rev. Lett.* 87:192302 (2001); *Nucl. Phys.* A717:268 (2003)

248. Lappi T. *Phys. Rev. C* 67:054903 (2003); Krasnitz A, Nara Y, Venugopalan R. *Nucl. Phys.* A727:427 (2003)

249. Armesto N, Pajares C. *Int. J. Mod. Phys.* A15:2019 (2000)

250. Kharzeev D, Nardi M. *Phys. Lett.* B507:121 (2001); Kharzeev D, Levin EM. *Phys. Lett.* B523:79 (2001); Kharzeev D, Levin EM, Nardi M. arXiv:hep-ph/0111315

251. Kharzeev D, Levin EM, McLerran LD. *Phys. Lett.* B561:93 (2003)

252. Krasnitz A, Nara Y, Venugopalan R. *Phys. Lett.* B554:21 (2003)

253. Huovinen P. arXiv:nucl-th/0305064

254. Hirano T, Nara Y. *Nucl. Phys.* A743:305 (2004)

255. *The eRHIC Zeroth Order Design Report*, ed. M Farkhondeh, V Ptitsyn. BNL Report C-A/AP/142, March (2004)

256. Merminga L, Derbenev Y. *AIP Conf. Proc.* 698:811 (2004)

257. Koop I, et al. *Electron Cooling for RHIC.* Presented at the IEEE Particle Accelerator Conference (PAC2001), Chicago, June 2001; *Proc. IEEE Particle Accelerator Conference (PAC2001)*, ed. P Lucas, S Webber, Chicago, June (2001); Parkhomchuk VV, Ben-Zvi I. *Electron Cooling for RHIC Design Report*, BNL Report C-A/AP/47 (2001)

258. Ben-Zvi I, et al. *Nucl. Instrum. Meth.* A532:177 (2004)

259. Farkhondeh M. Polarized electron sources for a linac-ring electron-ion collider. In *Proc. EIC Accelerator Workshop*, ed. M Davis, A Deshpande, S Ozaki, R Venugopalan. BNL Report-52663-V.1 (2002)

260. Abramowicz H, Caldwell A. *Rev. Mod. Phys.* 71:1275 (1999)

Annu. Rev. Nucl. Part. Sci. 2005. 55:229–70
doi: 10.1146/annurev.nucl.55.090704.151502

LITTLE HIGGS THEORIES

Martin Schmaltz

Physics Department, Boston University, Boston, Massachusetts 02215;
email: schmaltz@bu.edu

David Tucker-Smith

Department of Physics, Williams College, Williamstown, Massachusetts 01267;
email: dtuckers@williams.edu

Key Words   Electroweak symmetry breaking

■ **Abstract**   Recently there has been renewed interest in the possibility that the Higgs particle of the Standard Model is a pseudo-Nambu-Goldstone boson. This development was spurred by the observation that if certain global symmetries are broken only by the interplay between two or more coupling constants, then the Higgs mass-squared is free from quadratic divergences at one loop. This collective symmetry breaking is the essential ingredient in little Higgs theories, which are weakly coupled extensions of the Standard Model with little or no fine tuning, describing physics up to an energy scale ∼10 TeV. Here we give a pedagogical introduction to little Higgs theories. We review their structure and phenomenology, focusing mainly on the $SU(3)$ theory, the Minimal Moose, and the littlest Higgs as concrete examples.

CONTENTS

1. INTRODUCTION

A few years before the start of the Large Hadron Collider (LHC) program, electroweak symmetry breaking remains poorly understood. The detailed quantitative fit of Standard Model predictions to precision electroweak data strongly suggests that electroweak symmetry is broken by one or more weakly coupled Higgs doublets. However, fundamental scalar particles suffer from a radiative instability in their masses, leading us to expect additional structure (such as compositeness, supersymmetry, little Higgs, etc.) near the weak scale.

Interestingly, we can turn this problem into a prediction for the LHC. The argument goes as follows: Let us assume that precision electroweak data are telling us that there are no new particles beyond the Standard Model (with the exception of possible additional Higgs doublets) with masses at or below the weak scale. Then physics at the weak scale may be described by an effective Standard Model, which has the particle content of the Standard Model and in which possible new physics is parametrized by higher-dimensional operators suppressed by the new physics scale $\Lambda \gtrsim$ TeV. All renormalizable couplings are as in the Standard Model. If there are additional Higgs fields, then more complicated Higgs self-couplings as well as Yukawa couplings are possible. Because no Higgs particles have been discovered so far, the effects of additional Higgs fields can be parametrized by effective operators for the Standard Model fields.

The higher-dimensional operators can be categorized by the symmetries that they break. The relevant symmetries are baryon and lepton number (B and L), CP and flavor symmetries, and custodial $SU(2)$ symmetry. The wealth of indirect experimental data can then be translated into bounds on the scale suppressing the operators (1–4). Examples of such operators and the resulting bounds are summarized in Table 1.

TABLE 1 Lower bounds on the scale that suppresses higher-dimensional operators that violate approximate symmetries of the Standard Model

Broken symmetry	Operators	Scale Λ
B, L	$(QQQL)/\Lambda^2$	10^{13} TeV
Flavor (1,2nd family), CP	$(\bar{d}s\bar{d}s)/\Lambda^2$	1000 TeV
Flavor (2,3rd family)	$m_b(\bar{s}\sigma_{\mu\nu}F^{\mu\nu}b)/\Lambda^2$	50 TeV
Custodial $SU(2)$	$(h^\dagger D_\mu h)^2/\Lambda^2$	5 TeV
None (S-parameter)	$(D^2h^\dagger D^2h)/\Lambda^2$	5 TeV

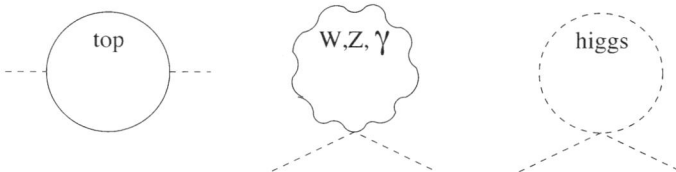

Figure 1   The most significant quadratically divergent contributions to the Higgs mass in the Standard Model.

The bounds imply that physics at the TeV scale has to conserve B and L, flavor, and CP to a very high accuracy, and that violations of custodial symmetry and contributions to the S-parameter should also be small.

The question then becomes if it is possible to add new physics at the TeV scale to the SM that stabilizes the Higgs mass but does not violate the above bounds. To better understand the requirements on this new physics, we must look at the source of the Higgs mass instability. The three most dangerous radiative corrections to the Higgs mass in the Standard Model come from one-loop diagrams with top quarks, $SU(2)$ gauge bosons, and the Higgs itself running in the loop (Figure 1).

All other diagrams give smaller contributions because they involve small coupling constants. Assuming that the Standard Model remains valid up to a cutoff of order the LHC center-of-mass energy, $\Lambda \sim 10$ TeV, the three diagrams give

$$\text{top loop} \quad -\frac{3}{8\pi^2}\lambda_t^2\Lambda^2 \sim -(2\text{ TeV})^2$$

$$SU(2)\text{ gauge boson loops} \quad \frac{9}{64\pi^2}g^2\Lambda^2 \sim (700\text{ GeV})^2$$

$$\text{Higgs loop} \quad \frac{1}{16\pi^2}\lambda^2\Lambda^2 \sim (500\text{ GeV})^2.$$

The total Higgs mass-squared includes the sum of these loop contributions and a tree-level mass-squared parameter.

To obtain a weak-scale expectation value for the Higgs without invoking a fine tuning of parameters except at the 10% level, the top, gauge, and Higgs loops must be cut off at scales satisfying

$$\Lambda_{top} \lesssim 2\text{ TeV} \quad \Lambda_{gauge} \lesssim 5\text{ TeV} \quad \Lambda_{Higgs} \lesssim 10\text{ TeV}. \qquad 1.$$

We see that the Standard Model with a cutoff near the maximum attainable energy at the Tevatron ($\sim$1 TeV) is natural, and we should not be surprised that we have not observed any new physics. However, the Standard Model with a cutoff of order the LHC energy would be fine tuned, and so we should expect to see new physics at the LHC.

More specifically, we expect new physics that cuts off the divergent top loop at or below 2 TeV. In a weakly coupled theory this implies that there are new particles with masses at or below 2 TeV. These particles must couple to the Higgs, giving rise to a new loop diagram that cancels the quadratically divergent contribution from the top loop. For this cancellation to be natural, the new particles must be related to the top quark by some symmetry, implying that the new particles have similar quantum numbers to top quarks. Thus naturalness arguments predict a new multiplet of colored particles with mass below 2 TeV, particles that would be easily produced at the LHC. In supersymmetry these new particles are of course the top squarks.

Similarly, the contributions from $SU(2)$ gauge loops must be canceled by new particles related to the Standard Model $SU(2)$ gauge bosons by symmetry, and the masses of these particles must be at or below 5 TeV for the cancellation to be natural. Finally, the Higgs loop requires new particles related to the Higgs itself at or below 10 TeV. Given the LHC's 14 TeV center-of-mass energy, these predictions are very exciting, and encourage us to explore different possibilities for what the new particles could be.

The new particles may be the superpartners predicted by the Minimal Supersymmetric Standard Model (MSSM) (5). In this case the quadratic divergence of the top quark loop is canceled by a corresponding loop with top squarks: Supersymmetry (SUSY) predicts the necessary relationship between the top and stop coupling constants. Alternatively, the Higgs may be a composite resonance at the TeV scale as in technicolor (6) or composite Higgs models (7–9). Or perhaps extra dimensions are lurking at the TeV scale, with new mechanisms to stabilize the Higgs mass (10).

Here we explore a different possibility, that the Higgs is a pseudo-Nambu-Goldstone boson. This idea was first suggested in References (11, 12) and was recently revived by Arkani-Hamed, Cohen, and Georgi when they constructed the first successful little Higgs model (13) and thereby started an industry of little model building (14–34). As we will see, in these theories the Higgs mass is protected from one-loop quadratic divergences by approximate global symmetries under which the Higgs field shifts. New particles must be introduced to ensure that the global symmetries are not broken too severely, and these are the states that cut off the quadratically divergent top, gauge, and Higgs loops.

The outline of the rest of the article is as follows: In the next section we review some basic aspects of Nambu-Goldstone bosons. In Section 3 we describe how to construct a little Higgs theory, using the $SU(3)$ model of References (19, 20, 30), as an illustrative example. In Section 4 we discuss the prototype product-group models of References (15, 16), along with several variations. Finally, in Section 5 we discuss precision electroweak constraints on little Higgs theories and the prospects for discovering the new particles they predict at the LHC.

2. NAMBU-GOLDSTONE BOSONS

Nambu-Goldstone bosons (NGBs) arise whenever a continuous global symmetry is spontaneously broken. If the symmetry is exact, the NGBs are exactly massless and have only derivative couplings.

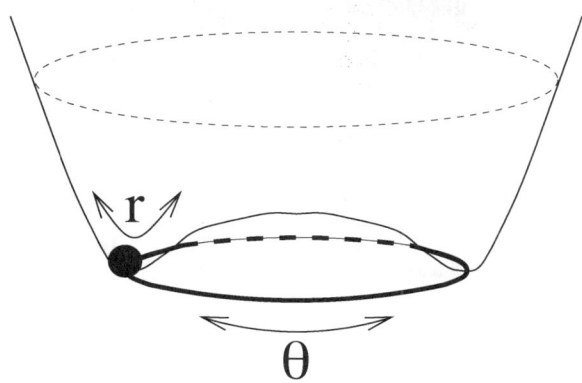

Figure 2 The "Mexican hat" potential for Φ. The black dot represents the vacuum expectation value f, r is the radial mode, and θ is the Nambu-Goldstone boson.

$U(1)$ example: Consider for example a theory with a single complex scalar field ϕ with potential $V = V(\phi^*\phi)$. The kinetic energy term $\partial_\mu \phi^* \partial^\mu \phi$ and the potential are invariant under the $U(1)$ symmetry transformation

$$\phi \to e^{i\alpha}\phi. \qquad\qquad 2.$$

If the minimum of the potential is not at the origin but at some distance f away as in the famous "wine bottle" or "Mexican hat" potential (Figure 2), then the $U(1)$ symmetry is spontaneously broken in the vacuum. When we expand the field for small fluctuations around its vacuum expectation value (VEV), it takes the form

$$\phi(x) = \frac{1}{\sqrt{2}}(f + r(x))e^{i\theta(x)/f}, \qquad\qquad 3.$$

where f is the VEV, $r(x)$ is the massive radial mode, and $\theta(x)$ is the NGB. The factor of $1/\sqrt{2}$ ensures canonical kinetic terms for the real fields r and θ.

The radial field r is invariant under the $U(1)$ symmetry transformation of Equation 2, whereas the NGB field θ shifts

$$\theta \to \theta + \alpha \qquad\qquad 4.$$

under $U(1)$ transformations. We say that the $U(1)$ symmetry is nonlinearly realized. Suppose that we now integrate out the massive field r. We can be sure that the resulting effective Lagrangian for the NGB $\theta(x)$ will not include a mass term for θ, or any potential terms for that matter, because the shift symmetry forbids all nonderivative couplings of θ.

Non-Abelian examples: In the generalization to spontaneously broken non-Abelian symmetries, we find one NGB for every broken symmetry generator. For example, suppose we break $SU(N) \to SU(N-1)$ with the VEV of a single fundamental

ϕ of $SU(N)$. The number of broken generators is the total number of generators of $SU(N)$ minus the number of unbroken generators, i.e.,

$$[N^2 - 1] - [(N-1)^2 - 1] = 2N - 1. \qquad 5.$$

The NGBs are conveniently parametrized by writing

$$\phi = \exp\left\{\frac{i}{f}\left(\begin{array}{c|c} & \begin{matrix} \pi_1 \\ \vdots \\ \pi_{N-1} \end{matrix} \\ \hline \pi_1^* \cdots \pi_{N-1}^* & \pi_0/\sqrt{2} \end{array}\right)\right\}\begin{pmatrix} 0 \\ \vdots \\ 0 \\ f \end{pmatrix} \equiv e^{i\pi/f}\phi_0. \qquad 6.$$

The field π_0 is real, whereas the the fields $\pi_1 \cdots \pi_{N-1}$ are complex. The last equality defines a convenient shorthand notation, which we will employ whenever the precise form of π and ϕ_0 is clear from the context.

Another example of symmetry breaking and NGBs that has found applications in little Higgs model building is

$$SU(N) \rightarrow SO(N). \qquad 7.$$

Here the number of NGBs is the number of fields in the adjoint of $SU(N)$ minus the number of fields in the adjoint of $SO(N)$ (antisymmetric tensor), i.e.,

$$[N^2 - 1] - \frac{N(N-1)}{2} = \frac{N(N+1)}{2} - 1. \qquad 8.$$

For even N we also have

$$SU(N) \rightarrow SP(N). \qquad 9.$$

and the number of NGBs is the number of fields in the adjoint of $SU(N)$ minus the number of fields in the adjoint of $SP(N)$ (symmetric tensor), i.e.,

$$[N^2 - 1] - \frac{N(N+1)}{2} = \frac{N(N-1)}{2} - 1. \qquad 10.$$

Finally, for

$$SU(N) \times SU(N) \rightarrow SU(N), \qquad 11.$$

the number of NGBs is

$$2[N^2 - 1] - [N^2 - 1] = N^2 - 1. \qquad 12.$$

In this last case the symmetry breaking can be achieved by a VEV that transforms as a bi-fundamental under the two $SU(N)$ symmetries. Denoting transformation matrices of the two $SU(N)$ as L and R respectively, we have

$$\phi \rightarrow L\phi R^\dagger. \qquad 13.$$

The symmetry-breaking VEV is proportional to the unit matrix

$$\langle \phi \rangle \equiv \phi_0 = f \begin{pmatrix} 1 & & 0 \\ & \ddots & \\ 0 & & 1 \end{pmatrix}. \qquad 14.$$

This VEV is left invariant under vector transformations for which $L = R \equiv U$,

$$\phi_0 \rightarrow U\phi_0 U^\dagger = \phi_0, \qquad 15.$$

whereas other symmetry generators (the axial generators) are broken. The corresponding NGBs can be parametrized as

$$\phi = \phi_0 e^{i\pi/f} = f e^{i\pi/f}, \qquad 16.$$

where π is a Hermitian traceless matrix with $N^2 - 1$ independent components.

2.1. How Do NGBs Transform?

We now show how NGBs transform under the broken and unbroken symmetries in the example of $SU(N) \rightarrow SU(N-1)$, which is often denoted in more mathematical notation as $SU(N)/SU(N-1)$. The NGBs can be parametrized as $\phi = e^{i\pi}\phi_0$ as in Equation 6. Let's consider first the unbroken $SU(N-1)$ transformations. Under these transformations we have

$$\phi \rightarrow U_{N-1}\phi = \left(U_{N-1} e^{i\pi/f} U_{N-1}^\dagger \right) U_{N-1}\phi_0 = e^{i(U_{N-1}\pi U_{N-1}^\dagger)/f}\phi_0, \qquad 17.$$

where in the second equality we used the fact that the symmetry-breaking ϕ_0 is invariant under the unbroken U_{N-1} transformations. Therefore the NGBs transform in the usual linear way under $SU(N-1)$, $\pi \rightarrow U_{N-1}\pi U_{N-1}^\dagger$. Explicitly, the unbroken $SU(N-1)$ transformations are

$$U_{N-1} = \begin{pmatrix} \hat{U}_{N-1} & 0 \\ 0 & 1 \end{pmatrix}. \qquad 18.$$

The single real NGB π_0 transforms as a singlet, whereas the $N-1$ complex NGBs transform as

$$\left(\begin{array}{c|c} 0 & \vec{\pi} \\ \hline \vec{\pi}^\dagger & 0 \end{array} \right) \rightarrow U_{N-1} \left(\begin{array}{c|c} 0 & \vec{\pi} \\ \hline \vec{\pi}^\dagger & 0 \end{array} \right) U_{N-1}^\dagger = \left(\begin{array}{c|c} 0 & \hat{U}_{N-1}\vec{\pi} \\ \hline \vec{\pi}^\dagger \hat{U}_{N-1}^\dagger & 0 \end{array} \right), \qquad 19.$$

where we have used a vector notation $\vec{\pi}$ to represent the $N-1$ complex NGBs as a column vector. We see that $\vec{\pi} \rightarrow \hat{U}_{N-1}\vec{\pi}$, i.e. $\vec{\pi}$ transforms in the fundamental representation of $SU(N-1)$.

Under the broken symmetry transformations we have

$$\phi \rightarrow Ue^{i\pi}\phi_0 = \exp\left\{ i \begin{pmatrix} 0 & \vec{\alpha} \\ \vec{\alpha}^\dagger & 0 \end{pmatrix} \right\} \exp\left\{ \frac{i}{f} \begin{pmatrix} 0 & \vec{\pi} \\ \vec{\pi}^\dagger & 0 \end{pmatrix} \right\} \phi_0$$

$$\equiv \exp\left\{\frac{i}{f}\begin{pmatrix} 0 & \vec{\pi}' \\ \vec{\pi}'^\dagger & 0 \end{pmatrix}\right\} U_{N-1}(\vec{\alpha}, \vec{\pi})\phi_0$$

$$= \exp\left\{\frac{i}{f}\begin{pmatrix} 0 & \vec{\pi}' \\ \vec{\pi}'^\dagger & 0 \end{pmatrix}\right\} \phi_0, \qquad 20.$$

where in the second equality we used the fact that any $SU(N)$ transformation can be written as the product of a transformation in the coset $SU(N)/SU(N-1)$ times an $SU(N-1)$ transformation (35). The $U_{N-1}(\vec{\alpha}, \vec{\pi})$ transformation, which depends on $\vec{\alpha}$ and $\vec{\pi}$, leaves ϕ_0 invariant and can therefore be removed. Equation 20 defines the transformed field $\vec{\pi}'$, and in general, $\vec{\pi}'$ is a complicated function of $\vec{\alpha}$ and $\vec{\pi}$. To linear order the transformation is simple,

$$\vec{\pi} \to \vec{\pi}' = \vec{\pi} + f\vec{\alpha}, \qquad 21.$$

showing that the NGBs shift under the nonlinearly realized symmetry transformations. As in the $U(1)$ case, the shift symmetry ensures that NGBs can only have derivative interactions.

2.2. Effective Lagrangian for NGBs

Our goal for this section is to write the most general effective Lagrangian involving only the massless NGB fields that still respects the full $SU(N)$ symmetry. This is where the utility of the exponentiated fields ϕ becomes obvious: Although the full $SU(N)$ transformations on the πs are complicated, the ϕs transform very simply. To get the low energy effective Lagrangian, we expand in powers of ∂_μ/Λ and write the most general possible $SU(N)$-invariant function of $\phi = e^{i\pi/f}\phi_0$ at every order. With no derivatives we can form two basic gauge invariant objects, $\phi^\dagger\phi = f^2$ and $\epsilon^{a_1\cdots a_N}\phi_{a_1}\phi_{a_2}\cdots\phi_{a_N} = 0$. Thus the most general invariant contribution to the potential is simply a constant, so that the most general term that can be written at quadratic order is a constant times $|\partial_\mu\phi|^2$, and therefore,

$$\mathcal{L} = \text{const.} + |\partial_\mu\phi|^2 + \mathcal{O}(\partial^4), \qquad 22.$$

where we normalized the coefficient of the second-order term such that the π fields have canonical kinetic terms. The kinetic term for ϕ expanded to higher-order in the π fields contains interactions that determine the scattering of arbitrary numbers of π's at low energies, in terms of the single parameter f.

3. CONSTRUCTING A LITTLE HIGGS MODEL: $SU(3)$

Now that we know how to write a Lagrangian for NGBs, we would like to use this knowledge to write a model where the Higgs is an NGB. The explicit model we are going to construct in this section is the simplest little Higgs (19, 20, 30).

Consider the symmetry-breaking pattern $SU(3)/SU(2)$, with NGBs

$$\pi = \left(\begin{array}{cc|c} -\eta/2 & 0 & \\ 0 & -\eta/2 & h \\ \hline h^\dagger & & \eta \end{array} \right).$$ 23.

Note that h is a doublet under the unbroken $SU(2)$, as required for the Standard Model Higgs, and it is also an NGB—it shifts under broken $SU(3)$ transformations. The field η is an $SU(2)$ singlet, which we will ignore for simplicity in most of the following. To see what interactions we get for h, we expand

$$\phi = \exp\left\{ \frac{i}{f} \left(\begin{array}{cc} 0 & h \\ h^\dagger & 0 \end{array} \right) \right\} \left(\begin{array}{c} 0 \\ f \end{array} \right) = \left(\begin{array}{c} 0 \\ f \end{array} \right) + i \left(\begin{array}{c} h \\ 0 \end{array} \right) - \frac{1}{2f} \left(\begin{array}{c} 0 \\ h^\dagger h \end{array} \right) + \cdots,$$ 24.

and therefore obtain

$$|\partial_\mu \phi|^2 = |\partial_\mu h|^2 + \frac{|\partial_\mu h|^2 h^\dagger h}{f^2} + \cdots,$$ 25.

which contains the Higgs kinetic term as well as interactions suppressed by the symmetry-breaking scale f.

Because the Lagrangian contains nonrenormalizable interactions, it can only be an effective low-energy description of physics. To determine the cutoff Λ at which the theory becomes strongly coupled, we can compute a loop and ask at what scale it becomes as important as a corresponding tree-level diagram. The simplest example is the quadratically divergent one-loop contribution to the kinetic term that stems from contracting $h^\dagger h$ into a loop in the second term in Equation 25. Cutting the divergence off at Λ we find a renormalization of the kinetic term proportional to

$$\frac{1}{f^2} \frac{\Lambda^2}{16\pi^2},$$ 26.

and therefore $\Lambda \lesssim 4\pi f$.

Summarizing, we now have a theory that produces a "Higgs" doublet transforming under an exactly preserved (global) $SU(2)$. This "Higgs" is an NGB and therefore exactly massless. It has nonrenormalizable interactions suppressed by the scale f, which become strongly coupled at $\Lambda = 4\pi f$. Because of the shift symmetry, no diagrams, divergent or not, can give rise to a mass for h. Of course, at this point the theory is still very far from what we want: An NGB can only have derivative interactions, which means no gauge interactions, no Yukawa couplings, and no quartic potential. Any of these interactions explicitly break the shift symmetry $h \to h + \text{const}$. In the following subsections we discuss how to add these interactions without reintroducing one-loop quadratic divergences.

3.1. Gauge Interactions

Let us try to introduce the $SU(2)$ gauge interactions for h (we ignore hypercharge for the moment—it will be easy to add later). To do so we simply follow our nose and see where it leads us. We will arrive at the right answer after a few unsuccessful attempts.

First attempt: Let's simply couple h to $SU(2)$ gauge bosons in the usual way, by adding to the Lagrangian of Equation 25 the term

$$|gW_\mu h|^2,$$

27.

along with another term with one derivative and one $SU(2)$ gauge boson W_μ, as required by gauge invariance. These terms lead to quadratically divergent Feynman diagrams (Figure 3) that generate a mass term

$$\frac{g^2}{16\pi^2}\Lambda^2 h^\dagger h.$$

28.

Note that these diagrams are exactly the quadratically divergent Standard Model gauge loops that we set out to cancel. We have apparently gained nothing: We started with a theory in which the Higgs was protected by a nonlinearly realized $SU(3)$ symmetry (under which h shifts), but then we added the term of Equation 27, which completely and explicitly breaks the symmetry. Of course, we necessarily have to break the shift symmetry in order to generate gauge interactions for h, but we must break the symmetry in a subtler way to avoid quadratic divergences in the Higgs mass.

Second attempt: Let's couple h to gauge fields through the more $SU(3)$-symmetric looking expression

$$\left|g\begin{pmatrix} W_\mu \\ & 0 \end{pmatrix}\phi\right|^2,$$

29.

where W_μ contains the three $SU(2)$ gauge bosons. (Really, we write $|D_\mu\phi|^2$, where the covariant derivative involves only $SU(2)$ gauge bosons. The two-gauge

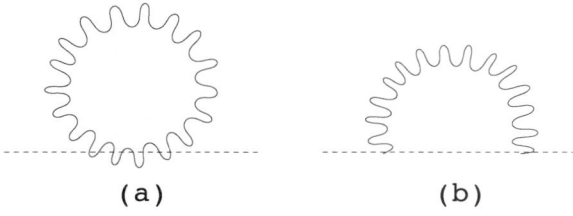

(a)   (b)

Figure 3 Quadratically divergent gauge loop contributions to the Higgs mass.

boson-coupling is then Equation 29). This still generates a quadratically divergent contribution to the Higgs mass. The diagram is the same as before except with external ϕ fields, and gives

$$\frac{g^2}{16\pi^2}\Lambda^2\phi^\dagger \begin{pmatrix} 1 & & \\ & 1 & \\ & & 0 \end{pmatrix} \phi = \frac{g^2}{16\pi^2}\Lambda^2 h^\dagger h + \cdots, \qquad 30.$$

where the projection matrix $\text{diag}(1, 1, 0)$ arises from summing over the three $SU(2)$ gauge bosons running in the loop. Not surprisingly, we got the same answer as before because we added the same interactions, just using a fancier notation.

Third attempt: Let us preserve $SU(3)$ by gauging the full $SU(3)$ symmetry, i.e., by adding $|D_\mu\phi|^2$, where now the covariant derivative contains the 8 gauge bosons of $SU(3)$. Again we can write the same quadratically divergent diagram and find

$$\frac{g^2}{16\pi^2}\Lambda^2\phi^\dagger \begin{pmatrix} 1 & & \\ & 1 & \\ & & 1 \end{pmatrix} \phi = \frac{g^2}{16\pi^2}\Lambda^2 f^2, \qquad 31.$$

which has no dependence on $\bar{h}$. The quadratic divergence contributes a constant term to the vacuum energy, but no mass for the "Higgs" doublet h! Unfortunately, we have also lost h: The NGBs are eaten by the heavy $SU(3)$ gauge bosons corresponding to the broken generators, i.e., they become the longitudinal components of the gauge bosons.

We have now exhausted all possible ways of adding $SU(2)$ gauge interactions to our simple toy model for h. The lesson is that we can avoid the quadratically divergent contribution to the mass of h by writing $SU(3)$-invariant gauge interactions; the problem that remains is that our "Higgs" was eaten. But this is easy to fix.

Fourth attempt (successful): We use two copies of NGBs, ϕ_1 and ϕ_2, and add $SU(3)$ invariant covariant derivatives for both. We expect no quadratic divergence for either of the NGBs, and only one linear combination will be eaten. To see how this works in detail we parametrize

$$\phi_1 = e^{i\pi_1/f}\begin{pmatrix} f \end{pmatrix} \qquad \phi_2 = e^{i\pi_2/f}\begin{pmatrix} f \end{pmatrix}, \qquad 32.$$

where we have assumed aligned VEVs for ϕ_1 and ϕ_2 and, for simplicity, identical symmetry-breaking scales $f_1 = f_2 = f$. The Lagrangian is

$$\mathcal{L} = |D_\mu\phi_1|^2 + |D_\mu\phi_2|^2. \qquad 33.$$

The two interaction terms produce two sets of quadratically divergent one-loop diagrams similar to those of the previous attempt (Figure 4a), which give

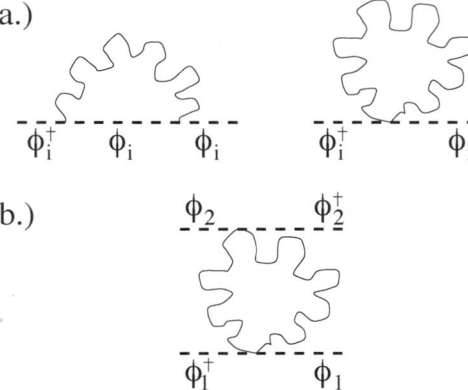

Figure 4 (*a*) Quadratically divergent gauge loop contributions that do not contribute to the Higgs potential, (*b*) log-divergent contribution to the Higgs mass.

$$\frac{g^2}{16\pi^2}\Lambda^2\left(\phi_1^\dagger\phi_1 + \phi_2^\dagger\phi_2\right) = \frac{g^2}{16\pi^2}\Lambda^2(f^2 + f^2), \qquad 34.$$

i.e., no potential for any of the NGBs. Moreover, only one linear combination of π_1 and π_2 is eaten as there is only one set of hungry massive $SU(3)$ gauge bosons. A simple way to understand this result is to notice that each set of diagrams involves only one of the ϕ fields. Therefore the diagrams are the same as in the theory with only one ϕ, where all NGBs were eaten, and so neither ϕ_1 nor ϕ_2 can get a potential.

This reasoning does not apply once we consider diagrams involving both ϕ_1 and ϕ_2. For example, the diagram in Figure 4*b* gives

$$\frac{g^4}{16\pi^2}\log\left(\frac{\Lambda^2}{\mu^2}\right)\left|\phi_1^\dagger\phi_2\right|^2, \qquad 35.$$

which does depend on h but is not quadratically divergent. To calculate the Higgs dependence we choose a convenient parametrization

$$\phi_1 = \exp\left\{\frac{i}{f}\begin{pmatrix} & k \\ k^\dagger & \end{pmatrix}\right\}\exp\left\{\frac{i}{f}\begin{pmatrix} & h \\ h^\dagger & \end{pmatrix}\right\}(f) \qquad 36.$$

$$\phi_2 = \exp\left\{\frac{i}{f}\begin{pmatrix} & k \\ k^\dagger & \end{pmatrix}\right\}\exp\left\{-\frac{i}{f}\begin{pmatrix} & h \\ h^\dagger & \end{pmatrix}\right\}(f). \qquad 37.$$

The field k can be removed by an $SU(3)$ gauge transformation, and it corresponds to the eaten NGBs, whereas h cannot simultaneously be removed from ϕ_1 and ϕ_2, and it is physical. In the following we will work in the unitary gauge for $SU(3)$, where k is rotated away. Then we have

$$\phi_1^\dagger \phi_2 = \begin{pmatrix} 0 & f \end{pmatrix} \exp\left\{ -\frac{2i}{f} \begin{pmatrix} & h \\ h^\dagger & \end{pmatrix} \right\} \begin{pmatrix} 0 \\ f \end{pmatrix}$$

$$= \left[f^2 \begin{pmatrix} 1 & \\ & 1 \end{pmatrix} - 2fi \begin{pmatrix} & h \\ h^\dagger & \end{pmatrix} - 2 \begin{pmatrix} hh^\dagger & \\ & h^\dagger h \end{pmatrix} + \cdots \right]_{33}$$

$$= f^2 - 2h^\dagger h + \cdots, \qquad\qquad\qquad 38.$$

and we see that Equation 35 contains a mass-squared for h equal to

$$g^4/(16\pi^2) \log\left(\frac{\Lambda^2}{\mu^2}\right) f^2, \qquad\qquad 39.$$

which is $\sim M_{weak}^2$ for g equal to the $SU(2)$ gauge coupling and $f \sim$ TeV.

To summarize, the theory of two complex triplets that both break $SU(3) \rightarrow SU(2)$ automatically produces a Higgs doublet pseudo-NGB that does not receive quadratically divergent contributions to its mass. There are log-divergent and finite contributions, and from these the natural size for the Higgs mass is $f/4\pi \sim M_{weak}$. Thus our theory has three relevant scales that are separated by loop factors, $\Lambda \sim 4\pi f \sim (4\pi)^2 M_{weak}$.

3.2. Symmetry Argument, Collective Breaking

Let us understand the absence of one-loop quadratic divergence in the mass of h using symmetries. The lesson we will learn is valuable as it generalizes to other couplings, and so provides a general recipe for constructing little Higgs theories.

Without gauge interactions, our theory would consist of two nonlinear sigma fields, each representing the spontaneous breaking of an $SU(3)$ global symmetry to $SU(2)$. The coset is thus $[SU(3)/SU(2)]^2$. There are 10 spontaneously broken generators for this coset, and therefore 10 NGBs. The gauge couplings explicitly break some of the global symmetries. For example, the two-gauge boson–two-scalar coupling

$$\mathcal{L} \sim |gA_\mu \phi_1|^2 + |gA_\mu \phi_2|^2 \qquad\qquad 40.$$

breaks the two previously independent $SU(3)$ symmetries to the diagonal (gauged) $SU(3)$. Thus only one of the spontaneously broken symmetries is exact, and only one set of exact NGBs arises, the eaten ones. The other linear combination, corresponding to the explicitly broken axial $SU(3)$, gets a potential from loops.

However, as we saw before, there is no quadratically divergent contribution to the potential. This is easy to understand by considering the symmetries left invariant by each of the terms in Equation 40 separately. Imagine setting the gauge coupling of ϕ_2 to zero. Then the Lagrangian has 2 independent $SU(3)$ symmetries, one acting on ϕ_1 (and A_μ) and the other acting on ϕ_2. Thus we now have two spontaneously broken $SU(3)$ symmetries and therefore 10 exact NGBs (5 of which are eaten). Similarly, if the gauge coupling of ϕ_1 is set to zero, there are again two spontaneously broken $SU(3)$'s. Only in the presence of gauge couplings for both

ϕ_1 and ϕ_2 are the two $SU(3)$ symmetries explicitly broken to one $SU(3)$, and only then can h develop a potential. Therefore any diagram that contributes to the h mass must involve the gauge couplings for both ϕ_1 and ϕ_2. But there are no quadratically divergent one-loop diagrams involving both couplings.

This is the general mechanism employed by little Higgs theories (13):

> The little Higgs is a pseudo-Nambu-Goldstone boson of a spontaneously broken symmetry. This symmetry is also explicitly broken but only collectively, i.e., the symmetry is broken when two or more couplings in the Lagrangian are nonvanishing. Setting any one of these couplings to zero restores the symmetry and therefore the masslessness of the little Higgs.

We now know how to construct a theory with a naturally light scalar doublet that couples to $SU(2)$ gauge bosons. To turn this into an extension of the Standard Model we still need (*a*) Yukawa couplings, (*b*) hypercharge and color, and (*c*) a Higgs potential with a quartic coupling.

3.3. Top Yukawa Coupling

The numerically most significant quadratic divergence stems from top quark loops. Thus the cancellation of the quadratic divergence associated with the top Yukawa is the most important. Let us construct a sector that guarantees this cancellation. The crucial trick is to introduce $SU(3)$ symmetries into the Yukawa couplings that are only broken collectively. First, we enlarge the quark doublets into triplets $\Psi \equiv (t, b, T)$ transforming under the $SU(3)$ gauge symmetry. The quark singlets remain the same, t^c and b^c, except that we also need to add a Dirac partner T^c for T. Note that we are using a notation in which all quark fields are left-handed Weyl spinors, and the Standard Model Yukawa couplings are of the form $h^\dagger Q t^c$. Let us change notation slightly to reflect the fact that t^c and T^c mix, and call them t_1^c and t_2^c. We can now write two terms that both look like they contribute to the top Yukawa coupling,[1]

$$\mathcal{L}_{yuk} = \lambda_1 \phi_1^\dagger \Psi t_1^c + \lambda_2 \phi_2^\dagger \Psi t_2^c. \qquad 41.$$

To see what couplings for the Higgs arise, we substitute the parametrization of Equation 37 and expand in powers of h. For simplicity, let us also set $\lambda_1 \equiv \lambda_2 \equiv \lambda/\sqrt{2}$. This will reduce the number of terms we encounter because it preserves a parity $1 \leftrightarrow 2$, but the main points here are independent of this choice. We find

$$\mathcal{L} \sim \frac{\lambda}{\sqrt{2}} \left[fT\left(t_2^c + t_1^c\right) + ih^\dagger Q\left(t_2^c - t_1^c\right) - \frac{1}{2f} h^\dagger h T\left(t_2^c + t_1^c\right) + \cdots \right]$$

$$= \lambda f \left(1 - \frac{1}{2f^2} h^\dagger h \right) T T^c + \lambda h^\dagger Q t^c + \cdots, \qquad 42.$$

---

[1]We do not write the couplings $\phi_1^\dagger \Psi t_2^c$ and $\phi_2^\dagger \Psi t_1^c$ as they would reintroduce quadratic divergences. They can be forbidden by global $U(1)$ symmetries and are therefore not generated by loops.

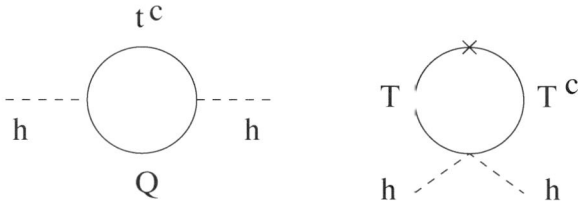

Figure 5   The quadratically divergent contribution to the Higgs mass from the top loop is canceled by the T loop.

where the second line is written in terms of the linear combinations $T^c = (t_2^c + t_1^c)/\sqrt{2}$ and $t^c = i(t_2^c - t_1^c)/\sqrt{2}$.

The last term of the second line in Equation 42 is the top Yukawa coupling, and so we identify $\lambda = \lambda_t$. The Dirac fermion T, T^c has a mass $\lambda_t f$ and a coupling to two Higgs fields with coupling constant $\lambda_t/(2f)$. The couplings and masses are related by the underlying $SU(3)$ symmetries. To see how the new fermion and its couplings to the Higgs cancel the quadratic divergence from the top quark loop, we compute the fermion loops including interactions to order λ^2. The two relevant diagrams (Figure 5) give

$$\frac{\lambda_t^2}{16\pi^2}\Lambda^2 h^\dagger h + \frac{\lambda_t^2 f^2}{16\pi^2}\left(1 - \frac{h^\dagger h}{f^2}\right)\Lambda^2 + \mathcal{O}(h^4) = \text{const.} + \mathcal{O}(h^4).   43.$$

The quadratically divergent contribution to the Higgs mass from the top and T loops cancel![2]

Although this computation allowed us to see explicitly that the quadratic divergence from t and T cancel, the absence of a quadratic divergence to the Higgs mass is much more naturally understood by analyzing the symmetries of the Lagrangian for the ϕ_i fields, Equation 41. First note that the Yukawa coupling Lagrangian preserves one $SU(3)$ symmetry, the gauge symmetry. The term proportional to λ_1 forces symmetry transformations of ϕ_1 and Ψ to be aligned, and the term proportional to λ_2 also forces ϕ_2 to transform like Ψ. Thus, in the presence of both terms the global symmetry-breaking pattern is only $SU(3)/SU(2)$, with 5 NGBs, which are all eaten by the heavy $SU(3)$ gauge bosons. However, if we set either of the λ_i to zero, the symmetry of Equation 41 is enhanced to $SU(3)^2$ because the ϕ_i can now rotate independently. Thus, with either of the λ_i turned off, we expect two sets of NGBs. One linear combination is eaten and the other is the little Higgs.

To understand radiative stability of this result we observe that a contribution to the Higgs potential can only come from a diagram that involves both λ_i. The lowest-order fermion diagram that involves both λ_i is the loop shown in Figure 6, which is proportional to $|\lambda_1\lambda_2|^2$. It is clear that one cannot produce a diagram

---

[2]In order for the two cutoffs for the two loops to be identical, the new physics at the cutoff must respect the $SU(3)$ symmetries. This is analogous to the situation in SUSY where the boson-fermion cancellation also relies on a supersymmetric regulator/cutoff.

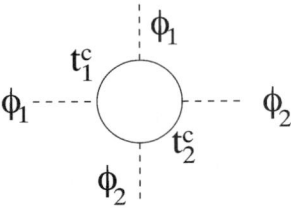

Figure 6   A log divergent contribution to the Higgs mass from the top and T loops proportional to $|\lambda_1\lambda_2|^2$.

that contributes to the Higgs potential and is proportional to only a single power of $\lambda_1\lambda_2$. This also follows from an argument using spurious symmetries: assign t_1^c charge 1 under a $U(1)_1$ symmetry while all other fields are neutral. The symmetry is broken by the Yukawa coupling λ_1, but it can be formally restored by assigning the spurion λ_1 charge -1. Any effective operators that may be generated by loops must be invariant under this symmetry. In particular, operators that contribute to the Higgs potential and do not contain the fermion field t_1^c can depend on the spurion λ_1 only through $|\lambda_1|^2$. Of course, the same argument shows that the dependence on λ_2 is through $|\lambda_2|^2$ only. A contribution to the Higgs potential requires both couplings λ_1 and λ_2 to appear and therefore the potential is proportional to at least $|\lambda_1\lambda_2|^2$, i.e., it has at least 4 coupling constants. But a one-loop diagram with 4 coupling constants can at most be logarithmically divergent and therefore does not destabilize the Higgs mass.

In the explicit formulae above, we assumed for simplicity that $f_1 = f_2 = f$ and $\lambda_1 = \lambda_2 = \lambda_t/\sqrt{2}$. In the general case we find

$$m_T = \sqrt{\lambda_1^2 f_1^2 + \lambda_2^2 f_2^2} \qquad\qquad 44.$$

$$\lambda_t = \lambda_1\lambda_2\frac{\sqrt{f_1^2 + f_2^2}}{m_T}. \qquad\qquad 45.$$

Note that the top Yukawa coupling goes to zero as either of the λ_i is taken to zero, as anticipated from the $SU(3)$ symmetry arguments. Furthermore, the mass of the heavy T quark can be significantly lower than the larger of the two f_i if the corresponding λ_i is smaller than 1. This is a nice feature because it allows us to take the heavy gauge boson's masses large ($\gtrsim$ few TeV as required by the precision electroweak constraints) while keeping the T mass near a TeV. Keeping the T mass as low as possible is desirable because the quadratic divergence of the top loop in the Standard Model is cut off at the mass of T.

3.4. Other Yukawa Couplings

The other up-type Yukawa couplings may be added in exactly the same way. We enlarge the $SU(2)$ quark doublets into triplets because of the gauged $SU(3)$. Then we add two sets of Yukawa couplings that couple the triplets to ϕ_1 and ϕ_2 and quark singlets q_1^c and q_2^c.

In the Standard Model, Yukawa couplings for down-type quarks arise from a different operator where the $SU(2)$ indices of the Higgs doublet and the quark doublets are contracted using an epsilon tensor (or, equivalently, the conjugate Higgs field $h^c = i\sigma_2 h^\dagger$ is used). Before explicitly constructing this operator from the quark and ϕ_i fields, note that even the bottom Yukawa coupling is too small to give a significant contribution to the Higgs mass. The quadratically divergent one loop diagram in the Standard Model yields

$$\frac{\lambda_b^2}{16\pi^2}\Lambda^2 \approx (30\ \text{GeV})^2 \qquad 46.$$

for $\Lambda \sim 10$ TeV. Therefore, we need not pay attention to symmetries and collective breaking when constructing the down-type Yukawa couplings. The Standard Model Yukawa is

$$\lambda_b \epsilon_{ij} h_i Q_j b^c. \qquad 47.$$

To obtain the epsilon contraction from an $SU(3)$-invariant operator we write the Lagrangian term

$$\frac{\lambda_b}{f}\epsilon_{ijk}\phi_1^i \phi_2^j \Psi_Q^k b^c. \qquad 48.$$

The ϵ_{ijk} contraction breaks both $SU(3)$ symmetries (acting on the two scalar triplets ϕ_1 and ϕ_2) to the diagonal subgroup, and therefore this operator does lead to a quadratic divergence. But the quadratic divergence is harmless because of the small size of the bottom Yukawa coupling.

3.5. Color and Hypercharge

Color is added by simply adding $SU(3)_{\text{color}}$ indices where we expect them from the Standard Model. $SU(3)_{\text{color}}$ commutes with all the symmetry arguments given above, and therefore nothing significant changes.

Hypercharge is slightly more complicated. The VEVs $\phi_i \propto (0, 0, 1)$ break the $SU(3)_{\text{weak}}$ gauge group to $SU(2)$, i.e., no $U(1)$ hypercharge candidate is left. Therefore, we gauge an additional $U(1)_X$. In order for the hypercharge of the Higgs to come out correctly, we assign it $SU(3) \times U(1)_X$ quantum numbers

$$\phi_i = 3_{-1/3}. \qquad 49.$$

The combination of generators that is unbroken by $\phi_i \sim (0, 0, 1)$ is

$$Y = \frac{-1}{\sqrt{3}}T^8 + X \quad \text{where} \quad T^8 = \frac{1}{2\sqrt{3}}\begin{pmatrix} 1 & & \\ & 1 & \\ & & -2 \end{pmatrix}, \qquad 50.$$

and X is the generator corresponding to $U(1)_X$. This uniquely fixes the $U(1)_X$ charges of all quarks and leptons once their $SU(3)$ transformation properties are chosen.

For example, the covariant derivative acting on ϕ_i is

$$D_\mu \phi = \partial_\mu \phi - \frac{1}{3} i g_X A_\mu^X \phi + i g A_\mu^{SU(3)} \phi. \qquad 51.$$

The $U(1)_X$ generator commutes with $SU(3)$, and the $U(1)_X$ gauge interactions do not change any of the symmetry arguments that we used to show that the Higgs mass is not quadratically divergent at one loop.

There are now three neutral gauge bosons corresponding to the generators T^3, T^8, X. These gauge bosons mix, and the mass eigenstates are the photon, the Z, and a Z'. The Z' leads to interesting modifications of predictions for precision electroweak observables, as discussed in Section 5.

3.6. Quartic Higgs Coupling

To generate a quartic Higgs coupling we want to write a potential $V(\phi_1, \phi_2)$ that (*a*) contains no mass at order f for the Higgs, (*b*) contains a quartic coupling, and (*c*) preserves the collective symmetry breaking of the $SU(3)$'s, i.e., the quartic coupling is generated by at least two couplings in V, and if one sets either one of them to zero the Higgs becomes an exact NGB. This last property guarantees radiative stability, that is, no Λ^2 contributions to the Higgs mass.

Writing down a potential that satisfies these properties appears to be impossible for the pure $SU(3)$ model. To see why it is not straightforward, note that $\phi_1^\dagger \phi_2$ is the only nontrivial gauge invariant that can be formed from ϕ_1 and ϕ_2. ($\phi_1^\dagger \phi_1 = $ const $= \phi_2^\dagger \phi_2$ and $\epsilon^{ijk} \phi_i \phi_j \phi_k = 0$.) But the $\phi_1^\dagger \phi_2$ invariant is a bad starting point because it breaks the two $SU(3)$'s to the diagonal, and it is not surprising that generic functions of $\phi_1^\dagger \phi_2$ always contain a mass as well as a quartic. For example, we have

$$\phi_1^\dagger \phi_2 \sim f^2 - h^\dagger h + \frac{1}{f} (h^\dagger h)^2 + \cdots, \qquad 52.$$

so that

$$\frac{1}{f^{2n-4}} (\phi_1^\dagger \phi_2)^n \sim f^4 - f^2 h^\dagger h + (h^\dagger h)^2 + \cdots. \qquad 53.$$

By dialing the coefficient of this operator we can either get a small enough mass term or a large enough quartic coupling, but not both. Of course, we could try to tune two terms with different powers n such that the mass terms cancel between them, but that tuning is not radiatively stable.

There are two different solutions to the problem in the literature. Both require enlarging the model and symmetry structure. One solution, suggested by Kaplan and Schmaltz (20), involves enlarging the gauge symmetry to $SU(4)$ and introducing four ϕ fields, each transforming as a **4** of $SU(4)$. The four ϕ fields break $SU(4) \rightarrow SU(2)$, yielding 4 $SU(2)$ doublets. Two of them are eaten, and the other

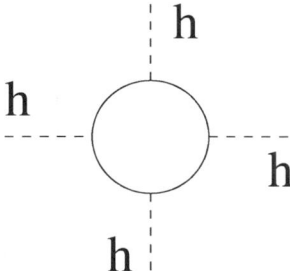

Figure 7   The top loop contribution to the Higgs quartic coupling.

two are little Higgs fields with a quartic potential similar to the quartic potential in SUSY.

The other solution, suggested by Skiba and Terning (22), keeps the $SU(3)$ gauge symmetry the same but enlarges the global $SU(3)^2$ symmetry to $SU(3)^3$, which is then embedded in an $SU(9)$. The larger symmetry also leads to two little Higgs doublets for which a quartic coupling can be written. Both of these solutions spoil some of the simplicity of the $SU(3)$ model, but they allow a large quartic coupling for the Higgs fields with natural electroweak symmetry breaking. We refer the reader to the original papers for details on these models.

A third option (30) is simply to add a potential with a very small coefficient. The resulting quartic coupling is then also very small, but as in the MSSM, radiative corrections from the top loop give a contribution that can raise the Higgs mass above the experimental bound of 114 GeV. Explicitly, below the T mass the cancellations in the top sector no longer occur, and the diagram in Figure 7 gives a quartic-term contribution

$$\frac{3\lambda_t^4}{16\pi^2} \log\left(\frac{m_T^2}{m_t^2}\right)(h^\dagger h)^2, \qquad\qquad 54.$$

which is too small by itself but does give successful electroweak symmetry breaking when combined with a small tree-level contribution. Because the tree-level term also contributes to the Higgs mass-squared, a moderate amount of tuning ($\sim$10%) is required. Although this is not completely satisfactory, it is better than most other models of electroweak symmetry breaking and certainly better than the MSSM with gauge coupling unification, which requires tuning at the few % level or worse.

3.7. The Simplest Little Higgs

To emphasize the simplicity of the model, we summarize the field content and Lagrangian of the simplest little Higgs (30), the $SU(3)$ model in which the Higgs quartic coupling is predominantly generated from the top loop.

The model has an $SU(3)_{\text{color}} \times SU(3)_{\text{weak}} \times U(1)_X$ gauge group with three generations transforming as

$$\Psi_Q = (3,3)_{\frac{1}{3}} \quad \Psi_L = (1,3)_{-\frac{1}{3}}$$

$$d^c = (\bar{3},1)_{\frac{1}{3}} \quad e^c = (1,1)_1$$

$$2 \times u^c = (\bar{3},1)_{-\frac{2}{3}} \quad n^c = (1,1)_0. \qquad 55.$$

The triplets Ψ_Q and Ψ_L contain the Standard Model quark and lepton doublets, the singlets are u^c, d^c, e^c, n^c.[3] The $SU(3)_{weak} \times U(1)_X$ symmetry is broken by expectation values for scalar fields $\phi_1 = \phi_2 = (1,3)_{-1/3}$.

The Lagrangian of the model contains the usual kinetic terms, Yukawa couplings and a tree-level Higgs potential

$$\mathcal{L}_{kin} \sim \Psi_Q^\dagger \slashed{D} \Psi_Q + \cdots + |D_\mu \phi_1|^2 + \cdots \qquad 56.$$

$$\mathcal{L}_{yuk} \sim \lambda_1^u \phi_1^\dagger \Psi_Q u_1^c + \lambda_2^u \phi_2^\dagger \Psi_Q u_2^c + \frac{\lambda^d}{f}\phi_1\phi_2\Psi_Q d^c$$

$$+ \lambda^n \phi_1^\dagger \Psi_L n^c + \frac{\lambda^e}{f}\phi_1\phi_2\Psi_L e^c \qquad 57.$$

$$\mathcal{L}_{pot} \sim \mu^2 \phi_1^\dagger \phi_2. \qquad 58.$$

Substituting the parametrization for the NGBs

$$\phi_1 = e^{i\Theta\frac{f_2}{f_1}}\begin{pmatrix} 0 \\ 0 \\ f_1 \end{pmatrix}, \quad \phi_2 = e^{-i\Theta\frac{f_1}{f_2}}\begin{pmatrix} 0 \\ 0 \\ f_2 \end{pmatrix}, \quad \Theta = \frac{\eta}{\sqrt{2}f} + \frac{1}{f}\begin{pmatrix} 0 & 0 & h \\ 0 & 0 & h \\ h^\dagger & 0 \end{pmatrix} \qquad 59.$$

where $f^2 = f_1^2 + f_2^2$, we can solve for the spectrum of heavy new gauge bosons $W^{+\prime}$, $W^{0\prime}$, Z', fermions T, U, C and scalar η (30).

4. PRODUCT-GROUP MODELS

4.1. The Minimal Moose

In this section, we describe two little Higgs theories whose structures are quite different than that of the $SU(3)$ model. Taken together, the three models illustrate different approaches to implementing the little Higgs mechanism economically.

---

[3]This fermion content is anomalous under the extended electroweak gauge group. Anomalies may be canceled by additional fermions, which can be as heavy as Λ. There are also charge assignments for which anomalies cancel among the fields $\Psi_Q, \Psi_L, u^c, d^c, e^c, n^c$ alone (30, 36, 37).

First we describe the Minimal Moose model presented in Reference (15). The coset on which this model is constructed bears some similarity to that of the chiral Lagrangian used to describe the low-energy dynamics of quantum chromodynamics (QCD). In the case of QCD, an approximate $SU(3)_L \times SU(3)_R$ chiral symmetry, which becomes exact in the limit of vanishing light quark masses, is spontaneously broken to its vector subgroup, so the coset is $(SU(3)_L \times SU(3)_R)/SU(3)_V$. There is an octet of pseudo-NGBs associated with this spontaneous breaking, and these are understood to be the light mesons (π^0, $\pi^\pm$, K^0, $\bar{K}^0$, $K^\pm$, η^0).

The coset used for the Minimal Moose is $[SU(3)_L \times SU(3)_R/SU(3)_V]^4$, so there are four sets of NGB octets that appear in four sigma fields Σ_i, $i = 1, 2, 3, 4$. The sigma fields transform under the global symmetries as

$$\Sigma_i \to L_i \Sigma_i R_i^\dagger, \qquad\qquad 60.$$

where L_i and R_i are $SU(3)$ matrices. The sigma fields can be parametrized in terms of the NGB's π_i^a, as

$$\Sigma_i = e^{2i\pi_i^a T_a/f}, \qquad\qquad 61.$$

where the T_a are the generators of $SU(3)$, normalized so that $\mathrm{tr}(T_a T_b) = \frac{1}{2}\delta_{ab}$.

The idea is to identify some of these NGBs as Higgs doublets responsible for electroweak symmetry breaking (this will turn cut to be a two-Higgs-doublet model). As with the $SU(3)$ model, the challenge will be to give these fields gauge interactions, Yukawa couplings to fermions, and self interactions, in such a way that no quadratically divergent contributions to their masses squared arise at one loop.

4.1.1. GAUGE INTERACTIONS   In this model, the electroweak group descends from a gauged $SU(3) \times SU(2) \times U(1)$ subgroup of the original $SU(3)^8$ global symmetry. The sigma fields transform identically under this gauge group: Under an $SU(3)$ gauge transformation U and an $SU(2) \times U(1)$ transformation V, we have

$$\Sigma_i \to U \Sigma_i V^\dagger. \qquad\qquad 62.$$

Here, $SU(2) \times U(1)$ gauge transformations are embedded in the 3×3 matrix V in such a way that the $SU(2)$ transformation lives in the upper left 2×2 space of V, whereas the gauged $U(1)$ generator is proportional to $\mathrm{diag}(1, 1, -2)$. We can take the Standard Model fermions to be charged under $SU(2) \times U(1)$ with their ordinary quantum numbers and neutral under $SU(3)$. Then, to allow for couplings between the sigma fields and fermions described below, the first two columns of Σ are assigned $U(1)$ charge $-1/6$, whereas the last has charge $1/3$.

The gauge structure of the model can be depicted by the diagram of Figure 8, in which there are two sites corresponding to the $SU(3)$ and $SU(2) \times U(1)$ gauge groups, and links representing the sigma fields that transform under the gauge groups of both sites. The Standard Model fermions live on the $SU(2) \times U(1)$ site.

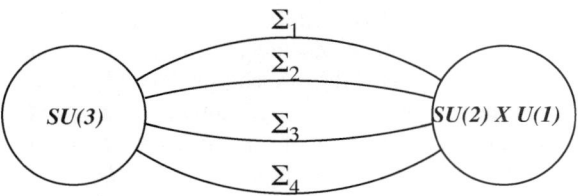

Figure 8   The Minimal Moose.

A simple modification of this model is to have $SU(2) \times U(1)$ gauge groups at both sites, and we will return to this possibility in Section 4.3.

The sigma fields spontaneously break the $SU(3) \times SU(2) \times U(1)$ gauge symmetry to $SU(2)_W \times U(1)_Y$, where, as the subscripts suggest, the unbroken group is identified as the gauge symmetry of the standard electroweak theory. Similarly to what we did for the $SU(3)$ theory, we can assemble the NGBs in each sigma field into a matrix and make their decomposition under $SU(2)_W \times U(1)_Y$ manifest:

$$\pi_i = \pi_i^a T^a = \begin{pmatrix} \phi_i + \eta_i/(2\sqrt{3}) & h_i/\sqrt{2} \\ h_i^\dagger/\sqrt{2} & -\eta_i/\sqrt{3} \end{pmatrix}, \qquad 63.$$

where the real scalar η_i is a singlet under $SU(2)_W \times U(1)_Y$, the real scalars $\phi_i = \phi_i^a \sigma^a/2$ transform as $\mathbf{3_0}$, and the complex doublet h_i transforms like the Standard Model Higgs, $\mathbf{2_{1/2}}$, once we fix the normalization of $U(1)_Y$ charge appropriately.

By expanding the covariant derivatives in the kinetic terms of the nonlinear sigma model,

$$\mathcal{L}_{kin} = \frac{f^2}{4} \sum_{i=1}^{4} \mathrm{tr}[(D_\mu \Sigma_i)^\dagger (D^\mu \Sigma_i)], \qquad 64.$$

one can calculate the masses of the octet of gauge bosons that become heavy when the gauge symmetry is spontaneously broken: These are $\sqrt{g_3^2 + g_2^2} f$ for an electroweak triplet, $g_3 f$ for a pair of doublets, and $\sqrt{g_3^2 + g_1^2/3} f$ for a singlet. Here g_1, g_2, and g_3 are the $SU(3)$, $SU(2)$, and $U(1)$ gauge couplings, which determine the gauge couplings of the electroweak theory according to

$$g' = \frac{g_1 g_3}{\sqrt{g_3^2 + g_1^2/3}} \qquad 65.$$

$$g = \frac{g_2 g_3}{\sqrt{g_2^2 + g_3^2}}. \qquad 66.$$

Assuming that $f \sim$ TeV, we expect that the masses of these heavy gauge bosons will be somewhere near the TeV scale.

The gauge couplings explicitly break the large $SU(3)^8$ global symmetry that we began with, and so the π_i^a fields are exact NGBs only in the limit of vanishing gauge couplings. The only global symmetry that remains exact when the couplings are turned on is $SU(3) \times SU(2) \times U(1)$ itself, and the octet of NGBs associated with the spontaneous breaking of this symmetry to $SU(2)_W \times U(1)_Y$ is eaten: Its members become the longitudinal components of the heavy gauge bosons. The eaten fields appear in the linear combination of π_is that shifts under $SU(3) \times SU(2) \times U(1)$ transformations associated with the broken generators. Because these transformations act identically on the various sigma fields, it is clear that $\pi_1 + \pi_2 + \pi_3 + \pi_4$ is the linear combination eaten (it shifts, whereas the linear combinations orthogonal to it do not).

The uneaten NGBs acquire potentials from loops, but none receives quadratically divergent contributions to its mass at one loop. In the limit that $g_3 \to 0$, an exact $SU(3)^4 \times SU(2) \times U(1)$ global symmetry is recovered. This is spontaneously broken to $SU(2) \times U(1)$, ensuring a total of four exactly massless octets of NGBs (including the ones eaten). If instead g_2 and g_1 are turned off, an $SU(3)^5$ global symmetry is recovered. This is spontaneously broken to $SU(3)$, so all of the NGBs are massless once again. Only when g_3 and at least one of g_1 and g_2 are turned on can these fields acquire mass, so a one-loop contribution to their masses must involve both g_3 and g_2, or both g_3 and g_1. Contributions of this sort are at most logarithmically divergent.

The absence of one-loop quadratic divergences depends crucially on the enlarged gauge symmetry. If only the diagonal $SU(2) \times U(1)$ is gauged, then there are no global symmetries to protect the masses of the scalars, and so they receive quadratically divergent contributions at one loop. This is not surprising given that, in this case, there are no extra particles around to cut off the divergences from ordinary gauge loops. Once the extra $SU(3)$ is gauged, the heavy gauge bosons associated with the spontaneous breaking down to $SU(2)_W \times U(1)_Y$ appear in one-loop diagrams whose quadratic divergences cancel those from diagrams with massless gauge bosons running the loops.

4.1.2. THE QUARTIC COUPLING   A quartic coupling for the Higgs doublets can be generated by adding the following terms to the Lagrangian:

$$\mathcal{L}_\Sigma = \left(\frac{f}{2}\right)^4 \text{tr}\left(c_1 \Sigma_1 \Sigma_2^\dagger \Sigma_3 \Sigma_4^\dagger\right) + \left(\frac{f}{2}\right)^4 \text{tr}\left(c_2 \Sigma_1 \Sigma_4^\dagger \Sigma_3 \Sigma_2^\dagger\right) + \text{h.c.,} \qquad 67.$$

where for now we can imagine that c_1 and c_2 are numbers of order unity, which we take to be real for simplicity. Precisely where these operators come from is an interesting question, but there is no need to specify their ultraviolet origin to work out their consequences in the effective theory. In Reference (15) it was shown that a particular arrangement of couplings of the sigma fields to fermions does generate these operators.

Let's consider the effects of $\mathcal{L}_\Sigma$, forgetting for the moment about the gauge couplings. If only c_1, and not c_2, is turned on, the original $SU(3)^8$ global symmetry is explicitly broken to $SU(3)^4$: Only those transformations satisfying $R_1 = R_2, R_3 = R_4, L_1 = L_4$, and $L_2 = L_3$ leave $\mathcal{L}_\Sigma$ invariant. This $SU(3)^4$ global symmetry is spontaneously broken to $SU(3)$ by the sigma fields, giving three octets of massless NGBs, one of which is eaten. Because we started out with four octets, we expect c_1 to give a potential to one of them.

It is useful to define the linear combinations

$$x = \frac{1}{\sqrt{2}}(\pi_1 - \pi_3) \qquad\qquad 68.$$

$$y = \frac{1}{\sqrt{2}}(\pi_2 - \pi_4) \qquad\qquad 69.$$

$$z = \frac{1}{2}(\pi_1 - \pi_2 + \pi_3 - \pi_4), \qquad\qquad 70.$$

and set the fourth linear combination, the eaten one, to zero. The field z is the linear combination that does not shift under any $SU(3)^4$ transformation preserved by c_1. For instance, under infinitesimal symmetry transformations $R_1 = R_2 = 1 - 2\sqrt{2}i\alpha/f$ and $L_1 = L_4 = 1 + 2\sqrt{2}i\beta/f$, the changes in the fields are

$$\delta x = \alpha + \beta + \cdots$$

$$\delta y = \alpha - \beta + \cdots$$

$$\delta z = \frac{i}{2f}\left([\alpha + \beta, y] - [\alpha - \beta, x]\right) + \cdots, \qquad\qquad 71.$$

where terms of order x^2, y^2, and z have been neglected. Both x and y shift, which tells us that no potential can be generated for them. On the other hand, z does not transform at zeroth order in the fields, and so it can acquire a potential. Expanding $\mathcal{L}_\Sigma$ for $c_2 = 0$, we find

$$\mathcal{L}_\Sigma = -c_1 f^2 \mathrm{tr}\left(z - \frac{i}{2f}[x, y]\right)^2 + \cdots, \qquad\qquad 72.$$

where we have kept terms only up to order z, x^4, and y^4 in the interactions (it is easy to check that this expression is indeed invariant under the transformations of Equation 71). As anticipated, only the z octet acquires a mass-squared $\sim c_1 f^2$, whereas x and y remain massless. In fact, after integrating out the z octet, the potential for x and y vanishes completely: There is an exact cancellation between the $\mathrm{tr}[x, y]^2$ term already in Equation 72, and an identical term generated by the $\mathrm{tr}(z[x, y])$ coupling through exchange of the heavy z octet. This is all as it must be based on the symmetry considerations above.

When both c_1 and c_2 are turned on, we have

$$\mathcal{L}_\Sigma = -c_1 f^2 \mathrm{tr}\left(z - \frac{i}{2f}[x, y]\right)^2 - c_2 f^2 \mathrm{tr}\left(z + \frac{i}{2f}[x, y]\right)^2. \qquad 73.$$

The crucial sign difference in the second term relative to the first arises because the two terms in $\mathcal{L}_\Sigma$ are related by $\Sigma_2 \leftrightarrow \Sigma_4$, which amounts to $\pi_2 \leftrightarrow \pi_4$, or $y \to -y$, with x and z left invariant. Because of this sign difference, it is no longer true that the potential for x and y vanishes once z is integrated out. This makes sense: Once c_1 and c_2 are both turned on, only a global $SU(3)^2$ symmetry is preserved, with each L the same and each R the same, and the only massless NGBs associated with the spontaneous breaking of this $SU(3)^2$ to $SU(3)$ are the eaten ones. Thus, neither x nor y are exact NGBs once c_1 and c_2 are both turned on, and a potential for them is possible at tree level.

In particular, after integrating out z we find the quartic term

$$\mathcal{L}_{quartic} = \frac{c_1 c_2}{c_1 + c_2} \mathrm{tr}[x, y]^2. \qquad 74.$$

Because this term arises from an interplay between the c_1 and c_2 couplings, it vanishes if either is turned off. The tree-level potential produced for x and y does not include mass terms, because neither c_1 nor c_2 can give mass to x or y by themselves, and there is no tree-level diagram with intermediate z that generates mass terms. Nor do the masses of the x and y fields receive quadratically divergent contributions at one loop, as there is no one-loop quadratically divergent diagram involving both c_1 and c_2. So the setup described here generates a tree-level quartic coupling for our pseudo-NGBs while keeping their masses-squared loop-suppressed relative to f^2. This is exactly what we want: If the quartic coupling suffered the same loop suppression as the mass terms, the hierarchy between f and v could not be realized without fine tuning.

Note that this mechanism of generating a tree-level quartic term for the NGBs while protecting their masses would not work with fewer than four sigma fields. For example, if we try using just three, and write down

$$\mathcal{L}_\Sigma = \left(\frac{f}{2}\right)^4 \mathrm{tr}\left(c_1 \Sigma_1 \Sigma_2^\dagger \Sigma_1 \Sigma_3^\dagger\right) + \left(\frac{f}{2}\right)^4 \mathrm{tr}\left(c_2 \Sigma_1 \Sigma_3^\dagger \Sigma_1 \Sigma_2^\dagger\right) + \mathrm{h.c.} \qquad 75.$$

to generate a quartic coupling, we find that c_1 by itself breaks the global symmetry down to $SU(3)^2$. When this is spontaneously broken to $SU(3)$, it yields only the single octet of massless NGBs that is eaten.

4.1.3. THE TOP YUKAWA COUPLING   We demand that the top Yukawa coupling arises in such a way that no quadratically divergent contributions to the masses-squared of the Higgs doublets are induced, and the strategy will again be to ensure that each coupling preserves enough global symmetry to protect these masses. The Standard Model fermions are not charged under $SU(3)$, so coupling the sigma fields to them requires us to work with $SU(3)$-singlet combinations, such as $\Sigma_1^T \Sigma_2^*$. Now consider the Lagrangian terms

$$\mathcal{L}_t = \lambda f \begin{pmatrix} 0 & 0 & u_3^{c\prime} \end{pmatrix} \Sigma_1^T \Sigma_2^* \chi + \lambda' f U^c U + \text{h.c.} \qquad 76.$$

Here, the third-generation electroweak-doublet and -singlet quarks $q_3 = (u_3, d_3)$ and $u_3^{c\prime}$ are joined by an extra vector-like electroweak-singlet quark (U, U^c), and we have grouped some of these fermions into a triplet $\chi = (d_3, -u_3, U)$ (the ordering of d_3 and u_3 in χ is appropriate because it ensures that an $SU(2)_W$ transformation V, which sends $h \to Vh$, acts on the upper two components of χ as $\chi \to V^*\chi$, as required for $\mathcal{L}_t$ to be gauge invariant).

As we will see shortly, these terms generate a Yukawa coupling for the top quark. Moreover, they do so without generating quadratically divergent contributions to the x and y masses-squared at one loop, as can be seen by an argument similar to that used in discussing the top Yukawa coupling in the $SU(3)$ model: In the absence of λ', the term with coefficient $\mathcal{L}_t$ has an $SU(3)$ global symmetry under which χ transforms as a triplet, and this symmetry prevents x and y from acquiring mass. So both couplings are required to generate masses for x and y, and the rephasing symmetries of $\mathcal{L}_t$ indicate that these couplings must appear as $|\lambda|^2$ and $|\lambda'|^2$ in any radiatively generated operators involving only Σ. Thus any one-loop diagram that contributes to the x and y masses must involve four coupling constants, and can be at most logarithmically divergent.

The absence of quadratic divergences can also be verified explicitly using the Coleman-Weinberg potential. At one loop, the effective potential generated for x and y by fermion loops can be calculated by turning on background values for the sigma fields and calculating the fermion mass matrix $M(\Sigma)$ in this background. The quadratically divergent piece of the effective potential is then proportional to $\text{tr} M(\Sigma)^\dagger M(\Sigma)$. In our case, we have

$$\mathcal{L}_t = \begin{pmatrix} 0 & 0 & u_3^{c\prime} & U^c \end{pmatrix} M(\Sigma)\chi + \text{h.c.}, \qquad 77.$$

where the mass matrix $M(\Sigma)$ has the form

$$M(\Sigma) = P \begin{pmatrix} \lambda f \Sigma_1^T \Sigma_2^* \\ 0 \quad 0 \quad \lambda' f \end{pmatrix}, \qquad 78.$$

with $P = \text{diag}(0, 0, 1, 1)$. The unitarity of the Σ's and the cyclic property of the trace then lead to

$$\text{tr} M(\Sigma)^\dagger M(\Sigma) = f^2(|\lambda|^2 + |\lambda'|^2). \qquad 79.$$

There is no dependence on x or y, and so as expected, no quadratically divergent contributions to their masses at one loop. This technique also allows one to check explicitly that the gauge interactions and $\mathcal{L}_\Sigma$ do not generate quadratically divergent contributions at one loop either; see Reference (15) for details.

To explore the coupling of the light scalars to fermions, we expand x and y in terms of $h_{x,y}$, $\eta_{x,y}$, and $\phi_{x,y}$ as in Equation 63, and find that $\mathcal{L}_t$ includes the terms

$$\lambda' f U^c U + \lambda f u_3^{c'} U + i \lambda u_3^c q_3 (h_x - h_y) + \text{h.c.} \qquad 80.$$

One linear combination of $u_3^{c'}$ and U^c marries U to make a Dirac fermion with mass $f \sqrt{|\lambda|^2 + |\lambda'|^2}$, whereas the orthogonal combination, u_3^c, is massless in the absence of electroweak symmetry breaking. Integrating out the heavy fermion, we are left with the Yukawa coupling

$$\lambda_t u_3^c q_3 (h_x - h_y)/\sqrt{2}, \qquad 81.$$

where the coefficient is

$$\lambda_t = \frac{\sqrt{2} \lambda \lambda'}{\sqrt{|\lambda|^2 + |\lambda'|^2}}. \qquad 82.$$

Thus an unsuppressed top Yukawa coupling is generated, without quadratically divergent contributions to the Higgs mass at one loop. The Yukawa couplings for the remaining Standard Model fermions are small enough that their quadratically divergent contributions to the Higgs mass-squared are numerically unimportant for the relatively low cutoff $\Lambda \sim 10$ TeV we have in mind. This means that to give masses to these fermions, we are free to write down couplings similar to the first term in Equation 76, except without introducing extra vector-like fermions analogous to (U, U^c).

Note that $\mathcal{L}_t$ breaks the original $SU(3)^8$ global symmetry to $SU(3) \times [SU(2) \times U(1)]^2$, which is spontaneously broken to $SU(3)^2 \times SU(2) \times U(1)$, giving three full octets of massless NGBs plus one (η, ϕ) set. One full octet is eaten, and another becomes massive at tree level owing to $\mathcal{L}_\Sigma$, leaving a full octet plus an additional (η, ϕ) set as light fields that acquire no potential from $\mathcal{L}_t$. This tells us that of the fields x and y, top loops contribute to the mass of only one combination of Higgs doublets $(h_x - h_y)$, and do not contribute to any η or ϕ masses.

4.1.4. ELECTROWEAK SYMMETRY BREAKING   All of the light scalars in x and y receive positive contributions to their masses-squared once one-loop contributions induced by $\mathcal{L}_\Sigma$ and gauge interactions are taken into account. As just discussed, the triplets and singlets do not receive contributions to their potentials owing to $\mathcal{L}_t$, but the linear combination $h_x - h_y$ does receive a one-loop contribution to its mass from the coupling to fermions, and this contribution is negative. It is possible that this negative contribution is large enough to force a vev for $h_x - h_y$, causing electroweak symmetry breaking.

To address the important question of the stability of electroweak symmetry breaking, let us expand the quartic coupling of Equation 74 in terms of the Higgs doublets. We find

$$\mathcal{L}_{quartic} = \frac{c_1 c_2}{4(c_1 + c_2)} \left((h_x^\dagger h_y - h_y^\dagger h_x)^2 + \text{tr}(h_y h_x^\dagger - h_x h_y^\dagger)^2 \right), \qquad 83.$$

where we have set $\eta_{x,y}$ and $\phi_{x,y}$ to zero. Defining new doublets $h_1 = (h_y - i h_x)/\sqrt{2}$ and $h_2 = (h_x - i h_y)/\sqrt{2}$, these terms become

$$\mathcal{L}_{quartic} = -\frac{c_1 c_2}{4(c_1 + c_2)} \left(\left(h_1^\dagger h_1 - h_2^\dagger h_2 \right)^2 + \mathrm{tr}\left(h_1 h_1^\dagger - h_2 h_2^\dagger \right)^2 \right), \qquad 84.$$

which, when restricted to neutral components, has the same form as the quartic potential of the supersymmetric Standard Model. This quartic potential has a flat direction along $h_1 = h_2$, which must be stabilized by the quadratic terms in the potential. These are of the form $m_1^2(|h_x|^2 + |h_y|^2) + m_2^2|h_x - h_y|^2$, where $m_2^2 < 0$ comes from top loops. Written in terms of h_1 and h_2, the quadratic terms become

$$\left(m_1^2 + m_2^2 \right)\left(|h_1|^2 + |h_2|^2 \right) + m_2^2 \left(h_1^\dagger h_2 + h_2^\dagger h_1 \right). \qquad 85.$$

Now, for electroweak symmetry to occur at all, we need the origin to be unstable, which requires

$$\left(m_1^2 + m_2^2 \right)^2 - \left(m_2^2 \right)^2 < 0. \qquad 86.$$

Because m_2^2 is negative, this condition is easily satisfied. But if it is satisfied, then the quadratic terms in the potential do not stabilize the $h_1 = h_2$ flat direction, and the potential is unbounded from below.

This means that the sigma field couplings of Equation 67 need to be modified. The modification suggested in Reference (15) is to give a nontrivial matrix structure to the couplings c_1 and c_2: $c_1 \rightarrow c_1 \mathbf{I} + \epsilon_1 \mathbf{T_8}$ and $c_2 \rightarrow c_2 \mathbf{I} + \epsilon_2 \mathbf{T_8}$. The coupling ϵ_1 breaks the global symmetry of the term with coefficient c_1 from $SU(3)^4$ to $SU(2) \times U(1) \times SU(3)^3$. It is true that because this global symmetry is spontaneously broken to $SU(2) \times U(1)$, the ϵ_1 term by itself would still leave three massless octets of NGBs. The point, though, is that the three octets protected when c_1 alone is turned on are not exactly the same as the three octets protected when ϵ_1 alone is turned on, and when both are turned on, some of the states contained in x and y acquire mass. By explicit calculation, one finds the Lagrangian term

$$\frac{\sqrt{3}}{8} f^2 \Im(\epsilon_1 - \epsilon_2) i (h_y^\dagger h_x - h_x^\dagger h_y), \qquad 87.$$

which is proportional to $|h_1|^2 - |h_2|^2$. Now that the masses-squared for h_1 and h_2 are split, it is possible to have stable electroweak symmetry breaking. For this term to have the right size, $\Im(\epsilon_1 - \epsilon_2)$ should be small, $\sim 10^{-2}$. However, because ϵ_1 and ϵ_2 encode the effects of additional sources of symmetry breaking, it is natural for them to be suppressed.

To summarize, the Minimal Moose has two Higgs doublets in its scalar sector, along with a complex triplet and a complex singlet, all with masses near the weak scale. Around the TeV scale, there is an additional octet of scalars, along with an electroweak-singlet vector-like quark, and an octet of heavy gauge bosons. This particle content is sufficient for cutting off the one-loop quadratic divergences associated with the Higgs doublets' gauge, Yukawa, and self couplings at around the TeV scale.

4.2. The Littlest Higgs

The third and final model that we will discuss in detail is the littlest Higgs model of Reference (16), which is constructed using an $SU(5)/SO(5)$ coset. We can imagine that the $SU(5)$ global symmetry is broken by the vacuum expectation value of a scalar Σ transforming as a two-index symmetric tensor, or **15** of $SU(5)$. Let's take this VEV to have the form

$$\langle \Sigma \rangle = \begin{pmatrix} & & \mathbb{1} \\ & 1 & \\ \mathbb{1} & & \end{pmatrix}, \qquad 88.$$

where $\mathbb{1}$ is the 2×2 identity matrix, and entries left blank vanish. Under an $SU(5)$ transformation $U = e^{i\theta_a T_a}$, we have $\Sigma \rightarrow U\Sigma U^T$, which means that the ten unbroken generators satisfy

$$\overline{T}_a \langle \Sigma \rangle + \langle \Sigma \rangle \overline{T}_a^T = 0. \qquad \text{(unbroken)} \qquad 89.$$

These are the 10 generators of $SO(5)$. The 14 remaining broken generators satisfy

$$\hat{T}_a \langle \Sigma \rangle - \langle \Sigma \rangle \hat{T}_a^T = 0, \qquad \text{(broken)} \qquad 90.$$

and in constructing the nonlinear sigma model, we keep only the fluctuations of Σ around its VEV in these broken directions:

$$\Sigma = e^{i\pi_a \hat{T}_a/f} \langle \Sigma \rangle e^{i\pi_a \hat{T}_a^T/f} = e^{2i\pi_a \hat{T}_a/f} \langle \Sigma \rangle, \qquad 91.$$

where the last equality follows from Equation 90.

It is straightforward to show that, in light of Equation 90, the matrix of NGBs may be written as

$$\pi = \pi_a \hat{T}_a = \begin{pmatrix} \chi + \eta/(2\sqrt{5}) & h^*/\sqrt{2} & \phi^\dagger \\ h^T/\sqrt{2} & -2\eta/\sqrt{5} & h^\dagger/\sqrt{2} \\ \phi & h/\sqrt{2} & \chi^T + \eta/(2\sqrt{5}) \end{pmatrix}, \qquad 92.$$

where $\chi = \chi^a \sigma^a/2$ is a Hermitian, traceless 2×2 matrix, η is a real singlet, h is a complex doublet, and ϕ is a 2×2 symmetric matrix. The various coefficients are chosen so that the kinetic term of the nonlinear sigma model,

$$\mathcal{L}_{kin} = \frac{f^2}{8} \text{tr}[(D_\mu \Sigma)^\dagger (D^\mu \Sigma)], \qquad 93.$$

yields canonically normalized kinetic terms for the NGBs.

4.2.1. GAUGE INTERACTIONS   As was done in the previous models, we now gauge a subgroup of the global symmetry, and as before, we do this in such a way that each gauge coupling by itself preserves enough of the global symmetry to ensure that the Higgs doublet (a single doublet in this model) remains an exact NGB. The gauge group is taken to be $[SU(2) \times U(1)]^2$, embedded in $SU(5)$ such that the gauged generators for the two $SU(2)$'s are

$$Q_1^a = \left(-\sigma^{a*}/2\right) \quad \text{and} \quad Q_2^a = \left(\sigma^a/2\right), \qquad 94.$$

so the first $SU(2)$ acts on the first two indices, and the second acts on the last two. The gauged generators for the two $U(1)$'s are $Y_1 = \text{diag}(-3, -3, 2, 2, 2)/10$ and $Y_2 = \text{diag}(-2, -2, -2, 3, 3)/10$, respectively. Because the trace of the product of these generators is nonvanishing, kinetic mixing between the two $U(1)$ gauge bosons will arise at loop level. Rediagonalizing the gauge kinetic terms modifies the couplings of the W and Z to fermions. But aside from unobservable rescalings of gauge couplings, physical effects of the kinetic mixing vanish in the limit of unbroken Standard Model gauge symmetry. Therefore contributions to precision electroweak observables are suppressed by $1/(16\pi^2)(M_Z/M_{Z'})^2 \sim 10^{-4}$, evading all constraints.

The linear combination $Q_1^a + Q_2^a$ satisfies Equation 89 and generates the unbroken symmetry that we identify as $SU(2)_W$ of the Standard Model. Similarly, the linear combination $Y_1 + Y_2$ also satisfies Equation 89 and generates the unbroken symmetry that we identify as $U(1)_Y$. The orthogonal combinations are broken, and by expanding $\mathcal{L}_{kin}$ one can check that the heavy $SU(2)$ and $U(1)$ gauge bosons have masses $M_{W'} = f\sqrt{g_1^2 + g_2^2}/2$ and $M_{B'} = f\sqrt{(g_1'^2 + g_2'^2)/20}$, respectively. Here g_1 and g_2 are the gauge couplings of the two $SU(2)$'s, and g_1' and g_2' are the gauge couplings of the two $U(1)$'s. These determine the Standard Model gauge couplings

$$g = \frac{g_1 g_2}{\sqrt{g_1^2 + g_2^2}} \qquad 95.$$

$$g' = \frac{g_1' g_2'}{\sqrt{g_1'^2 + g_2'^2}}, \qquad 96.$$

where the $U(1)$ charge is normalized so that $Y = 1/2$ for the doublet h (i.e., $Y = Y_1 + Y_2$). Note that the mass of the B' is somewhat suppressed. As discussed in Section 5, the effects of this particle modify M_W/M_Z and the couplings of the Z to fermions, so the relative smallness of $m_{B'}/f$ leads to important constraints on the model.

When $[SU(2) \times U(1)]^2$ is broken to its diagonal subgroup, the η and χ fields of the π matrix in Equation 92 are eaten, leaving only h and ϕ, which transform as $\mathbf{2}_{1/2}$ and $\mathbf{3}_1$ under $SU(2)_W \times U(1)_Y$. What can we say about the potential for these fields generated by gauge loops? Let's first imagine that all of the gauge couplings vanish, except for g_2. In this case, the original $SU(5)$ global symmetry is explicitly broken to $SU(3) \times SU(2) \times U(1)$, where the $SU(3)$ acts on the first three indices and the $SU(2)$ acts on the last two. This is spontaneously broken to the

$SU(2) \times U(1)$ of the electroweak group, and so in this limit, there are eight NGBs: These include the four that are eaten when the full $[SU(2) \times U(1)]^2$ is gauged and the four that make up the Higgs doublet h. The Higgs doublet h shifts under part of the $SU(3)$, and so it is forbidden from picking up a potential. The six real scalars in ϕ, on the other hand, are not protected by the global symmetry and are allowed to pick up a potential. The same symmetry argument applies when only g_2' is turned on. Finally, when g_1 or g_1' is turned on, a different $SU(3) \times SU(2) \times U(1)$ is preserved (this time with the $SU(3)$ acting on the last three indices), and this symmetry is also enough to keep the Higgs massless. The fact that more than one coupling is required to break enough of the global symmetry to let h acquire a potential tells us that even when all couplings are turned on, the Higgs does not receive quadratically divergent contributions to its mass-squared at one loop.

On the other hand, ϕ does pick up a quadratically divergent mass-squared at one loop, and after integrating ϕ out, a tree-level quartic potential is generated for h. This is very similar to what happened in the Minimal Moose model with the introduction of $\mathcal{L}_\Sigma$, except that in the present model, no extra terms for Σ are required: The gauge interactions by themselves generate a quartic coupling for the Higgs. This can be verified by studying the quadratically divergent piece of the Coleman-Weinberg potential, $V = \Lambda^2 \text{tr} M^2(\Sigma)/(16\pi^2) \sim f^2 \text{tr} M^2(\Sigma)$, where $\text{tr} M^2(\Sigma)$ is the trace of the gauge boson mass-squared matrix in the presence of a background value for Σ. It is calculated from $\mathcal{L}_{kin}$ as

$$\text{tr} M^2(\Sigma) = \frac{f^2}{2} \sum_{i=1,2} \left(g_i'^2 \text{tr}\big[(\Sigma^\dagger Y_i)^*(\Sigma Y_i)\big] + g_i^2 \sum_c \text{tr}\big[(Q_i^a \Sigma)^*(Q_i^a \Sigma)\big] \right), \quad 97.$$

where terms that do not depend on Σ have been dropped. Expanding this expression gives

$$\text{tr} M^2(\Sigma) = (g_1^2 + g_1'^2)\text{tr}(K_-{}^\dagger K_-) + (g_2^2 + g_2'^2)\text{tr}(K_+{}^\dagger K_+) + \cdots, \quad 98.$$

where $K_\pm = \phi \pm \frac{i}{2f} h h^T$. As claimed, ϕ picks up a quadratically divergent mass-squared at one loop, but h does not. In fact, the potential for h vanishes entirely once ϕ is integrated out, if only the first term, or only the second term, is present. In the presence of both terms, however, integrating out ϕ generates the quartic term

$$\mathcal{L}_{quartic} = -c \frac{(g_1^2 + g_1'^2)(g_2^2 + g_2'^2)}{g_1^2 + g_1'^2 + g_2^2 + g_2'^2} |h h^\dagger|^2, \quad 99.$$

where c is a coefficient of order unity that encodes the details of how the quadratic divergences are cut off in the full UV-completed theory.

4.2.2. THE TOP YUKAWA COUPLING   In this model, there are a number of approaches to generating a top Yukawa coupling while protecting the Higgs mass. Here we describe a setup that introduces the minimal number of additional fermions. To avoid generating quadratically divergent contributions to the Higgs mass-squared at one loop, we require that the coupling of the top quark to Σ respect one of the $SU(3)$ global symmetries under which h shifts. Suppose that we take the third generation quark doublet $q_3 = (u_3, d_3)$ to transform under $SU(2)_1$. Then to achieve an $SU(3)$-symmetric coupling, we proceed as we did for the Minimal Moose: We add an extra electroweak-singlet vector-like fermion (U, U^c), form a triplet $\chi = (d_3, -u_3, U)$, and couple this triplet to Σ and the electroweak-singlet $u_3^{c\prime}$ in an $SU(3)$-symmetric fashion.

Consider the Lagrangian terms

$$\mathcal{L}_t = \lambda f \epsilon_{ijk} \chi_i \Sigma_{j4} \Sigma_{k5} u_3^{c\prime} + \lambda' f U U^c + \text{h.c.} \qquad 100.$$

The indices i, j, k run over the values 1, 2, 3, so the first term is an $SU(3)$-invariant antisymmetric contraction of three triplets. It is also invariant under the gauged $SU(2)_2$, as required: For each i and j, the combination $\Sigma_{i4}\Sigma_{j5} - \Sigma_{i5}\Sigma_{j4}$ appears after the sum is carried out, and this is an $SU(2)_2$-invariant antisymmetric contraction of two doublets. The $U(1)_1 \times U(2)_2$ charges of the fermions are chosen so that the above terms are neutral. For example, if we take $(1/6, 0)$ to be the charges of q_3, then we have $u_3^c(-7/15, -1/5)$ and $U(2/3, 0)$. Anomalies associated with the broken generators of the extended gauge group are assumed to be canceled by heavy fermions.

In the presence of λ alone, a linear combination of u_3 and U remains massless even after electroweak symmetry breaking, but the second term makes this mode heavy. When we expand $\mathcal{L}_t$ we find

$$\mathcal{L}_t = i\sqrt{2}\lambda q_3 h u_3^{c\prime} + f U \left(\lambda u_3^{c\prime} + \lambda' U^c\right) + \text{h.c.}, \qquad 101.$$

where we have only kept terms up to linear order in h. Inspection of the second term shows that U marries a linear combination of $u_3^{c\prime}$ and U^c to become a Dirac particle with mass $f\sqrt{|\lambda|^2 + |\lambda'|^2}$. Once this particle is integrated out, the orthogonal linear combination, u_3^c, appears in the Yukawa coupling

$$\mathcal{L}_t = \frac{\sqrt{2}\lambda\lambda'}{\sqrt{|\lambda|^2 + |\lambda'|^2}} q_3 h u_3^c + \text{h.c.} \qquad 102.$$

As discussed earlier, the Yukawa couplings for the remaining Standard Model fermions are small enough that we need not worry about the quadratic divergences they generate. We can produce these couplings by writing down terms similar to the first term in Equation 100, except without introducing extra vector-like fermions analogous to (U, U^c).

We can verify that the couplings in $\mathcal{L}_t$ do not generate quadratically divergent contributions to the Higgs mass-squared at one loop by calculating the quadratically divergent piece of the Coleman-Weinberg potential. Defining $\phi_i = \epsilon_{ijk}\Sigma_{j4}\Sigma_{k5}$, the fermion mass matrix in a Σ background is

$$M(\Sigma) = f \begin{pmatrix} \lambda\phi_1 & 0 \\ \lambda\phi_2 & 0 \\ \lambda\phi_3 & \lambda' \end{pmatrix}. \qquad 103.$$

The quadratically divergent piece of the Coleman-Weinberg potential is then proportional to

$$\text{tr}M(\Sigma)^\dagger M(\Sigma) = f^2\left(|\lambda|^2 \sum_i |\phi_i|^2 + |\lambda'|^2\right) = -|\lambda|^2 \text{tr}(K_-{}^\dagger K_-), \qquad 104.$$

where for the last equality we have kept only terms up to order ϕ^2 and h^4. This term has the same form as the second term in Equation 98, which should come as no surprise given that λ and g_2 respect the same global symmetry. When we integrate out ϕ, the contributions to the potential arising from Equation 104 will modify the quartic coupling obtained in Equation 99: In the end, the quartic coupling depends on unknown order-one coefficients that are dependent on UV physics, along with the gauge couplings and λ.

Although $\mathcal{L}_t$ does not generate a quadratically divergent contribution to the Higgs mass-squared at one loop, it does generate a logarithmically divergent contribution at this level. Moreover, this contribution is negative, and if its magnitude is sufficiently large, it can overcome the positive contributions from gauge and self-interactions and cause electroweak symmetry breaking.

We have seen that the $SU(5)/SO(5)$ coset allows for a very economical implementation of the little Higgs mechanism. In this model, the Higgs doublet of the Standard Model is the only anomalously light pseudo-NGB. The other scalars, which form an electroweak triplet, have a mass $\sim f \sim$ TeV. The other new states with masses around this scale are one electroweak-singlet vector-like quark (for the model of the top Yukawa coupling presented here), and weak-triplet and singlet heavy gauge bosons. An especially interesting feature of this model is that the gauge interactions of the nonlinear sigma field are by themselves sufficient for generating a quartic coupling for the Higgs doublet.

4.3. Other Models

In the next section we will discuss various effects that give rise to precision electroweak constraints on little Higgs theories. An order-v^2/f^2 modification of M_W/M_Z arises from tree-level exchange of the heavy electroweak-singlet gauge boson. Depending on the theory, corrections to M_W/M_Z may also come from a triplet VEV, or from dimension-six operators in the expansion of the nonlinear sigma model kinetic term. Finally, integrating out the heavy gauge bosons also modifies the couplings of the light Standard Model gauge bosons to fermions and generates four-fermion operators in the effective theory. Because these effects impose serious constraints on the models described so far, we now briefly describe alternative models in which the constraints are less severe.

4.3.1. LOSING THE TRIPLET   A variation of the littlest Higgs model was constructed on the coset $SU(6)/Sp(6)$ in Reference (18). This breaking pattern gives rise to 14 NGBs: 4 are eaten owing to the gauge symmetry breaking $[SU(2) \times U(1)]^2 \to SU(2) \times U(1)$, 8 appear in two Higgs doublets, and 2 appear in a neutral complex singlet. The singlet's mass-squared receives quadratically divergent contributions at one loop, while the doublets remain light, and integrating out the singlet generates a quartic coupling for the Higgs doublets. Having an extra singlet instead of an extra triplet allows for smaller values of f, as discussed in the following section.

4.3.2. MODELS WITH CUSTODIAL SYMMETRY   Consider the 2×2 matrix Σ, whose first column is $i\sigma_2 h^*$ and whose second column is h. In the limit of vanishing gauge couplings, the Standard Model Higgs sector is invariant under

$$\Sigma \to L\Sigma R^\dagger, \qquad\qquad 105.$$

where L and R and are independent $SU(2)$ global symmetry transformations. The group $SU(2)_L \times SU(2)_R$ is broken to the diagonal subgroup by the Higgs VEV $\langle \Sigma \rangle \propto \mathbb{1}$, and the unbroken group is called custodial symmetry, or $SU(2)_C$. Under $SU(2)_C$, the three NGBs from Σ transform as a triplet. These become the longitudinal components of the $W^\pm$ and Z. When the Standard Model gauge couplings are turned on, the Higgs doublet couples to the $SU(2)_W$ gauge fields through currents j^a_μ, and these transform as a triplet under $SU(2)_C$ just as the NGBs do. The fact that $SU(2)_C$ breaking in the Higgs and gauge boson sectors arises only from the hypercharge gauge coupling leads to the tree-level relation $M_W = M_Z \cos\theta$.

Given a modified Higgs sector, there is no guarantee that this relation will be preserved, unless there is an unbroken custodial $SU(2)$ under which the NGBs and the $SU(2)_W$ currents both transform as triplets. The heavy $U(1)$ gauge bosons in little Higgs models couple as T^3_R, that is, in an $SU(2)_C$-violating fashion, and integrating out these particles generates dangerous operators such as $(h^\dagger D_\mu h)^2$. Violation of $SU(2)_C$ also may arise from higher-order terms in the nonlinear sigma model kinetic term. Although these effects are typically quite constraining, they can be minimized by building $SU(2)_C$ in as an approximate symmetry of the theory, as was done in References (21, 23).

Reference (21) presented a modification of the Minimal Moose in which the $SU(3)$ global symmetries are replaced by $SO(5)$ global symmetries, and the $SU(3) \times SU(2) \times U(1)$ gauge group is replaced by $SO(5) \times SU(2) \times U(1)$. The $SO(5)$ global symmetry is large enough to contain $SU(2)_C$, so the $SU(2)_C$-violating contributions from the nonlinear sigma model structure are automatically absent. The $SU(2)_C$ violating contributions from exchange of the heavy $U(1)$ gauge boson are partially canceled because this particle is now joined by other heavy gauge bosons with which it forms a triplet of $SU(2)_C$. In the limit where the $SO(5)$ gauge coupling becomes large, these states become nearly degenerate, which suppresses the $SU(2)_C$-violating effects produced when the triplet is integrated out.

One complication is that the model has two Higgs doublets, with a potential that requires their VEVs to be misaligned. This misalignment generates its own source of $SU(2)_C$ violation, so that even when the triplet becomes degenerate, integrating it out still yields an $SU(2)_C$-violating term. In Reference (23), this complication is avoided by constructing a single-Higgs-doublet model with an

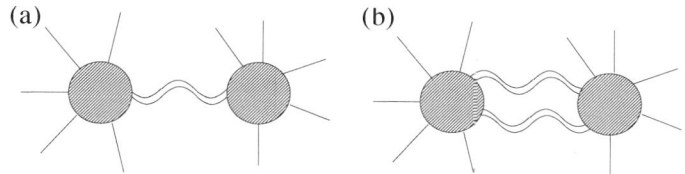

Figure 9   (*a*) Tree-level exchange of T-odd particles is forbidden.
(*b*) Loop processes are allowed.

approximate $SU(2)_C$, an extension of the littlest Higgs with coset $SO(9)/(SO(5)\times SO(4))$.

4.3.3. T-PARITY   A particularly interesting class of models (25, 28, 31) incorporates a discrete symmetry called T-parity, under which the heavy particles are odd and the Standard Model fields are even. With this symmetry in place, no effective operators involving just light fields are generated by tree-level exchange of heavy fields, because an even number of heavy fields are required at each vertex (see Figure 9). As a result, precision electroweak constraints are weakened. This is similar to the way in which R-parity is helpful for preventing excessive corrections to precision electroweak observables in supersymmetric theories.

Reference (25) implemented T-parity in a variation of the Minimal Moose model. To avoid $SU(2)_C$-violation from higher-order terms in the nonlinear sigma model kinetic term, the global symmetry is enlarged to contain products of $SO(5)$ as in Reference (21), rather than products of $SU(3)$. If we further modify the Minimal Moose by gauging $[SU(2) \times U(1)]^2$ instead of $SU(3) \times SU(2) \times U(1)$, and set the two $SU(2)$ couplings and the two $U(1)$ couplings equal, then the theory possesses a Z_2 symmetry that exchanges the gauge fields of the two $SU(2) \times U(1)$ groups and sends $\Sigma \rightarrow \Sigma^\dagger$, if we neglect the fermions for the time being. This is apparent from the reflection symmetry that exchanges the two sites in Figure 8. Denoting one of the gauge fields of the first site by A_1 and the analogous gauge field from the second site by A_2, the linear combination $A_1 + A_2$ is even under this Z_2, and the orthogonal combination $A_1 - A_2$ is odd. But $A_1 - A_2$ is precisely the linear combination that becomes heavy when the gauge symmetry is broken to the diagonal subgroup: As desired, the heavy gauge boson is odd under T-parity.

Fermions must be introduced into the picture in a way that preserves the Z_2, and so identical copies are introduced on the two sites, i.e., one copy transforms under one $SU(2) \times U(1)$, and the other transforms under the second $SU(2) \times U(1)$. To make one linear combination of these fermions heavy, a third site is introduced, with mirror fermions that marry one linear combination. Once this is done, it is no longer true that all heavy gauge bosons are T-parity odd, but taking the gauge couplings on the third site to be large makes these T-parity-even gauge bosons extra heavy and suppresses their couplings to Standard Model fermions, which still live on the other two sites.

References (28, 31) explore the potential role of T-parity in little Higgs models further, for instance by elucidating a systematic approach to incorporating fermions in models with T-parity and by implementing T-parity in the littlest Higgs model and in variants of the littlest Higgs. It turns out that it is not possible to implement T-parity in models with a simple gauge group, such as the $SU(3)$ model.

5. LITTLE HIGGS PHENOMENOLOGY

5.1. Precision Electroweak Constraints

As discussed in the introduction and alluded to in the previous section, indirect constraints from precision electroweak measurements on new physics at the TeV scale are severe. The new physics predicted in little Higgs theories is no exception, and it is important to explore whether there are regions of parameter space in which these models are consistent with precision electroweak data. Of course, agreement with data can always be achieved by increasing f, and thus the masses of the new particles, but if f becomes too large, this comes at the price of reintroducing fine tuning into the theory[4].

There are several sources of corrections to precision electroweak observables in little Higgs theories. The heavy gauge bosons couple to fermions and Higgs doublets through the currents j_F and j_H. After integrating out these gauge bosons one obtains operators of the form $j_F j_F$, $j_F j_H$, and $j_H j_H$. The $j_F j_F$ terms correspond to new four-fermion operators, which are most strongly constrained by limits from LEPII and from measurements of atomic parity violation, whereas the $j_F j_H$ terms lead to modifications of the couplings of Standard Model gauge boson to fermions, and so are constrained by Z-pole data. The $j_H j_H$ terms include the $SU(2)_C$-violating operator $(h^\dagger D_\mu h)^2$, which is generated by heavy $U(1)$ gauge bosons, but not by heavy $SU(2)$-triplet gauge bosons.

Another potential source of $SU(2)_C$ violation comes from expanding the kinetic term of the nonlinear sigma model. This is not the case for the littlest Higgs or the simplest little Higgs, but it is true, for example, for the Minimal Moose. If the theory contains triplet scalars that acquire VEVs, as the littlest Higgs does, these will also contribute to $SU(2)_C$ violation. All of the effects discussed so far are tree-level effects that can be analyzed using the type of general effective field theory analysis given in Reference (4). Finally, there are also $SU(2)_C$-violating loop contributions coming from the extended top sector, and depending on the model, from an extended Higgs sector.

For generic choices of parameters, these effects typically require little Higgs models to have fairly large values of f. For the littlest Higgs, for instance, it was found in Reference (38) that constraints from precision electroweak data imply

---

[4]Although constraints from precision electroweak data are often expressed as lower bounds on f, one should keep in mind that the particle masses of the TeV-scale states are a better indicator of how finely tuned the theory is.

$f > 4$ TeV, which in turn puts a lower bound on the mass of the heavy quark equal to 5.7 TeV. An even slightly stronger bound was obtained in Reference (39), by combining indirect constraints with constraints from direct production of the heavy $U(1)$ gauge boson at the Tevatron. Using the fact that the heavy quark cuts off the quadratic divergence in the top-loop contribution to the Higgs mass-squared, the bound $m_U > 5.7$ TeV was estimated to correspond to fine tuning at roughly the percent level.

Simple modifications of the littlest Higgs improve the situation without spoiling the stabilization of the electroweak scale. If we let the light fermions be charged equally under both $U(1)$'s, rather than just under $U(1)_1$, then taking $g_1' = g_2'$ decouples the heavy $U(1)$ gauge boson from j_F and j_H simultaneously, and the precision electroweak problems associated with this particle go away. Alternatively, one can just gauge $U(1)_Y$, rather than a product of $U(1)$s, because the quadratically divergent contribution to the Higgs mass-squared caused by hypercharge interactions is numerically small for $\Lambda \sim 10$ TeV. In Reference (40) it was shown that if only $U(1)_Y$ is gauged, there exists allowed parameter space with $f \sim$ TeV, with the main constraint coming from the VEV of the triplet scalar. In the model built on $SU(6)/Sp(6)$, these regions expand due to the absence of the triplet (40, 41).

Precision electroweak constraints were also studied for the Minimal Moose model in Reference (44), where it was found that by taking the gauge group to be $[SU(2) \times U(1)]^2$ rather than $SU(3) \times SU(2) \times U(1)$—again, this allows one to choose couplings that effectively decouple the heavy $U(1)$ gauge boson from both j_F and j_H—it is possible to find regions in parameter space that are consistent with precision electroweak constraints, without severe fine tuning. Finally, Reference (40) also analyzed contributions to precision electroweak observables in the $SU(3)$ model, and found the constraint $f > 4$ TeV; however, in this case, the mass of the heavy fermion may be well below f, and so it is possible to have only mild fine tuning even with large values of f.

For more on the indirect effects of new particles in little Higgs theories, we refer the reader to the literature (42–54).

Although there is parameter space in which the simplest little Higgs theories are consistent with precision electroweak data, it is interesting to consider alternative models in which the most dangerous effects are automatically absent. The T-parity conserving models discussed in the previous section are one example. Because tree-level effects associated with the T-odd particles are removed, bounds from precision electroweak data can be satisfied even with f below 1 TeV (21, 23). The models incorporating $SU(2)_C$, also already discussed, are another example. These provide an alternative to gauging only $U(1)_Y$ by making the heavy $U(1)$ gauge boson a member of an $SU(2)_C$ triplet.

5.2. Direct Production of Little Higgs Partners

The spectrum of new particles varies somewhat from one little Higgs model to another, but all of them predict at least one vector-like quark at the TeV scale, along with extra gauge bosons and scalars. This is guaranteed given that the little-Higgs mechanism arranges for quadratic divergences to cancel between states of the same statistics. Here we summarize the prospects for these particles to be discovered at the LHC.

Let us focus on theories with a product group structure. The collider phenomenology of a heavy $U(1)$ gauge boson is certainly interesting, but because this particle is associated with the most stringent precision electroweak constraints, and because it is not essential for stabilizing the weak scale, most phenomenological studies have concentrated on the heavy $SU(2)$ gauge bosons.

At the LHC, these gauge bosons would be produced in pp collisions predominantly through quark-antiquark annihilation. The production rate depends on $\cot\theta = g_1/g_2$, where g_1 is the gauge coupling of the $SU(2)$ under which the Standard Model fermion doublets transform, and g_2 is the coupling of the other $SU(2)$; the fermion couplings to the heavy gauge bosons are proportional to $g\cot\theta$. For $\cot\theta = 1$, the cross section for producing a 3 TeV neutral Z_H at the LHC is ~ 100 fb (55, 56), corresponding to tens of thousands of events. A clean discovery channel would arise from the decay of Z_H to pairs of highly energetic leptons, and the discovery reach for $\cot\theta = 1$ is roughly 5 TeV. Precision electroweak data prefer smaller values, $\cot\theta \lesssim .2$, but even for these values, the discovery reach is still well into the multi-TeV region. After ameliorating or eliminating the precision electroweak constraints associated with a heavy $U(1)$ gauge boson, the lower bound on the mass of the heavy $SU(2)$ gauge bosons is roughly 2 TeV, leaving plenty of room for discovery.

The phenomenology of the heavy neutral gauge boson of the $SU(3)$ model (simplest little Higgs) is similar to the phenomenology of the Z_H in product gauge group models. On the other hand, the detection of the heavy $SU(2)$-doublet gauge bosons of the $SU(3)$ model is complicated by a v/f suppression of the coupling of these states to light quarks.

Even if heavy gauge bosons are discovered, there is still the question of whether they arise from a little Higgs model. Studying the littlest Higgs, the authors of Reference (55) pointed out that the partial width of the Z_H to fermions is proportional to $\cot^2\theta$, whereas the partial width into boson pairs Zh and W^+W^- is proportional to $\cot^2 2\theta$. This feature was proposed as an interesting test of the little Higgs structure, because in an alternative theory with the Higgs charged under just one $SU(2)$, the partial width into bosons is proportional to $\cot^2\theta$, just like for the fermions.

The production of the heavy quark (U, U^c) in the littlest Higgs model was also studied in References (56, 57). These can be pair-produced via their coupling to gluons, but owing to the mixing in the top quark sector, they can also be produced singly via Wb fusion, $W^+b \to U$ (56). This mode, whose rate depends on the ratio of the λ and λ' couplings that appear in the top Yukawa sector of the model, tends to dominate for larger values of the mass, $m_U \gtrsim 1$ TeV. The heavy quark branching fractions satisfy $\Gamma(U \to th) \approx \Gamma(U \to tZ) \approx \frac{1}{2}\Gamma(U \to bW^+)$, and all three decay modes lead to identifiable signatures. The discovery reach can be estimated to be roughly 3 TeV.

The parameters λ and λ' determine the top Yukawa coupling, and along with f, they also determine the mass of U and its coupling to light states. The decay constant f can be determined by measurements of the properties of the heavy gauge bosons (55), and the known top Yukawa coupling gives one equation constraining λ and λ'. By measuring m_U one would obtain a second, independent equation. Reference (57) considered whether it might be possible to test the structure of the littlest Higgs top sector by combining the result of the measurement of m_U with a measurement of the U production cross section, which depends on λ and λ' in yet another way. This would be another interesting experimental probe of the little Higgs mechanism for canceling quadratic divergences, this time in the top sector. The authors concluded that such a test might be feasible if the uncertainty in the bottom-quark parton distribution function were reduced.

The prospects for discovering extra scalar particles are quite model dependent. For the littlest Higgs, Reference (56) pointed out the possibility that the doubly charged scalar belonging to the triplet ϕ would mediate a resonant contribution to longitudinal WW scattering, possibly giving a signal above background. In other little Higgs theories, for example $SU(6)/Sp(6)$, the triplet is replaced by a neutral scalar, whose discovery prospects look grim (that model has two Higgs doublets, so the Higgs phenomenology would still be different than in the Standard Model).

Finally, we should mention that the phenomenology of little Higgs theories is drastically altered if the theory conserves T-parity, in which case there are missing-energy signals that are similar to those in supersymmetric theories with R-parity conservation (58). Other work on little Higgs phenomenology appears in References (59–70).

6. CONCLUSIONS

Little Higgs theories are a compelling possibility for physics beyond the Standard Model. These theories feature weakly coupled new physics at the TeV scale, which stabilizes the Higgs potential even with a cutoff as large as ~ 10 TeV. The key ingredient is that the Higgs is a pseudo-Nambu-Goldstone boson of a spontaneously broken approximate global symmetry, with the explicit breaking of this symmetry collective in nature, that is, more than one coupling at a time must be turned on for the symmetry to be broken. The collective breaking ensures that no quadratically divergent contributions to the Higgs potential arise at one loop. New TeV-scale particles cancel the quadratic divergences of Standard Model fields with the same statistics, and some of these new particles should be revealed at the LHC if they play a role in stabilizing the weak scale.

For little Higgs theories with T-parity or any other unbroken discrete symmetry at the TeV scale, the lightest particle charged under the symmetry is stable, and might play an important role in cosmology. If it is electrically neutral it may well be a good cold dark matter candidate (71).

Little Higgs theories are effective field theories valid up to a cutoff $\Lambda \sim 4\pi f$. An important question that this review has not addressed is what lies beyond the cutoff Λ. This question has not been explored extensively in the literature, but we will conclude here by mentioning a few possibilities. One possibility is that the global symmetry is broken by a weakly coupled scalar, and this scalar's potential is protected by its own little Higgs mechanism—it is a pseudo-NGB of a different symmetry. By building a structure with a single iteration of this type, one can raise the cutoff to $\Lambda \sim 100\,\text{TeV}$ (29), and more ambitiously, one can attempt to construct a theory with many iterations, with a much higher cutoff (33). A different possibility is that the global symmetry is broken by strong dynamics that give rise to fermion condensation. In this case the Higgs is a composite particle. An explicit little Higgs UV completion of this type was constructed in Reference (26), which employs soft supersymmetry breaking at $\sim 10\,\text{TeV}$ to generate the four-fermion operators required to give masses to the Standard Model fermions. Finally, theories have been constructed in five-dimensional Anti-de Sitter space that can be interpreted as holographic duals of composite Higgs theories (24, 32, 34), and these theories can also be thought of as UV completions that involve strong dynamics at the scale Λ.

ACKNOWLEDGMENTS

MS is supported by the Outstanding Junior Investigator Award DE-FG02-91ER 40676 and an Alfred P. Sloan Research Fellowship. DT-S is supported by a Research Corporation Cottrell College Science Award.

**The *Annual Review of Nuclear and Particle Science* is online at
http://nucl.annualreviews.org**

LITERATURE CITED

1. Eidelman S, et al. (Particle Data Group). *Phys. Lett.* B592:1 (2004)
2. TLE Group, TS Electroweak Group, HF Group (OPAL Collab.). arXiv:hep-ex/0412015
3. D'Ambrosio G, Giudice GF, Isidori G, Strumia A. *Nucl. Phys.* B645:155 (2002)
4. Han Z, Skiba W. arXiv:hep-ph/0412166
5. Haber H. Contribution to "Physics in d $\geq$ 4 (TASI 2004)." Singapore: World Sci. (In press); Luty M. Contribution to "Physics in d $\geq$ 4 (TASI 2004)." Singapore: World Sci. (In press)
6. Chivukula S, Simmons E. Contribution to "Physics in d $\geq$ 4 (TASI 2004)." Singapore: World Sci. (In press)
7. Kaplan DB, Georgi H. *Phys. Lett.* B136:83 (1984)
8. Kaplan DB, Georgi H, Dimopoulos S. *Phys. Lett.* B136:187 (1984)
9. Georgi H, Kaplan DB. *Phys. Lett.* B145:216 (1984).
10. Sundrum R. Contribution to "Physics in d $\geq$ 4 (TASI 2004)." Singapore: World Sci. (In press); Csaki C. Contribution to "Physics in d $\geq$ 4 (TASI 2004)." Singapore: World Sci. (In press); Kribs G. Contribution to "Physics in d $\geq$ 4 (TASI 2004)." Singapore: World Sci. (In press)
11. Georgi H, Pais A. *Phys. Rev. D* 10:539 (1974)
12. Georgi H, Pais A. *Phys. Rev. D* 12:508 (1975)

13. Arkani-Hamed N, Cohen AG, Georgi H. *Phys. Lett.* B513:232 (2001)
14. Arkani-Hamed N, Cohen AG, Gregoire T, Wacker JG. *JHEP* 0208:020 (2002)
15. Arkani-Hamed N, Cohen AG, Katz E, Nelson AE, Gregoire T, Wacker JG. *JHEP* 0208:021 (2002)
16. Arkani-Hamed N, Cohen AG, Katz E, Nelson AE. *JHEP* 0207:034 (2002)
17. Gregoire T, Wacker JG. *JHEP* 0208:019 (2002)
18. Low I, Skiba W, Smith D. *Phys. Rev. D* 66:072001 (2002)
19. Schmaltz M. *Nucl. Phys. Proc. Suppl.* 117:40 (2003)
20. Kaplan DE, Schmaltz M. *JHEP* 0310: 039 (2003)
21. Chang S, Wacker JG. *Phys. Rev. D* 69: 035002 (2004)
22. Skiba W, Terning J. *Phys. Rev. D* 68: 075001 (2003)
23. Chang S. *JHEP* 0312:057 (2003)
24. Contino R, Nomura Y, Pomarol A. *Nucl. Phys.* B671:148 (2003)
25. Cheng HC, Low I. *JHEP* 0309:051 (2003)
26. Katz E, Lee JY, Nelson AE, Walker DGE. arXiv:hep-ph/0312287
27. Birkedal A, Chacko Z, Gaillard MK. *JHEP* 0410:036 (2004)
28. Cheng HC, Low I. *JHEP* 0408:061 (2004)
29. Kaplan DE, Schmaltz M, Skiba W. *Phys. Rev. D* 70:075009 (2004)
30. Schmaltz M. *JHEP* 0408:056 (2004)
31. Low I. *JHEP* 0410:067 (2004)
32. K Agashe, Contino R, Pomarol A. arXiv:hep-ph/0412089
33. Batra P, Kaplan DE. arXiv:hep-ph/0412267
34. Thaler J, Yavin I. arXiv:hep-ph/0501036
35. Callan CG, Coleman SR, Wess J, Zumino B. *Phys. Rev.* 177:2247 (1969)
36. Kong OCW. arXiv:hep-ph/0307250
37. Kong OCW. *Phys. Rev. D* 70:075021 (2004)
38. Csaki C, Hubisz J, Kribs GD, Meade P, Terning J. *Phys. Rev. D* 67:115002 (2003)
39. Hewett JL, Petriello FJ, Rizzo TG. *JHEP* 0310:062 (2003)
40. Csaki C, Hubisz J, Kribs GD, Meade P, Terning J. *Phys. Rev. D* 68:035009 (2003)
41. Gregoire T, Smith DR, Wacker JG. *Phys. Rev. D* 69:115008 (2004)
42. Chivukula RS, Evans NJ, Simmons EH. *Phys. Rev. D* 66:035008 (2002)
43. Huo W, Zhu S. *Phys. Rev. D* 68:097301 (2003)
44. Kilic C, Mahbubani R. *JHEP* 0407:013 (2004)
45. Chen MC. Dawson S. *Phys. Rev. D* 70: 015003 (2004)
46. Casalbuoni R, Deandrea A, Oertel M. *JHEP* 0402:032 (2004)
47. Kilian W, Reuter J. *Phys. Rev. D* 70:015004 (2004)
48. Yue CX, Wang W. *Nucl. Phys.* B683:48 (2004)
49. Choudhury SR, Gaur N, Joshi GC, McKellar BHJ. arXiv:hep-ph/0408125
50. Lee JY. *JHEP* 0412:065 (2004)
51. Buras AJ, Poschenrieder A, Uhlig S. arXiv:hep-ph/0410309
52. Choudhury SR, Gaur N, Goyal A, Mahajan N. *Phys. Lett.* B601:164 (2004)
53. Buras AJ, Poschenrieder A, Uhlig S. anXiv:hep-ph/0501230
54. Marandella G, Schappacher C, Strumia A. arXiv:hep-ph/0502096
55. Burdman G, Perelstein M, Pierce A. *Phys. Rev. Lett.* 90:241802 (2003); Erratum. *Phys. Rev. Lett.* 92:049903 (2004)
56. Han T, Logan HE, McElrath B, Wang LT. *Phys. Rev. D* 67:095004 (2003)
57. Perelstein M, Peskin ME, Pierce A. *Phys. Rev. D* 69:075002 (2004)
58. Hubisz J, Meade P. arXiv:hep-ph/0411264
59. Han T, Logan HE, McElrath BE, Wang LT. *Phys. Lett.* B563:191 (2003) Erratum. *Phys. Lett.* B603:257 (2004)
60. Dib C, Rosenfeld R, Zerwekh A. arXiv:hep-ph/0302068
61. Sullivan Z. arXiv:hep-ph/0306266
62. Yue CX, Wang SZ, Yu DQ. *Phys. Rev. D* 68:115004 (2003)
63. Chang S, He H-J. *Phys. Lett.* B586:95 (2004)
64. Azuelos G, et al. arXiv:hep-ph/0402037

65. Logan HE. *Phys. Rev. D* 70:115003 (2004)
66. Gonzalez-Sprinberg GA, Martinez R, Rodriguez JA. arXiv:hep-ph/0406178
67. Cho GC, Omote A. arXiv:hep-ph/0408270
68. Kilian W, Rainwater D, Reuter J. *Phys. Rev. D* 71:015008 (2005)
69. Yue CX, Wang W. *Phys. Rev. D* 71:015002 (2005)
70. Park SC, Song J. *Phys. Rev. D* 69:115010 (2004)
71. Birkedal-Hansen A, Wacker JG. *Phys. Rev. D* 69:065022 (2004)

Annu. Rev. Nucl. Part. Sci. 2005. 55:271–310
doi: 10.1146/annurev.nucl.55.090704.151526

PHYSICS OF ULTRA-PERIPHERAL NUCLEAR COLLISIONS

Carlos A. Bertulani,[1] Spencer R. Klein,[2] and Joakim Nystrand[3]

[1]*Department of Physics, University of Arizona, Tucson, Arizona 85721;
email: bertulani@physics.arizona.edu*
[2]*Nuclear Science Division, Lawrence Berkeley National Laboratory, Berkeley, California,
94720; email: srklein@lbl.gov*
[3]*Department of Physics and Technology, University of Bergen, 5007 Bergen, Norway;
email: joakim.nystrand@ift.uib.no*

Key Words   nuclear collisions, relativistic, heavy ions, virtual photons, particle production, nuclear fragmentation, two-photon reactions, photonuclear reactions

■ **Abstract**   Moving highly-charged ions carry strong electromagnetic fields that act as a beam of photons. In collisions at large impact parameters, hadronic interactions are not possible, and the ions interact through photon-ion and photon-photon collisions known as ultra-peripheral collisions (UPCs). Hadron colliders like the Relativistic Heavy Ion Collider (RHIC), the Tevatron, and the Large Hadron Collider (LHC) produce photonuclear and two-photon interactions at luminosities and energies beyond that accessible elsewhere; the LHC will reach a γp energy ten times that of the Hadron-Electron Ring Accelerator (HERA). Reactions as diverse as the production of anti-hydrogen, photoproduction of the ρ^0, transmutation of lead into bismuth, and excitation of collective nuclear resonances have already been studied. At the LHC, UPCs can study many types of new physics processes.

CONTENTS

0163-8998/05/1208/0271$20.00

1. INTRODUCTION

In 1924, Enrico Fermi, then 23 years old, submitted a paper "On the Theory of Collisions Between Atoms and Elastically Charged Particles" to Zeitschrift für Physik (1). This paper does not appear in his "Collected Works." Nevertheless, it is said that this was one of Fermi's favorite ideas and that he often used it later in life (2). In this publication, Fermi devised a method known as the equivalent (or virtual) photon method, where he treated the electromagnetic fields of a charged particle as a flux of virtual photons. Ten years later, Weizsäcker and Williams extended this approach to include ultra-relativistic particles, and the method is often known as the Weizsäcker-Williams method (3).

A fast-moving charged particle has electric field vectors pointing radially outward and magnetic fields circling it. The field at a point some distance away from the trajectory of the particle resembles that of a real photon. Thus, Fermi replaced the electromagnetic fields from a fast particle with an equivalent flux of photons. The number of photons with energy ω, $n(\omega)$, is given by the Fourier transform of the time-dependent electromagnetic field. The virtual photon approach used in quantum electrodynamics (QED) to describe, e.g., atomic ionization or nuclear excitation by a charged particle, can be simply described using Fermi's approach.

When two nuclei collide, two types of electromagnetic processes can occur. A photon from one ion can strike the other, Figure 1a, or, photons from each nucleus can collide, in a photon-photon collision, as in Figure 1b.

Ultra-peripheral hadron-hadron collisions will provide unique opportunities for studying electromagnetic processes. At the Large Hadron Collider (LHC), photon-proton collisions will occur at center of mass energies an order of magnitude higher than are available at existing accelerators, and photon-heavy ion collisions will reach 30 times the energies available at fixed target accelerators. The electromagnetic fields of heavy ions are very strong, so reactions involving multi-photon excitations can be studied.

Ultra-relativistic heavy-ion interactions have been used to study nuclear photoexcitation (e.g., to a Giant Dipole Resonance) and photoproduction of hadrons. Coulomb excitation is a traditional tool in low-energy nuclear physics. The strong electromagnetic fields from a heavy ion allow for the study of multi-photon excitation of nuclear targets. This can produce high-lying states in nuclei, e.g., the double-giant resonance (4, 5). Multiple, independent interactions among a single ion pair are also possible. Reactions like multiple vector meson production can

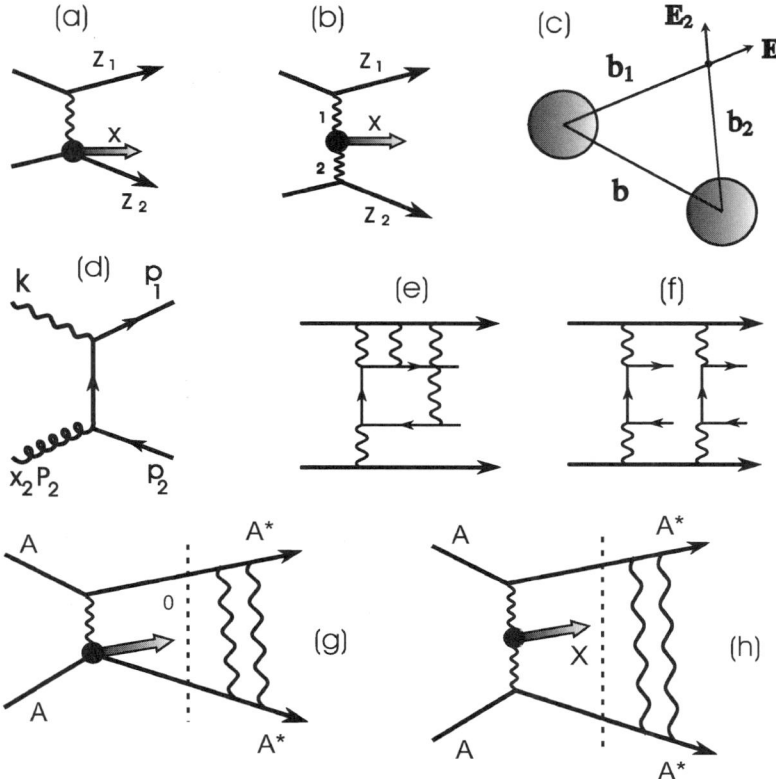

Figure 1   (*a*) One-photon and (*b*) two-photon processes in heavy ion collisions. (*c*) Geometrical representation of the photon fluxes at a point outside nuclei 1 and 2, in a collision with impact parameter *b*. The electric field of the photons at that point are also shown. (*d*) Feynman diagram for $q\bar{q}$ production through photon-gluon fusion to leading order. (*e, f*) Example of higher order corrections to pair production: (*e*) Coulomb distortion, and (*f*) production of multiple pairs. (*g*) The dominant diagram for Au+Au → Au*+ Au*+ ρ^0 and (*h*) for Au+Au → Au*+Au* + e^+e^- or a meson *X*. The dotted lines in panels (*g*) and (*h*) show how the mutual Coulomb nuclear excitation factorizes from the particle production.

be used for studies involving polarized photons. The high photon energies can be used to study the gluon density in heavy nuclei (6) at low Feynman-*x*.

The cross section for photoproduction of a state *x* is

$$\sigma_X = \int d\omega \frac{n(\omega)}{\omega} \sigma_X^\gamma(\omega),$$ 1.

where $\sigma_X^\gamma(\omega)$ is the photonuclear cross section.

Photon-photon (or two-photon) processes have long been studied at e^+e^- colliders. They are an excellent tool for many aspects of meson spectroscopy and tests of QED. At hadron colliders, they are also used to study atomic physics processes, often involving electrodynamics in strong fields. One striking success was the production of antihydrogen atoms at CERN's[1] LEAR[2] (7) and at the Fermilab Tevatron (8). At the highest energy colliders, reactions like $\gamma \gamma \to X$ may be used to probe the quark content and spin structure of meson resonances. Production of meson or baryon pairs can also probe the internal structure of hadrons. At the LHC, electroweak processes such as $\gamma \gamma \to W^+ W^-$ may be probed. The cross section for two-photon processes is (9)

$$\sigma_X = \int d\omega_1 \, d\omega_2 \frac{n(\omega_1)}{\omega_1} \frac{n(\omega_2)}{\omega_2} \sigma_X^{\gamma \gamma}(\omega_1, \omega_2), \qquad 2.$$

where $\sigma_X^{\gamma \gamma}(\omega_1, \omega_2)$ is the two-photon cross section.

Fermi's method has found application beyond the realms of QED. It has been extended to strong interactions between nuclei in peripheral collisions. These interactions are mainly mediated by pion exchange, and an equivalent pion method has been applied to describe subthreshold pion production in nucleus-nucleus collisions (10). Feshbach used the term nuclear Weiszsäcker-Williams method to describe excitation processes induced by the nuclear interaction in peripheral collisions of heavy ions (11). More recently, a non-Abelian Weiszsäcker-Williams field was used to describe the boosted gluon distribution functions in nuclear collisions (12).

Since Fermi's original work, much progress has been achieved in this field, especially with the advent of relativistic heavy-ion accelerators like the Bevalac accelerator at Lawrence Berkeley National Laboratory (LBNL). Intermediate energy processes have been explored at heavy-ion accelerators at NSCL/MSU, GANIL, RIKEN, and GSI[3]; these facilities have explored the collective excitation and electromagnetic fragmentation of nuclei and studied many reactions that occur in the sun, supernovae, and the big bang. Experimental studies of higher-energy processes have recently begun at Brookhaven's RHIC. In the next few years, CERNs LHC will begin operations, allowing for the study of heavy mesons, measurements of gluon distributions in nuclei, and searches for a host of new physics processes.

This review will discuss these experiments, their theoretical interpretation, and some future possibilities in this field. UPCs have been previously reviewed by a number of authors (9, 13–16).

---

[1]European Organization for Nuclear Research (Conseil Européenne pour la Recherche Nucléaire).
[2]Low Energy Antiproton Ring.
[3]NSCL/MSU: National Science Cyclotron Laboratory at Michigan State University, GANIL: Grand Accelerateur National d'Ions Lourds in Caen/France, RIKEN: The Institute of Physical and Chemical Research, Wako, Saitama/Japan, GSI: Gesellschaft fuer Schwerionenforschung, Darmstadt/Germany.

1.1. The Photon Flux

The flux of equivalent photons from a charged particle is determined from the Fourier transform of the electromagnetic field of the moving charge. The fields of a relativistic particle Lorentz contract toward a co-moving pancake (see Figure 2). The photon energy spectrum depends on the time a target particle spends in this pancake, i.e., on the minimum distance between the target and the charge and on the projectile velocity; the minimum photon wavelength is the width of the pancake at the target. At an ion-ion separation (impact parameter) b, the interaction time is $\Delta t \sim b/(\gamma v)$. In the lab frame, the maximum photon energy is

$$\omega^{\mathrm{max}} = \frac{\hbar}{\Delta t} \sim \frac{\gamma \hbar v}{b},$$

3.

where γ is the Lorentz factor of the particle, $\gamma = (1 - v^2/c^2)^{-1/2}$. In the target frame, this equation applies, as long as γ is taken as the boost to go from the frame of one nucleus to the other ($\gamma = 2\gamma^2_{collider} - 1$).

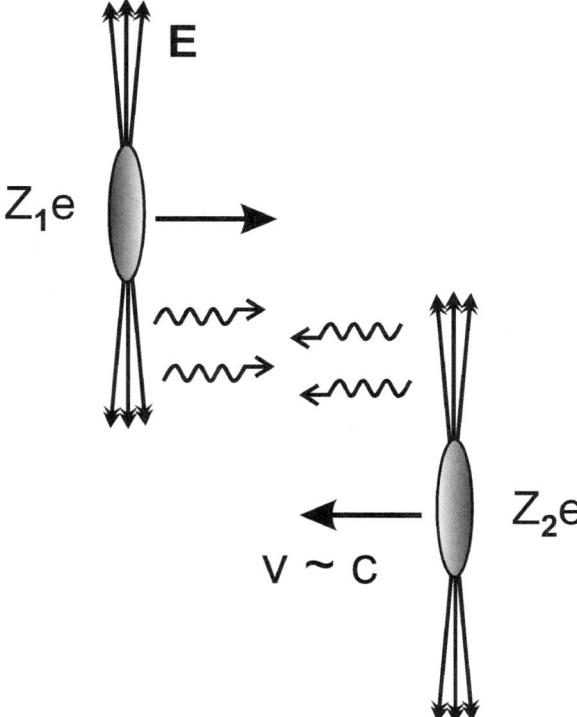

Figure 2   Highly energetic charges have Lorentz contracted electric (and magnetic) fields. The interaction between these fields can be replaced by the interaction of real (or quasi-real) photons.

For a grazing collision, where the two nuclei barely touch, we can take $b_{min} = 2R_A$, and the maximum photon energy is $\gamma \hbar v / 2R_A$ (R_A = nuclear radius). The maximum photon energy is about $\hbar/(2R_A A m_p c)$ of the ion energy. Here, $A m_p$ is the ion mass. For heavy ions, $R_A \approx 7$ fm so ω^{max} is about $0.03/A$ of the ion energy. For protons, R_A is not well defined, but taking ω^{max} to be 10% of the proton energy is a reasonable rule of thumb. RHIC (see Table 1) can reach photon-gold center of mass energies up to about 30 GeV per nucleon, and photon-proton center of mass energies up to 300 GeV. These energies are slightly higher than are available at fixed target accelerators and at HERA respectively. At the LHC, the corresponding energies are 1 TeV and 10 TeV respectively, more than an order of magnitude higher than is available elsewhere.

The equivalent (or virtual) photon flux per unit area (the relation between $n(\omega)$ and $N(\omega, b)$) is $n(\omega) = \int N(\omega, b) \, d^2 b$) is (1, 3, 17)

$$N(\omega, b) = \frac{Z^2 \alpha \omega^2}{\pi^2 \gamma^2 \hbar^2 \beta^2 c^2} \left(K_1^2(x) + \frac{1}{\gamma^2} K_0^2(x) \right). \qquad 4.$$

where $x = \omega b / \gamma \beta \hbar c$, Z is the ion charge, $\alpha = 1/137$, βc is the particle velocity, and K_0 and K_1 are modified Bessel functions. The first term $(K_1(x)^2)$ gives the flux of photons transversely polarized to the ion direction and the second is the flux for longitudinally polarized photons. The transverse polarization dominates for ultra-relativistic particles ($\gamma \gg 1$). The photon flux is exponentially suppressed when $\omega > \gamma \beta \hbar c / b$, justifying the estimates in the beginning of this section.

These photons are almost real, with virtuality $-q^2 < (\hbar/R_A)^2$. Except for the production of $e^+ e^-$ pairs, the photons can usually be treated as real photons.

The usable photon flux depends on the geometry. Most UPC reactions lead to final states with a handful of particles. These final states will be overwhelmed by any

TABLE 1 Some ion species, maximum energy and luminosity for several accelerators (170). Also shown are the maximum effective γp and $\gamma \gamma$ energies. For photon beams, the maximum effective photon energy here is 10% of the proton energy; there is some flux at higher energies. The CERN SPS is a fixed target accelerator; the effective luminosity depends on the target thickness. Not mentioned here are lower-energy accelerators, where photon exchange processes have also been studied.

Accelerator	Ions	Max. Energy per nucleon pair (CM)	Luminosity	Max. γp	Max. $\gamma \gamma$ energy
CERN SPS	Pb+Pb	17 GeV	—	3.1 GeV	0.8 GeV
RHIC	Au+Au	200 GeV	4×10^{26} cm^{-2} s^{-1}	24 GeV	6.0 GeV
RHIC	p+p	500 GeV	6×10^{30} cm^{-2} s^{-1}	79 GeV	50 GeV
LHC	Pb+Pb	5.6 TeV	10^{27} cm^{-2} s^{-1}	705 GeV	178 GeV
LHC	p+p	14 TeV	10^{34} cm^{-2} s^{-1}	3130 GeV	1400 GeV
Tevatron	p+p̄	20 TeV	5×10^{31} cm^{-2} s^{-1}	320 GeV	200 GeV

hadronic interactions between the fast-moving ion and the target. Thus, the useful photon flux is that for which the ions do not overlap, i.e., when the impact parameter $b = |b_1 - b_2|$ is greater than twice the nuclear radius ($2R_A$) (see Figure 1c). Usually, we can take $R_A = 1.2 A^{1/3}$ fm, where A is the atomic number. The $b > 2R_A$ requirement treats the nuclei as hard spheres; it is accurate for heavy nuclei, but less appropriate for lighter ions.

The photons can interact with a target nucleus in a one-photon process (when $b_1 < R_A$) or with its electromagnetic field in a two-photon process (when $b_1 > R_A$ and $b_2 > R_A$). In a photonuclear (one-photon) interaction, the usable photon flux is obtained by integrating Equation 4 over $b > b_{min} = 2R_A$:

$$n(\omega) = \frac{2Z^2\alpha}{\pi\beta^2}\left[\xi K_0(\xi)K_1(\xi) - \frac{\xi^2}{2}\left(K_1^2(\xi) - K_0^2(\xi)\right)\right]$$ 5.

where $\xi = \omega b_{min}/\gamma\beta\hbar c = 2\,\omega\,R_A/\gamma\beta\hbar c$.

For two-photon exchange processes, the equivalent photon numbers in Equation 2 must account for the electric field orientation of the photon fluxes with respect to each ion (see Figure 1b), obeying the ion non-overlap criteria $b_1, b_2 > R_A$ (14). The effects of orientation are also not included in Equation 5. For instance, symmetry properties dictate that $J^\pi = 0^+$ (scalar) particles originate from configurations such that $E_1 \parallel E_2$, whereas 0^- (pseudo-scalar) particles originate from $E_1 \perp E_2$ (18, 19). If one uses Equation 5 for $n(\omega_1)$ and $n(\omega_2)$, the total photoproduction cross section obtained from Equation 2 is higher than in a more detailed calculation, and the difference increases with increasing particle masses (18). Even more detailed calculations can be done by replacing the sharp-cutoff, $b_1, b_2 > R_A$, criterion with integrals over b_1 and b_2, which are weighted by the hadronic non-interaction probability. Asymmetric collisions (especially pA and dA) are also of interest; the higher-Z nucleus is likely to be the photon emitter, so the photon direction is known.

Low-energy processes, e.g., nuclear excitation, are also sensitive to the electromagnetic multipolarity involved. Equations 4 and 5 are only appropriate for electric dipole (E1) excitations. Equations for higher multipolarities are described in Reference (9).

For protons, the hard sphere approximation is inadequate. Instead, the proton size is included by the use of a form factor. With a dipole form factor, the flux is (20)

$$n(\omega) = \frac{\alpha}{2\pi z}[1 + (1 - z)^2]\left(\ln\chi - \frac{11}{6} + \frac{3}{\chi} - \frac{3}{2\chi^2} + \frac{1}{3\chi^3}\right)$$ 6.

where

$$\chi = 1 + \frac{0.71\ \text{GeV}^2}{Q_{min}^2 c^2}$$ 7.

accounts for the proton structure and $z = W^2/s$, with W the γp center of mass energy, and s the squared ion-ion center of mass energy per-nucleon. Here, Q_{min}

is the minimum momentum transfer possible in the reaction. For proton-proton collisions, the form factor has an effect similar to imposing a requirement $b_{min} = 0.7$ fm (21).

For protons and light nuclei, the weak electromagnetic interactions introduce another complication. The momentum transfer due to elastic scattering, $\Delta p = 2\eta\hbar/b$, with $\eta = Z_1 Z_2 \alpha \ll 1$ is small enough that the impact parameter is not a well-defined observable because $\Delta p \Delta b \sim \hbar$, leading to $\Delta b > b$ for $\eta < 1$ (22). This does not affect the total photon flux. However, it might affect the component of the photon flux that is unaccompanied by hadronic interactions. The uncertainty may also affect the probabilities for multiple interactions discussed in Section 5.

Equation 6 is valid when the proton remains intact. When photon emission with proton excitation, such as to the Δ resonance, is included, then the flux increases about 30% (23). At very high photon energy ($z \to 1$), the magnetic form factor of the proton can also become important (24).

1.2. Experimental Characterization

Ultra-peripheral collisions look very different from the more conventional hadronic interactions. The final state multiplicity is much smaller, and usually the events are fully reconstructed. Because the photon p_T are both small (roughly, $p_T \approx \omega/\gamma c$) in two-photon interactions, the final state p_T will be small. Photonuclear interactions that involve coherent scattering from the target nucleus (such as vector meson production) also have a very small p_T: $p_T < \hbar/R_A$. This gives the events a distinctive experimental signature, greatly simplifying detection (25).

UPCs are studied at a variety of accelerators. The characteristics of some relevant accelerators are given in Table 1. Each accelerator can accelerate many different species; Table 1 gives only a few candidates. The CERN Super Proton Synchrotron (SPS) has produced results on lead-to-bismuth transmutation and $e^+ e^-$ pair production in ion-ion collisions.

Although RHIC only began taking data in 2000, it has already released UPC results on ρ^0 photoproduction and on $e^+ e^-$ pair production. RHIC has enough energy and luminosity to photoproduce a wide variety of mesons, including the J/ψ. However, because it is a collider, detection of very low p_T particles is difficult, complicating the study of $e^+ e^-$ pairs and other atomic phenomena.

Although it is exclusively a $\overline{p}p$ collider, the Fermilab Tevatron is an interesting place to study UPCs. Antihydrogen was produced there using the process $\gamma\gamma \to e^+ e^-$, with the positron bound to an antiproton (8). Photoproduction of the J/ψ (21) may have been observed by the CDF collaboration (26).

The Large Hadron Collider (LHC), scheduled to begin operation in 2007, will search for physics beyond the standard model. A UPC program at the LHC can contribute to this search. Especially for pp collisions, where $\gamma\gamma$ and γp energies up to about 10% of the beam energy are accessible, UPCs may be an attractive place to search for new physics. With ion beams, the photon energies are lower, but $W^{\pm}$, Z, and heavy quark physics may be studied.

2. LOW-ENERGY PHOTONUCLEAR INTERACTIONS

Relativistic Coulomb Excitation (RCE) is now a popular tool to investigate the intrinsic nuclear dynamics and structure of the colliding nuclei. It is especially important in reactions involving radioactive nuclear beams (27–33), and has been used for many decades in low-energy nuclear collisions to study nuclear structure (34). However, nuclear-induced processes may also contribute to the reactions being studied.

RCE may involve single or multiple photon exchange between the projectile and the target. In the first case, perturbation theory directly relates the data to the matrix elements of electromagnetic transitions. These matrix elements are clean probes of the nuclear structure, and RCE can be used to study short-lived unstable nuclei that cannot be probed with real photons or electron scattering (28, 31, 33).

Radiative capture processes ($b + c \rightarrow a + \gamma$) play a major role in astrophysical sites, e.g., in a pre-supernova (35, 36). Some reactions of interest for astrophysics, e.g., $^7\text{Be}(p, \gamma)^8\text{B}$, can be studied via the inverse photo-dissociation reaction $^8\text{B}(\gamma, p)^7\text{Be}$ (37) using relativistic Coulomb collisions. The Coulomb breakup reaction $a + A \rightarrow b + c + A$ is useful to obtain the corresponding γ-induced cross section $\gamma + a \rightarrow b + c$. Using detailed balance, this cross section can be related to the radiative capture cross section $b + c \rightarrow a + \gamma$, of astrophysical interest (37). The radiative capture cross sections are often expressed in terms of the astrophysical S-factor: $S(E) = \sigma(E) \exp(-2\pi Z_b Z_c e^2 / \hbar v_{bc})/E$, where $E \equiv E_{rel} = m_{bc} v_{bc}^2/2$ is the relative energy between b and c. In this equation v_{bc} is the relative velocity and $m_{bc} = m_b m_c/(m_b + m_c)$ is the reduced mass of $b + c$. Because the Coulomb penetration factor is explicitly factored out, the S-factor is a much flatter function of E than $\sigma(E)$, allowing a better extrapolation of the measurements.

As an example, Figure 3a shows the result of an experiment performed at the GSI laboratory in Darmstadt, Germany (38) for the Coulomb dissociation of ^8B. Data on the reaction $^7\text{Be} + p \rightarrow \gamma + ^8\text{B}$ is important for understanding the structure of our sun. The decay of ^8B is responsible for the high-energy neutrinos observed by earth-bound detectors. The measured S-factor (S_{17}, $1 = \text{proton}$, $7 = {}^7\text{Be}$) is shown in Figure 3a as solid circles. The solid curve is a fit using a theoretical model for $S_{17}(E)$. Some of the data shown in the figure are from direct capture experiments (39).

Other b(c, γ) radiative capture reactions are planned or have already been studied with the Coulomb dissociation method (40). These processes may occur in the sun, supernovae, or during the Big Bang. Most of these reactions cannot be directly studied because the Coulomb barrier leads to very small values for the cross section, beyond the reach of present experimental techniques.

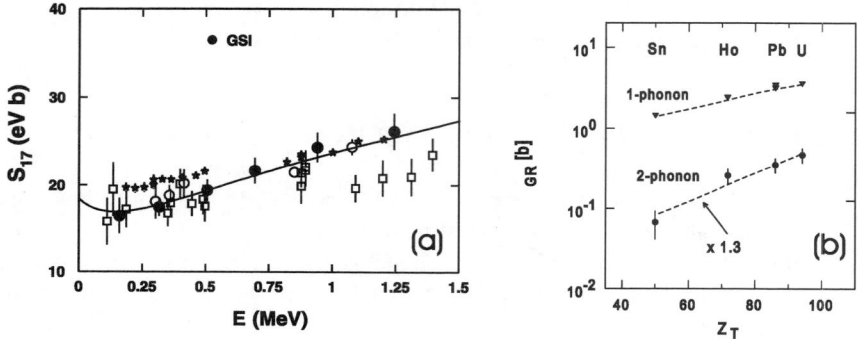

Figure 3 (*a*) S-factors (S_{17}) for the ^{7}Be(p, γ)^{8}B reaction. The GSI data was obtained using the Coulomb dissociation method (38). The other data are from direct capture measurements (39). (*b*) Cross sections for the excitation of the GDR (1-phonon) and the DGDR (2-phonon) in ^{208}Pb projectiles incident on different targets. The dashed curves are theoretical calculations.

Giant dipole resonances (GDR) occur in nuclei at energies of 10–20 MeV. Their gross properties are well described in terms of an out-of-phase collective motion (oscillation) of protons against neutrons in a nucleus (41, 42). If this oscillation is harmonic, with excitation energy $\hbar\omega$, then higher excitation modes, with energies equal to $N\hbar\omega$, also occur. These modes are interpreted as double, triple, . . . , giant dipole resonances. The double giant dipole resonances (DGDR) are thus two giant dipole vibrations superimposed in one nucleus, with about twice the energy of the GDR (4, 9, 29, 30). In the harmonic model, the RCE cross section for all multiphonon states can be calculated exactly (4).

A series of experiments at the GSI laboratory obtained the energy spectra, cross sections, and angular distribution of fragments following the decay of the DGDR (43–49). The experimental cross sections are about 30% bigger than the theoretical ones. This is shown in Figure 3*b*, where the dashed lines are the result of theoretical calculations (29, 30, 50–54). These experiments are promising for the studies of the nuclear response in very collective states. The giant resonances are still poorly understood, even with the best current microscopic approaches. The study of the width of the DGDRs will be help improve this scenario (30).

In heavy-ion colliders, the mutual Coulomb excitation of the ions (leading to their simultaneous fragmentation) is a useful tool for beam monitoring (55). A recent measurement at RHIC (56), using the Zero Degree Calorimeters to measure the neutron decay of the reaction products, has proved the feasibility of the method.

DGDRs constitute only 10% of the total fragmentation cross section induced by Coulomb excitation in UPC. The dominant contribution is the excitation of a single GDR, which then decays mostly by neutron emission. This is also a major source of beam losses in relativistic heavy-ion colliders (57), and an important fragmentation mode of relativistic nuclei in cosmic rays.

Another useful reaction is deuteron photodissociation in d+A collisions—a photon from a heavy ion photodissociates a deuteron (58). The reaction has a large cross section, 1.38 b for d+Au at RHIC, and 2.49 b for d+Pb at the LHC (59), and has been used as a "standard candle" for luminosity monitoring (60). d+A collisions are studied because they are technically simpler than p+A collisions.

3. PHOTOPRODUCTION AT HADRON COLLIDERS

The main interest in photoproduction at hadron colliders is derived from the possibility it offers of a direct determination of the gluon distribution in nucleons and nuclei. Examples of interactions in which the gluon distribution can be probed are exclusive production of heavy vector mesons, photoproduction of heavy quark-anti-quark pairs, and photoproduction of jets. The gluon distributions are not directly accessible in deep inelastic scattering, because the gluons carry neither electrical nor weak charge.

Measuring the nuclear shadowing using heavy-ion beams is particularly interesting. The nuclear gluon density can, as a first approximation, be written as the nucleon gluon distribution, $g(x, Q^2)$, multiplied by the number of nucleons (A):

$$G^A(x, Q^2) = Ag(x, Q^2).$$ 8.

Here, x is the fraction of the projectile momentum carried by the gluon, and Q^2 is the 4-momentum transfer squared.

Results from deep inelastic scattering of electrons on nuclear targets have, however, showed deviations from such a simple scaling for the structure function, $F_2(x_2, Q^2)$. Depending on x, suppressions (shadowing) of up to $\sim 30\%$ and enhancements (anti-shadowing) of up to $\sim 10\%$ have been observed. The effects of shadowing on $G^A(x, Q^2)$ are hard to determine directly. The current best estimates of the modification to the gluon distribution in nuclei are obtained from the Q^2 evaluation of $F_2(x, Q^2)$ (61, 62) and from studies of diffractive interactions (63). Photoproduction at heavy-ion colliders may provide a more direct measurement of $G^A(x, Q^2)$.

The particle production in photon-hadron or photon-nucleus interactions can be exclusive, when the protons or nuclei remain in their ground state or are only internally excited, or inclusive, when at least one of the nucleons or nuclei breaks up. Exclusive production will be discussed first. When the momentum transfer is small compared with its inverse nucleon/nuclear size, $Q \sim \hbar c/R$, the fields couple coherently to the entire target. The kinematics of coherent, exclusive interactions is very similar to that of two-photon interactions, which will be discussed in section 4.

3.1. Exclusive Particle Production

The dominant coherent interaction leading to the production of a hadronic final state is the exclusive production of vector mesons,

$$A + A \rightarrow A + A + V. \qquad\qquad 9.$$

In these reactions, a photon from the electromagnetic field of one of the projectiles interacts coherently with the nuclear field of the other (target), producing the vector meson.

Exclusive vector meson photoproduction on proton and nuclear targets has been studied since the mid-1960s using photon beams (64) and since 1992 at the HERA electron-proton accelerator (65). The first results from a heavy-ion collider on exclusive ρ^0 production (Au+Au $\rightarrow$ Au+Au + ρ^0) were recently published by the STAR collaboration at RHIC (25).

The total vector meson cross section in p+p or A+A interactions can be calculated from Equation 1. By differentiating and changing the variable from ω to y, the rapidity of the produced vector meson, one obtains

$$\frac{d\sigma(A + A \rightarrow A + A + V)}{dy} = n(\omega)\sigma_{\gamma A \rightarrow VA}(\omega) \qquad\qquad 10.$$

where the photon energy, ω, is related to y through $\omega = (M_V c^2/2) \exp(y)$ and M_V is the mass of the vector meson. If the photon flux is known, the differential cross section, $d\sigma/dy$, is thus a direct measure of the vector meson photoproduction cross section for a given photon energy.

The bulk of the photon-hadron cross section can be explained by the photon first fluctuating to a $q\bar{q}$ pair, which interacts with the target through the strong nuclear force. Since the photon has quantum numbers $J^{PC} = 1^{--}$, it preferentially fluctuates to a vector meson. The lifetime of the fluctuation is determined by the uncertainty principle. For a photon of virtuality Q fluctuating to a state of mass M_V the lifetime is of order

$$\Delta t \approx \frac{\hbar}{\sqrt{M_V^2 c^4 + Q^2 c^2}} \approx \frac{\hbar}{M_V c^2}. \qquad\qquad 11.$$

The last approximation is always true at hadron colliders because of the low virtuality of the photons. The photon wave function is written as a Fock decomposition (66):

$$|\gamma> = C_{bare}|\gamma_{bare} > + C_\rho|\rho > + C_\omega|\omega > + C_\phi|\phi > + \cdots + C_q|q\bar{q} >. \qquad 12.$$

Here $C_{bare} \approx 1$ and $C_V \sim \sqrt{\alpha_{em}}$ ($V = \rho, \omega, \phi, \ldots$). The coefficients C_V are related to the photon-vector meson coupling, f_V, through

$$C_V = \frac{\sqrt{4\pi\alpha_{em}}}{f_V}. \qquad\qquad 13.$$

The numerical values of the couplings f_V are usually determined from the vector meson leptonic decay widths, $\Gamma\ (V \rightarrow e^+ e^-)$.

According to the Generalized Vector Meson Dominance Model (GVMD), the scattering amplitude for the process $\gamma + A \rightarrow B$ is the sum over the corresponding vector meson scattering amplitudes,

$$A_{\gamma+A\rightarrow B}(s, t) = \sum_V C_V A_{V+A\rightarrow B}(s, t).$$ 14.

For elastic scattering, $\gamma + A \rightarrow V + A$, the cross-terms, i.e., $V' + A \rightarrow V + A$, are usually small (67) and are often neglected. The cross section is then

$$\frac{d\sigma(\gamma + A \rightarrow V + A)}{dt} = C_V^2 \frac{d\sigma(V + A \rightarrow V + A)}{dt},$$ 15.

where t is the momentum transfer from the target nucleus squared and $d\sigma/dt = |A|^2$.

The momentum transfer of the elastic scattering is determined by a hadronic form factor, $F(t)$,

$$\frac{d\sigma}{dt} = \frac{d\sigma}{dt}\bigg|_{t=0} |F(t)|^2.$$ 16.

For proton targets, the form factor is well represented by an exponential function, $|F(t)|^2 = \exp(-b|t|)$ with a slope $b \approx 10\,\text{GeV}^{-2}\,c^2$ for the light vector mesons (ρ, ω) and $b \approx 4\,\text{GeV}^{-2}\,c^2$ for the J/Ψ. The form factor for nuclear targets is peaked at much smaller momentum transfers because of the larger size of the target.

The form factor reflects the size and shape of the target. It can, in principle, be calculated if the spatial distribution is known. The dynamical information is contained in the forward scattering amplitude, $d\sigma/dt(t = 0)$. The optical theorem relates this to the total vector meson cross section, $\sigma_{tot}(VA)$:

$$\frac{d\sigma}{dt}\bigg|_{t=0} = C_V^2 \frac{\sigma_{tot}^2(VA)}{16\pi\hbar^2}(1 + \eta^2)$$ 17.

Here, η is the ratio of the real to the imaginary part of the scattering amplitude.

In Reference (68), data on vector meson photoproduction with proton targets were used to extract the total vector meson nucleon cross section, $\sigma_{tot}(VN)$. This result was then used to calculate the total vector meson nucleus cross section, $\sigma_{tot}(VA)$, from the nuclear geometry. This gave the vector meson production cross sections for heavy-ion interactions at RHIC and the LHC shown in Table 2. For heavier vector mesons, like the J/ψ, gluon shadowing may reduce the cross section (69).

In the Glauber Model (70), the elastic scattering amplitude is given by the two-dimensional Fourier transform of the nuclear profile function, $\Gamma(b)$:

$$\frac{d\sigma(\gamma + A \rightarrow V + A)}{dt} = \frac{\pi}{\hbar^2}\left|\int e^{i\mathbf{p}_T \cdot \mathbf{b}/\hbar}\Gamma(\mathbf{b})d^2\mathbf{b}\right|^2$$ 18.

TABLE 2 Cross sections for exclusive vector meson production in Au+Au and Pb+Pb interactions at RHIC and the LHC, respectively (68)

Meson	Au+Au, RHIC σ [mb]	Pb+Pb, LHC σ [mb]
ρ^0	590	5200
ω	59	490
ϕ	39	460
J/Ψ	0.29	32

$\Gamma(b)$ is a function of the distribution of matter inside the nucleus, $\rho(b, z)$, and the vector meson-nucleon forward scattering amplitude, f_{VN} (which can be related to the total vector meson-nucleon cross section through Equation 17):

$$\Gamma(b) = 1 - \exp\left[\frac{2i\pi\hbar c}{\omega} \int \rho(b, z') f_{VN}(0) \, dz'\right]. \qquad 19.$$

This approach only works for high photon energies, when $c\gamma\beta\Delta t > R_A$ so the interaction is longitudinally coherent over the entire nucleus. At lower ω, the loss of coherence reduces the cross section. The Glauber model is discussed in References (71, 72).

A Glauber model calculation of the coherent ρ^0 production cross section in Au+Au collisions at RHIC gave a total cross section of 934 mb (73). This is about 50% higher than the result in Reference (68) (cf. Table 2). The main reason for the difference is that in Reference (68), the total vector meson-nucleus cross section was calculated assuming that $\sigma_{tot}(\rho A) \approx \sigma_{inel}(\rho A)$. The calculation in Reference (73) furthermore includes the contribution from off-diagonal elements corresponding to $\rho' + Au \rightarrow \rho + Au$ scattering, as well as a non-zero real part of the forward scattering amplitudes (η in Equation 17). For a discussion of the $\rho' + A \rightarrow \rho + A$ contribution, see also Reference (67).

The measured ρ^0 production cross section at RHIC is $\sigma(Au + Au \rightarrow Au + Au + \rho^0) = 460 \pm 220 \pm 110$ mb at 130 A GeV (25). This can be compared with the Glauber Model calculations, which give $\sigma = 490$ mb at this energy (74). The corresponding number from the method used in Reference (68) is $\sigma = 350$ mb.

The rapidity distribution for coherent ρ^0 production measured by the STAR collaboration in Au+Au interactions at 200 A GeV is shown in Figure 4a (75). This is for ρ^0 production in coincidence with mutual Coulomb breakup of the beam nuclei (cf. Section 5). The rapidity distribution and cross section are in excellent agreement with the distribution obtained from the Monte Carlo model based on the calculations in Reference (76), corrected for the experimental acceptance. The reconstructed invariant $\pi^+\pi^-$ mass is shown in Figure 4b. The shape is well described by the sum of a relativistic Breit-Wigner function and a Söding interference term for direct $\pi^+\pi^-$ production (77). The STAR collaboration has very recently presented the first preliminary data on ρ^0 production in deuteron-gold

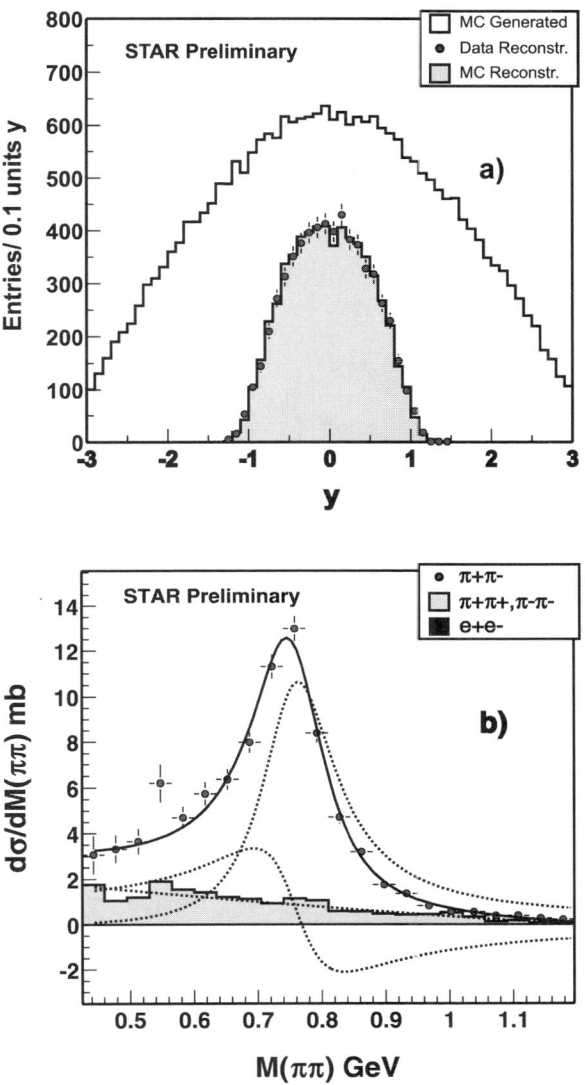

Figure 4   Rapidity (*a*) and invariant mass (*b*) distributions for coherent ρ^0 production in Au+Au interactions accompanied by mutual Coulomb breakup at $\sqrt{s} = 200$ A GeV, by the STAR collaboration. The dashed curves in (*b*) correspond to a relativistic Breit-Wigner function and a Söding interference term; the solid curve is the sum of the two. The dash-dotted curve describes the background from incoherent interactions (75).

interactions (the gold nucleus acts as photon emitter) (78) and on coherent production of $\pi^+\pi^-\pi^+\pi^-$ (79); the latter may be attributed to ρ^* photoproduction. PHENIX has shown indications of coherent J/Ψ and e^+e^--pair production in Au+Au interactions at RHIC (60, 80).

The forward scattering amplitude for heavy vector mesons has been calculated from two-gluon exchange in QCD. To leading-order (81) it was found that

$$\left.\frac{d\sigma(\gamma p \to V p)}{dt}\right|_{t=0} = \frac{\alpha_s^2 \hbar^2 \Gamma_{ee}}{3\alpha M_V^5 c^6} 16\pi^3 \left[x g\left(x, M_V^2/4\right)\right]^2. \qquad 20.$$

Here, x is the fraction of the proton or nucleon momentum carried by the gluons, and the gluon distribution, $g(x, Q^2)$, is evaluated at a momentum transfer $Q^2 = (M_V/2)^2$. This approach has been developed further by including relativistic wave functions and off-diagonal parton distributions (82, 83). The result is a total vector meson nucleon cross section which grows rapidly with increasing photon-proton center-of-mass energy, $W_{\gamma p}$. For Υ production, $\sigma \propto W_{\gamma p}^{1.7}$ is expected.

The dependence of $d\sigma/dt$ on $[g(x)]^2$ makes exclusive vector meson production a very sensitive probe of the proton and nuclear gluon distributions. An Υ-meson produced at mid-rapidity at the LHC would come from gluons with $x \approx 7 \cdot 10^{-4}$ and $x \approx 2 \cdot 10^{-3}$ in p+p and Pb+Pb interactions, respectively.

Figure 5 shows the predicted $d\sigma/dy$ for heavy vector mesons in nucleus-nucleus and proton-proton collisions. The calculations are based on parameterizations of

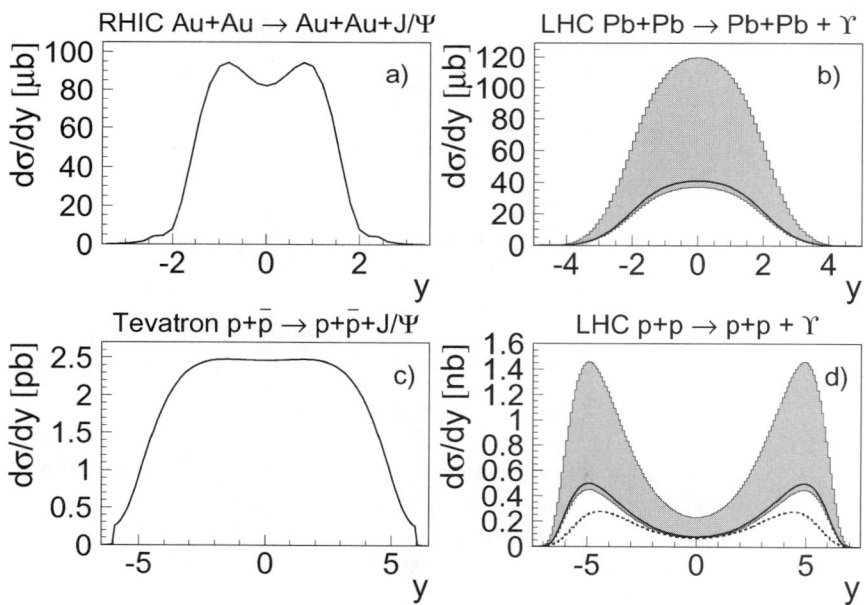

Figure 5   Rapidity distributions for exclusive J/ψ and Υ production in nucleus-nucleus and proton-proton collisions. Adapted from (21) and (68).

the photon-proton cross sections derived from measurements at HERA and from QCD-based models (21, 68). The photoproduction cross sections rise rapidly with energy. With lead at the LHC, the ρ^0 cross section is comparable to the hadronic cross section. With calcium beams at full luminosity, the LHC will produce about 230,000 ρ^0, 15,000 ϕ and 800 J/ψ per second. These rates are comparable to dedicated e^+e^- colliders, qualifying the LHC as a meson factory!

3.2. Interference in Exclusive Vector Meson Production

When a single vector meson is produced through a coherent photonuclear inter-action in a nucleus-nucleus or proton-(anti-)proton collision, it is in general not possible to determine which projectile acted as target and which was the photon-emitter. The two possibilities are indistinguishable, and under certain conditions they will interfere quantum mechanically (84). Because of this interference, it is incorrect to add the cross sections for the two possibilities.

The cross section is given by adding the corresponding amplitudes A_1 and A_2

$$\frac{\hbar d\sigma}{dy dp_T} = \int_{b>2R} |A_1 \pm A_2|^2 d^2\mathbf{b}.$$ 21.

The interference is maximal at mid-rapidity, where symmetry requires that $A_1 = A_2$. For ion-ion and proton-proton collisions, the interference is destructive be-cause of the negative parity of the vector meson; exchanging the position of the two nuclei or the two protons is equivalent to a reflection of the spatial coordinates, i.e., a parity transformation. For $p\overline{p}$ collisions, as at the Fermilab Tevatron, exchanging the proton and antiproton involves a charge-parity (CP) trans-formation. Since CP is positive for vector mesons, the interference is constructive at $p\overline{p}$ colliders (21).

The amplitudes A_1 and A_2 depend on the photon flux (and thus on rapidity) and on the photonuclear cross sections. Their p_T dependence comes from the convolu-tion of the photon p_T spectrum and the p_T from the photon-nucleus scattering. The former is given by the equivalent photon spectrum (19, 85), and the latter comes from the form factor of the target.

If the outgoing vector meson is treated as a plane wave (appropriate for a distant observer), at mid-rapidity, $A_1 = A_2$ and the square of the sum of the amplitudes is

$$|A_1 \pm A_2|^2 = 2A_0^2 \left(1 \pm \cos\left(\frac{\mathbf{p} \cdot \mathbf{b}}{\hbar}\right)\right).$$ 22.

For very low momenta, $p_T \ll \hbar/\langle b\rangle$, $\cos(\mathbf{p} \cdot \mathbf{b}/\hbar) \approx 1 - (\mathbf{p} \cdot \mathbf{b}/\hbar)^2/2$ and, as $p_T \to 0$, the interference is complete; emission disappears in ion-ion collisions, but doubles for $p\overline{p}$ colliders. Interference is significant for $p_T < 20$ MeV/c for the ρ^0 at RHIC (84), and $p_T < 250$ MeV/c for the J/ψ at the Tevatron (21). When $b \gg \hbar/p_T$, the cosine term oscillates rapidly as b varies, and the interference dis-appears. In this regime, the cross section reduces to the sum of cross sections for the two photon directions.

Away from mid-rapidity, $|A_1| \neq |A_2|$ because the photon energies for the two possibilities are different: $\omega_{1,2} = (M_V c^2/2) \exp(\pm y)$. Both the photon flux and photonuclear cross sections will be different. The interference will thus be reduced. A_1 and A_2 could also have slightly different phases, adding a phase factor δ to the cosine term in Equation 22. However, $\delta = 0$ in the standard Pomeron models (84), and a significant phase difference seems unlikely.

The interference in exclusive vector meson production is of particular interest because the vector mesons have very short lifetimes compared to the typical impact parameters. The median impact parameter for ρ^0 production in Au+Au collisions at RHIC, for example, is 46 fm, much larger than the lifetime of the ρ^0 ($\tau = 1.3$ fm/c). The vector meson cannot, on the other hand, be produced more than ~ 1 fm away from one of the two nuclei because of the short range of the nuclear force. Observing the expected interference pattern would thus prove that the wave function of the vector meson is preserved long after it has decayed. This is an example of the Einstein-Podolsky-Rosen paradox (84).

Preliminary data from the STAR collaboration on ρ^0 production in Au+Au at $\sqrt{s_{nn}} = 200$ GeV seem to confirm the presence of interference (86). The measured t spectrum is shown in Figure 6. The data are for interactions where both gold nuclei Coulomb dissociate. The coincident Coulomb dissociation selects events with smaller impact parameters compared to exclusive production (cf. Section 5); with Coulomb dissociation, the median impact parameter is only 18 fm (76).

The data are fit to a function,

$$\frac{dN}{dt} = a \exp(-b|t|) \left[1 + c(R(t) - 1) \right], \qquad 23.$$

with three parameters. These correspond to a normalization constant (a), the width of the nuclear form factor ($b \approx R_A^2/\hbar^2$), and a parameter to quantify the magnitude of the interference (c). The function $R(t)$ is the ratio of the $d\sigma/dt$ from the Monte Carlo calculated with and without interference. This functional form separates the interference from the nuclear form factor. No interference would correspond to $c = 0$, while complete interference according to the calculations above would correspond to $c = 1$. A fit to the data finds $c = 1.01 \pm 0.08$ ($0.1 \leq y \leq 0.5$) and $c = 0.78 \pm 0.13$ ($0.5 \leq y \leq 1.0$) for the two ranges in rapidity.

3.3. Inclusive Photoproduction

The high photon flux at hadron colliders and the large total photon-hadron cross sections lead to high rates for photonuclear interactions. In Au+Au collisions at RHIC, the total photonuclear cross section for photon-nucleon center-of-mass energies above 4 GeV is about 2 barns, or nearly 1/3 of the total hadronic Au+Au cross section. The majority of these interactions are resolved interactions, i.e., they are preceded by a fluctuation of the photon to a $q\bar{q}$ state. They therefore

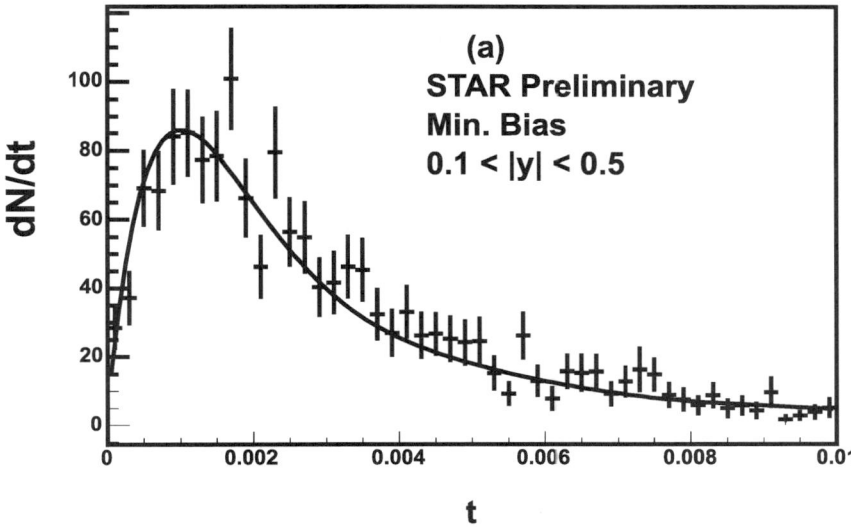

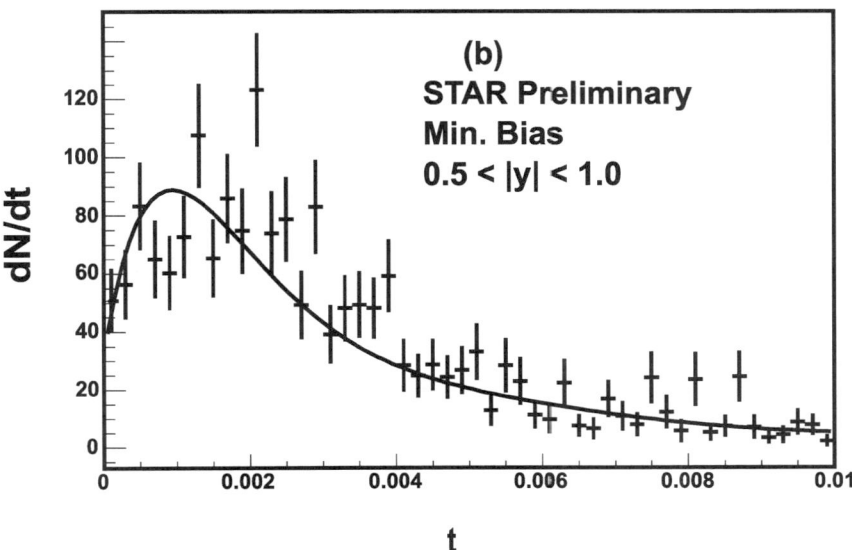

Figure 6 Efficiency corrected $t_\perp$ spectrum $t_\perp = p_{T^2}$ for ρ^0 production in Au+Au collisions at RHIC with Coulomb breakup. The points are the data, and the solid curve is a fit to Equation 23. Interference causes the dip at low $t \le 0.001$ (GeV/c)^2 (86).

resemble inelastic hadron-nucleon/nucleus collisions. Because the photon energies are much lower than the beam energies, the kinematics is similar to that of fixed target interactions.

Despite the large cross section, particle production in resolved photon-nucleon/nucleus interactions at hadron colliders has so far attracted relatively little interest. See, however, Reference (87). Understanding the kinematics of these interactions is nevertheless essential, because they are a significant background to other ultra-peripheral processes, particularly at the trigger level (88).

Considerably more interest has been devoted to direct photon interactions, in particular the production of heavy $q\bar{q}$-pairs (6, 89–91). Recently, the cross section for photoproduction of heavy quark pairs in pp collisions has been calculated (92). In these interactions, a photon interacts with a parton in the target and the partonic cross section can be calculated from QCD.

The leading order contribution to the photoproduction of a $q\bar{q}$-pair corresponds to photon-gluon fusion, as is illustrated in the Feynman diagram in Figure 1d. The cross section for the partonic sub-process is (93, 94)

$$\sigma_{\gamma g \to q\bar{q}}(W_{\gamma g}) = \frac{\pi e_q^2 \alpha_{em} \alpha_s(Q^2) \, \hbar^2 c^2}{W_{\gamma g}^2} \left[(3 - \beta^4) \ln\left(\frac{1+\beta}{1-\beta}\right) - 2\beta(2 - \beta^2) \right].$$

24.

Here, m_q and e_q are the quark mass and electric charge, respectively, $\beta = (1 - 4m_q^2 c^4 / W_{\gamma g}^2)$ and $W_{\gamma g}$ is the photon-gluon center-of-mass energy. If the gluon carries a fraction x of the nucleon momentum, then $W_{\gamma g}^2 = 2\omega x \sqrt{s}$. The strong coupling constant, $\alpha_s(Q^2)$, is evaluated to one loop at scale $Q^2 = m_q^2 c^2 + p_T^2$, where p_T is the quark transverse momentum.

The total photoproduction cross section $\sigma(A[\gamma]A \to Aq\bar{q}X)$ is obtained by convoluting the partonic cross section with the equivalent photon flux, $n(\omega)$, and the nuclear/nucleon gluon distribution, $G^A(x_2, Q^2)$, i.e.,

$$\sigma(A[\gamma]A \to Aq\bar{q}X) = \int \int \frac{n(\omega)}{\omega} G^A(x, Q^2) \sigma_{\gamma g}(W_{\gamma g}) \, \Theta\left(W_{\gamma g} - 2m_q c^2\right) d\omega \, dx.$$

25.

This equation is the equivalent of Equation 2 for two-photon interactions with the photon flux from one nucleus replaced by the gluon distribution, $G^A(x, Q^2)$. The final state $q\bar{q}$ rapidity depends on the photon energy and the gluon x. The rapidity distributions of bottom and top quarks produced in Pb+Pb and O+O collisions at the LHC are shown in Figure 7. The kinematics are discussed in more detail elsewhere (91, 95). The $q\bar{q}$ production cross section is peaked near threshold, $W_{\gamma g} \approx 4m_q^2$. Mid-rapidity production of $c\bar{c}$- and $b\bar{b}$-pairs therefore mainly probes x-values of $x \sim 1 \cdot 10^{-3}$ ($c\bar{c}$) and $x \sim 3 \cdot 10^{-3}$ ($b\bar{b}$) in heavy-ion collisions at the LHC. The corresponding numbers at RHIC are $x \sim 2 \cdot 10^{-2}$ and $x \sim 1 \cdot 10^{-1}$. (For a $q\bar{q}$-pair with invariant mass $W_{\gamma g}$ and pair-rapidity y, $x = (W_{\gamma g}/\sqrt{s})e^y$.)

The total cross sections for $c\bar{c}$ and $b\bar{b}$ production in various systems at RHIC and the LHC are listed in Table 3. The calculations without shadowing are compared

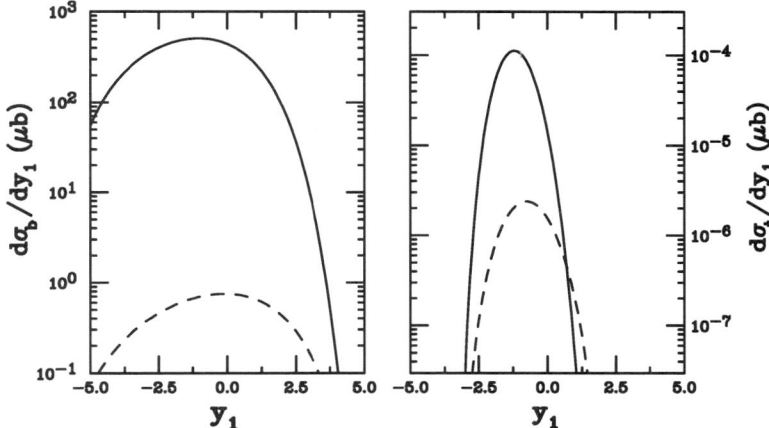

Figure 7   Rapidity distributions of (*left*) and (*right*) quarks produced in photonuclear collisions at the LHC. The solid and dashed curves are for Pb+Pb and O+O interactions, respectively. Here, the photon is emitted by the nucleus with positive rapidity; the complete cross section is the sum of this curve plus its mirror image. Shadowing is included using the parameterizations of Reference (62). Adapted from References (90, 91).

with two calculations that include nuclear modifications. As expected, shadowing has the largest effect on the production of lighter quarks ($c\bar{c}$) using heavy nuclei. The cross section for producing a $c\bar{c}$-pair in Pb+Pb interactions at the LHC is of the order of 1 b, nearly 1/6 of the total hadronic cross section.

Quark pairs can also be produced in anomalous interactions, where a parton from the resolved photon interacts with a parton in the target, or in two-photon interactions. The contribution from these processes is small compared with the direct production cross section (91). The anomalous cross sections are 1–20% of the direct cross sections, depending on quark flavor and collision energy. The two-photon contribution is usually less than 1% of the anomalous cross section.

In addition to probing the nuclear gluon distribution, photon-gluon fusion reactions are of interest as a means to determine the electric charge of the top quark. The top quark measurements are all consistent with a standard model top with electric charge $q_t = +2/3$. However, since the correlation between the decay products $W^+ \leftrightarrow b$ or $W^- \leftrightarrow \bar{b}$ are never measured, q_t is unconstrained (96). The data can still accommodate an exotic particle t' with $q_t = 4/3$, which decays via $t' \rightarrow W^- + b$. Because the cross section for $\gamma + g \rightarrow q + \bar{q}$ is proportional to q_t^2, $q_t = -4/3$ would quadruple the cross section.

There have so far been no experimental measurements of heavy quark pair photoproduction in heavy-ion or pp collisions. Some of the experimental techniques

TABLE 3 Cross sections for $q\bar{q}$ photoproduction through direct photon-gluon fusion in heavy-ion interactions. The numbers in columns 4 and 5 include nuclear gluon shadowing from References (62) and (97), respectively.

		$q\bar{q}$ cross sections in heavy-ion collisions		
Colliding system	**Flavor**	σ [mb] **No shadowing**	σ [mb] **EKS98**	σ [mb] **FGS**
RHIC Au+Au	$c\bar{c}$	15.8	17.4	17.6
RHIC Au+Au	$b\bar{b}$	$2.9 \cdot 10^{-3}$	$3.0 \cdot 10^{-3}$	$3.0 \cdot 10^{-3}$
RHIC Si+Si	$c\bar{c}$	0.196	0.203	0.20
RHIC Si+Si	$b\bar{b}$	$1.07 \cdot 10^{-4}$	$1.13 \cdot 10^{-4}$	$1.14 \cdot 10^{-4}$
LHC Pb+Pb	$c\bar{c}$	1250	1050	850
LHC Pb+Pb	$b\bar{b}$	4.9	4.7	4.4
LHC Ar+Ar	$c\bar{c}$	16.3	14.3	12.3
LHC Ar+Ar	$b\bar{b}$	0.073	0.070	0.066

that could be used to separate this signal from hadronic backgrounds are discussed in Section 6.

3.4. Dijets, Compton Scattering, Vector Boson Production and Other Processes

When the final state quarks in the process shown in Figure 1d have high p_T, the final state is two roughly back-to-back jets (98). The cross section is calculated in a manner similar to that of Equation 24, except that light quarks are included. Jets may also be produced to leading order through the process $\gamma + q \rightarrow g + q$. The cross section for dijet production is sensitive to the gluon distribution in the target; because of the simplicity of the reaction, there are fewer systematic uncertainties than in other processes such as vector meson production. However, it may be more difficult to isolate dijet photoproduction from backgrounds such as hadronic production of dijets and diffractive hadronic production.

Since the gluon-contributing nucleus does not stay intact, the experimental signature for this process is two jets, accompanied by a single rapidity gap between the jets and the photon-emitting nucleus. The two jets may have very different rapidities and it may be difficult to reconstruct the entire event. Calculations have considered the case where a single jet is detected, with $|\eta| < 1$ (98). Without shadowing, the rate to photoproduce jets with energies above 21 GeV in lead-lead collisions at the LHC is 0.015 Hz. In a 10^6 s run, jets up to 80 GeV should be detectable.

A closely related process is the production of a photon + jet final state; this is essentially Compton scattering. The rates for this process are about two orders of magnitude below that for dijet production (98).

The strong Coulomb fields may also dissociate hadrons into jets. For example, a proton may fragment into three quarks, leading to a reaction such as $\gamma p \to 3$ jets; this would be a distinctive signature in pA collisions (99). One photon can also dissociate another, leading to reactions such as $\gamma\gamma \to 2$ jets (99, 100).

$W^\pm$ and Z^0 can be photoproduced in ultra-peripheral collisions. A Z^0 can be produced when a photon fluctuates to a $q\bar{q}$ pair, scatters from a target nucleus and emerges as a Z^0. In the high-energy limit, the cross section for $\gamma p \to Z^0 p$ is about 0.01 pb (101). Unfortunately, even with a coherent photon beam, the cross section seems too low to be observable.

$W^\pm$ pairs can be produced directly from a photon fluctuation, $\gamma \to W^+W^-$. One of the Ws can interact with the target nucleus, leading to a hadronic jet plus a real W. Unfortunately, this process has not been studied in detail.

4. TWO-PHOTON PROCESSES

4.1. Production of Free- and Bound-Pairs

Between 1933 and 1937, Furry, Carlson, Landau, Lifshitz, Bhabha, Racah, Nishina, Tomonaga, and several others performed calculations of e^+e^- production in relativistic collisions of fast particles (cosmic rays) (102–106). The purpose was to test the newly born Dirac theory for the positron. Starting with the Dirac equation for the electron and its antiparticle, they found (106),

$$\sigma = \frac{28}{27\pi}\sigma_0[L^3 - 2.198L^2 + 3.821L - 1.632], \qquad 26.$$

where $\sigma_0 = (Z_1 Z_2 \alpha^2 \hbar/m_e c)^2$, $L = \ln(\gamma_1 \gamma_2)$, and γ_i is the Lorentz factor of ion i in the laboratory system. The first term of this equation can be simply obtained from Equation 1 and the cross sections for $\gamma\gamma$ pair production. The production cross sections for heavy lepton pairs ($\mu^+\mu^-$, or $\tau^+\tau^-$) can be obtained similarly. The production of $\mu^+\mu^-$ pairs using hadron beams was first observed in 63 GeV pp collisions at CERNs Intersecting Storage Rings (107).

For meson pairs like $\pi\pi$, neglecting internal substructure, as is done for Equation 26, may be appropriate near threshold. However, at higher pair masses, the quark substructure of the mesons becomes important, and the cross section for $\gamma\gamma \to \pi^0\pi^0$ becomes comparable to that for $\pi^+\pi^-$ (108). In fact, studies of $\gamma\gamma$ production of mesons pairs are interesting probes of meson structure. Baryon-antibaryon pairs are also of interest, because the cross section is sensitive to the baryon internal structure.

Because the cross sections depend on the inverse of the square of the particle mass, production of heavier pairs ($\mu^+\mu^-$, $\tau^+\tau^-$) is much smaller than for e^+e^-. Their Compton wavelength, $\lambda_i = \hbar/m_i c$, is smaller than the nuclear radius R. This requires the replacement $L \to \mathcal{L} = \ln(\gamma_1\gamma_2\delta/m_i cR)$ in Equation 26, where $\delta = 0.681\ldots$ is a number related to Euler's constant.

Bound particles, such as positronium or $q\bar{q}$ mesons, are also produced in two-photon interactions. The cross section is given by Equation 2. The cross section for $\gamma_1\gamma_2 \rightarrow X$ depends on the particle's decay width to two photons, $\Gamma_{\gamma\gamma}$ (109). Because decay and $\gamma\gamma$ production use the same matrix elements, only the phase-space factors and polarization summations are distinct. One finds (109)

$$\sigma(\omega_1, \omega_2) = 8\pi^2(2J + 1)\frac{\Gamma_{\gamma\gamma}}{Mc^2}\,\delta\!\left(4\omega_1\omega_2 - M^2c^4\right) \qquad 27.$$

where J, M, and $\Gamma_{M\rightarrow\gamma\gamma}$ are the spin, mass, and two-photon decay width of the meson, respectively. The delta-function imposes energy conservation.

Using Equation 27, the production of mesons with mass M in HI colliders is (9):

$$\sigma = \frac{128}{3}\,(Z_1 Z_2 \alpha)^2\,\frac{\hbar\Gamma_{\gamma\gamma}}{M^3c^5}[\mathcal{L}^3 + \cdots]. \qquad 28.$$

This equation is obtained using Equation 2 and the high-energy limit ($\gamma \gg 1$) of the equivalent photon number $n(\omega)$ (for more details, see Reference (110)).

A more detailed account of the space geometry of the two-photon collision is necessary (18) and will be discussed in Section 4.4. Because spin 1 particles cannot couple to two real photons (111), only spin 0 and spin 2 particles should be produced.

The treatment of bound states in quantum field theory (QFT) is a very complex subject (for reviews, see References (112, 113)). In the case of positronium production by two photons (para-positronium) and by three photons (ortho-positronium), standard QFT techniques allow a simple and accurate way to calculate the cross sections from first principles (114, 115). The para-positronium production cross sections are quite large, 19.4 mb and 116 mb, for RHIC (Au+Au) and LHC (Pb+Pb), respectively (115). However, Coulomb corrections reduce these values by as much as 43% for RHIC and 27% for LHC (114). The cross sections for the production of ortho-positronium, which requires three-photon exchange, are also large: 11.2 mb and 35 mb, for RHIC and LHC, respectively (114). Even the ortho-positronium cross sections correspond to production rates of 4 and 35 per second respectively. If the positronium could be extracted from the interaction points, they could be used to test interesting properties of QFT for bound states. Relativistic positronium has an unusually large transparency in thin layers (see Reference (116) and references therein).

The same diagrams for the calculation of the positronium apply for production of bound $q\bar{q}$ pairs (mesons) in UPC (115). However, proper accounting for the color degrees of freedom is needed (117).

4.2. Production of Free e^+e^- Pairs

Due to experimental difficulties, Equation 26 (and its newer counterparts) has never been fully tested. With the construction of RHIC and the LHC, interest in this process has grown. For heavy ions, the e^+e^- production probabilities are close to one, and lowest-order perturbative calculations of the cross sections violate unitarity (i.e., $d^2\sigma/d^2b > 1$) (9).

This observation led to more detailed calculations (19, 118–122), involving high-order processes, such as the exchange of multiple photons (Coulomb distortion) and the production of multiple pairs, as shown in Figures 1e, f. These processes are important for collisions at small impact parameters. Diverse theoretical methods have been considered. Perturbative calculations are simple to write down, but they involve rather complicated integrals, especially for low-energy electrons, due to Coulomb distortion and relativistic effects on the continuum electronic wavefunction (9). A general sum of the contribution of diagrams like those in Figure 1e, f and unitarity corrections (involving the production of virtual e^+e^- pairs) was obtained in Reference (123). To account for Coulomb distortions, one needs to add to Equation 26 a term of the form [see eq. 7.3.10 of Reference (9)],

$$\sigma_C = -\frac{28}{9\pi} \left[f(Z_1\alpha) + f(Z_2\alpha) \right] \sigma_0 L^2, \qquad 29.$$

where

$$f(x) = x^2 \sum_{n=1}^{\infty} \frac{1}{n(n^2 + x^2)} \qquad 30.$$

is the Bethe-Maximon correction. Equation 29 has been correctly derived in Reference (124). Their result was later confirmed by independent calculations in Reference (125). For Pb+Pb collisions at LHC, the Coulomb distortion correction reduces the pair-production cross section by 14%. Other unitarity corrections further reduce the cross sections by 3% (123).

The calculation of the production of multiple pairs, as shown in Figure 1e, f, is directly connected with the unitarity problem. It is possible to interpret $d^2\sigma/d^2b$ as the mean number of pairs produced at a given impact parameter. For Ca-Ca collisions at the LHC ($Z\alpha = 0.15$), $\sigma_{2-pairs} = 0.11$ b (123), or about 27,000 $e^+e^-e^+e^-$ events per second. In the literature one finds different methods to calculate the cross section for the production of $n > 2$ pairs. Reference (123) does a simple fit to numerical calculations. Reference (117) is based on the expression for the probability $P(b)$ of the pair production taken from Reference (9). Reference (122) claims that this expression is wrong. They derived expressions for this probability using two different methods. Reference (123) obtains $\sigma_n \propto L^n$, whereas References (118) and (124) obtain $\sigma_n \propto L^{2n}$ and $\sigma_n \propto L^{3n}$, respectively. The calculations differ in the method used to include Coulomb and unitarity corrections. The production of multiple pairs has been studied with a variety of different theoretical approaches (14, 123, 125, 127, 133–137).

The calculation of multiple photon exchange can be considerably simplified in the ultra-relativistic limit. In this limit, the electromagnetic field of the ions is squeezed in the plane perpendicular to its trajectory (i.e., it can be approximated by a delta-function along this plane). In the appropriate gauge, the Coulomb potential is two-dimensional and the time-dependent Dirac equation may be solved exactly (130, 138, 139). This should be equivalent to an all-orders perturbative calculation.

This approach yields good results as long as $\omega b/\gamma\hbar c \leq 0.1$ (140). Above this value, the delta-function approximation breaks down. Because the most important impact parameters for this process are of order $b \simeq \hbar/m_e c$ (9), the calculation can be separated into two regions: (a) $b \simeq \hbar/m_e c$ where the approximation is valid; and (b) large impact parameters, for which perturbative calculations are accurate (130, 138, 139). This method describes well the differential cross sections for e^+e^- pair production up to energies of order 0.1 $\gamma m_e c^2$, above which the delta-function approximation breaks down for the same reason as above (140). The initial calculations using this technique found results that matched the lowest order perturbation theory without Coulomb corrections (131, 132). This was inconsistent with both theoretical expectations and with data (132). However, regularization of the integrals was critical; with this regularization, the Dirac approach reproduced the lowest order result, with Coulomb corrections (126, 141). This technique allows for the calculation of cross section for free e^+e^- pair production to all orders in $Z\alpha$ (141).

Electron-positron pair production has been studied at RHIC in combination with mutual Coulomb excitation (142). As will be discussed in the next section, the mutual Coulomb excitations were required to trigger on these events, and also had the effect of selecting events with $b < 30$ fm, where non-perturbative effects were strongest. The cross section, pair mass, rapidity and pair p_T distributions were all in accord with the predictions of lowest order perturbation theory (136). The pair p_T distribution deviated from the Weiszaecker-Williams virtual photon approach, showing that the photon mass was important in that kinematic regime.

The STAR study suffered from low statistics. Earlier experiments on e^+e^- production in sulfur-platinum (and sulfur-nuclei) collisions at the CERN SPS had higher statistics, but lower beam energies; they also found good agreement with lowest order calculations (143, 144).

It remains disappointing that these 70-year-old QED calculations are still not fully tested. Although many aspects of QED have been tested to high precision, studies involving strong fields are much less advanced; pair production with relativistic heavy ions (with $Z\alpha \sim 1$) is one important example.

4.3. Pair Production with Capture and Antihydrogen

An important phenomenon occurs when the electron is captured in an atomic orbit of the projectile or of the target (9). At RHIC and the LHC, this is an important source of beam particle loss (110). The produced beam of single-electron ions carries considerable power (145); at the LHC, at full luminosity, the produced $^{+81}$Pb beam carries sufficient power to quench the superconducting accelerator magnets; this limits the LHC luminosity with heavy ions (146).

One striking application of this process was the recent production of antihydrogen atoms using relativistic antiproton beams (7). Here the positron is produced and captured in an orbit of the antiproton. The expression

$$\sigma = 3.3\pi \frac{Z_1^2 Z_2^6 \alpha^8 \hbar^2}{m_e^2} \frac{1}{\exp(2\pi Z_2 \alpha) - 1} (L - 2.051) \qquad 31.$$

for pair production with electron capture in the nucleus with charge Z_2 is obtained in first order perturbation theory (9). Although Equation 31 works reasonably well for explaining antihydrogen production, it is only valid for small Z_i ($Z_i \leq 15$) (7, 147, 148). For large Z_i, as with the experiments at RHIC and LHC, non-perturbative calculations may be necessary (119, 130, 149–152). Equation 31 includes higher order effects related to the electron capture, but is not a complete all-orders result. The additional higher order corrections are apparently small, and Equation 31 should be usable for most purposes. The fraction with the exponential term is due to the distortion of the positron wavefunction. It accounts for the reduction of the magnitude of the positron (continuum) wavefunction near the nucleus where the electron is localized (bound).

Equation 31 shows that the cross section depends on energy as $\sigma = A \ln \gamma_1 \gamma_2 + B$, where the coefficients A and B depend on the nuclear charges. This scaling was confirmed in numerical calculations of Reference (130) and was used in the analysis of the experiment in Reference (153), shown in Figure 8a. The comparison between theory and the data of Reference (153) is not completely

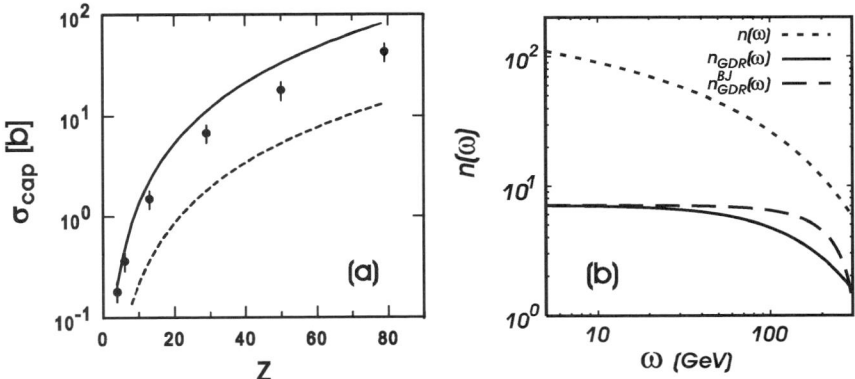

Figure 8 (a) Pair production with capture for Pb^{82+} (33 TeV) beams on several targets. The solid and dashed curves are theoretical calculations (153). See text for more details. (b) The tagged total photon flux (accompanied by a single giant dipole resonance excitation) for a complete calculation (*solid line*) and a simplified "box" integration (*dashed line*), compared to the untagged flux (*dotted line*), for gold at RHIC. Although tagging reduces the low-energy flux by an order of magnitude, at high energies, the difference is much smaller. From Reference (22).

valid because atomic screening was not included. When screening is present, the cross sections are smaller by at least a factor of 2-4 (see Equation 7.4.3 of Reference (9)).

Similar reactions occur with hadron pairs. Processes like pion pair production with capture (or, similarly, photoexcitation of a Δ resonance, which decays by π^+ emission) increase the nuclear charge by 1 and can turn lead into bismuth, plus a π^+. This change cannot occur electromagnetically, because e^+ will not bind to lead. These transmutations have been studied at the SPS/CERN for ^{208}Pb at 160 GeV/nucleon (154). The data can be described quantitatively with electromagnetic excitation calculations (87, 155). For high-Z nuclei, the dominant contribution to nuclear-charge pickup is due to electromagnetic production of π^- by virtual photons. This contribution is completely negligible for a similar experiment at an energy of 10.6 GeV per nucleon (156).

4.4. Two-Photon Production of Mesons

For the production of composite particles, one can use Equation 28 as a first guess. However, even the lightest mesons (π^0) require photons of relatively large energy (≥ 70 MeV). Mesons are produced primarily in collisions with relatively small impact parameters (compared to $2R_A$) where hard photons are more abundant. Substantial changes to Equation 28 are required to account for the collision geometry (18, 157).

One can rewrite Equation 2 more conveniently as

$$\sigma_X = \int ds \frac{d\mathcal{L}(W_{\gamma\gamma})}{dW_{\gamma\gamma}} \, \sigma_{\gamma\gamma \to X}(W_{\gamma\gamma}), \qquad\qquad 32.$$

where $W_{\gamma\gamma} = 4\omega_1\omega_2$ is the square of the center-of-mass energy of the two photons, $\sigma_{\gamma\gamma \to X}(W_{\gamma\gamma})$ is the two-photon production of particle X, and $d\mathcal{L}/dW_{\gamma\gamma}$ is the "photon-photon luminosity." $d\mathcal{L}/dW_{\gamma\gamma}$ can be multiplied by the ion beam luminosities, yielding an effective two-photon luminosity $d\mathcal{L}_{eff}/dW_{\gamma\gamma}$ which can be directly compared to other two-photon luminosities, such as at e^+e^- or pp colliders (158). With heavy-ion beams, the LHC two-photon luminosities are much higher than are available elsewhere, either with proton beams at the LHC or at the e^+e^- LEP-II collider for energies up to $\sqrt{W_{\gamma\gamma}} \approx 500$ GeV (14).

Table 4 (115) shows the cross sections for the production of $C = even$ mesons for the RHIC and LHC colliders. Other calculations were done in References (88, 100, 159). The cross section corresponds to a $\gamma\gamma \to \pi^0$ rate of 30 events/second with lead beams at the LHC. For heavier mesons, like η_c, the rate is still large, of the order of 1 per minute.

For mesons of comparable mass, the two-photon cross sections in Table 4 are about two orders of magnitude lower than the cross sections for photonuclear vector meson production (Table 2). This difference stems from the different coupling strengths of the strong and electromagnetic interactions, $\alpha_s \sim 1$ vs. $\alpha_{em} \approx 1/137$.

TABLE 4 Cross sections for two-photon production of ($C = even$) mesons at RHIC (Au+Au) and at LHC (Pb+Pb) (115)

Meson	Mass [MeV]	σ^{RHIC} [mb]	σ^{LHC} [mb]
π_0	134	4.9	28
η	547	1.0	16
η'	958	0.75	21
$f_2(1270)$	1275	0.54	22
$a_2(1320)$	1318	0.19	8.2
η_c	2981	3.3×10^{-3}	0.61
χ_{0c}	3415	0.63×10^{-3}	0.16
χ_{2c}	3556	0.59×10^{-3}	0.15

Two-photon meson spectroscopy is thus greatly complicated by the large background from photonuclear interactions. For example, with lead beams at the LHC, the rate of J/ψ photoproduction followed by $J/\psi \rightarrow \gamma \eta_c$ is about 2.5 per minute, higher than the $\gamma\gamma \rightarrow \eta_c$ rate.

Although it may be possible to separate the different event classes with cuts on meson p_T, rapidity, and final state particles, the vector meson background seems daunting to most efforts (68).

4.5. Searches for New Physics

The LHC will reach high enough energies that two-photon interactions will be an attractive place to search for some types of new physics. Many early calculations focused on the search for the Higgs particle (163–165). Other examples include supersymmetric particle pairs, magnetic monopoles, and possible extra spatial dimensions (14). The LHC will also be able to probe vector boson couplings through reactions like $\gamma\gamma \rightarrow W^+ W^-$.

The two-photon production rate for the Higgs is small enough that, for most models, it is likely to be discovered in hadronic interactions. However, for standard model Higgs masses under ≈ 200 GeV, with medium ion beams, the $\gamma\gamma$ channel should produce a handful of events per year (14). In some supersymmetric scenarios, the production of the Higgs in UPCs could be significantly enhanced (159). The $\gamma\gamma$ production channel could also be studied in pp collisions; the greatly increased luminosity and running time will more than compensate for the smaller production cross section. However, in pp collisions, there is a considerable background due to diffractive interactions. It may be possible to separate $\gamma\gamma$ from diffractive interactions by studying the p_T of the scattered protons. This could be done by placing small detectors, known as Roman pots, inside the beampipe, to detect protons scattered at very small angles (160).

Supersymmetric particle pairs are a similar story. If present, they are likely to be discovered in hadronic interactions. However, if supersymmetry is correct, a large number of new particles are likely to be present, and the $\gamma\gamma$ production of sparticle pairs is likely to provide significant new information; two-photon production is sensitive to significant regions of phase space (159). Two-photon interactions are sensitive to sparticles that do not participate in the strong interaction. Photonuclear interactions may also be useful for studying supersymmetry.

Real or virtual magnetic monopoles can be produced by two-photon fusion. Investigators searched for the process $\gamma\gamma \to \gamma\gamma$ at the Fermilab Tevatron, and mass limits were set on magnetic monopoles with various charges (161). The LHC will be able to do far better.

The presence of extra dimensions could be detected via the two-photon production of gravitons. The cross section to produce a graviton, $\gamma\gamma \to G$ increases in the presence of compact dimensions (162). There are unresolved theoretical issues regarding the cross section, but one calculation finds that, for 2 extra compact dimensions, at the LHC, the cross section is of order 1 nb for lead and 10 pb for calcium (162). Rates are not given for proton beams, but for calcium, and likely protons, a few events would be produced each month. The experimental signature of graviton production has not been worked out in detail.

Most of these new physics channels involve relatively high p_T particles and so should be within the purview of the planned trigger setup for the ATLAS and CMS detectors. This may not be true for supersymmetric final states; two charged sleptons that do not interact hadronically will challenge any trigger.

5. MULTIPLE INTERACTIONS BETWEEN A SINGLE ION PAIR

Because heavy ions have such large charges, a single ion pair can undergo multiple electromagnetic reactions in a single interaction. Even though the reactions may be independent, the geometry introduces correlations between the photon energies and polarizations. Multiple interactions are also a key experimental tool, allowing for many cross checks under different triggering conditions. One example of such a reaction is the photoproduction of a ρ^0 meson, accompanied by the mutual Coulomb excitation of both nuclei:

$$Au + Au \to Au^* + Au^* + \rho^0. \qquad 33.$$

This reaction was studied by the STAR collaboration (25). This process occurs predominantly via 3-photon exchange, as is shown in Figure 1g.

Investigators have also observed four photon reactions, such as the production of an e^+e^- pair accompanied by mutual Coulomb excitation (142), as shown in Figure 1h.

In a multi-photon process, each photon emission may be treated independently if the energy lost by the nucleus is not significant. As long as the photon emission does not excite the emitter, the reactions may be treated as completely independent. The cross section is calculated in impact parameter space

$$\sigma = \int d^2 b \, P(b) \qquad 34.$$

where $P(b)$ is the probability for the reaction to occur at impact parameter b. This is

$$P(b) = \int \frac{d\omega}{\omega} N(\omega, b) \sigma_{\gamma A}(\omega). \qquad 35.$$

When the cross section for a reaction is very large (as with $e^+ e^-$ production or GDR excitation), the naive $P(b)$ calculated in Equation 35 may exceed 1. $P(b)$ should then not be interpreted as a probability but rather as the mean number of produced particles at that impact parameter. The actual number of produced particles follows a Poisson distribution with this mean. The generalization to 2 (or more) photon exchanges is obvious. Calculations using this approach accurately predicted the cross section and kinematic distributions for Reaction 33 (25).

This factorization only holds if several conditions are satisfied. Photon emission must not excite the emitting nucleus, and the photons must be emitted independently. As long as the fractional energy loss of the nuclei is small, this is valid (168). Finally, the excitation must not change the nuclear form factor significantly on the time scale of the reaction (the frozen nucleus approximation). As long as these strictures are satisfied, the ordering of the subprocesses is unimportant. These conditions hold for heavy-ion collisions. For proton beams, with $\eta = Z_1 Z_2 e^2 \alpha / \beta < 1$, the factorization is on weaker ground because of the poorly defined impact parameter (22).

It can be convenient to treat one reaction as a trigger (or selector) for a range of impact parameters. Picking events with mutual Coulomb excitation, for example, selects events with small impact parameters. The reason can be seen in Equation 34. In the low-energy limit ($\omega \ll \gamma \hbar c / R_A$), for a fixed ω, $P(b) \approx N(\omega, b) \approx 1/b^2$. For a two-photon reaction, $P_1(b) P_2(b) \approx 1/b^4$. The mean impact parameter b_n for an n-photon interaction is (22)

$$\bar{b}_n = \frac{\int d^2 b \, b \, P_1(b) \ldots P_n(b)}{\int d^2 b \, P_1(b) \ldots P_n(b)}. \qquad 36.$$

Here, $b_{min} = 2 R_A$ is the minimum impact parameter, and $b_{max} = \gamma \hbar / R_A$. For $n = 1$ the result $\bar{b} = (b_{max} - b_{min}) / \ln(b_{max}/b_{min})$ is not so useful. However, for 2 or more photon exchanges

$$\bar{b}_{n>1} = 2 R_A \frac{2n - 2}{2n - 3}; \qquad 37.$$

b_{max} drops out, leaving a simple result. For $n = 2$, this reduces to $\bar{b}_2 = 2 R_A$; for larger n, $\bar{b}$ is even smaller. At heavy-ion colliders, mutual Coulomb excitation is an effective trigger for selecting low-impact parameter events. Detailed calculations of the median impact parameter in 1 and 3 photon interactions find a similar scaling (76). Reducing $\bar{b}$ is very helpful in studying interference in vector meson production, by increasing the p_T range over which the interference is visible.

This selection can also be viewed in momentum space. In the low-energy limit, the photon flux $n(\omega)$ (Equation 5) scales as $1/\omega$. However, when an additional photon is present, the spectrum becomes much harder. The spectrum for photons that are accompanied by Coulomb excitation is:

$$n(\omega) \approx \int_{2R_A}^{\gamma \hbar c/\omega} d^2b N(\omega, b) \frac{\text{Const.}}{b^2} \approx \text{Const.} \qquad 38.$$

The extra photon line adds a $1/b^2$ weighting, and the resultant flux is independent of the photon energy.

Figure 8b compares the spectra with and without tagging. By selecting reactions with additional accompanying photon interactions, experimenters can tune their photon beam, hardening or softening the spectrum. This tuning allows many cross-checks. For example, in vector meson production, $y = \ln(2\omega/M_V c^2)$, and there is a two-fold ambiguity over which nucleus emitted the photon. By comparing vector meson production with and without mutual Coulomb excitation, it is possible to account for this ambiguity and find the production cross section as a function of photon energy.

The coupling is also very useful in experimental triggering. A simple reaction like multiple Coulomb excitation can be used to trigger on small-impact-parameter collisions; the remainder of the event can then be studied without experimental bias.

For two-photon final states, like e^+e^- pairs, the situation is more complex because the particles are produced outside the nuclei. However, two-photon reactions are also enhanced at small nucleus-nucleus impact parameters. The STAR collaboration used mutual Coulomb excitation to study low-mass e^+e^- pair production at RHIC, because it was not possible to trigger on the e^+e^- pair itself (142). The presence of the mutual excitation also significantly hardens the pair mass spectrum.

In multiple photonuclear interactions, the photon polarizations are collinear. Photons are linearly polarized along the electric field of the emitting nucleus. In photonuclear interactions, the electric field vector follows the impact parameter vector. When a single nucleus emits multiple photons, these photons all have the same linear polarization. When the other nucleus emits a photon, it will have the opposite polarization.

When the final states are sensitive to the photon polarization, then angular correlations will be present. In ρ^0 decay, in the plane perpendicular to the ρ^0 direction, the angle between the photon polarization vector and the π^+ (or π^-) direction is distributed following a $\cos\theta$ distribution. For the case of two independent ρ^0 decays, with uncorrelated photon polarization, the angle between the two π^+, $\Delta\phi$ is evenly distributed between 0 and 2π. With the polarization correlation, the angular correlation is

$$N(\Delta\phi) \approx 1 + \frac{1}{2}\cos(2\Delta\phi).$$ 39.

A similar distribution is also expected for neutrons from giant dipole resonance decay. Correlations between neutron p_T in ρ^0 production and GDR excitation(s) should also be measurable at RHIC. In the longer term, GDR excitation neutrons could be used to tag photons according to their polarization direction, allowing for studies with polarized photons. Similar polarization should also occur for medium-energy nuclear reactions.

In addition to the correlations due to the geometry, multiple interactions may be a place to study Bose-Einstein correlations, such as in $\rho^0\rho^0$ production by two independent photonuclear interactions. When the two ρ^0 are produced on the same nucleus, the production should be bosonically enhanced (exhibit super-radiance) when the ρ have (in the nuclear target rest frame) a momentum difference $\Delta p < \hbar/R_A$.

6. EXPERIMENTAL POSSIBILITIES AND LIMITATIONS FOR ULTRA-PERIPHERAL COLLISIONS

The experimental study of electromagnetic interactions at high-energy colliders is quite new. Since the characteristics of these interactions are very different from the more common hadronic interactions, most existing and planned detector systems are not optimized to study them. This section will discuss some of the general experimental possibilities and limitations at current and future colliders.

With hadron colliders (unlike electron beams) it is generally not possible to detect the outgoing projectiles following an electromagnetic interaction. This is because of the small momentum transfers involved. Tagging of nuclei will never be possible because the angular deflection following the coherent emission of a photon is smaller than the angular dispersion of the beam. Proton tagging has been proposed, but requires extremely high resolution Roman pots (160).

For exclusive particle production, characterized by the emission of only a few final state particles, good signal to background ratios can be achieved by selecting events with small total transverse momentum when the event is reconstructed (88). The total event transverse momentum is the sum of the momentum transfer from each projectile, which is determined by the form factors and can be calculated accurately. The method works best for heavy nuclei. This is illustrated in Figure 9, which shows the p_T distribution for exclusive ρ^0 production in Au+Au interactions at RHIC. The coherent peak for events with exactly one reconstructed π^+ and one reconstructed π^- can be clearly seen with $p_T < 100$ MeV/c. The incoherent background can be estimated by measuring events with two reconstructed pions

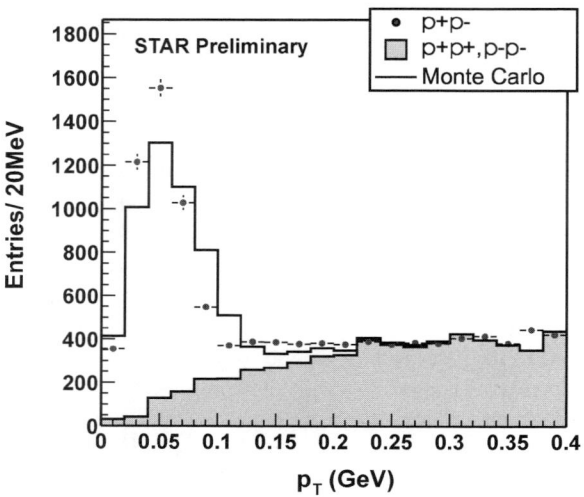

Figure 9 Transverse momentum distribution for events with exactly 2 reconstructed charged pions. Data from the STAR collaboration (75). The Monte Carlo calculation is based on Reference (76).

with the same charge (dashed histogram in the figure). The technique of using the total event p_T to identify electromagnetic interactions works only for coherent and exclusive particle production. Other photonuclear interactions require different methods. Incoherent photon-induced interactions are characterized by a gap in rapidity, void of particles, between the rapidity of the photon-emitting projectile and the rapidity of the produced state. This distinguishes the electromagnetic interactions from ordinary hadronic interactions.

In hadronic interactions, the probability of having a rapidity gap of width Δy void of particles decreases exponentially with increasing size of the interval:

$$p(n = 0) = \exp(-\langle dn/dy \rangle \Delta y) \qquad 40.$$

where $\langle dn/dy \rangle$ is the mean multiplicity per unit of rapidity. Requiring a rapidity gap of width $\Delta y = 2$ leads to a hadronic rejection factor of 10^{-2} to 10^{-3} for single nucleon-nucleon interactions at RHIC and LHC energies. Additional rejection of hadronic events can be obtained if the fragmentation of the beam nuclei is detected, for example in forward (zero-degree) calorimeters. In hadronic interactions, both nuclei normally break up, whereas in photoproduction, the photon emitter normally remains intact. Further discussions of the experimental aspects can be found in Reference (91). The conclusion is that sufficient rejection can be obtained with current and planned detectors at heavy-ion colliders to study, e.g., the production of heavy quark pairs.

One significant background to electromagnetic diffraction is hadronic diffraction. Pomeron exchange can also produce rapidity gaps, and double-Pomeron interactions produce isolated systems in a central detector, with rapidity gaps on each side (100). This is not an issue for vector meson production, because double-Pomeron interactions do not produce $J^{PC} = 1^{--}$ final states, but these backgrounds could be problematic for other reactions. Because the strong force has a short range ($\leq$1 fm), Pomeron exchange between nuclei and double-Pomeron interactions can only occur between surface nucleons in grazing collisions; this limits the cross sections. Unfortunately, quantitative estimates are lacking. However, for pp collisions, Pomeron interactions may constitute a significant background.

Another experimental challenge is finding an efficient trigger with adequate background rejection. Because the outgoing beams are not tagged, it is necessary to trigger on the particles emitted from the final state X in order to study the reaction A+A $\rightarrow$ A+A+X. These particles are usually produced at or near mid-rapidity. Most of the current and planned heavy-ion experiments at RHIC and the LHC have triggers that are optimized for hard, high p_T interactions. These triggers are difficult to adapt to low multiplicity low-p_T final states. STAR at RHIC is an exception, but faces challenges because of the low allowable trigger rates (169). Backgrounds from beam-gas interactions, grazing hadronic collisions, ambient neutrons, and other beam-related backgrounds are serious concerns to any experiment.

In combination with the large probability for multiple Coulomb interactions with heavy beams, factorization has made triggering on neutrons emitted in the forward direction following single or double Coulomb excitation an attractive alternative. This reduces the photon flux, but provides some control of the impact parameter distribution, as discussed in Section 5.

The experimental feasibility of studying ultra-peripheral collisions is demonstrated through the data presented earlier. However, the considerations above limit the types of processes that can be studied. The study of two-photon production of multiple e^+e^--pairs seems difficult because of the extremely low p_T of the emitted electrons. A dedicated experiment would do better, but that is likely to be a difficult proposition at a high-energy physics laboratory. The study of meson spectroscopy at the LHC also seems problematic because of the lack of triggers for low momentum charged particles around mid-rapidity.

7. CONCLUSIONS

We have presented the formalisms for studying photoproduction and two-photon reactions at hadron colliders, and discussed some of the more interesting applications of these techniques. Low-energy nuclear physics has used these techniques for many years, with good results. However, only recently have higher-energy machines like RHIC and the Tevatron begun to study particle production in very strong fields.

The amount of experimental data on UPCs is still rather limited. Despite this, the field has developed tremendously over the past decade, with much theoretical progress. The new data from RHIC is helping to focus new theoretical work on the channels that are most readily accessible. With the new results from RHIC and the coming LHC startup, UPCs are now ready to make substantial contributions to many areas of physics.

The LHC will produce photonuclear interactions with 10 times the energy available at other accelerators like HERA. This will open a huge window to search for new physics processes (some of them not accessible in hadronic collisions), measure gluon densities at very low Feynman-x, and perform a host of other measurements. At the same time, UPC reactions are very important for machine operations - e^+e^- production with e^- capture will limit the LHC luminosity with heavy ions.

ACKNOWLEDGMENTS

We thank Ramona Vogt for providing Figure 7, Gerhard Baur and Kai Hencken for beneficial discussions, and Heather Gray for helpful comments on the manuscript. This work was supported by the U.S. Department of Energy under grants No. DE-FG02-04ER41338 and DE-AC-03076SF00098.

**The *Annual Review of Nuclear and Particle Science* is online at
http://nucl.annualreviews.org**

LITERATURE CITED

1. Fermi E. *Z. Physik* 29:315 (1924)
2. Marciano W, White S, eds. *Electromagnetic Probes of Fundamental Physics.* Singapore: World Sci. (2003)
3. Weizsäcker CF. *Z. Physik* 88:612 (1934); Williams EJ. *Phys. Rev.* 45:729 (1934)
4. Baur G, Bertulani CA. *Phys. Lett.* B174:23 (1986)
5. Ritmann JL, et al. *Phys. Rev. Lett.* 70:533 (1993); *Phys. Rev. Lett.* 70:2659 (1993); Schmidt R, et al. *Phys. Rev. Lett.* 70:1767 (1993)
6. Goncalves VP, Bertulani CA. *Phys. Rev. C* 65:054905 (2002)
7. Baur G, et al. *Phys. Lett.* B368:251 (1996)
8. Blanford G, et al. *Phys. Rev. Lett.* 80:3040 (1998)
9. Bertulani CA, Baur G. *Phys. Rep.* 163:299 (1988)
10. Pirner HJ. *Phys. Rev. C* 22:1962 (1980)
11. Feshbach H, Zabek M. *Ann. Phy.* (*NY*) 107:110 (1977); Feshbach H. *Theoretical Nuclear Physics: Nuclear Reactions.* New York: Wiley-InterScience (1993)
12. McLerran L, Venugopalan R. *Phys. Rev. D* 50:2225 (1994)
13. Krauss F, Greiner M, Soff G. *Prog. Nucl. Part. Phys.* 39:503 (1997)
14. Baur G, et al. *Phys. Rep.* 364:359 (2002)
15. Baur G, et al. hep-ex/0201034
16. Baur G, Hencken K, Trautmann D. *J. Phys. G* 24:1657 (1998)
17. Jackson JD. *Classical Electrodynamics*, 2nd ed. New York: Wiley (1975)
18. Baur G, Ferreira Filho LG. *Nucl. Phys.* A518:786 (1990)
19. Vidović M, Greiner M, Best C, Soff G. *Phys. Rev. C* 47:2308 (1993)
20. Drees M, Zeppenfeld D. *Phys. Rev. D* 39:2536 (1989)
21. Klein S, Nystrand J. *Phys. Rev. Lett.* 92:142003 (2004)
22. Baur G, et al. *Nucl. Phys.* A729:787 (2003)
23. Ohnemus J, Walsh TF, Zerwas PM. *Phys. Lett. B* 328:369 (1994)
24. Kniehl BA. *Phys. Lett.* B254:267 (1991)
25. Adler C, et al. (STAR Collaboration). *Phys. Rev. Lett.* 89:272302 (2002)

26. Wyatt A. Presented at "Small-x and Diffraction 2003," Fermilab, IL. Transparencies are available at http://conferences.fnal.gov/smallx/new_program.htm

27. Bertulani CA, Canto LF, Hussein MS. *Phys. Rep.* 226:282 (1993)

28. Glasmacher T. *Annu. Rev. Nucl. Part. Sci.* 48:1 (1998)

29. Aumann T, Bortignon PF, Emling H. *Annu. Rev. Nucl. Part. Sci.* 48:351 (1998)

30. Bertulani CA, Ponomarev V Yu. *Phys. Rep.* 321:139 (1999)

31. Hansen PG, Jensen AS, Jonson B. *Annu. Rev. Nucl. Part. Sci.* 45:591 (1995)

32. Bertulani CA, Hussein M, Muenzenberg G. *Physics of Radioactive Beams.* Hauppage, NY: Nova Science (2002)

33. Jonson B. *Phys. Rep.* 389:1 (2004)

34. Alder K, et al. *Rev. Mod. Phys.* 28:432 (1956)

35. Clayton DD. *Principles of Stellar Evolution and Nucleosynthesis.* New York: McGraw-Hill (1968)

36. Rolfs C, Rodney WS. *Cauldrons in the Cosmos* Chicago: Univ. Chicago Press (1988)

37. Baur G, Bertulani CA, Rebel H. *Nucl. Phys.* A458:188 (1986)

38. Schuemann F, et al. *Phys. Rev. Lett.* 90:232501 (2003)

39. Hammache F, et al. *Phys. Rev. Lett.* 86:3985 (2001); Junghans AR, et al. *Phys. Rev. Lett.* 88:041101 (2002); Baby LT, et al. *Phys. Rev. Lett.* 90:022501 (2003); *Phys. Rev. C* 67:065805 (2003)

40. Baur G, Rebel H. *Annu. Rev. Nucl. Part. Sci.* 46:321 (1996); Baur G., Hencken K, Trautmann D. *Prog. Part. Nucl. Phys.* 51:487 (2003)

41. Goldhaber M, Teller E. *Phys. Rev.* 74:1046 (1948)

42. Steinwedel H, Jensen JHD. *Z. Naturforsch.* 5a:413 (1950)

43. Ritman J, et al. *Phys. Rev. Lett.* 70:533 (1993)

44. Schmidt R, et al. *Phys. Rev. Lett.* 70:1767 (1993)

45. Aumann T, et al. *Phys. Rev. C* 47:1728 (1993)

46. Boretzky K, et al. *Phys. Lett.* B384:30 (1996)

47. Grünschloss A, et al. *Phys. Rev. C* 60:051601 (1999)

48. Boretzky K, et al. *Nucl. Phys.* A649:235c (1999)

49. Ilievski S, et al. *Nucl. Phys.* A687:178c (2001)

50. Bertulani CA, Canto LF, Hussein MS, de Toledo Piza AFR. *Phys. Rev. C* 53:334 (1996)

51. Hussein MS, de Toledo Piza AFR, Vorov OK. *Phys. Rev. C* 59:R1242 (1999)

52. de Toledo Piza AFR, et al. *Phys. Rev. C* 59:3093 (1999); Carlson BV, et al. *Phys. Rev. C* 59:2689 (1999); Carlson BV, Hussein MS. *Phys. Rev. C* 59:R2343 (1999); Carlson BV, Hussein MS, de Toledo Piza AFR, Canto LF. *Phys. Rev. C* 60:014604 (1999)

53. de Paula DT, et al. *Phys. Rev. C* 64:064605 (2001)

54. Pshenichnov IA, et al. *Phys. Rev. C* 64:024903 (2001)

55. Baltz AJ, et al. *Phys. Rev. E* 54:4233 (1996); Baltz AJ, Chasman C, White SN. *Nucl. Instrum. Meth. A*, 417:1 (1998); White SN. *Nucl. Instrum. Meth. A* 409:618 (1998)

56. Chiu M, et al. *Phys. Rev. Lett.* 89:012302 (2002)

57. Bertulani CA, Baur G. *Physics Today* 22: March (1994)

58. Hoffman B, Baur G. *Phys. Rev. C* 30:247 (1984)

59. Klein S, Vogt R. *Phys. Rev. C* 68:017902 (2003)

60. White SN. nucl-ex/0501004.

61. Gousset T, Pirner HJ. *Phys Lett. B* 375:349 (1996)

62. Eskola KJ, Kolhinen VJ, Ruuskanen PV. *Nucl. Phys. B* 535:351 (1998)

63. Frankfurt L, Strikman M. *Eur. Phys. J.* A5:293 (1999)

64. Bauer TH, Spital RD, Yennie DR. *Rev. Mod. Phys.* 50:261 (1978)

65. Crittenden JA. *Exclusive Production of Neutral Vector Mesons at the Electron-Proton Collider HERA.* Berlin: Springer-Verlag (1997)

66. Schuler GA, Sjöstrand T. *Nucl. Phys.* B 407:539 (1993)

67. Pautz A, Shaw G. *Phys. Rev. C* 57:2648 (1998)

68. Klein S, Nystrand J. *Phys. Rev. C* 60: 014903 (1999)

69. Frankfurt L, Strikman M, Zhalov M. *Phys. Lett.* B540:220 (2002)

70. Glauber RJ. In *Lectures in Theoretical Physics,* ed. WE Britten, LG Dunham. New York: Interscience (1959)

71. Grammer G, Sullivan JD. In *Electromagnetic Interactions of Hadrons,* Vol. 2, ed. A Donnachie, G Shaw. New York and London: Plenum (1978)

72. Alberi G, Goggi G. *Phys. Rep.* 74:1 (1981)

73. Frankfurt L, Strikman M, Zhalov M. *Phys. Lett.* B537:51 (2002)

74. Frankfurt L, Strikman M, Zhalov M. *Phys. Rev. C* 67:034901 (2003)

75. Meissner F (for the STAR Collaboration). *Nucl. Phys.* A715:522c (2003)

76. Baltz AJ, Klein SR, Nystrand J. *Phys. Rev. Lett.* 89:012301 (2002)

77. Söding P. *Phys. Lett.* 19:702 (1966)

78. Timoshenko SL (for the STAR Collaboration). nucl-ex/0501010

79. Ogawa A (for the STAR Collaboration). Presented at DIS2004, April 14–18, 2004, Strbske Pleso, Slovakia

80. Silvermyr D (for the PHENIX Collaboration). Presented at DNP04, Meeting of the Nucl. Phys. Div. of the APS, Chicago, Oct. 27–30, 2004. http://www.phenix.bnl.gov/conferences.html

81. Ryskin MG. *Z. Phys. C* 57:89 (1993)

82. Frankfurt LL, McDermott MF, Strikman M, *J. High Energy Physics* 02:002 (1999)

83. Martin AD, Ryskin MG, Teubner T *Phys. Lett.* B454:339 (1999)

84. Klein SR, Nystrand J. *Phys. Rev. Lett.* 84: 2330 (2000); *Phys. Lett.* A308:323 (2003)

85. Baur G, Ferreira-Filho LG. *Phys. Lett.* B 254:30 (1991)

86. Klein SR (for the STAR Collaboration). nucl-ex/0310020; nucl-ex/0402007

87. Pshenichnov IA, et al. *Phys. Rev. C* 60: 044901 (1999)

88. Nystrand J, Klein S. *nucl-ex/9811007* (1998), *Proc. Workshop on Photon Interactions and the Photon Structure,* eds. G. Jarlskog, T. Sjöstrand, Lund, Sweden, Sept. (1998)

89. Greiner M, et al. *Phys. Rev. C* 51:911 (1995)

90. Klein SR, Nystrand J, Vogt R. *Eur. Phys. J. C* 21:563 (2001)

91. Klein SR, Nystrand J, Vogt R. *Phys. Rev. C* 66:044906 (2001)

92. Goncalves VP, Machado MVT. *Phys. Rev. D* 71:014025 (2005); *Eur. Phys. J.* C40:519 (2005)

93. Jones LM, Wyld HW. *Phys. Rev. D* 17:759 (1978)

94. Fritzsch H, Streng KH. *Phys. Lett.* B72:385 (1978)

95. Smith J, van Neerven WL. *Nucl. Phys.* B 374:36 (1992)

96. Baur U, Buice M, Orr LH, *Phys. Rev. D* 64:094019 (2001)

97. Frankfurt L, Guzey V, Strikman M. hep-ph/0303022

98. Vogt R. hep-ph/0407298

99. Frankfurt L, Strikman M. *Phys. Rev. D* 67:017502 (2003)

100. Eggert K, et al. *FELIX Letter of Intent, CERN/LHCC 97/45;* Ageev, et al. *J. Phys. G* 28:R117 (2002); Engel R, Ranft J., Roesler S. *Phys. Rev. D* 55:6957 (1997)

101. Pumplin J. hep-ph/9612356

102. Furry WH, Carlson JF. *Phys. Rev.* 44:238 (1933)

103. Landau LD, Lifshitz EM. *Phys. Zs. Sowjet* 6:244 (1934)

104. Bhabha HJ. *Proc. R. Soc. London Ser. A* 152:559 (1935)

105. Nishina Y, Tomonaga S, Kobayashi M. *Sci. Pap. Phys. Chem. Res.* 27:137 (1935)

106. Racah G. *Nuovo Cimento* 14:93 (1937)

107. Antreasyan D, et al. preprint CERN-EP/80–82 (1990)

108. Morgan D, Pennington MR, Whalley MR, *J. Phys. G* 20:A1 (1994)
109. Low FE. *Phys. Rev.* 120:582 (1960)
110. Baur G, Bertulani CA. *Z. Phys. A* 330:77 (1988); Bertulani CA, Baur G. *Nucl. Phys.* A505:835 (1989)
111. Yang CN. *Phys. Rev.* 77:242 (1950); Wolfenstein L and Ravenhall DG. *Phys. Rev.* 88:279 (1952)
112. Bodwin GT, Yennie DR, Gregorio MA *Rev. Mod. Phys.* 57:723 (1985)
113. Sapirstein J, Yennie DR. In *Quantum electrodynamics*, ed. T. Kinoshita. Singapore: World Sci. (1990)
114. Kotkin GL, Kuraev EA, Schiller A, Serbo VG. *Phys. Rev. C* 59:2734 (1999)
115. Bertulani CA, Navarra F. *Nucl. Phys.* A703:861 (2002)
116. Nemenov LL. *Yad. Fiz.* 51:444 (1990); *Sov. J. Nucl. Phys.* 51:284 (1990); Lyuboshitz VL, Podgoretsky MI. *Zh. Eksp. Teor. Fiz.* 81:1556 (1981)
117. Appelquist T, Politzer HD. *Phys. Rev. Lett.* 34:43 (1975)
118. Baur G. *Phys. Rev. D* 41:3535 (1990); *Phys. Rev. A* 42:5736 (1990)
119. Bottcher C, Strayer MR. *Phys. Rev. D* 39:1330 (1989); *J. Phys. G* 16:975 (1990); *Phys. Lett.* B237:175 (1990)
120. Rhoades-Brown MJ, Weneser J. *Phys. Rev. A* 44:330 (1991)
121. Best C, Greiner W, Soff G. *Phys. Rev. A* 46:261 (1992)
122. Hencken K, Trautmann D, Baur G. *Phys. Rev. A* 51:998 (1995); *Phys. Rev. A* 51:1874 (1995)
123. Lee RN, Milstein AI, Serbo VG. *Phys. Rev. A* 65:022102 (2002)
124. Güçlü MC. *Nucl. Phys.* A668:149 (2000)
125. Ivanov DY, Schiller A, Serbo VG. *Phys. Lett. B* 454:15 (1999)
126. Lee RN, Milstein AI. *Phys. Rev. A* 61:032103 (2000); *Phys. Rev. A* 64:032106 (2001)
127. Aste A, et al. *Eur. Phys. J. C* 23:545 (2002)
128. Bartos E, Gevorkyan SR, Kuraev EA, Nikolaev NN. *Phys. Lett.* B538:45 (2002)
129. Eichler J, Meyerhof W. *Relativistic Atomic Collisions*. San Diego: Academic (1995)
130. Baltz AJ. *Phys. Rev. Lett.* 78:1231 (1997)
131. Baltz AJ, McLerran L. *Phys. Rev. C* 58:1679 (1998)
132. Segev B, Wells JC. *Phys. Rev. A* 57:1849 (1998); *Phys. Rev. C* 58:1697 (1998)
133. Eichmann U, Reinhardt J, Schramm S, Greiner W *Phys. Rev. A* 59:1223 (1999)
134. Hencken K, Trautmann D, Baur G. *Phys. Rev. A* 59:841 (1999)
135. Eichmann U, Reinhardt J, Greiner W. *Phys. Rev. A* 61:062710 (2000); *Phys. Rev. C* 61:064901 (2000)
136. Hencken K, Baur G, Trautmann D. *Phys. Rev. C* 69:054902 (2004)
137. Baltz AJ, Gelis F, McLerran L, Peshier A. *Nucl. Phys. A* 695:395 (2001)
138. Baltz AJ, Rhoades-Brown MJ, Weneser J. *Phys. Rev. A* 44:5569 (1991); *Phys. Rev. A* A48:2002 (1993); *Phys. Rev. A* 47:3444 (1993); *Phys. Rev. A* 50:4842 (1994)
139. Baltz AJ. *Phys. Rev. A* 52:4970 (1995)
140. Bertulani CA. *Phys. Rev. A* 63:062706 (2001)
141. Baltz AJ. nucl-th/0409044
142. Adams J, et al. *Phys. Rev. C* 70:031902 (2004)
143. Baur R, et al. *Phys. Lett.* B332:471 (1994)
144. Vane CR, et al. *Phys. Rev. A* 50:2313 (1994)
145. Klein S. *Nucl. Instrum. Meth. A* 459:51 (2001)
146. Brandt D. *LHC Project Report 450.* www.cern.ch (2000)
147. Bertulani CA, Baur G. *Phys. Rev. D* 58:034005 (1998)
148. Bertulani CA, Dolci D. *Nucl. Phys.* A683:635 (2001)
149. Anholt R, Becker U. *Phys. Rev. A* 36:4628 (1987)
150. Becker U. *J. Phys. B* 20:6563 (1987)
151. Bottcher C, Strayer MR. *J. Phys. G* 16:975 (1990); *Phys. Lett.* B237:175 (1990)
152. Meier H, et al. *Phys. Rev. A* 63:032713 (2001)

153. Krause HF, et al. *Phys. Rev. Lett.* 80:1190 (1998)

154. Scheidenberger C, et al. *Phys. Rev. Lett.* 88:042301 (2002); *Phys. Rev. C* 70:014902 (2004)

155. Pshenichnov IA, et al. *Phys. Rev. C* 57:1920 (1998); *Phys. Rev. C* 64:024903 (2001)

156. Geer LY, et al. *Phys. Rev. C* 52:334 (1995)

157. Cahn RN, Jackson JD. *Phys. Rev. D* 42:3690 (1990)

158. Khoze VA, Martin AD, Ryskin MG. *Eur. Phys. J. C* 23:311 (2002)

159. Vidovic M, Greiner M, Soff G. *Phys. Rev. C* 47:2288 (1993)

160. Piotrzkowski K. *Phys. Rev. D* 63:071502 (2001)

161. Abbott B, et al. *Phys. Rev. Lett.* 81:524 (1998)

162. Ahern SC, Norbury JW, Poyser WJ. *Phys. Rev. D* 62:116001 (2000)

163. Papageorgiu E. *Phys. Rev. D* 40:92 (1989)

164. Grabiak M, Müller B, Greiner W, Koch P. *J. Phys. G* 15:L25 (1989)

165. Drees M, Ellis J, Zeppenfeld D. *Phys. Lett.* B223:454 (1989)

166. Nikulin V. Presented at 2nd Workshop on Ultra-Peripheral Heavy Ion Collisions, Oct. 11–12, 2002, CERN, Geneva

167. Baur G, Hencken K, Trautmann D, Sadovsky S and Kharlov Y. e-Print Archive hep-ph/990436

168. Gupta SN. *Phys. Rev.* 99:1015 (1955)

169. Bieser FS, et al. *Nucl. Instrum. Meth. A* 499:766 (2003)

170. S. Eidelman, et al. (Particle Data Group), *Phys. Lett.* B592:1 (2004)

Annu. Rev. Nucl. Part. Sci. 2005. 55:311–55
doi: 10.1146/annurev.nucl.55.090704.151558

LEPTOGENESIS AS THE ORIGIN OF MATTER

W. Buchmüller,[1] R.D. Peccei,[2] and T. Yanagida,[3]

[1]*Deutsches Elektronen-Synchrotron DESY, 22603 Hamburg, Germany;*
email: buchmuwi@mail.desy.de
[2]*Department of Physics and Astronomy, University of California at Los Angeles,*
Los Angeles, California 90095; email: peccei@physics.ucla.edu
[3]*Department of Physics, University of Tokyo, Tokyo 113-0033, Japan;*
email: tsutomu@hep-th.phys.s.u-tokyo.ac.jp

Key Words Leptogenesis, Baryogenesis, matter-antimatter asymmetry, seesaw
mechanism, neutrino mass and mixing, neutrino oscillations, sphaleron, inflation,
reheating temperature, dark matter, axion, supersymmetry, LSP, gravitino

■ **Abstract** We explore in some detail the hypothesis that the generation of a primordial lepton-antilepton asymmetry (Leptogenesis) early on in the history of the
Universe is the root cause for the origin of matter. After explaining the theoretical
conditions for producing a matter-antimatter asymmetry in the Universe we detail
how, through sphaleron processes, it is possible to transmute a lepton asymmetry—or,
more precisely, a $(B - L)$-asymmetry—into a baryon asymmetry. Because Leptogenesis depends in detail on properties of the neutrino spectrum, we review briefly existing
experimental information on neutrinos as well as the seesaw mechanism, which offers
a theoretical understanding of why neutrinos are so light. The bulk of the review is
devoted to a discussion of thermal Leptogenesis, and we show that for the neutrino
spectrum suggested by oscillation experiments, one obtains the observed value for the
baryon to photon density ratio in the Universe, independently of any initial boundary
conditions. In the latter part of the review we consider how well Leptogenesis fits with
particle physics models of dark matter. Although axionic dark matter and Leptogenesis
can be very naturally linked, there is a potential clash between Leptogenesis and models
of supersymmetric dark matter because the high temperature needed for Leptogenesis
leads to an overproduction of gravitinos, which alter the standard predictions of Big
Bang Nucleosynthesis. This problem can be resolved, but it constrains the supersymmetric spectrum at low energies and the nature of the lightest supersymmetric particle
(LSP). Finally, as an illustration of possible other options for the origin of matter, we
discuss the possibility that Leptogenesis may occur as a result of non-thermal processes.

CONTENTS

0163-8998/05/1208-0311$20.00

1. INTRODUCTION

Our understanding of the Universe has deepened considerably in the last 25 years, so much so that a standard cosmological model has emerged (1). In this model, after the Big Bang, a period of inflationary expansion (2) ensued that effectively set the Universe's curvature to zero. After inflation, the Universe's expansion continued, not in an exponential fashion but with the rate of expansion being determined by which component of the Universe's energy density dominated the total energy density.

In the present epoch this energy density is dominated by a so-called dark energy component whose negative pressure causes the Universe's expansion to accelerate (3). Dark energy now accounts for approximately 70% of the total energy density, with the other 30% of the remaining energy density of the Universe dominated by some kind of nonluminous (dark) matter. In detail, the angular distribution of the temperature fluctuations of the microwave background radiation measured by the Wilkinson Microwave Anisotropy Probe (WMAP) collaboration (4) determines the various components of the ratio of the Universe's energy density now ρ_o to the critical energy density ρ_c, $\Omega = \rho_o/\rho_c$.[1] The results are: $\Omega_{\text{dark energy}} = 0.73 \pm 0.04$; $\Omega_{\text{matter}} = 0.27 \pm 0.04$; and $\Omega_{\text{B}} = 0.044 \pm 0.004$. Here Ω_{B} is the contribution of baryonic matter to Ω_{matter}, confirming that about 85% of Ω_{matter} is indeed contributed by dark matter. The contribution of neutrinos and photons is at a few per mil, or below, and is negligible.

Although the standard cosmological model sketched above provides an accurate description of the present Universe and its evolution, deep questions remain to be answered. What exactly constitutes the dark energy? Is it just a cosmological constant? But if that is so, why is the energy scale associated with the corresponding vacuum energy density [$\rho_{\text{cc}} = E_o^4; E_o \simeq 2 \times 10^{-3}$ eV] so small? Equally mysterious is the nature of the dark matter, although in this case there are at least some particle physics candidates that may be the source for this component of the Universe's energy density.

A further mystery is associated with the observed baryon energy density. This number can be used to infer the ratio of the number density of baryons to photons in the Universe, a quantity that is measured independently from the primordial nucleosynthesis of light elements. The WMAP results (4) are in agreement with the most recent nucleosynthesis analysis of the primordial Deuterium abundance, but there are discrepancies with both the inferred ^{4}He and ^{7}Li values (5). These latter values, however, may have an underestimated error (6). Averaging the WMAP result only with that coming from the primordial abundance of Deuterium gives:

$$\frac{n_B}{n_\gamma} \equiv \eta_B = 6.1 \pm 0.3 \times 10^{-10}. \qquad 1.$$

Why does this ratio have this value?

In this review, we will principally try to address this last question, which, as we shall see, is intimately related to the existence of a primordial matter-antimatter asymmetry. Nevertheless, we shall, when germane, try to connect our discussion with the broader issues of what constitutes dark energy and dark matter.

There is good evidence that the Universe is mostly made up of matter, although it is possible that small amounts of antimatter exist (7). However, antimatter certainly does not constitute one of the dominant components of the Universe's energy density. Indeed, as Cohen, de Rujula, and Glashow (8) have compellingly argued, if there were to exist large areas of antimatter in the Universe, they could only be at a cosmic distance scale from us. Thus, along with the question of why n_B/n_γ has the value given in Equation 1, there is a parallel question of why the Universe is predominantly composed of baryons rather than antibaryons.

In fact, these two questions are interrelated. If the Universe had been matter-antimatter symmetric at temperatures of O(1 GeV), as the Universe cools

---

[1]The critical density ρ_c is the density that corresponds to a closed Universe now, $\rho_c = 3H_o^2/8\pi G_N$. Here H_o is the value of the Hubble parameter now. Inflation predicts that $\rho_o = \rho_c$, so that $\Omega = 1$.

further and the inverse process $2\gamma \to B + \bar{B}$ becomes ineffective because of the Boltzmann factor, the number density of baryons and antibaryons relative to photons would have been reduced dramatically as a result of the annihilation process $B + \bar{B} \to 2\gamma$. A straightforward calculation gives, in this case, (9):

$$\frac{n_B}{n_\gamma} = \frac{n_{\bar{B}}}{n_\gamma} \simeq 10^{-18}. \qquad\qquad 2.$$

Thus, in a symmetric Universe, the question is really why observationally n_B/n_γ is so large!

It is very difficult to imagine processes at temperatures below a GeV that could enhance the ratio of the number density of baryons relative to that of photons much beyond the value this quantity attains when baryon-antibaryon annihilation occurs.[2] Thus, because Equation 2 does not agree with the observed value given in Equation 1, one is led to the interesting conclusion that a primordial matter-antimatter asymmetry must have existed at temperatures of $\mathcal{O}(1\,\text{GeV})$ in the Universe. The observed value for n_B/n_γ and the lack of antimatter in the Universe are manifestations of this primordial asymmetry. Hence, in reality, the ratio η_B is, in effect, a measure of the number density of matter minus that of antimatter relative to the photon number density:

$$\eta_B = \frac{n_B - n_{\bar{B}}}{n_\gamma} = 6.1 \pm 0.3 \times 10^{-10}. \qquad\qquad 3.$$

It is interesting to consider the physical origins of this primordial matter-antimatter asymmetry. From the seminal work of Sakharov (11) one knows that, under certain conditions which we will amplify later on in this article, this asymmetry can be generated by physical processes. In this review we will focus on Leptogenesis—the creation of a primordial lepton-antilepton asymmetry—as the root source for the observed baryon-antibaryon asymmetry of Equation 3 (12). In our view, Leptogenesis provides the most compelling scenario for generating the observed baryon asymmetry in the Universe. In particular, because Leptogenesis is closely linked with parameters in the neutrino sector that can be eventually determined experimentally, this scenario can be tested and can be either confirmed or ruled out by data.

The plan of this review is as follows. In Section 2 we discuss the theoretical conditions necessary for producing a primordial matter-antimatter asymmetry in the Universe and explain how, through a mechanism first discussed by Kuzmin, Rubakov, and Shaposhnikov, (13) it is possible to turn a lepton asymmetry into a baryon asymmetry. In Section 3 we review existing experimental information on the neutrino sector, as well as the seesaw mechanism (14) that provides a theoretical framework for understanding this data. Section 4 discusses thermal Leptogenesis

---

[2]An exception is provided by some versions of Affleck-Dine Baryogenesis (10) where a baryon excess is produced by the decay of a scalar field very late in the history of the Universe, which reheats the Universe to temperatures of the $\mathcal{O}(100\,\text{MeV})$.

and contains the main quantitative results of the review. In particular, we show in this section that the observed value for η_B obtained from Leptogenesis significantly constrains low energy neutrino properties, and vice versa. In Section 5 we turn to dark matter and discuss how supersymmetric candidates for dark matter are significantly constrained if thermal Leptogenesis is the source of the observed baryon asymmetry in the Universe. The constraint arises from the overproduction of gravitinos. Section 6 discusses nonthermal Leptogenesis and other nonthermal processes that can lead to Baryogenesis. Finally, we present our conclusion and summary of results in Section 7.

2. THEORETICAL FOUNDATIONS

2.1. Sakharov's Conditions for Baryogenesis

In 1967, Sakharov (11) considered the consequences of the hypothesis that the observed expanding Universe originated from a superdense initial state with temperature of order the Planck mass, $T_i \sim M_P$. Since he could not imagine how, starting from such an initial state, one could obtain a macroscopic separation of matter and antimatter, he concluded that our Universe contains today only matter. That is, the Universe evolved from an initial state even under charge conjugation to a state odd under charge conjugation today. He then realized that in an expanding Universe, a matter-antimatter asymmetry could be generated dynamically, if C, CP, and baryon and lepton number were violated, and these processes were out of thermal equilibrium.

Sakharov also described a concrete model for Baryogenesis. He proposed as the origin for the baryon and lepton asymmetry the CP-violating decays of maximons, hypothetical neutral spin zero particles with mass of order the Planck mass. Their existence leads to a departure from thermal equilibrium already at temperatures $T \sim M_P$, where a small matter-antimatter asymmetry is then generated. An unavoidable consequence of this model is that protons are unstable and decay. However, the proton lifetime in Sakharov's model turned out to be unobservably long, $\tau_p > 10^{50}$ years.

During the past four decades many models of Baryogenesis have been proposed, demonstrating that the conditions Sakharov spelled out to allow Baryogenesis to take place are quite readily satisfied in the Standard Model of particle physics and its extensions. There are, however, significant differences among the various mechanisms suggested for producing the baryon asymmetry. Grand Unified Theories (GUTs) have been of particular importance for the development of realistic models of Baryogenesis (15). These theories provide natural heavy particle candidates, whose decays can be the source of the baryon asymmetry. However, in general, the simplest GUT models based on $SU(5)$ lead to a creation of a (B + L)-asymmetry, with a vanishing asymmetry for B − L. As will be made clear below, a (B + L)-asymmetry generated at the GUT scale eventually gets erased by sphaleron processes. In Leptogenesis, heavy Majorana neutrinos required by the seesaw mechanism (14) serve to trigger Baryogenesis. Because B − L is violated, the erasure present in GUTs is avoided. In principle, Electroweak Baryogenesis (16) is an attractive possibility, as the relevant parameters could then be tested in collider experiments. However, in general, the electroweak phase transition is not sufficiently out of equilibrium to generate an asymmetry of the magnitude observed in the Universe. Finally, in supersymmetric theories, the baryon and lepton number stored in scalar expectation values can also lead to Baryogenesis, through the so-called Affleck-Dine mechanism (10), which will be discussed in some detail in Section 6.

2.2. $B + L$ Violation in the Standard Model

Due to the chiral nature of the electroweak interactions, baryon and lepton number are not conserved in the Standard Model (17). The divergence of the B and L currents,

$$J_\mu^B = \frac{1}{3} \sum_{generations} \left(\overline{q_L} \gamma_\mu q_L + \overline{u_R} \gamma_\mu u_R + \overline{d_R} \gamma_\mu d_R \right), \qquad 4.$$

$$J_\mu^L = \sum_{generations} \left(\overline{l_L} \gamma_\mu l_L + \overline{e_R} \gamma_\mu e_R \right), \qquad 5.$$

is given by the triangle anomaly,

$$\partial^\mu J_\mu^B = \partial^\mu J_\mu^L$$

$$= \frac{N_f}{32\pi^2} \left(-g^2 W_{\mu\nu}^I \widetilde{W}^{I\mu\nu} + g'^2 B_{\mu\nu} \widetilde{B}^{\mu\nu} \right). \qquad 6.$$

Here N_f is the number of generations, and W_μ^I and B_μ are, respectively, the $SU(2)$ and $U(1)$ gauge fields with gauge couplings g and g'.

As a consequence of the anomaly, the change in baryon and lepton number is related to the change in the topological charge of the gauge field,

$$B(t_f) - B(t_i) = \int_{t_i}^{t_f} dt \int d^3x \, \partial^\mu J_\mu^B$$

$$= N_f [N_{cs}(t_f) - N_{cs}(t_i)], \qquad 7.$$

where

$$N_{cs}(t) = \frac{g^3}{96\pi^2} \int d^3x \, \epsilon_{ijk} \epsilon^{IJK} W^{Ii} W^{Jj} W^{Kk}. \qquad 8.$$

For vacuum to vacuum transitions W^{Ii} is a pure gauge configuration and the Chern-Simons numbers $N_{cs}(t_i)$ and $N_{cs}(t_f)$ are integers.

In a non-abelian gauge theory there are infinitely many degenerate ground states, which differ in their value of the Chern-Simons number, $\Delta N_{cs} = \pm 1, \pm 2, \ldots$.

The correponding points in field space are separated by a potential barrier whose height is given by the so-called sphaleron energy E_{sph} (18). Because of the anomaly, jumps in the Chern-Simons number are associated with changes of baryon and lepton number,

$$\Delta B = \Delta L = N_f \Delta N_{cs}.$$
                                                                              9.

Obviously, in the Standard Model the smallest jump is $\Delta B = \Delta L = \pm 3$.

In the semiclassical approximation, the probability of tunneling between neighboring vacua is determined by instanton configurations. In the Standard Model, $SU(2)$ instantons lead to an effective 12-fermion interaction

$$O_{B+L} = \prod_{i=1...3} (q_{Li}q_{Li}q_{Li}l_{Li}),$$
                                                                             10.

which describes processes with $\Delta B = \Delta L = 3$, such as

$$u^c + d^c + c^c \rightarrow d + 2s + 2b + t + \nu_e + \nu_\mu + \nu_\tau.$$
                                                                             11.

The transition rate is determined by the instanton action and one finds (17)

$$\Gamma \sim e^{-S_{inst}} = e^{-\frac{4\pi}{\alpha}}$$

$$= \mathcal{O}\left(10^{-165}\right).$$
                                                                             12.

Because this rate is extremely small, $(B + L)$-violating interactions appear to be completely negligible in the Standard Model. However, this picture changes dramatically when one is in a thermal bath.

2.3. Sphalerons and the KRS Mechanism

As emphasized in the seminal paper of Kuzmin, Rubakov, and Shaposhnikov (13), in the thermal bath provided by the expanding Universe one can make transitions between the gauge vacua not by tunneling, but through thermal fluctuations over the barrier. For temperatures larger than the height of the barrier, the exponential suppresion in the rate provided by the Boltzmann factor disappears completely. Hence (B + L)-violating processes can occur at a significant rate and these processes can be in equilibrium in the expanding Universe.

The finite-temperature transition rate in the electroweak theory is determined by the sphaleron configuration (18), a saddle point of the field energy of the gauge-Higgs system. Fluctuations around this saddle point have one negative eigenvalue, which allows one to extract the transition rate. The sphaleron energy is proportional to $v_F(T)$, the finite-temperature expectation value of the Higgs field, and one finds

$$E_{sph}(T) \simeq \frac{8\pi}{g} v_F(T).$$
                                                                             13.

Taking translational and rotational zero-modes into account, one obtains for the transition rate per unit volume in the Higgs phase (19)

$$\frac{\Gamma_{B+L}}{V} = \kappa \frac{M_W^7}{(\alpha T)^3} e^{-\beta E_{sph}(T)}, \qquad 14.$$

where $\beta = 1/T$, $M_W = g^2 v_F(T)/2$ and κ is some constant.

Extrapolating this semiclassical formula to the high-temperature symmetric phase, where $v_F(T) = 0$, and using for M_W the thermal mass, $M_W \sim g^2 T$, one expects in this phase $\Gamma_{B+L}/V \sim (\alpha T)^4$. However, detailed studies have shown that this naive extrapolation from the Higgs to the symmetric phase is not quite correct. The relevant spatial scale for nonperturbative fluctuations is the magnetic screening length $\sim 1/(g^2 T)$, but corresponding time scale turns out to be $1/(g^4 T \ln g^{-1})$, which is larger for small coupling (20,21). As a consequence, one obtains for the sphaleron rate in the symmetric phase

$$\Gamma_{B+L}/V \sim \alpha^5 \ln \alpha^{-1} T^4. \qquad 15.$$

It turns out that the dynamics of low-frequency gauge fields can be described by a remarkably simple effective theory, derived by Bödeker (21). The color magnetic and electric fields satisfy the equation of motion

$$\vec{D} \times \vec{B} = \sigma \vec{E} - \vec{\zeta}. \qquad 16.$$

Here $\vec{\zeta}$ is Gaussian noise, a random vector field with variance

$$\langle \zeta_i(x) \zeta_j(x') \rangle = 2\sigma \delta_{ij} \delta^4(x - x'). \qquad 17.$$

These equations define a stochastic three-dimensional gauge theory. The parameter σ is the 'color conductivity,' $\sigma = m_D^2/(3\gamma)$, where $m_D \sim gT$ is the Debye screening mass and $\gamma \sim g^2 T \ln(1/g)$ is the hard gauge boson damping rate. To leading-log accuracy one has $1/\sigma \sim \ln g^{-1}$. A next-to-leading order analysis yields for the sphaleron rate (22)

$$\frac{\Gamma_{B+L}}{V} = (10.8 \pm 0.7) \left(\frac{gT}{m_D}\right)^2 \alpha^5 T^4 \left[\ln\left(\frac{m_D}{\gamma}\right) + 3.041 + \left(\frac{1}{\ln(1/g)}\right)\right]. \qquad 18.$$

The overall coefficient has been determined by a numerical lattice simulation (23). From Equation 18 one easily obtains the temperature range where sphaleron processes are in thermal equilibrium:

$$T_{EW} \sim 100\,\text{GeV} < T < T_{sph} \sim 10^{12}\,\text{GeV}. \qquad 19.$$

The effective theory describing topological fluctuations of the gauge field in the high-temperature phase is valid for small coupling, $g \ll 1$. Yet for $T_{EW} < T < T_{sph} \sim 10^{12}$ GeV one has $g = \mathcal{O}(1)$. This implies that the electric screening length $1/(gT)$ and the magnetic screening length $1/(g^2 T)$ are not well separated and that nonperturbative corrections to the sphaleron rate, Equation 18, may be large. This will modify the temperature range given in Equation 19, but one expects that the qualitative picture of fluctuations in baryon and lepton number in the high-temperature phase of the Standard Model will not be affected.

2.4. Electroweak Baryogenesis and Its Experimental Constraints

An important ingredient in the theory of Baryogenesis is related to the nature of the electroweak transition from the high-temperature symmetric phase to the low-temperature Higgs phase. Because in the Standard Model baryon number, C, and CP are not conserved, it is conceivable that the cosmological baryon asymmetry could have been generated at the electroweak phase transition (13), provided that this transition is of first-order, because then there is also the necessary departure from thermal equilibrium. This possibility has stimulated much theoretical activity during the past years to determine the phase diagram of the electroweak theory.

Electroweak Baryogenesis requires that the baryon asymmetry generated during the phase transition is not erased by sphaleron processes afterwards.[3] This leads to a condition on the jump of the Higgs vacuum expectation value $v_F = \sqrt{H^\dagger H}$ at the critical temperature (24):

$$\frac{\Delta v_F(T_c)}{T_c} > 1. \qquad\qquad 20.$$

The strength of the electroweak transition has been studied by numerical and analytical methods as function of the Higgs boson mass. For the $SU(2)$ gauge-Higgs model one finds from lattice simulations as well as perturbative calculations that the lower bound of Equation 20 is violated for Higgs above 45 GeV (25).[4] Because the present lower bound from LEP on the Higgs mass is 114 GeV (26), it is clear that the electroweak transition in the Standard Model is too weak for Baryogenesis. However, for special choices of parameters or by adding singlet fields, in certain circumstances supersymmetric extensions of the Standard Model have a sufficiently strong first-order phase transition to allow Electroweak Baryogenesis to take place (27).

For large Higgs masses, the nature of the electroweak transition is dominated by nonperturbative effects of the $SU(2)$ gauge theory at high temperatures. At a critical Higgs mass $m_H^c = \mathcal{O}(M_W)$, an intriguing phenomenon occurs: The first-order phase transition turns into a smooth crossover (28–30), as expected on general grounds (25). At the endpoint of a critical line of first-order transitions, which is reached for $m_H = m_H^c$, the phase transition is of second order (31).

The value of the critical Higgs mass can be estimated by comparing the W-boson mass M_W in the Higgs phase with the magnetic mass m_{SM} in the symmetric phase. This yields $m_H^c \simeq 74\,\text{GeV}$ (32). Numerical lattice simulations have determined the precise value $m_H^c = 72.1 \pm 1.4\,\text{GeV}$ (33). The analytic estimate of the critical Higgs mass can be generalized to supersymmetric extensions of the Standard

---

[3]The produced asymmetry will be erased if, after the phase transition, (B + L)-violating processes are in equilibrium.

[4]For Higgs masses below 50 GeV, the Higgs model provides a good approximation for the full Standard Model.

Model, where one finds $m_h^c < 130 \ldots 150$ GeV (34), which is still compatible with the present experimental lower bound.

2.5. The Relation Between Baryon and Lepton Asymmetries

In a weakly coupled plasma, one can assign a chemical potential μ to each of the quark, lepton, and Higgs fields. In the Standard Model, with one Higgs doublet H and N_f generations, one then has $5N_f + 1$ chemical potentials.[5] For a non-interacting gas of massless particles, the asymmetry in the particle and antiparticle number densities is given by

$$n_i - \bar{n}_i = \frac{gT^3}{6} \begin{cases} \beta\mu_i + \mathcal{O}((\beta\mu_i)^3), & \text{fermions,} \\ 2\beta\mu_i + \mathcal{O}((\beta\mu_i)^3), & \text{bosons.} \end{cases} \qquad 21.$$

The following analysis is based on these relations for $\beta\mu_i \ll 1$. However, one should keep in mind that the plasma of the early Universe is very different from a weakly coupled relativistic gas, owing to the presence of unscreened non-abelian gauge interactions, where nonperturbative effects are important in some cases.

Quarks, leptons, and Higgs bosons interact via Yukawa and gauge couplings and, in addition, via the nonperturbative sphaleron processes. In thermal equilibrium all these processes yield constraints between the various chemical potentials (35). The effective interaction of Equation 10 induced by the $SU(2)$ electroweak instantons implies

$$\sum_i \left(3\mu_{qi} + \mu_{li}\right) = 0. \qquad 22.$$

One also has to take the $SU(3)$ Quantum Chromodynamics (QCD) instanton processes into account (36), which generate an effective interaction between left-handed and right-handed quarks. The corresponding relation between the chemical potentials reads

$$\sum_i \left(2\mu_{qi} - \mu_{ui} - \mu_{di}\right) = 0. \qquad 23.$$

A third condition, valid at all temperatures, arises from the requirement that the total hypercharge of the plasma vanishes. From Equation 21 and the known hypercharges one obtains

$$\sum_i \left(\mu_{qi} + 2\mu_{ui} - \mu_{di} - \mu_{li} - \mu_{ei} + \frac{2}{N_f}\mu_H\right) = 0. \qquad 24.$$

The Yukawa interactions, supplemented by gauge interactions, yield relations between the chemical potentials of left-handed and right-handed fermions,

---

[5]In addition to the Higgs doublet, the two left-handed doublets q_i and ℓ_i and the three right-handed singlets u_i, d_i, and e_i of each generation each have an independent chemical potential.

$$\mu_{qi} - \mu_H - \mu_{dj} = 0, \quad \mu_{qi} + \mu_H - \mu_{uj} = 0, \quad \mu_{li} - \mu_H - \mu_{ej} = 0. \quad 25.$$

These relations hold if the corresponding interactions are in thermal equilibrium. In the temperature range $100\,\text{GeV} < T < 10^{12}\,\text{GeV}$, which is of interest for Baryogenesis, this is the case for gauge interactions. On the other hand, Yukawa interactions are in equilibrium only in a more restricted temperature range that depends on the strength of the Yukawa couplings. In the following we shall ignore this complication, which has only a small effect on our discussion of Leptogenesis.

Using Equation 21, the baryon number density $n_B \equiv gBT^2/6$ and the lepton number densities $n_{L_i} \equiv L_i gT^2/6$ can be expressed in terms of the chemical potentials:

$$B = \sum_i \left(2\mu_{qi} + \mu_{ui} + \mu_{di}\right), \quad 26.$$

$$L_i = 2\mu_{li} + \mu_{ei}, \quad L = \sum_i L_i. \quad 27.$$

Consider now the case where all Yukawa interactions are in equilibrium. The asymmetries $L_i - B/N_f$ are then conserved and we have equilibrium between the different generations, $\mu_{li} \equiv \mu_l$, $\mu_{qi} \equiv \mu_q$, etc. Using also the sphaleron relation and the hypercharge constraint, one can express all chemical potentials, and therefore all asymmetries, in terms of a single chemical potential that may be chosen to be μ_l,

$$\mu_e = \frac{2N_f + 3}{6N_f + 3}\mu_l, \quad \mu_d = -\frac{6N_f + 1}{6N_f + 3}\mu_l, \quad \mu_u = \frac{2N_f - 1}{6N_f + 3}\mu_l,$$

$$\mu_q = -\frac{1}{3}\mu_l, \quad \mu_H = \frac{4N_f}{6N_f + 3}\mu_l. \quad 28.$$

The corresponding baryon and lepton asymmetries are

$$B = -\frac{4N_f}{3}\mu_l, \quad L = \frac{14N_f^2 + 9N_f}{6N_f + 3}\mu_l. \quad 29.$$

This yields the important connection between the B, B-L and L asymmetries (37)

$$B = c_s(B - L); \quad L = (c_s - 1)(B - L), \quad 30.$$

where $c_s = (8N_f + 4)/(22N_f + 13)$. The above relations hold for temperatures $T \gg v_F$. In general, the ratio $B/(B - L)$ is a function of v_F/T (38).

The relations (30) between B-, (B − L)-, and L-number suggest that (B − L)-violation is needed in order to generate a B-asymmetry.[6] Because the

---

[6]In the case of Dirac neutrinos, which have extremely small Yukawa couplings, one can construct Leptogenesis models where an asymmetry of lepton doublets is accompanied by an asymmetry of right-handed neutrinos such that the total L-number is conserved and the

(B − L)-current has no anomaly, the value of B − L at time t_f, where the Leptogenesis process is completed, determines the value of the baryon asymmetry today,

$$B(t_0) = c_s(B - L)(t_f).$$ 31.

On the other hand, during the Leptogenesis process the strength of (B − L)- and therefore L-violating interactions can only be weak. Otherwise, because of Equation 30, they would wash out any baryon asymmetry. As we shall see in the following, the interplay between these conflicting conditions leads to important constraints on the properties of neutrinos.

3. EXPERIMENTAL AND THEORETICAL INFORMATION ON THE NEUTRINO SECTOR

3.1. Results from Oscillation Experiments

The search for neutrino mass has a long history (40). Positive results are now provided by neutrino oscillation experiments. The allowed values for the mass-squared differences Δm_{ij}^2 and the mixing angles θ_{ij} at the 3σ level for three generations of neutrinos are summarized below (26):

$$\sin^2 2\theta_{23} = 0.92 - 1, \quad \left|\Delta m_{23}^2\right| = (1.2 - 4.8) \times 10^{-3} \, \text{eV}^2 (\text{atmospheric } \nu)$$ 32.

$$\sin^2 2\theta_{12} = 0.70 - 0.95, \left|\Delta m_{12}^2\right| = (5.4 - 9.5) \times 10^{-5} \, \text{eV}^2 (\text{solar } \nu).$$ 33.

The CHOOZ experiment (41) gives only an upper limit on the remaining mixing angle θ_{13}:

$$\sin^2 2\theta_{13} = 0 - 0.17 (\text{CHOOZ } \nu).$$ 34.

The LSND experiment (42) reports neutrino oscillations from $\bar{\nu}_\mu$ to $\bar{\nu}_e$. The mixing angle and mass-squared difference inferred from this experiment are $\sin^2 2\theta = 0.003 - 0.03$ and $|\Delta m^2| = 0.2 - 2 \, \text{eV}^2$. This parameter region is almost excluded by negative results from a comparable experiment by the KARMEN collaboration (43), but there still remains a narrow region allowed at the 90% CL.

It is very difficult to explain all the above data from neutrino oscillation experiments within the three neutrino framework. Indeed, to accommodate the LSND data one must introduce at least one sterile neutrino (44), or make the radical assumption that CPT is not conserved (45). In this review, for simplicity we will disregard the data from the LSND experiment.

(B − L)-asymmetry vanishes (39).

3.2. Information from β-Decay, 2β-Decay, and Cosmology

The oscillation experiments discussed in the previous subsection are only sensitive to mass-squared differences. In this subsection we quote results from direct searches for the absolute values of neutrino masses.

The direct laboratory limits on the neutrino masses are summarized as follows (26):

$$m_{\nu_e} < 2.5 \text{ eV};\qquad\qquad 35.$$

$$m_{\nu_\mu} < 170 \text{ keV};\qquad\qquad 36.$$

$$m_{\nu_\tau} < 18 \text{ MeV}.\qquad\qquad 37.$$

The ν_e mass measurements use the decay of tritium, $^3\text{H} \rightarrow {}^3\text{He} + \bar{\nu}_e + e^-$, which has a small Q value, $Q = 18.6$ keV, and looks at the electron spectrum near the end point in the Kurie plot. The limit on the ν_μ mass is obtained from the two-body kinematics of the pion decay, $\pi^+ \rightarrow \mu^+ + \nu_\mu$. Finally, the limit on the ν_τ mass is obtained from measurements of the invariant mass distribution of 3π and 5π systems in the τ decays, $\tau \rightarrow 3(5)\pi + \nu_\tau$.

Neutrinoless double β decay experiments provide a bound on an element of the Majorana mass matrix, $m_{\nu_{ee}}$ (46). The best limit comes from the ^{76}Ge results. Because the calculation of the double β decay rate is model dependent, we quote a range for this bound:

$$m_{\nu_{ee}} < 0.3 - 0.8 \text{ eV}.\qquad\qquad 38.$$

One can derive a stringent upper limit on the sum of neutrino masses from cosmology. In the early Universe, neutrinos were in thermal equilibrium with radiation, and one can infer their number density today to be $n_i \simeq 110 \text{ cm}^{-3}$ for each neutrino species. Although the contribution of massless neutrinos to the Universe's energy density is negligible, the contribution of massive neutrinos could be important. By requiring that the energy density of massive neutrinos does not exceed that of dark matter, one obtains the bound:

$$\sum_i m_i \leq 30h^2 \text{ eV},\qquad\qquad 39.$$

where h is the Hubble constant in units of $100 \text{ km s}^{-1}\text{Mpc}^{-1}$ [$h = 0.71^{+0.04}_{-0.03}$ (4)].

Even if neutrinos satisfy the above limit, massive neutrinos would affect the formation of cosmic structure, because the free streaming of neutrinos suppresses density fluctuations at small scales. The normalization of large- and small-scale fluctuations constrains the contribution from neutrinos. Recent detailed analyses (47) lead to the bound:

$$\sum_i m_i < 0.65 \text{ eV}.\qquad\qquad 40.$$

It is interesting that a similar constraint, $\sum_i m_i < 2.0\,\text{eV}$, has been obtained by using the cosmic microwave background data alone (48).

3.3. The Seesaw Mechanism

If there are right-handed neutrinos, then neutrinos can have a Dirac mass, much as quarks and charged leptons do. However, if this is the only source for their mass, it is not easy to find a natural reason for the very small mass of neutrinos. With only a Dirac mass term, the smallness of neutrino masses needs to be ascribed to having very tiny Yukawa coupling constants h ($h \sim 10^{-13}$ gives a neutrino mass $m_\nu \simeq 0.01$ eV). Although it is possible to imagine mechanisms that result in very small Dirac masses for neutrinos,[7] the seesaw mechanism, which entails Majorana masses, provides a natural explanation for the smallness of neutrino masses in theories of unification at ultra-high energy scales (14).

To appreciate this point, we should first note that the Standard Model does not require neutrinos to be massive, because right-handed neutrinos are not necessary for the electroweak theory. Without right-handed neutrinos, these particles may acquire their mass only from what are called irrelevant operators—operators such as $\ell\ell HH$ with dimension greater than four. These operators can give rise to a Majorana mass for neutrinos in theories with a cutoff. However, in the limit of an infinite cutoff, neutrino masses vanish. It should be noted that if one adopts the Planck scale as the Wilson cutoff for the Standard Model, one finds neutrino masses to be at most 10^{-5} eV. Thus, the observed mass $\sqrt{\Delta m_{\text{atm}}^2} \simeq 0.05$ eV is unable to be explained within the Standard Model.

If one considers possible extensions of the Standard Model gauge group, it is natural to consider a gauge group G whose rank is at least 5, because the rank of the Standard Model group, $SU(3) \times SU(2) \times U(1)_Y$, is 4. This new group G may contain an extra $U(1)$ as a subgroup, in addition to the Standard Model gauge group. The simplest candidate for the extra $U(1)$ is a B-L gauge symmetry, because we know that the global $U(1)_{B-L}$ does not have any anomaly due to the Standard Model gauge interactions. However, when one gauges this B-L symmetry, the theory does have a self-anomaly of the B-L interactions. That is, the triangle anomaly of $[U(1)_{B-L}]^3$ is nonvanishing. A crucial point, however, is that this anomaly is cancelled by introducing right-handed neutrinos. This also cancels the mixed gravitational/B-L anomaly. Thus, right-handed neutrinos are required for consistency of the theory! A famous example, which includes $U(1)_{B-L}$, is provided by SO(10) grand unification. But our argument is more general. For instance, the string brane world predicts many $U(1)$'s, and it is quite natural to consider some of them to be anomaly free and survive as low-energy (compared with the string scale) gauge symmetries.

---

[7] At the end of this subsection we outline a recent approach that may explain how such extremely small Yukawa coupling constants might arise in a higher dimensional theory.

It is usually assumed that the unification group G is broken down to the Standard Model group at high energies. Then, the right-handed neutrinos naturally obtain large Majorana masses, because they are singlets of the Standard Model and there is no unbroken symmetry to protect them from acquiring a large Majorana mass. In this article we shall denote heavy Majorana neutrinos as N. Then, the masses of the neutrinos written as a matrix take a simple form:

$$\begin{pmatrix} 0 & m \\ m^T & M \end{pmatrix}.$$ 41.

Here m is the Dirac mass matrix between the left-handed neutrinos ν and the heavy Majorana neutrinos N, which is of order the electroweak scale, while M is the Majorana mass matrix of the heavy neutrinos N. For three generations, both M and m are 3×3 matrices. Integrating out the heavy Majorana neutrinos N leads to a small neutrino mass via the seesaw mechanism (14). For one neutrino generation, one simply has that:

$$m_\nu \simeq \frac{m^2}{M}.$$ 42.

We see from the above that a small neutrino mass is a reflection of the ultra-heavy mass of the heavy neutrino N. The observed neutrino mass $\sqrt{\Delta m_{\text{atm}}^2} \simeq 0.05$ eV implies $M \simeq 10^{15}$ GeV, which is very close to the Grand Unification scale. Thus, effectively, the small neutrino masses provide a window to new physics at an ultra-high energy scale.

One may question why the unification group G, or the $U(1)_{B-L}$ symmetry, should be broken at such a high energy scale. Perhaps the answer to this question, as we shall amplify in this review, is because otherwise, the baryon number in the Universe would be too small for us to exist! It turns out that if the Universe's baryon number were to be two orders of magnitude below the present observed value, galaxies would not be formed (49). One of the purposes of this review article is to explain in some detail this fundamental point.

3.3.1. SMALL NEUTRINO YUKAWA COUPLINGS FROM HIGHER DIMENSIONAL THEOR-IES

To explain how one can generate small Dirac masses for neutrinos, consider a theory described in $(4 + 1)$-dimensional space-time, while our world is on a $(3 + 1)$-dimensional hyperplane—a so-called D3 brane. The Einstein action of gravity in five-dimensional space-time is given by

$$S = \frac{M_*^3}{16\pi} \int d^4x \int dy \sqrt{-g_5} \mathcal{R}_5,$$ 43.

where M_* is the gravitational scale in five-dimensional space-time, and g_5 and $\mathcal{R}_5$ are the metric and the scalar curvature, respectively. We assume that the fifth dimension is compactified to a space of radius L, and consider the metric to be

$$d^2s = g_{\mu\nu}dx^\mu dx^\nu - dy^2,$$ 44.

where $g_{\mu\nu}$ is the metric in four-dimensional space-time. The integration over dy leads to the four-dimensional action

$$S_4 = \frac{M_*^3 L}{16\pi} \int d^4x \sqrt{-g_4}\, \mathcal{R}_4. \qquad 45.$$

Then, the Planck scale, $M_P \simeq 1.2 \times 10^{19}$ GeV, in four-dimensional space-time, is given by

$$M_P^2 = M_*^3 L. \qquad 46.$$

One can get the observed value for M_P even for $M_* = 1$ TeV by taking a very large L. The weakness of gravity in these theories is the result of having a large compactification scale in the fifth dimension (50).

Let us now assume that all the Standard Model particles reside on a D3 brane at the boundary $y = 0$ and that the right-handed neutrino lives in the five-dimensional bulk (51). Then the action involving the right-handed neutrinos is given by

$$S = M_* \int d^4x \int_0^L dy \sqrt{-g_5}\, \bar{N}_{Ri}\partial N_R + \int d^4x \int_0^L dy \sqrt{-g_5} h \bar{N}_R \ell_L H \delta(y) + h.c.$$
$$47.$$

Here, ℓ_L and H are $SU(2)_L$ doublets of the left-handed leptons and of the Higgs boson, respectively. One obtains the action in four dimensions by integrating over dy, and one finds:

$$S_4 = M_* L \int d^4x \sqrt{-g_4}\, \bar{N}_{Ri}\partial N_R + \int d^4x \sqrt{-g_4} h \bar{N}_R \ell_L H + h.c. \qquad 48.$$

After renormalizing the wave function of N_R so that it has a canonical kinetic term, one finds that the Yukawa coupling constant is suppressed by $1/\sqrt{M_* L} = M_*/M_P$. Thus in this model one obtains an effective Yukawa coupling constant $h_{\text{eff}} \simeq 10^{-13}$ (corresponding to a neutrino Dirac mass $m_\nu \simeq 0.01$ eV) for $h = 1$ and $M_* \simeq 10^3$ TeV.

This model, however, has a serious drawback: There is no symmetry to protect the right-handed neutrinos from acquiring a Majorana mass, because they are singlets of the Standard Model gauge group. A solution to this problem may be found by imposing a $U(1)_{B-L}$ gauge symmetry in the five-dimensional bulk. As long as the B-L gauge symmetry is exact, the right-handed neutrinos cannot have a Majorana mass because this mass carries a nonvanishing B-L charge. If this symmetry is exact, the corresponding gauge boson is completely massless. However, this may not cause any phenomenological difficulties at low-energies, because the B-L gauge coupling constant must also be extremely suppressed.[8]

---

[8]The B-L gauge coupling constant α_{B-L} is constrained to be $\alpha_{B-L} < 10^{-21}\alpha_{\text{em}}$. This bound comes from the empirical limits of the electromagnetic charges for the neutron and the neutrino. That is, $Q_n = (-0.4 \pm 1.1) \times 10^{-21}$ (52) and $Q_\nu = (0.5 \pm 2.9) \times 10^{-21}$ (53).

But, as we explained in Section 2, if the B-L gauge symmetry is exact, it is very difficult to account for the baryon-number asymmetry in the present Universe.

3.4. CP-Violating Phases at Low and High Energies in the Lepton Sector

In the Standard Model, the Lagrangian for the lepton sector, augmented by including right-handed neutrinos, is given by

$$\mathcal{L} = \bar{\ell}_{Li} i \partial \ell_{Li} + \bar{e}_{Ri} i \partial e_{Ri} + \bar{N}_{Ri} i \partial N_{Ri}$$

$$+ f_{ij} \bar{e}_{Ri} \ell_{Lj} H^\dagger + h_{ij} \bar{N}_{Ri} \ell_{Lj} H - \frac{1}{2} M_{ij} N_{Ri} N_{Rj} + h.c, \qquad 49.$$

where $i, j = \{1 - 3\}$ are the family-number indices. We adopt, without losing generality, a basis where the matrices f_{ij} and M_{ij} are diagonal. The Yukawa matrix h_{ij} in this basis is in general complex and thus has CP-violating phases. Because for three families the matrix h_{ij} has 9 complex parameters, we have 9 possible CP-violating phases. However, three of these phases can be absorbed into the wave function of ℓ_L and hence 6 CP-violating phases remain physically relevant. These are known as high-energy phases, because they enter in the full theory.[9]

Let us now discuss the CP-violating phases at low energies. To do that we first need to integrate out the heavy Majorana neutrinos, N_i. Doing so the effective Lagrangian for the lepton sector reduces to:

$$\mathcal{L}_{\text{eff}} = \bar{\ell}_{Li} i \partial \ell_{Li} + \bar{e}_{Ri} i \partial e_{Ri} + f_{ii} \bar{e}_{Ri} \ell_{Li} H^\dagger + \frac{1}{2} \sum_k h_{ik}^T h_{kj} \ell_{Li} \ell_{Lj} \frac{H^2}{M_k} + h.c. \quad 50.$$

The last term can be rewritten as

$$-\frac{1}{2} m_{\nu_{ij}} \ell_{Li} \ell_{Lj} \frac{H^2}{\langle H \rangle^2}, \qquad 51.$$

so that all low-energy phases appear in the mass matrix of light neutrinos. Because the neutrino mass matrix is symmetric, it has 6 complex parameters, and hence one has 6 possible CP-violating phases. However, as before, 3 of these 6 phases can be absorbed into the wave function of ℓ_L. Therefore, there remain only three physical low-energy CP-violating phases (54). Because they are different in number, it is unfortunately very difficult to establish a direct link between the low-energy and the high-energy CP-violating phases (55).

Furthermore, in practice, it is not possible to measure all three low-energy phases. One of these three phases can be measured by neutrino oscillation

---

The present model suggests $\alpha_{B-L} \simeq (\frac{M_*}{M_P})^2 \times \alpha_{\text{em}} \simeq 10^{-25} \times \alpha_{\text{em}}$, which may be in an interesting region for future experiments.

[9]In particular, as we will show in the next section, the phase contributing to the generation of the lepton asymmetry in the decay of N_1 is a combination of these high-energy phases, given by $\sum \text{Im}[(hh^\dagger)_{1i}]^2$.

experiments, while neutrinoless double β decay, if it were to be observed, would provide information on another phase. However, the remaining phase is undetermined. In other words, one cannot perform a complete experiment to determine the neutrino mass matrix. Nevertheless, if the neutrino mass matrix were to have an extra constraint, one may be able to determine all matrix elements of m_ν. This constraint must be independent of the frame of the family basis. One example of such a constraint is the requirement that $\det(m_\nu) = 0$. In this case, we have only 7 independent physical parameters including the phases in the neutrino mass matrix (56), which can be determined in principle in future experiments.[10]

4. THERMAL LEPTOGENESIS

4.1. Lepton Number Violation and Leptogenesis

As we discussed above, lepton number violation is most simply realized by adding right-handed neutrinos to the Standard Model. Their existence is predicted by all extensions of the Standard Model containing B-L as a local symmetry and allows for an elegant explanation of the smallness of the light neutrino masses via the seesaw mechanism (14).

The most general Lagrangian for couplings and masses of charged leptons and neutrinos is given in Equation 49. The vacuum expectation value of the Higgs field, $\langle H \rangle = v_F$, generates Dirac masses m_e and m_D for charged leptons and neutrinos, $m_e = f v_F$ and $m_D = h v_F$, which are assumed to be much smaller than the Majorana masses M. This yields the light and heavy neutrino mass eigenstates

$$\nu \simeq V_\nu^T \nu_L + \nu_L^c V_\nu^*, \quad N \simeq N_R + N_R^c, \qquad 52.$$

with masses

$$m_\nu \simeq -V_\nu^T m_D^T \frac{1}{M} m_D V_\nu, \quad m_N \simeq M. \qquad 53.$$

In a basis where the charged lepton mass matrix m_e and the Majorana mass matrix M are diagonal, V_ν is the mixing matrix in the leptonic charged current.

The right-handed neutrinos can efficiently erase any pre-existing lepton asymmetry at temperatures $T > M$, but they can also generate a lepton asymmetry by means of their out-of-equilibrium decays at temperatures $T < M$. This asymmetry is then partially transformed into a baryon asymmetry by sphaleron processes. This is the Leptogenesis mechanism proposed by Fukugita and Yanagida (12).

The decay width of the heavy neutrino N_i at tree level reads,

$$\Gamma_{Di} = \Gamma\left(N_i \rightarrow H + \ell_L\right) + \Gamma\left(N_i \rightarrow H^\dagger + \ell_L^\dagger\right) = \frac{1}{8\pi}(hh^\dagger)_{ii} M_i. \qquad 54.$$

---

[10]As will be shown in Section 6, Affleck-Dine Leptogenesis suggests a constraint, $m_{\nu_1} \simeq 10^{-9}$ eV and hence $\det(m_\nu) \simeq 0$.

Once the temperature of the universe drops below the mass M_1, the heavy neutrinos are not able to follow the rapid change of the equilibrium distribution. Hence, the necessary deviation from thermal equilibrium ensues as a result of having a too large number density of heavy neutrinos, compared to the equilibrium density. Eventually, however, the heavy neutrinos decay, and a lepton asymmetry is generated owing to the presence of CP-violating processes. The CP asymmetry involves the interference between the tree-level amplitude and the one-loop vertex and self-energy contributions. In a basis where the right-handed neutrino mass matrix M is diagonal, one obtains (57) for the CP asymmetry parameter ϵ_1 assuming hierarchical heavy neutrino masses ($M_1 \ll M_2, M_3$):

$$\varepsilon_1 \simeq \frac{3}{16\pi} \frac{1}{(hh^\dagger)_{11}} \sum_{i=2,3} \mathrm{Im} \left[(hh^\dagger)^2_{i1}\right] \frac{M_1}{M_i}. \qquad 55.$$

In the case of mass differences of order the decay widths, one obtains a significant enhancement from the self-energy contribution (58), although the influence of the thermal bath on this effect is presently unclear.

The CP asymmetry of Equation 55 can be obtained in a very simple way by first integrating out the heavier neutrinos N_2 and N_3 in the leptonic Lagrangian. This yields

$$\mathcal{L}_\nu^{eff} = h_{1j} \overline{N_{R1}} \ell_{Lj} H - \frac{1}{2} M_1 \overline{N^c_{R1}} N_{R1} + \frac{1}{2} \eta_{ij} \ell_{Li} H \ell_{Lj} H + \text{h.c.,} \qquad 56.$$

with

$$\eta_{ij} = \sum_{k=2}^{3} h^T_{ik} \frac{1}{M_k} h_{kj}. \qquad 57.$$

The asymmetry ε_1 is then obtained from the interference of the Born graph and the one-loop graph involving the cubic and the quartic couplings (see Figure 1). This includes automatically both vertex and self-energy corrections (59) and yields an expression for ε_1 directly in terms of the light neutrino mass matrix:

$$\varepsilon_1 \simeq -\frac{3}{16\pi} \frac{M_1}{(hh^\dagger)_{11} v_F^2} \mathrm{Im} \left(h^* m_\nu h^\dagger\right)_{11}. \qquad 58.$$

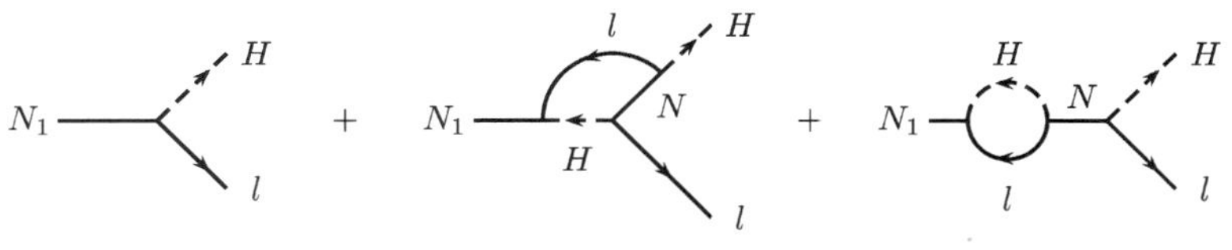

Figure 1   Tree level and one-loop diagrams contributing to heavy neutrino decays whose interference leads to Leptogenesis.

The CP asymmetry then leads to a $(B - L)$-asymmetry (12),

$$Y_{B-L} \simeq -Y_L = -\frac{n_L - n_{\bar{L}}}{s} = -\kappa \frac{\varepsilon_1}{g_*}.$$ 59.

Here s is the entropy and, in the present epoch $s = 7.04 \, n_\gamma$, whereas $g_* \sim 100$ is the number of degrees of freedom in the plasma. The factor $\kappa < 1$ in the above takes into account the effect of washout processes. As we shall discuss below, in order to determine κ one has to solve the Boltzmann equations.

Early studies of Leptogenesis were partly motivated by trying to find alternatives to Electroweak Baryogenesis, which did not seem to produce a big enough asymmetry. Some extensions of the Standard Model were considered and, in particular, in the simple case of hierarchical heavy neutrino masses, the observed value of the baryon asymmetry is naturally obtained with $(B - L)$ broken at the unification scale, $M_{GUT} \sim 10^{15}$ GeV. The corresponding light neutrino masses are then very small, $m_{1,2} < m_3 \sim 0.1$ eV, and the typical parameters for the necessary CP asymmetry and the Baryogenesis temperature are $\varepsilon_1 \simeq 10^{-6}$ and $T_B \sim M_1 \sim 10^{10}$ GeV, respectively (60, 61).[11] Subsequently, researchers realized that such small neutrino masses are consistent with the small mass differences inferred from the solar and atmospheric neutrino oscillations. This fact has given rise to a strong interest in Leptogenesis in recent years, and a large number of interesting models have been suggested (63).

4.2. Departure from Thermal Equilibrium

Leptogenesis takes place at temperatures $T \sim M_1$. For a decay width small compared to the Hubble parameter, $\Gamma_1(T) < H(T)$, heavy neutrinos are out of thermal equilibrium, otherwise they are in thermal equilibrium (64). The borderline between the two regimes is given by $\Gamma_1 = H|_{T=M_1}$, which is equivalent to the condition that the effective neutrino mass

$$\tilde{m}_1 = \frac{(m_D m_D^\dagger)_{11}}{M_1}$$ 60.

is equal to the "equilibrium neutrino mass"

$$m_* = \frac{16\pi^{5/2}}{3\sqrt{5}} g_*^{1/2} \frac{v_F^2}{M_P} \simeq 10^{-3} \text{ eV}.$$ 61.

Here we have used the Hubble parameter $H(T) \simeq 1.66 \, g_* T^2/M_P$ where $g_* = g_{SM} = 106.75$ is the total number of degrees of freedom and $M_P = 1.22 \times 10^{19}$ GeV is the Planck mass.

It is quite remarkable that the equilibrium neutrino mass m_* is close to the neutrino masses suggested by neutrino oscillations, $\sqrt{\Delta m_{\text{sol}}^2} \simeq 8 \times 10^{-3}$ eV and

---

[11] For early work based on SO(10), see (62).

$\sqrt{\Delta m_{\text{atm}}^2} \simeq 5 \times 10^{-2}$ eV. This encourages one to think that it may be possible to understand the cosmological baryon asymmetry via Leptogenesis as a process close to thermal equilibrium. Ideally, $\Delta L = 1$ and $\Delta L = 2$ processes should be strong enough at temperatures above M_1 to keep the heavy neutrinos in thermal equilibrium and weak enough to allow the generation of an asymmetry at temperatures below M_1.

In general, the generated baryon asymmetry is the result of a competition between production processes and washout processes that tend to erase any generated asymmetry. Unless the heavy Majorana neutrinos are partially degenerate, $M_{2,3} - M_1 \ll M_1$, the dominant processes are decays and inverse decays of N_1 and the usual off-shell $\Delta L = 1$ and $\Delta L = 2$ scatterings (65, 66).

The Boltzmann equations for Leptogenesis are[12]

$$\frac{dN_{N_1}}{dz} = -(D + S)\left(N_{N_1} - N_{N_1}^{\text{eq}}\right), \qquad 62.$$

$$\frac{dN_{B-L}}{dz} = -\varepsilon_1 D\left(N_{N_1} - N_{N_1}^{\text{eq}}\right) - W N_{B-L}, \qquad 63.$$

where $z = M_1/T$. The number density N_{N_1} and the amount of B-L asymmetry, N_{B-L}, are calculated in a portion of comoving volume that contains one photon at the onset of Leptogenesis, so that the relativistic equilibrium N_1 number density is given by $N_{N_1}^{\text{eq}}(z \ll 1) = 3/4$. Alternatively, one may normalize the number density to the entropy density s and consider $Y_X = n_X/s$. If entropy is conserved, both normalizations are related by a constant.

There are four classes of processes that contribute to the different terms in the above equations: decays, inverse decays, $\Delta L = 1$ scatterings, and $\Delta L = 2$ processes mediated by heavy neutrinos. The first three processes all modify the N_1 abundance and try to push it towards its equilibrium value $N_{N_1}^{\text{eq}}$. Denoting by H the Hubble expansion rate, the term $D = \Gamma_D/(Hz)$ accounts for decays and inverse decays, whereas the scattering term $S = \Gamma_S/(Hz)$ represents the $\Delta L = 1$ scatterings. Decays also yield the source term for the generation of the B-L asymmetry, the first term in Equation 63, whereas all other processes contribute to the total washout term $W = \Gamma_W/(Hz)$, which competes with the decay source term. The dynamical generation of the N_1 abundance and the B-L asymmetry is shown in Figure 2 (see color insert) for typical parameters.

4.3. Decays and Inverse Decays

It is very instructive to consider first a simplified picture in which decays and inverse decays are the only processes that are effective.[13] For consistency, in this approximation, the real intermediate state contribution to the $2 \rightarrow 2$ processes has

---

[12]We use the conventions of (67). We have also summed over the three lepton flavors neglecting the dependence on the lepton Yukawa couplings (68).

[13]This section follows closely (69).

to be included. In the kinetic Equations 62 and 63, one then has to replace $D + S$ by D and W by W_{ID}, respectively, where W_{ID} is the contribution of inverse decays to the washout term. The solution for N_{B-L} in this case is the sum of two terms (64),

$$N_{B-L}(z) = N^i_{B-L}e^{-\int_{z_i}^{z} dz' W_{ID}(z')} - \frac{3}{4}\varepsilon_1\kappa(z; \tilde{m}_1).$$ 64.

Here the first term accounts for an initial asymmetry which is partly reduced by washout, and the second term describes B-L production from N_1 decays. It is expressed in terms of the *efficiency factor* κ (68), which does not depend on the CP asymmetry ε_1,

$$\kappa(z) = \frac{4}{3}\int_{z_i}^{z} dz' D\left(N_{N_1} - N^{eq}_{N_1}\right) e^{-\int_{z'}^{z} dz'' W_{ID}(z'')}.$$ 65.

As we shall see, decays and inverse decays are sufficient to describe qualitatively many properties of the full problem.

We will first study in detail the regimes of weak and strong washout. If just decays and inverse decays are taken into account, these regimes correspond, respectively, to the limits $K \ll 1$ and $K \gg 1$ of the decay parameter

$$K = \frac{\Gamma_D(z = \infty)}{H(z = 1)} = \frac{\tilde{m}_1}{m_*},$$ 66.

introduced in the context of ordinary GUT Baryogenesis (64). Based on the insight into the dynamics of the non-equilibrium process gained from these limiting cases, one can then obtain analytic interpolation formulas that describe rather accurately the entire parameter range.

To proceed, let us first recall some basic definitions and formulas. The decay rate is given by the formula (70),

$$\Gamma_D(z) = \Gamma_{D1}\left\langle\frac{1}{\gamma}\right\rangle,$$ 67.

where the thermally averaged dilation factor is given by the ratio of the modified Bessel functions K_1 and K_2,

$$\left\langle\frac{1}{\gamma}\right\rangle = \frac{K_1(z)}{K_2(z)}.$$ 68.

For the decay term D, one then obtains

$$D(z) = Kz\left\langle\frac{1}{\gamma}\right\rangle.$$ 69.

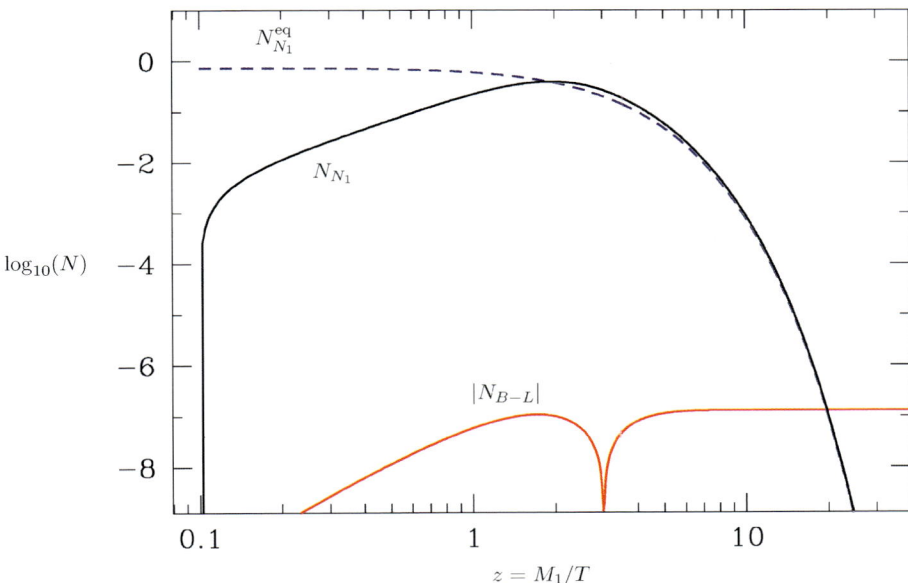

Figure 2 The evolution of the N_1 abundance and the $B-L$ asymmetry for a typical choice of parameters, $M_1 = 10^{10}$ GeV, $\varepsilon_1 = 10^{-6}$, $\tilde{m}_1 = 10^{-3}$ eV, and $\bar{m} = 0.05$ eV. From (67).

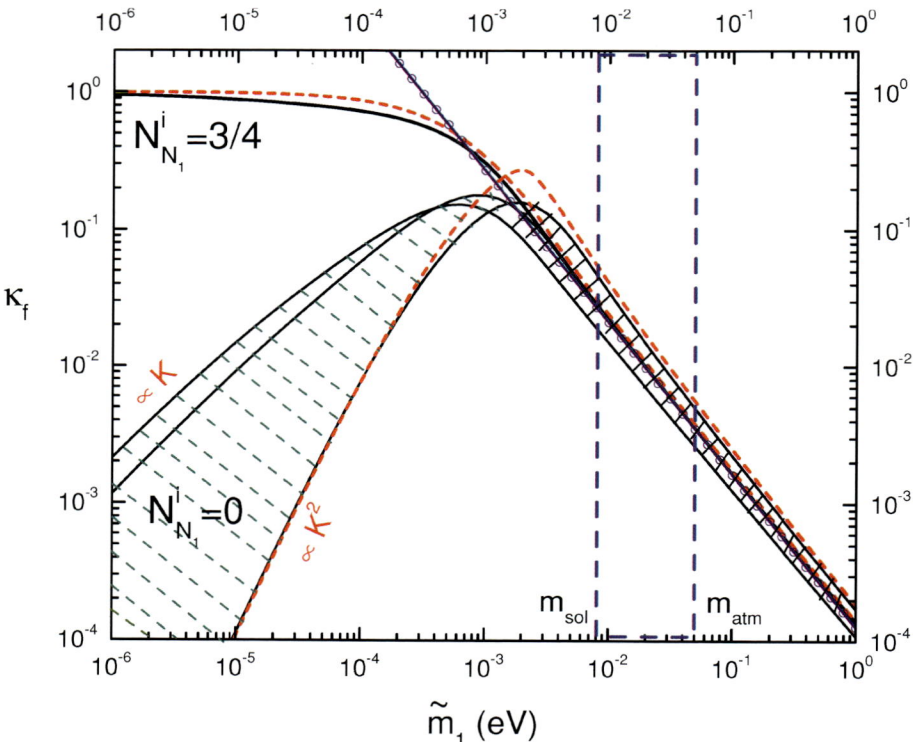

Figure 3   Final efficiency factor when the washout term ΔW is neglected. From (69).

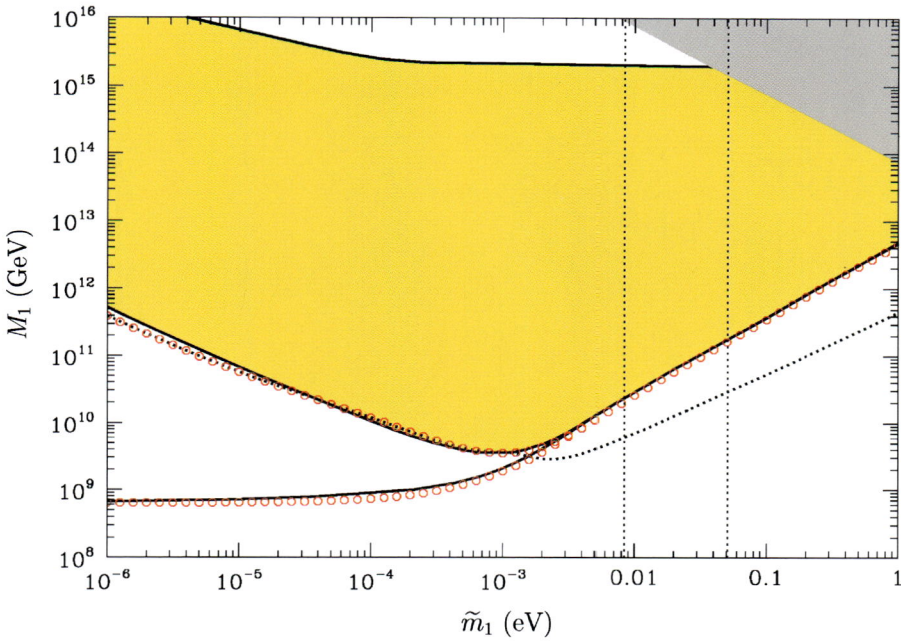

Figure 4 Analytical lower bounds on M_1 (circles) and T_i (dotted line) for $m_1 = 0$, $\eta_B^{CMB} = 6 \times 10^{-10}$ and $m_{atm} = 0.05$ eV. The analytical results for M_1 are compared with the numerical ones (solid lines). Upper and lower curves correspond to zero and thermal initial N_1 abundance, respectively. The vertical dashed lines indicate the range (m_{sol}, m_{atm}). The gray triangle at large M_1 and large m_1 is excluded by theoretical consistency. From (69).

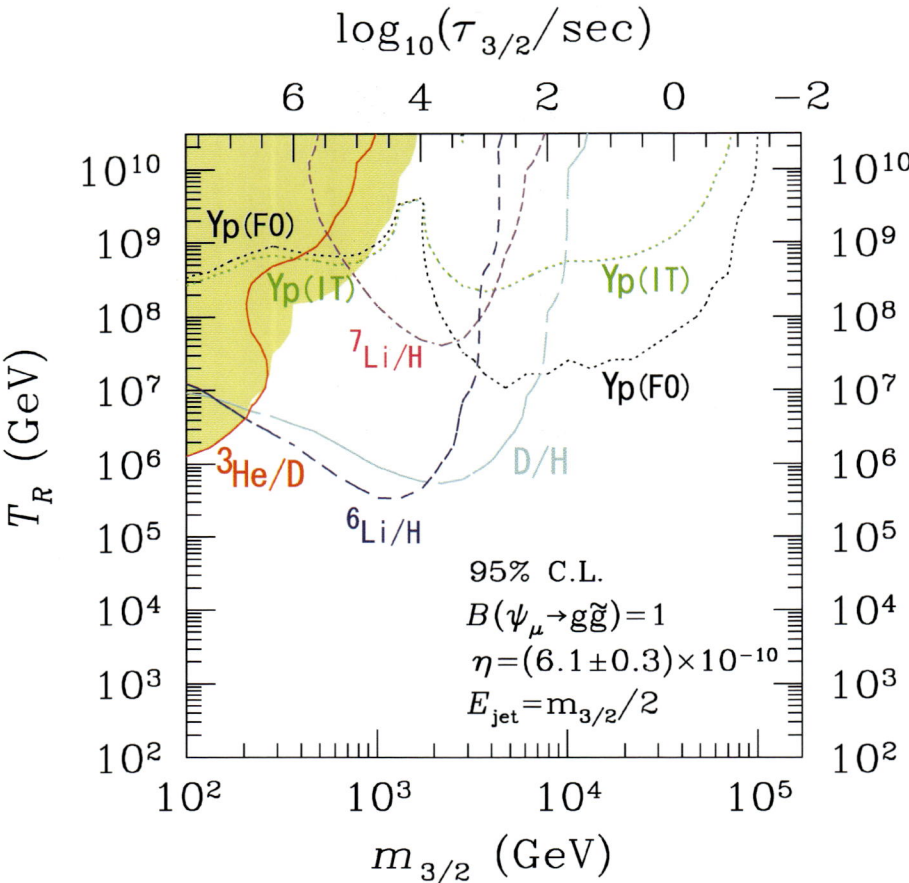

Figure 5 Upper bounds on the reheating temperature as function of the gravitino mass for the case where the gravitino dominantly decays into a gluon-gluino pair. From (90).

The inverse decay rate is related to the decay rate by

$$\Gamma_{ID}(z) = \Gamma_D(z)\frac{N_{N_1}^{eq}(z)}{N_l^{eq}}, \qquad 70.$$

where N_l^{eq} is the equilibrium density of lepton doublets. Because the number of degrees of freedom for heavy Majorana neutrinos and lepton doublets is the same, $g_{N_1} = g_l = 2$, one has

$$N_{N_1}^{eq}(z) = \frac{3}{8}z^2 K_2(z), \qquad N_l^{eq} = \frac{3}{4}. \qquad 71.$$

This yields for the contribution of inverse decays to the washout term W:

$$W_{ID}(z) = \frac{1}{2}D(z)\frac{N_{N_1}^{eq}(z)}{N_l^{eq}}. \qquad 72.$$

All relevant quantities are given in terms of the Bessel functions K_1 and K_2, which can be approximated by simple analytical expressions.

In the regime "far out of equilibrium," $K \ll 1$, decays occur at very small temperatures, $z \gg 1$, and the produced (B–L)-asymmetry is not reduced by washout effects. In this case, using Equation 62 with $S = 0$, the integral for the efficiency factor given in Equation 65 becomes simply,

$$\kappa(z) \simeq \frac{4}{3}\left(N_{N_1}^i - N_{N_1}(z)\right). \qquad 73.$$

The final value of the efficiency factor $\kappa_f = \kappa(\infty)$ is proportional to the initial N_1 abundance. If $N_1^i = N_1^{eq} = 3/4$, then $\kappa_f = 1$. But if the initial abundance is zero, then $\kappa_f = 0$ as well. Therefore, in this region there is the well known problem that one has to invoke some external mechanism to produce the initial abundance of neutrinos. Moreover, an initial (B − L)-asymmetry is not washed out. Thus in the regime $K \ll 1$, the results strongly depend on the initial conditions and there is little predictivity.

In order to obtain the efficiency factor in the case of "vanishing initial N_1-abundance," $N_{N_1}(z_i) \equiv N_{N_1}^i \simeq 0$, one has to calculate how heavy neutrinos are dynamically produced by inverse decays. This requires solving the kinetic equation Equation 62 with the initial condition $N_{N_1}^i = 0$.

Let us define a value z_{eq} by the condition

$$N_{N_1}(z_{eq}) = N_{N_1}^{eq}(z_{eq}). \qquad 74.$$

Then Equation 62 implies that the number density reaches its maximum at $z = z_{eq}$. For $z > z_{eq}$ the efficiency factor is always the sum of two contributions,

$$\kappa_f(z) = \kappa^-(z) + \kappa^+(z). \qquad 75.$$

Here $\kappa^-(z)$ and $\kappa^+(z)$ correspond to the integration domains $[z_i, z_{eq})$ and $[z_{eq}, z)$, respectively.

Consider first the case of "weak washout," $K \ll 1$, which implies $z_{eq} \gg 1$. One then finds,

$$N_{N_1}(z_{eq}) \simeq \frac{9\pi}{16} K.$$ 76.

It turns out that to first order in K, there is a cancellation between κ^+ and κ^-, yielding for the final efficiency factor

$$\kappa_f(K) \simeq \frac{9\pi^2}{64} K^2.$$ 77.

Note that Equation 77 does not hold for $K > 1$, because in this case z_{eq} becomes small, and washout effects change the result.

In the case of "strong washout," $K \gg 1$, we can neglect the negative contribution κ^-, because the asymmetry generated at high temperatures is efficiently washed out. Now the neutrino abundance tracks closely the equilibrium behavior. Because $D \propto K$, one can solve Equation 62 systematically in powers of $1/K$, which yields

$$D\left(N_{N_1}(z) - N_{N_1}^{eq}(z)\right) = \frac{3}{2Kz} W_{ID}(z) + \mathcal{O}\left(\frac{1}{K}\right),$$ 78.

where we have used properties of the Bessel functions. From Equations 65 and 78, one obtains for the efficiency factor[14]

$$\kappa(z) = \frac{2}{K} \int_{z_{eq}}^{z} dz' \frac{1}{z'} W_{ID}(z') e^{-\int_{z'}^{z} dz'' W_{ID}(z'')}.$$ 79.

The integral is dominated by the contribution from a region around the value z_B where the integrand has a maximum, which is determined by the condition

$$W_{ID}(z_B) = \left\langle \frac{1}{\gamma} \right\rangle^{-1} (z_B) - \frac{3}{z_B}.$$ 80.

For $K \gg 1$ one has $z_B \gg 1$, and the condition (80) becomes approximately $W_{ID}(z_B) \simeq 1$, with $W_{ID}(z) > 1$ for $z < z_B$ and $W_{ID}(z) < 1$ for $z > z_B$. This means that the asymmetry produced for $z < z_B$ is essentially erased, whereas for $z > z_B$, washout is negligible. Hence, the expression of Equation 79 is a good approximation for the final efficiency factor.

One finds that a rather accurate expression for $z_B(K)$ is given by

$$z_B(K) \simeq 1 + \frac{1}{2} \ln\left(1 + \frac{\pi K^2}{1024}\left[\ln\left(\frac{3125\pi K^2}{1024}\right)\right]^5\right).$$ 81.

---

[14]Because κ^- does not contribute, we can take the lower limit below as z_i.

The integral of Equation 79, which gives the final efficiency factor in terms of $z_B(K)$, is well approximated by

$$\kappa_f(K) \simeq \frac{2}{z_B(K)K}\left(1 - e^{-\frac{1}{2}z_B(K)K}\right). \qquad 82.$$

Both equations can also be extrapolated into the regime of weak washout, $K \ll 1$, where one obtains $\kappa_f = 1$ corresponding to thermal initial abundance, $N_{N_1}^i = N_{N_1}^{eq} = 3/4$. At $K \simeq 3$ a rapid transition takes place from strong to weak washout. Even here, analytical and numerical results agree within 30%. For the case of zero initial N_1 abundance, one obtains an interpolation formula $\kappa_f(K)$ analogous to Equation 82.

The above discussion of decays and inverse decays can be extended to include $\Delta L = 1$ and $\Delta L = 2$ scattering and washout processes. In the weak washout regime, $K \ll 1$, the main effect is that the efficiency factor of Equation 77 is enhanced to $\kappa_f \propto K$. Relevant effects include scattering processes involving gauge bosons (71, 72) and thermal corrections to the decay and scattering rates (72, 73). The range of different results is represented in Figure 3 (see color insert) by the hatched region. An additional uncertainty in the weak washout regime is due to the dependence of the final results on the initial N_1 abundance and a possible initial asymmetry created before the onset of Leptogenesis.

The situation is very different in the strong washout regime. Here the final efficiency factor is not sensitive to the neutrino production because a thermal neutrino distribution is always reached at high temperatures. For $\tilde{m}_1 > m_* \simeq 10^{-3}$ eV, the effect of $\Delta L = 1$ processes on the washout is not larger than about 50%, as indicated by the hatched region in Figure 3. Within these uncertainties, the final efficiency factor is given by the simple power law:

$$\kappa_f = (2 \pm 1)10^{-2}\left(\frac{0.01\,\text{eV}}{\tilde{m}_1}\right)^{1.1\pm0.1}. \qquad 83.$$

Both the scale of solar neutrino oscillations, $m_{sol} \equiv \sqrt{\Delta m_{sol}^2} \simeq 8 \times 10^{-3}$ eV, and the scale of atmospheric neutrino oscillations, $m_{atm} \equiv \sqrt{\Delta m_{atm}^2} \simeq 0.05$ eV, are larger than the equilibrium neutrino mass m_*. Hence, the range of neutrino masses, and therefore $\tilde{m}_1$, indicated by neutrino oscillations lies entirely in the strong washout regime where theoretical uncertainties are small and the efficiency factor is still large enough to allow for successful Leptogenesis.

4.4. Bounds on Neutrino Masses

The $\Delta L = 2$ processes with heavy neutrino exchange generate a contribution to the washout rate that depends on the absolute neutrino mass scale $\bar{m}^2 = m_1^2 + m_2^2 + m_3^2$,

$$\Delta W \propto \frac{M_P M_1 \bar{m}^2}{v_F^4}. \qquad 84.$$

As long as ΔW can be neglected, the efficiency factor is independent of M_1. With increasing $\bar{m}$ however, the washout rate ΔW becomes important and eventually prevents successful Leptogenesis. This leads to the upper bound on the absolute neutrino mass scale (67, 74).

One can also obtain a lower bound on the heavy neutrino masses (75), because the CP asymmetry ε_1 satisfies an upper bound (75–78), which is a function of M_1, $\widetilde{m}_1$ and $\bar{m}$. Since the rates entering the Boltzmann equations depend on the same quantities, there exists for arbitrary neutrino mass matrices a maximal baryon asymmetry $\eta_B^{\max}$,

$$\eta_B \leq \eta_B^{\max}(\widetilde{m}_1, M_1, \bar{m})$$
$$\simeq 0.01\varepsilon_1^{\max}(\widetilde{m}_1, M_1, \bar{m})\kappa(\widetilde{m}_1, M_1\bar{m}^2). \qquad 85.$$

Requiring the maximal baryon asymmetry to be larger than the observed one,

$$\eta_B^{\max}(\widetilde{m}_1, M_1, \bar{m}) \geq \eta_B^{CMB}, \qquad 86.$$

then yields a constraint on the neutrino mass parameters $\widetilde{m}_1$, M_1 and $\bar{m}$. For each value of $\bar{m}$, one obtains a domain in the $(\widetilde{m}_1 - M_1)$-plane, which is allowed by successful Baryogenesis. For $\bar{m} \geq 0.20$ eV this domain shrinks to zero. One can easily translate this bound into upper limits on the individual neutrino masses. In a similar way, one finds a lower bound on M_1, the smallest mass of the heavy Majorana neutrinos. The resulting upper and lower bounds are (77)

$$m_i < 0.1\,\text{eV}, \qquad M_1 > 4 \times 10^8\,\text{GeV}, \qquad 87.$$

where we have assumed thermal initial N_1 abundance. The upper bound on the light neutrino masses holds for a normal as well as an inverted hierarchy of masses. For zero initial N_1 abundance one obtains the more restrictive lower bound $M_1 > 2 \times 10^9$ GeV. For $\widetilde{m}_1 > m_*$, the baryon asymmetry is generated at the temperature $T_B \simeq M_1/z_B < M_1$. Hence the lower bound on the reheating temperature T_i is less restrictive than the lower bound on M_1. The results of a detailed analytical and numerical calculation are summarized in Figure 4 (see color insert). For the lower bound on the reheating temperature one finds $T_i > 2 \times 10^9$ GeV (69, 72).[15]

What is the theoretical error on the upper bound for the light neutrino masses? In order to answer this question one needs a full quantum mechanical treatment of Leptogenesis, a challenging problem! A possible starting point is the Kadanoff-Baym equations for which a systematic expansion around the Boltzmann equations can be constructed (59). One then has to calculate relativistic corrections, off-shell effects, "memory effects," higher order loop corrections, etc. One important effect is the running of neutrino masses between the Fermi scale and the energy scale of Leptogenesis (68, 81). Also relevant are the $\Delta L = 1$ scattering processes

---

[15]In the supersymmetric case the CP asymmetry is enhanced but also the washout processes are stronger. These two effects partly compensate each other (79), leading to the slightly less stringent bound $T_i > 1.5 \times 10^9$ GeV (80).

involving gauge bosons (71, 72). Conceptually interesting are thermal corrections at large temperatures, $T > M_1$, which correspond to loop corrections involving gauge bosons and the top quark (72). Their effect is large if thermal masses are treated as kinematical masses in the evaluation of scattering matrix elements. At sufficiently high temperatures the process $N_1 \rightarrow H\ell_L$ is then kinematically forbidden, whereas the process $H \rightarrow N_1\ell_L^\dagger$ is allowed by "phase space." On the contrary, thermal corrections are small if they are only included as propagator effects (73). It is important to clarify this issue for the treatment of non-equilibrium processes at high temperatures.

The analysis (72) leads to the upper bound on the light neutrino masses $m_i < 0.15$ eV. In (69) an upper bound of 0.12 eV has been obtained. About 0.02 eV of this difference is due to the different treatment of radiative corrections (81), the remaining 0.01 eV reflects differences in the treatment of thermal corrections. This discrepancy has to be compared with an uncertainty of about -0.02 eV due to "spectator processes" (82), which have not been taken into account in both analyses. Hence, within the minimal seesaw model and the present status of theoretical calculations, the upper bound on the light Majorana neutrino masses is now known rather precisely.

The main result of this section is summarized in Figure 3. For $\widetilde{m}_1 > m_*$, the efficiency factor, and therefore the baryon asymmetry η_B, is independent of the initial N_1 abundance. Furthermore, the final baryon asymmetry does not depend on the value of an initial baryon asymmetry generated by some other mechanism (77). Hence, the value of η_B is entirely determined by neutrino properties. In this way Leptogenesis singles out the neutrino mass range

$$10^{-3}\,\text{eV} < m_i < 0.1\,\text{eV}. \qquad\qquad 88.$$

The firm predictions of thermal Leptogenesis open a window into the physics of the early universe at temperatures $T_B = \mathcal{O}(10^{10}\,\text{GeV})$, and we can ask what the implications are for dark matter, cosmology, and particle physics.

4.5. Triplet Models and Resonant Leptogenesis

Measurements in neutrino physics determine the parameters of the neutrino mass matrix,

$$m_\nu = -m_D^T \frac{1}{M} m_D + m_\nu^{\text{triplet}}, \qquad\qquad 89.$$

which in general contains a contribution from $SU(2)$ triplet fields (83) in addition to the seesaw term generated by $SU(2)$ singlet heavy Majorana neutrinos. So far, we have only considered the minimal case, $m_\nu^{\text{triplet}} = 0$. Clearly, a dominant triplet contribution would destroy the connection between Leptogenesis and low energy neutrino physics.

The discovery of quasi-degenerate neutrinos with masses above the bound 0.1 eV would require significant modifications of minimal Leptogenesis and/or the seesaw mechanism. In this case $SU(2)$ triplet contributions to neutrino masses could be a possible way out (78, 84, 85). Clearly, one then has no upper bound on the light neutrino masses anymore. Yet Leptogenesis with right-handed neutrino decays can still work, yielding a slightly relaxed lower bound on the heavy neutrino masses. For instance, one may have $m_i \sim 0.35 \, \text{eV}$ with $M_1 > 4 \times 10^8 \, \text{GeV}$ (85).

Another way to reconcile quasi-degenerate light neutrinos with Leptogenesis makes use of the enhancement of the CP asymmetry in case of quasi-degenerate heavy neutrinos (86). For instance, to raise the upper bound from $0.1 \, \text{eV}$ to $0.4 \, \text{eV}$, a degeneracy of $\Delta M / M$ for the heavy neutrinos in the range $0.4 - 10^{-3}$ is required, depending on assumptions about the neutrino mass matrices (77, 78). In the extreme case of "resonant Leptogenesis" (71), CP asymmetries $\varepsilon = \mathcal{O}(1)$ are reached for degeneracies $\Delta M / M = \mathcal{O}(10^{-10})$. In this case the right-handed neutrino masses may be as small as $1 \, \text{TeV}$, which may lead to observable signatures at colliders. A number of models of this type have been constructed (87), some of which make use of the relative smallness of soft supersymmetry breaking terms (88).

4.6. Is the CP Violation in Leptogenesis Connected with the Low Energy CP Violation in the Neutrino Sector?

As was shown in Section 3, the seesaw model has 6 CP-violating phases in the Yukawa matrix h_{ij}. Leptogenesis depends on one combination of these 6 phases. However, there are only 3 CP-violating phases at low energies. Hence it is impossible to determine all 6 phases in the theory, even if one were to measure all 3 low-energy phases. Futhermore, as we discussed earlier, one of these low-energy phases remains undetermined by experiments feasible at low energies.

Nevertheless, the effective number of high energy CP-violating phases is reduced if one of the superheavy Majorana neutrinos N_i is extremely heavy and decouples from the seesaw system. In this case, the Yukawa couplings h_{ij} effectively are given by a 2×3 matrix that contains 6 complex parameters and hence 6 phases. Three of the 6 phases can be absorbed into the wave functions of ℓ_L and thus one is left with only 3 CP-violating phases at high energies. In this case, the 3 low-energy CP-violating phases that appear in the neutrino mass matrix m_ν are reduced to only two physical phases, because $\det(m_\nu) \simeq 0$. Although these 2 low-energy phases are, in principle, measurable in future experiments, this is still not enough to determine all three phases in the full theory. Therefore, even in this simplified example, one cannot establish any link between the sign of the Universe's baryon-number asymmetry with the observable CP-violating phases at low energies.

In the very special case where h_{ij} has two zeros, one has only one CP-violating phase. In this case the CP-violating phase in neutrino oscillations is connected with the phase in Leptogenesis or, equivalently, the sign of the baryon-number asymmetry in the Universe (89).[16] Thus, in this case, one may indeed test directly the idea of Leptogenesis. It is interesting that such a restricted model, where $h_{13} = h_{21} = 0$, is still consistent with data on neutrino oscillations.

5. DARK MATTER CONSIDERATIONS

It is certainly possible that the mechanism that generates a primordial matter-antimatter asymmetry in the Universe is not physically related to the existence of a nonluminous component of the energy density of the Universe, a component that now accounts for about 25% of the total energy density. However, it would be very interesting if these two phenomena, so central to the history of the Universe, were connected in some deep way. It turns out, as we shall see, that if Leptogenesis is the mechanism by which a primordial matter-antimatter asymmetry in the Universe is established, it considerably impacts what the dark matter in the Universe can be.

Of the three viable options for dark matter, from the point of view of particle physics, two are either linked or constrained by thermal Leptogenesis and the third has clear connections to nonthermal Leptogenesis. Before discussing these points in some detail, it is useful to briefly review the extant dark matter candidates motivated by particle physics.

5.1. PQ Symmetry and Axions

It is well known that QCD admits the presence of an additional CP violating term in its Lagrangian density (17),

$$\mathcal{L}_\theta = \frac{\alpha_3}{8\pi} \theta F_i^{\mu\nu} \tilde{F}_{i\mu\nu}, \qquad\qquad 90.$$

where $\tilde{F}_{i\mu\nu} = 1/2\epsilon_{\alpha\beta\mu\nu} F_i^{\alpha\beta}$. If θ is nonvanishing $\mathcal{L}_\theta$, which is C-even and P-odd, violates CP and T invariance. Because possible CP violating parameters of the strong interactions, like the electric dipole moment of the neutron, are very tightly bounded by experiment (91), the parameter θ must be very small ($\theta \leq 10^{-10}$) (92). The reason for this is a mystery and is known as the Strong CP Problem (93).

Probably the most "natural" solution suggested for the Strong CP Problem is to assume that the total Lagrangian for the strong and electroweak interactions is invariant under a global chiral $U(1)_{PQ}$ symmetry (94). Even though this symmetry is spontaneously broken, one can show (94) that as a result of the $U(1)_{PQ}$ symmetry, the parameter θ is driven to zero. In effect, what happens is that the CP violating Lagrangian term $\mathcal{L}_\theta$ is replaced by a CP conserving interaction between the CP

---

[16]The prediction of the sign of the CP-violating phase in neutrino oscillations depends on which heavy Majorana neutrino is responsible for Leptogenesis. This problem is solved in the inflaton-decay scenario in supersymmetry (SUSY) theories, because one choice is unable to produce enough lepton-number asymmetry due to the constraint on the reheating temperature $T_R < 10^7$ GeV (90).

odd pseudo-Goldstone boson[17] associated with the spontaneous breakdown of $U(1)_{PQ}$—the axion a—(95) and $F\tilde{F}$:

$$\mathcal{L}_\theta \rightarrow \frac{\alpha_3}{8\pi} \frac{a}{f_a} F_i^{\mu\nu} \tilde{F}_{i\mu\nu}. \qquad 91.$$

Originally, it was supposed (94) that the $U(1)_{PQ}$ symmetry was broken at the electroweak scale. Then $f_a \sim v_F \simeq 200$ GeV and the axion mass lies in the keV range. However, axions in this mass range, which are coupled with strength $1/f_a$, have been ruled out by experiment (93). Astrophysical considerations, however, impose very strong constraints on axions much lighter than a keV, as their emission from stars would significantly alter their properties. Only axions that are sufficiently weakly coupled (hence, with large enough f_a and thus a correspondingly small mass) avoid these constraints, and one finds the bound (96) $f_a \geq 10^{10}$ GeV. On the other hand, f_a cannot be arbitrarily large, because zero-momentum axion oscillations in the early Universe would carry enough energy density (proportional, approximately, to f_a) to overclose the Universe (97). Thus, for an appropriate value for f_a, axions can be the dark matter in the Universe. In particular, one finds (98) that $f_a \simeq 10^{12}$ GeV gives $\Omega_a \simeq 1$.

5.2. Dark Matter Candidates from Supersymmetry

Supersymmetry, a boson-fermion symmetry, has been invoked extensively as the solution of the so-called hierarchy problem. This problem is related to the fact that without some stabilizing influence, radiative corrections in the electroweak theory would naturally push the Fermi scale v_F to have the value of whatever cutoff delimits the validity of the theory. Typically, this cutoff is imagined to be at the Planck scale M_P, and why $v_F \ll M_P$ is the hierarchy problem. This problem is resolved if there is some low energy (spontaneously broken) supersymmetry. Due to the fermion-boson nature of supersymmetry, radiative corrections of parameters in the electroweak theory (like v_F) are now only logarithmically dependent on the cutoff, not quadratically dependent. Hence, effectively, if there is some low energy supersymmetry, one can contemplate having a hierarchy like $v_F \ll M_P$, because radiative shifts can only change v_F logarithmically.

In general, supersymmetric theories possess a discrete symmetry (R-symmetry) that distinguishes particles from their supersymmetric partners. As a result, the lightest supersymmetric particle (the LSP) is stable and, in principle, could be the source of the dark matter in the Universe. Indeed, it is known (64) that the energy density of particles of mass of $O(v_F)$, whose annihilation cross section is of electroweak strength, is of the order of the critical energy density that closes the Universe. With supersymmetric partners of ordinary particles having electroweak

---

[17]Axions are not true Goldstone bosons because the $U(1)_{PQ}$ symmetry is anomalous (95). In fact, the same effective potential for axions that serves to drive θ to zero gives axions a small mass. This mass is of order (93) $m_a^2 \sim m_q \Lambda_{QCD}^3/f_a^2$, where f_a is the scale where the $U(1)_{PQ}$ symmetry breaks down spontaneously, and m_q is the (light) quark mass.

scale masses and interactions, the LSP is therefore an ideal candidate for the dark matter in the Universe (99). In this review we will discuss both the cases of neutralinos (the SUSY partners of gauge and Higgs boson) and of gravitinos (the spin 3/2 partner of the graviton) as LSP candidates.

5.3. Extended Structures

Scalar fields are necessary ingredients of the standard electroweak model as well as its supersymmetric extension. It is well known that theories with scalar fields can lead to the formation of nontopological solitons. These extended structures, known as Q-balls (100), may be stable or unstable and arise when some scalar field carries a conserved $U(1)$ charge. For example, in supersymmetric theories, sleptons and squarks carry, respectively, lepton and baryon number.

In supersymmetric theories, more generally, Q-balls can develop along flat directions of the scalar potential (101). These Q-balls can, in a number of instances, carry baryon number. If the baryon number of the Q-ball is large enough, and its mass is small enough, the baryonic Q-balls are stable. Because of their stability, one can imagine that these Q-balls could be the dark matter in the Universe.[18] Typically (101), if stable Q-balls exist, they have both very large baryon number ($B \sim 10^{26}$) and are very massive ($M_Q \leq 10^{26}$ GeV). Unfortunately, this makes their detection very difficult, because their flux is very low (104).

5.4. Natural Connection of Axions with Leptogenesis

The scale of $U(1)_{PQ}$ breaking needed for axions to be the dark matter in the Universe ($f_a \simeq 0.3 \times 10^{12}$ GeV) is close enough to the mass of the lightest right-handed neutrino ($M_1 \simeq 10^{10}$ GeV) needed for Leptogenesis to seek for a common linkage. In fact, the existence of such a linkage was observed long ago by Langacker, Peccei, and Yanagida (105). What these authors observed was that if M_1 were due to the VEV of a scalar field σ, one could identify this field as carrying a PQ-symmetry rather that lepton number.

Let us examine this assertion in a bit more detail by looking at the Yukawa interactions of the quarks and leptons with the three Higgs fields[19] ϕ_1, ϕ_2 and σ.

---

[18]This, however, is not easily achieved because, in general, the squarks are unstable with their baryon number eventually residing on quarks. If the squarks are light enough, stability can be achieved. However, as Kasuya et al. point out (102), it is difficult to explain both the baryon asymmetry and the dark matter density simultaneously. Nevertheless, there are scenarios where unstable Q-balls are the source for both Baryogenesis and neutralino dark matter (103).

[19]For a PQ-symmetry to exist, one needs to have two $SU(2)$ doublet Higgs fields, ϕ_1 and ϕ_2, rather than just the single Higgs field of the Standard Model H (and its Hermitian adjoint $H^\dagger$).

Schematically, one has

$$\mathcal{L}_{\text{Yukawa}} = h_\sigma \sigma N_R N_R + h \bar{N}_R \phi_2 \ell_L + h_u \bar{u}_R \phi_2 q_L + f_d \bar{d}_R \phi_1 q_L + f \bar{e}_R \phi_1 \ell_L + h.c.. \qquad 92.$$

One sees that Equation 92 is invariant under a PQ-symmetry, where

$$\phi_1, \phi_2 \rightarrow e^{i\alpha} \phi_1, e^{i\alpha} \phi_2 \qquad 93.$$

$$N_R, l_R, u_R, d_R \rightarrow e^{i\alpha} N_R, e^{i\alpha} l_R, e^{i\alpha} u_R, e^{i\alpha} d_R, \qquad 94.$$

provided that

$$\sigma \rightarrow e^{-2i\alpha} \sigma. \qquad 95.$$

To allow $\langle \sigma \rangle = f_a \gg v_F$, as in all invisible axion models (106, 107), requires one fine tuning. In the above case, this requires the PQ-invariant term in the scalar potential

$$V = \kappa \sigma \phi_1 \phi_2 + h.c. \qquad 96.$$

to have the constant $\kappa \sim v_F^2 / f_a$, to allow electroweak symmetry breaking to occur at a scale much below the scale of $U(1)_{PQ}$ symmetry breaking ($v_F \ll f_a$).

The Yukawa couplings of this model guarantee that the mass of the lightest right-handed neutrino and f_a are related: $M_1 = 2(h_\sigma)_{11} f_a$. Thus, if axions are the source of the dark matter energy density in the Universe and the baryon aymmetry arises from Leptogenesis, because $\Omega_{\text{DM}} \sim f_a$ and $\Omega_B \sim M_1 \sim f_a$, their ratio is independent of the scale of $U_{PQ}(1)$ breaking. Hence it is perhaps not surprising that this ratio is of order unity.

5.5. The Gravitino Problem in Supersymmetric Theories

As we discussed in Section 4, for Leptogenesis to be effective, the mass of the lightest right-handed neutrino has to be greater than 2×10^9 GeV. This bound, in turn, means that thermal Leptogenesis must have occurred at temperatures above 2×10^9 GeV. Hence, if the Universe went through an inflationary period, as all evidence seems to suggest (4), the reheating temperature after inflation T_R must have been greater than 2×10^9 GeV for Leptogenesis to be the source of the matter-antimatter asymmetry in the Universe. This high reheating temperature is problematic for supersymmetric theories because it leads to an overproduction of light states, like the gravitino, with catastrophic consequences for the evolution of the Universe after inflation. Unless these observational inconsistencies can be avoided, it appears that Leptogenesis in supersymmetric theories cannot produce the desired baryon asymmetry in the Universe.

The production of gravitinos after inflation has been studied in some detail (108). The thermal production of gravitinos produced by the strong interactions of quarks, squarks, gluons, and gluinos is governed by the Boltzmann equation (109)

$$\frac{dn_{3/2}}{dt} + 3Hn_{3/2} = C_{3/2}(T),$$ 97.

where

$$C_{3/2}(T) = \frac{3\zeta(3)\alpha_3(T)}{\pi^2} \frac{T^6}{M_P^2} \left(1 + \frac{m_{\tilde{g}}^2}{3m_{3/2}^2}\right) F(T).$$ 98.

Here $F(T)$ is a thermal factor of O(10) and $m_{\tilde{g}}$ and $m_{3/2}$ are, respectively, the gluino and the gravitino masses. Integrating Equation 97 to a reheating temperature T_R, the resulting relic density of produced gravitinos is given by

$$\Omega_{3/2}h^2 \simeq 0.44\alpha_3(T_R)\left[1 + \frac{1}{3}\left(\frac{\alpha_3(T_R)}{\alpha_3(\mu)}\right)^2\left(\frac{m_{\tilde{g}}(\mu)}{m_{3/2}}\right)^2\right]\left(\frac{T_R}{10^{10}\,\text{GeV}}\right)\left(\frac{m_{3/2}}{100\,\text{GeV}}\right),$$
99.

where h is the scaled Hubble parameter and $\mu \sim M_Z$.

If gravitinos are stable (i.e. they are the LSP), the WMAP constraint on the amount of dark matter in the Universe (4)

$$\Omega_{DM}h^2 = 0.1126^{+0.0161}_{-0.0181}$$ 100.

constrains $\Omega_{3/2}h^2$ to be below this value and, for any given reheating temperature T_R and gravitino mass $m_{3/2}$, gives a bound on the gluino mass. If, on the other hand, the gravitinos are not stable, their rate of production for $T_R > 2 \times 10^9$ GeV is so large that subsequent gravitino decays completely alter the standard Big Bang Nucleosynthesis (BBN) scenario. Thus, in either case, there are severe constraints imposed on supersymmetric dark matter, which we will discuss in detail below.

If the gravitino is unstable, it has a long lifetime and decays during or after BBN for an interesting range of the gravitino mass, $m_{3/2} \simeq 100\,\text{GeV} - 10\,\text{TeV}$. The gravitino decay products destroy the light elements produced by the BBN and hence the relic abundance of gravitinos is constrained from above to keep the success of the BBN. This leads to an upper bound of the reheating temperature T_R after inflation, because the abundance of gravitinos is proportional to the reheating temperature. A recent detailed analysis derived a stringent upper bound $T_R < 10^{6-7}$ GeV when the gravitino decay has hadronic modes [see Figure 5 (see color insert) (90)]. This upper bound is much lower than the temperature for Leptogenesis, $T_R > 2 \times 10^9$ GeV (69, 72). Therefore, thermal Leptogenesis seems difficult to reconcile with low energy supersymmetry if gravitino masses lie in the range $m_{3/2} \simeq 100\,\text{GeV} - 10\,\text{TeV}$—a natural range for Supergravity (SUGRA) models.

5.6. Solutions to the Gravitino Problem in Thermal Leptogenesis

There have been several attempts to solve the gravitino problem in thermal Leptogenesis. Here we will briefly review a number of these proposed solutions.

One possibility has been proposed by Pilaftsis, who considers quasi-degenerate heavy Majorana neutrinos ($M_1 \simeq M_2$) (111). In this model the lepton-asymmetry parameter ε is enhanced by a factor of $M_1/(M_1 - M_2)$ and hence the decays of both N_1 and N_2 may produce enough asymmetry even for $T_R < 10^{6-7}$ GeV. However, it is difficult to find a compelling justification for having such a degeneracy in the heavy neutrino spectrum.

Another proposal was made by Bolz, Buchmüller, and Plümacher (112), who consider the case where the gravitino is the stable lightest SUSY particle (LSP). In this case the next lightest supersymmetric particle (NLSP) is the subject of the cosmological constraint, because its decay products may destroy the light elements created by the BBN, much like the unstable gravitino. Detailed analyses show that this scenario favors the $\tilde{\tau}$ NLSP compared to the neutralino NLSP and gravitino masses below $m_{3/2} \simeq 100$ GeV (113, 114).[20] In general, the gravitino production can be dominated by NLSP decays (115) or by thermal processes (116).

A third proposed solution makes use of gauge-mediation model of supersymmetry breaking in which the gravitino is the stable LSP with a mass $m_{3/2} < 1$ GeV. It turns out that, if the gravitino mass is $m_{3/2} < 16$ eV (117), then there is no gravitino problem. However, in this case the gravitino cannot be the dark matter. It must be something else, perhaps the axion. For the range of gravitino masses $m_{3/2} \simeq 100$ keV $- 1$ GeV, there is the interesting possibility that late-time entropy production in a class of gauge mediation models can naturally make the gravitino the dominant component of dark matter (118). In this scenario the reheating temperature can be as high as $T_R \simeq 10^{13}$ GeV. A light axino with mass $\mathcal{O}(1$ keV$)$ as LSP and a gravitino as NLSP would also solve the gravitino problem (119).

Finally, another possible solution arises if supersymmetry breaking effects are transmitted to the Standard Model sector through the scale anomaly, resulting in very heavy gravitino masses $m_{3/2} > 100$ TeV. In this case the gravitino decays before the time of BBN (110) and hence there is no cosmological problem. However, the gravitino decay modes contain always one LSP and hence the relic abundance of the gravitino must be constrained from above so that the density of the nonthermal LSP produced by the gravitino decays does not exceed the dark-matter density. This condition leads to $T_R < 10^{11}$ GeV (120, 121), which is consistent with thermal Leptogenesis.

We stress here that each of the above "solutions" predicts distinct particle spectra at the TeV scale, which may be testable in future collider experiments at the Large Hadron Collider (LHC). If one discovers supersymmetry but all the above possibilities are excluded experimentally, this would argue strongly against thermal Leptogenesis, although nonthermal Leptogenesis could still be viable. On the other hand, if some of these scenarios are confirmed experimentally, thermal Leptogenesis will become much more compelling.

---

[20]The gravitino mass is even less constrained if the LSP is a scalar neutrino or the gluino. For a class of supergravity models, an upper bound of 5×10^9 GeV on the reheating temperature has been obtained (114).

5.7. Leptogenesis and Lepton Flavor Violation in SUSY Models

Lepton flavor violation is another area in which thermal Leptogenesis and super-symmetry may have some linkages. If the neutrinos have a mass, lepton flavor violation (LFV) processes such as $\mu \to e + \gamma$ decay can occur. In nonsupersymmetric theories, these processes are strongly suppressed by a factor of $(m_\nu/M_W)^2$ in rate and hence are unmeasurable physically. However, this is not the case in the SUSY Standard Model if the seesaw mechanism is effective.

In the SUSY Standard Model, scalar quarks and leptons are assumed to have a universal SUSY-breaking soft mass, m_0, at the Planck scale. Otherwise one would have too large flavor-changing neutral currents (FCNC). However, even then quantum corrections resulting from Yukawa interactions of the quarks generate a violation of the universality of the soft masses for scalar quarks, which induces FCNC. In the lepton sector the Yukawa couplings of the superheavy Majorana neutrinos N_i also generate non-universal masses for scalar leptons that serve as a source of LFV (122).[21] If the relevant Yukawa couplings are of $O(1)$, or equivalently $M_3 \simeq 10^{15}$ GeV, one predicts a branching ratio for $\mu \to e + \gamma$ decay that may be testable in future experiments. However, an accurate prediction for LFV processes is very difficult because it hinges on unknown Yukawa coupling constants (124). In particular, the constraint coming from Leptogenesis that $M_1 \simeq 10^{10}$ GeV is not strong enough to suggest that LFV processes have potentially testable rates.

6. NONTHERMAL LEPTOGENESIS

Supersymmetry is an important symmetry for the unification of all interactions and all matter, and the SUSY Standard Model is considered as a plausible scenario for producing new physics at the TeV scale. Thus, it is quite interesting to consider theories where supersymmetry is spontaneously broken in a hidden sector connected to ordinary matter by gravitational strength interactions—the SUGRA framework. The seesaw mechanism is easily incorporated into this framework. However, as we discussed in some detail in Section 5, the gravitino problem argues against thermal Leptogenesis, particularly in SUGRA.

A possible solution to this problem may be provided by nonthermal Leptogenesis (10, 125–129), where one does not have a strong constraint on the reheating temperature. We will discuss here specifically nonthermal Leptogenesis via

---

[21] The Yukawa couplings h_{ij} of the Higgs to the N_is and leptons induce flavor dependent soft masses for scalar leptons. At the one-loop level the induced mass is given by $(m_{\tilde{\ell}}^2)_{ij} \simeq -6m_0^2/(4\pi)^2 h_{ik}^\dagger h_{kj} ln(M_P/M_k)$, where m_0 is the universal soft mass for scalar leptons at the Planck scale M_P. Thus, one may obtain information on the high-energy Yukawa coupling h_{ij} and the heavy neutrino masses M_k by measuring directly the mass matrix for the scalar leptons. However, the phases in $h^\dagger h$ are different from the phases contributing to Leptogenesis (123).

inflaton decay (125, 126), which we consider an interesting scenario. In the next subsection, we present general arguments for this scenario and show that it suggests a lower bound on the mass of the heaviest light neutrino $m_3 > 0.01$ eV. In the subsequent subsection, we will also discuss the Affleck-Dine mechanism (10) for Leptogenesis, which, specifically in supersymmetric theories, is also an interesting mechanism to generate the matter-antimatter asymmetry.

6.1. Nonthermal Leptogenesis via Inflaton Decay

Inflation early on in the history of the Universe is one of the most attractive hypotheses in modern cosmology because it not only solves long-standing problems in cosmology, like the horizon and the flatness problems (130), but it also accounts for the origin of density fluctuations (131). In this subsection we discuss the hypothesis that the inflaton Φ decays dominantly into a pair of the lightest heavy Majorana neutrinos, $\Phi \rightarrow N_1 + N_1$. We assume, for simplicity, that other decay modes including those into pairs of N_2 and N_3 are energetically forbidden. The produced N_1 neutrinos decay subsequently into $H + \ell_L$ or $H^\dagger + \ell_L^\dagger$. If the reheating temperature T_R is lower than the mass M_1 of the heavy neutrino N_1, then the out-of-equilibrium condition (11) is automatically satisfied.

The above two channels for N_1 decay have different branching ratios when CP conservation is violated. Interference between tree-level and one-loop diagrams generates a lepton-number asymmetry (57). Following our discussion in Section 4, the lepton asymmetry parameter ε can be written as (128, 129, 132)[22]

$$\varepsilon = -\frac{3}{8\pi} \frac{M_1}{\langle H \rangle^2} m_3 \delta_{\text{eff}}, \qquad\qquad 101.$$

where the effective CP-violating phase δ_{eff} is given by

$$\delta_{\text{eff}} = \frac{\text{Im}\left[h_{13}^2 + \frac{m_2}{m_3} h_{12}^2 + \frac{m_1}{m_3} h_{11}^2 \right]}{|h_{13}|^2 + |h_{12}|^2 + |h_{11}|^2}. \qquad\qquad 102.$$

Numerically, one obtains for the ε parameter

$$\varepsilon \simeq -2 \times 10^{-6} \left(\frac{M_1}{10^{10}\text{GeV}} \right) \left(\frac{m_3}{0.05\text{eV}} \right) \delta_{\text{eff}}. \qquad\qquad 103.$$

The chain decays $\Phi \rightarrow N_1 + N_1$ and $N_1 \rightarrow H + \ell_L$ or $H^\dagger + \ell_L^\dagger$ reheat the Universe, producing not only the lepton-number asymmetry but also entropy for the thermal bath. The ratio of the lepton number to entropy density after reheating is estimated to be (126)

---

[22]Because of supersymmetry, the asymmetry parameter ε below is a factor of 2 larger than that given in Equation 58.

$$\frac{n_L}{s} \simeq -\frac{3}{2}\varepsilon\frac{T_R}{m_\Phi}$$

$$\simeq 3 \times 10^{-10} \left(\frac{T_R}{10^6\,\text{GeV}}\right)\left(\frac{M_1}{m_\Phi}\right)\left(\frac{m_3}{0.05\,\text{eV}}\right), \qquad 104.$$

where m_Φ is the inflaton mass and we have taken $\delta_{\text{eff}} = 1$. This lepton-number asymmetry is converted into a baryon-number asymmetry through the sphaleron effects and one obtains (35)

$$\frac{n_B}{s} \simeq -\frac{8}{23}\frac{n_L}{s}. \qquad 105.$$

We should stress an important merit of the inflaton-decay scenario: It does not require a reheating temperature $T_R \sim M_1$, but it requires only $m_\Phi > 2M_1$. On the other hand, for thermal Leptogenesis to work it is necessary that $T_R \sim M_1$, which necessitates higher reheating temperature for Leptogenesis to produce enough matter-antimatter asymmetry.

If one assumes that $T_R < 10^7$ GeV to satisfy the cosmological constraint on the gravitino abundance (90) discussed earlier and uses $m_\Phi > 2M_1$, the observed baryon number to entropy ratio (4) gives a constraint on the heaviest light neutrino:

$$m_3 > 0.01\,\text{eV}. \qquad 106.$$

It is very interesting that the neutrino mass suggested by atmospheric neutrino oscillation experiments, $\sqrt{\Delta m_{\text{atm}}^2} \simeq 0.05$ eV, just satisfies the above constraint. However, to get this bound, we assumed that the inflaton decays dominantly into a pair of N_1s. If this branching ratio is only 10%, the lower bound on the neutrino mass exceeds the observed neutrino mass $\sqrt{\Delta m_{\text{atm}}^2} \simeq 0.05$ eV.

A variety of models have been considered to restore the bound of Equation 106 by imposing a symmetry. However, it is perhaps most interesting to consider that the scalar partner of the heavy Majorana neutrino N_1 is the inflaton itself (127), and the inflaton decay into a lepton plus a Higgs boson gives an effective branching ratio of 100%. In this model, one must assume that the initial value of the scalar partner of N_1 is much larger than the Planck scale to cause inflation [chaotic inflation (133)]. However, chaotic inflation is not easily realized in SUGRA, because the minimal supergravity potential has an exponential factor, $\exp(\phi^*\phi/M_G^2)$, that prevents any scalar field ϕ from having a value larger than the reduced Planck scale $M_G \simeq 2.4 \times 10^{18}$ GeV. Ref. (134) uses a restricted form of the Kahler potential.

6.2. Affleck-Dine Leptogenesis

In the SUSY Standard Model, in the limit of unbroken supersymmetry, some combinations of scalar fields do not enter the potential, constituting so-called flat directions of the potential. Since the potential is almost independent of these fields, they may have large initial values in the early Universe. Such flat directions receive soft masses in the SUSY-breaking vacuum. When the expansion rate H_{exp} of the

Universe becomes comparable to their masses, the flat directions begin to oscillate around the minimum of the potential. If the flat directions are made of scalar quarks and carry baryon number, the baryon-number asymmetry can be created during these coherent oscillations. This is the Affleck-Dine (AD) mechanism for Baryogenesis (10).

QCD corrections, however, make the potential of the AD fields milder than $|\phi|^2$. This allows nontopological soliton solutions (Q-balls) (135) to form in the early Universe, as a result of the coherent oscillations in the flat directions. Because Q-balls have long lifetimes, their decays produce a huge amount of entropy at late times. To avoid this problem one must choose parameters in the SUSY theory so that the density of the lightest SUSY particle (LSP) does not exceed the dark matter density in the present Universe (135). Although this may not be a problem, it is much safer to consider flat directions without QCD interactions, because such directions most likely do not have Q-ball solutions.

The most interesting candidate (128) for such a flat direction is

$$\phi_i = (2H\ell_i)^{1/2},$$ 107.

where ℓ_i is the lepton doublet field of the i-th family. Here, H and ℓ_i represent the scalar components of the corresponding chiral multiplets. The Yukawa interactions of H make the potential of ϕ_i steeper than the mass term and hence there is no instability of the coherent oscillation (i.e. there are no Q-ball solutions). Because this flat direction carries lepton number, a lepton asymmetry will be created during the coherent oscillation (AD Leptogenesis) (128). Sphaleron processes then transmute, in the usual fashion, this lepton asymmetry into a baryon asymmetry.

The seesaw mechanism induces a dimension-five operator in the superpotential for the theory,[23]

$$W = \frac{m_v}{2|\langle H \rangle|^2}(\ell H)^2,$$ 108.

where we have used a basis in which the neutrino mass matrix is diagonal. With this superpotential we have a SUSY-invariant potential for the flat direction ϕ given by

$$V_{\text{SUSY}} = \frac{m_v^2}{4|\langle H \rangle|^4}|\phi|^6.$$ 109.

In addition to the SUSY-invariant potential we have a SUSY-breaking potential,

$$\delta V = m_\phi^2|\phi|^2 + \frac{m_{\text{SUSY}}m_v}{8|\langle H \rangle|^2}\left(a_m\phi^4 + h.c.\right).$$ 110.

Here, a_m is a complex number. We take $m_\phi \simeq m_{\text{SUSY}} \simeq 1$ TeV and $|a_m| \sim 1$. The second term in δV is very important, because it gives rise to the lepton-number generation.

---

[23]For ease of notation we have dropped the subscript i below.

We assume that the flat direction ϕ acquires a negative (mass)^2 induced by the inflaton potential and rolls down to the point balanced by the SUSY-invariant potential V_{SUSY} during inflation. Thus, the AD field ϕ has an initial value of $\sqrt{H_{inf}|\langle H \rangle|^2/m_\nu}$, where H_{inf} is the Hubble constant (the expansion rate) during inflation. ϕ decreases in amplitude gradually after inflation, and begins to oscillate around the potential minimum when the Hubble constant H_{exp} of the Universe becomes comparable to the SUSY-breaking mass m_ϕ. At the beginning of the oscillation, the AD field has the value $|\phi_0| \simeq \sqrt{m_\phi|\langle H \rangle|^2/m_\nu}$, which, as shown below, is an effective initial value for Leptogenesis.

Let us consider now lepton-number generation in this scenario. The evolution of the AD field ϕ is described by

$$\frac{\partial^2 \phi}{\partial t^2} + 3H_{exp}\frac{\partial \phi}{\partial t} + \frac{\partial V}{\partial \phi^*} = 0, \qquad 111.$$

where $V = V_{SUSY} + \delta V$. Because the lepton number is given by

$$n_L = i\left(\frac{\partial \phi^*}{\partial t}\phi - \phi^*\frac{\partial \phi}{\partial t}\right), \qquad 112.$$

the evolution of n_L is given by

$$\frac{\partial n_L}{\partial t} + 3H_{exp}n_L = \frac{m_{SUSY}m_\nu}{2|\langle H \rangle|^2}\text{Im}(a_m^* \phi^{*4}). \qquad 113.$$

The motion of ϕ in the phase direction generates the lepton number. This is predominantly created just after the AD field ϕ starts its coherent oscillation, at a time $t_{osc} \simeq 1/H_{osc} \simeq 1/m_\phi$, because the amplitude $|\phi|$ damps as t^{-1} during the oscillation. Thus, we obtain for the lepton number

$$n_L \simeq \frac{m_{SUSY}m_\nu}{2|\langle H \rangle|^2}\delta_{eff}|a_m\phi_0^4| \times t_{osc}, \qquad 114.$$

where $\delta_{eff} = \sin(4\text{arg}\phi + \text{arg}a_m)$ represents an effective CP-violating phase. Using $m_{SUSY} \simeq m_\phi$, $|\phi_0| \simeq \sqrt{m_\phi|\langle H \rangle|^2/m_\nu}$ and $t_{osc} \simeq 1/m_\phi$, we find

$$n_L \simeq \delta_{eff}m_\phi^2\frac{|\langle H \rangle|^2}{2m_\nu}. \qquad 115.$$

After the end of inflation, the inflaton begins to oscillate around the potential minimum and n_L/ρ_{inf} stays constant until the inflaton decays. Here ρ_{inf} is the energy density of the inflaton. The inflaton decay reheats the Universe producing entropy s. Because $\rho/s = 2T_R/4$, we find for the lepton-number asymmetry the expression

$$\frac{n_L}{s} \simeq \left(\frac{\rho_{inf}}{s}\right)\left(\frac{n_L}{\rho_{inf}}\right) \simeq \delta_{eff}\frac{3T_R}{4M_G}\frac{|\langle H \rangle|^2}{6m_\nu M_G}. \qquad 116.$$

Here we have used $\rho_{\text{inf}} \simeq 3m_\phi^2 M_G^2$ at the begining of the AD field oscillation (when most of the lepton number is generated). This lepton-number asymmetry is converted to a baryon-number asymmetry by the KRS mechanism. In this way one obtains for the baryon-number asymmetry

$$\frac{n_B}{s} \simeq \frac{1}{23} \frac{|\langle H \rangle|^2 T_R}{m_\nu M_G^2}. \qquad 117.$$

The observed ratio $n_B/s \simeq 0.9 \times 10^{-10}$ implies $m_\nu \simeq 10^{-9}$ eV for $T_R \simeq 10^6$ GeV. This small mass corresponds to the mass of the lightest neutrino. We should note that for such a low reheating temperature one may neglect the effects due to thermal mass for the AD field ϕ (136).

7. CONCLUSIONS AND SUMMARY OF RESULTS

In this article we have discussed the physical mechanism responsible for the origin of matter in the Universe. Both the rather large observed value for the ratio of baryons to photons, η_B, in the present epoch and the absence of antimatter are the consequences of a primordial asymmetry between matter and antimatter generated early on in the Universe. Although a variety of mechanisms have been proposed for producing this primordial asymmetry, in this review we have focused on Leptogenesis as the origin of matter. In our view, this is the most appealing scenario for the origin of matter, for at least three reasons:

1. Explicit lepton number violation is very natural once one includes right-handed neutrinos in the Standard Model. Furthermore, the lightness of the observed neutrinos strongly suggests, through the seesaw mechanism, the presence of superheavy neutrinos, whose decays can produce a lepton-antilepton asymmetry.

2. Quantum mechanically, through the KRS mechanism, one can automatically turn a leptonic asymmetry into a baryonic asymmetry. Indeed, because of the existence of these sphaleron processes, the origin of matter is linked to phenomena in the early Universe that result in the establishment of a (B − L)-asymmetry, like Leptogenesis.

3. If neutrino masses lie in the range 10^{-3} eV $< m_i < 0.1$ eV, as suggested by neutrino oscillation experiments, the leptonic asymmetry produced in thermal Leptogenesis is both independent of the abundance of heavy neutrinos and of any pre-existing asymmetry and has the right magnitude to yield the observed value for η_B.

Because, in the final analysis, η_B is just one number, it is important to ask if the particular mechanism proposed for the origin of matter has other consequences. Thus in this review we examined in some detail how Leptogenesis fit with ideas proposed to explain the dark matter that constitutes about 25% of the Universe's energy density.[24] We pointed out that axionic dark matter is perfectly compatible

---

[24]We did not try to examine models of dark energy in the light of Leptogenesis, because our understanding of dark energy is still in its infancy.

with Leptogenesis. Indeed, it is possible to very naturally link the scale of the heavy neutrinos with that of $U(1)_{PQ}$ breaking f_a, so that the ratio Ω_{DM}/Ω_B is independent of these large scales. The situation, however, is more complex in the case of supersymmetric dark matter.

To be effective, thermal Leptogenesis needs to occur at high temperatures, above $T = 2 \times 10^9$ GeV. This means that the Universe after inflation must have reheated to at least this temperature. However, in supersymmetric theories such a high reheating temperature is problematic as it leads to an overproduction of gravitinos. When they decay, gravitinos of such abundances completely alter the primordial abundance of light elements produced in Big Bang Nucleosynthesis. The gravitino problem, however, is not fatal, as there are a number of ways to mitigate the overproduction of gravitinos. Nevertheless, if Leptogenesis is at the root of the origin of matter, the supersymmetric spectrum at low energies and the nature of the LSP are quite constrained. Thus, in a sense, Leptogenesis is also quite predictive in this context.

Although much of our review, very naturally, focused on thermal Leptogenesis, we also discussed two examples where matter originated through a leptonic asymmetry produced in nonthermal processes. These models, although much more speculative, illustrate some of the possible other options for the origin of matter. Naturally, in this case some of the specific predictivity is lost.

ACKNOWLEDGMENTS

In our work on the topics discussed in this review we have benefitted from the insight of many colleagues. W.B. and T.Y. owe special thanks to M. Bolz, A. Brandenburg, P. Di Bari, K. Hamaguchi, M. Ibe, K. Izawa, M. Moroi, M. Plümacher, and M. Ratz. The work of T.Y. has been supported in part by a Humboldt Research Award. R.D.P.'s work was supported in part by the Department of Energy under Contract No. FG03-91ER40662, Task C.

The *Annual Review of Nuclear and Particle Science* is online at http://nucl.annualreviews.org

LITERATURE CITED

1. Seljak U, et al. *Phys. Rev. D* 71:103515 (2005)
2. Guth AH. *Phys. Rev. D* 23:347 (1981); Linde AD. *Phys. Lett.* B129:177 (1982); Albrecht A, Steinhardt PJ. *Phys. Rev. Lett.* 48:1220 (1982)
3. Tonry JL, et al. *Astrophys. J.* 594:1 (2003); Knop RA, et al. *Astrophys. J.* 598:102 (2003); Barris BJ, et al. *Astrophys. J.* 602:571 (2004); Riess AG, et al. *Astrophys. J.* 607:665 (2004); Perlmutter S, Schmidt B. In *Supernovas and Gamma Ray Bursts*, ed. K Weller. Berlin: Springer Verlag (2003)
4. WMAP Collaboration, Bennett CL, et al. *Astrophys. J.* Suppl. 148:1 (2003); Spergel DN, et al. *Astrophys. J.* Suppl. 148:175 (2003)
5. Kneller JP, Steigman G. *New J. Phys.* 6:117 (2004)

6. Olive KE, Skillman ED. *Astrophys. J.* 617:29 (2004)
7. Dolgov A, Silk J. *Phys. Rev. D* 47: 4244 (1993); Khlopov MY, Rubin MG, Sakharov AS. *Phys. Rev. D* 62:083505 (2000)
8. Cohen A, De Rujula A, Glashow S. *Astrophys. J.* 495:539 (1998)
9. Turner MS. In *Intersection Between Particle Physics and Cosmology, Jerusalem Winter School for Theoretical Physics*, ed. T Piran, S Weinberg, vol. 1, p. 99. Singapore: World Sci. (1986)
10. Affleck I, Dine M. *Nucl. Phys.* B249:361 (1985)
11. Sakharov AD. *JETP Lett.* 5:24 (1967)
12. Fukugita M, Yanagida T. *Phys. Lett* B174:45 (1986)
13. Kuzmin VA, Rubakov VA, Shaposhnikov MA. *Phys. Lett.* B155:36 (1985)
14. Yanagida T. *Proc. of the Workshop on "The Unified Theory and the Baryon Number in the Universe,"* Tsukuba, Japan, Feb. 13–14, p. 95, eds. O. Sawada and S. Sugamoto, (KEK Report KEK-79-18, 1979, Tsukuba) (1979) *Progr. Theor. Phys.* 64:1103 (1980); Ramond P. Talk given at the Sanibel Symposium, Palm Coast, Fla., Feb. 25-Mar. 2, preprint CALT-68-709 (1979)
15. Yoshimura M. *Phys. Rev. Lett.* 41:281 (1978); Yoshimura M. *Phys. Rev. Lett.* 42:746E (1979); Toussaint D, Treiman SB, Wilczek F, Zee A. *Phys. Rev. D* 19:1036 (1979); Weinberg S. *Phys. Rev. Lett.* 42:850 (1979); Dimopoulos S, Susskind L. *Phys. Rev. D* 18:4500 (1978).
16. Rubakov VA, Shaposhnikov MS. *Phys. Usp.* 39:461 (1996)
17. t'Hooft G. *Phys. Rev. Lett.* 37:8 (1976); t'Hooft G. *Phys. Rev. D* 14:4332 (1976)
18. Klinkhammer FR, Manton NS. *Phys. Rev. D* 30:2212 (1984)
19. Arnold P, McLerran L. *Phys. Rev. D* 36:581 (1987)
20. Arnold P, Son D, Yaffe L. *Phys. Rev. D* 55:6264 (1997); Arnold P, Son D, Yaffe L. *Phys. Rev. D* 59:105020 (1999)
21. Bödeker D. *Phys. Lett.* B426:351 (1998); Bödeker D. *Nucl. Phys.* B559:502 (1999)
22. Arnold P, Yaffe L. *Phys. Rev. D* 62: 125014 (2000)
23. Bödeker D, Moore GD, Rummukainen K. *Phys. Rev. D* 61:056003 (2000)
24. Shaposhnikov ME. *JETP Lett.* 44:465 (1986)
25. Jansen K. *Nucl. Phys. (Proc. Supp.)* B47:196 (1996)
26. Particle Data Group Eidelman S, et al. *Phys. Lett* B592:1 (2004)
27. Prokopec T, Kainulainen K, Schmidt MG, Weinstock S. *Strong and Electoweak Matter 2002*, ed. M.G. Schmidt. Singapore: World Sci. (2003); Carena M, Megevand A, Quirós M, Wagner CEM. *Nucl. Phys.* B716:319 (2005)
28. Buchmüller W, Philipsen O. *Nucl. Phys.* B443:47 (1995)
29. Kajantie K, Laine M, Rummukainen K, Shaposhnikov M. *Phys. Rev. Lett.* 77:2887 (1996)
30. Bergerhoff B, Wetterich C. In *Current Topics in Astrofundamental Physics*, eds. N Sanchez, A Zichichi, p. 162. Singapore: World Sci. (1997)
31. Rummukainen K, Tsypin M, Kajantie K, Laine M, Shaposhnikov M. *Nucl. Phys.* B532:283 (1998)
32. Buchmüller W, Philipsen O. *Phys. Lett.* B397:112 (1997)
33. Fodor Z. *Nucl. Phys. (Proc. Supp.)* B83–84:121 (2000)
34. de Carlos B, Espinosa JR. *Nucl. Phys.* B503:24 (1997)
35. Harvey JA, Turner MS. *Phys. Rev. D* 42:3344 (1990)
36. Mohapatra RN, Zhang X. *Phys. Rev. D* 45:2699 (1992)
37. Khlebnikov SY, Shaposhnikov ME. *Nucl. Phys.* B308:885 (1988)
38. Laine M, Shaposhnikov ME. *Phys. Rev. D* 61:117302 (2000)
39. Dick K, Lindner M, Ratz M, Wright D. *Phys. Rev. Lett.* 84:4039 (2000)

40. Bilenky SM. Report at the Nobel Symposium on Neutrino Physics, Haga Slott, Enkoping, Sweden, August 19–24, 2004, hep-ph/0410090, *Phys. Scripta* (In press)

41. CHOOZ Collaboration, Apollonio M, et al. *Phys. Lett* B420:397 (1998); CHOOZ Collaboration, Apollonio M, et al. *Phys. Lett* B466:415 (1999); Palo Verde Collaboration, Boehm F, et al. *Phys. Rev. D* 64:112001 (2001)

42. LSND Collaboration, Aguilar A, et al. *Phys. Rev. D* 64:112007 (2001)

43. KARMEN Collaboration, Armbruster B, et al. *Phys. Rev. D* 65:112001 (2002)

44. Cirelli M, Marandella G, Strumia A, Vissani F. *Nucl. Phys.* B708:215 (2005)

45. Barenboim G, Beacom JF, Borissov L, Kayser B. *Phys. Lett* B537:227 (2002)

46. Kayser B. In *Neutrino Mass*, eds. G Altarelli, K Winter, p. 1. Berlin: Springer Tracts in Modern Physics (2003)

47. Hannestad S, Raffelt G. *J. Cosmol. Astropart. Phys.* 0404:008 (2004)

48. Ichikawa K, Fukugita M, Kawasaki M. *Phys. Rev. D* 71:043001 (2005)

49. Tegmark M, Rees M. *Astrophys. J. B* 499:526 (1998)

50. Arkani-Hamed M, Dimopoulos S, Dvali G. *Phys. Lett* B429:263 (1998)

51. Dienes RK, Dudas E, Gherghetta T. *Nucl. Phys.* B557:25 (1999); Arkani-Hamed N, Dimopoulos S, Dvali G, March-Russell J. *Phys. Rev. D* 65:024032 (2002); Grossman Y, Neubert M. *Phys. Lett* B474:361 (2000); Gherghetta T. *Phys Rev. Lett.* 92:161601 (2004)

52. Baumann J, Gahler R, Kalus J, Mampe W. *Phys. Rev. D* 37:3107 (1988)

53. Dylla HF, King JG. *Phys. Rev. A* 7:1224 (1973)

54. Bilenky SM, Hosek J, Petcov ST. *Phys. Lett* B94:495 (1980); Schechter J, Valle JWF. *Phys. Rev. D* 22:2227 (1980)

55. Branco GC, Rebelo MN. *New J. Phys.* 7:86 (2005)

56. Branco GC, Felipe RG, Joaquim FR, Yanagida T. *Phys. Lett* B562:265 (2003)

57. Fukugita M, Yanagida T. See Ref. 12; Flanz M. Paschos EA, Sarkar U. *Phys. Lett.* B345:248 (1995); Covi L, Roulet E, Vissani F. *Phys. Lett.* B384:169 (1996); Buchmüller W, Plumacher M. *Phys. Lett.* B431:354 (1998)

58. Pilaftsis A. *Int. J. Mod. Phys.* A14:1811 (1999)

59. Buchmüller W, Fredenhagen S. *Phys. Lett* B483:217 (2000)

60. Buchmüller W, Plümacher M. *Phys. Lett* B389:73 (1996)

61. Buchmüller W, Yanagida T. *Phys. Lett.* B445:399 (1999)

62. Gherghetta T, Jungman G. *Phys. Rev. D* 48:1546 (1993)

63. Mohapatra RN, Nasri S, Yu H. *Phys. Lett.* B615:231 (2005); Xing Z-Z. *Phys. Rev. D* 70:071302 (2004); Cosme N. *JHEP* 0408:027 (2004); Rodejohann W, *Eur. Phys. J.* C32:235 (2004); Grimus W, Lavoura L. *J. Phys.* G30:1073 (2004); Barger V, Dicus DA, He H-J, Li T. *Phys. Lett* B583:173 (2004); Chankowski PH, Turzyński K. *Phys. Lett* B570:198 (2003); Velasco-Sevilla L, *JHEP* 0310:035 (2003); Akhmedov EKh, Frigerio M, Smirnov A Yu. *JHEP* 0309:021 (2003); Branco GC, González Felipe R. Joaquim FR, Masina I, Rebelo MN, Savoy CA. *Phys. Rev. D* 67:07025 (2003)

64. Kolb EW, Turner MS. *The Early Universe*. New York: Addison Wesley (1990)

65. Luty MA. *Phys. Rev. D* 45:455 (1992)

66. Plümacher M, *Z. Phys.* C74:549 (1997)

67. Buchmüller W, Di Bari P, Plümacher M. *Nucl. Phys.* B643:367 (2002)

68. Barbieri R, Creminelli P, Strumia A, Tetradis N. *Nucl. Phys.* B575:61 (2000)

69. Buchmüller W, Di Bari P, Plümacher M. *Ann. Phys.* 315:303 (2005)

70. Kolb EW, Wolfram S. *Nucl. Phys.* B172:224 (1980); Kolb EW, Wolfram S. *Nucl. Phys.* B195:542(E) (1982)

71. Pilaftsis A, Underwood TEJ. *Nucl. Phys.* B692:303 (2004)

72. Giudice GF, Notari A, Raidal M, Riotto A, Strumia A. *Nucl. Phys.* B685:89 (2004)

73. Covi L, Rius N, Roulet E, Vissani F. *Phys. Rev. D* 57:93 (1998)

74. Buchmüller W, Di Bari P, Plümacher M. *Phys. Lett.* B547:128 (2002)

75. Davidson S, Ibarra A. *Phys. Lett.* B535:25 (2002)

76. Hamaguchi K, Murayama H, Yanagida T. *Phys. Rev. D* 65:043512 (2002)

77. Buchmüller W, Di Bari P, Plümacher M. *Nucl. Phys.* B665:445 (2003)

78. Hambye T, Lin Y, Notari A, Papucci M, Strumia A. *Nucl. Phys.* B695:169 (2004)

79. Plümacher M. *Nucl. Phys.* B530:207 (1998)

80. Di Bari P. hep-ph/0406115

81. Antusch S, Kersten J, Lindner M, Ratz M. *Nucl. Phys.* B674:401 (2003)

82. Buchmüller W, Plümacher M. *Phys. Lett.* B511:74 (2001)

83. Lazarides G, Shafi Q, Wetterich C. *Nucl. Phys.* B181:287 (1981); Mohapatra RN, Senjanović G. *Phys. Rev. D* 23:165 (1981); Wetterich C. *Nucl. Phys.* B187:343 (1981)

84. Hambye T, Senjanović G. *Nucl. Phys.* B582:73 (2004); Rodejohann W. *Phys. Rev. D* 70:073010 (2004); Gu P-H, Bi X-J. *Phys. Rev. D* 70:063511 (2004); D'Ambrosio G, Hambye T, Hektor A, Raidal M, Rossi A. *Phys. Lett* B604:199 (2004)

85. Antusch S, King SF. *Phys. Lett.* B597:199 (2004)

86. Flanz M, et al. and Covi L, et al. See Ref. (57)

87. Dar S, Huber S, Senoguz VN, Shafi Q. *Phys. Rev D* 69:077701 (2004); Albright CH, Barr SM. *Phys. Rev. D* 70:033013 (2004); Pilaftsis A. hep-ph/0408103

88. Grossman Y, Kashti T, Nir Y, Roulet E. *Phys. Rev. Lett.* 91:251801 (2003); D'Ambrosio G, Giudice GF, Raidal M. *Phys. Lett.* B575:75 (2003); Hambye T, March-Russell J, West SM, *JHEP* 0407:070 (2004); Boubekeur L, Hambye T, Senjanović G. *Phys. Rev. Lett.* 93:111601 (2004); Grossman Y, Kashti T, Nir Y, Roulet E. *JHEP* 0411:080 (2004);

Chen MC, Mahanthappa KT. *Phys. Rev. D* 70:113013 (2004); Allahverdi R, Drees M. *Phys. Rev. D* 79:123522 (2004)

89. Frampton PH, Glashow SL, Yanagida T, *Phys. Lett.* B548:119 (2002); Endoh T, et al. *Phys. Rev. Lett.* 89:231601 (2002)

90. Kawasaki M, Kohri K, Moroi T. *Phys. Rev. D* 71:083502 (2005)

91. Harris PG, et al. *Phys. Rev. Lett.* 82:904 (1999)

92. Baluni V. *Phys. Rev. D* 19:2227 (1979); Crewther RJ, di Vecchia P, Veneziano G, Witten E. *Phys. Lett.* B88:123 (1979); Crewther RJ, di Vecchia P, Veneziano G, Witten E. *Phys. Lett.* B91:487(E) (1980)

93. Peccei RD. In *CP Violation*, ed. C Jarlskog, p. 503. Singapore: World Sci. (1989)

94. Peccei RD, Quinn HR. *Phys. Rev. Lett.* 38:1440 (1977); Peccei RD, Quinn HR. *Phys. Rev. D* 16:1791 (1977)

95. Weinberg S. *Phys. Rev. Lett.* 40:223 (1978); Wilczek F. *Phys. Rev. Lett.* 40:271 (1978)

96. Raffelt G. *Stars as Laboratories for Fundamental Physics*. Chicago: University of Chicago Press (1996)

97. Preskill J, Wise M, Wilczek F. *Phys. Lett.* B120:127 (1983); Abbott L, Sikivie P. *Phys. Lett.* B120:133 (1983); Dine M, Fischler W. *Phys. Lett.* B120:137 (1983)

98. Sikivie P, *Nucl. Phys. Proc. Suppl.* B87:41 (2000)

99. Olive KA. In *Proc. 5th Int. Heidelberg Conf. on Dark Matter in Astro and Particle Physics*. London: Springer-Verlag (In press). hep-ph/0412054

100. Coleman S. *Nucl. Phys.* B262:263 (1985)

101. Dine M, Kusenko A. *Rev. Mod. Phys.* 76:1 (2004)

102. Kasuya S, Kawasaki M, Takahashi F. *Phys. Rev. D* 68:023501 (2003)

103. Fujii M, Yanagida T. *Phys. Lett.* B542:80 (2002)

104. Arafune J, et al. *Phys. Rev. D* 62:105013 (2000)

105. Langacker P, Peccei RD, Yanagida T. *Mod. Phys. Lett.* A9:541 (1986)

106. Dine M, Fischler W, Srednicki M. *Phys. Lett.* B104:199 (1981); Zhitnitski A. *Sov. Jour. Nucl. Phys.* 31:260 (1980)

107. Kim JE. *Phys. Rev. Lett.* 43:103 (1979); Shifman MA, Vainshtein AI, Zakharov VI. *Nucl. Phys.* B166:493 (1980)

108. Khlopov ML, Linde AD. *Phys. Lett.* B138:265 (1984); Ellis JR, Kim JE, Nanopoulos DV. *Phys. Lett.* B145:181 (1984)

109. Bolz M, Brandenburg A, Buchmüller W. *Nucl. Phys.* B606:518 (2001)

110. Folomkin IV, et al. *Nuovo Cim.* A79:193 (1984) [*Yad. Fiz.* 39:990 (1984)]

111. Pilafsis A. *Phys. Rev. D* 56:5431 (1997); Ellis J, Raidal M, Yanagida T. *Phys. Lett.* B546:228 (2002)

112. Bolz M, Buchmüller W, Plümacher M. *Phys. Lett.* B443:209 (1998)

113. Fujii M, Ibe M, Yanagida T. *Phys. Lett.* B579:6 (2004); Ellis JR, Olive KA, Santoso Y, Spanos VC. *Phys. Lett.* B588:7 (2004); Feng JL, Su S, Takayama F. *Phys. Rev. D* 70:075019 (2004)

114. Roszkowski L, Ruiz de Austri R. hep-ph/0408227

115. Feng JL, Rajaraman A, Takayama F. *Phys. Rev. Lett.* 91:011302-1 (2003)

116. Buchmüller W, Hamaguchi K, Ratz M. *Phys. Lett.* B574:156 (2003)

117. Viel M, et al. *Phys. Rev. D* 71:063534 (2005)

118. Fujii M, Yanagida T. *Phys. Lett.* B549:273 (2002); Fujii M, Ibe M, Yanagida T. *Phys. Lett.* B579:6 (2004)

119. Asaka T, Yanagida T. *Phys. Lett.* B494:297 (2000)

120. Gherghetta T, Giudice GF, Wells JD. *Nucl. Phys.* B559:27 (1999)

121. Ibe M, Kitano R, Murayama H, Yanagida T. *Phys. Rev. D* 70:075012 (2004)

122. Borzumati F, Masiero A. *Phys. Rev. Lett.* 57:961 (1986); Hisano J, Moroi T, Tobe K, Yamaguchi M, Yanagida T. *Phys. Lett.* B357:579 (1995)

123. Davidson S, Ibarra A. *JHEP* 0109:013 (2001)

124. Masina I, Savoy CA. hep-ph/0410382

125. Lazarides G, Shafi Q. *Phys. Lett.* B258:305 (1991); Lazarides G, Panagiotakopoulos C, Shafi Q. *Phys. Lett.* B315:325 (1993)

126. Kumekawa K, Moroi T, Yanagida T. *Progr. Theor. Phys.* 92:437 (1994); Asaka T, Hamaguchi K, Kawasaki M, Yanagida T. *Phys. Lett.* B464:12 (1999); Asaka T, Hamaguchi K, Kawasaki M, Yanagida T. *Phys. Rev. D* 61:083512 (2000); Giudice GF, Peloso M, Riotto A, Tkachev I. *J. High Energy Phys.* 08:014 (1999)

127. Murayama H, Suzuki H, Yanagida T, Yokoyama J. *Phys. Rev. Lett.* 70:1912 (1993); Ellis JR, Raidal M, Yanagida T. *Phys. Lett.* B581:9 (2004)

128. Murayama H, Yanagida T. *Phys. Lett.* B322:349 (1994); Dine M, Randall L, Thomas S. *Nucl. Phys.* B458:291 (1996)

129. Hamaguchi K, Murayama J, Yanagida T, *Phys. Rev. D* 65:043512 (2002)

130. Guth AH. *Phys. Rev. D* 23:347 (1981)

131. Guth AH, Pi S-Y. *Phys. Rev. Lett.* 49:1110 (1982); Hawking S. *Phys. Lett.* B115:295 (1982); Starobinsky AA. *Phys. Lett.* B117:175 (1982)

132. Covi L, et al. See Ref. (57).

133. Linde A. *Phys. Lett.* B129:177 (1983); Linde A. *Phys. Scripta* T117:40 (2005)

134. Goncharov AS, Linde AD. *Phys. Lett.* B139:27 (1984); Murayama H, Suzuki H, Yanagida T, Yokoyama J. *Phys. Rev. D* 50:R2356 (1994)

135. Kusenko A, Shaposhnikov M. *Phys. Lett.* B418:46 (1998)

136. Asaka T, Fujii M, Hamaguchi K, Yanagida T. *Phys. Rev. D* 62:123514 (2000); Fujii M, Hamaguchi K, Yanagida T. *Phys. Rev. D* 63:123513 (2001)

Annu. Rev. Nucl. Part. Sci. 2005. 55:357–402
doi: 10.1146/annurev.nucl.55.090704.151533

FEMTOSCOPY IN RELATIVISTIC HEAVY ION COLLISIONS: Two Decades of Progress

Michael Annan Lisa

Department of Physics, The Ohio State University, Columbus, Ohio 43210; email: lisa@mps.ohio-state.edu

Scott Pratt

Department of Physics and Astronomy, Michigan State University, East Lansing, Michigan 48824; email: pratts@pa.msu.edu

Ron Soltz

N-Division, Livermore National Laboratory, 7000 East Avenue, Livermore, California 94550; email: soltz1@llnl.gov

Urs Wiedemann

Theory Division, CERN, Geneva, Switzerland; email: urs.wiedemann@cern.ch

Key Words   HBT, intensity interferometry, heavy ion collisions, femtoscopy

■ **Abstract**   Analyses of two-particle correlations have provided the chief means for determining spatio-temporal characteristics of relativistic heavy ion collisions. We discuss the theoretical formalism behind these studies and the experimental methods used in carrying them out. Recent results from RHIC are put into context in a systematic review of correlation measurements performed over the past two decades. The current understanding of these results is discussed in terms of model comparisons and overall trends.

CONTENTS

0163-8998/05/1208-0357$20.00

357

1. INTRODUCTION

The study of nucleus-nucleus collisions at ultra-relativistic energies aims to characterize the dynamical processes by which matter at extreme densities is produced and the fundamental properties that this matter exhibits. In nucleus-nucleus collisions, how do partonic equilibration processes proceed? For how long, over which spatial extension, and at which density is a QCD equilibration state approached, and what are its properties? Particle densities attained during a heavy ion collision are expected to exceed significantly the inverse volume of a hadron. This implies that the high temperature phase of QCD, the Quark Gluon Plasma, comes within experimental reach. Chiral symmetry restoration and deconfinement phase transition are testable in heavy ion collisions. However, the experimental study of QCD at high temperatures and densities is complicated by the short lifetime and mesoscopic extension of the produced system. Femtoscopy, the spatio-temporal characterization of the collision region on the femtometer scale, is needed to frame any discussion of dynamical equilibration processes.

The Relativistic Heavy Ion Collider (RHIC) just completed the first part of a dedicated experimental heavy ion program. Center of mass energies ($\sqrt{s_{\text{NN}}} = 200$ GeV) exceeded those of previous fixed target experiments by a factor 10. The current discussion of RHIC data focuses mainly on several qualitatively novel phenomena that all support the picture that dense and rapidly equilibrating QCD matter is produced in the collision region (1–4). In particular, identified single inclusive hadron spectra appear to emerge from a common flow field whose size and dependence on transverse momentum and azimuth is consistent with expectations that the produced matter is a locally equilibrated, almost ideal fluid of very small viscosity. Moreover, high-p_T hadron spectra show a centrality dependent, strong suppression in Au+Au collisions, but not in a d+Au control experiment, indicating that even the hardest partons produced in the collision participate significantly in equilibration processes. In short, experiments at RHIC have demonstrated already that heavy ion collisions produce dense and equilibrating matter, and that controlled experimentation of this matter is possible using a large variety of probes.

Despite these successes, numerous questions remain concerning the state of the matter produced in these collisions. Most notably, the equation of state is far from being determined, and issues concerning chiral symmetry restoration are largely unresolved. Addressing these fundamental questions about bulk matter requires a detailed understanding of the dynamics and chemistry of the collision, which can only be acquired by thorough and coordinated analyses of data and theory. In particular, spatio-temporal aspects of the reaction need to be experimentally addressed. The small size, $\sim 10^{-14}$ m, and transient nature, $\sim 10^{-22}$ seconds, of the reactions preclude direct measurement of times or positions. Instead, femtoscopy must exploit measurements of asymptotic momenta. Correlations of two final-state particles at small relative momentum provide the most direct link to the size and lifetime of subatomic reactions (5–12). Because correlations from either interactions or from identity interference are stronger for smaller separations in space-time, spatio-temporal information can be most easily extracted for small sources, unlike the limitations of microscopes and telescopes.

The interference of two particles emitted from chaotic sources was first applied by Hanbury-Brown and Twiss (13, 14), where photons were exploited to determine source sizes for both laboratory and stellar sources in the 1950s and 1960s. Correlations of identical pions were shown to be sensitive to source dimensions in proton-antiproton collisions by Goldhaber, Goldhaber, Lee, and Pais in 1960 (15). In the 1970s, these methods were refined by Kopylov and Podgoretsky (16–19), Koonin (20), and Gyulassy (21), and other classes of correlations were shown to be useful for source-size measurements, such as strong and Coulomb interactions. Bevalac analyses showed that interferometry was truly capable of quantitatively determining spatial and temporal source dimensions (22, 23) and providing a stringent test of dynamical models (24). Throughout the last 25 years this phenomenology has developed into a precision tool for heavy ion collisions. Whereas hadronic sources are short lived and one measures correlations of the momentum of outgoing particles, stars are long lived and require experimental filters to enforce the approximate simultaneity of the two photons. Although the theory for these two classes of measurement are very different (25, 26), the heavy-ion community often uses the term *HBT*, in reference to Hanbury-Brown and Twiss's original work with photons, to refer to any type of analysis related to size and shape that uses two particles at small relative momentum. To some, however, the term HBT refers only to identical-particle interferometry. Following Lednicky, we will employ the term *femtoscopy* (27, 28) to denote any measurement that provides spatio-temporal information, including coalescence analyses.

Femtoscopic measurements from truly relativistic heavy ion collisions were first reported almost twenty years ago. Since then, measurements have been performed for collisions at energies spanning two orders of magnitude. In a double sense, then, this review examines recent RHIC results within the larger context of two decades' worth of femtoscopy. The theory and phenomenology of correlation femtoscopy are reviewed in the next section, with particular emphasis on describing the approximations used to derive the connection between spatio-temporal aspects of the emission function and correlations constructed from final-state momenta. Experimental methods and techniques are correspondingly reviewed in Section 3. Section 4 presents experimental results, with an emphasis on describing the systematics of source dimensions and lifetimes as a function of beam energy, system size, particle species and a particle's momentum. In addition to source dimensions, results for phase space density and entropy are presented. Comparisons of experimental results and transport models are presented in Section 5, with an emphasis on explaining the "HBT puzzle," i.e., the fact that dynamic descriptions that incorporate a phase transition to a new state of matter with many degrees of freedom significantly over-predict observed source sizes. Section 6 summarizes the current state of the field and lists new directions and challenges for future theoretical and experimental analyses.

2. THEORY AND PHENOMENOLOGY BASICS

2.1. Formalism

Two-particle correlation functions are constructed as the ratio of the measured two-particle inclusive and single-particle inclusive spectra,

$$C^{ab}(\mathbf{P}, \mathbf{q}) = \frac{dN^{ab}/(d^3 p_a d^3 p_b)}{(dN^a/d^3 p_a)(dN^b/d^3 p_b)},$$

$$P \equiv p_a + p_b, \quad q^\mu = \frac{(p_a - p_b)^\mu}{2} - \frac{(p_a - p_b) \cdot P}{2P^2} P^\mu. \qquad 1.$$

The theoretical analysis of (1) aims at relating this experimentally measured correlation to the space-time structure of the particle emitting source (6–8, 11). Two forms are common for connecting the measured correlation function to the space-time emission function $s(p, x)$ through a convolution with the wave function ϕ. In the first form (29),

$$C^{ab}(\mathbf{P}, \mathbf{q}) = \frac{\int d^4 x_a d^4 x_b s_a(p_a, x_a) s_b(p_b, x_b) |\phi(\mathbf{q}', \mathbf{r}')|^2}{\int d^4 x_a s_a(p_a, x_a) \int d^4 x_b s_b(p_b, x_b)}. \qquad 2.$$

In calculations of two-particle correlation functions, the squared relative two-particle wave function $|\phi|^2$ serves generally as a weight, and the emission function $s(p, x)$ contains all space-time information about the source because it describes the probability of emitting a particle with momentum p from a space-time point x. Here, and throughout this section, primes denote quantities in the center-of-mass frame, i.e., the frame where $\mathbf{P} = 0$. The source function s_a is evaluated at the momentum $\bar{\mathbf{p}}_a = m_a \mathbf{P}/(m_a + m_b)$, $\bar{p}_a^0 = E_a(\bar{\mathbf{p}}_a)$.

The second form, which is equally justified as Equation 2 by the approximations described further below, is,

$$C^{ab}(\mathbf{P}, \mathbf{q}) - 1 = \int d^3 r' \mathcal{S}_{\mathbf{P}}(\mathbf{r}') \left[|\phi(\mathbf{q}', \mathbf{r}')|^2 - 1 \right],$$

$$\mathcal{S}_{\mathbf{P}}(\mathbf{r}') \equiv \frac{\int d^4 x_a d^4 x_b s_a(\bar{p}_a, x_a) s_b(\bar{p}_b, x_b) \delta \left(\mathbf{r}' - \mathbf{x}'_a + \mathbf{x}'_b \right)}{\int d^4 x_a d^4 x_b s_a(\bar{p}_a; x_a) s_b(\bar{p}_b, x_b)}. \qquad 3.$$

This expression allows one to consider $|\phi|^2$ as a kernel with which one can transform from the coordinate-space basis to the relative-momentum basis. It also emphasizes the limitation of correlation functions, that they can provide, at best, the function $\mathcal{S}_{\mathbf{P}}(\mathbf{r}')$, the distribution of relative positions of particles with identical velocities and total momentum $\mathbf{P}$ as they move in their asymptotic state. Thus, correlations do not measure the size of the entire source. Instead, they address the dimensions of the "region of homogeneity," a term coined by Sinyukov (30), i.e., the size and shape of the phase space cloud of outgoing particles whose velocities have a specific magnitude and direction. If the collective expansion of the produced matter is strong, as is the case in central collisions, then the region of homogeneity is significantly smaller than the entire source volume. In the following, we discuss the assumptions on which Equations 2 and 3 are based.

We start from explicit expressions for the one- and two-particle spectra in terms of T-matrix elements. For one-particle emission,

$$E \frac{dN}{d^3 p} = \int d^4 x s(p, x) = \sum_{F'} \left| \int d^4 x T_{F'}(x) e^{-ip \cdot x} \right|^2, \qquad 4.$$

$$s(p, x) = \sum_{F'} \int d^4 \delta x T_{F'}^*(x + \delta x / 2) T_{F'}(x - \delta x / 2) e^{ip \cdot \delta x}. \qquad 5.$$

Here, F' refers to the state of all other particles in the system. All interactions with the residual system are incorporated into the T matrix. However, there is a choice as to whether mean-field interactions are included in the T matrix or are instead incorporated into the evolution matrix (31–35). For example, one can include the Coulomb interactions with the residual system by replacing the phase factor $e^{ip \cdot x}$ in Equation 4 with an outgoing Coulomb wave function. This can be quantitatively important, in particular for slow particles. It becomes difficult when the two particles interact with one another through the strong or Coulomb force, as this represents a quantum three-body problem.

Assumption 1: Higher order (anti)symmetrization can be neglected. Equation 4 implies that all particles with asymptotic momentum p must have had their last interaction with the source at some point x. For distinguishable particles, this is indeed the case and Equation 4 does not represent an assumption. However, if there are $N_a > 1$ particles of the same type a, then one must consider $T_a(x_1 \cdots x_{N_a})$. The evolution matrix is then no longer a simple phase factor but includes $N_a!$ interference terms. The single-particle probability is then obtained by integrating over the other $N_a - 1$ momenta. This can be accomplished explicitly for simple source functions (36–40). The distortion to the single-particle spectra and to the two-particle correlation function were found to be important when the phase space density approached unity. Otherwise, Equation 4 is well justified.

Assumption 2: The emission process is initially uncorrelated. In writing Equation 4, one requires that two-particle matrix elements factorize, $T_{F''}(x_a, x_b) = T_{F'_a}(x_a)T_{F'_b}(x_b)$, i.e., that the emission is independent. If multi-particle symmetrization can be neglected, the two-particle evolution operator factorizes into a center-of-mass and a relative operator. One has $U(x_a, x_b; p_a, p_b) = u_{q'}(x'_a - x'_b)e^{iP \cdot (E'_a x_a/M_{inv} + E'_b x_b/M_{inv})}$ for non-identical particles, whereas for identical particles $U = e^{iP \cdot (x_a + x_b)/2}(u_{q'}(x'_a - x'_b) \pm (u_{q'}(x'_b - x'_a))/\sqrt{2}$. This is illustrated in Figure 1 (see color insert). Then, the two-particle probability can be expressed in terms of one-particle source functions,

$$E_a E_b \frac{dN_{ab}}{d^3 p_a d^3 p_b} = \int d^4 x_a d^4 x_b d^4 \tilde{q} s_a((E'_a/M_{inv})P + \tilde{q}, x_a)s_b((E'_b/M_{inv})P$$
$$- \tilde{q}, x_b)d^4 \delta r' e^{i\tilde{q} \cdot \delta r'} u^*_{q'}(x'_a - x'_b + \delta r'/2)u_{q'}(x'_a - x'_b - \delta r'/2).$$
$$6.$$

Assumption 3: Smoothness approximation (41–43). Equation 6 is difficult to evaluate as it requires knowledge of the source function evaluated off-shell. For the special case where the particles do not interact aside from identical particle interference, $u_q(x_a - x_b) = [e^{iq \cdot (x_a - x_b)} \pm e^{iq \cdot (x_b - x_a)}]/\sqrt{2}$, the integrals over $\tilde{q}$ and $\delta r'$ can be performed analytically,

$$E_a E_b \frac{dN_{ab}}{d^3 p_a d^3 p_b} = \int d^4 x_a d^4 x_b \{s(p_a, x_a)s(p_b, x_b)$$
$$\pm s(P/2, x_a)s(P/2, x_b)\cos[(p_a - p_b) \cdot (x_a - x_b)]\}.$$
$$7.$$

The source functions in the interference term are evaluated off-shell for non-zero q, $P_0/2 \neq E(\mathbf{P}/2)$. The smoothness approximation replaces $s(P/2, x_a)s(P/2, x_b)$ with either $s(E(\mathbf{P}/2), \mathbf{P}/2, x_a)s(E(\mathbf{P}/2), \mathbf{P}/2, x_b)$, which leads to Equation 3, or with $s(p_a, x_a)s(p_b, x_b)$, which leads to Equation 2. If the first approximation is performed, one should also make the same approximation for the denominator. The smoothness approximation has been checked for expanding thermal sources, where it was found to be very reasonable for large (RHIC-like) sources, but quite questionable for smaller sources such as those found in pp or e^+e^- collisions (43).

Assumption 4: Equal time approximation. For the general case where the evolution operator incorporates Coulomb or strong interactions, deriving Equations 3 and 2 from Equation 6 is more complicated. First, the smoothness assumption amounts here to neglecting the $\tilde{q}$ dependence in the product of the source functions in Equation 6. This assumption is somewhat more stringent in the presence of final state interactions because the relevant range of $\tilde{q}$ extends beyond q. With this assumption, one obtains a δ-function constraint for $\delta r'$, and the integrand of (6) is proportional to the squared evolution matrix $|u'_q(x'_a - x'_b)|^2$. This evolution matrix has non-zero time components, which must be neglected if one is to identify it with the relative wave function. Because the relative motion in the pair rest frame is small, one expects this approximation to be reasonable, but it has not yet been tested in model studies.

The above formalism is semi-classical in the sense that a quantum-mechanical particle emission probability, defined by the T-matrix elements, is usually approximated by classical source functions. As a consequence, quantum uncertainty limits the applicability of Equations 3 and 2. To illustrate this limitation, source functions have been evaluated by convoluting the emission function with wave packets (44–47) of spatial width σ. This leads to a broadening of spatial distributions by $(\Delta R)^2 \sim 0.5 - 1.0 \, \text{fm}^2$. Because quantum smearing may already be incorporated into some of the semi-classical treatments through the choice of the initial density distribution, these calculations should be regarded as indicative of the theoretical uncertainty. Because this error affects the size in quadrature, it is negligible for large sources, but might be significant for sources near 1 fm in size with strong space-time correlations (42, 43, 45, 46, 48). In particular, analyses of $\pi\pi$ correlations from e^+e^- collisions, which result in source sizes of less than 1.0 fm (49, 50), are difficult to interpret in the above formalism.

2.2. Identical-Particle Interference

In the absence of strong and electromagnetic final state interactions, the wave function of an identical particle pair in Equation 3 becomes

$$|\phi(\mathbf{q}', \mathbf{r}')|^2 - 1 = \pm \cos(2\mathbf{q}' \cdot \mathbf{r}'). \qquad 8.$$

The distribution of separations in coordinate space $\mathcal{S}_\mathbf{P}(\mathbf{r}')$ can then be determined by performing a three-dimensional Fourier transform of $C(\mathbf{q}') - 1$. It is instructive to consider the properties of this inversion in more detail. The curvature of $C(\mathbf{q})$ at $\mathbf{q} = 0$ can be related to the mean-square separation of the three-dimensional shape of $\mathcal{S}_\mathbf{P}(\mathbf{r})$ (we neglect the $\mathbf{P}$ labels in C and $\mathcal{S}$),

$$-\frac{d^2 C(\mathbf{q})}{dq_i' dq_j'} \bigg|_{q=0} = \langle r_i r_j \rangle = \int d^3 r \, \mathcal{S}(\mathbf{r}) r_i r_j. \qquad 9.$$

This relation has been useful to qualitatively illustrate the relation between specific space-time information and specific features of the correlator. However, applying the identity quantitatively requires careful consideration of pions from longer-lived resonances which can dominate the calculation of $\langle r^2 \rangle$ if not accounted for.

2.3. Correlations from Coulomb and Strong Interactions

Compared to the case for non-interacting identical particles, where the transformation between $\mathcal{S}_\mathbf{P}(\mathbf{r})$ and $C(\mathbf{P}, \mathbf{q})$ is a Fourier transform, analyzing the experimentally measured correlation function with Equation 3 to determine the source function is more complicated. Understanding the resolving power of the kernel $|\phi(\mathbf{q}, \mathbf{r})|^2$ requires a detailed understanding of the relative wave function. Once one averages over spins, the squared relative wave function is a function of q, r and $\cos \theta_{qr}$. In relativistic collisions, correlation analyses are usually confined to light, singly charged hadrons. Coulomb-induced correlations are then weak and must be analyzed at small q, where quantum effects become important ($qr/\hbar \sim 1$). The relative two-particle wave function in the presence of Coulomb interactions can then be written as a function of $qr/\hbar$, r/a_0 and $\cos\theta_{qr}$.

$$\phi = \Gamma(1 + i\eta)e^{-\pi\eta/2}e^{i\mathbf{q}\cdot\mathbf{r}}\left\{1 + \sum_{n=1}^{\infty} h_n \cdot (r/a_0)^n\right\}, \qquad 10.$$

where a_0 is the Bohr radius, $h_1 = 1$ and $h_n = \frac{n-1-i\eta}{-in\eta}h_{n-1}$. Here, $\eta \equiv \mu e^2/\hbar q$ is independent of r, and for small r/a_0 the correlation function behaves as the Gamow factor, $G(\eta) \equiv e^{-\pi\eta}|\Gamma(1 + i\eta)|^2 = 2\pi\eta/(e^{2\pi\eta} - 1)$. Thus, the Coulomb kernels have little resolving power for $\pi\pi$ correlations where $a_0 = 387$ fm, but have greater resolving power for pK correlations where $a_0 = 84$ fm.

For $r \ll a_0$, the effects of Coulomb interactions can be removed easily from the correlation function because η is independent of r (21),

$$|\phi|^2 \approx G(\eta)[1 + \cos(2qr\cos\theta_{qr})]. \qquad 11.$$

For realistic source sizes, the order r/a_0 corrections are of the order of 10% for $\pi\pi$ correlations and are larger for heavier pairs (41, 51). Significant effort has been invested in "correcting" experimental correlation functions to remove Coulomb effects to all orders, but such corrections are model-dependent. The safest method for determining the source function is either inverting the full kernel (52–57) or fitting $C(\mathbf{q})$ to some parameterized form for $\mathcal{S}(\mathbf{r})$, which is convoluted with the full kernel. Neither of these tactics are computationally prohibitive.

Strong interactions can also be exploited to provide size and shape information. If the size of the source is much larger than the range of the potential between the two particles, the kernel ($|\phi|^2 - 1$) can be determined entirely from knowledge of the phase shifts. Pairs that have a resonant interaction are especially useful, because the resonance will lead to a peak whose height is inversely proportional to the source volume, if $qR \gg \hbar$. At small q the kernel is determined by the scattering length, a (58), and the height of the correlation at $q = 0$ becomes

$$C(\mathbf{q} = \mathbf{0}) - 1 = \left\langle -\frac{2a}{r} + \frac{a^2}{r^2}\right\rangle, \qquad 12.$$

where the averaging is performed using $\mathcal{S}_P(\mathbf{r})$ as a weight. Of course, the effects of strong interactions, Coulomb interference, and identical-particle interference can all combine as is the case for pp correlations. The pp kernel has been analyzed in depth by Brown and Danielewicz, where the kernel was inverted and applied to experimental pp data. Evidence was seen for significant non-Gaussian behavior in the resulting source functions (56, 57). Strong interactions also provide angular resolving power (59) which can be understood from the perspective of classical trajectories. Even s-wave scattering can be exploited to discern information about shape.

Strong and Coulomb-induced correlations apply to both identical and non-identical particles. For non-identical particles, the wave function is not symmetrized and $|\phi(\mathbf{q}, \mathbf{r})|^2 \neq |\phi(\mathbf{q}, -\mathbf{r})|^2$, which results in odd components of the correlations function, $C(\mathbf{q}) \neq C(-\mathbf{q})$, if there are odd components of $\mathcal{S}(\mathbf{r})$ as is the case for non-identical particle pairs. This asymmetry can be experimentally exploited to investigate the spatio-temporal differences between the emission functions of different particle species (60, 61); this requires, however, sufficient statistics to select on the angle between the total and relative momentum in the pair center of mass $\angle(\mathbf{q}, \mathbf{P})$ (62).

2.4. Coordinate Systems

Correlation functions depend on two three-dimensional momenta, $\mathbf{P}$ and $\mathbf{q}$. For high-energy collisions, analyses are usually performed in the longitudinally co-moving system (LCMS), a rest frame moving along the longitudinal (beam) direction such that $P_z = 0$. Axes are usually chosen according to the out-side-long prescription. The longitudinal axis is chosen parallel to the beam, while the outward axis points in the direction of $\mathbf{P}$, which is perpendicular to the beam axis. The sideward axis is chosen perpendicular to the other two. If the system is boost-invariant, observables expressed in the LCMS variables are independent of P_z and the source has a reflection symmetry about the $r_{\text{long}} = 0$ plane. If the collision is central, there is also a reflection symmetry about the $r_{\text{side}} = 0$ plane. Any four-vector V can be expressed in this coordinate system using the four-momentum P to project out the components (63–65),

$$V_{\text{long}} = (P_0 V_z - P_z V_0)/M_T,$$

$$V_{\text{out}} = (P_x V_x + P_y V_y)/P_T,$$ 13.

$$V_{\text{side}} = (P_x V_y - P_y V_x)/P_T,$$

where $M_T^2 = P_0^2 - P_z^2$ and $P_T^2 = P_x^2 + P_y^2$. Dimensions of the source function are typically quoted in this coordinate system. One could also perform a second boost to the pair frame, in which the transverse components of the total momentum are zero. Then,

$$V'_{\text{out}} = \frac{M_{\text{inv}}}{M_T} \frac{(P_x V_x + P_y V_y)}{P_T} - \frac{P_T}{M_T M_{\text{inv}}} P \cdot V,$$ 14.

where $M_{\text{inv}}^2 = P^2$. Relative wave functions are more conveniently expressed in the frame of the pair. For instance, a sharp resonant peak is no longer sharp if the correlation is viewed away from the pair frame. For pairs where the correlation is influenced by Coulomb and strong interactions, most analyses are conducted in the pair frame.

For non-zero impact parameters, azimuthal symmetry is lost and source functions also depend on the azimuthal direction of the pair's total momentum (66–68). Also, if boost-invariance is broken, the pair's rapidity needs to be specified. In this more general case, reflection symmetries are broken and the choice of the coordinate axes becomes somewhat arbitrary. One could orient the axes according the event's impact parameter, or one could rotate the coordinate system so that in the new frame cross terms such as $\langle xy \rangle$ vanish (illustrated in Figure 2, see color insert). One would then specify the Euler angles as part of the description of the source function.

2.5. Gaussian Parameterizations

To gain a physical understanding of the three-dimensional spatio-temporal source distribution, it is useful to summarize its size and shape with a few parameters. This motivates the study of Gaussian parameterizations for the source $S_P(\mathbf{r}')$ and the two-particle correlator. Realistic sources deviate from Gaussians, e.g., by exponential tails caused by resonance decay contributions. The extracted Gaussian source parameters may thus depend on details of the fitting procedures. These shortcomings can be overcome with imaging methods, or more complicated forms for the fitting. A more general three-dimensional analysis of correlations would involve decomposing both the correlation and source functions in either spherical or Cartesian harmonics (69, 70). Although the detailed non-Gaussian aspects of the correlation are important, the extra information can also cloud the main trends in the data. In practice, Gaussian parameterizations provide the standard minimal description of experimental data.

2.5.1. THE GENERAL CASE   The most general form for a Gaussian source is $\exp\{-A_{\alpha\beta}(x_\alpha - \bar{x}_\alpha)(x_\beta - \bar{x}_\beta)\}$, where x_α refers to the four dimensions x, y, z, t, and A is a 4×4 real symmetric matrix. The most general form has 14 parameters, 10 parameters for A and four more parameters for the offsets $\bar{x}_\alpha$. Reflection symmetries can be used to eliminate certain cross terms and some of the offsets (71). Furthermore, if the particles are identical or have the same phase space distributions, all the offsets can be set to zero. Because the source function for the second species b might also have 14 parameters, there could be up to 28 Gaussian parameters in describing both s_a and s_b. However, because $S(\mathbf{r})$ depends only on the distribution of relative spatial coordinates in the pair frame, only nine Gaussian parameters are required to describe the most general S for a given $\mathbf{P}$. Three of these nine parameters can be identified with Gaussian widths, three can be identified with offsets, and the last three can either be identified with cross terms or with the three Euler angles describing the orientation of the three-dimensional ellipse.

Using the reflection symmetries for mid-rapidity sources in a symmetric central collision, a Gaussian parameterization of the emission function for particle species a in the out-side-long coordinate system reads

$$s_a(p, x) \sim \exp\left\{-\frac{(x_{\text{out}} - \bar{x}_{a,\text{out}} - V_{s,a}(t - \bar{t}_a))^2}{2R_{a,\text{out}}^2} - \frac{(x_{\text{side}})^2}{2R_{a,\text{side}}^2} - \frac{(x_{\text{long}})^2}{2R_{a,\text{long}}^2} - \frac{(t - \bar{t}_a)^2}{2(\Delta\tau_a)^2}\right\}.$$

15.

The symmetries forbid any cross terms in the exponential involving x_{side} or x_{long}, such as $x_{side}x_{out}$ or $x_{long}t$, but do not forbid a cross term between outward position x_{out} and time t. Here, this cross term is taken into account by allowing the source to move in the outward direction with a velocity V_s. The symmetries also forbid offsets in the sideward or longitudinal direction. These other offsets have also been addressed for non-central collisions (72, 73).

The correlation function is determined by the phase space density of the final state, Equation 26. The resulting phase space density is

$$f_a(\mathbf{p}, \mathbf{r}, t) \sim \exp \left\{ -\frac{\left[x_{out} - \bar{X}_a(t) \right]^2}{2 \left[R_{a,out}^2 + (V_{s,a} - V_\perp)^2 (\Delta \tau_a)^2 \right]} - \frac{x_{side}^2}{2 R_{a,side}^2} - \frac{x_{long}^2}{2 R_{a,long}^2} \right\},$$

$$\bar{X}_a(t) = \bar{x}_{a,out} + V_\perp(t - \bar{t}_a). \qquad \qquad 16.$$

Here, $V_\perp$ is the velocity of the pair in the LCMS frame.

The correlation function is determined by the relative distance function $\mathcal{S}_\mathbf{P}(\mathbf{r}')$, see Equation 3,

$$\mathcal{S}_\mathbf{P}(\mathbf{r}') \sim \exp \left\{ -\frac{\left[r'_{out} - \bar{X}_{out} \right]^2}{4 \gamma_\perp^2 R_{out}^2} - \frac{r_{side}'^2}{4 R_{side}^2} - \frac{r_{long}'^2}{4 R_{long}^2} \right\}$$

$$R_{out}^2 = \frac{1}{2} \left[R_{a,out}^2 + R_{b,out}^2 + (V_{s,a} - V_\perp)^2 (\Delta \tau_a)^2 + (V_{s,b} - V_\perp)^2 (\Delta \tau_b)^2 \right],$$

$$R_{side}^2 = \frac{1}{2} \left[R_{a,side}^2 + R_{b,side}^2 \right], \qquad R_{long}^2 = \frac{1}{2} \left[R_{a,long}^2 + R_{b,long}^2 \right],$$

$$\bar{X}_{out} = \bar{x}'_{a,out} - \bar{x}'_{b,out}, \qquad \qquad 17.$$

where $\gamma_\perp \equiv (1 - V_\perp)^{-1/2}$. Thus, there are four measurable parameters, R_{out}, R_{side}, R_{long}, and $\bar{X}_{out}$. In the absence of any symmetry, there are five more terms: three cross terms ($R_{out,side}^2$, $R_{out,long}^2$ and $R_{side,long}^2$), and two more offsets (X_{side} and X_{long}). For identical particles, all the offsets are zero.

2.5.2. SENSITIVITY TO LIFETIME   Given the symmetries used to derive Equation 17, the most experiment can provide is the determination of the four parameters, R_{out}, R_{side}, R_{long}, and $\bar{X}_{out}$. The only way to independently determine the lifetime is to assume that the two transverse spatial sizes are approximately equal (74). After assuming $R_{a,out}^2 + R_{b,out}^2 \approx R_{a,side}^2 + R_{b,side}^2$, Equation 17 yields for identical particles ($V_{a,s} = V_{b,s} = V_s$, $\Delta \tau_a = \Delta \tau_b = \Delta \tau$),

$$(V_\perp - V_s)^2 (\Delta \tau)^2 \approx R_{out}^2 - R_{side}^2. \qquad \qquad 18.$$

In general, however, the source velocity is not precisely known and outward and sideward spatial dimensions are not exactly equal; these factors result in a significant systematic error when applying Equation 18, especially because temporal effects enter in quadrature. The R_{out}/R_{side} ratio only provides a reliable estimate of the lifetime when $(V_\perp - V_s)\Delta\tau$ is much larger than the transverse size.

2.5.3. GAUSSIAN CROSS TERMS   In addition to the three Gaussian parameters, R_{out}, R_{side} and R_{long}, that describe the dimensions of a Gaussian source, one needs, in the general non-symmetric case, three more parameters to describe the angular orientation of the principal axes. Figure 2 displays how the principal axes might differ from the out-side-long axes once the collisions are off center. These three Euler angles combined with three sizes can also be related to the parameters A_{ij} in the general form for a three-dimensional Gaussian, $\exp(-A_{ij}r_ir_j)$, where A is a symmetric matrix with six independent components (66, 68, 71, 75).

For pairs of identical particles, the correlation function for Gaussian sources is also Gaussian,

$$C(\mathbf{q}) = 1 + \exp(-(\mathcal{R}^2)_{ij}Q_iQ_j),\qquad\qquad 19.$$

where $Q = 2q$. The six experimentally determined parameters, $(\mathcal{R})^2_{ij}$, can be related to the moments of $\mathcal{S}(\mathbf{r}')$ (76),

$$\langle r_ir_j\rangle = (\mathcal{R}^2)_{ij}.\qquad\qquad 20.$$

For central collisions, the source sizes $(R^2)_{ij}$ depend only on the longitudinal pair momentum P_L and on the modulus of the transverse pair momentum $|\vec{P}_T|$. This is different for non-central collisions for which the azimuthal direction $\phi_{pair} = \angle(\vec{P}_T, \hat{b})$ of the transverse pair momentum with respect to the impact parameter direction $\hat{b}$ matters. The ϕ_{pair}-dependences of $(R^2)_{out}$, $(R^2)_{side}$, and $(R^2)_{out-side}$ then characterize the degree to which the initially out-of-plane-extended source geometry has expanded to the point where it becomes in-plane-extended (66, 67). The out-longitudinal and side-longitudinal cross terms contain information about the extent to which the main axis of the emission ellipsoid is tilted with respect to the beam axis (67). We return to this topic in Section 4.2.

The Yano-Koonin parameterization (71, 77–79) provides an alternative basis for describing the out-long cross term. The Yano-Koonin form is based on the assumption that one can boost along the beam axis to a source frame where the correlation function has a simple form.

$$C(\mathbf{P}, \mathbf{Q}) = 1 + \exp\left\{-\tilde{Q}_\perp^2 R_\perp^2 - \tilde{Q}_{||}^2 R_{||}^2 - \tilde{Q}_0^2 R_0^2\right\},\qquad 21.$$

where $\tilde{Q}$ is the momentum difference defined in the source frame which has rapidity y_{YK}. In that frame R_0 is the Gaussian lifetime and $R_\perp$ and $R_{||}$ are the dimensions of the source. This can be transposed to the out-long-side frame by boosting $\tilde{Q}$ along the beam axis to the frame where $P_z = 0$, i.e., the LCMS frame, then using the fact that $Q_0 = Q_{out}P_\perp/P_0 \equiv Q_{out}V_\perp$ in the new frame. This yields $\tilde{Q}_0 = \cosh(y_{\pi\pi} - y_{YK})Q_{out}V_\perp - \sinh(y_{\pi\pi} - y_{YK})Q_{long}$, and $\tilde{Q}_{||} = \cosh(y_{\pi\pi} - y_{YK})Q_{long} - \sinh(y_{\pi\pi} - y_{YK})Q_{out}V_\perp$, where $y_{\pi\pi}$ is the rapidity of the LCMS frame. Substituting these expressions into Equation 21 yields a cross term in the exponential equal to $Q_{\text{out}}Q_{\text{long}}(R_0^2 + R_{||}^2)\sinh 2(y_{\pi\pi} - y_{YK})$, which disappears when $y_{YK} = y_{\pi\pi}$. By fitting y_{YK} as a function of the pair rapidity, aspects of boost invariance can be tested. Given that the distribution of source rapidities should fall off for large rapidities, one expects y_{YK} to lag $y_{\pi\pi}$, because pions of a given rapidity would more likely have been emitted from sources with smaller rapidities (78).

2.6. The λ Factor

Many pions measured in experiment come from long-lived decays. Pions from weak decays may or may not, depending on the experiment, be identified and removed from the analysis, because their decay vertices are typically a few centimeters from the reaction center. Decays from η or η' mesons occur a few thousand fm away from the center of the collision. At these distances, they are effectively uncorrelated with other particles but cannot be identified with experiment. If a fraction λ of the pairs originate from the spatio-temporal region relevant for correlations, the correlation is muted by the factor λ (21). If the source function is divided into two contributions, $\mathcal{S}(\mathbf{r}) = \lambda\mathcal{S}_{\text{local}} + (1 - \lambda)\mathcal{S}_{\infty}$, where both $\mathcal{S}_{\text{local}}$ and $\mathcal{S}_{\infty}$ integrate to unity, the resulting correlation is

$$C(\mathbf{q}) = (1 - \lambda) + \lambda \int d^3r' \mathcal{S}_{\text{local}}(\mathbf{r}')[|\phi(\mathbf{q}', \mathbf{r}')|^2 - 1]. \qquad 22.$$

If the experimental sample includes a contamination from weak decays, η, or misidentified particles of a fraction f, the lambda factor becomes $(1 - f)^2$. It is not uncommon for this contamination factor to be near 30%, which results in λ near one half.

Certainly, this division is somewhat artificial, as there are non-Gaussian tails, or halos (80), to $\mathcal{S}_{\text{local}}$ due to such causes as the exponential fall-off of the source function in the longitudinal direction or semi-long-lived resonances such as the ϕ, whose lifetime is 40 fm/c. Non-Gaussian behavior is a subject for either imaging (52–57) or for more complicated parameterizations.

The λ factor is often referred to as an incoherence factor, the name being motivated by the properties of a coherent state, $\exp\{i \int d\mathbf{p}\eta(\mathbf{p})[a(\mathbf{p})+a^\dagger(\mathbf{p})]\}|0\rangle$, which for identical particles leads to no correlation. Coherent states represent highly correlated emissions caused by phase coherence and thus violate the assumption of incoherence or uncorrelated emission implied by Assumption 2 as described early in this section. The question of whether an observation of $\lambda \neq 1$ is due to coherence or due to contamination from particles from far outside the source volume can be tested by analyzing three-particle correlations (81, 82). Such analyses of data at both SPS and RHIC have been consistent with the incoherence conjecture (83–85). Microscopic model calculations at the AGS reproduce the excitation function of λ when resonances contributions are included (86).

2.7. Collective Flow and Blast-Wave Models

Both longitudinal and radial collective expansion reduce the size of the region of homogeneity, i.e., the relevant volume for particles of a given velocity. For an infinite volume, the size of this region is set by the length one must move before collective velocity overcomes the thermal velocity, $R \sim V_{\text{therm}}/(dv/dz)$.

2.7.1. LONGITUDINAL FLOW   The Gaussian lifetime $\Delta\tau$ described in Equation 18 represents the spread of the emission times. A small value of $\Delta\tau$ does not imply that particles were emitted early, but that they were emitted suddenly. Inferring the mean time at which particles are emitted requires a different assumption. For instance, at RHIC, the initial nuclei are Lorentz contracted by a factor of 100, and if there were no subsequent expansion, R_{long} would be less than a Fermi. If one assumes that the system expands along the beam axis with no longitudinal acceleration, the collective velocity becomes

$$V_{\text{coll},z} = z/t. \qquad 23.$$

If emission then comes from sources moving over a large range of rapidities (a boost-invariant expansion), the dimension along the beam axis for the source emitting zero-rapidity particles is determined by the distance one can move before the collective velocity overwhelms the thermal velocity to force the emission function back to zero. The size can then be expressed as:

$$R_{\text{long}} \approx \frac{V_{\text{therm}}}{dv/dz} = V_{\text{therm}}\langle t \rangle. \qquad 24.$$

Whereas $R_{\text{out}}/R_{\text{side}}$ gives information about the suddenness of emission, R_{long} provides insight into the mean time at which emission occurs given an estimate of the thermal velocity.

For a thermal source with relativistic motion, the thermal velocity along the beam axis is determined by the temperature and the transverse mass, $m_T = \sqrt{m^2 + p_T^2}$ (63). For large m_T the thermal velocity in the longitudinal direction becomes non-relativistic, $V_{\text{therm}} = \sqrt{T/m_T}$, and the source size falls as $1/\sqrt{m_T}$ which is referred to as m_T scaling (87). This is illustrated in Figure 3 (see color insert). However, this assumes all particles are emitted with the same Bjorken time τ_B and temperature, independent of the transverse mass. Because particles with high m_T are probably emitted at lower τ_B, and because the temperature roughly behaves at $\tau_B^{-4/3}$, the longitudinal size could fall even more quickly than $m_T^{-1/2}$.

In a boost invariant expansion, emission is a function of the Bjorken time $\tau_B = \sqrt{t^2 - z^2}$, not the time t, and because $t = \sqrt{\tau_B^2 + z^2}$, those particles emitted with small z have a head start. This is sometimes referred to as an inside-outside cascade. The transverse shape of $\mathcal{S}(\mathbf{r})$ is then affected non-trivially by the expansion along the beam. The resulting correlation function can be calculated analytically in the case of pure identical-particle correlations (88, 89).

Boost invariance is incorporated into blast-wave models with transverse expansion and assumed for many hydrodynamic models. The finite size of the system would alter the results for two reasons. First, if the distribution of sources covers only a finite range in η, the tails of the distribution $S(\mathbf{r})$ are chopped off. Assuming the distribution in η is Gaussian rather than uniform.

$$\frac{1}{R_{\text{long}}^2} \sim \frac{1}{V_{\text{therm}}^2 \tau_B^2} + \frac{1}{\eta_G^2 \tau_B^2}, \qquad\qquad 25.$$

where η_G is the range of rapidities over which the sources are distributed. If η_G were 1.5 units of rapidity, the extracted values of τ_B from boost-invariant pictures would be underestimating τ_B by $\sim 10\%$.

A second shortcoming of boost-invariant models is that they ignore acceleration in the longitudinal direction. Accounting for this acceleration would alter the relation between the time and the velocity gradient. Neglecting this acceleration could also lead to a modest underestimate of τ_B (180).

2.7.2. TRANSVERSE COLLECTIVE FLOW   Because transverse collective flow is intimately related to the pressure and viscosity, it is of central importance. Blast-wave models are based on pictures of thermal sources superimposed onto the transverse and longitudinal collective velocity profiles. Simple forms are then chosen for the profiles. Only two parameters are important for analyzing spectra, the temperature and the transverse velocity. Because heavier particles are more sensitive to flow than are light particles, the two parameters can be adjusted to fit the spectra of several species.

In addition to the temperature and transverse velocity, correlation measurements are also sensitive to the space-time parameters of the blast wave. In a minimal parameterization this would include the lifetime τ_B and the transverse size R. More sophisticated models would also include a spread in lifetime $\Delta\tau$ and a surface diffuseness ΔR. Additional parameters ensue when one considers sensitivity to the reaction plane. Then, two parameters are needed to describe the transverse size, and two parameters are required to describe the transverse collective velocity. These parameters can then depend on the azimuthal direction of $\mathbf{P}$ (90).

Choosing a blast-wave parameterization involves a number of choices about the form of the parameterization (91). Chemical potentials and temperatures might be chosen to vary with the transverse position r (92, 93) or might be chosen to be uniform. A wide variety of parameterizations have been employed for the transverse velocity profile, which might choose linear profiles for either v, γv or the transverse rapidity $asinh(\gamma v)$. In some parameterizations, the velocity profile has been chosen to rise quadratically with r (94, 95). Although hydrodynamics has been invoked as justification for different parameterizations, profiles from hydrodynamics vary according to the equation of state.

For particles moving much faster than the surface velocity, transverse flow manifests itself by constraining particles to an increasingly small fraction of the blast-wave volume for the same reason that R_{long} falls with m_T owing to longitudinal expansion (96). For large m_T, this leads to both R_{side} and R_{out} falling as $1/\sqrt{m_T}$ (71, 92, 93, 97). The fact that transverse dimensions fall with m_T might also result from the dynamics of cooling, superimposed with a growing fireball.

This correlates high-energy particles with earlier times when the fireball was both smaller and at a higher temperature.

Non-identical particles are of special interest in a blast-wave. For particles moving faster than the surface of the blast wave, there is a stronger tendency for heavier particles to be more confined to the region of the surface owing to their slower thermal velocities (72). This results in heavy particles being ahead of lighter particles of the same asymptotic velocity, and leads to a non-zero $\Delta\mathbf{r}$ illustrated in Figure 3. As discussed in Section 2.3, these displacements are accessible through measuring odd components of the correlation function (60–62).

2.8. Generating Correlations Functions from Hydrodynamics and from Microscopic Simulations

Any model that predicts final-state space-time and momentum information of emitted particles can be used to predict correlation functions. This information may be extracted from both microscopic simulations or from hydrodynamic calculations.

Microscopic simulations model the collision by evolving particles along straight-line classical trajectories which are punctuated by collisions that are programmed to be consistent with free-space cross-sections. When the modeling is done on a one-to-one basis, the simulations are referred to as cascades. Boltzmann simulations are similar but employ an oversampling by a factor N_s accompanied by a scaling down of the cross-sections by the same factor. These are then consistent with the Boltzmann equation and become local and relativistically covariant in the large N_s limit (98, 99). To generate correlation functions from either class of simulation, there are essentially two methods which are equally justified within the smoothness approximation. Method I is motivated by Equation 3. This involves first creating two lists, one for each species, of the space-time coordinates x_a and x_b of all those particles that were emitted with momenta $m_a\mathbf{P}/(m_a + m_b)$ and $m_b\mathbf{P}/(m_a + m_b)$. From these lists, one generates $\mathcal{S}_\mathbf{P}(\mathbf{r})$ by sampling the distributions of $x_a - x_b$. This list is then convoluted with $|\phi(\mathbf{q}, \mathbf{r})|^2$ to generate $C(\mathbf{P}, \mathbf{q})$ for all $\mathbf{q}$. In Method II, one samples pairs randomly without regard to their momenta. The numerator of the correlation function is then calculated by generating pairs with the same weight as one expects to observe experimentally and applying a weight given by the square of the relative wave function. The denominator would be calculated in a similar manner, but without the weight from the wave function. This method reflects the description of the correlation function in Equation 2. Acceptance effects or kinematic cuts can then be performed exactly as they would be performed for real particles. Method II has an advantage in that it is easier to accurately incorporate acceptance effects or tight kinematic cuts. Method I makes for a much quicker calculation because the procedure does not require sampling particles for irrelevant momenta (42).

Given the equation of state and the initial energy density, hydrodynamics provides the means for solving for the space-time development of the stress-energy tensor which can be used to make predictions for correlations (100–102). Viscous effects can also be incorporated and are non-negligible (103, 104). Generating source functions from the output of hydrodynamic calculations is not as straight-forward as it might seem. The Cooper-Frye prescription (105) conserves energy and momentum if the equation of state is one of free particles, but it suffers from the fact that the particles that cross backwards across the surface into the hydro-dynamic volume enter the source function as a negative emission probability. If the relative velocity of the surface, as measured by an observer in the matter's rest frame, is not much faster than the thermal velocity, a different prescription is required. Numerous prescriptions have been proposed to address these issues (106–108).

Hydrodynamic models, even those that incorporate viscosity, cannot be justified once the system expands to the point that the mean free path is similar to the characteristic size of the system. However, Boltzmann descriptions or cascades are well justified at lower densities. Several efforts have thus focused on coupling the two approaches (109–112). Because the final-state trajectories are established in the Boltzmann part of the prescription, one can apply either of the methods mentioned above.

2.9. Phase Space Density, Entropy and Coalescence

Because phase space density depends on both the momentum $\mathbf{p}$ and the position $\mathbf{r}$, a measurement of the phase space density must specify the spatial region over which it is determined. In practice, spatial information from two-particle correlation functions is instrumental to this end. For identical particle pairs, the "area" under the correlation function determines the average phase space density (113). Substituting the final phase space density for the time-integrated source function,

$$\int^{t_f} \frac{dx_0}{E_p(2\pi)^3} s(p, x)|_{p_0=E_p} = f(\mathbf{p}, \mathbf{x}, t_f),$$ 26.

and inserting into Equation 3 with $Q = 2q$ leads to

$$\int d^3Q\,[C(\mathbf{Q}) - 1] = (2\pi)^3 \frac{\int d^3r |f(\mathbf{P}/2, \mathbf{r}, t_f)|^2}{[\int d^3r f(\mathbf{P}/2, \mathbf{r}, t_f)]^2} = \frac{\bar{f}(\mathbf{P}/2)}{dN/d^3p},$$

$$\bar{f}(\mathbf{p}) = \frac{(2\pi)^3}{(2S+1)} \frac{E_p}{m} \frac{dN}{d^3p} \mathcal{S}_{\mathbf{P}=2\mathbf{p}}(\mathbf{r}' = 0).$$ 27.

Equation 27 applies also for the case of non-identical particles of the same phase space density. We note that $\bar{f}(\mathbf{p})$ is the phase space density averaged over coordinate space for a specific momentum using the phase space density itself as the weight. Unless $f(\mathbf{p}, \mathbf{r}, t)$ is a constant within a fixed volume, the average phase space density will fall below the maximum phase space density (106). For instance, if $f(\mathbf{p}, \mathbf{r}, t)$ has a Gaussian profile in coordinate space, the average phase space density will be $2^{-3/2}$ of the maximum phase space density for that momentum.

For a Gaussian source,

$$\bar{f}(\mathbf{p}) = \frac{\pi^{3/2}}{(2S+1)mR_{\text{inv}}^3} E\frac{dN}{d^3p}, \qquad 28.$$

where $R_{\text{inv}}^3 = (E/m)R_{\text{out}}R_{\text{side}}R_{\text{long}}$ is the product of the three radii as measured in the frame of the pair. The phase space density is determined by combining a source size measurement with the spectra. Entropy can be related to the phase space density in the standard way (114)

$$S = (2S+1)\int \frac{d^3r d^3p}{(2\pi)^3}\left[-f\ln f \pm (1\pm f)\ln(1\pm f)\right], \qquad 29.$$

$$dS/dy \approx \int d^2p_T E\frac{dN}{d^3p}\left[\frac{5}{2} - \ln(2^{3/2}\bar{f}(\mathbf{p})) \pm \bar{f}(\mathbf{p})/2\right]. \qquad 30.$$

Here, Equation 30 ignores higher powers of $\bar{f}$.

The average phase space density is also straightforward to determine by constructing ratios of spectra with species that can either bind or form a resonance. If species a and b can bind to form species c, thermal arguments would state

$$f_c(\mathbf{r}, t) = f_a(\mathbf{P}m_a/m_c, \mathbf{r}, t)f_b(\mathbf{P}m_b/m_c, \mathbf{r}, t)e^{B/T}, \qquad 31.$$

where B is the binding energy, or the excitation energy if the resonance is unstable. Coalescence arguments, which give the same expression but without the binding energy (116, 117), are identical if the binding energy is small compared to the temperature, as is the case for nucleon coalescence. The average phase space density for a or b can be determined by inserting Equation 31 into Equation 28,

$$\bar{f}_{a,b}(\mathbf{p}) = e^{-B/T}\frac{m_{b,a}(2S_{b,a}+1)}{m_c(2S_c+1)}\frac{E_c dN_c/d^3P}{E_{b,a}dN_{b,a}/d^3p}. \qquad 32.$$

Here, the binding energy needs to be expressed in the frame of the thermal bath. For the case where B is small, the assumption of a thermal bath can be neglected. Two examples where particles with similar phase space densities form low-energy resonances or bound states are $pn \to d$ and $\phi \to K^+K^-$.

By comparing the expression for $\bar{f}$ from the ratio of spectra in Equation 32, with Equations 27 or 28, one can determine either $\mathcal{S}_{\mathbf{P}}(\mathbf{r}' = 0)$ or the Gaussian parameters from ratios of spectra.

$$\mathcal{S}_{\mathbf{P}}(\mathbf{r}' = 0) = e^{-B/T}\frac{m_a m_b(2S_a+1)(2S_b+1)}{(2\pi)^3 m_c(2S_c+1)}\frac{E_c dN_c/d^3P}{E_a dN_a/d^3p_a \cdot dN/d^3p_b}, \qquad 33.$$

$$R_{\text{inv}}^3(\mathbf{P}) = \frac{1}{(4\pi)^{3/2}\mathcal{S}_{\mathbf{P}}(\mathbf{r}' = 0)}. \qquad 34.$$

Coalescence analyses can provide powerful measurements of volumes, but they only provide a single number, $\mathcal{S}(\mathbf{r}' = 0)$, and they cannot provide any insight into either the shape or the r dependence.

3. FEMTOSCOPIC MEASUREMENTS

Experimental techniques have developed considerably in response to significant improvements in both the theory and the quantity and quality of experimental data. In this section we discuss the general experimental approach for defining and analyzing femtoscopic correlations and their systematic dependence on global and kinematic quantities.

3.1. Correlation Function Definition

In practice, the formal definition of the correlation function in Equation 1 is seldom used in heavy ion physics. Instead the correlation of two particles, a and b for a given pair momentum $\mathbf{P}$ and relative momentum $\mathbf{q}$, is nominally given by

$$C_{\mathbf{P}}^{ab}(\mathbf{q}) = \frac{A_{\mathbf{P}}^{ab}(\mathbf{q})}{B_{\mathbf{P}}^{ab}(\mathbf{q})} \cdot \xi_{\mathbf{P}}(\mathbf{q}), \qquad\qquad 35.$$

where $A_{\mathbf{P}}^{ab}(\mathbf{q})$ is the signal distribution, $B_{\mathbf{P}}^{ab}(\mathbf{q})$ is the reference or background distribution which is ideally similar to A in all respects except for the presence of femtoscopic correlations, and $\xi_{\mathbf{P}}(\mathbf{q})$ is a correction factor introduced to compensate for non-femtoscopic correlations present in the signal that are not fully accounted for in the background as well as artifacts resulting, e.g., from finite resolution and contamination.

3.2. Signal Construction

The signal $A_{\mathbf{P}}^{ab}(\mathbf{q})$ refers to the relative momentum distribution of particles a and b for a given range of pair momenta, $\mathbf{P}$, and a given set of event characterizations. Although not all analyses proceed in exactly the following fashion, the mechanics of constructing the signal and background are most easily understood if one considers as separate steps:

1. Event quality cuts and event-class binning;
2. Single-track (including particle identification) cuts and single-particle binning; and
3. Two-particle pairing, two-track cuts, and pair momenta binning.

Here the term event class refers to both physics observables, such as collision centrality and reaction plane orientation, and detector considerations, such as event vertex position and the condition of the detector when the event was recorded, usually keyed by run number. The latter considerations are relevant to the proper construction of the background. For event class, particle, and pair bin, the final signal is usually stored as a set of 3D-histograms in the canonical relative momentum variables.

Single-particle acceptances divide out with a properly constructed background, but 2-track acceptances can have a large effect on the correlation function. For this reason the analysis of 2-track cuts is the dominant consideration in the signal construction for most analyses. The cuts and terminology are different for Time Projection Chamber (TPC) experiments with near continuous hit distributions and Drift Chamber experiments with projective geometry, but the goals are the same. Split-tracks (118, 119) and ghost-tracks both refer to single tracks which are incorrectly reconstructed as a pair of tracks with very low relative momenta. Even after the event-reconstruction algorithms (which generate a list of individual tracks) have been optimally tuned, small traces of these false pairs remain and must be removed from the analysis with identical pairwise cuts. Usually only a tiny fraction of tracks are split, and this effect may be ignored in essentially all experimental analyses except femtoscopic ones. Various methods have been developed for identifying likely split tracks, usually based on the number (119) or topology (118) of space-points associated with the track.

Pairwise effects usually also result in the loss of pairs at low relative momentum, because two tracks with very similar trajectories tend to be reconstructed as a single track. (Note that in tracking detectors, this is not a problem if one or both of the particles is a topologically identified neutral particle. In that case, the decay daughters may be well-separated even if their parents have identical momentum.) Such merging issues are usually resolved by pairwise cuts that remove merged pairs. Developing efficient and appropriate cuts can be a subtle exercise, and it requires good knowledge of the detector and event reconstruction software. In most cases these cuts are supported by simulations, but final determinations are nearly always based upon data.

What does it mean to cut out merged pairs? After all, if the tracks have merged, then the pair is lost anyway. The point is twofold. First, the pair efficiency usually does not drop from 100% to 0% sharply as a function of any variable. Thus, the cuts are usually tuned to exclude all (86, 118) or most (120–122) of the inefficient region. If regions with less than perfect efficiency remain in the analysis, a 2-track efficiency correction based on Monte Carlo simulations must be applied, typically leading to systematic uncertainties of a few percent. The second reason for the cut is that it is applied equally to the signal and to the background distribution $B_{\mathbf{P}}^{ab}(\mathbf{q})$ (123). Thus, if some fraction f of pairs is lost at some relative momentum $\mathbf{q}$ in $A_{\mathbf{P}}^{ab}(\mathbf{q})$, the same fraction is lost in $B_{\mathbf{P}}^{ab}(\mathbf{q})$, and the ratio in Equation 35 is robust against the effect.

3.3. Background Construction

For reasons described above, all cut-imposed effects on the signal pair distribution A^{ab} must be applied to B^{ab}. This often means identifying which pairs would have been removed by merging, splitting, or other cuts, had the particles come from in the same event.

The ideal background should be identical to the signal in all respects except for the presence of femtoscopic correlations. Therefore, the global event characteristics, single particle distributions, and acceptances should match those of the signal. A simple and straightforward way to construct such a background is to form pairs from different events within a single event class. This event-mixing technique (124) has gained wide acceptance in relativistic heavy ion collisions where violations to energy-momentum conservation are negligible in the high multiplicity environment. This technique will be described in detail in what follows. However, other methods have also been used, especially if one considers femtoscopy in other systems.

For elementary-particle collisions or in low-multiplicity events, event mixing can violate total energy-momentum conservation, especially when exclusive final states or jet-axes must be preserved; thus, the correlation function would reflect non-femtoscopic in addition to femtoscopic correlations. In these cases, the most common techniques form a background from unlike-signed pairs, with resonance regions excluded with cuts (125) or normalized with a correlation of like- to unlike-signed pairs from a Monte Carlo (126). Other experiments have constructed a background using only Monte Carlo generated pairs (127). A few experiments have investigated backgrounds formed by swapping (128) or reversing momentum components relative to a jet-axis (125), but these methods are not widely used. For detectors with symmetrical acceptance, such as the STAR TPC (129), momentum conservation effects may be eliminated by mixing pairs from the same event, with the lab momentum of one particle flipped (130). Backgrounds constructed from single-particle distributions as formally defined by Equation 1 have been used for heavy ion collisions at lower energies and shown to be consistent with the more commonly used event-mixing technique (131).

In order to avoid inducing artificial structure in the correlation function, the particles forming pairs in the background distribution should originate from parent events with the same event characteristics. The parent events should have similar vertex positions to within the experimental resolution. Because detector acceptances can vary with time (e.g., components may fail for some runs), parent events should have been measured close in time to each other; this is usually easiest in any case, because event mixing is done on the fly as time-ordered data is read sequentially.

Parent events whose particles are mixed should also have the same single-particle momentum distributions. Thus, they should have similar centralities and orientations of the reaction plane. For example, mixing particles from events with very different p_T slopes or directions of preferred emission (elliptic flow) would produce differences between $A^{ab}(\mathbf{q})$ and $B^{ab}(\mathbf{q})$ even in the absence of physical correlations. Because almost all analyses to date have ignored these potential biases, it is comforting that they make little difference in practice (118).

The list of event classes given here is by no means exhaustive. One can expect future analyses to incorporate the orientation of high p_T particles (jet axis) or any other event-related observable.

The procedure for deciding how many events to mix remains something of an art and involves optimizing over the range of data runs, bin width, and statistics. In order to minimize statistical errors, one typically forms approximately ten times the number of pairs in the background as in the signal. For the special case when all possible combinations are formed, the variance of a particular relative momentum bin is proportional to $n^{\frac{3}{4}}$, where n is the number of entries in the bin (22). However, as the number of pairs formed is reduced, the variance per bin approaches the $n^{\frac{1}{2}}$ value expected for Poisson statistics (121). It is possible that non-Poisson fluctuations persist in the co-variance between different bins, but this has not yet been investigated.

Once the pairs have been mixed, the background must be subject to the same 2-track cuts that have been applied to the signal. For example, the exact same track merging cuts or minimum separation on a detector must be applied to both signal and background.

3.4. Corrections

Corrections to the correlation function fall into three categories: finite resolution effects, mis-identified particle contamination, and compensation for deficiencies in the background.

The first category concerns single-track momentum resolution, and reaction-plane resolution. We consider finite momentum resolution corrections first. Typically, momentum resolutions are on the order of 1%. One approach is to correct for momentum resolution by a double ratio of the ideal correlation function generated from a Monte Carlo simulation with perfect momentum resolution divided by a Monte Carlo correlation function with momentum resolution turned on. The femtoscopic weights are inserted into the simulations iteratively until the fitted radii converge (86, 118, 119, 123). A second approach is similar, but corrects only the Coulomb interaction term, which is most greatly affected by momentum resolution effects (122). In both cases, the corrections change the fitted radii by only $<5\%$.

As discussed in Section 4.2, azimuthally sensitive analyses (132–134) measure oscillations in correlations as a function of emission angle with respect to the reaction plane. Finite resolution effects in the reaction plane angle (135) artificially reduce the oscillation strengths. Methods have been developed to correct the distributions $A^{ab}(\mathbf{q})$ and $B^{ab}(\mathbf{q})$ for these effects (68, 136).

Another type of correction accounts for the inclusion of mis-identified and secondary-particle contamination. For example, electrons may be mistakenly identified as π^- mesons. It is usually assumed that the mis-identified or secondary particles are uncorrelated with other particles, so the net effect on the correlation function is to damp all structure uniformly in $\mathbf{q}$. For purely Gaussian correlations (see Section 3.5), this effect is absorbed wholly into the λ factor discussed in Section 2.6; homogeneity lengths—derived from the width of the Gaussian correlation—are unaffected by the reduction in its strength. In many cases, however, the homogeneity length is extracted largely from the strength of the correlation, and so contamination effects must be removed. In the general case, for which the purity ρ depends on the relative momentum, the correlation function is corrected according to $C^{\text{true}}(\mathbf{q}) = (C^{\text{raw}}(\mathbf{q}) - 1)/\rho(\mathbf{q}) + 1$ (62, 118).

It is more difficult to correct for correlated contamination. For example, if cuts cannot completely distinguish primary protons from those coming from Λ decay, then measured $p - \Lambda$ correlations will contain contributions from $\Lambda - \Lambda$ correlations. Unlike the white-noise contamination discussed above, this introduces structure into $C^{p\Lambda}(\mathbf{q})$ that can be accounted for only with detailed simulations. Such corrections will become more important at RHIC due to copious resonance production, and especially for baryon correlation measurements, in which the heavy daughter carries most of the momentum of the parent resonance.

The last category of corrections are applied to fix deficiencies in the background distribution. This includes corrections to account for two-particle inefficiencies, which have been discussed in the previous section. A second correction of this type deals with the residual signal correlation that is present in all backgrounds derived from events that contribute to the signal. The residual correlation arises because femtoscopic correlations can modify the single particle distributions. This is especially true for small-aperture spectrometers. This effect can be removed with an iterative procedure (22, 137); however, for many large experiments the induced error is often 1% or less, and it is easier to fold this into the systematic errors (122).

3.5. Fitting

After the application of all cuts and corrections, the correlation is formed according to Equation 35 and then fit to determine spatial parameters. As described in Section 2, there are three approaches to fitting the correlation function: fitting to a simplified Gaussian form with strong and Coulomb interactions neglected or factored out, fitting to a convolution of the full kernel convoluted to a parameterization of $S(\mathbf{r})$, and inverting the kernel to fit a source image. The simplified Gaussian fits based on Equation 19 are limited to correlations of identical pions, kaons, and photons, but it has been the most widely used method to date because of the computational demands of the other methods. We expect its use to continue for the large systematic studies in which binning in centrality, reaction plane, and k_T leads to fits of more than one hundred separate correlation functions for a single colliding system.

However, the functional form of the Gaussian parameterization used by experimentalists has evolved over the years. Before reviewing the most recent functional forms, it is necessary to review the treatment of the Coulomb interaction and the fraction of pairs coming from the source that contribute to the femtoscopic correlations. Both were first introduced into the literature by Zajc (22) in the form of the Gamow factor given in Section 2 and an empirical parameter λ to account for the observation that not all pairs exhibit femtoscopic correlations. With steady improvements in data quality and CPU speed, the Gamow factor has been replaced with a calculation of an squared unsymmetrized Coulomb wave for a finite Gaussian source. The improvements in data quality have also led to a self-consistent treatment of λ with respect to both Coulomb and Gaussian components of the fit function (138, 139). For this to be accurate, we must assume that the source is fully chaotic, an assumption that has recently been verified with three-pion correlations

(81, 85). The non-femtoscopic pairs consist of mis-identified particles and particles that emanate from too far from the source for the correlation to be resolved experimentally. The region far from the source has been referred to the source halo, to differentiate it from the core. The correlation fit function is therefore given by Equation 36,

$$C(\mathbf{q}) = N \left[\lambda G(\mathbf{q}) F(\mathbf{q}) + (1 - \lambda) \right], \qquad 36.$$

where N is the overall normalization, F is the Coulomb component, and G is the Gaussian form for the un-damped correlation function, Equation 19 for out-long-side coordinates, or Equation 21 for Yano-Koonin variables.

Figure 4 shows projections of a $\pi^- - \pi^-$ correlation function measured by the STAR collaboration (118). The filled symbols are the measured correlation function corrected for momentum resolution only and fit with Equation 36. The open symbols have been overcorrected by applying to all pairs the Coulomb correction for the fitted source dimensions. Depending on the shape of the correlation and degree of experimental contamination, extracted homogeneity lengths may vary by up to ~15% if the correlation function is overcorrected.

For proton-proton correlations and non-identical particle correlations, direct fits are performed by convoluting the full kernel with a parameterized source. For

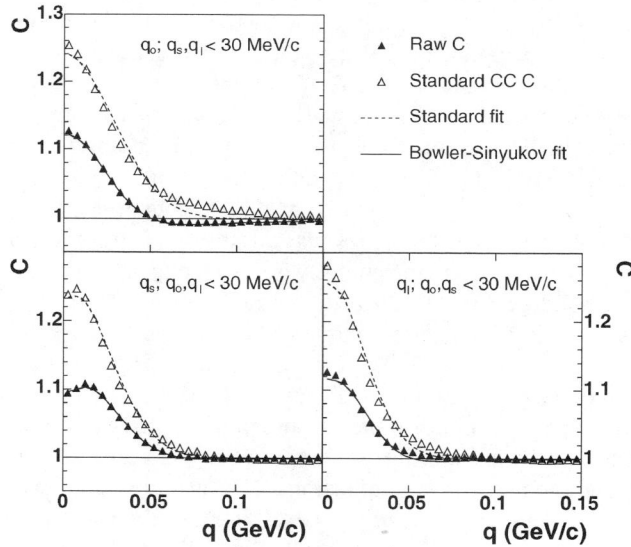

Figure 4 Projections of a three-dimensional correlation function (integrated over 0-30 MeV/c in orthogonal components) for low-k_T π^- pairs for 200 GeV central Au+Au collisions (118) (filled symbols) with fit function. Open symbols include correction for Coulomb interaction among all pairs. Projections were generated according to the prescription described in (140).

2

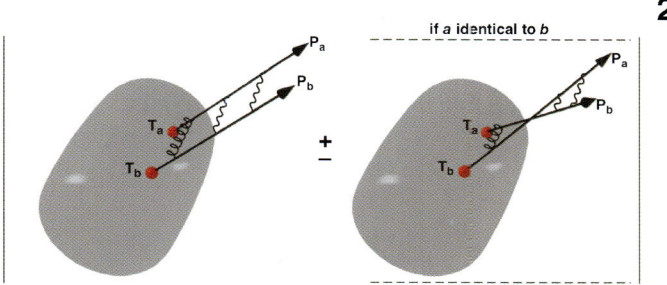

Figure 1   Schematic representation of the squared emission amplitude for two particles emitted independently from the grey-shaded source region and interacting with each other in the final states. For identical bosons $(+)$ and fermions $(-)$, correlation also involves interference between the paths.

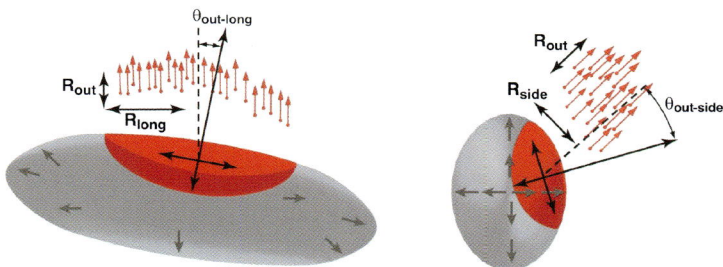

Figure 2   For non-central collisions the principal axes describing the orientation of the region of homogeneity can differ from the out-side-long axes. By viewing the source distribution from the perspective where the beam axis is oriented horizontally (*left panel*) and from peering down the beam pipe (*right panel*), the orientations leading to out-long and out-side cross terms are illustrated.

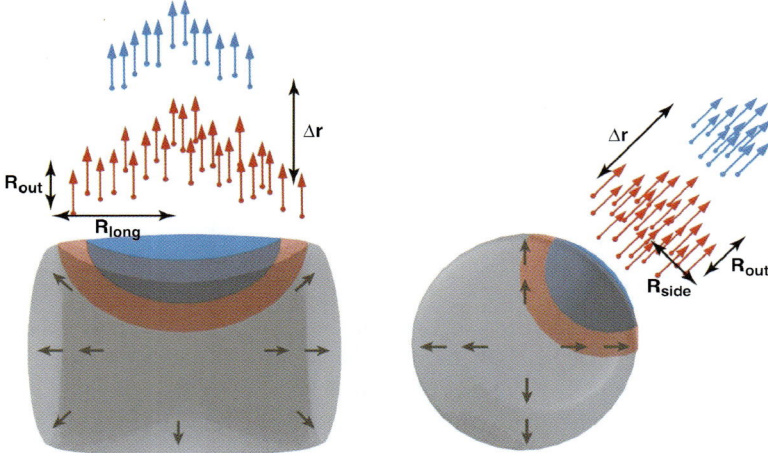

Figure 3   Because particles with heavier masses have smaller thermal velocities, their source volumes are more strongly confined by collective flow. For longitudinal flow (*left panel*) this results in smaller values of R_{long} for particles with higher $m_T = \sqrt{m^2 + p_T^2}$. For radial flow (*right panel*) this confines heavier particles toward the surface, which results in both a reduced volume and an offset Δr in the outward direction.

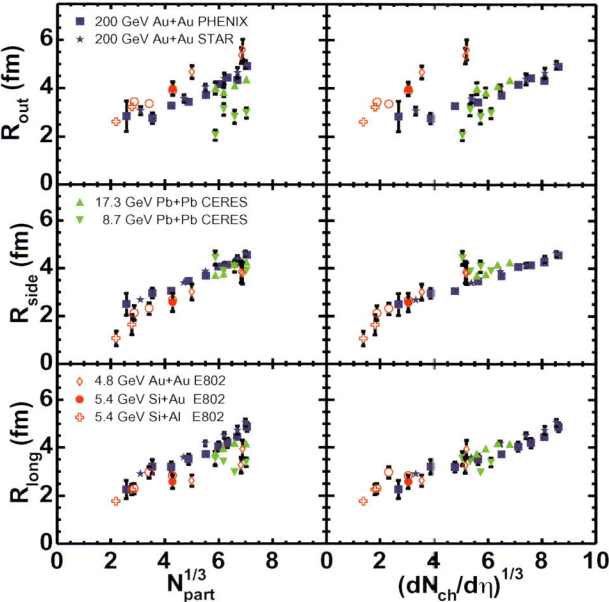

Figure 5   Pion source radius dependence on number of participants (*left*) and on charged particle multiplicity (*right*). Data are for for Au+Au (Pb+Pb) collisions at several values of $\sqrt{s_{NN}}$ and also for Si+A collisions at the lowest energy. Average transverse momentum $\langle k_T \rangle \sim 450$ MeV/c for the PHENIX data and ~390 MeV/c for the others. Data from (118, 121, 122, 149).

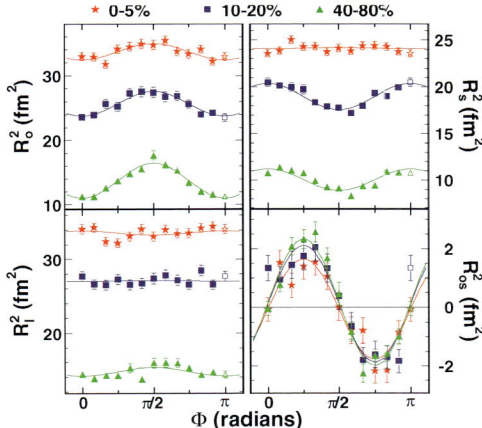

Figure 6   Squared source radii measured at mid-rapidity in Au+Au collisions at RHIC, as a function of pair emission angle relative to the reaction plane. Data for three centralities are shown. Figure from (134).

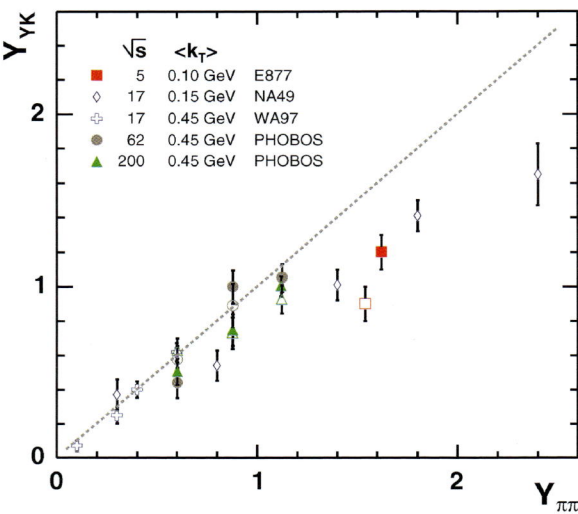

Figure 8 The Yano-Koonin source rapidity is shown as a function of the pion pair rapidity for central Au(Pb)+Au(Pb) collisions over a broad range of energies (open symbols are for $\pi^-\pi^-$, closed symbols for $\pi^+\pi^+$). Both quantities are in the center-of-mass frame of the colliding system. Data from (176–179).

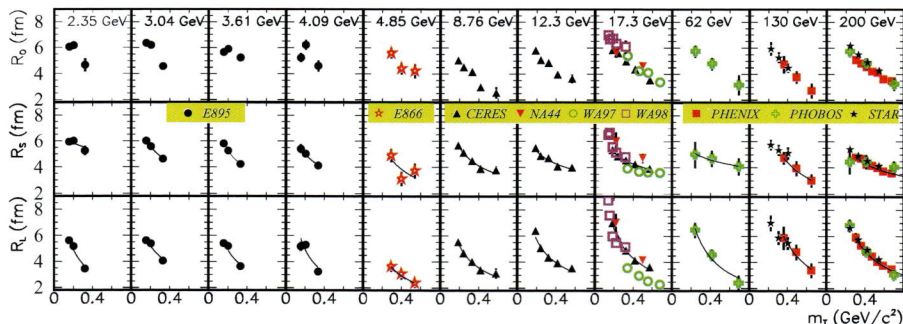

Figure 9 World data set of published m_T dependence of pion Bertsch-Pratt radii near mid-rapidity from Au+Au (Pb+Pb) collisions. Centrality selection is roughly top 10% of cross-section, but varies somewhat with experiment. Data from (86, 118, 121, 122, 137, 142, 149, 178, 179, 181, 182). Lines represent parameterized fits; see text for details.

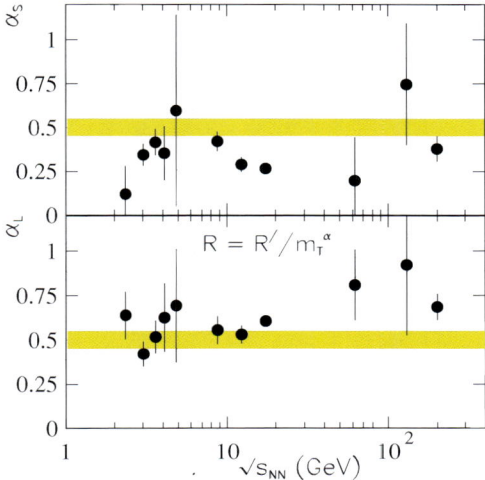

Figure 10   The excitation function of m_T power-law fall-off of pion source radii $(R_i(m_T) \sim m_T^{-\alpha_i})$. Shaded regions show $\alpha_i = 0.5 \pm 0.05$.

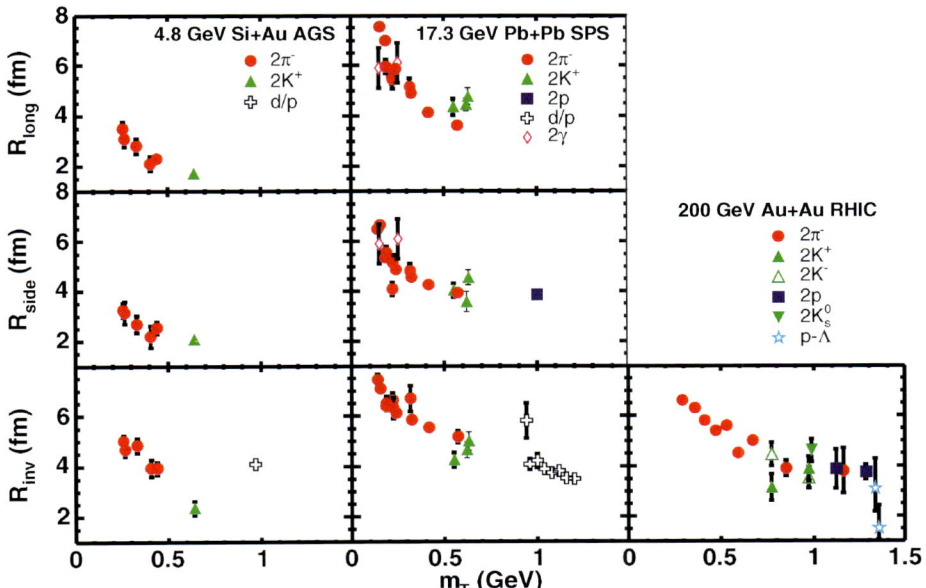

Figure 12   Transverse mass dependence of homogeneity lengths from correlations between particles of (nearly) identical mass.

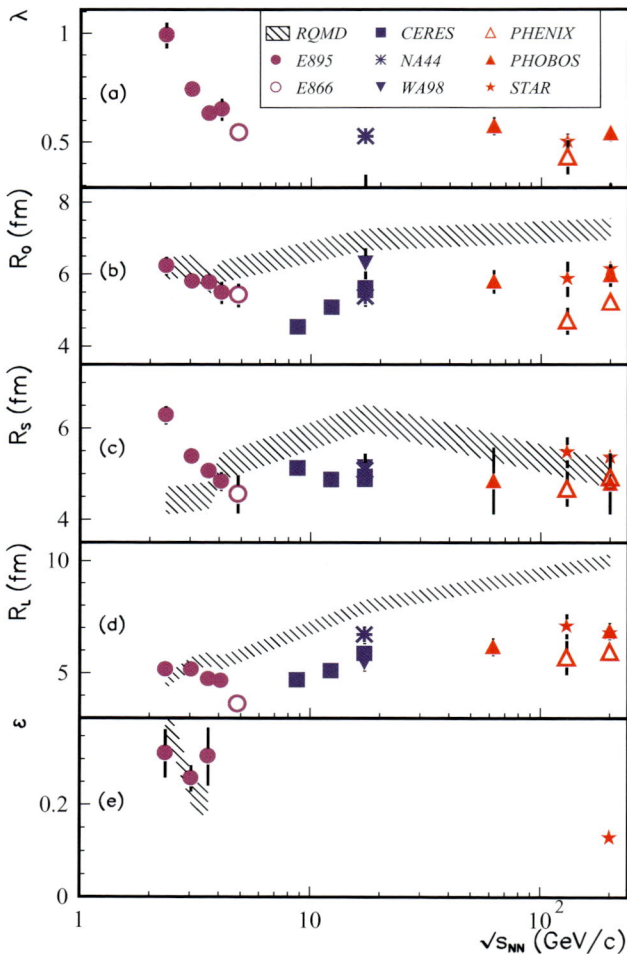

Figure 13 Panels (*a–d*): Excitation function of π^- source parameters at mid-rapidity and low k_T (~ 0.17 GeV/c) in central Au+Au (Pb+Pb) collisions. PHENIX data are for $k_T \sim 0.26$ GeV/c and so fall somewhat below the trend. Panel (*e*): Transverse freeze-out anisotropy parameter from non-central ($|\vec{b}| \sim 8$ fm) Au+Au collisions, estimated from the azimuthal dependence of source radii. Data are from (86, 118, 121, 122, 132, 134, 137, 142, 149, 179, 181, 182). Also shown are calculations (86, 132, 137, 211) at several energies with the RQMD model (212); hashed region at other values of $\sqrt{s_{NN}}$ interpolates between these calculations.

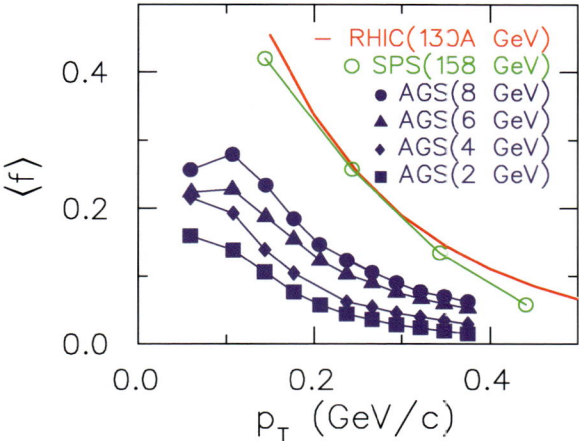

Figure 14   Average pionic phase space densities for central Au+Au and Pb+Pb collisions from the AGS to RHIC rise with beam energy but seem to plateau at SPS energies. Values were calculated with Equation 28.

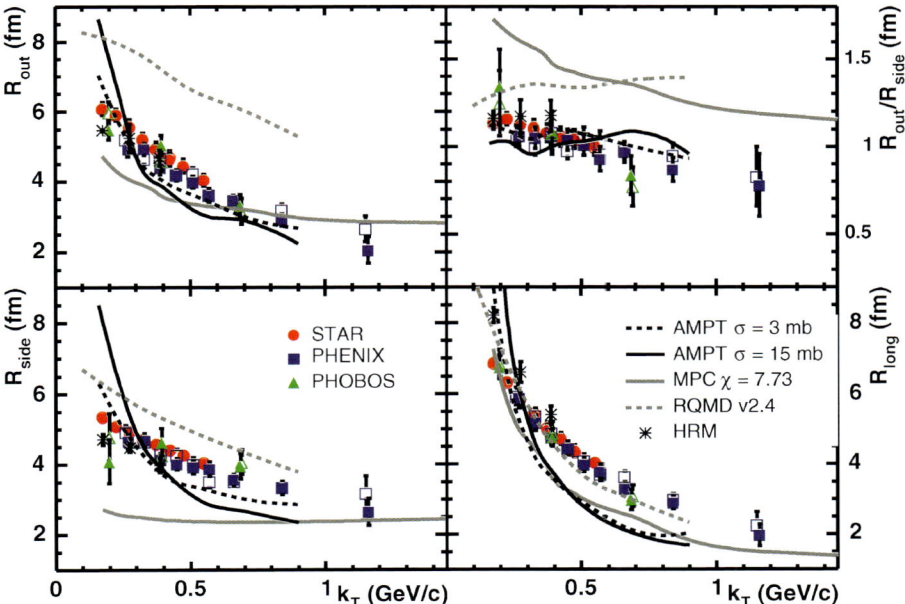

Figure 15   Pion radius parameters from four Boltzmann/cascade models are compared to experimental RHIC data. For Molnar's Parton Cascase calculations, the reported radii are simply space-time variances. For the other models, radii are obtained from Gaussian fits to correlation functions generated according to the methods described in Section 2.8. Measured data for positive and negative pions are indicated by closed and open symbols, respectively.

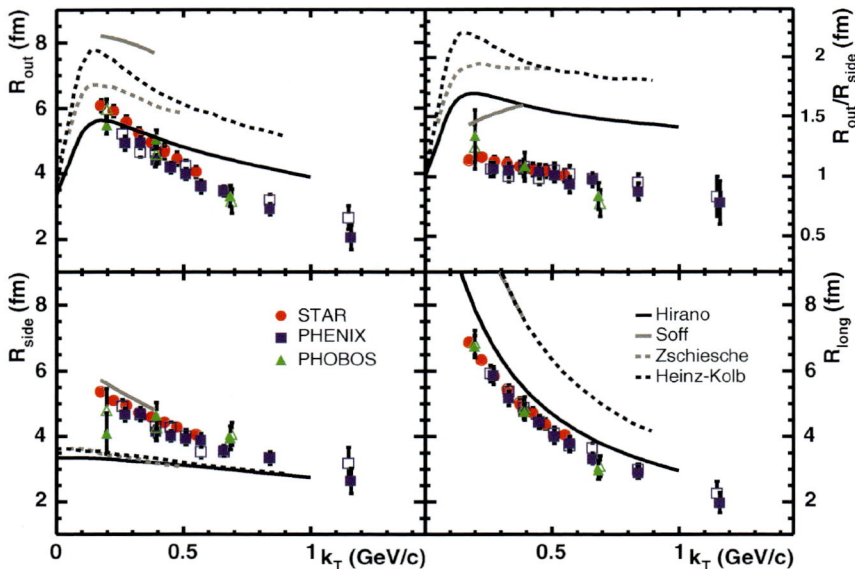

Figure 16   Hydrodynamic (Zschiesche, Hirano, and Kolb) and hybrid hydrodynamic/cascade (Soff) models calculations in comparison to RHIC data. Data are for $2\pi^-$ (open symbols) and $2\pi^+$ (closed symbols) source radii.

these analyses, the paucity of statistics has been more of a limitation than the relatively modest demands in CPU power. The examples given in Section 4 are all for one-dimensional analyses, but recent data from RHIC will soon be analyzed in multi-dimensions.

The ability to image the source by inverting the kernel is a relatively recent development, but one with very general applications. Because the source is parameterized by a series of B-splines, it is a very general form which is sensitive to non-Gaussian shapes. To date, source imaging has been performed only with one-dimensional correlations, but like with the direct fits, a multi-dimensional kernel will soon be possible (69, 141).

Non-Gaussian effects were reported in the first pion correlation measurement at RHIC (142). With higher statistics, the STAR Collaboration has studied the issue in greater detail (118), performing a functional expansion (the so-called Edgeworth expansion (143)) about a Gaussian shape. Although significant non-Gaussian contributions were reported, the dominant length scales were already extracted in the purely Gaussian fits.

3.5.1. MINIMIZATION   A simple chi-squared test is inappropriate for fitting correlation functions because the ratio of two Poisson distributions is not itself Poisson distributed, especially when taking the ratio of small numbers. For this reason, a log-likelihood fit function of the form given in Equation 37 is preferred.

$$\chi^2_{PML} = -2 \left[A \ln \left(\frac{C(A+B)}{A(C+1)} \right) + B \ln \left(\frac{A+B}{B(C+1)} \right) \right], \qquad 37.$$

where A, B, and C were introduced in Equation 35. This equation was derived from the principle of maximum likelihood assuming that both signal and background are Poisson distributed (121). The full derivation of Equation 37 and comparison to earlier log-likelihood functions is given in (121).

4. MEASURED FEMTOSCOPIC SYSTEMATICS

The first systematic study to compare femtoscopic measurements across several systems and experiments was performed almost 20 years ago with data from intermediate-energy heavy ion collisions at the Bevalac (144). The data, taken from experiments with different acceptances, triggers, and analysis techniques, were sufficient to demonstrate a crude $A^{1/3}$ scaling of the one-dimensional radii, indicating that spatial scales were indeed being probed. The first femtoscopic measurements for relativistic heavy ion collisions were presented by the NA35 Collaboration at the Quark Matter meeting in Nordkirchen (145, 146). More detailed measurements followed with the availability of sulphur and silicon beams at the SPS (123) and AGS (147, 148).

Since then, increasingly sophisticated experiments at the AGS, SPS, and RHIC have performed femtoscopic measurements corresponding to a wide range of control parameters. The experimental community performing the measurements has reached critical mass and matured substantially; a common language and knowledge base have developed concerning sometimes subtle details in performing and interpreting femtoscopic measurements. The result of this effort is a striking degree of consistency across experiments in regions of phase space where acceptances overlap and meaningful generation of systematics across experiments. Large-statistics data sets routinely allow three-dimensional correlation measurements with small statistical error bars. Systematic errors, which now dominate the experimental errors, have been reduced to the level of $\sim 5\%$, or ~ 0.25 fm for most measurements. It is no exaggeration to state that femtoscopic measurements have become a precision tool.

Here, we cover the most important systematics of femtoscopic measurements from the AGS, SPS, and RHIC. We discuss only generally the physics probed by a given systematic, appealing to intuitive schematic models such as the blast wave (72). Full interpretations and comparisons to dynamic models are given in Section 5.

4.1. System Size: N_{part} and Multiplicity

As discussed earlier, femtoscopic radii probe homogeneity regions, and not the entire source (hereafter, the term *source* will be used to refer to the entire source of particle emission). Nevertheless, the claim that two-particle correlations probe spatial scales would be given little credence if the radii did not exhibit a strong, positive correlation with system size. Therefore, measuring the systematic variation of the radii vs. system composition and centrality represents the most basic test of both theoretical and experimental femtoscopic techniques.

Coalescence studies (150) and two-proton measurements at the AGS (151) and SPS (152) unambiguously demonstrate that nucleon homogeneity lengths increase with decreasing impact parameter and/or increasing projectile mass, continuing the trend mapped at lower energies (153, 154), where directional cuts have allowed measurement of the shape of the homogeneity region (155–157). More detailed information comes from pion correlations at relativistic energies, for which three-dimensional analyses allow partial isolation of purely geometrical effects. The centrality dependence of Bertsch-Pratt source radii are shown in Figure 5 (see color insert) for a wide range of collision energies. The left panels show the dependence on the number of participating nucleons, N_{part}, a generalization of the $A^{1/3}$ linear scaling of nuclear radii used to approximate the initial overlap geometry. All of the radii exhibit a linear scaling in $N_{\text{part}}^{1/3}$, most with finite intercepts. Only the slope of the R_{long} dependence shows a significant increase from the AGS to RHIC, consistent with a lifetime that increases with both centrality and $\sqrt{s_{\text{NN}}}$. The trend of increasing R_{long} with increasing $\sqrt{s_{\text{NN}}}$ is reversed for $\sqrt{s_{\text{NN}}} < 5$ GeV (86).

The right panels of Figure 5 show the same radii as a function of $(dN_{\text{ch}}/d\eta)^{1/3}$. The primary motivation for exploring the $(dN_{\text{ch}}/d\eta)^{1/3}$ dependence is its relation to the final state geometry through the density at freeze-out. However, the two scaling quantities are highly correlated. In fact, the values of $dN_{ch}/d\eta$ shown on the right side of Figure 5 were derived from N_{part} using the N_{part}^{α} parameterizations given in (158), and conversely, the N_{part} values are often calculated from multiplicity distributions using a Glauber model. Given this caveat, the R_{side} and R_{long} values exhibit a linear dependence on $(dN_{ch}/d\eta)^{1/3}$, again with finite intercepts. The strong uniformity from $\sqrt{s_{NN}}$ of 5 to 200 GeV leads one to believe that the approximate N_{part} scaling (initial overlap geometry) is a result of the scaling with multiplicity (final freeze-out geometry) and not the other way around.

The parameter R_{out}, which mixes spatial and temporal information (see Section 2.5), increases with multiplicity at each given collision energy, but does not follow a universal curve. However, the strikingly $\sqrt{s_{NN}}$-independent multiplicity scaling of the geometric radii R_{side} and R_{long} strongly suggests that the observed increase of these radii with collision energy for $\sqrt{s_{NN}} > 5$ GeV (see Section 5.1) is due simply to the rise of multiplicity with collision energy. This trend, as well as its violation at $\sqrt{s_{NN}} < 5$ GeV, has been interpreted in terms of changing chemical composition of the source as the system evolves with energy from baryon to meson dominance (210).

We note that the systematics in system size represent an initial sanity check for the femtoscopic technique. The obvious direct connection of the radii to the source geometry estimated in two ways refutes suggestions (159) that smaller, non-geometric length scales dominate experimentally extracted transverse radii.

4.2. Source Shape: Pair Emission Angle Relative to $\hat{b}$

The variation of femtoscopic radii with the pair emission angle relative to $\vec{b}$ (ϕ_{pair}) can be used to probe the three-dimensional shape of the source (66–68, 72, 75, 160–162). The anisotropic shape transverse to the beam direction—the coordinate-space analog to the elliptic flow characterizing momentum-space—gives rise to $\cos(n\phi_{pair})$ (n even) oscillations in the squared transverse source radii R_{out}^2, R_{side}^2, $R_{out,side}^2$ (66, 68).

Just as one expects the source (and homogeneity regions) to be larger for decreasing $|\vec{b}|$, one also expects it to be rounder, reflected by small oscillations of the radii. Figure 6 (see color insert) for mid-rapidity pions from Au+Au collisions at RHIC confirms this expectation. As $|\vec{b}|$ increases, the oscillations indicate a transverse source increasingly elongated out of the reaction plane (134).[1]

The strong in-plane expansion (163) does not fully convert the initial out-of-plane (overlap) geometry into an in-plane-extended source at freeze-out. This suggests a rather short evolution time; in essence, the system did not have time to reverse its deformation. However, this is only a hint, and a full dynamical transport calculation is required to extract physical timescales (90).

---

[1]The out-of-plane nature of the elongation may be read directly from Figure 6. Ignoring collective flow or opacity effects (e.g., 67) an out-of-plane-extended source would produce $R_{side}^2(\phi_{pair} = 0°) > R_{side}^2(\phi_{pair} = 90°)$, as seen in Figure 6. Collective flow effects complicate this picture (66, 72), but the sign of the oscillations are determined by geometric, not dynamic, effects for realistic sources at RHIC (72, 90).

Whereas at the highest RHIC energy, the freeze-out anisotropy is $\sim\frac{1}{3}$ of the initial (134), at low AGS energies, the final anisotropy is consistent with that of the initial overlap region (132), or perhaps slightly lower. Because elliptic flow vanishes—changes sign—at these energies (164), these trends make intuitive sense and suggest an underlying connection to the evolution dynamics. It would be desirable to map the source anisotropy at intermediate (AGS and SPS) energies, for which there may be interesting changes in the space-time systematics. At these energies, there have been intriguing hints of asymmetries in the homogeneity regions for pions (165–168) and protons (169), and in the proton-pion separation (73, 166), although they have not been finalized.

If the impact parameter direction $\hat{b}$—not simply the 2nd-order event-plane angle (unambiguous only over a range $[0, \pi]$)—is known, then more detailed information may be obtained. In the left panel of Figure 2, the source is tilted with respect to the beam axis, toward $\hat{b}$. Just as anisotropic azimuthal geometry in the transverse plane is related to the structure of elliptic flow (90, 100), a tilted geometry can reveal important information on the underlying nature of directed flow (67, 132, 170, 171). The structure in $R^2_{o,l}$ and $R^2_{s,l}$ shown in Figure 7 is generated by this tilt.

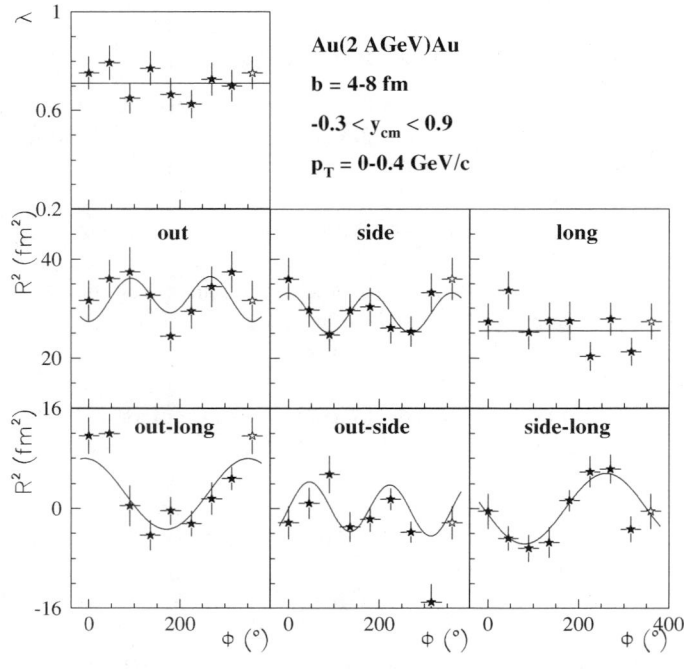

Figure 7   Squared source radii measured near mid-rapidity in mid-central Au+Au collisions at lowest-energy AGS settings, as a function of pair emission angle relative to the reaction plane. Figure from (132).

The spatial tilt has been measured only at low AGS energies (132); a measurement at RHIC might reveal exotic geometric configurations generated by quark gluon plasma (QGP) formation (171, 172) and would impact the important issue of boost-invariance at mid-rapidity.

4.3. Boost Invariance : $Y_{\pi\pi}$

In high-energy hadronic collisions, the initial parton distribution is expected to be approximately flat in rapidity. This momentum rapidity distribution may correspond to producing matter that initially exhibits a boost-invariant Hubble-type scaling correlation between longitudinal flow velocity and space-time points, $v_{L,flow} = z/t$ (173, 174). Relativistic hydrodynamics preserve boost-invariance of the initial conditions throughout its dynamical evolution (174). The combination of the above arguments underlies expectations that in ultra-relativistic heavy ion collisions, particle production emerges from boost-invariant longitudinal flow, and that dN/dy exhibits an approximately boost-invariant plateau around mid-rapidity. However, an extended plateau has never been observed from AGS through RHIC energies (3).

Because correlation measurements access spatio-temporal information, the question arises (78, 175) whether they allow us to test the relation $v_{L,flow} = z/t$ between the space-time rapidity and the momentum rapidity of the source. For boost-invariant sources, one can show that the pair momentum rapidity is equal to the Yano-Koonin source velocity, which is directly obtained from the Gaussian radius parameters, $Y_{\pi\pi} \approx Y_{L,flow} \equiv \tanh^{-1} v_{L,flow}$. However, even if the source density distribution shows significant deviations from boost-invariance, this relation still holds approximately as long as the velocity profile is boost invariant, and k_T is sufficiently large (78).

Figure 8 (see color insert) reveals a roughly universal dependence of Y_{YK} on $Y_{\pi\pi}$ for pions from central collisions, depending weakly, if at all, on $\sqrt{s_{NN}}$ (see Section 2.5.3). This trend is particularly striking given the very different center-of-mass projectile rapidities (~ 1.55 and 5.5 for $\sqrt{s_{NN}} = 5$ and 200 GeV, respectively) and corresponding widths of the pion distributions dN/dy.

By way of caution, we note that the results from RHIC are limited to the region $Y_{\pi\pi} < 1.2$, and that the deviations from boost-invariance are mostly in the lower energy data. Extending the RHIC results to more forward rapidities would provide an important test for both the velocity scaling at RHIC and the energy-independence that is exhibited in Figure 8.

For central collisions, the roughly universal behavior approximately obeys the boost-invariant consistency relationship discussed above. Moreover, $Y_{L,flow}$ shows a significant k_T dependence and falls below the linear relation in particular for small k_T (178). Qualitatively, this is consistent with blast-wave models in which a boost-invariant longitudinal flow is superimposed on a source density distribution of finite longitudinal width. However, a full dynamical understanding of the dependence is missing so far. The flat Y_{YK} dependence on $Y_{\pi\pi}$ measured at the SPS (178) for the most peripheral collisions is counter-intuitive, and requires further study.

The question of whether the source has boost-invariant space-time structure is an important one. There are many reports of very short evolution timescales (lifetimes) based on fits to the data with Equation 24, which is based upon an assumption of boost-invariance (87). Relaxation of that assumption might lead to considerably larger estimates (180).

4.4. Collective Dynamics: k_T and Particle Mass

As discussed in Section 2.7.2, the dynamic substructure of the source is encoded in space-momentum (**xp**) correlations. Longitudinal **xp** correlations, encoded in $R_{\text{long}}(k_T)$, are generally acknowledged (72, 87) to reflect longitudinal flow. Because all transverse correlations are generated in the collision itself, considerably more attention has generally been paid to the transverse substructure than to the longitudinal flow discussed in Section 4.3.

The most common explanation for transverse **xp** correlations is collective transverse flow (96). These correlations have mostly been studied through pion correlations, but transverse flow implies also a systematic trend as the particle mass is varied.

4.4.1. k_T DEPENDENCE OF PION RADII   Collective flow generates a characteristic fall-off of the pion source radii with k_T, which is ubiquitously observed in data. Final results for the k_T-dependence of Gaussian radii from central Au+Au (Pb+Pb) collisions exist at the AGS (86, 121), SPS (137, 149, 177, 178, 181), and RHIC (118, 122, 134, 142, 179, 182). As is clear from Figure 9 (see color insert), aside from a small variation in overall scale (discussed later), the k_T dependence is startlingly similar for all energies.

Figure 10 (see color insert) quantifies the evolution of the m_T-dependence of the pion source radii with $\sqrt{s_{\text{NN}}}$, using fits to $R_i(m_T) \sim m_T^{-\alpha_i}$. As discussed in Section 2.7.1, $\alpha_i = 0.5$ would represent expectations for instantaneous thermal emission for a three-dimensionally expanding fireball in the limit of large m_T.

The similarity persists as N_{part} is varied. In Au+Au collisions at RHIC, $R_i(m_T)$ (i = out, side, long) simply scale as $N_{\text{part}}^{1/3}$, with perhaps some flattening for $N_{\text{part}} <$ 100 (118, 122). Very similar k_T dependence for different N_{part} is also observed in Pb+Pb collisions at SPS (149) and for Si+Au and Au+Au collisions at the AGS (121).

In a flow-dominated freeze-out scenario, the fall-off of transverse radii with m_T increases as flow increases and/or temperature decreases (e.g., 72). Blast-wave fits to spectra (183) indicate that freeze-out flow and temperature vary significantly with $\sqrt{s_{\text{NN}}}$ for $\sqrt{s_{\text{NN}}} \lesssim 10$ GeV. The overall approximate $\sqrt{s_{\text{NN}}}$-independence of the α_i parameters may reflect the fact that significantly changing the slope of $R_i(m_T)$ requires very large changes in flow and temperature; on the other hand, it could be that the compensating effects of smaller (larger) homogeneity lengths generated by larger flow (temperature) cancel almost exactly in nature. Although $R_i(m_T)$ almost certainly reflects strong collective flow, determining the strength of that flow requires other information, such as particle spectra (72, 94).

Because the radii fall off roughly as $1/\sqrt{m_T}$ (see Figure 10) and such a dependence has been discussed frequently in the literature (e.g., 184, 185), it is interesting to look at the overall scale parameter from a single-parameter fit to $R_i'/\sqrt{m_T}$. The $\sqrt{s_{NN}}$-dependence of R_{side}' and R_{long}' are shown in Figure 11. The scale of the longitudinal homogeneity length grows significantly with $\sqrt{s_{NN}}$, consistent with an increase of the system evolution time. However, R' varies only very weakly with $\sqrt{s_{NN}}$.

4.4.2. SYSTEMATICS WITH PARTICLE MASS   Systematic studies for different mass particles provide additional controls probing the space-time evolution of the source. In particular for kaons, the interpretation may be simplified owing to reduced effects of resonance feed-down (186) and a reduced scattering cross-section for K^+ in nuclear matter, raising the possibility that kaon correlations could peer farther back to earlier stages of the collision (187). Indeed, the first kaon measurements (176, 188–190) reported smaller source radii for kaons. However, the observation that radii for K^+ and K^- were very similar (189) was an early experimental indication that different cross-sections were not the driving physics behind these smaller radii. This was supported by model calculations (191), which suggested that K^+ and K^- in fact scattered roughly equally in the dense medium created in heavy ion collisions. In this case, the smaller radii for kaons result from their increased mass in a flow field, not different cross-sections.

If indeed flow is generated in matter sufficiently dense that individual cross-sections are unimportant, then all particles participate equally in collective

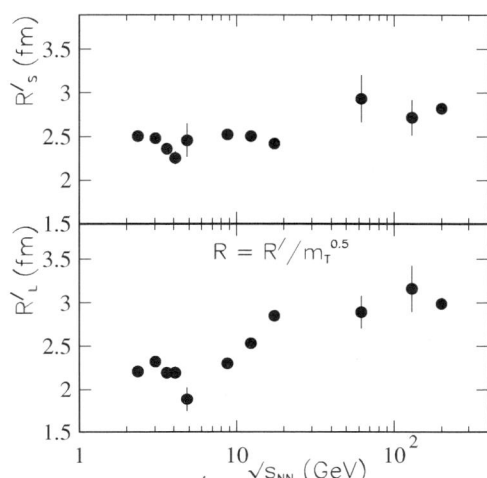

Figure 11   The excitation function of scale parameters R_i' from fits of the data in Figure 9 to $R_i(m_T) = R_i'/\sqrt{m_T}$. Note: this is a different functional form than that used in Figure 10.

transverse flow. In this case, their source radii should approximately follow a common m_T scaling (8, 30, 159, 192). Within uncertainties, first results on kaon interferometry by NA44 at the SPS in S+Pb (184) and Pb+PB (193) collisions were consistent with a $1/\sqrt{m_T}$ scaling expected for isotropic Hubble flow (30, 159, 192). In some more recent analyses (194), the common m_T systematic for the transverse radii is less steep than for $R_{\text{long}}(m_T)$ (see Figure 10), as might be expected for more boost-invariant (non-isotropic) flowing systems (8, 72).

Figure 12 (see color insert) collects the m_T dependence of homogeneity lengths for several energies. The left panels show results for Si+Au collisions at $\sqrt{s_{NN}} = 5.4$ GeV, measured by E802 for pions (121) and kaons (188, 190). Femtoscopic radii for pions (137, 149, 181, 194), kaons (193), protons (195), and photons (196) measured in Pb+Pb collisions at the SPS are shown in the center panels. The right panel shows the one-dimensional radius parameter R_{inv} measured at RHIC for pions, charged kaons, and protons (197), neutral kaons (198), and with $\Lambda - p$ correlations (199).[2] To compare across energies, R_{inv} results are included for the AGS and also for the SPS, where the R_{inv} values were calculated from the 3D fit results by accounting for the boost along the outwards direction from the LCMS (in which $P_z = 0$) to the PCOM (pair center-of-mass, in which $|P| = 0$) frame, $R_{\text{inv}}^2 = R_{\text{long}}^2 + R_{\text{side}}^2 + \gamma^2 R_{\text{out}}^2$, where γ is given by m_T/m of the pair. Note that for massless particles, such as photons, γ is given by k_T/Q_{inv}. For a given orientation of the photon pair momentum R_{inv} can be related to 3D radii in the LCMS frame through the following expression for the exponent of the Gaussian source term,

$$Q_{\text{inv}}^2 R_{\text{inv}}^2 = Q_{\text{inv}}^2 \left(R_{\text{long}}^2 \cos^2\theta + R_{\text{side}}^2 \sin^2\theta \sin^2\phi + k_T^2/Q_{\text{inv}}^2 R_{\text{out}}^2 \sin^2\theta \cos^2\phi \right)$$

$$= k_T^2 R_{\text{out}}^2 \sin^2\theta \cos^2\phi + Q_{\text{inv}}^2 \left(R_{\text{long}}^2 \cos^2\theta + R_{\text{side}}^2 \sin^2\theta \sin^2\phi \right). \qquad 38.$$

Thus R_{out} drops completely out of the source term and enters into the λ coefficient (196), but as a function of k_T. For this reason the 1D Q_{inv} fits from WA98 are plotted alongside values of R_{side} and R_{long} instead.

This consistency between different particle types may carry an important message. It calls into question theoretical scenarios which appear to explain $R_i(m_T)$ for particular particle types only (33). Further, the consistency with emission from a common flow-dominated source may also support freeze-out scenarios in which the last scattering in the dense phase determines the homogeneity region, instead of milder rescatterings in the more dilute stage, which are dominated by particle species-dependent cross-sections (200, 201).

---

[2]Protons and Λ baryons are not identical particles, of course. However, their masses are sufficiently close to try including $R_{\text{inv}}^{p-\Lambda}$ on a m_T-scaling plot. Adherence to the scaling is consistent with a flow-dominated scenario in which homogeneity lengths depend only on mass.

4.5. New Possibilities: Non-Identical Particle Correlations

Almost all femtoscopic measurements to date have been done through correlations of identical particles, usually charged pions. With the availability of high-statistics data sets and new theoretical ideas (202), however, experiments are beginning to make full multi-dimensional analyses of non-identical particle correlations. These correlations are being used to test and refine the treatment of Coulomb effects in identical-particle correlation analyses, to explore violations of flow-induced universal m_T scaling, and to extract qualitatively new information on the space-time substructure of the source.

4.5.1. OPPOSITE-SIGN PION CORRELATIONS   The most common non-identical particle femtoscopy measurements have involved $\pi^+ - \pi^-$ correlations. Here, the natural assumption is that the homogeneity regions of these particles coincide. The primary interest is in whether the homogeneity volume is extracted from $\pi^\pm - \pi^\pm$ correlations—driven by quantum statistics and Coulomb final state interactions (with minimal strong FSI effects)—is consistent with that extracted from $\pi^+ - \pi^-$ correlations, determined only by Coulomb and strong FSI (118). This issue has obvious implications for the FSI weight $F(\mathbf{q})$ discussed in Section 3.5.

At the SPS, $\pi^+ - \pi^-$ correlations from reactions with sulphur (203) and lead (177) beams were consistent with emission from a homogeneity region with spatial scale roughly the same as the radii from identical-particle correlations. Preliminary analysis of opposite-sign pion correlations from Au+Au collisions at the AGS (73, 176) similarly find consistency with scales extracted from like-sign pion correlations. Recently, STAR reported (118) strong consistency at RHIC energies as well. Using large-statistics data sets, they further show that contributions to the $\pi^+ - \pi^-$ correlation function from strong-force interactions, though small, are nevertheless important in explaining the data in detail.

4.5.2. OTHER NON-IDENTICAL PARTICLE CORRELATIONS IN $|\vec{q}|$ For statistical reasons, most non-identical particle correlations are measured in one-dimensional $|\vec{q}|$-space, and thus probe only the average spatial separation between the two particles at freeze-out in the pair center-of-momentum frame (see Section 2.3). Because they are sensitive to the size of each particles emission source and this separation between them, it is often necessary to rely on the identical particle correlations for a proper interpretation of results.

Despite these ambiguities, one-dimensional non-identical particle correlations may be used to test existing systematics and expectations of the freeze-out scenario. Conversely, if the freeze-out geometry may be taken as given, non-identical particle correlations may place constraints on hadronic scattering parameters, e.g., to measure the squared relative wave-function of a pion and a Ξ baryon. First studies along these lines are underway at RHIC (204).

Wang and Pratt (58) suggested the measurement of $p - \Lambda$ correlations, which may be more sensitive to large structure than $p - p$ correlations, and which have a higher two-track reconstruction efficiency for some experiments. Published results from Au+Au collisions at the AGS (55) suggest a $p - \Lambda$ separation distribution with a width similar to the proton homogeneity length and roughly consistent with m_T scaling expectations (cf Section 4.4.2). Preliminary results at SPS (205) and RHIC (204, 206) give similar conclusions. Because the $p - \Lambda$ potential is not unambiguously known theoretically (58), it is unclear whether possible statistically marginal violations of m_T scaling (55) or differences between baryon and anti-baryon emission regions (204) are meaningful. To first order, existing $p - \Lambda$ correlations confirm existing systematics.

With the high-statistics and high-quality data sets at the largest collision energies, truly exotic correlation studies are possible. Preliminary results (206) from RHIC on $\pi - \Xi$ correlations look particularly promising. Here, the well-measured pion emission distribution may be used to study the strange baryon freeze-out configuration. It may also provide information on the $\pi - \Xi$ final state interaction and scattering cross-section which in turn can be used to constrain our understanding of the sources of collective flow. Blast-wave calculations reproduce the preliminary $\pi - \Xi$ correlations, suggesting that the Ξ flow is determined by its mass, not its quark content.

4.5.3. NEW INFORMATION: NON-IDENTICAL PARTICLE CORRELATIONS WITH DIRECTIONAL CUTS ON $\vec{q}$ Non-identical particle correlation analyses as a function of— or with cuts on—the relative direction of $\vec{q}$ and $\vec{P}$ reveal qualitatively new information (60, 61, 202, and Section 2). In particular, in addition to the root mean square (RMS) width of the separation distribution, the direction and size of the average separation between the particles is probed, although offsets in time and space cannot be disentangled; this is shown as Δr in Figure 3. The correlation functions selected for $\vec{q} \parallel \vec{P}$ and $\vec{q} \nparallel \vec{P}$ differ if $\Delta r \neq 0$. Furthermore, collective flow will induce position-momentum correlations detectable with directionally selected non-identical particle correlations (61, 72).

These correlations are statistically challenging, and few results are available. At RHIC, STAR has reported (62) asymmetries in $K^{\pm} - \pi^{\pm}$ correlations measured in central Au+Au collisions. Blast-wave calculations with transverse flow roughly adjusted to reproduce other observations at RHIC (72) describe the data semi-quantitatively. Preliminary studies of $p - \pi$ correlations at the SPS (205) and RHIC (206) exhibit similar mass-ordered spatial asymmetries in the transverse plane. A preliminary study at the AGS (73) reported very large ($\Delta r_{\text{long}} \approx 10$ fm) average $p - \pi$ separations in the longitudinal (beam) and impact parameter direction for forward-moving particles, suggesting strong longitudinal flow; however, this result was never confirmed.

We expect full three-dimensional analyses of a wide range of non-identical particle combinations to be available in the near future from RHIC experiments. Sophisticated analyses may probe non-trivial geometric substructure when selecting on reaction plane, including the sideward shift predicted by blast-wave calculations (72) when anisotropic flow structure is present.

5. INTERPRETATIONS OF EXPERIMENTAL RESULTS

In this section, we ask what we can learn from the spectrum of results just presented. Beginning with the broadest, least detailed observations, we move to two fundamental quantities that may be directly extracted from the data, and finally on to comparisons with specific models of heavy ion collisions.

5.1. General Conclusions from Systematic Trends

One of the first messages to take away from the discussion of Section 4 is that the results are stable across detector and method; experimental systematics and uncertainties are under control. Whatever difficulties we may have in interpreting measurements, we may be confident that they do not have their origin in experimental artifact.

The size and shape inferred from two-particle correlations tracked with collision geometry as anticipated. Kinematic and mass dependences of femtoscopic measurements showed the expected clear signatures of strong collective flow in the beam direction and perpendicular to it.

At a generic level (ignoring quantitative predictions), we are first given pause at the jejune nature of the $\sqrt{s_{NN}}$-dependence of femtoscopic parameters. The most common example discussed is the pion source radii excitation function shown in Figure 13 (see color insert). In Section 4 we explored the trends in considerably greater detail, but the figure conveys the right message: Scanning $\sqrt{s_{NN}}$ through a range of two orders of magnitude changes final-state geometry little.

Based on rather generic arguments of soft points in the equation of state or entropy generation during a phase transition, there had been hopes for non-trivial structure in the excitation function, as the energy threshold for quark gluon plasma (QGP) creation was crossed (63, 207–209). General expectations were for long (relative to the explosion of a purely hadronic system) system evolution time scales if QGP was formed. Blast-wave analyses of data (e.g., 72) appear to rule out systems with lifetimes in the neighborhood of 20 fm/c or higher. However, owing to dynamic effects, careful comparison with a dynamical model is required to extract detailed evolution information; this is discussed in Section 5.3.

As discussed in Section 4.1, the weak increase with R_{out} of the femtoscopic radii for central collisions shown in Figure 13 is mostly due simply to increasing multiplicity. The CERES collaboration (210) has suggested that this increase may be understood in terms of an energy-independent mean free path of 1 fm at freeze-out. Interestingly, this explanation also appears to describe the decrease of radii with $R_{out} < 5$ GeV, when the different cross-sections and abundances of pions and protons are accounted for. This observation emphasizes the importance of chemistry in the determination of the freeze-out geometry. However, it neglects the dynamical structure of the source (e.g.. flow) and the importance of the six-dimensional phase space density discussed in Section 2.9. We discuss this fundamental quantity next.

5.2. Phase Space Density and Entropy

As shown in Section 2.9, average phase space densities, $\bar{f}(\mathbf{p})$, can be calculated by combining source-size measurements with spectra. Pionic phase space densities have been estimated for $130A$ GeV collisions at RHIC (114), at the top SPS energies, and for several AGS energies (213). Figure 14 (see color insert) shows results for all three regions. In each case, the phase space densities were calculated via Equation 28. For the SPS case, results were generated by applying Equation 28 to published spectra (214) and source-size measurements (215). We note that our calculations for the SPS are higher than previously published values at low p_t (216). This discrepancy is likely due to the fact that in Reference (216), analytic parameterizations were used that differ significantly from published spectra at low p_t.

When applying Equation 28, an issue arises as to whether one should subtract the contribution from resonances to the spectra. Indeed, if the pions are created by decays so far outside the source volume that they do not contribute to the correlation function, they should not be considered as pions when calculating the phase space density. There are two strategies to correct for such pions. First, one could use spectra where such particles are subtracted and apply Equation 28 literally. As a second option, one could use the spectra without subtractions, but then multiply the expression for $\bar{f}$ by $\sqrt{\lambda}$. Because most published spectra have been purged of the products of weak decays, the first method is usually applied. However, published spectra still include the contribution from η's, which decay thousands of fm away from the source. The η contribution was accounted for in the SPS calculation for Figure 14 by reducing the spectra by 5%.

The phase space densities in Figure 14 show a steady rise with beam energy that seems to plateau at SPS energies. Because the displayed phase space densities have been averaged over coordinate space, the peak values are higher (91, 106), by a factor of $2\sqrt{2}$, if the spatial profiles are Gaussian. For a breakup temperature of 110 MeV, this requires a rather high chemical potential, near or above 80 MeV (217–219).

Ratios of particle yields are consistent with chemical freeze-out at temperatures near 170 MeV (220–222). This suggests an interpretation where at higher densities the system is so strongly interacting that yields equilibrate until the system reaches this temperature and are then frozen during the subsequent freeze-out. At AGS energies, the increase in pion production brought on by increased beam energy results in more pions being pushed into a given amount of phase space as realized in Figure 14 by the rising phase space density. However, one would expect that as the excitation energy surpasses the threshold for a 170 MeV temperature, the phase space density would reach a limiting value. Because the average phase space density is preserved in an isentropic expansion with fixed particle number (not exactly true if several masses are mixed together), one would expect phase space densities to saturate once energy densities reached this value. Indeed, the behavior in Figure 14 is consistent with this scenario.

As shown in Section 2.9, entropy can be calculated from average phase space densities using Equation 30. Spectra and source-size measurements for baryons and mesons were used to estimate phase space densities and entropy for $130A$ GeV Au+Au collisions at RHIC. The total entropy in the central unit of rapidity was estimated at $dS/dy = 4450 \pm 10\%$ (114). In a hydrodynamic expansion, entropy is conserved though viscosity, and shock waves might result in a roughly 10% increase during the evolution. Thus, this measurement provides an upper bound for the entropy at $\tau \sim 1$ fm/c when thermalization first occurs. At $\tau = 1$ fm/c, the volume for particles in this rapidity slice is determined by the geometric cross-sectional areas of the overlapping gold nuclei multiplied by $c\tau$, thus an upper bound for the entropy density can be determined $s \leq (dS/dy)/(\tau \pi R^2)$. Knowing the energy density at $\tau = 1$ fm/c would then provide a point in the equation of state, s vs. ϵ. The value of dS/dy estimated in (114) is consistent with lattice calculations if $\epsilon(\tau = 1$ fm/c$) \sim 7$ GeV/fm^3. Estimates of the original energy density from the final-state measurement of $dE_t/d\eta$ are in the range of 4.5 GeV/fm^3 (223), but because these estimates neglect losses from longitudinal work or the energy from longitudinal thermal motion, the 7 GeV/fm^3 value seems reasonable.

5.3. Dynamic Models and Their Comparison with Data

It is increasingly recognized that the comparison of dynamic models of heavy ion collisions to data is only insightful if it involves a sufficiently large variety of experimental data. A comparison of dynamic models to femtoscopic measurements alone (or to any other class of measurements alone) is of limited value, simply because for realistic models, the number of possible model-dependent parameter choices then tends to exceed the number of experimental constraints. In fact, all the model results that we review in the current subsection remain unsatisfactory with this respect: They either deviate significantly from femtoscopic data, or they reproduce these data at the price of missing other important experimental information. In particular, there is so far no dynamically consistent model that reproduces quantitatively both the systematic trends discussed in Section 4 and the corresponding single inclusive spectra. In this situation, the scope of this subsection is somewhat limited. We want to explain why a dynamical understanding of femtoscopic measurements is important. We shall also discuss the key physics input that enters current attempts of dynamic modeling and the uncertainties resulting from it. However, we shall try to bypass as far as possible model-dependent details and rather focus on the question of which qualitative changes in the underlying dynamics result in characteristic changes of femtoscopic data.

Correlation measurements provide a snapshot of the geometrical distribution of particles at the time they decouple from the reaction. This geometrical distribution provides a unique test of the dynamical evolution of the produced matter at the late stage. Because the spatial extension, dynamical evolution, and lifetime of the produced system determines phase space density and thus particle reaction rates, any dynamic model for the latter has to be consistent with femtoscopic information. So far, correlation analyses have focused mainly on Boltzmann (or cascade) models, on hydrodynamic models, or on combinations of both (hybrid models). These model classes correspond to rather different equations of state.

The equations of state represented by cascade models tend to be stiff unless they incorporate large number of resonant scatterings. If particles collide via $2 \rightarrow 1 \rightarrow 2$ processes where the intermediate state has a finite lifetime, the equation of state can be softened (224, 225). A prominent example of a cascade model is RQMD (191), Relativistic Quantum Molecular Dynamics; it is the only one which has been compared to data at AGS, SPS, and RHIC. Results for RQMD are shown in Figure 13. However, for RHIC energies, the dynamic consistency of RQMD is questionable because for most of its evolution, the model uses hadronic degrees of freedom, although RQMD simulations for RHIC yield an energy density which stays above that of normal nuclear matter for a significant duration ($\sim$5 fm/c). As can be seen in Figure 15 (see color insert), RQMD, which models the expansion as a hadronic cascade, overpredicts R_{out} and R_{long} at RHIC despite the fact that it underpredicts multiplicities. Another hadron cascade model, the Hadronic Rescattering Model (226, 227), based solely on hadronic rescattering with sudden collisions, gives smaller sources, which come closer to the data. Part of the difference between the two hadronic cascades may derive from RQMD's treatment of scatterings as resonant interactions with finite lifetimes, which differs from the instantaneous collisions employed in the Hadronic Scattering Model. Figure 15 also displays results for Molnar's Parton Cascase (MPC) (228), which aims to provide a transparent partonic toy model by modeling the collision of light partons which undergo a one-to-one hadronization to pions. MPC, which has only instantaneous two-to-two scatterings, should have the stiffest effective equation of state and underestimates the source radii.

One of the most complicated transport codes available is A Multi-Phase Transport (AMPT) model. AMPT (229, 230) aims at a realistic description of all aspects of the reaction dynamics and includes a partonic cascade coupled to a hadronic cascade employing a full list of resonant interactions. AMPT provides a good fit to experimental radii, though the k_T fall-off is stronger than that found in the data, which fall off $\sim m_T^{-1/2}$. The more rapid decrease of radius parameters with respect to k_T may be due to the continuous surface-like emission characteristic of microscopic models, which when combined with cooling gives a higher relative weight for high-energy particles to have been emitted earlier in time, before the reaction volume has reached its full spatial extent. Any modification that would reduce this type of emission should improve agreement with data. Cautiously, one would conclude that the results in Figure 15 favor models with a stiff, but not too stiff, equation of state and no latent heat.

This conclusion is further supported if one compares the results of cascade models with those of hydrodynamic simulations or hybrid models. Figure 16 (see color insert) compares results from RHIC to a three-dimensional hydrodynamic model of Hirano et al. (231) that investigates the effects of resonance decays on chemical composition, the two-dimensional model of Heinz and Kolb (232), and a two-dimensional chiral model by Zschiesche et al. (233) that performs calculations for both first order and cross-over transitions. In all cases, the more favorable calculations are compared with the data: partial chemical freeze-out for Hirano, and cross-over transition for Zschiesche, but large discrepancies between the models and the data still remain. The results from Zschiesche were taken from the cross-over transition for the lowest critical temperature, 80 MeV. Substituting a first order phase transition or increasing the critical temperature to 130 MeV increased the value of R_{out} relative to R_{side}, thereby increasing the differences with the data. Hirano also reported a calculation under complete chemical equilibrium at freeze-out; this improves the agreement with v_2, but at the expense of a much larger value of R_{out}/R_{side}.

In general, these models invoke equations of state that are typically softer than those used in cascades and Boltzmann calculations, and they often have latent heats to accompany the transition from the partonic phase. As a consequence, lifetime and emission duration of the produced matter are significantly larger than what one finds in cascade models, and such models often significantly over-predict R_{long} and R_{out}/R_{side}. The fact that R_{side} comes out smaller than the data in Figure 16 is mainly due to an attempt to compensate within the available model-parameter range the very large time-scales as much as possible. Purely hydrodynamic models can vary overall source volumes by adjusting the break-up criteria, but doing so can make it difficult to fit the three source dimensions, and their m_T dependence, as well as other observables. Several prescriptions have been applied to improve the modeling of the breakup in the late stage to depend on microscopic considerations determined by free-space cross-sections without having an extra adjustable parameter, e.g., break-up density (106–108). An alternative to improve the description of breakup is the use of hybrid models, in which hydrodynamic evolution in the early stage is combined with cascading in the late stage. In Figure 16, we compare results of one such hybrid model, Ultrarelativistic Quantum Molecular Dynamics (URQMD) (109, 110), to results of hydrodynamic calculations. Similar results were obtained by Teaney et al. (111, 112). Compared to hydrodynamic simulations, hybrid descriptions do not seem to notably reduce the overpredicted lifetimes. They tend to emit most of the pions at a time near or above 15 fm/c and significantly overpredict R_{out}/R_{side} ratios, whereas blast-wave parameterizations favor breakup times near or slightly below 10 fm/c. However, firm conclusions that the relative failure of hybrid models derives from the chosen equations of state cannot be made until comparisons are made between Boltzmann and hybrid calculations that use the same equation of state. Until such an analysis is performed, other issues will cloud the interpretation, such as whether viscous effects or details of the hydrodynamic/Boltzmann interface dominate results and might even invalidate the hydrodynamic approach.

Entropy and pressure are intimately related, in that knowing the entropy density as a function of the energy density determines the pressure as a function of energy density. Whereas one can intuitively understand why a lower pressure would lead to longer lifetimes and larger source dimensions, the manifestations of changing the entropy are less transparent. One way to understand the effects of changing the entropy is to associate higher entropy with a larger effective number of light degrees of freedom. For instance, if one were to hadronize via a one-to-one parton-to-pion scheme, the volume per identical particle would change by a factor of the number of degrees of freedom. If the r.m.s. momentum is the same before and after hadronization, this would imply a change in entropy per particle equal to the logarithm of the ratio of effective degrees of freedom, and if $\sim$40 light partonic degrees of freedom were immediately replaced by three pionic degrees of freedom, the system would lose $\ln(40/3)$ units of entropy per particle. To conserve entropy, a system must expand its volume by a similar ratio which implies an increase in radius parameters. Hydrodynamic models, which manifestly conserve entropy, use the energy stored in the latent heat to provide the heat necessary to preserve entropy during hadronization. Entropy conservation is more difficult to enforce in microscopic approaches. This underlines the challenges involved in applying microscopic simulations in an environment of strongly interacting matter with ill-defined degrees of freedom, and it re-emphasizes the importance of gaining a better understanding of the validity of hydrodynamic calculations in such manifestly finite situations. It is peculiar that the entropy extracted from source-size and spectral measurements is consistent with the lattice-inspired equation of state (111, 112), whereas the source sizes extracted from hybrid models incorporating similar equations of state significantly overpredict source sizes. Part of this contradiction can be explained by increases in the populations of baryons, which, because they inherently have more entropy per particle than do pions, can account for much of the missing entropy (114). Thus, the HBT puzzle does not necessarily imply an entropy puzzle.

The HBT puzzle is not so much that radius parameters cannot be fit by models, but that our most sophisticated models, which incorporate a phase transition, fail to reproduce the data. The very gradual evolution of extracted source sizes as beam energies traverse a large range of energies is remarkable and puzzling in its own right. A simple explanation is that the equations of state do not dramatically change as the energy density changes from hadronic to super-hadronic densities, i.e., there is not even a hint of a latent heat. In fact, the failure of hydrodynamical models to reproduce femtoscopic measurements might be largely due to unrealistically large latent heats needed to reproduce elliptic flow data (115). However, as emphasized above, a host of unresolved issues prevent more quantitative conclusions. These qualifiers cannot be lifted until much more thorough analyses of models are performed that entail a systematic exploration of the sensitivity of model predictions to both parameters and assumptions. This would necessitate a tremendous commitment from the community, but without it, many conclusions about the matter created in relativistic heavy ion collisions will remain vague.

6. SUMMARY

Twenty-five years ago, the goal of femtoscopy was to demonstrate that one could measure a hadronic length scale with correlations, and if a result was on the order of a few Fermi, the analysis was deemed a success. In contrast, with the improved accuracy of measurements, the enormous increase in statistics, and the simultaneous development of phenomenology and theory, femtoscopy is now considered a precision measurement. Ten percent deviations between theory and experiment are now taken seriously as evidence that the spatio-temporal description of a model is significantly flawed. At RHIC energies, all six dimensions of the correlation function have been exploited to provide truly three-dimensional insight into the phase space cloud for particles of any momentum with any direction.

For relativistic heavy ion collisions, there were expectations that a transition from hadronic to partonic matter might be accompanied by a large latent heat, which would bring about a dramatic change in the dynamics as beam energies traversed the range for exploring the mixed phase. Within this range, it was expected that the latent heat and the associated softening of the equation of state would manifest itself by slowing the explosion with lifetimes approaching or exceeding 20 fm/c. The signal of the phase transition would have been an increase of the effective lifetime for a range of beam energies followed by a return to more explosive and shorter-lived reactions at even higher energies.

Extended lifetimes were not observed. Increasing beam energies from AGS to RHIC indeed causes larger energy densities and higher multiplicities, which push toward increasing the source volumes. However, much of this increase in multiplicities is absorbed by higher phase space densities rather than larger source sizes. Combined with the increasing strength of radial flow, which provides smaller regions of homogeneity relative to the overall source volume, the result is that the effective dimensions change remarkably little over a wide range of beam energies. Furthermore, it appears that lifetimes of the reaction never leave the neighborhood of 10 fm/c. Not only does this represent a lack of evidence for a latent heat, it also suggests that there is no such latent heat.

These conclusions remain only modestly guarded. Theory and phenomenology are progressing, but improvements in such aspects as mean-field effects or accounting for the smoothness approximation are not expected to change conclusions by more than 10%. As with the improvements in including Coulomb effects in Section 4, removing some of the distortions and aberrations from the analyses is likely to be significant for fine-tuning models, but should not alter the conclusion that there is no large latent heat associated with the reaction.

The theory of modeling, i.e., generating the source functions, has the greatest need for progress. Because femtoscopic measurements are determined solely by the geometry of breakup, changes in chemical or kinetic evolution may have a significant impact on extracted source dimensions. The next generation of transport theories should be more flexible and will probably incorporate numerous effects such as in-medium mass changes, in-medium reduction of scattering cross-sections, viscous effects, and dynamical solutions for chemical rates.

Despite the progress listed above, measurement has only begun to address the rich expanse of information available in correlations. Nearly all the three-dimensional analyses have been focused on identical-pion correlations. The huge data sets of the recent and upcoming runs at RHIC make it possible to analyze source functions for many pairs of particles in six full dimensions. In addition to providing important verification of identical-pion measurements, these analyses address other issues, such as whether all species flow and break up together.

This is not a field for the complacent. As emphasized above, efforts at RHIC are just beginning to explore wholly new classes of correlations. Energy densities at the LHC might surpass those at RHIC by the same factor that those at RHIC surpassed the AGS. Just as our visions of the future from twenty years ago proved largely naive, we should be prepared to be surprised with the femtoscopy of the next 25 years.

ACKNOWLEDGMENTS

This work was supported by U.S. Department of Energy, Grants DE-FG02-03ER41259 and W-7405-ENG-48, and by U.S. National Science Foundation Grant PHY-0355007. The authors thank David Brown, Ulrich Heinz, Sergei Panitkin, Nu Xu, and Kacper Zalewski for sharing insights. For providing and explaining data, we thank Harry Appelshäuser, Giuseppe Bruno, Mike Heffner, Mercedes López-Noriega, Michael Murray and the experimental collaborations who conscientiously post data in numerical form and respond to querries. We thank Mark Meamber of the Physics Art Team at Lawrence Livermore National Laboratory for creating the illustrative figures.

The *Annual Review of Nuclear and Particle Science* is online at
http://nucl.annualreviews.org

LITERATURE CITED

1. Adams J, et al. nucl-ex/0501009
2. Adcox K, et al. nucl-ex/0410003
3. Back BB, et al. nucl-ex/0410022
4. Arsene I, et al. nucl-ex/0410020
5. Boal DH, Gelbke CK, Jennings BK. *Rev. Mod. Phys.* 62:553 (1990)
6. Bauer W, Gelbke CK, Pratt S. *Annu. Rev. Nucl. Part. Sci.* 42:77 (1992)
7. Heinz UW, Jacak BV. *Annu. Rev. Nucl. Part. Sci.* 49:529 (1999)
8. Wiedemann UA, Heinz UW. *Phys. Rep.* 319:145 (1999)
9. Csorgo T. *Heavy Ion Phys.* 15:1 (2002)
10. Weiner RM. *Phys. Rep.* 327:249 (2000)
11. Tomasik B, Wiedemann UA. hep-ph/0210250
12. Alexander G. *Rep. Prog. Phys.* 66:481 (2003)
13. Hanbury-Brown R, Twiss RQ. *Nature* 178:1046 (1956)
14. Hanbury-Brown R, Twiss RQ. *Philos. Mag.* 45:663 (1954)
15. Goldhaber G, Goldhaber S, Lee W-Y, Pais A. *Phys. Rev.* 120:300 (1960)
16. Kopylov GI, Podgoretsky MI. *Sov. J. Nucl. Phys.* 15:219 (1972)
17. Kopylov GI, Lyuboshits VL, Podgoretsky MI. JINR-P2-8069 (1974)
18. Kopylov GI, Podgoretsky MI. *Sov. J. Nucl. Phys.* 18:336 (1974)
19. Kopylov GI. *Phys. Lett.* B50:472 (1974)
20. Koonin SE. *Phys. Lett.* B70:43 (1977)
21. Gyulassy M, Kauffmann SK, Wilson LW. *Phys. Rev. C* 20:2267 (1979)
22. Zajc WA, et al. *Phys. Rev. C* 29:2173 (1984)
23. Fung SY, et al. *Phys. Rev. Lett.* 41:1592 (1978)
24. Humanic TJ. *Phys. Rev. C* 34:191 (1986)
25. Baym G. *Acta Phys. Pol.* B29:1839 (1998)

26. Kopylov GI, Podgoretsky MI. *Sov. Phys.-JETP* 42:211 (1976)
27. Lednicky R, Lyuboshits VL. In *Proc. Particle Correlations and Interferometry in Nuclear Collisions*, p. 42. Nantes, France: World Sci. (1990)
28. Lednicky R. nucl-th/0212089
29. Lednicky R, Lyuboshits VL. *Sov. J. Nucl. Phys.* 35:770 (1982)
30. Akkelin SV, Sinyukov YM. *Phys. Lett.* B356:525 (1995)
31. Barz HW. *Phys. Rev. C* 59:2214 (1999)
32. Barz HW. *Phys. Rev. C* 53:2536 (1996)
33. Cramer JG, Miller GA, Wu JMS, Yoon J-H. nucl-th/0411031
34. Chu MC, Gardner S, Matsui T, Seki R. *Phys. Rev. C* 50:3079 (1994)
35. Kapusta JI, Li Y. nucl-th/0503075
36. Pratt S. *Phys. Lett.* B301:159 (1993)
37. Heinz UW, Scotto P, Zhang QH. *Ann. Phys.* 288:325 (2001)
38. Zhang QH, Scotto P, Heinz UW. *Phys. Rev. C* 58:3757 (1998)
39. Csorgo T, Zimanyi J. *Phys. Rev. Lett.* 80:916 (1998)
40. Zimanyi J, Csorgo T. *Heavy Ion Phys.* 9:241 (1999)
41. Anchishkin D, Heinz UW, Renk P. *Phys. Rev. C* 57:1428 (1998)
42. Zhang QH, Wiedemann UA, Slotta C, Heinz UW. *Phys. Lett.* B407:33 (1997)
43. Pratt S. *Phys. Rev. C* 56:1095 (1997)
44. Aichelin J. *Nucl. Phys.* A617:510 (1997)
45. Wiedemann UA, et al. *Phys. Rev. C* 56:614 (1997)
46. Wiedemann UA, Ferenc D, Heinz UW. *Phys. Lett.* B449:347 (1999)
47. Padula SS, Roldao CG. *Phys. Rev. C* 58:2907 (1998)
48. Martin M, Kalechofsky H, Foka P, Wiedemann UA. *Eur. Phys. J. C* 2:359 (1998)
49. Barate R, et al. *Phys. Lett.* B478:50 (2000)
50. Abreu P, et al. *Phys. Lett.* B471:460 (2000)
51. Pratt S. *Phys. Rev. D* 33:72 (1986)
52. Brown DA, Danielewicz P. *Phys. Rev. C* 64:014902 (2001)
53. Brown DA, Wang F, Danielewicz P. *Phys. Lett.* B470:33 (1999)
54. Brown DA, Danielewicz P. *Phys. Lett.* B398:252 (1997)
55. Chung P, et al. *Phys. Rev. Lett.* 91:162301 (2003)
56. Verde G, et al. *Phys. Rev. C* 65:054609 (2002)
57. Verde G, et al. *Phys. Rev. C* 67:034606 (2003)
58. Wang F, Pratt S. *Phys. Rev. Lett.* 83:3138 (1999)
59. Pratt S, Petriconi S. *Phys. Rev. C* 68:054901 (2003)
60. Lednicky R, Lyuboshits VL, Erazmus B, Nouais D. *Phys. Lett.* B373:30 (1996)
61. Voloshin S, Lednicky R, Panitkin S, Xu N. *Phys. Rev. Lett.* 79:4766 (1997)
62. Adams J, et al. *Phys. Rev. Lett.* 91:262302 (2003)
63. Pratt S. *Phys. Rev. D* 33:1314 (1986)
64. Bertsch GF. *Nucl. Phys.* A498:C173 (1989)
65. Csorgo T, et al. *Phys. Lett.* B241:301 (1990)
66. Wiedemann UA. *Phys. Rev. C* 57:266 (1998)
67. Lisa MA, Heinz UW, Wiedemann UA. *Phys. Lett.* B489:287 (2000)
68. Heinz UW, Hummel A, Lisa MA, Wiedemann UA. *Phys. Rev. C* 66:044903 (2002)
69. Danielewicz P, Pratt S. nucl-th/0501003
70. Chajecki Z, Gutierrez TD, Lisa MA, Lopez-Noriega M. nucl-ex/0505009
71. Chapman S, Nix JR, Heinz UW. *Phys. Rev. C* 52:2694 (1995)
72. Retiere F, Lisa MA. *Phys. Rev. C* 70:044907 (2004)
73. Miskowiec D. nucl-ex/9808003
74. Heinz UW, Tomasik B, Wiedemann UA, Wu YF. *Phys. Lett.* B382:181 (1996)
75. Voloshin SA, Cleland WE. *Phys. Rev. C* 53:896 (1996)
76. Wiedemann UA, Heinz UW. *Phys. Rev. C* 56:610 (1997)
77. Yano FB, Koonin SE. *Phys. Lett.* B78:556 (1978)

78. Wu YF, Heinz UW, Tomasik B, Wiedemann UA. *Eur. Phys. J. C* 1:599 (1998)

79. Chapman S, Scotto P, Heinz UW. *Phys. Rev. Lett.* 74:4400 (1995)

80. Csorgo T, Lorstad B, Zimanyi J. *Z. Phys. C* 71:491 (1996)

81. Heinz UW, Zhang QH. *Phys. Rev. C* 56:426 (1997)

82. Heinz UW, Sugarbaker A. *Phys. Rev. C* 70:054908 (2004)

83. Boggild H, et al. *Phys. Lett.* B455:77 (1999)

84. Bearden IG, et al. *Phys. Lett.* B517:25 (2001)

85. Adams J, et al. *Phys. Rev. Lett.* 91:262301 (2003)

86. Lisa MA, et al. *Phys. Rev. Lett.* 84:2798 (2000)

87. Makhlin AN, Sinyukov YuM. *Z. Phys. C* 39:69 (1988)

88. Kolehmainen K. *Phys. Lett.* B180:203 (1986)

89. Padula SS, Gyulassy M. *Nucl. Phys.* B339:378 (1990)

90. Heinz UW, Kolb PF. *Phys. Lett.* B542:216 (2002)

91. Tomasik B, Heinz UW. *Phys. Rev. C* 65:031902 (2002)

92. Chojnacki M, Florkowski W, Csorgo T. *Phys. Rev. C* 71:044902 (2005)

93. Csorgo T, et al. *Phys. Rev. C* 67:034904 (2003)

94. Lee KS, Heinz UW, Schnedermann E. *Z. Phys. C* 48:525 (1990)

95. Schnedermann E, Sollfrank J, Heinz UW. *Phys. Rev. C* 48:2462 (1993)

96. Pratt S. *Phys. Rev. Lett.* 53:1219 (1984)

97. Sinyukov YuM, Akkelin SV, Xu N. *Phys. Rev. C* 59:3437 (1999)

98. Molnar D, Gyulassy M. *Phys. Rev. C* 62:054907 (2000)

99. Cheng S, et al. *Phys. Rev. C* 65:024901 (2002)

100. Kolb PF, Heinz U. nucl-th/0305084

101. Huovinen P, el al. *Phys. Lett.* B503:58 (2001)

102. Hirano T. nucl-th/0410017

103. Heinz UW, Kolb PF. *Nucl. Phys.* A702:269 (2002)

104. Teaney D. *Phys. Rev. C* 68:034913 (2003)

105. Cooper F, Frye G, Schonberg E. *Phys. Rev. D* 11:192 (1975)

106. Tomasik B, Wiedemann UA. *Phys. Rev. C* 68:034905 (2003)

107. Csernai LP, et al. hep-ph/0406082

108. Sinyukov YuM, Akkelin SV, Hama Y. *Phys. Rev. Lett.* 89:052301 (2002)

109. Soff S, Bass SA, Dumitru A. *Phys. Rev. Lett.* 86:3981 (2001)

110. Bass SA, Dumitru A. *Phys. Rev. C* 61:064909 (2000)

111. Teaney D, Lauret J, Shuryak EV. *Nucl. Phys.* A698:479 (2002)

112. Teaney D, Lauret J, Shuryak EV. nucl-th/0110037

113. Bertsch GF. *Phys. Rev. Lett.* 72:2349 (1994)

114. Pal S, Pratt S. *Phys. Lett.* B578:310 (2004)

115. Huovinen P. nucl-th/0505036

116. Danielewicz P, Bertsch GF. *Nucl. Phys.* A533:712 (1991)

117. Llope WJ, et al. *Phys. Rev. C* 52:2004 (1995)

118. Adams J, et al. *Phys. Rev. C* 71:044906 (2005)

119. Lisa MA, et al. nucl-ex/0503017

120. Boggild H, et al. *Phys. Lett.* B349:386 (1995)

121. Ahle L, et al. *Phys. Rev. C* 66:054906 (2002)

122. Adler SS, et al. *Phys. Rev. Lett.* 93:152302 (2004)

123. Boggild H, et al. *Phys. Lett.* B302:510 (1993)

124. Kopylov GI. *Phys. Lett.* B50:472 (1974)

125. Abreu P, et al. *Phys. Lett.* B286:201 (1992)

126. Abbiendi G, et al. *Eur. Phys. J. C* 16:423 (2000)

127. Uribe Duque J. *Phys. Rev. D* 49:4373 (1994). UMI-93-29679

128. Avery P, et al. *Phys. Rev. D* 32:2294 (1985)

129. Anderson M, et al. *Nucl. Instrum. Methods A* 499:659 (2003)

.t al. *Nukleonica* 49:S23

130. Stavinskiy
(2004) , WG, Gelbke CK, Lynch
131. Lisa M. *C* 44:2865 (1991)
WG. *Phys. Lett.* B496:1 (2000)
132. Lis? thesis. Ohio State Univ.
W. 02)
133. W. l. *Phys. Rev. Lett.* 93:012301

134 M, Voloshin SA. *Phys. Rev.*
(1998)
V, Ollitrault JY. *Phys. Rev. C*
7.34, 5 (2004)
IG, et al. *Phys. Rev. C* 58:1656
ett.
MG. *Phys. Lett.* B270:69 (1991)
ov Yu, et al. *Phys. Lett.* B432:248
6)
nacon AD, et al. *Phys. Rev. C* 43:2670
(1991)

141. Brown DA, Danielewicz P, Heffner M, Soltz R. nucl-th/0404067

142. Adler C, et al. *Phys. Rev. Lett.* 87:082301 (2001)

143. Csorgo T, Hegyi S, Zajc WA. *Eur. Phys. J. C* 36:67 (2004)

144. Bartke J. *Phys. Lett.* B174:32 (1986)

145. Humanic TJ, et al. *Z. Phys. C* 38:79 (1988)

146. Bamberger A, et al. *Phys. Lett.* B203:320 (1988)

147. Abbott T, et al. *Phys. Rev. Lett.* 69:1030 (1992)

148. Barrette J, et al. *Phys. Lett.* B333:33 (1994)

149. Adamova D, et al. *Nucl. Phys.* A714:124 (2003)

150. Barrette J, et al. *Phys. Rev. C* 50:1077 (1994)

151. Barrette J, et al. *Phys. Rev. C* 60:054905 (1999)

152. Boggild H, et al. *Phys. Lett.* B458:181 (1999)

153. Lisa MA, et al. *Phys. Rev. Lett.* 70:3709 (1993)

154. Kotte R, et al. *Eur. J. Phys. A* 23:271 (2005)

155. Lisa MA, et al. *Phys. Rev. Lett.* 71:2863 (1993)

156. Lisa MA, et al. *Phys. Rev. C* 49:2788 (1994)

157. Kotte R, et al. *Z. Phys. A* 359:47 (1997)

158. Adler SS, et al. *Phys. Rev. C* 71:034908 (2005)

159. Csorgo T, Lorstad B. *Phys. Rev. C* 54:1390 (1996)

160. Voloshin SA, Cleland WE. *Phys. Rev. C* 54:3212 (1996)

161. Heiselberg H. *Phys. Rev. Lett.* 82:2052 (1999)

162. Heiselberg H, Levy A-M. *Phys. Rev. C* 59:2716 (1999)

163. Ackermann KH, et al. *Phys. Rev. Lett.* 86:402 (2001)

164. Pinkenburg C, et al. *Phys. Rev. Lett.* 83:1295 (1999)

165. Miskowiec D. *Nucl. Phys.* A590:C473 (1995)

166. Filimonov K, et al. *Nucl. Phys.* A661:198 (1999)

167. Nishimura S, et al. *Proc. 15th Winter Workshop on Nuclear Dynamics, Park City, UT* (1999)

168. Aggarwal MM, et al. *Nucl. Phys.* A663:729 (2000)

169. Panitkin SY. nucl-ex/9905007

170. Csernai LP, Rohrich D. *Phys. Lett.* B458:454 (1999)

171. Brachmann J, et al. *Phys. Rev. C* 61:024909 (2000)

172. Magas VK, Csernai LP, Strottman DD. hep-ph/0101125

173. Shuryak EV. *Phys. Rep.* 61:71 (1980)

174. Bjorken JD. *Phys. Rev. D* 27:140 (1983)

175. Heinz UW. *Nucl. Phys.* A610:C264 (1996)

176. Miskowiec D, et al. *Nucl. Phys.* A610:C227 (1996)

177. Appelshauser H, et al. *Eur. Phys. J. C* 2:661 (1998)

178. Antinori F, et al. *J. Phys. G* 27:2325 (2001)

179. Back BB, et al. nucl-ex/0409001

180. Renk T. *Phys. Rev. C* 69:044902 (2004)

181. Aggarwal MM, et al. *Phys. Rev. C* 67:014906 (2003)

182. Adcox K, et al. *Phys. Rev. Lett.* 88:192302 (2002)
183. Xu N, Kaneta M. *Nucl. Phys.* A698:306 (2002)
184. Beker H, et al. *Phys. Rev. Lett.* 74:3340 (1995)
185. Csorgo T, Lorstad B. *Phys. Rev. C* 54: 1390 (1996)
186. Gyulassy M, Padula SS. *Phys. Rev. C* 41: 21 (1990)
187. Schnetzer S, et al. *Phys. Rev. Lett.* 49:989 (1982)
188. Akiba Y, et al. *Phys. Rev. Lett.* 70:1057 (1993)
189. Beker H, et al. *Z. Phys. C* 64:209 (1994)
190. Cianciolo V. *Nucl. Phys.* A590:C459 (1995)
191. Sorge H, Stocker H, Greiner W. *Nucl. Phys.* A498:C567 (1989)
192. Wiedemann UA, Scotto P, Heinz UW. *Phys. Rev. C* 53:918 (1996)
193. Bearden IG, et al. *Phys. Rev. Lett.* 87: 112301 (2001)
194. Afanasiev SV, et al. *Phys. Lett.* B557:157 (2003)
195. Appelshauser H, et al. *Phys. Lett.* B467:21 (1999)
196. Aggarwal MM, et al. *Phys. Rev. Lett.* 93: 022301 (2004)
197. Heffner M. *J. Phys. G* G30:S1043 (2004)
198. Bekele S. *J. Phys. G* G30:S229 (2004)
199. Renault G. hep-ex/0404024
200. Kapusta JI, Li Y. *J. Phys. G* 30:S1069 (2004)
201. Wong C-Y. hep-ph/0403025
202. Lednicky R. nucl-th/0112011
203. Alber T, et al. *Z. Phys. C* 73:443 (1997)
204. Renault G. hep-ex/0406066
205. Blume C, et al. *Nucl. Phys.* A715:55 (2003)
206. Kisiel A. *J. Phys. G* 30:S1059 (2004)
207. Bertsch G, Gong M, Tohyama M. *Phys. Rev. C* 37:1896 (1988)
208. Rischke DH, Gyulassy M. *Nucl. Phys.* A608:479 (1996)
209. Harris JW, Müller ʙ. Part. Sci. 46:71 (1996). *Rev. Nucl.*
210. Adamova D, et al. *Rev. Lett.* 90:022301 (2003)
211. Hardtke D, Voloshin SA. *Lett.* 024905 (2000)
212. Sorge H. *Phys. Rev. C* 52:3251.
213. Lisa MA, et al. *Nucl. Phys* (2002)
214. Afanasiev SV, et al. *Phys. R* 054902 (2002)
215. Kniege S, et al. nucl-ex/0403034
216. Ferenc D, et al. *Phys. Lett.* B4 (1999)
217. Greiner C, Gong C, Muller B. *Phys.* B316:226 (1993)
218. Pratt S, Haglin K. *Phys. Rev. C* 59:3ɔ (1999)
219. Akkelin SV, Sinyukov YuM. *Phys. Rev. C* 70:064901 (2004)
220. Braun-Munzinger P, Redlich K, Stachel J. nucl-th/0304013
221. Braun-Munzinger P, Magestro D, Redlich K, Stachel J. *Phys. Lett.* B518:41 (2001)
222. Andronic A, Braun-Munzinger P, Redlich K, Stachel J. *Phys. Lett.* B571:36 (2003)
223. Adcox K, et al. *Phys. Rev. Lett.* 87:052301 (2001)
224. Danielewicz P, Pratt S. *Phys. Rev. C* 53: 249 (1996)
225. Larionov AB, Effenberger M, Leupold S, Mosel U. *Phys. Rev. C* 66:054604 (2002)
226. Humanic TJ. *Nucl. Phys.* A715:641 (2003)
227. Humanic TJ. nucl-th/0301055
228. Molnar D, Gyulassy M. *Phys. Rev. Lett.* 92:052301 (2004)
229. Lin ZW, Ko CM, Pal S. *Phys. Rev. Lett.* 89:152301 (2002)
230. Lin ZW, Ko CM. *J. Phys. G* 30:S263 (2004)
231. Hirano T, Tsuda K. *Phys. Rev. C* 66: 054905 (2002)
232. Heinz UW, Kolb PF. hep-ph/0204061
233. Zschiesche D, Stocker H, Greiner W, Schramm S. *Phys. Rev. C* 65:064902 (2002)

Annu. Rev. Nucl. Part. Sci. 2005. 55:403–65
doi: 10.1146/annurev.nucl.53.041002.110615
Copyright © 2005 by Annual Reviews. All rights reserved

SMALL-x PHYSICS: From HERA to LHC and Beyond

Leonid Frankfurt

*School of Physics and Astronomy, Tel Aviv University, 69978 Tel Aviv, Israel;
email: frankfur@lev.tau.ac.il*

Mark Strikman

*Department of Physics, Pennsylvania State University, University Park,
Pennsylvania 16802; email: strikman@phys.psu.edu*

Christian Weiss

Theory Group, Jefferson Lab, Newport News, Virginia 23606; email: weiss@jlab.org

Key Words   diffraction, hadronic final states, high-energy scattering, quantum chromodynamics

PACS Codes: 11.80.La, 12.40.Gg, 12.40.Pp, 25.40.Ve, 27.75. + r

■ **Abstract**   We summarize the lessons learned from studies of hard scattering processes in high-energy electron-proton collisions at HERA and antiproton-proton collisions at the Tevatron, with the aim of predicting new strong interaction phenomena observable in next-generation experiments at the Large Hadron Collider (LHC). Processes reviewed include inclusive deep-inelastic scattering (DIS) at small x, exclusive and diffractive processes in DIS and hadron-hadron scattering, as well as color transparency and nuclear shadowing effects. A unified treatment of these processes is outlined on the basis of factorization theorems of quantum chromodynamics, and using the correspondence between the "parton" picture in the infinite-momentum frame and the "dipole" picture of high-energy processes in the target rest frame. The crucial role of the three dimensional quark and gluon structure of the nucleon is emphasized. A new dynamical effect predicted at high energies is the unitarity, or black disk, limit (BDL) in the interaction of small dipoles with hadronic matter, owing to the increase of the gluon density at small x. This effect is marginally visible in diffractive DIS at HERA and will lead to the complete disappearance of Bjorken scaling at higher energies. In hadron-hadron scattering at LHC energies and beyond (cosmic ray physics), the BDL will be a standard feature of the dynamics, with implications for (*a*) hadron production at forward and central rapidities in central proton-proton and proton-nucleus collisions, in particular events with heavy particle production (Higgs), (*b*) proton-proton elastic scattering, and (*c*) heavy-ion collisions. We also outline the possibilities for studies of diffractive processes and photon-induced reactions (ultraperipheral collisions) at LHC, as well as possible measurements with a future electron-ion collider.

CONTENTS

1. INTRODUCTION

In understanding the nature of strong interactions, progress has mostly come from the investigation of certain "extreme" kinematic regions, in which the dynamics simplifies. One such region is high-energy hadron-hadron scattering, in which

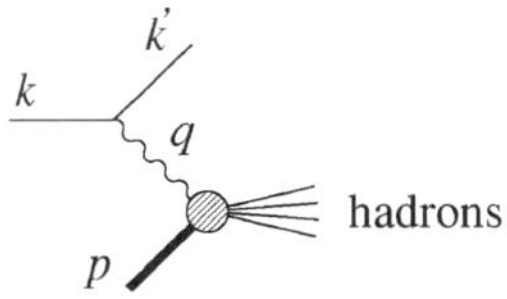

$$Q^2 = -q^2$$
$$W^2 = (p+q)^2$$
$$x = \frac{Q^2}{W^2 + Q^2}$$

Figure 1   The kinematics of deep-inelastic lepton-hadron scattering (DIS). The interaction proceeds by exchange of a virtual photon, whose four-momentum is given by the difference of the lepton momenta, $q = k' - k$. The hadronic scattering process is characterized by two kinematic invariants, the photon virtuality, $q^2 \equiv -Q^2 < 0$, and the photon-proton center-of-mass energy, W, or, alternatively, the Bjorken scaling variable, x. (We neglect the target mass.)

the center-of-mass energy is significantly larger than the masses of the hadronic systems in the initial and final state. Historically, this was the first area in which powerful mathematical methods, such as dispersion relations and Reggeon calculus, could be applied to strong interaction phenomena. They are based on the general principles of unitarity of the scattering matrix (conservation of probability) and analyticity of scattering amplitudes (causality). These methods have given us important insights into general properties of high-energy processes, such as the increase of the radius of interaction with energy predicted by Gribov (1, 2), the Froissart bound for the growth of total hadronic cross sections with energy (3), and the Pomeranchuk theorem of asymptotic equality of particle-particle and particle-antiparticle cross sections.

Further progress came with the study of "hard" scattering processes, characterized by a momentum transfer significantly larger than the typical mass scale associated with hadron structure, μ (a reasonable numerical value for this scale is the ρ meson mass). Such processes can be described in quantum chromodynamics (QCD), the field theory of interacting quarks and gluons, the fundamental property of which is the smallness of the effective coupling constant in small space-time intervals (asymptotic freedom) (4, 5). Hard processes happen so "rapidly" that they do not significantly change the environment of the interacting quarks and gluons inside the hadrons. This allows one to calculate their amplitudes using a technique termed factorization—a systematic separation into a hard quark-gluon scattering process and certain functions describing the distribution of quarks and gluons in the participating hadrons. The simplest such process is deep-inelastic lepton-hadron scattering (DIS) in the so-called Bjorken limit, $Q^2 \sim W^2 \gg \mu^2$ (Figure 1). Historically, the observation of scaling behavior in the structure functions of inclusive DIS (6) gave the first indication of the presence of quasi-free, pointlike constituents in the proton (7). Another class of processes for which factorization is possible are certain hard processes in hadron-hadron scattering, such as the production of jets with large transverse momenta or large-mass dilepton pairs.

A particularly interesting region of strong interactions are hard scattering processes in the region where the center-of-mass energy becomes large compared to the momentum transfer, $W^2 \gg Q^2 \gg \mu^2$. In DIS this limit corresponds to values of the Bjorken variable $x \ll 1$ (Figure 1), whence this field is known as "small-x physics." On one hand, because of the large momentum transfer, such processes probe the quark and gluon degrees of freedom of QCD. On the other hand, they share many characteristics with high-energy hadron-hadron scattering, such as a large spatial extension of the interaction region along the collision axis (this will be explained in detail in Section 2.2). The treatment of such processes generally requires a combination of the methods of QCD factorization and "pre-QCD" methods of high-energy hadron-hadron scattering for modeling the dynamics of the hadronic environment of the quarks and gluons participating in the hard process. From the point of view of QCD, the high-energy (small-x) region corresponds to a greatly increased phase space for gluon radiation as compared to $x \gtrsim 0.3$. QCD predicts a fast rise of the gluon density in the nucleon with decreasing x, and thus a strong increase of the DIS cross section with energy (8, 9). A challenging question, which is presently being addressed in different approaches, is the role of unitarity of the scattering matrix in such processes at high energies. More generally, one hopes to eventually understand the quark-gluon dynamics underlying such general high-energy phenomena as the growth of the radius of interaction with energy, and the Froissart bound. This dynamics would correspond to a strongly interacting quark-gluon system at small coupling constant, and represent a fascinating new form of "QCD matter" that could be produced in the laboratory.

An intuitive understanding of the dynamics of hard scattering processes and the basis for QCD factorization can be developed by following the space-time evolution of the reactions in certain reference frames. At high energies (small x), the space-time evolution of DIS can be discussed in two complementary ways. In a frame where the proton is fast-moving one obtains the well-known parton picture of hard processes, in which the hard scattering process involves quarks and gluons carrying a certain fraction of the proton's momentum. In the proton rest frame, on the other hand, the DIS process takes the form of the scattering of a quark-antiquark dipole from the target, with the dipole formed a long time before reaching the target, and having a distribution of transverse sizes extending down to values $\sim 1/Q$. This representation reveals a close relation between DIS at small x and the so-called "color transparency" phenomenon—the transparency of hadronic matter to the propagation of spatially small color-singlet configurations, as observed, e.g., in the suppression of the interaction of heavy quarkonia with hadronic matter. The correspondence between the "parton" and the "dipole" picture of small-x processes is a powerful tool for analyzing the dynamics of strong interactions in this regime. (A pedagogical introduction to these concepts will be given in Section 2.)

The experimental investigation of small-x processes became possible with the advent of high-energy colliders (colliding beam facilities). Extensive studies of DIS at small x have been performed at the HERA electron-proton (ep) collider at DESY. Measurements of inclusive cross sections have spectacularly confirmed the rise of the gluon density in the proton at small x, as predicted by QCD, down to values $x \sim 10^{-4}$. Measurements of exclusive processes in DIS, such as heavy and light vector meson production (J/ψ, ρ), provide information about the spatial distribution of partons in the transverse plane ("generalized parton distributions") and allow us to construct a three dimensional image of the quark and gluon structure of the nucleon. Finally, measurements of diffractive processes in DIS, in which the produced hadronic system is separated from the target remnants by a large rapidity gap, make it possible to probe the interaction of various small-size color-singlet configurations with the proton in much more detail than inclusive DIS.

Another—potentially much more powerful—laboratory for studying small-x physics are high-energy proton-proton (pp) and antiproton-proton ($\bar{p}p$) colliders, such as LHC at CERN (under construction) and the Tevatron at Fermilab. QCD factorization can be applied to $pp/\bar{p}p$ collisions with hard processes, such as the production of dijets with large transverse momenta or heavy particles ($W^{\pm}$ bosons, Higgs bosons, etc.), which originate from binary collisions of quarks and gluons in the two colliding hadrons. At LHC, such processes can probe parton distributions down to values of $x \sim 10^{-7}$. Even higher energies are reached in collisions of cosmic-ray particles near the Greisen-Zatsepin-Kuzmin cutoff (10) with atmospheric nuclei. In pp scattering, as compared to ep, one is dealing with collisions of two objects with a complex internal structure. This results, e.g., in a high probability of multiple hard scattering processes at high energies, and a much richer spectrum of soft hadronic interactions. Thus, although QCD factorization can still be applied to certain hard processes in $pp/\bar{p}p$ collisions, the modeling of the hadronic environment of the quarks and gluons participating in the hard process (or processes) is generally much more challenging than in ep scattering.

This review is an attempt to summarize what has been learned about small-x physics from experiments at HERA and the Tevatron and related theoretical studies, and use this information to make predictions for new QCD phenomena observable at LHC. With LHC about to be commissioned, and the HERA program nearing completion, this is a timely exercise. It is not our aim to give a comprehensive overview of the existing collider experiments and their numerous implications for our understanding of QCD. Rather, we identify certain specific "lessons" that are of particular importance in making the transition from HERA to LHC:

- *Black-disk limit in dipole-hadron interactions.* Because of the fast rise of the gluon density at small x, the strength of interaction of small color-singlet configurations (dipoles) with hadronic matter can approach the maximum value allowed by s-channel unitarity. We quantify this effect by formulating an optical model of dipole-hadron scattering, in which the unitarity limit corresponds to the scattering from a "black disk" whose radius increases with energy (black-disk limit, or BDL). We argue that the onset of the BDL regime can be seen in the diffractive DIS data at the upper end of the HERA energy range, as well as in elastic $pp/\bar{p}p$ scattering at Tevatron energies. In DIS at higher energies, the BDL leads to a breakdown of Bjorken scaling. In hadron-hadron collisions at LHC energies and beyond (cosmic ray physics), the dynamics will be deep inside the BDL regime, with numerous consequences for the hadronic final states.

■ *Small transverse area of leading partons.* Studies of hard exclusive processes at HERA and Fermilab show that partons in the nucleon with $x > 10^{-2}$ and significant transverse momenta are concentrated in a small transverse area, $\ll 1 \, \text{fm}^2$, substantially smaller than the area associated with the nucleon in soft (hadronic) interactions at high energies. The resulting "two-scale picture" of the transverse structure of the nucleon is essential for modeling the hadronic environment of the colliding partons in high-energy pp collisions with hard processes.

On the basis of these observations, we make several predictions for new strong interaction phenomena observable in pp, pA (proton-nucleus), and AA (nucleus-nucleus) collisions at LHC:

■ *Hard processes as a trigger for central pp collisions.* In pp scattering at LHC, hard QCD processes involving binary collisions of partons with momentum fractions $x_{1,2} > 10^{-2}$ occur practically only in pp events with small impact parameters (central collisions). This makes it possible to trigger on central pp events by requiring the presence of a hard dijet (or double dijet) at small rapidities.

■ *Black-disk limit in central pp/pA/AA collisions at LHC.* The approach to the BDL at high energies will strongly affect the dynamics of central $pp/pA/AA$ collisions at LHC energies and above. A crucial point is that in hadron-hadron scattering at such energies one is dealing mostly with gluon-gluon dipoles, whose cross section for scattering from hadronic matter is $9/4$ times larger than that of the quark-antiquark dipoles dominating in ep scattering. We argue that, as a consequence of the approach to the BDL, the leading partons in central pp collisions will acquire large transverse momenta ($p_\perp^2 \sim$ several $10 \, \text{GeV}^2$) and fragment independently, resulting in the disappearance of leading hadrons with small transverse momenta at forward/backward rapidities, increased energy loss, and increased soft particle production at central rapidities. These observable effects allow for experimental studies of this fascinating new regime of "strong gluon fields" at LHC. We also outline the role of the BDL in heavy-ion collisions and cosmic ray physics.

■ *Diffraction in high-energy pp collisions.* LHC offers the possibility to study a wide variety of diffractive processes in high-energy pp scattering, which probe the interaction of small-size color singlets with hadronic matter and can be used to map the gluon distribution in the proton. Such processes involve a delicate interplay between hard (partonic) and soft (hadronic) interactions. A crucial ingredient in understanding the dynamics is the information about the transverse spatial distribution of gluons obtained from exclusive vector meson production in DIS at HERA.

We also comment on the potential of LHC for parton distribution measurements at small x, and for studies of small-x dynamics via photon-induced reactions in ultraperipheral $pp/pA/AA$ collisions. Finally, we discuss the opportunities for studies of small-x dynamics provided by the planned electron-ion collider.

The primary purpose of LHC is the search for new heavy particles (Higgs bosons, supersymmetry) in high-energy pp collisions. The small-x phenomena we describe here directly impacts this program. Heavy particles are produced in hard partonic collisions. For the reason described above, the production of heavy particles in inelastic pp collisions happens predominantly in central collisions, which are strongly affected by the approach to the BDL, and the strong interaction background may be completely different from what one would expect on the basis of the naive extrapolation of existing data (Tevatron). Likewise, the search for Higgs bosons in diffractive pp events depends crucially on the understanding of the strong interaction dynamics in these processes.

2. QCD FACTORIZATION AND THE SPACE-TIME EVOLUTION OF SMALL-x SCATTERING

2.1. QCD Factorization of Hard Processes

We begin by introducing the basic concept of QCD factorization of hard processes, and outlining the space-time evolution of small-x scattering processes at high energies. Our main point is the correspondence between the "parton" picture of hard processes in a frame in which the nucleon is moving fast, and the "dipole" picture of high-energy processes in the target rest frame. In this section we illustrate this correspondence using as an example the simplest high-energy process, inclusive DIS at small x. Below we shall apply these results to analyze the HERA DIS data (Section 3), and generalize them to processes with exclusive (Section 4) and diffractive (Section 5) final states. The correspondence between the two pictures also plays a crucial role in formulating the approach to the unitarity limit at high energies (Section 6), and for understanding the dynamics of high-energy hadron-hadron collisions (Sections 7 and 8).

DIS is essentially the scattering of a virtual photon (γ^*) from a hadronic target (Figure 1). By the optical theorem of scattering theory, the total γ^*p cross section is given by the imaginary part of the forward scattering amplitude (virtual Compton amplitude) (Figure 2a). We consider the Bjorken limit, in which both the photon virtuality and the γ^*p center-of-mass energy become large compared to the typical hadronic mass scale, $Q^2 \sim W^2 \gg \mu^2$. As a consequence of the asymptotic freedom of QCD, DIS in this limit can be described as the scattering of the virtual photon from quasi-free quarks (and antiquarks) in the proton. In the simplest approximation, one neglects the interactions of the quarks altogether, and considers the scattering from a free quark (Figure 2b). This is equivalent to the space-time picture of DIS expressed in the parton model (7). Its basic assumption

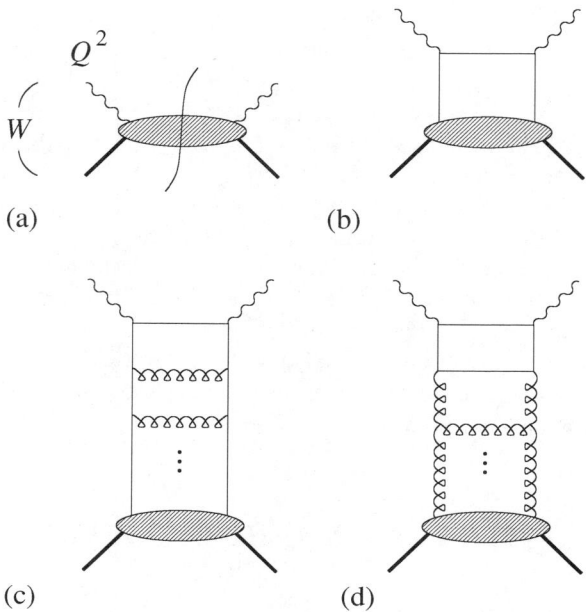

Figure 2   (*a*) The total cross section for $\gamma^* p$ scattering is given by the imaginary part of the forwad scattering amplitude. (*b*) The total cross section in the parton model (Bjorken scaling). (*c*, *d*) QCD radiative corrections, giving rise to the leading scaling violations in $\alpha_s \ln(Q^2/Q_0^2)$. Diagrams of type (*c*) dominate at large *x* and diagrams of type (*d*) dominate at small *x*.

is that, in a reference frame where the proton moves along the collision axis with a large momentum, the interaction of the γ^* with the quarks ("partons") is instantaneous compared to the characteristic time of their internal motion in the proton. In this picture, the total cross section for $\gamma^* p$ scattering in the Bjorken limit is given by

$$\sigma^{\gamma^* p \to X}(Q^2, W) = \frac{4\pi^2 \alpha_{\mathrm{em}}}{Q^2(1-x)} F_2(x), \quad F_2(x) = \sum_f e_f^2 \, x \left[q_f(x) + \bar{q}_f(x) \right],$$

1.

where α_{em} is the electromagnetic fine structure constant, e_f are the quark charges ($f = u, d, s \ldots$ labels the quark flavor), and $q_f(x)$ and $\bar{q}_f(x)$ are the parton densities, describing the number density of quarks and antiquarks carrying a fraction, *x*, of the fast-moving proton's momentum. The transverse momenta of the quarks and antiquarks are of the order $k_\perp^2 \sim \mu^2$, and are integrated over. Equation 1 exhibits the famous property of Bjorken scaling (6), i.e., the structure function, F_2, depends on the kinematic invariants characterizing the initial state only through the dimensionless Bjorken variable (we neglect the nucleon mass),

$$x \equiv \frac{Q^2}{W^2 + Q^2}. \qquad\qquad 2.$$

This is a direct consequence of the scattering from pointlike, quasi-free particles.

The parton model assumption of widely different timescales for the γ^*-quark interaction and the internal motion of the quarks in the proton becomes invalid in quantum field theory, where the ultraviolet divergences introduce a scale larger than Q^2. At its most elementary, this is the reason why Bjorken scaling is violated in quantum chromodynamics—an argument originally owing to Gribov. The leading scaling violations in $\alpha_s \ln(Q^2/Q_0^2)$ arise from gluon bremsstrahlung, as described by the ladder-type Feynman diagrams shown in Figures 2c and d, and can be summed up in closed form. Here, α_s is the strong coupling constant, and Q_0^2 is an arbitrarily chosen initial scale in the region of approximate Bjorken scaling. The result can be expressed in the form of a Q^2-dependence of the parton densities, governed by a differential equation, the Dokshitzer-Gribov-Lipatov-Altarelli-Parisi (DGLAP) evolution equation (9, 11, 12). This formulation allows one to retain the basic space-time picture of the parton model while incorporating QCD radiative corrections by way of a Q^2-dependence of the parton densities. Note that through evolution the gluon density in the proton effectively enters into the structure functions of γ^*p scattering (Figure 2d).

The basic structure of the γ^*p total cross section in the Bjorken limit in QCD is that of a product of a "hard" photon-parton cross section (involving virtualities $\sim Q^2$) and a "soft" matrix element (involving virtualities $\sim \mu^2$), describing the distribution of partons in the proton. QCD radiative corrections can be incorporated by a systematic redefinition of the "hard" and "soft" factors. This property is referred to as factorization. Factorization for the γ^*p total cross section in QCD has been formally demonstrated in several different approaches, including operator methods, in which the parton densities appear as nucleon matrix elements of certain non-local light-ray operators, and the evolution equations coincide with the QCD renormalization group equations for these operators. The calculations have also been extended to sum up next-to-leading (NLO) corrections in $\alpha_s \ln(Q^2/Q_0^2)$. We shall see below that the basic technique of factorization can be applied also to exclusive (Section 4) and diffractive (Section 5) final states in γ^*p scattering, as well as to certain hard processes in hadron-hadron scattering (Section 7).

More generally, QCD factorization allows one to perform an asymptotic expansion of the DIS structure functions in the Bjorken limit. The contribution from the diagram of Figure 2b, Equation 1, determines the leading power behavior at large Q^2, with an additional logarithmic dependence appearing owing to radiative corrections (Figures 2c and d). In the context of operator methods this is known as the leading-twist approximation. Power corrections of the order μ^2/Q^2 (higher-twist corrections) arise from taking into account the effect of the quark transverse momentum on the hard scattering process and the interaction of the "struck" quark with the non-perturbative gluon field in the proton; the two effects are intimately related because of gauge invariance in QCD (13).

From a mathematical perspective, Bjorken scaling of the moments of the DIS structure functions can be seen as a consequence of the conformal invariance of the QCD Lagrangian. The ultraviolet divergences associated with radiative corrections give rise to anomalous representations of the conformal group, with a logarithmic scale dependence. Later we shall see that at high energies a new dynamical scale appears in QCD, related to the gluon density in the nucleon and its transverse area, which breaks the conformal invariance, and thus leads to the complete disappearance of Bjorken scaling—the black-disk limit (BDL) (Section 6).

2.2. Space-Time Evolution of Small-x Scattering in the Target Rest Frame

We now turn to DIS at high energies, $W^2 \gg Q^2 \gg \mu^2$, which corresponds to values of the Bjorken variable $x \ll 1$. Though this process can be discussed within the standard QCD factorization approach described above, one faces the practical question at which point higher-twist $(1/Q^2-)$ corrections enhanced at small x, or radiative corrections beyond the DGLAP approximation giving rise to factors $\ln(1/x)$, become important. These and other questions can be addressed in a transparent way by considering the time evolution of DIS in the target rest frame, where the process takes the form of the scattering of a small-size $q\bar{q}$ dipole from a hadronic target. More generally, this formulation suggests a new understanding of QCD factorization, closely related to the so-called "color transparency" phenomenon observed in diffractive processes in hadron-hadron scattering.

QCD factorization in DIS and the DGLAP approximation have been formulated using the covariant language of Feynman diagrams. A typical Feynman diagram relevant at small x is shown in Figure 3a. In order to arrive at a space-time interpretation one needs to perform the integration over the "energy" variable using the residue theorem. It is this step that actually introduces the dependence of the amplitudes on the reference frame. Alternatively, one may directly trace the space-time evolution using the language of time-ordered perturbation theory. In this formulation, the time scales of the processes are determined by the energy denominators associated with the various transitions, via the energy-time uncertainty relation, $\Delta t = 1/\Delta E$.

In the target rest frame, the virtual photon in DIS at small x moves with 3-momentum $P \approx Q^2/(2m_N x)$. Consider its conversion into a $q\bar{q}$ pair with longitudinal momenta zP and $(1-z)P$ and transverse momenta $\pm k_\perp$ ("longitudinal" and "transverse" are defined relative to the direction of motion of the photon). The energy denominator for this transition is

$$\Delta E = \frac{M_{q\bar{q}}^2 + Q^2}{2P}, \qquad\qquad 3.$$

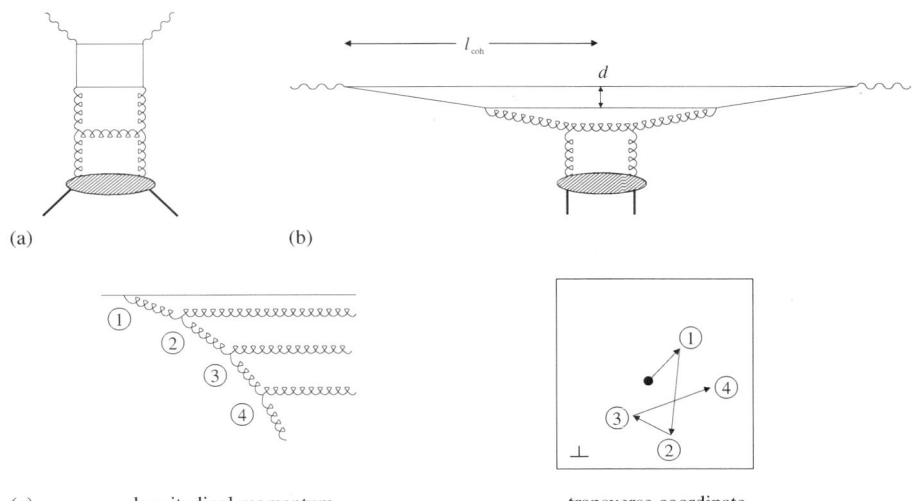

Figure 3   (a) A typical Feynman diagram for inclusive γ^*p scattering at small x. (b) Time evolution of small-x scattering in the target rest frame. (c) The decay of a fast parton in the $q\bar{q}$ dipole. The degradation of longitudinal momenta is accompanied by a random walk in the transverse coordinate.

where $M_{q\bar{q}}^2 \equiv (m_q^2 + k_\perp^2)/[z(1-z)]$ is the invariant mass of the $q\bar{q}$ pair (m_q is the quark mass, $k_\perp \equiv |\boldsymbol{k}_\perp|$). For a longitudinally polarized photon, the dominant contribution to the cross section comes from values $z \sim 1/2$ and $k_\perp \sim Q$, for which $M_{q\bar{q}}^2 \sim Q^2$ (this will be explained in more detail in Section 3.2). For such values the time associated with the $\gamma^* \to q\bar{q}$ transition is

$$\Delta t = 1/\Delta E \sim Q^2/P \approx 1/(2m_N x). \qquad 4.$$

At small x, the photon converts into a $q\bar{q}$ pair long before reaching the target, as illustrated in Figure 3b. Both the quark and antiquark move essentially with the speed of light. The distance between the point of their creation and the target, the so-called coherence length, is given by (we use units in which the velocity of light is $c = 1$)

$$l_{\text{coh}} = \Delta t. \qquad 5.$$

It is important to realize that the $q\bar{q}$ wave packet remains well localized in the longitudinal direction as it travels towards the target, and that the transverse separation between the quark and antiquark (at a given time) is a meaningful concept. It is generally of the order $d \sim 1/M_{q\bar{q}} \sim 1/Q$, and thus small if Q^2 is sufficiently large.

In short, high-energy γ_L^*-hadron scattering in the target rest frame is essentially the scattering of a small-size $q\bar{q}$ dipole from the target hadron. In this formulation, the quantity describing the strong interaction effects is the cross section for dipole-hadron scattering. It can be computed using methods of QCD factorization, with the dipole size, $d \ll \mu$ acting as the factorization scale (Section 3.2). Without going into details, we can immediately state an important property of the cross section, namely that in the limit $d \to 0$ it vanishes as

$$\sigma^{q\bar{q}-\mathrm{hadron}} \propto d^2, \qquad\qquad 6.$$

up to logarithmic corrections in d. Equation 6 reflects a fundamental property of QCD as a gauge theory—the interaction of a small-size color singlet object with hadronic matter is small ("color transparency"). In this understanding of QCD factorization, high-energy γ^*-hadron scattering exhibits a close relation to the interaction of heavy quarkonia with hadronic matter and a number of other color transparency phenomena in hadron-hadron scattering.

To make the dipole picture quantitative, one has to take into account the effects of QCD radiation. In particular, this is necessary in order to determine the coefficient in Equation 6 with logarithmic accuracy. The importance of different types of radiation can again be studied using the language of time-ordered perturbation theory in the target rest frame. The characteristic time for the quark to radiate a gluon with fraction x_g and transverse momentum $k_{\perp,g}$, relative to the time the $q\bar{q}$ pair spends between its creation and "hitting" the target, (4), is (see Figure 3b)

$$\frac{\Delta t_1}{\Delta t} = \left[Q^2 + \frac{m_q^2 + k_\perp^2}{z(1-z)} \right] \bigg/ \left[Q^2 + \frac{m_q^2 + k_\perp^2}{1-z} + \frac{m_q^2 + (k_\perp - k_{\perp g})^2}{z - x_g} + \frac{k_{\perp g}^2}{x_g} \right].$$
$$7.$$

If x is sufficiently small, and for average values of z, the emission process can be repeated several times before the evolved system reaches the target.

There exist several kinematic domains where gluon emission during the propagation of the $q\bar{q}$ wave packet is likely because of a large phase space at small x. One is the emission of partons with transverse momenta smaller than $k_\perp$ of the parent parton. Each such emission contributes a factor $\alpha_s \ln(Q^2/Q_0^2)$ in the amplitude, where the logarithm arises from the integration over the phase volume of the radiated gluon. In the standard QCD description of DIS, these are the radiative corrections summed up by the DGLAP evolution equations for the parton distributions described above (9,11,12; see also Reference 14). In the context of the dipole picture at small x, the summation of these corrections in leading order (LO) corresponds to a dipole-hadron cross section of the form (15–17)

$$\sigma^{q\bar{q}-\mathrm{hadron}}(x, d^2) = \frac{\pi^2}{4} F^2 d^2 \alpha_s(Q_{\mathrm{eff}}^2) \, xG(x, Q_{\mathrm{eff}}^2). \qquad 8.$$

Here $F^2 = 4/3$ is the Casimir operator of the fundamental representation of the $SU(3)$ gauge group (for gg dipoles in the adjoint representation, $F^2 = 3$). Furthermore, $\alpha_s(Q_{\mathrm{eff}}^2)$ is the LO running coupling constant and $G(x, Q_{\mathrm{eff}}^2)$ the LO gluon density in the target. They are evaluated at a scale $Q_{\mathrm{eff}}^2 \propto d^{-2}$. The coefficient of proportionality is not fixed within the LO approximation, and needs to be determined from NLO calculations or from phenomenological considerations (Sections 3.2 and 4).[1]

Equation 8 actually quotes a simplified expression for the dipole-hadron cross section. The original expression involves an integral over the gluon momentum fractions, which is concentrated in a narrow range above x. Also neglected in Equation 8 is the contribution proportional to the quark/antiquark distribution in the target, which at small x is suppressed compared to the gluon distribution. This contribution would lead to a flavor dependence of the dipole-nucleon cross section (19).

Another large contribution, specific to small x, comes from the large phase space in rapidity ($\propto z_g$) for emission of gluons without strong degrading of transverse momenta in the leading approximation. Such emissions give rise to factors $\alpha_s(N_c/2\pi)\Delta y$, where $N_c = 3$ is the number of colors in QCD, and $\Delta y = (y_i - y_{i+1})$ is the difference in rapidities between successive partons in the ladder. In terms of x, this corresponds to corrections proportional to $\ln(x_0/x)$, where $x_0 \sim 0.1$ accounts for the fact that nucleon fragmentation enters in the definition of the gluon density in the nucleon and does not produce a logarithm in x. If the rapidity interval for emissions (i.e., the lifetime of the quark-gluon system) becomes very large, one needs to sum these logarithms in addition to the $\alpha_s \ln(Q^2/Q_0^2)$ terms (20) (Section 3.3).

QCD radiation generally leads to an increase of the transverse size of the "dressed" dipole with decreasing x, and thus to an increase of the radius of the dipole-hadron interaction with energy. Each individual emission shifts the transverse coordinate of the radiating parton by $\Delta\rho \sim 1/k_\perp$ (Figure 3c). If there are n successive emissions with comparable, randomly oriented $k_\perp$ (this is the case in the limit of large $\ln x$), the overall shift is (2)

$$\Delta\rho^2 = n/k_\perp^2 = y/(\Delta y k_\perp^2), \qquad 9.$$

where y is the rapidity of the initial parton. A similar diffusion mechanism for soft partons was discussed by Gribov as a model for the increase of the radius of soft hadronic interactions with energy (1, 2). In the case of hard processes such as γ^*-hadron scattering, in the region where the DGLAP approximation is valid, the rate of expansion with energy is much smaller than for soft interactions, because of the larger transverse momenta of the emitted partons and the larger rapidity intervals between the emissions. This manifests itself e.g., in a much weaker energy dependence of the t-slope of hard exclusive processes as compared to elastic hadron-hadron scattering (21) (Section 4).

---

[1]There is an approach to high-energy scattering in which the projectile particle is represented as a superposition of eigenstates of the scattering matrix (e.g., Reference (18) and references therein). Equation 8 implies that states with different transverse size, d, should be orthogonal. However, the extension of Equation 8 to the case of elastic scattering indicates that transitions between configurations with different d are allowed for finite t. This suggests that the eigenstate model should be a reasonable approximation only for small values of t.

For transversely polarized virtual photons the space-time picture of the interaction is more complicated than in the longitudinal case. Owing to the different spin structure of the $\gamma_T^* \to q\bar{q}$ vertex, configurations of very different size—from hadronic size to $1/Q$—contribute to the interaction. The hadronic size configurations correspond to $z \sim 1$ or 0, and $k_\perp \sim \Lambda_{QCD}$. They are dual to two jets aligned along the virtual photon direction and are referred to as aligned jet configurations. They are expected to interact with the target with typical hadronic cross sections, giving the dominant contribution to the structure function F_2 at $Q^2 \sim$ few GeV^2 and $x \sim 10^{-2}$ (Section 3.2). Also, such configurations can easily scatter elastically from the target, and thus are an important source of diffractive scattering (Section 5).

3. INCLUSIVE $\gamma^* p$ SCATTERING AT SMALL x

3.1. DGLAP Evolution and the HERA Data

We start our discussion of $\gamma^* p$ scattering at high energies with inclusive DIS. Inclusive DIS is the main source of information about the parton distributions in the nucleon at small x. Because of the relatively simple structure of QCD factorization, it is also the main testing ground for higher-order QCD calculations and resummation approaches.

The validity of QCD factorization and DGLAP evolution for inclusive DIS have extensively been tested in fixed-target experiments, probing the quark/antiquark densities in the nucleon at values $x > 10^{-2}$ (see Reference (22) for a review). Going to smaller x, DGLAP evolution produces a fast increase of the parton densities, related to the fact that the gluon has spin 1 (8, 9), which implies a fast increase of the DIS cross section with energy. This prediction has spectacularly been confirmed by the measurements with the HERA ep collider. Figure 4 (see color insert) shows a summary of the F_2 proton structure function data taken by H1 and ZEUS compared to a QCD fit based on NLO DGLAP evolution (23). The data clearly support the interrelation of the x- and Q^2-dependence as predicted by DGLAP evolution. The analysis of the data found that effects of next-to-next-to-leading order (NNLO) terms of the form of α_s^2 multiplied by a function of $\alpha_s \ln(Q^2/Q_0^2)$ generally appear to be small. It is remarkable that the DGLAP approximation, which does not account for all potentially large terms containing $\ln(1/x)$, describes the presently available high-energy data so well.

More detailed insights into the "workings" of the DGLAP approximation can be gained by studying the effective power behavior in x of the structure function and the individual parton distributions in the NLO fit,

$$F_2 \propto x^{-\lambda_2}, \quad xG(x) \propto x^{-\lambda_g}, \quad \sum e_f^2 x\bar{q}_f(x) \propto x^{-\lambda_q} \quad (x < 10^{-2}), \qquad 10.$$

where the exponents depend on Q^2 (Figure 5, see color insert). At low Q^2, $\lambda_2 \approx 0.1$, reflecting the energy dependence expected for the cross section of soft hadronic processes. Starting from $Q^2 \approx 0.5\,\mathrm{GeV}^2$ λ_2 grows, reaching a value of ~ 0.4 at $Q^2 \sim 100\,\mathrm{GeV}^2$ (A. Levy, private communication). For $Q^2 > 3\,\mathrm{GeV}^2$, one observes that $\lambda_g \approx \lambda_2$, indicating that in this Q^2-region the x-dependence of the structure function is indeed driven by the gluon distribution. For lower Q^2, however, λ_g is significantly different from λ_2, becoming even negative at $Q^2 \approx 2\,\mathrm{GeV}^2$. Thus, although the NLO DGLAP approximation formally describes the x-dependence of the structure function even at low Q^2, the price to be paid is the lack of a smooth matching of the x-dependence of the gluon distribution to the soft regime. This indicates the presence of significant corrections to the leading-twist description of DIS at small x for $Q^2 \leq 3\,\mathrm{GeV}^2$. The dynamical origin of these corrections will be discussed in Section 3.2.

The data show that the deviation from the soft energy dependence of F_2 starts at surprisingly low scales, $Q^2 \ll 1\,\mathrm{GeV}^2$. Within the DGLAP approximation this behavior can be explained by the presence of a large non-perturbative gluon density in the nucleon at moderate x at a low scale (24). This is principally consistent with the idea of spontaneous chiral symmetry breaking in QCD, according to which most of the nucleon mass resides in gluon fields.

3.2. Space-Time Picture of Inclusive DIS

Many of the observed features of inclusive DIS at small x can be understood within the space-time picture in the target rest frame (Section 2.2). In this formulation, corrections to the leading-twist approximation at low Q^2 appear because of the contribution from large dipole sizes. This allows us to quantify the region of validity of the leading-twist approximation, and develop an "interpolating" approximation valid in a wide range of Q^2.

Following the logic outlined in Section 2.2, one can express the total $\gamma^* p$ cross section at small x as a superposition of $q\bar{q}$ dipole cross sections, characterized by the longitudinal momentum fraction of the quark, z, and the dipole size, d:

$$\sigma_{L,T}(x, Q^2) = \int_0^1 dz \int d^2 d \; \sigma^{q\bar{q}-\mathrm{hadron}}(z, d, x) \left| \psi_{L,T}^{\gamma}(z, d, Q^2) \right|^2 , \qquad 11.$$

where $\psi_{L,T}^{\gamma}(z, d)$ denotes the light-cone wave function of the $q\bar{q}$ component of the virtual photon, calculable in quantum electrodynamics. An important question is which dipole sizes dominate in the integral. For a longitudinally polarized photon, the modulus squared of the wave function is given by

$$\left| \psi_L^{\gamma}(z, d, Q^2) \right|^2 = \frac{6\alpha_{\mathrm{em}} Q^2}{\pi^2} \sum_{f=1}^{N_f} e_f^2 \left[z(1-z) K_0(\epsilon d) \right]^2 , \qquad 12.$$

where K_0 is the modified Bessel function and $\epsilon^2 = z(1-z)Q^2 + m_f^2$ (25). One can verify by direct calculation that in this case the contributions from large dipole sizes are suppressed at large Q^2, if the integral in Equation 11 is evaluated with the LO expression for the dipole-nucleon cross section (Equation 8) (21). In fact,

Equations 11, 12 and 8 are formally equivalent to the LO DGLAP approximation in QCD (see the discussion below). The effective scale in the gluon distribution entering in the dipole cross section can be determined by comparing Equation 11 with the LO DGLAP expression; one finds $Q^2_{\text{eff}} \approx 9/d^2$ for HERA kinematics (26). Note that the factor $xG(x, Q^2_{\text{eff}})$ in Equation 8 results in a fast increase of the cross section with energy, in contrast to the two-gluon exchange model of References (27–29), where the cross section is energy-independent.

When applying Equation 11 to transversely polarized photons, the distribution of dipole sizes is significantly wider than in the longitudinal case. At $Q^2 \sim$ few GeV^2, the transverse cross section receives sizable contributions from dipole sizes for which the perturbative approximation for the dipole-nucleon cross section becomes invalid (see Equation 8). Still, at large Q^2 the perturbative contribution should dominate, because of the faster increase with energy of the parton distribution for the smaller-size quark-gluon configuration. The contribution from large-size $q\bar{q}$ configurations is strongly suppressed by Sudakov form factors; it is actually represented by large-size $q\bar{q}g, \dots$ configurations.

Equation 11 can serve as the basis for an "interpolating" model that describes $\gamma^* p$ interactions over a wide range of Q^2 for both transverse and longitudinal polarizations (30). There is ample evidence—e.g., from studies of γN and πN elastic scattering—that real photons in high-energy reactions have transverse sizes comparable to pions. A way to ensure this within the $q\bar{q}$ dipole description is to introduce a dynamical quark mass of $\sim$300 MeV, which is consistent with the phenomenology of spontaneous chiral symmetry breaking (31). The cross section for the scattering of such a "hadronic-size" dipole with the target can then be inferred from the πN scattering data. For small dipoles, $d \leq 0.4$ fm, the cross section can be calculated perturbatively. When evaluating the leading-twist expression, it is important to accurately treat the kinematic limits of the integral over the gluon momentum fractions, as this leads to an additional dependence of the dipole cross section on Q^2. A dipole cross section obtained by matching the two prescriptions is shown in Figure 6. This function is then averaged with the photon wave function for massive quarks (see Equation 11). This model reproduces well the HERA F_{2p} data for $Q^2 \geq 0.1$ GeV^2, and correctly predicts σ_L (30).

In order to make contact with the analysis of Section 3.1, we need to state more precisely how the dipole picture is related to the DGLAP approximation in QCD. In LO, it has been demonstrated explicitly that Equations 8 and 11 can be obtained by rewriting the LO DGLAP expression for the $\gamma^* p$ cross section (17). This simple relation appears because in the leading logarithmic approximation the separation of the process according to time in the target rest frame—transition of the virtual photon into a $q\bar{q}$ pair (photon wave function), and interaction of the pair with the target (dipole cross section)—coincides with the separation of transverse momenta in $k^2_\perp \sim Q^2$ and $k^2_\perp \ll Q^2$ in the partonic ladder. Beyond the leading order, one needs to explicitly include $q\bar{q}g$ components of the photon wave function, and the distinction between the wave function and the dipole interaction with the target becomes more delicate. Although in principle the leading-twist dipole picture

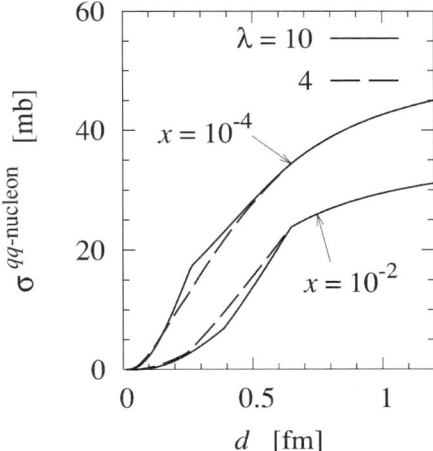

Figure 6   The dipole-nucleon cross section in the "interpolating" model of Reference (30). Shown are the results corresponding to two different values of the parameter, λ, determining the effective scale, Q^2_{eff}.

should be equivalent to the DGLAP approximation in any order of the expansion, in practice the problem of formulating a consistent dipole picture in NLO has not been solved yet. Also, it is worth emphasizing that the correspondence between parton distributions and amplitudes of physical processes is not always direct. In particular, the parton model-type contribution, which naturally leads to diffractive processes, is hidden in the boundary condition for the QCD evolution of the parton distributions.

In spite of the lack of an explicit NLO analysis in the dipole picture, it still seems to be of relevance that the region of Q^2 where the NLO DGLAP analysis leads to a gluon exponent λ_g dropping below the soft value (Figure 5) corresponds in the dipole model to contributions from $d > 0.4$ fm, where nonperturbative effects become important. Thus, it seems that the "anomalous" behavior of λ_g is a consequence of the leading-twist DGLAP approximation trying to mock up higher-twist corrections at low Q^2. The dipole picture allows us to quantify the region of applicability of the leading-twist approximation at low Q^2, and suggests a natural way to incorporate non-perturbative effects.

Equations 8 and 11 are valid also within the leading $\alpha_s \ln(x_0/x)$ approximation. Furthermore, they can be derived from the eikonal model expression for the propagation of a heavy quarkonium through a hadronic medium (32).

3.3. Breakdown of the DGLAP Approximation at Very Small *x*

The observation of the fast increase of parton densities at small *x* has stimulated theoretical discussions of the stability of the DGLAP approximation at small *x*.

In fact, in the kinematic limit of fixed Q^2 and $x \to 0$ the effective parameter of the perturbative QCD expansion is multiplied by a factor $\ln(x_0/x)$, which arises because of gluon emission in multi-Regge kinematics (rapidity distance between adjacent gluons $\gg 2$), and the hierarchy of dominant terms is changed as compared to the DGLAP approximation. The constant x_0 is determined by the typical momentum fraction in the initial parton distributions; usually $x_0 \sim 0.1$.

A simple kinematic estimate shows that in typical HERA kinematics the DGLAP approximation is still reliable. The rapidity span at HERA is approximately $\ln(Q/xm_N) \approx 10$ for $x = 10^{-4}$ and $Q = 2.5\,\text{GeV}$. To obtain a significant $\ln(x_0/x)$ term, the distance in rapidity between adjacent partons in the ladder should be $\gg 2$. Thus, the number of radiated gluons in multi-Regge kinematics at HERA is $\ll (10 - 4)/2 - 1 \approx 2$, where we took into account that each of the fragmentation regions occupies at least two units in rapidity. This simple estimate agrees well with a numerical study of NLO QCD evolution, which indicates that the average change of x in the HERA region does not exceed 10, corresponding to $\Delta y = \ln(x_0/x) \approx 2$, provided that $Q_0^2 \geq 1\,\text{GeV}^2$ (33). Because one (two) logarithms of x are effectively taken into account by the NLO (NNLO) approximation, there is no need for a special treatment of $\ln(x_0/x)$ effects at HERA kinematics. A similar estimate shows that at LHC kinematics the radiation of 5–6 gluons is permitted. Thus, at LHC energies and above the resummation of $\ln(x_0/x)$ terms becomes a practical issue.

The program of resumming leading $\alpha_{\text{em}} \ln(x_0/x)$ terms started in quantum electrodynamics (20). In QCD, the reggeization of gluons slows down the energy dependence of amplitudes of high-energy processes (34, 35).[2] In the leading $\alpha_s \ln(x_0/x)$ approximation (Balitsky-Fadin-Kuraev-Lipatov, or BFKL, approximation) (36), where energy-momentum conservation and the running of the coupling constant are neglected, the reggeization of gluons is canceled by contributions from multigluon radiation. NLO corrections to a large extent subtract kinematically forbidden contributions, leading to a large negative contribution to the structure functions (37, 38). Another feature of this approach is the lack of an unambiguous separation between perturbative and nonperturbative QCD effects (37). Thus, this approximation seems to be limited to the description of single-scale hard processes where DGLAP evolution is unimportant in a wide kinematic range, such as $\gamma^*(Q^2) + \gamma^*(Q^2) \to$ hadrons, or two-body processes where the hardness is controlled by proper choice of final state like, such as $\gamma + \gamma \to \Upsilon + \Upsilon$.

The resummation approaches of References (39, 40) predict a significantly slower increase of amplitudes with energy than the LO BFKL approximation, and possibly even oscillations in the energy dependence. Most of the reduction is due to the better account of energy-momentum conservation in these approaches, and

---

[2]The high-energy behavior of two-body amplitudes with color-octet quantum numbers in the crossed channel in QCD is given by the Regge pole formula, $(1/x)^{\beta(t)}$, where $\beta(t)$ decreases with increase of $-t$. In the lowest order of α_s QCD gives $\beta(t) = 1$. Thus, gluons in QCD (as well as quarks) are reggeons.

account of the running of the coupling constant. At extremely small x (beyond the reach of LHC) much of the LO BFKL results reappear, but with a slower dependence on x. For the parton densities in the nucleon, where $x_0 \approx 0.1$ is a reasonable value for the constant in the $\ln(x_0/x)$ factor, resummation effects should be small for $x \geq 10^{-4}$, that is, for the whole HERA range above $Q^2 \geq 2\,\mathrm{GeV}^2$. At smaller x, the result of the resummed evolution is close to that of NLO DGLAP evolution down to $x \sim 10^{-6}$, but differs strongly from NNLO (41). This suggests that NLO DGLAP evolution could be a good guess for the parton densities down to the very small x values probed at LHC, even though the underlying dynamics may change significantly at $x \leq 10^{-4}$.

4. EXCLUSIVE PROCESSES IN $\gamma^* p$ SCATTERING AT SMALL x

4.1. QCD Factorization for Hard Exclusive Processes

The concept of QCD factorization can be extended to certain exclusive channels in $\gamma^* p$ scattering, namely processes of the type

$$\gamma_L^*(q) + N(p) \to \text{``Meson''}(q + \Delta) + \text{``Baryon''}(p - \Delta),\qquad 13.$$

at large virtuality, $Q^2 \equiv -q^2$, and center-of-mass energy, $W^2 \equiv (p + q)^2$, with fixed $x = Q^2/(W^2 + Q^2)$, and fixed small invariant momentum transfer, $t \equiv \Delta^2$. Examples include the production of light vector mesons (ρ, ρ') (21), heavy vector mesons (J/ψ, ψ', Υ) (21), and real photons (deeply virtual Compton scattering, DVCS) (42–48). Closely related to these processes are certain hadron-induced reactions, such as the diffractive dissociation of pions, $\pi + T \to 2\,\mathrm{jets} + T$, where T denotes a hadronic target (nucleon or nucleus) (16). These exclusive processes probe the interaction of small-size color singlets with hadronic matter in much more detail than inclusive DIS. They also provide new information about the transverse spatial structure of the nucleon, contained in the so-called generalized parton distributions.

The basis for the analysis of exclusive processes (Equation 13) is the QCD factorization theorem (49), which extends the initial analysis of Reference (21) for the small-x limit. It states that the amplitude can be represented as a convolution of three functions, as depicted in Figure 7:

$$A^{\gamma_L^* N \to M+B} = \sum_{i,j} \int_0^1 dz \int dx_1\ f_{i/p}(x_1, x - x_1, t; \mu_{\mathrm{Fact}})$$

$$\times H_{ij}(x_1, x, z, Q^2; \mu_{\mathrm{Fact}})\, \phi_j^M(z, \mu_{\mathrm{Fact}}) + \text{power corrections.}\qquad 14.$$

Here, f is the generalized parton distribution (GPD), which describes the amplitude for the nucleon to "emit" and "absorb" a parton with longitudinal momentum fractions x_1 and $x_2 = x_1 - x$, respectively, accompanied by an invariant momentum transfer, t, and, possibly, a transition to another baryonic state. μ_{Fact} is the factorization scale. At zero momentum transfer, $x_1 = x_2$ and $t = 0$,

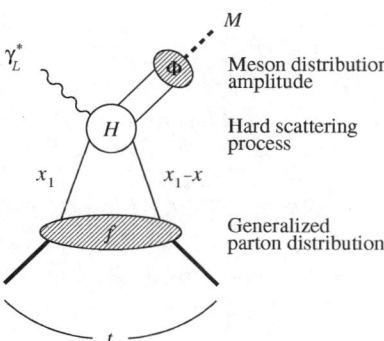

Figure 7   Factorization of the amplitude of hard exclusive meson production (Equation 14).

the GPD's coincide with the usual parton densities measured in inclusive DIS. For recent reviews of the properties of GPD's and their applications, see References (50, 51). Furthermore, ϕ^M is the distribution amplitude describing the conversion of a $q\bar{q}$ pair with relative longitudinal momentum fraction z to the produced meson (or photon). Finally, H_{ij} denotes the amplitude of the hard partonic scattering process, which is calculable in powers of $\alpha_s(Q^2)$. The indices i, j label the different parton species. The contribution of diagrams in which the hard scattering process involves more than the minimum number of partons is suppressed by $1/Q^2$. An important consequence of factorization is that the t-dependence of the amplitude rests entirely in the GPD. Thus, different processes probing the same GPD at similar x and Q^2 should exhibit the same t-dependence.

4.2. Space-Time Picture of Hard Exclusive Processes

The physics of hard exclusive processes at small x becomes most transparent when following the space-time evolution in the target rest frame. As in the case of inclusive scattering, this approach allows one to expose the limits of the leading-twist approximation, and to quantify power corrections related to the finite transverse size of the produced meson.

In exclusive vector meson production, $\gamma_L^* + N \rightarrow V + N$, one can identify three distinct stages in the time evolution in the target rest frame (21). The virtual photon dissociates into a $q\bar{q}$ dipole of transverse size $d \sim 1/Q$ at a time $\tau_i = l_{\text{coh}} \approx 1/(m_N x)$ before interacting with the target (see Equation 5). The $q\bar{q}$ dipole then scatters from the target, and "lives" for a time $\tau_f \gg \tau_i$ before forming the final state vector meson. The difference in the time scales is due to the smaller transverse momenta (virtualities) allowed by the meson wave function as compared to the virtual photon.

In the leading logarithmic approximation in $\ln(Q^2/\Lambda_{\text{QCD}}^2)$, the effects of QCD radiation can again be absorbed in the amplitude for the scattering of the small-size dipole off the target. It can be shown by direct calculation of Feynman diagrams that the leading term for small dipole sizes is proportional to the generalized gluon distribution, $G(x_1, x_2, t; Q_{\text{eff}}^2)$, where $Q_{\text{eff}}^2 \propto d^{-2}$ (17). A simpler approach is to infer the result for the imaginary part of the amplitude from the expression for the cross section, Equation 8, via the optical theorem. The imaginary part is proportional to the generalized gluon distribution at $x_1 = x$ and $x_2 = 0$. At sufficiently large Q^2, the generalized gluon distribution at small x_1 and x_2 can be calculated by perturbative evolution, starting from the "diagonal" generalized gluon distribution, $x_1 = x_2 \gg x$, at a low scale (52–54). In applications to vector meson production at HERA, where the effective scale is of the order $Q_{\text{eff}}^2 \sim$ few GeV^2, the "skewness" effects induced by the evolution are not very substantial, and one may approximate the generalized gluon distribution by the diagonal one at the scale Q_{eff}^2. It is convenient to separate the t-dependence and write the diagonal generalized gluon distribution in the form

$$G\left(x, x, t; Q_{\text{eff}}^2\right) = G\left(x, Q_{\text{eff}}^2\right) F_g\left(x, t; Q_{\text{eff}}^2\right), \qquad 15.$$

where $G(x, Q_{\text{eff}}^2)$ is the usual gluon density and F_g is the "two-gluon form factor" of the target, which satisfies $F_g(x, t = 0; Q_{\text{eff}}^2) = 1$. Altogether, one obtains for the dipole-nucleon scattering amplitude in this approximation

$$A^{q\bar{q}-N}(x, d^2, t) = 2\pi i\, F^2\, W^2 d^2\, \alpha_s\left(Q_{\text{eff}}^2\right) x G\left(x, t, Q_{\text{eff}}^2\right) F_g\left(x, t, Q_{\text{eff}}^2\right). \qquad 16.$$

The amplitude for the hadronic process (13) is then given by the convolution of Equation 16 with the light-cone wave function of the virtual photon, Equation 12, and that of the produced vector meson, ψ^V. In coordinate representation,

$$A^{\gamma^*+N \to V+N} = \int_0^1 dz \int d^2d\; \psi_L^{\gamma}(z, d)\, A^{q\bar{q}-N}(x, d^2, t)\, \psi^V(z, d), \qquad 17.$$

where the integration is over the quark longitudinal momentum fraction, z, and the transverse dipole size, d.

In Equation 17, the wave function of the vector meson of transverse size $1/m_V$ is convoluted with the wave function of the virtual photon of significantly smaller transverse size, $1/Q$. One may say that the meson in this process is "squeezed," i.e., forced to couple in a configuration much smaller than its natural hadronic size. In the leading-twist approximation one neglects the spatial variation of the vector meson wave function and substitutes it by the distribution amplitude,

$$\psi^V(z, d) \to \psi^V(z, 0) \equiv \phi^V(z) \qquad 18.$$

(in momentum representation, the distribution amplitude is the integral of the wave function over transverse momenta). The integral over transverse sizes can then be performed explicitly, using Equations 12 and 16. After restoring the real part of the amplitude using its analyticity properties, the differential cross section is obtained as (21)

$$
\frac{d\sigma_L^{\gamma^*+N \to V+N}}{dt} = \frac{3\pi^3 \Gamma_V m_V \eta_V^2}{N_c^2 \alpha_{\text{em}} Q^6}
$$

$$
\times \; \alpha_s^2(Q_{\text{eff}}^2) \; \left| \left(1 + \frac{i\pi}{2}\frac{d}{d\ln x}\right) xG\left(x; Q_{\text{eff}}^2\right) \right|^2 \; F_g^2\left(x, t; Q_{\text{eff}}^2\right). \quad 19.
$$

Here, Γ_V is the leptonic width of the vector meson, which defines the normalization of the meson wave function, and

$$
\eta_V \equiv \frac{1}{2} \int_0^1 dz \frac{\phi^V(z)}{z(1-z)} \Bigg/ \int_0^1 dz \phi^V(z); \quad\quad 20.
$$

$\eta_V \to 1$ at asymptotically large Q^2. These expressions apply to production by a longitudinally polarized photon. For transverse polarization, the nonperturbative contribution is suppressed only by a Sudakov-type form factor, similar to the case of $F_2(x, Q^2)$ in inclusive γ^*p scattering. This contribution originates from highly asymmetric $q\bar{q}$ pairs ($z \sim 0$ or 1) in the γ_T^* wave function, which have transverse size similar to that of hadrons. We note that elastic photo/electroproduction of J/ψ mesons has been evaluated also within the LO BFKL approximation (55). The function $xG(x, Q^2)$ that enters there has no relation to the conventional DGLAP gluon distribution, which is defined within the DGLAP approximation only.

Equation 19 is based on the leading logarithmic approximation in $\ln(Q^2/\Lambda_{\text{QCD}}^2)$, as well as on the leading-twist approximation, Equation 18. Though it already exhibits many qualitative features seen in the data (see below), two important effects need to be taken into account before a quantitative comparison can be attempted. First, because the wave function of the vector meson in Equation 17 is significantly broader than that of the γ_L^*, the effective dipole sizes in the meson production amplitude, Equation 16, are substantially larger than in σ_L, Equation 11 (Figure 8). As a result, the effective scale in the gluon distribution, Q_{eff}^2, is smaller in vector meson production than in σ_L (Figure 8) (26, 56). This effect slows the x- (energy) dependence of the cross section compared to the naive estimate, $Q_{\text{eff}}^2 = Q^2$. Second, numerical studies using model wave functions show that retaining the full d-dependence of the vector meson wave function in the convolution integral (17) results in a substantial decrease of the absolute cross section at moderate Q^2 as compared to the leading-twist approximation, Equation 18, as well as in a slower Q^2 dependence (26, 56). These higher-twist effects, related to the finite size of the vector meson, limit the region of validity of the leading-twist approximation (18) and need to be taken into account in quantitative estimates at low Q^2.

4.3. Vector Meson Production at HERA

With proper choice of the effective scale, Q_{eff}^2, and inclusion of higher-twist effects due to the finite transverse size of the meson, one can quantitatively compare the results of the leading logarithmic approximation, Equations 16 and 17, with the

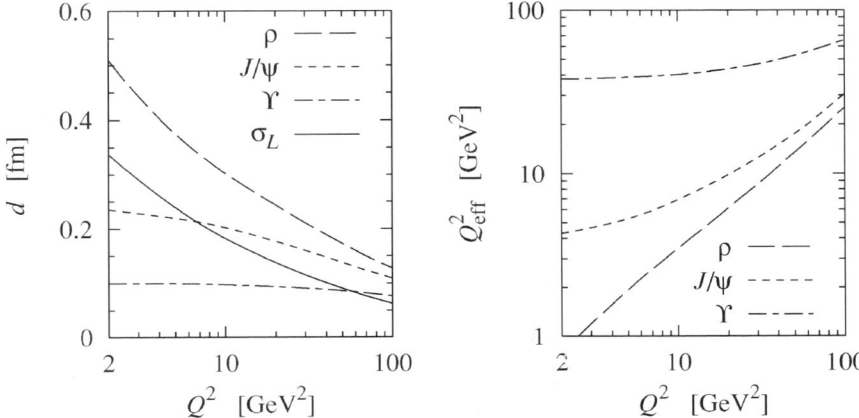

Figure 8   The average dipole size, d, (*left*) and the effective scale, Q^2_{eff}, (*right*) in exclusive vector meson production (ρ, J/ψ, Υ) by longitudinally polarized photons, as a function of Q^2 (26, 56). Also shown are the average values of d in the integrand of the expression for the inclusive cross section, σ_L.

HERA data on heavy and light vector meson production. The data confirm in particular the following predictions of this picture:

- *Increase of cross sections with energy.* Equation 19 implies that $d\sigma/dt(t = 0)$ grows with energy as $[xG(x, Q^2_{\text{eff}})]^2$, with Q^2_{eff} estimated to be $\sim 3\,\text{GeV}^2$. When combined with the LO gluon density obtained from fits to DIS data, this implies a growth $\propto W^{0.8}$. Such behavior has been observed for ρ production at $Q^2 = 10–20\,\text{GeV}^2$, and for J/ψ production starting from $Q^2 = 0$ (57). The later onset of the hard regime for ρ electroproduction is due to the rather slow "squeezing" of the $q\bar{q}$ configuration in the ρ meson; it reaches a size comparable to that of the J/ψ only at $Q^2 \sim 20\,\text{GeV}^2$ (Figure 8).[3] The naive choice $Q^2_{\text{eff}} = Q^2$ would imply a too fast growth (Figure 5). For soft interactions, on the other hand, $d\sigma/dt(t = 0) \propto W^{0.32}$, and the growth is even smaller for the cross section integrated over t.

- *Decrease of cross sections with Q^2.* The decrease with Q^2 of σ_L for ρ-meson production, and of the total cross section for J/ψ production, is slower than $1/Q^6$, owing to the Q^2-dependence of $\alpha_s G$ in Equation 19, as well as finite-size (higher-twist) effects. This is best observed in J/ψ electroproduction, where the model of Reference (55), which neglects finite-size effects, predicts

---

[3]In the case of ρ-meson production initiated by transverse photons, the squeezing is generated by the Sudakov form factor, as well as by the more rapid increase with energy of the small size contribution. The observed behavior of σ_L/σ_T can be fitted within the current models (58).

a decrease of the cross section by a factor of ~5 faster than observed in the kinematic region covered by the H1 experiment.

■ *Absolute cross sections.* The absolute cross sections for vector meson production are well reproduced, provided that higher-twist effects related to the finite size of the vector meson are taken into account (26, 56).

■ *Dominance of longitudinal cross section.* The data on ρ production indicate $\sigma_L \gg \sigma_T$ for $Q^2 \gg m_V^2$, in agreement with QCD factorization.

■ *Universality of t-dependence.* Comparison of ρ and J/ψ electroproduction data clearly show the universality of the t-dependence at large Q^2, where the vector mesons are "squeezed," and the t-dependence originates solely from the two-gluon form factor (Figure 9, see color insert).

■ *Flavor symmetry.* Because the interaction of the $q\bar{q}$ dipole with the gluon distribution is flavor-blind, one expects the restoration of $SU(3)$ flavor symmetry in vector meson production for $Q^2 \gg m_V^2$. For example, $\phi : \rho = (2 : 9)$ in the flavor symmetry limit. The violation of $SU(3)$ flavor symmetry owing to the increase of the wave function of the vector meson at small distances with increasing quark mass leads to an enhancement of this ratio by a factor ~1.2.

A new situation is encountered in the photoproduction of Υ mesons. In this case, the approximation of the generalized gluon distribution by the usual gluon density becomes invalid (large "skewness" and large Q_{eff}^2), and the real part of the amplitude becomes significant. Together, these effects increase the predicted cross section by a factor of approximately 4 (59, 60). For Υ production $Q_{\text{eff}}^2 \approx 40\,\text{GeV}^2$, leading to an energy dependence of the cross section as $d\sigma/dt(t = 0) \propto W^{1.7}$.

Closely related to vector meson production is the production of real photons (deeply virtual Compton scattering, DVCS). This process has been the subject of intense theoretical study in the region of moderate x, accessible in fixed-target experiments (HERMES at DESY, COMPASS at CERN, Jefferson Lab), and is considered the main tool for probing the generalized quark distributions in the nucleon (43, 45, 46, 61). At small x, the DVCS amplitude has been computed in the leading $\ln(Q^2/\Lambda_{\text{QCD}}^2)$ approximation outlined in Section 4.2, and found to be substantially enhanced as compared to the forward $\gamma^* p \rightarrow \gamma^* p$ amplitude (47). The DVCS cross section reported by the HERA experiments is in reasonable agreement with these predictions, as well as with the color dipole model of Reference (62); see Reference (63) and references therein. The HERA data at small x are also well described by an NLO QCD analysis (64, 65), in which the modeling of the input GPD's is a much more challenging problem than in LO (see Reference (65) for details). DVCS at small x and the closely related process of production of Z bosons, $\gamma + p \rightarrow Z + p$, were also studied within the leading $\alpha_s \ln(x_0/x)$ approximation (42).

To summarize, the HERA data on exclusive electroproduction of vector mesons clearly show the transition to the perturbative QCD regime for $Q^2 \geq 10\text{–}20\,\text{GeV}^2$. This conclusion is consistent with the observation of color transparency phenomena in several other processes (see Section 4.5). It establishes the study of exclusive processes (x, Q^2 and t-dependence of the cross section) as a way to extract detailed information about the interaction of small dipoles with hadrons, as well as about the generalized parton distribution in the nucleon.

4.4. Transverse Spatial Distribution of Gluons in the Nucleon

An important aspect of hard exclusive processes at small x is that they provide information about the transverse spatial distribution of gluons in the nucleon. It is contained in the Fourier transform of the two-gluon form factor, Equation 15,

$$F_g\left(x, \rho; Q^2_{\text{eff}}\right) \equiv \int \frac{d^2\Delta_\perp}{(2\pi)^2}\, e^{i(\Delta_\perp \rho)}\, F_g\left(x, t = -\Delta^2_\perp; Q^2_{\text{eff}}\right),\qquad 21.$$

where ρ is a transverse coordinate variable. The function $F_g(x, \rho; Q^2_{\text{eff}})$ is positive definite (66) and describes the spatial distribution of gluons in the transverse plane, $\int d^2\rho F_g(x, \rho; Q^2_{\text{eff}}) = 1$.

The convergence of the t-slopes of ρ and J/ψ production at large Q^2 (Figure 9) demonstrates that the t-dependence of the differential cross section is dominated by the two-gluon form factor. The two-gluon form factor can thus be extracted from the J/ψ photoproduction data ($Q^2_{\text{eff}} \approx 3\,\text{GeV}^2$), with small corrections ($\sim 10\%$) due to the finite transverse size of the J/ψ meson. This process has been measured over a wide range of energies; see References (67, 68) for an overview of the data. At fixed-target energies, $x \sim 10^{-1}$, the t-dependence of the data is well described by a two-gluon form factor of dipole form,

$$F_g = \left(1 - t/m^2_g\right)^{-2},\quad m^2_g = 1.1\,\text{GeV}^2\quad (x \sim 10^{-1}),\qquad 22.$$

where the parameter, m_g, is close to that in the dipole fit to the axial form factor of the nucleon. This corresponds to a narrow spatial distribution of gluons in the transverse plane, with an average transverse radius $\langle\rho^2\rangle = 8/m^2_g \approx 0.28\,\text{fm}^2$ (Figure 10). At HERA energies, $x \sim 10^{-2}$–10^{-3}, the average radius is larger, $\langle\rho^2\rangle \approx 0.35\,\text{fm}^2$. It also exhibits a slow growth with $\ln(1/x)$, with a slope, α', significantly smaller than the value for soft interactions. The J/ψ photoproduction data from H1 give $\alpha'_{\text{hard}} = 0.08 \pm 0.17\,\text{GeV}^{-2}$ (69), the ZEUS electroproduction data $\alpha'_{\text{hard}} = 0.07 \pm 0.05(\text{stat})^{+0.03}_{-0.04}(\text{syst})\,\text{GeV}^{-2}$ (70), which should be compared to $\alpha' \approx 0.25\,\text{GeV}^{-2}$ for pp elastic scattering. The difference reflects the suppression of Gribov diffusion for partons with large virtualities (see the discussion in Section 2.2).

The change of the nucleon's average transverse radius between $x \sim 10^{-1}$ and 10^{-2} can naturally be explained by chiral dynamics. Pions in the nucleon wave function carry momentum fractions of the order m_π/m_N. For $x > m_\pi/m_N$ the pion cloud does not contribute to the gluon distribution, and the two-gluon form factor is similar to the nucleon axial form factor, which also does not receive contributions from the pion cloud. For $x \ll m_\pi/m_N$, the pion cloud contributes to the gluon distribution and leads to an increase of $\langle\rho^2\rangle$ by 20–30% (Figure 10) (71).

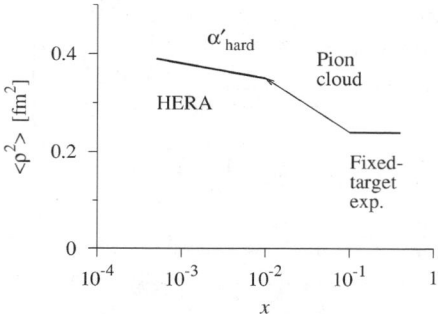

Figure 10   The average squared transverse radius of the gluon distribution in the nucleon, $\langle \rho^2 \rangle = \int d^2 \rho \rho^2 F_g(x, \rho; Q^2_{\text{eff}})$, as a function of x, as extracted from J/ψ photoproduction data ($Q^2_{\text{eff}} = 3\,\text{GeV}^2$) at various energies.

It is worth emphasizing that for smaller x the increase of the transverse size should continue owing to Gribov diffusion. Indeed, a hard probe can interact with a parton of the soft ladder, responsible for the growth of the soft radius, if the soft parton's momentum fraction is sufficiently large compared to x. At very small x and fixed Q^2 the rate of the growth should thus be comparable to that in the soft case (72). No such effect is present in the BFKL model where the interaction of two small dipoles is considered.

The change of the transverse spatial distribution of gluons in the nucleon with the scale, Q^2_{eff}, due to DGLAP evolution should generally be small (73). For Q^2_{eff} sufficiently large compared to the transverse spatial resolution, the parton decays happen essentially locally in transverse position. For fixed x, one finds that the transverse spatial distribution shrinks with increasing scale, because the distribution becomes sensitive to the input distribution (at the initial scale) at higher values of x, where it is concentrated at smaller transverse distances.

4.5. Color Transparency in Hard Processes with Nuclei

QCD predicts that the spatially small quark-gluon wave packets formed in hard γ^*-induced scattering processes interact weakly with hadronic matter, because of the color neutrality of the photon. At sufficiently small x, where the cross section is proportional to the gluon density (see Equation 8) one expects the ratio of the cross sections for γ^* scattering from a nucleus and a single nucleon to be equal to the ratio of the respective gluon densities, a property known as generalized color transparency (16, 21). Because with increasing Q^2 gluon shadowing at fixed x disappears (see the discussion in Section 5.4), one further expects that

$$\sigma^{\gamma^* A}_{\text{tot}} / \left(A \sigma^{\gamma^* N}_{\text{tot}} \right) \to 1 \quad (Q^2 \to \infty;\ x \text{ fixed, small}), \qquad 23.$$

which is referred to as color transparency proper. Conversely, at fixed Q^2 and decreasing x, the ratio in Equation 23 should decrease owing to the more important

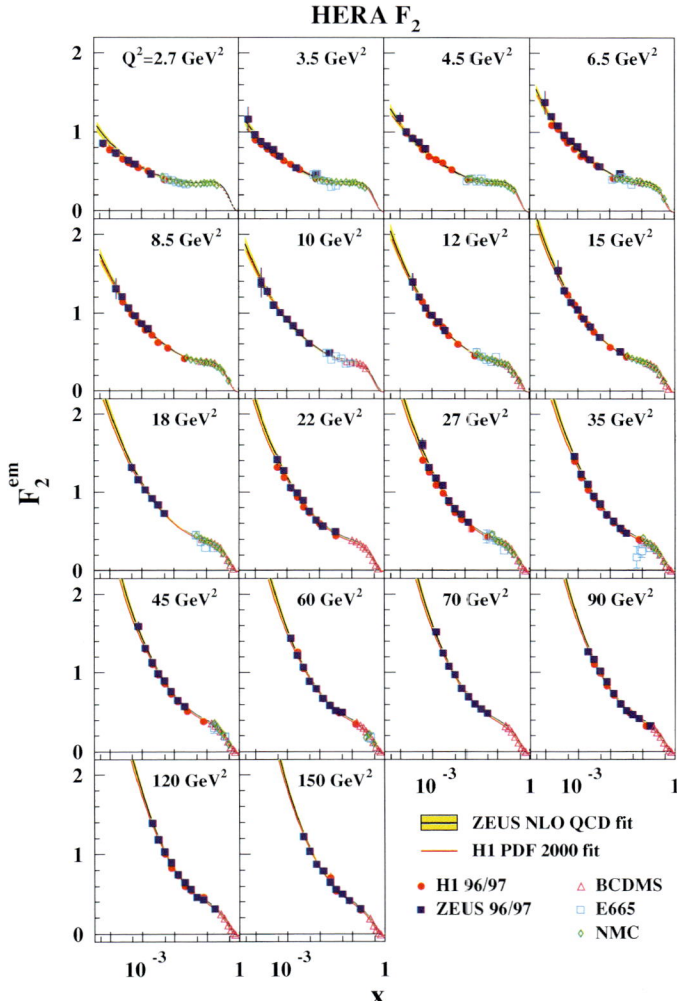

Figure 4 The proton structure function, $F_2(x)$, as measured by the H1 and ZEUS experiments at HERA (23). Also included are data from fixed-target experiments. The lines show a QCD fit based on the NLO DGLAP approximation.

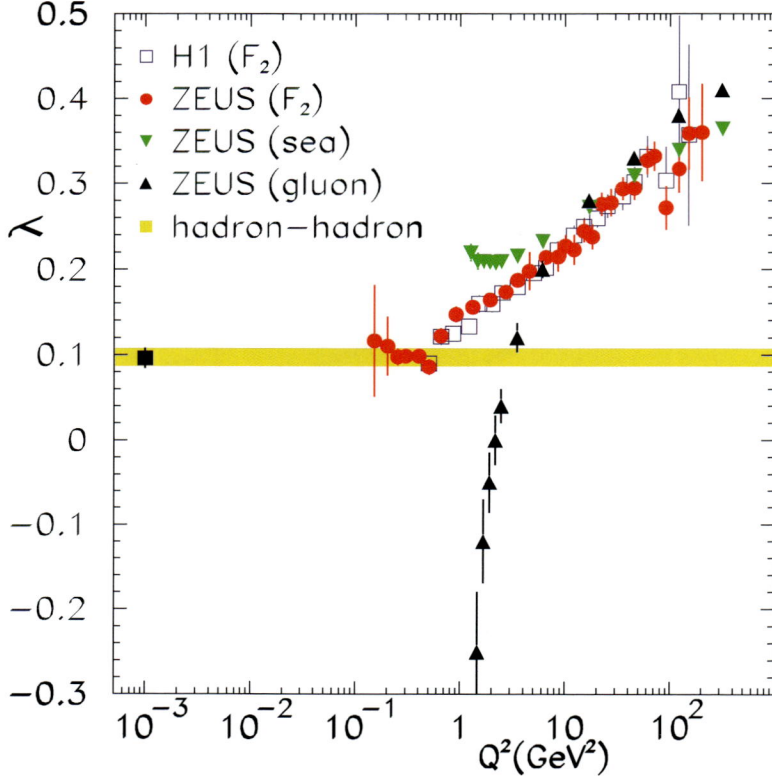

Figure 5   The exponents characterizing the x-dependence of F_2, λ_2, the gluon distribution, λ_g (*black squares*), and the sea quark distributions, λ_q (*green triangles*), see Equation 10, as extracted from the NLO DGLAP fit to the H1 and ZEUS data (A. Levy, private communication).

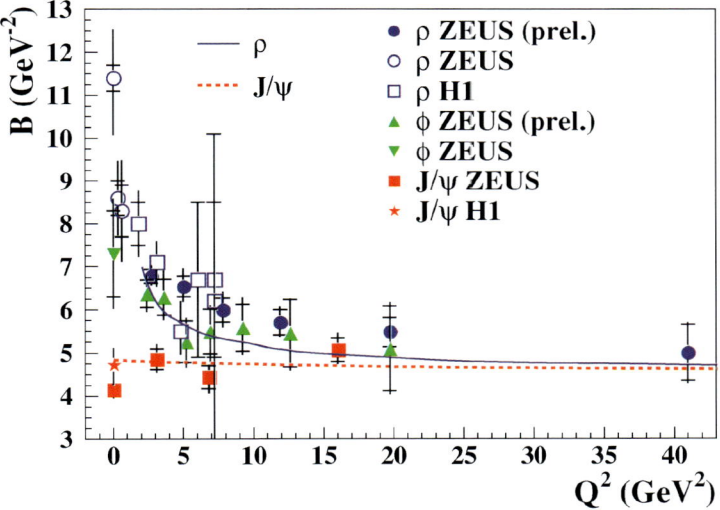

Figure 9 The HERA H1 and ZEUS data for the *t*-slopes of the differential cross sections for the exclusive electroproduction of ρ, ϕ and J/ψ mesons, as a function of Q^2. The convergence of the different slopes at large Q^2 indicates the dominance of small-size configurations in the production process ("squeezing"). The solid line shows the Q^2-dependence obtained in the calculation of Reference (26). The data are from Reference (57).

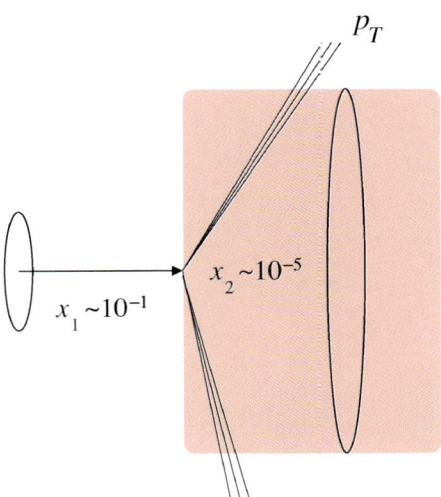

Figure 16 The black-disk limit (BDL) in central *pA* collisions: Leading partons in the proton, $x_1 \sim 10^{-1}$, interact with a dense medium of small-x_2 gluons in the nucleus (*shaded area*), acquiring a large transverse momentum, $p_\perp$.

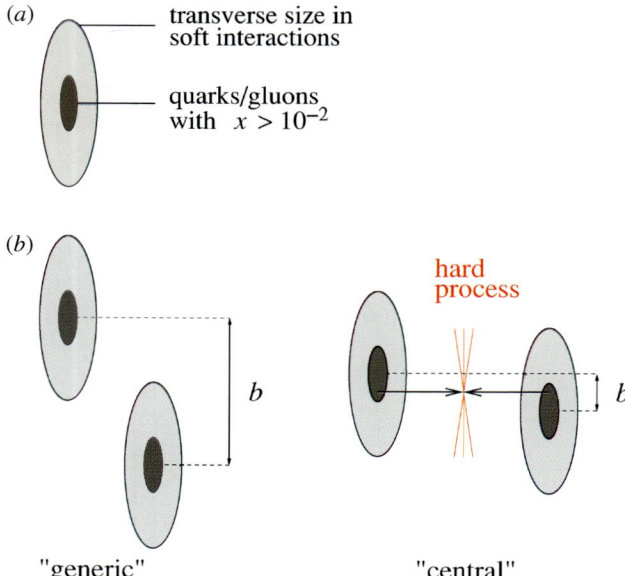

Figure 14   (*a*) The two-scale picture of the transverse structure of the nucleon in high-energy collisions. (*b*) The resulting classification of $pp/\bar{p}p$ events in "generic" and "central" collisions.

role of nuclear shadowing, and color transparency phenomena should completely disappear at very small x, where QCD factorization breaks down. This is in contrast to the two-gluon exchange model of References (74, 75), which neglects the space-time evolution of the dipole. In this model nuclear shadowing is obtained from exchanges of additional gluon between the current and target fragmentation regions, which is a higher-twist effect ($\propto 1/Q^2$) and disappears at large Q^2.

The color transparency phenomenon has been directly observed in three experiments:

- The total cross section for $\gamma^* A$ scattering increases with the atomic number as A^α with $\alpha \approx 1$, faster than the cross section for a hadronic projectile (see Reference (76) for a review of the experimental data).

- The cross section for coherent photoproduction of J/ψ mesons from nuclei increases with A much faster than that for coherent ρ meson production. The Fermilab E691 experiment (77) observed $\sigma^{\gamma^*+A \to J/\psi+A} \propto A^{1.46}$ at $E_\gamma = 150\,\text{GeV}$. Color transparency predicts that the coherent cross section integrated over t is $\propto (A^2/R_A^2) \approx A^{4/3}$. This A-dependence corresponds to the coherent sum of collisions from independent nucleons without absorption. A somewhat faster A-dependence emerges because of the contribution of incoherent diffractive processes (19).

- The A-dependence of the cross section for coherent dijet production in pion-nucleus scattering is anomalously large, as predicted in Reference (16). The Fermilab E791 experiment (78) observed a dependence $\propto A^{1.54}$ at $E_\pi \approx 600\,\text{GeV}$, similar to that in coherent J/ψ production. Note that the conventional Glauber approximation predicts $A^{1/3}$. Furthermore, the observed dependence of the cross section on the pion momentum fraction and the jet transverse momentum is well consistent with the perturbative QCD prediction of Reference (19). Notwithstanding the fact that the absolute cross section has not been measured, this is probably the first experimental observation of the high-momentum tail of the pion wave function as due to one-gluon exchange.

To summarize, there exists strong experimental evidence for color transparency in high-energy scattering. This phenomenon could be the basis for new "nondestructive" methods of investigating the microscopic structure of hadrons and nuclei in the future.

5. DIFFRACTION IN $\gamma^* p$ SCATTERING

5.1. QCD Factorization for Hard Diffractive Processes

Measurements of DIS at HERA have established the existence of a class of events in which the proton is observed in the final state, with a small invariant momentum transfer, t, and a hadronic system of invariant mass $M_X^2 \ll W^2$ is produced with a rapidity gap relative to the proton. In a frame in which the nucleon is fast-moving (i.e., in parton model kinematics) such processes are characterized by the fractional energy loss of the proton, $x_{I\!P} = (E_p^i - E_p^f)/E_p^i$, and the transverse momentum transfer, $\Delta_\perp$, with $t = -(\Delta_\perp^2 + x_{I\!P}^2 m_N^2)/(1 - x_{I\!P})$. In analogy with the corresponding phenomenon in hadronic collisions one refers to such processes as diffractive, although a priori the dynamics is not governed by soft physics.

Following suggestions of earlier works, a formal QCD factorization theorem was proved in References (79, 80) for hard processes of the type

$$\gamma^* + p \rightarrow h + (\text{rapidity gap}) + X, \qquad\qquad 24.$$

where X is either an inclusive state, or a state with extra hard activity (dijet production, heavy quark production, etc.). Similar to inclusive DIS, processes (Equation 24) with a given hadron h in the target fragmentation region are characterized by so-called conditional parton distribution functions, $f_j^h(\beta, Q^2, x_h, t)$, which are independent of the hard process and satisfy the same DGLAP evolution equations for fixed x_h and t. Here $\beta \equiv x/(1 - x_h) = Q^2/(Q^2 + M_X^2)$ is the fraction of the light-cone momentum of the target available for hard interactions in hadron h. Most of the current studies focus on diffractive kinematics, where $h = p$ and $1 - x_h = x_{I\!P} \leq 0.01$. The conditional parton distribution functions in this case are referred to as diffractive parton distribution functions (dPDF's), and denoted as f_j^D (Figure 11). In current data analysis it is usually assumed that the dependence of the dPDF's on $x_{I\!P}$, t and β, Q^2 can be factorized as (81)

$$f_j^D(\beta, Q^2, x_{I\!P}, t) = f_{I\!P/p}(x_{I\!P}, t)\, f_j^D(\beta, Q^2), \quad f_{I\!P/p}(x_{I\!P}, t) = f(t)x_{I\!P}^{-2\alpha_{I\!P}(t)+2}.$$
$$25.$$

This assumption is inspired by the soft Pomeron exchange model (which does not follow from the QCD factorization theorem) and referred to as Regge factorization (Figure 11). An additional term can be added to Equation 25 in analogy to non-vacuum exchange in soft physics; it gives a small contribution below $x_{I\!P} \sim 0.01$ and dominates at $x_{I\!P} \geq 0.05$.

Extensive studies of hard diffractive channels have been performed at HERA. The inclusive diffractive cross section was measured both integrated over t (the so-called diffractive structure function, $F_2^{D(3)}$), and as a function of t for a limited range of $x_{I\!P}$. These measurements are mostly sensitive to the quark dPDF, whereas the gluon dPDF enters through scaling violations. Diffractive dijet production for real and virtual photons, as well as diffractive charm production, primarily probe the gluon dPDF. The analysis of these data on the basis of QCD evolution equations has led to the following conclusions:

- The data at $Q^2 \geq 4\,\text{GeV}^2$ are described by universal dPDFs, consistent with the factorization theorem.
- $f_g(\beta, Q_0^2) \gg \sum_q f_q(\beta, Q_0^2)$ for the studied range of β. This conclusion was initially based on the weak scaling violation for $F_2^{D(3)}(\beta, Q^2)$ for large β, and was later confirmed by the studies of diffractive dijet production and charm

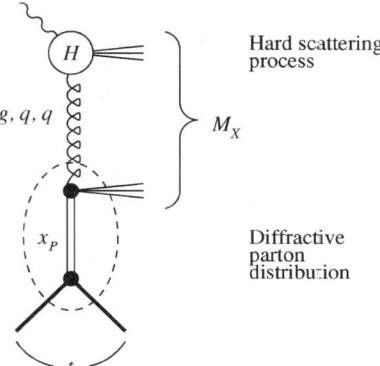

Figure 11 Factorization in diffractive DIS. The amplitude in the dashed blob, multiplied with its complex conjugate and summed over final states, defines the diffractive parton distribution. The internal structure of the dashed blob illustrates the assumption (25), which does not follow from the QCD factorization theorem.

production. However, the latter processes have so far been treated only in the LO approximation, and one should await the NLO analysis before drawing final conclusions.

- The data are consistent with Regge factorization, Equation 25, although the $x_{I\!P}$ dependence is faster than in soft physics. The analysis of ZEUS and H1 diffractive data finds $\alpha_{I\!P}(t = 0) = 1.2 \pm 0.07$ and increasing with Q^2, which should be compared with the energy dependence expected in soft hadronic collisions, $\alpha_{I\!P} \approx 1.1$.

- The absolute probability of diffraction in $\gamma^* p$ scattering is of the order of 10% for moderate Q^2, and thus of the same magnitude as in soft pion-nucleon collisions. However, the rate of increase of the diffractive cross section with energy for fixed M_X^2 and Q^2 is significantly faster than that of the total DIS cross section.

Another interesting characteristic of diffractive DIS is the probability of diffractive scattering depending on the type of parton coupling to the hard probe (72),

$$P_j(x, Q^2) = \int dt \int dx_{I\!P} \ f_j^D(x/x_{I\!P}, Q^2, x_{I\!P}, t) \Big/ f_j(x, Q^2). \qquad 26.$$

This ratio cannot exceed the value 0.5, which corresponds to the unitarity limit (BDL) (see the discussion in Section 6). Using the H1 fit to the diffractive DIS data (Figure 12), we find $P_g \gg P_q$, and $P_g(x \sim 10^{-3}) \approx 0.4(0.3)$ for $Q^2 = 4(10)\,\mathrm{GeV}^2$. That is, quark-induced diffraction is small, whereas gluon induced diffraction is close to the maximum value allowed by unitarity. We shall return to this point in our discussion of the profile function for the dipole-nucleon interaction in Section 6. Note that the H1 fit is based on the data at $x \geq 10^{-4}$. The fact that it

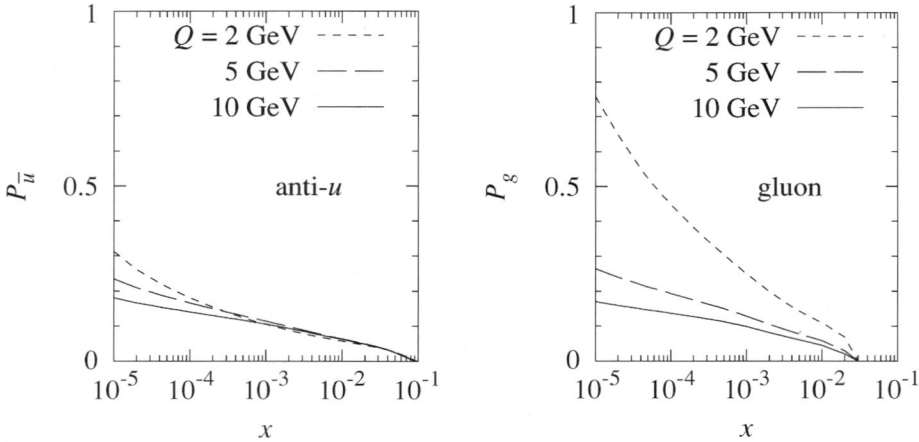

Figure 12 Probability of diffractive scattering from anti-u quarks (*left*) and gluons (*right*), extracted from a fit to the H1 data (82, updated by V. Guzey).

leads to $P_g \geq 0.5$ at smaller x indicates that the H1 parameterization should break down near the upper end of the HERA energy range.

5.2. Space-Time Picture of Hard Diffractive Processes

To understand the observed pattern of hard diffraction, it is instructive to consider the space-time evolution of such processes in the target rest frame. Such studies reveal new information about the interaction of small-size $q\bar{q}$ as well as $q\bar{q}g \ldots g$ configurations with hadronic matter. In particular, the ratio of the diffractive to the total cross section probes the interactions of such configurations with the target without reference to the probe that created them.

An immediate consequence of the QCD factorization theorem is that, in the target rest frame, the number of components in the photon wave function evolves with Q^2. Although at low Q^2 an approximation by a few components in the photon wave function may be reasonable, it is definitely inappropriate for large Q^2. This point is illustrated by the following example: Diffractive processes induced by longitudinally polarized photons are a leading-twist effect. If, however, all but the $q\bar{q}$ component of the photon wave function were neglected, one would erroneously conclude that diffraction is a higher-twist effect in this channel, because the transverse size of the longitudinal photon is $d^2 \propto 1/Q^2$. The proper Q^2 dependence is restored by the $q\bar{q}g \ldots g$ configurations in the photon wave function.

At low Q^2, aligned jet model-type configurations of large transverse size dominate in the wave function of the projectile photon (Section 2.2). Such configurations interact with a hadronic cross section, $\sim\sigma_{\text{tot}}(\pi N)$, and thus have a significant probability to rescatter elastically. If Q^2 increases, these configurations cannot be effectively produced without emission of gluons. Because these gluons are predominantly emitted collinearly, they do not change the transverse size of the diffracting system, and hence the probability of elastic rescattering. Assuming smooth matching between the strength of interaction in the perturbative and nonperturbative regimes, models reasonably describe the data on hard diffraction in ep scattering if $P_g > P_q$ at the initial scale, Q_0^2 (83, 84). Configurations of size $d^2 \geq \lambda/Q_0^2$, where $\lambda \approx 9$ (see the discussion in Section 3.2) should be included in the definition of the dPDF at the initial scale. For $Q_0^2 = 4\,\mathrm{GeV}^2$, this includes rather small transverse sizes, for which the cross section increases with energy significantly faster than at the soft scale, which is consistent with the trend of the HERA data mentioned in Section 5.1. Also, because the $q\bar{q}g$ configurations have masses considerably larger than $q\bar{q}$ configurations, they should manifest themselves in diffraction at relatively small $\beta < 0.5$. Indications of diffraction into non-aligned jet final states were indeed found in a number of HERA experiments (85). To summarize, it appears that hard diffraction at HERA with $Q^2 \sim 4\,\mathrm{GeV}^2$ represents the border between the high-Q^2 region, where leading-twist QCD dominates, and the low-Q^2 region, where higher-twist effects become important.

A quantitative analysis of the HERA data within the gluon dipole picture indicates that the interaction of gluon dipoles at HERA energies is rather close to the BDL (72) (see the discussion in Section 6). However, owing to our inability to build an effective trigger on the interaction of gluon dipoles with $d \geq 0.3\,\mathrm{fm}$ it is difficult to observe this effect directly in the experiments.

To conclude this discussion, we briefly want to comment on the assumption of Regge factorization, Equation 25. Because strong deviations of the energy dependence of diffraction from the soft regime are observed, there is a priori no reason for the validity of this assumption. Several effects are likely to contribute to the breakdown of Regge factorization: (a) different energy dependence of the cross section for diffraction of configurations of different transverse size, (b) emission of gluons by $q\bar{q}$ dipoles, which at smaller x occurs at large coherent lengths (see the discussion in Section 2.2), and (c) soft screening effects, which become more important with increasing energy (these effects were observed in soft diffraction (86, 87)), and which should be different for the various diffractive configurations, as they interact with different strengths.

One way to probe the degree of validity of Regge factorization is to check that the hadrons produced in the photon fragmentation region do not "talk" with hadrons in the target fragmentation region, i.e., that there are no long-range correlations in rapidity. Such a factorization ignores the existence of color fluctuations, which lead to processes in which the proton also dissociates. If Regge factorization were valid, the probability of dissociation would not depend on the properties of the state into which the virtual photon has diffracted. However, if screening effects are present and non-universal (for example, due to the different strength of interaction in quark and gluon induced diffraction), Regge factorization should be broken.

5.3. Diffraction and Leading-Twist Nuclear Shadowing

A direct relation between diffraction in high-energy hadron-hadron collisions and the nuclear shadowing effect in hadron-nucleus collisions was derived by Gribov (88, 89), in the approximation where the nucleon radius is considerably smaller than the mean internucleon distance in nuclei. The same reasoning in conjunction with the leading-twist approximation for hard diffractive processes allows one to calculate nuclear shadowing of PDF's in light nuclei (72).

Application of the Abramovsky-Gribov-Kancheli cutting rules (90), which relate the shadowing phenomenon for the total cross section of high energy processes to that of partial cross sections (such as for diffraction, multiparticle production, etc.) and are valid in perturbative QCD (see Reference (91) for a recent discussion), explicitly demonstrates that the interference of the amplitudes of diffraction from the proton and the neutron leads to a decrease of the total cross section for γ^*-deuteron scattering. When combined with the factorization theorem for inclusive diffraction, one can calculate the modification of nuclear PDF's at low values of Bjorken x (72). In the limit of low nuclear thickness, the nuclear shadowing corrections to the nuclear parton densities are given by

$$\frac{f_{j/A}(x, Q^2)}{A} = f_{j/N}(x, Q^2) - \frac{1}{2} \int d^2b \int_{-\infty}^{\infty} dz_1 \int_{z_1}^{\infty} dz_2$$

$$\times \int_x^{x_0} dx_{I\!P} \, \cos\left[x_{I\!P} m_N (z_1 - z_2)\right]$$

$$\times \frac{1 - \eta^2}{1 + \eta^2} \, f_{j/N}^D(\beta, Q^2, x_{I\!P}, t)\big|_{k_\perp = 0} \, \rho_A(b, z_1) \, \rho_A(b, z_2), \quad 27.$$

where $f_{j/N}(x, Q^2)$ is the usual parton density in the proton, $f_{j/N}^D(\beta, Q^2, x_{I\!P}, t)$ the diffractive parton density (Section 5.1), and $\rho_A(r)$ is the nucleon density in the nucleus with atomic number A. The momentum transfer, t, is given by $-t = (k_\perp^2 + (x_{I\!P} m_N)^2)/(1 - x_{I\!P})$, where $k_\perp$ is the transverse component of the momentum transferred to the struck nucleon, and $\beta = x/x_{I\!P}$. In Equation 27, the factor $(1 - \eta^2)/(1 + \eta^2)$, where $\eta \equiv -\pi/2\partial \ln(\sqrt{f_{i/N}^D})/\partial \ln(1/x_{I\!P}) = \pi/2[\alpha_{I\!P}(t = 0) - 1]$, accounts for the real part of the amplitude of diffractive scattering (92). One can easily modify Equation 27 to include the dependence of the diffractive amplitude on t. Obviously, both the left- and right-hand side of Equation 27 satisfy the QCD evolution equations in all orders in α_s, and this relation does not depend on the renormalization scheme. These expressions represent the model-independent result for leading-twist nuclear shadowing in the low-density limit.

In the case of heavy nuclei one may with good accuracy neglect the fluctuations of the strength of interaction in the hadron component of the photon wave function. This approximation makes it possible to extend the above formulas to the case of heavy nuclei (72, 93). Numerical studies of shadowing using Equation 27 and the corresponding expression for the total DIS cross section for heavy nuclei found large leading-twist shadowing effects for quark and gluon distributions, with gluon shadowing being larger up to rather high values of Q^2. The latter effect can be traced to the higher probability of gluon-induced diffraction as compared to quarks (Figure 12).

The connection between diffraction and nuclear shadowing does not depend on the twist decomposition of the cross section, and was successfully applied also to data on nuclear shadowing of F_{2A} at intermediate Q^2 (see e.g., References (94, 95)). One can use this to estimate the relative importance of leading-twist and higher-twist nuclear shadowing at $Q^2 \leq 2\,\mathrm{GeV}^2$, using experimental information on the leading-twist contribution to the diffractive cross section at these values of Q^2. One finds that a significant higher-twist contribution to diffractive DIS originates from ρ-meson production. This implies that a significant fraction ($\sim$40%) of the nuclear shadowing observed in the experiments at CERN and Fermilab (see Reference (76) for a review) may be due to higher-twist effects (93).

Recently, leading-twist nuclear shadowing was included in the initial conditions for the small-*x* evolution in the McLerran-Vegnugopalan model (96). A distinctive feature of this model is that the invariant masses in the nuclear vertex of the BFKL ladder should be very large as compared to Q^2. That is, small β should dominate in the integral, in analogy to Equation 27. A numerical analysis of gluon shadowing using the HERA dPDF's finds that the region $\beta \leq 0.1$ becomes important only for $x \leq 10^{-4}$ (93). Thus, the assumption of the dominance of large diffractive masses may give rise to important dynamical effects at the next generation of accelerators.

5.4. Implications of Nuclear Shadowing for Heavy-Ion Collisions

The typical x values relevant for semihard production of hadrons in heavy-ion collisions decreases with energy as $x_A \sim 2p_\perp/\sqrt{s_{NN}}$ for central rapidities, and much faster, $\sim 4p_\perp^2/s_{NN}$, for the fragmentation regions (s_{NN} is the squared center-of-mass energy of the effective nucleon-nucleon collisions). For central rapidities and $p_\perp \geq 2\,\mathrm{GeV}$, gluon shadowing is still a small correction at RHIC. However, it will be a large effect at LHC for a wide range of $p_\perp$, because the relevant x_A are much smaller than 0.01. The expected suppression of jet production is given by the factor $[G_A(x_A, p_\perp^2)/AG(x_A, p_\perp^2)]^2$, where G_A and G are the gluon densities in the nucleus and the nucleon, respectively. This factor can be of the order of $1/4$ (82).

Because the current RHIC detectors have rather limited forward coverage, they have limited sensitivity to small-*x* phenomena. One exception is J/ψ production, which, if interpreted within perturbative QCD, probes x down to 0.003. The observed suppression of the J/ψ yield is consistent with the estimates of Reference (93) (see Reference (97) for a review). The A-dependence of forward high-$p_\perp$ hadron production was studied by BRAHMS (98). Although at large rapidities small x contribute to the high $p_\perp$ spectra, the QCD analysis indicates that average x are $\sim$0.03 (99). Consequently, the yields are practically not sensitive to shadowing effects, or, more generally, to any initial-state modifications of the nucleus wave function consistent with present DIS data. Final-state interaction effects, which could explain the data, are nonperturbative contributions to the production of leading hadrons, due to coalescence of spectator partons and the relatively small energy losses in the initial and final state (on the scale of 3%).

6. BLACK-DISK LIMIT IN DIPOLE-HADRON INTERACTIONS

6.1. Violation of the Leading-Twist Approximation at Small x

A fundamentally new dynamical effect expected at high energies is the unitarity limit, or black-disk limit (BDL), in the interaction of a small dipole with hadronic matter. We now describe and quantify this effect, using the information gathered in our studies of inclusive, exclusive and diffractive DIS in Sections 3, 4 and 5.

A simple argument shows that the twist expansion for the cross sections of hard processes breaks down at sufficiently small x. QCD factorization predicts that the total cross section for DIS at fixed Q^2 increases with decreasing x as $\sigma_{tot} \propto xG(x)/Q^2$. At the same time, the cross section of elastic dipole-hadron scattering (which corresponds to the production of diffractive states with masses $M_X \propto 1/d \propto Q$) grows much faster, $\sigma_{diff} \propto (xG)^2/(BQ^4)$, where B is the t-slope of the corresponding differential cross section (21). Clearly, at sufficiently small x there is a contradiction—the total cross section should always be larger than that for any particular channel (44). The resolution of this paradox is that the decomposition of hard amplitudes in powers of $1/Q^2$ becomes inapplicable at sufficiently small x.

In order to quantify the onset of the new regime, it is instructive to consider the effects of unitarity (conservation of probability) on a purely theoretical scattering process, namely the scattering of a $q\bar{q}$ (quark-antiquark) or gg (gluon-gluon) dipole of small transverse size, d, from a hadronic target. Neglecting other constituents in the dipole is justified in a wide kinematic range by the smallness of the coupling constant; large terms $\propto \alpha_s \ln(x_0/x)$ arise only from interactions at large rapidity intervals. The invariant amplitude for dipole-proton elastic scattering is a function of the invariants $s \equiv W^2$, and t. We write it as a Fourier integral over the dipole-proton impact parameter, b,

$$A^{dp}(s, t) = \frac{is}{4\pi} \int d^2b \; e^{-i(\mathbf{\Delta}_\perp \mathbf{b})} \; \Gamma^{dp}(s, b) \quad (t = -\mathbf{\Delta}_\perp^2), \qquad 28.$$

where $\Gamma^{dp}(s, b)$ is the so-called profile function. Making use of unitarity, one can express the total, elastic, and inelastic (total minus elastic) cross sections in terms of the profile function as

$$\left.\begin{array}{r} \sigma_{tot}(s) \\ \sigma_{el}(s) \\ \sigma_{inel}(s) \end{array}\right\} = \int d^2b \; \times \; \left\{\begin{array}{l} 2 \, \text{Re} \; \Gamma^{dp}(s, b) \\ |\Gamma^{dp}(s, b)|^2 \\ \left[1 - |1 - \Gamma^{dp}(s, b)|^2\right]. \end{array}\right. \qquad 29.$$

In the situation where elastic scattering is the "shadow" of inelastic scattering, the profile function at a given b is restricted to values $|\Gamma^{dp}(s, b)| \le 1$. The value

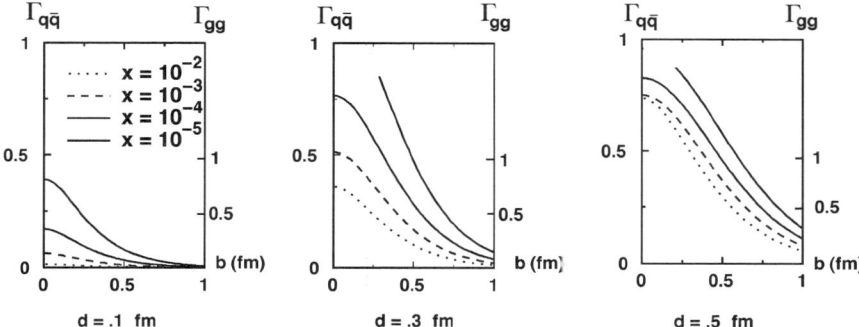

Figure 13   The profile function of dipole-nucleon scattering, Γ^{dp}, as a function of the impact parameter, b, for various values of the dipole size, d, and x, as obtained from a phenomenological estimate (see text). Shown are the results for $q\bar{q}$ (*left scale*) and gg dipoles (*right scale*), $\Gamma_{gg} = (9/4)\Gamma_{q\bar{q}}$.

$\Gamma^{dp}(s, b) = 1$ corresponds to complete absorption at a given impact parameter, the so-called black-disk limit (BDL).[4]

The proximity of $\Gamma^{dp}(s, b)$ to unity is an important measure of the strength of the interaction of the dipole with the proton. As outlined in Section 2.2, the analysis of $\gamma^* N$ scattering in the target rest frame allows one to determine with reasonable accuracy (LO approximation) the total cross section for the scattering of a $q\bar{q}$ dipole from the proton. Combining this with information on the transverse spatial distribution of gluons in the nucleon, obtained from measurements of the t-dependence of exclusive J/ψ photoproduction and other hard exclusive processes (Section 4.4), we can calculate the profile function for dipole-nucleon scattering (73). A sample of the results of Reference (100) is presented in Figure 13. Here, $x = Q^2/s$, where $Q^2 = \lambda/d^2 \approx 9/d^2$ is the characteristic virtuality corresponding to the dipole size. One sees that up to the top of the HERA energy range, $x \sim 10^{-4}$, the profile function for a $q\bar{q}$ dipole remains small for dipole sizes $d \leq 0.3$ fm, corresponding to $Q^2 \geq 4\,\text{GeV}^2$, the value usually used as a starting point for DGLAP evolution (101). Even for larger dipole sizes, $d \sim 0.5$ fm, the fraction of the cross section due to scattering with $\text{Re}\Gamma \geq 0.5$ remains small. This shows that the BDL is not reached in inclusive DIS at HERA energies, in agreement with what we observed in Section 3.1 and 3.2. However, the DGLAP evolution starting from a $q\bar{q}$ dipole generates gg dipoles, whose interaction at leading twist is larger by a factor of $9/4$ (see Equation 8) and thus approaches the BDL much earlier. The

---

[4]In non-relativistic quantum mechanics, the scattering of a particle from a completely absorptive sphere is referred to as the "black-body limit." In contrast, the high-energy limit of scattering amplitudes in QCD is essentially two dimensional, with the radius of interaction increasing with energy. It is thus natural to refer to the limit of complete absorption as the "black-disk limit."

theoretical estimate shown in Figure 13 indicates that the strength of interaction of gg dipoles is close to maximal at the top of the HERA energy range, for a wide range of b, and dipole sizes corresponding to $Q^2 \leq 4\,\text{GeV}^2$. This implies that in gluon-induced interactions at HERA at such Q^2 the probability of diffraction should be close to $1/2$—exactly as we found in the analysis of HERA diffractive data in Section 5.1.

To summarize, one may expect that the leading-twist approximation for DIS breaks down for $Q^2 \leq 4\,\text{GeV}^2$, especially in the gluon sector. Unfortunately, there are no readily available probes of gluons at low Q^2, except possibly the longitudinal cross section, σ_L. Without such measurements, it is impossible to determine whether the successful DGLAP fits to the HERA data down to $Q^2 \sim 1\,\text{GeV}^2$ are an artifact of using an essentially arbitrary gluon density at low Q^2 and low x; this function is practically not constrained by the data at larger Q^2, where it is dominated by DGLAP evolution from larger x. Also, it is worth emphasizing that although $\Gamma^{dp}(b)$ for gluon dipoles (and, at higher energies, also for quark dipoles) reaches values close to unity, the actual deviation of the nucleon structure function from the DGLAP fits may still be rather small, as the contributions from scattering at large b remain dominant.

6.2. Theoretical Issues in Describing the Black-Disk Limit

The analysis of the full DGLAP evolution equation and the resummation approaches shows that the cross section for the scattering of a small dipole increases effectively as x^{-n}, with $n \geq 0.2$ for $Q^2 \sim$ few GeV^2, and somewhat faster at larger Q^2. [At very large Q^2 and extremely high energies—well above the LHC range—the resummation approaches predict that $n \propto \alpha_s(Q^2)$, and thus $n \to 0$.] Within these approximations, the t-slope of the differential cross section for elastic scattering, B, increases with energy rather slowly (see the discussion in Section 4.4). Thus, the cross section for dipole-nucleon elastic scattering increases with energy faster than the total cross section, and unitarity is violated for this hard processes within the leading-twist approximation (see the discussion in Section 6.1). Probability conservation will be violated at fixed impact parameter, b, in a region corresponding to a disk of finite size. At large enough b, the interaction is too small to violate the leading-twist approximation. Note that in soft interactions the increase of the elastic cross section does not necessarily lead to a contradiction with unitarity, because soft interactions generate also α', and therefore give rise to an increase of the t-slope.

The conventional assumption is that, beyond the leading-twist approximation, taming of the growth of cross sections occurs due to the shadowing phenomenon (this follows, e.g., from Gribov's reggeon calculus). Specific to this phenomenon is that bare particles may experience only one inelastic collision, but any number of elastic interactions, without changing trajectory. The behavior of the amplitudes for high-energy processes in QCD differs from that given by the eikonal approximation in non-relativistic quantum mechanics because of the necessity to account for the non-conservation of the number of bare particles. Application of the Abramovsky-Gribov-Kancheli cutting rules (90) shows that the requirement of positive probabilities for total cross sections, single particle densities, etc., impose serious restrictions on the dynamics in the case of cross sections increasing with energy. To satisfy these requirements in a series of multiple rescatterings in which consecutive terms have alternating signs, the effective number of constituents in the dipole wave function should increase with energy. Thus, the increase of the cross section with energy leads to resolution of constituents in the wave functions of the colliding particles, and therefore to an evolution of final states. Evolution and gluon emission by dipoles are the key for generating multiple inelastic collisions without violation of causality and energy-momentum conservation. The evolution of a dipole in time manifests itself in the expansion of the system, emission of gluons, transitions between components containing different numbers of bare particles, change of impact parameters in the intermediate states, and the related effect of the cross section for inelastic diffraction exceeding the elastic cross section for the scattering of small dipoles (see the discussion in Section 5). In this regime, the concept of a parton density of the target cannot be defined in a model-independent way, because the parton distributions in the dipole and the target are intertwined and not restricted by probability conservation.

At energies where the dipole cross section becomes comparable or even larger than $2\pi R_N^2$, the whole picture of rescattering becomes inconsistent if the radius squared of the interaction is not proportional to the dipole cross section. For example, in this case the Glauber approximation for hadron-deuteron scattering would lead to negative total cross sections. Fortunately, in QCD the wave function of a fast projectile contains many partons. This fact, combined with the increase of the dipole cross sections with energy, is sufficient to ensure complete absorption for central collisions, without detailed knowledge of the hadron wave function (102). To illustrate the rapid onset of complete absorption for central collisions related to the increase of the number of constituents, we adopt here a simple approximation, namely that the perturbative QCD (pQCD) description of dipole-proton scattering works up to $\mathrm{Re}\Gamma^{dp}(b, x) = 1$, and that $\Gamma^{dp}(b, x) = 1$ if the pQCD formulas lead to $\mathrm{Re}\,\Gamma^{dp}(b, x) > 1$. That is,

$$\mathrm{Re}\,\Gamma^{dp}(b, x) = \mathrm{Re}\,\Gamma^{dp}(b, x)_{\mathrm{pQCD}}\,\Theta[1 - \mathrm{Re}\,\Gamma^{dp}(b, x)_{\mathrm{pQCD}}]$$
$$+ \Theta[\mathrm{Re}\,\Gamma^{dp}(b, x)_{\mathrm{pQCD}} - 1]. \qquad 30.$$

Many of the models currently discussed in the literature use the elastic eikonal approximation to describe the taming of the increase of the dipole-hadron cross section with energy as due to the shadowing phenomenon (see e.g., References (103, 104) for a nucleon target and Reference (105) for a heavy nuclear target). These models assume that taming becomes significant for $\Gamma^{dp}(b) \geq 0.5$, i.e., at significantly larger x than the values where unitarity is explicitly violated in the pQCD approximation. Early taming results in a slow approach to the unitarity limit, $\Gamma^{dp}(b) = 1$. Obviously, these conclusions are model dependent, as such models neglect most of the QCD effects mentioned above.

The condition of the BDL for dipole-hadron scattering, $\Gamma^{dp}(b) = 1$, expresses the complete "loss of memory" of the cross section on the structure of the projectile and the target, and of the value of the running coupling constant, in a finite region of transverse space. It reflects the breakdown of two dimensional conformal invariance (which is the basis of approximate Bjorken scaling in DIS) because of the appearance of a dynamical scale related to the high gluon density and the radius of the transverse distribution of gluons. The qualitative departure from pQCD dynamics in the BDL cannot be explained as a soft interaction effect. This can be understood when considering collisions of two small dipoles of same size near the BDL, e.g., $\gamma^*(Q^2) + \gamma^*(Q^2)$ scattering at sufficiently large Q^2, in which soft interaction effects are under control and negligible. An interesting question is whether, from a general perspective, the appearance of this new scale corresponds to a spontaneous breakdown of conformal symmetry, or is related to the conformal anomaly.

6.3. High-Energy Limit of Nuclear and Hadronic Structure Functions

The approach to the BDL in dipole-hadron scattering at high energies has interesting implications for the theoretical behavior of hadron and nuclear structure functions at extremely high energies, which is subject to the Froissart bound.

We consider the scattering of a virtual photon from a heavy nucleus (radius R_A) at high energies as a superposition of the scattering of dipoles of different sizes. The interaction at impact parameters $b \leq R_A$ will be black for dipoles with sizes larger than some critical size, $d > d(x)$, leading to a contribution to the cross section (see Equation 11)

$$\sigma^{\gamma^* A} \approx 2\pi R_A^2 \int d^2 d \int_0^1 dz \, |\psi^\gamma(d, z)|^2 \, \Theta[d - d(x)]. \qquad 31.$$

Because the profile function of dipole-nucleus scattering increases like a power of energy in the region where it is <1, one concludes that $d(x) \propto x^m$, with $m > 0$. When the nuclear radius significantly exceeds the essential impact parameters in $\gamma^* N$ collisions, one has

$$F_{2A}(x, Q^2) = \frac{Q^2}{12\pi^3} \left(\sum_f e_f^2 \right) (2\pi R_A^2) \ln \frac{x_0(Q^2)}{x}, \qquad 32.$$

where $x_0(Q^2)$ slowly decreases with increasing Q^2.[5] For illustrative purposes, we neglect here the contributions from peripheral collisions, which grow with the atomic number as $A^{1/3}$. Although formally these contributions increase with energy

---

[5]Because in the BDL the masses of the intermediate states are much larger than Q^2, the coherence length is much smaller than the naive estimate, $l_{coh} \sim 1/(2m_N x)$. If the gluon density in the approach to the BDL grows as $x^{-\lambda(Q^2)}$, one expects that $l_{coh} \propto 1/(m_N x^{1-\lambda})$.

faster than those from central collisions (which are nearly energy-independent), they are still comparatively small at all achievable energies. The gross violation of Bjorken scaling in Equation 32, $F_{2A} \propto Q^2 \ln(x_0/x)$, and the numerical coefficient follow from the normalization of the photon wave function to the Q^2-derivative of the photon polarization operator, as opposed to unity as for hadron wave functions. Equation 32 represents a QCD modification of the Gribov BDL (89), which assumed that all configurations in the virtual photon with masses $M^2 \leq 2m_N x/R_A$ interact with the heavy nucleus with maximum strength.

In DIS from a proton target, scattering at large impact parameters is always important in the regime where the pQCD interaction becomes strong. Indeed, on the basis of the studies of the transverse spatial distribution of gluons in hard exclusive processes (Section 4.4) one expects that $\Gamma^{dp}(s, b) \propto \exp(-\mu b)$ at large b, with $\mu \approx m_g$ for moderately small x (see Equation 22), and $\mu \to 2m_\pi$ in the limit of infinitely large energies. It follows from the unitarity bound, Equation 29, that the essential impact parameters increase with energy as $b^2 \propto \mu^{-2} \ln^2 \left[\sigma^{dp}(s, d)/(8\pi B) \right]$, where B is the t-slope of the differential cross section of dipole-nucleon scattering, which is almost energy-independent within the leading-twist approximation. The cross section of γ^*-nucleon scattering therefore increases with energy as (106)

$$\sigma^{\gamma^* p} \propto \mu^{-2} \ln^2 \frac{\sigma^{dp}(s, d)}{8\pi B}. \qquad 33.$$

In general, this behavior differs from the Froissart limit for soft hadronic interactions, because of the more complicated dependence of the dipole cross section, $\sigma^{dp}(s, d)$, on the energy, as described by the resummation approaches. To simplify the formulas, below we shall use the observation that effectively $\sigma^{dp}(s, d) \propto s^{n(d)}$, with $n \geq 0.2$ for small d. This approximation leads to the limiting behavior familiar from soft hadronic interactions (3). The leading asymptotic term in x for fixed Q^2 for the nucleon structure function (106) is

$$F_2(x, Q^2) = \frac{Q^2}{12\pi^3} \left(\sum_f e_f^2 \right) \sigma^{\gamma^* p} \ln \frac{s}{s_0} \propto \ln^3 \frac{s}{s_0}, \qquad 34.$$

where two logarithms originate from the dipole-nucleon cross section, and one from the integral over the photon wave function, similar to the case of scattering from nuclei.

In hard exclusive processes in $\gamma^* p$ scattering, the approach to the BDL implies that the t-slope increases with energy $\propto \ln^2(s/s_0)$ (see also Section 7.4). A promising strategy in searching for BDL effects would be to extract partial waves for small impact parameters from the cross sections of processes such as DVCS from the nucleon, ρ-meson production (107), coherent photoproduction of high $p_\perp$ dijets from the nucleon, and coherent photoproduction of $J/\psi, \psi', \psi''$ mesons. This would allow one to probe small b, where the gluon density is maximal and unitarity effects should manifest themselves early. Another possible strategy is to study the structure functions of heavy nuclei, where due to higher gluon densities the BDL effects are enhanced by a factor $A^{1/3}$ (108). Note that this enhancement is partly compensated by nuclear shadowing; see the discussion in Section 5.3.

It is interesting also to explore the behavior of nuclear structure functions at extremely high energies, where the radius of the $\gamma^* N$ interaction becomes comparable to or even exceeds the nuclear radius. In this case, first the edge of the nucleus contributes terms $\propto A^{1/3} \ln(x_0/x)^3$ to the cross section. Ultimately, for $s \to \infty$ and fixed Q^2, one would reach the universality regime where $F_{2A}(x, Q^2)/F_{2p}(x, Q^2) \to 1$ (102). However, the relevant scale is comparable to the gravitational scale.

6.4. Black-Disk Limit in Hard Diffractive Scattering from Heavy Nuclei

An important consequence of the BDL is that, in diffractive scattering at high energies, non-diagonal transitions between diffractive eigenstates are forbidden (89). This follows from the orthogonality of the eigenstates—if every configuration in the projectile interacts with the same strength, the relative proportion between different configurations remains the same. This implies that half of the nuclear DIS cross section should be due to diffraction, with the nucleus remaining intact, and a "jetty" diffractive final state resembling that of $e^+ e^- \to$ hadrons. In contrast, in the leading-twist approximation this cross section should be negligible.

At the onset of the BDL regime, where the contributions from configurations in the virtual photon interacting with the BDL strength and those for which pQCD is applicable are comparable, one can calculate the differential cross section for diffraction to final states of small mass, M_X, for which the interaction is already black (106),

$$\frac{dF_T^{\gamma_T^* \to X}(x, Q^2, M_X^2)}{dM_X^2 d\Omega_X} = \frac{\pi R_A^2}{12\pi^3} \frac{Q^2 M_X^2}{(M_X^2 + Q^2)^2} \frac{d\sigma(e^+ e^- \to X)/d\Omega_X}{\sigma(e^+ e^- \to \mu^+ \mu^-)}. \qquad 35.$$

This shows a much slower decrease with Q^2 than in the leading-twist approximation, and corresponds to "jetty" final states (mostly diffraction to $q\bar{q}$ and $q\bar{q}g$ jet states). The earliest signal for the change of the Q^2 dependence should be in ρ-meson production, where the Q^2 dependence of the dominant longitudinal cross section should change from $1/Q^6$ (Section 4.3) to $1/Q^2$.

Theoretical studies show that at HERA kinematics the fraction of the cross section due to diffraction should be much closer to 1/2 for nuclear targets than for the proton (109). The use of nuclear beams would greatly facilitate the exploration of the BDL regime. Possibilities for such measurements in ultraperipheral collisions at LHC will be discussed in Section 9.

7. SMALL-x DYNAMICS IN HADRON-HADRON COLLISIONS

7.1. Transverse Radius of Hard and Soft Interactions

Our studies of γ^*p scattering at HERA have taught us several important lessons about the gluon density at small x, the transverse spatial distribution of gluons, and about the interaction of small-size color singlets with hadrons (see Section 1 for a summary). We now explore the implications for the physics of $pp/p\bar{p}$ and pA collisions at high energies.

In hadron-hadron collisions, hard processes arise from binary collisions of partons in the colliding hadrons, in which a system of large invariant mass, $M^2 \gg \Lambda_{QCD}^2$, is produced. Examples are the production of dijets, dilepton pairs (Drell-Yan process), and the production of heavy particles such as Higgs bosons or SUSY particles. Such hard partonic processes are generally accompanied by a rich spectrum of soft interactions, which dominate the total hadronic cross section and determine the overall characteristics of hadron production in the final state. Understanding the interplay of hard and soft interactions is the main challenge in describing hadron-hadron collisions with hard processes.

A crucial observation in studies of hard exclusive processes in γ^*p scattering is that gluons with significant momentum fraction ($x > 10^{-2}$) are concentrated in a transverse area much smaller than that associated with the nucleon in pp elastic scattering at high energies, which is dominated by soft interactions. The difference between the two areas becomes more pronounced with increasing energy. When considering the production of a system of fixed mass, M^2, in collision of partons with $x_1 x_2 = M^2/s$, the transverse area of the hard partons grows with energy as $\langle \rho^2 \rangle = \alpha'_{hard} \ln s$, whereas the transverse area for soft interactions grows at a much faster rate, $\alpha'_{soft} \gg \alpha'_{hard}$. The cause of this difference is the suppression of Gribov diffusion for partons of large virtuality, as described in Section 2.2. Thus, in high-energy pp collisions one is dealing with an "onion-like" transverse structure of the nucleon (two-scale picture) (Figure 14a, see color insert).

The two-scale picture of the transverse structure of the nucleon implies a classification of $pp/\bar{p}p$ events in "central" and "generic" collisions, depending on whether the transverse areas occupied by the large-x partons in the two protons overlap or not (Figure 14b) (73). Generic collisions give the dominant contribution to the overall inelastic cross section. Hard processes, such as heavy particle production at central rapidities, practically happen in central collisions only. (Obviously, in these collisions multiparton interactions due to the small-x gluon fields are strongly enhanced, giving rise to the dynamical effects described in the following subsections.)

To quantify the distinction between generic and central collisions, we estimate the distribution of the probability for both types of events over the impact parameter of the pp collision, b. For generic collisions, the distribution is determined by the b-dependent probability of inelastic interaction, obtained via unitarity from the elastic pp amplitude in the impact parameter representation, analogous to Equation 29.

We define a normalized b-distribution as

$$P_{in}(s, b) = [2\,\text{Re}\,\Gamma^{pp}(s, b) - |\Gamma^{pp}(s, b)|^2]/\sigma_{in}^{pp}(s), \qquad 36.$$

where $\sigma_{in}^{pp}(s)$ is the inelastic cross section, which is given by the integral $\int d^2b$ of the expression in the numerator. For collisions with a hard process, on the other hand, the b-distribution is determined by the overlap integral of the distribution of hard partons in the two colliding protons (see Figure 14b),

$$P_2(b) \equiv \int d^2\rho_1 \int d^2\rho_2\, \delta^{(2)}(\boldsymbol{b} - \boldsymbol{\rho}_1 + \boldsymbol{\rho}_2)\, F_g(x_1, \rho_1)\, F_g(x_2, \rho_2). \qquad 37.$$

Numerical estimates can be performed with our parametrization of the transverse spatial distribution of hard gluons (Section 4.4), which takes into account the change of the distribution with x and the scale of the hard process. The two b-distributions are compared in Figure 15 for Tevatron and LHC energies. For the hard process we have taken the production of a dijet with transverse momentum $q_\perp = 25\,\text{GeV}$ at rapidity $y = 0$ in the center-of-mass frame; in the case of Higgs production at LHC the distribution $P_2(b)$ would be even narrower. The results clearly show that events with hard processes have a much narrower impact parameter distribution than generic inelastic events.

One expects that at LHC energies the rate of production of multiple dijets will be very high. It is interesting to consider the b-distribution also for the production of two dijets in two binary parton-parton collisions. Neglecting possible correlations between the partons in the transverse plane it is given by

$$P_4(b) \equiv [P_2(b)]^2 \Big/ \int\!\int d^2b'\, [P_2(b')]^2. \qquad 38.$$

Figure 15 shows that this distribution is significantly narrower than P_2, i.e., the requirement of two hard processes results in a further reduction of effective impact parameters.

Correlations in the transverse positions of partons can be probed by studying $pp/\bar{p}p$ events with two hard processes, involving two binary collisions of partons. At the Tevatron such a process—production of three jets and a photon—was studied by the CDF collaboration (111). The observed cross section is by a factor of 4 larger than the naive estimate based on the assumption that the partons are distributed uniformly in the transverse plane, over an area whose size was inferred from the proton electromagnetic form factor. The effect of correlations in the transverse position of partons (i.e., their localization in "spots" of significantly smaller size than the radius of their distribution within the nucleon) reduces this discrepancy by a factor of 2. This hints at the presence of significant correlations in the parton transverse positions for $x \geq 0.05$. A possible explanation of such correlations is the localization of the non-perturbative gluon fields in "constituent" quarks (and antiquarks), as suggested by the instanton vacuum model of chiral symmetry breaking in QCD and supported by a large body of information on

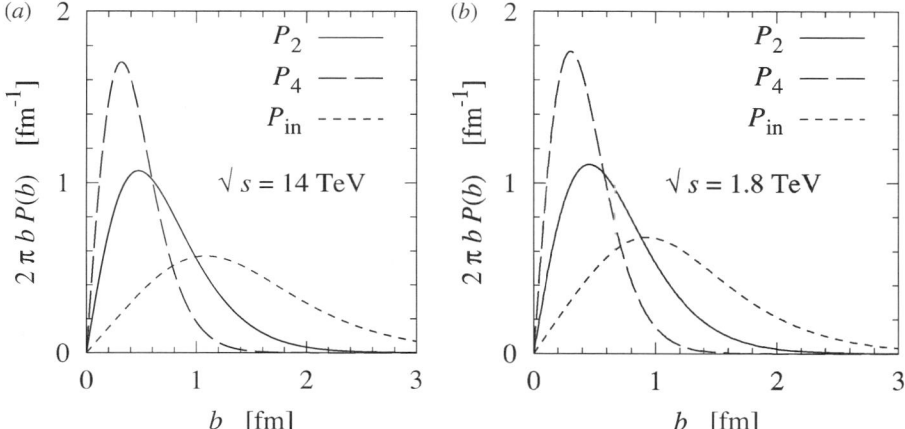

Figure 15   Solid lines: Impact parameter distributions in events with hard dijet production, $P_2(b)$, in pp collisions at LHC (*a*) and $\bar{p}p$ collisions at the Tevatron (*b*), for a back-to-back dijet at zero rapidity with transverse momentum $q_\perp = 25$GeV. Long dashed lines: Same for double dijet events, $P_4(b)$. Short dashed lines: Impact parameter distributions in generic inelastic collisions, obtained from the parameterization of the elastic pp amplitude of Islam et al. (110). The plots show the "radial" distributions in the impact parameter plane, $2\pi b P(b)$.

low-energy hadron structure (see Reference (31) for a review). We find that the parton correlations implied by this model indeed give rise to a further enhancement of the cross section for two hard processes by a factor ~ 2 (see References (73, 112) for details).

7.2. Black-Disk Limit in High-Energy pA and pp Collisions

A new effect encountered in high-energy pA and pp collisions is that the interaction of leading partons in the proton with the gluon field in the nucleus (or other proton) approaches the maximum strength allowed by s-channel unitarity, the BDL. This leads to certain qualitative modifications of the hadronic final state, which will be observable in central pp and pA collisions at LHC. In particular, this effect dramatically changes the strong interaction environment for new heavy particle production in central pp collisions at LHC.

In a central high-energy pA collision, consider a leading parton in the proton, with longitudinal momentum fraction $x_1 \sim 10^{-1}$ and typical transverse momentum of the order of the inverse hadron size (Figure 16, see color insert). Assuming a leading-twist two-body scattering process with transverse momentum $p_\perp$ in the final state, this leading parton can interact with partons in the nucleus of momentum fraction

$$x_2 = \frac{4p_\perp^2}{x_1 s} \qquad\qquad 39.$$

(x_2 and s here refer to the effective pN collision). If s becomes sufficiently large, x_2 can reach very small values even for sizable transverse momenta, $p_\perp^2 \gg \Lambda_{\mathrm{QCD}}^2$. For example, at LHC $x_2 \sim 10^{-6}$ is reached for $p_\perp \approx 2\,\mathrm{GeV}$. At such values of x_2, the gluon density in the nucleus becomes large. The leading parton can be thought of as propagating through a dense "medium" of gluons. In this situation, the probability for the leading parton to split into several partons and scatter inelastically approaches unity, corresponding to the scattering from a "black" object. As a result the leading parton effectively (via splittings) undergoes inelastic collisions, losing energy and acquiring transverse momentum, until its transverse momentum is so large that the interaction probability becomes small, and the nucleus no longer appears "black." To summarize, we can say that in central pA collisions the leading partons acquire transverse momenta of the order of the maximum transverse momentum, $p_{\perp,\mathrm{BDL}}$, for which their interaction with the nucleus at their respective x_1 is close to the BDL. This transverse momentum represents a new hard scale in high-energy hadron-hadron collisions, which appears due to the combined effect of the rise of the gluon density at small x and the unitarity condition.[6]

To estimate the maximum transverse momentum for interactions close to the BDL, we can treat the leading parton as one of the constituents of a small dipole scattering from the target. This "trick" allows us to apply the results of Section 6 to hadron-hadron scattering. In this analogy, the effective scale in the gluon distribution is $Q_{\mathrm{eff}}^2 = 4p_\perp^2$, corresponding to an effective dipole size of $d \approx 3/(2p_\perp)$. For simplicity, we first consider the case of central collisions of a proton with a large nucleus, which allows us to neglect the spatial variation of the gluon density in the target in the transverse direction. This amounts to approximating the transverse spatial distribution of gluons in the nucleus by

$$G_A(x, \rho; Q_{\mathrm{eff}}^2) \approx \frac{G_A(x; Q_{\mathrm{eff}}^2)}{\pi R_A^2} \qquad (\rho < R_A). \qquad\qquad 40.$$

As a criterion for the proximity to the BDL, we require that the profile function of the dipole-nucleus amplitude at zero impact parameter satisfy $\Gamma^{dA}(b = 0) > \Gamma_{\mathrm{crit}}$ (Figure 17a). For an estimate, we choose $\Gamma_{\mathrm{crit}} = 0.5$, corresponding to a probability for inelastic interaction of 0.75, reasonably close to unity. We then determine the maximum $p_\perp$ for which the criterion is satisfied. Figure 17 shows the result for

---

[6]The kinematics of the final state produced in the interaction of the large-x_1 parton with the small-x_2 gluon field resembles the backscattering of a laser beam from a high-energy electron beam. The large-x_1 parton gets a significant transverse momentum and loses a certain fraction of its longitudinal momentum, accelerating at the same time a small-x_2 parton.

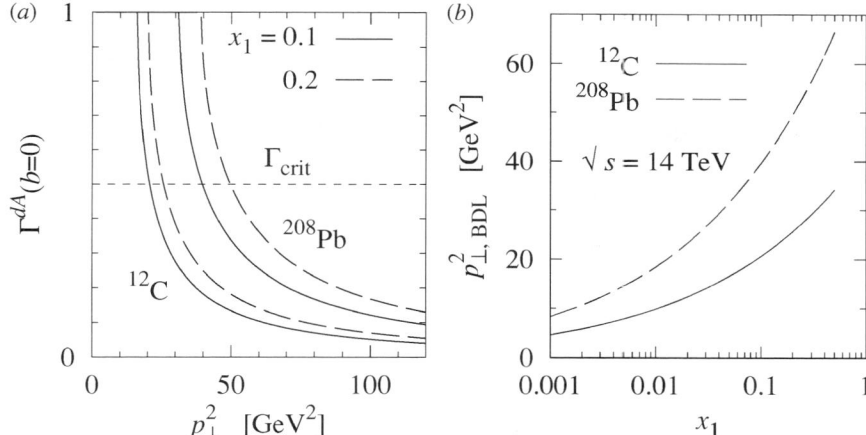

Figure 17   Black-disk limit in central *pA* collisions at LHC: (*a*) The profile function for the scattering of a leading gluon in the proton (regarded as a constituent of a *gg* dipole) from the nucleus at zero impact parameter, $\Gamma^{dA}(b = 0)$, as a function of the transverse momentum squared, $p_\perp^2$. (*b*) The maximum transverse momentum squared, $p_{\perp,\text{BDL}}^2$, for which the interaction of the leading gluon is "black," $\Gamma^{dA} > \Gamma_{\text{crit}}$, as a function of the gluon's momentum fraction, x_1. Here we assume $\sqrt{s} = 14\,\text{TeV}$ for the effective *NN* collisions, in order to facilitate comparison with the case of central *pp* collisions in Figure 16.

$p_{\perp,\text{BDL}}^2$ for a leading gluon, as a function of the gluon momentum fraction, x_1; for leading quarks, the result for $p_{\perp,\text{BDL}}^2$ is approximately 0.5 times the value for gluons. The numerical estimates show that leading partons indeed receive substantial transverse momenta when traversing the small-x_2 gluon medium of the nucleus. We emphasize that our estimate of $p_{\perp,\text{BDL}}$ applies equally well to the interaction of leading partons in the central region of *AA* collisions.

Turning now to *pp* collisions, we have to take into account the transverse spatial structure of the colliding hadrons. A crucial point is that high-energy interactions do not significantly change the transverse position of the leading partons, so that their interaction with the small-x_2 gluons is primarily determined by the gluon density at this transverse position. Because the leading partons in the "projectile" proton are concentrated in a small transverse area, and the small-x_2 gluon density in the "target" proton decreases with transverse distance from the center, it is clear that the maximum transverse momentum for interactions close to the BDL, $p_{\perp,\text{BDL}}^2$, decreases with the impact parameter of the *pp* collision, *b*. Figure 18 (upper row) shows the dependence of $p_{\perp,\text{BDL}}^2$ on *b*, as obtained with the parametrization of the transverse spatial distribution of gluons based on analysis of the HERA exclusive data (Section 4.4) (73). One sees that $p_{\perp,\text{BDL}} \sim$ several GeV in central collisions at LHC. Substantially smaller values are obtained at the Tevatron energy.

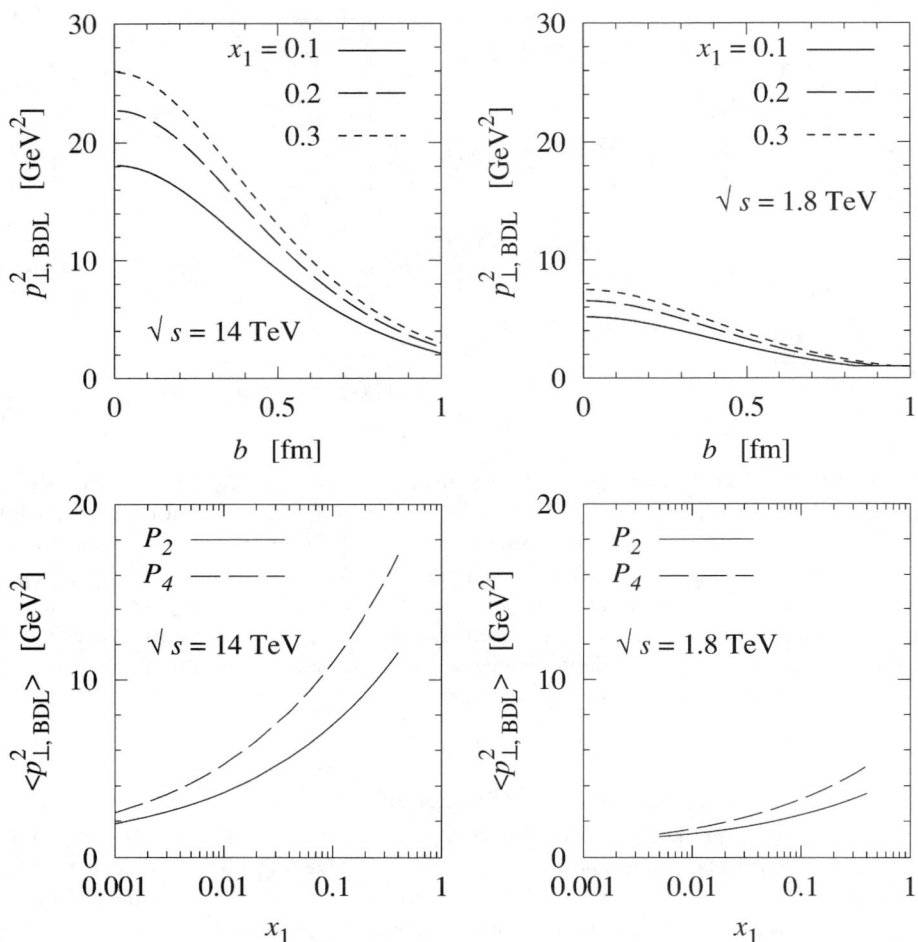

Figure 18   Upper row: The transverse momentum squared, $p^2_{\perp,\text{BDL}}$, acquired by a leading gluon (momentum fraction x_1) through interactions with the small-x_2 gluon field in the other proton near the BDL, as a function of the impact parameter of the pp collision, b. Shown are the estimates for LHC (*left panel*) and Tevatron energies (*right panel*). Lower row: Average values of $p^2_{\perp,\text{BDL}}$ in pp collisions with a single hard process (impact parameter distribution P_2) and two hard processes (distribution P_4) (see Figure 14). For leading quarks, the values of $\langle p^2_{\perp,\text{BDL}} \rangle$ are approximately half of those for gluons shown here.

To determine the typical transverse momenta of leading partons in events with new particle (or hard dijet) production, we need to average the results for $p^2_{\perp,\text{BDL}}$ over pp impact parameters, with the distribution implied by the hard production process, $P_2(b)$, Equation 37, or, in the case of four jet production, with $P_4(b)$, Equation 38. The resulting average values of $p^2_{\perp,\text{BDL}}$ are shown in Figure 18 (lower row). We find that the suppression of large impact parameters implied by the hard process, described in Section 7.1, is sufficient to keep $p_{\perp,\text{BDL}}$ above 1 GeV in more than 99% of events at LHC.

To summarize, our estimates show that in central pA and central (triggered) pp collisions at LHC the leading partons acquire substantial transverse momenta due to interactions near the BDL. A much weaker effect is found at the Tevatron energy. The origin of this difference is the increase in the gluon density due to the decrease of x_2 between Tevatron and LHC energies (see Equation 39).

7.3. Final State Properties in Central pp Collisions at LHC

The approach to the BDL in the interaction of leading partons implies certain qualitative changes in the hadronic final state in central pp and pA collisions. In particular, these effects will profoundly influence the strong interaction environment for the production of new heavy particles (Higgs boson, etc.) at LHC.

The main effect of the BDL is that the leading partons in the projectile acquire substantial transverse momenta, of the order $p_{\perp,\text{BDL}}^2$, when propagating through the dense medium of small-x_2 gluons in the target. As a result, the projectile becomes "shattered": The leading partons lose coherence and fragment independently over a wide range of rapidities close to the maximal rapidity, corresponding to hadron momentum fractions $z \sim x_1 \mu / p_{\perp,\text{BDL}}$ (μ is a typical hadronic mass scale). The differential multiplicity of leading hadrons, integrated over $p_\perp$, is approximately given by the convolution of the nucleon parton density, f_a, with the corresponding parton fragmentation function, $D_{h/a}$, at the scale $Q_{\text{eff}}^2 = 4 p_{\perp,\text{BDL}}^2$,

$$\frac{1}{N} \left(\frac{dN}{dz} \right)^{pp \to h+X} = \sum_{a=q,g} \int dx_1 \, x_1 \, f_a\!\left(x_1, Q_{\text{eff}}^2\right) D_{h/a}\!\left(z/x_1, Q_{\text{eff}}^2\right), \quad 41.$$

where N is total number of inelastic events (106, 113, 114). This corresponds to a very strong suppression of forward hadron production as compared to generic inelastic pp events. The suppression is particularly pronounced for nucleons; one expects that for $z \geq 0.1$ the differential multiplicity of pions should exceed that of nucleons. At the same time the transverse spectrum of the leading hadrons will be much broader, extending up to $p_{\perp,\text{BDL}} \gg 1$ GeV. Finally, the independent fragmentation mechanism implies that there will be no correlations between the transverse momenta of leading hadrons (some correlations will remain, however, because two partons produced in collisions of large-x_1 and small-x_2 partons may end up at similar rapidities).

In central pp collisions at LHC, where leading particles are suppressed in both the forward and backward direction, one expects a large fraction of events with no particles with $z \geq 0.02$–0.05 in both fragmentation regions. This amounts to the appearance of long-range rapidity correlations. Such events should show a large energy release at rapidities $y = 4$–6. However, similar to the situation with rapidity gap survival in pp scattering (see the discussion in Section 8), one should expect that there is a $\sim$10% probability for dijets to be produced in pp collisions at large impact parameters without additional BDL interactions between the constituents of the nucleons.

Another important effect of the BDL is a significantly increased energy loss of the leading partons, due to the larger probability of inelastic collisions, and the wider distribution of the propagating parton (dipole) over transverse momenta. In particular, a 3% energy loss would explain the suppression of forward pion production in deuteron-nucleus collisions at RHIC at $p_\perp \sim 4\,\text{GeV}$ (99). Studies of this effect would be possible both at RHIC and LHC, in particular if the forward capabilities of the current detectors were upgraded, as discussed in several proposals presently under consideration. Note that energy loss is neglected in Equation 41. This corresponds to the usual assumption of models in which parton propagation is treated as multiple elastic rescattering of the parton's accompanying gluon field from the medium. A consistent treatment of energy loss and transverse momentum broadening near the BDL remains a challenge for theory. We note that the pattern of energy loss in our approach is qualitatively different from models in which the leading partons scatter from a classical gluon field (in that case energy loss is negligible) (115).

The approach to the BDL has consequences also for hadron production in the central rapidity region. The multiple scattering of large-x_1 projectile partons from the small-x_2 gluons in the target shifts a large number of these gluons to larger rapidities, leaving numerous "holes" in the target wave function. Furthermore, multiple interactions of partons with moderately small $x_1 \sim x_2$ also occur with large probability. (Unitarity effects should be important for these interactions as well, but have not been studied so far.) Both effects lead to the creation of a substantial amount of color charge, which should result in an increase of soft particle multiplicities over a broad range of rapidities as compared to the situation far from the BDL. This increase should in fact be observable already at Tevatron energies, in central events selected by a trigger on two-jet or Z^0 production. An increase of the multiplicity at rapidities $|y| \leq 1.0$ in such events compared to minimum bias events was indeed reported in Reference (116). It would be extremely interesting to extend such studies to higher rapidities.

Our findings imply that new heavy particles at LHC will be produced in a much more "violent" strong-interaction environment than one would expect from the extrapolation of the properties of minimum bias events at the Tevatron. Even the extrapolation of properties of hard dijet events should not be smooth, as the transverse momenta acquired by leading partons are estimated to be substantially larger at LHC than at Tevatron (Figure 18).

7.4. Black-Disk Limit in Elastic pp Scattering

The assumption of the BDL in the interaction of leading partons, combined with the complex structure of the proton wave function in QCD, allows us to estimate the profile function of the pp elastic amplitude at small impact parameters. This simple estimate nicely explains the observed "blackness" of the phenomenological

pp profile function at $b = 0$ at the Tevatron energy, and allows us to extrapolate the profile function at small b to higher energies. It also raises the question whether the observed blackness could be explained on the basis of soft interactions.

In pp collisions at small impact parameters, leading quarks on average receive substantial transverse momenta when passing through the small-x gluon field of the other proton (Figure 18). When a single leading quark gets a transverse momentum, $p_\perp$, the probability for the nucleon to remain intact is approximately given by the square of the nucleon form factor, $F_N^2(p_\perp^2)$, which is ≤ 0.1 for $p_\perp > 1$ GeV, i.e., very small. One may thus conclude that the probability of survival, averaged over $p_\perp$, should be less than 1/2 (on average, half of the time the quark should receive a transverse momentum larger than the average transverse momentum, $p_\perp > 1$ GeV). Because there are six leading quarks (plus a number of leading gluons), the probability for the two protons to stay intact, $|1 - \Gamma^{pp}(s, b)|^2$ (see Equation 29), should go as a high power of the survival probability for the case of single parton removal, and thus be very small. This crude estimate shows that already at Tevatron energies $|1 - \Gamma^{pp}(s, b = 0)|^2$ should be close to zero owing to hard interactions. The conclusion that the small impact parameter hadron-hadron interactions should become black at high energies follows principally from the composite structure of the hadrons and does not depend on details of the dynamics. In particular, if taming effects stopped the growth of the dipole-nucleon interaction at a fraction of the BDL, our result would not change. Our conclusion that $\Gamma^{pp}(\sqrt{s} \geq 2\,\text{TeV}, b = 0) \approx 1$ agrees well with the current analysis of the Tevatron data (see e.g., Reference (110)).

One can estimate the maximum impact parameter, b_F, up to which hard interactions cause the pp interaction to be "black." The probability for a leading parton with $x_1 \sim 10^{-1}$ to experience a hard inelastic interaction increases with the collision energy at least as fast as dictated by the increase of the gluon density in the other proton at $x_2 = 4p_\perp^2/(sx_1)$ (see Equation 39 below). Because $x_2 G(x_2, Q^2) \propto x_2^{-n_h}$ with $n_h \geq 0.2$, the probability should grow as s^{n_h}.[7] The dipole parametrization of the transverse spatial distribution of gluons, Equation 22, suggests that the gluon density decreases with the distance from the center of the nucleon approximately as $\sim \exp[-m_g(x_2)\rho]$. If one neglects the transverse spread of the large-x_1 partons as compared to that of the small-x_2 gluons one arrives at an estimate of the energy dependence of b_F as due to hard interactions (102),

$$ b_F \approx \frac{n_h \ln(s/s_T)}{m_g(x_2)}, \qquad\qquad 42. $$

where $\sqrt{s_T} = 2\,\text{TeV}$ is the Tevatron energy. In principle, n_h may decrease at very large virtualities, which would become important at extremely high energies.

---

[7]The HERA data on dipole-nucleon scattering suggest that the taming of the gluon density starts only when the probability of inelastic interactions becomes large, $\geq 1/2$. However, for such probabilities of single parton interactions, multiparton interactions ensure that the overall interaction is practically black.

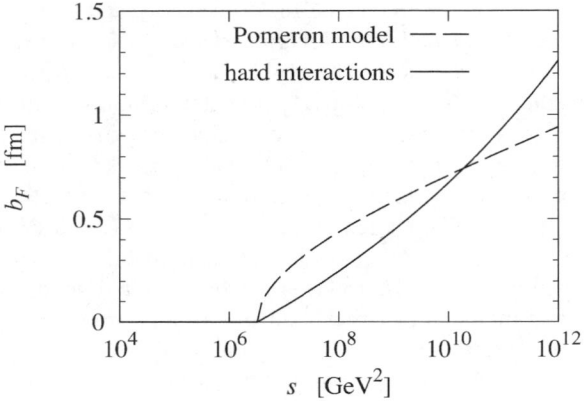

Figure 19 Energy dependence of the maximum impact parameter for "black" pp interactions, b_F. Solid line: Estimate based on the Pomeron model fit of pp elastic scattering of (117). Dashed line: Estimate based on hard interactions (Equation 42).

However, this effect is likely to be compensated by the increased number of constituents in the nucleon wave function affected by the BDL. The above estimate is consistent with the popular Pomeron model parameterization of the pp elastic amplitude (117). In this model $\Gamma^{pp}(0) \approx 1$ for $s = s_T$ and s-channel unitarity is violated at $b < b_F$ for $s > s_T$. The dependence of b_F on the energy in the Pomeron model is similar to our estimate (42) (Figure 19). This shows that the two-scale picture of the transverse structure of the nucleon—and the ensuing picture of hard and soft interactions—are self-consistent.

Overall, the analysis of Reference (102) summarized in this subsection suggests that at high energies hadron-hadron interactions should be "black" in a range of impact parameters growing approximately like $\ln s$, with a coefficient growing only very slowly with energy owing to decrease of $m_g(x_2)$ with energy. This corresponds to the Froissart regime, with interactions in the black region dominated by semi-hard interactions, and interactions at large impact parameters dominated by single Pomeron exchange. At the same time, this analysis indicates that the concept of summing multi-Pomeron exchanges, which should give the dominant contribution at small b, breaks down as the soft physics is gradually squeezed out to large b.

It is worth emphasizing here that, in principle, the BDL can emerge in hadron-hadron interactions already at the level of the soft interactions (118, 119). However, it is hardly possible to reconcile it with the pre-QCD Feynman parton model description of high energy processes, if one would require it to be valid both in the rest frame of the target and in the center-of-mass frame. Really, within the parton model one cannot generate complete absorption of the projectile in central collisions in the target rest frame, where the target consists of few partons (120). This puzzle is resolved in QCD, where radiation leads to the "blackening" of the hard interactions at central impact parameters. At extremely high energies, as a result of this effect, all memory of the colliding hadrons is lost. Hence the universal behavior of total cross sections, $\sigma_{tot} \propto \ln^2(s/s_0)$, with universal coefficient for all hadrons (nuclei) (102).[8]

7.5. Ion-Induced Quark-Gluon Implosion

The small-*x* phenomena outlined above—the approach to the BDL, and large leading-twist gluon shadowing—play an important role also in heavy ion collisions at LHC energies. Here we consider just one example, the so-called ion-induced quark-gluon implosion in the nucleus fragmentation region. For a review of other effects in the framework of the color glass condensate model, see Reference (115).

In generic central *AA* collisions at collider energies, in analogy to the central *pA* collisions discussed in Section 7.2, all the leading partons of the individual nucleons are stripped off "soft" partons and form a collection of quarks and gluons with large $p_\perp$. In the rest frame of the fragmenting nucleus, the incoming nucleus has a "pancake" shape with longitudinal length ~ 1 fm for soft partons, and $R_N(m_N/p_N)(x_V/x) \ll R_N$ for hard partons, where $x_V \sim 0.2$ is the average *x* value for the valence quarks. That is, the nucleons in the nucleus at rest at different locations along the collision axis are hit by the hard partons in the incoming nucleus one after another. In the BDL, no spectators are left. The hit partons are produced with practically the same *x* that they had in the nucleus (because the fractional energy loss is small), transverse momenta $\sim p_{\perp,BDL}$, and virtualities $\leq p^2_{\perp,BDL}$. The partons move in the direction of the projectile nucleus. Because they are emitted at finite angles, their longitudinal velocity is smaller than the speed of light, and they are left behind the projectile wave. However, because the emission angles are small, a shock wave is formed, compressing the produced system in the nucleus rest frame. In the frame co-moving with the shock wave, valence quarks and valence gluons are streaming in opposite directions. The resulting pattern of fragmentation of the colliding nuclei leads to an "implosion" of the quark and gluon constituents of the nuclei. The non-equilibrium state produced at the initial stage in the nucleus fragmentation region is estimated to have densities $\propto p^2_{\perp,BDL}$, which is ≥ 50 GeV/fm^3 at LHC, and probably ≥ 10 GeV/fm^3 at RHIC. It seems likely that the partons would rescatter strongly at the second stage, although much more detailed modeling is required to find out whether the system would reach thermal equilibrium. Such large-angle rescattering of partons would

---

[8]The increase of the interaction with energy, and the related increase of essential impact parameters, show that the theoretical description of high-energy hadronic collisions should be closer to classical mechanics than to quantum mechanics (V. Gribov, private communication to Yu. Dokshitzer). An example is the cross section for the scattering of a high-energy particle from a potential rapidly decreasing with impact parameter. The essential impact parameters—and therefore the cross section—are infinite within classical mechanics, but finite in quantum mechanics, whereas they increase with energy in QCD.

lead to production of partons at higher rapidities, and re-population of the cool region. In particular, two gluons from the pancake could have the right energies to produce near-threshold $c\bar{c}$ pairs and χ_c-mesons with small transverse momenta and $x_F(c\bar{c}) \sim 2x_g \sim 0.1$.

7.6. Cosmic Ray Physics Near the GZK Cutoff

An extensive program of studies of cosmic rays at energies close to the Greisen-Zatsepin-Kuzmin (GZK) cutoff (10), $E_{GZK} \simeq 6 \times 10^{10}$ GeV, is under way, using several cosmic ray detectors. These experiments detect cosmic rays indirectly, via the air showers induced when they enter the atmosphere. The properties of the primary particle need to be inferred from those of the observed shower. For this, a good understanding of the physics of high-energy interactions in the atmosphere is mandatory. The observed characteristics of the shower are predominantly sensitive to leading hadron production ($x_F \geq 10^{-2}$), which, according to our discussion above, at these energies probes small-x dynamics down to $x \sim 10^{-10}$, deep inside the regime affected by the approach to the BDL. First studies of these effects were performed in Reference (121). It was found that the steeper x_F-distribution of leading hadrons as compared to low-energy collisions, caused by the strong increase of the gluon densities at small x (see Section 7.4), leads to a reduction of the position of the shower maximum, X_{max}. Account of this effect in the models currently used for the interpretation of the data may shift fits of the composition of the cosmic ray spectrum near the GZK cutoff towards lighter elements. Also, it appears that the present data on $X_{max}(E)$ exclude the possibility that the prediction of a rapid growth of the critical x-value where the BDL becomes effective ($\sim 1/x^{0.3}$), which is compatible with RHIC and HERA data, would persist up to the GZK cutoff energy.

8. HARD DIFFRACTION AT HADRON COLLIDERS

8.1. Diffractive Proton Dissociation into Three Jets

LHC will offer an opportunity to study a variety of hard diffractive processes in pp and pA scattering. One interesting aspect of such processes is that they allow us to probe rare small-size configurations in the nucleon wave function. A proton in such a configuration can scatter elastically off the target and fragment into three jets, corresponding to the process

$$p + p(A) \rightarrow \text{jet1} + \text{jet2} + \text{jet3} + p(A). \qquad 43.$$

The cross section for the diffractive process (Equation 43) can be evaluated on the basis of the kind of QCD factorization theorem derived in Reference (19). The cross section is proportional to the square of the gluon density in the nucleon at $x \approx M^2(3 \text{ jets})/s$, and virtuality $Q^2 \sim (1-2)p_\perp^2$ (122). The distribution over the fractions of the proton longitudinal momentum carried by the jets is proportional to the square of the light-cone wave function of the $|qqq\rangle$ configuration. The numerical estimates suggest (123) that the process could be observed at the LHC energies provided one were able to measure jets with $p_\perp \sim 10\,\mathrm{GeV}$ at very high rapidities, $y_{\mathrm{jet}}(p_\perp = 10\,\mathrm{GeV}) \sim 6$, and with a large background from leading-twist hard diffraction. The latter will be suppressed in pA collisions, because the coherent 3-jet process has a much stronger A-dependence than the background due to soft and hard diffraction induced by strong interactions. The main background will be due to hard electromagnetic interactions of the proton with the Coulomb field of the nucleus. We note that the discussed mechanism of hard diffraction requires that the interaction of the spatially small three-quark color singlet configuration with the proton be far from the BDL at small impact parameters. Otherwise production at small impact parameters would be suppressed, leading to a dip in the t-dependence of the differential cross section for the production of three jets with moderate $p_\perp$.

The detection of the three-jet final state produced by diffractive scattering of a qqq configuration from a proton should be easier than that resulting from e^+e^- annihilation into $q\bar{q}g$, as in the former case all color charges are in the triplet representation, leading to less radiation between the jets. Finally, it would also be possible to study the process $pp \rightarrow pn + \text{two jets}$, which is similar to pion dissociation into two jets. Experimentally, this would require the measurement of jets at rapidities $y \sim 4$, together with the detection of a leading neutron by a zero-degree calorimeter, as is present in several of the LHC detectors (ALICE, ATLAS, CMS).

8.2. Exclusive Diffractive Higgs Production

Hard diffractive processes are also being considered in connection with the production of new heavy particles in pp collisions at LHC. In particular, the exclusive diffractive production of Higgs bosons,

$$p + p \rightarrow p + (\text{rapidity gap}) + H + (\text{rapidity gap}) + p, \qquad 44.$$

is regarded as a promising candidate for the Higgs search; see Reference (124) and references therein. From the point of view of strong interactions, this process involves a delicate interplay between "hard" and "soft" interactions, which can be described within our two-scale picture of the transverse structure of the nucleon (112). The Higgs boson is produced in a hard partonic process, involving the exchange of two hard gluons between the nucleons. The impact parameter distribution of the cross section for this process is described by the convolution of the squared transverse spatial distributions of gluons in the in and out states defined in Equation 37. (This distribution turns out to be numerically close to the distribution P_4, Equation 38, and we approximate it by P_4 in the numerical estimates below.) The scale here is of the order of the gluon transverse momentum squared, $\sim M_H^2/4$. In addition, the soft interactions between the spectator systems have to conspire in such a way as not to fill the rapidity gaps left open by the hard process. The probability for this to happen is approximately given by one minus the probability of an inelastic pp interaction at a given impact parameter, or $|1 - \Gamma^{pp}(s, b)|^2$. The product of the two probabilities, which determines the b-distribution for the total process, is shown in Figure 20a. At small b the probability for no inelastic interaction is very small $|1 - \Gamma^{pp}|^2 \approx 0$, leading to a strong suppression of small b in the overall distribution.

The so-called rapidity gap survival probability, which measures the "price" to be paid for leaving the protons intact, is given by the integral (112)

$$S^2 \equiv \int d^2b \, |1 - \Gamma^{pp}(s, b)|^2 \, P_4(b). \qquad 45.$$

Figure 20b shows our result for this quantity, with s ranging between Tevatron and LHC energies, for various values of the dipole mass in the two-gluon form factor of the nucleon, m_g^2, Equation 22. The survival probability decreases with s because the size of the "black" region at small impact parameters (in which inelastic interactions happen with high probability) grows with the collision energy. Note that the effective x values in the gluon distribution decrease with the energy (for fixed mass of the produced Higgs boson), resulting in smaller effective values of m_g^2. This makes the actual drop of the survival probability with energy slower than appears from the fixed-m_g^2 curves of Figure 20b. Our estimates of S^2 are in reasonable agreement with those obtained in Reference (125) in a multi-Pomeron model, as well as with those reported in Reference (126). In view of the different theoretical input to these approaches this is very encouraging.

Our results for the rapidity gap survival probability apply equally well to the production of two hard dijets instead of a Higgs boson. For this process, one expects a much larger cross section, and it would be possible to investigate experimentally the interplay of hard physics and absorptive effects, which leads to a rich, distinctive structure of the cross section as a function of the transverse momenta of the two

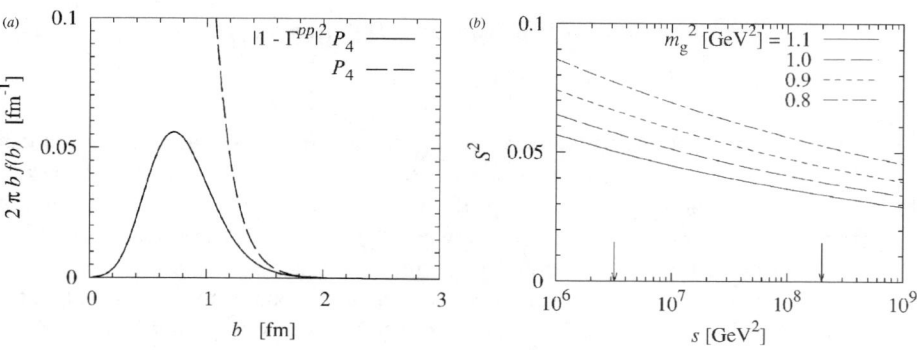

Figure 20 (*a*) The impact parameter distribution of the cross section for diffractive Higgs production at LHC ($\sqrt{s} = 14$ TeV). Dashed line: b-distribution of the hard process, $P_4(b)$, Equation 38 (see Figure 14). Solid line: b-distribution of the total process, $|1 - \Gamma^{pp}(s, b)|^2 P_4(b)$. Here, $m_g^2 = 1$ GeV^2. (*b*) The rapidity gap survival probability, S^2, Equation 45 (112). Shown is the result as a function of s, for various values of the mass parameter in the two-gluon form factor, m_g^2. The Tevatron and LHC energies are marked by arrows.

protons, $\mathbf{\Delta}_{1\perp}$ and $\mathbf{\Delta}_{2\perp}$ (127). This structure should also rather strongly depend on the rapidities of the jets, due to the x-dependence of the transverse spatial distribution of gluons (Section 4.4) (L. Frankfurt et al., in preparation).

8.3. Inclusive Hard Diffractive Processes

Inclusive hard diffractive processes, such as

$$p + p \rightarrow p + (\text{rapidity gap}) + 2\,\text{jets} + X,$$
$$p + p \rightarrow p + (\text{rapidity gap}) + 2\,\text{jets} + X + (\text{rapidity gap}) + p,$$

46.

offer a possibility to probe the "periphery" of the proton with hard scattering processes. The cross sections for these processes are again suppressed compared to the naive estimate based on the diffractive parton densities of the proton measured in ep scattering at HERA. As in the case of exclusive diffractive Higgs production, the cause of this is the very small probability for the protons not to interact inelastically at small impact parameters. The suppression factors can be estimated by generalizing the approach to the description of hard and soft interactions outlined in Section 8.2. Simple estimates along the lines of Equation 45 naturally reproduce the suppression factors of the order 0.1–0.2 observed at Tevatron. However, the results in this case are more sensitive to the details of the impact parameter dependence of the hard scattering process and the soft spectator interactions.

Measurements of inclusive hard diffractive processes at LHC would allow one to perform many interesting tests of the diffractive reaction mechanism. In particular, one could (*a*) investigate how the overall increase of the nucleon size with energy leads to a suppression of hard diffraction, (*b*) check how the rate of suppression depends on the x-value of the parton involved in the hard process, (*c*) look for the breakdown of Regge factorization, that is, the change of the diffractive parton distributions with $x_{I\!P}$.

9. SUMMARY AND OUTLOOK

9.1. From HERA to LHC

The HERA experiments and the theoretical investigations they stimulated have greatly advanced our knowledge of small-x dynamics. The key result of these studies are: (*a*) The rapid increase with energy of the cross section for the scattering of small $q\bar{q}$ wave packets from the nucleon. HERA energies are not sufficient to reach the BDL in the $q\bar{q}$ dipole-nucleon interaction in average configurations. The interaction of gluon dipoles in diffractive scattering appears to be close to the unitarity limit for $Q^2 \sim 4\,\text{GeV}^2$, but this can hardly be verified directly because of the lack of a trigger for such configurations. (*b*) The establishment of a three-dimensional picture of the partonic structure of the nucleon. The leading partons are concentrated in a much smaller transverse area than that associated with the nucleon in soft hadronic processes at high energies.

We have demonstrated that these elements of small-x dynamics are of utmost importance for building a realistic description of pp/pA collisions at LHC. The BDL will be commonplace in central pp/pA collisions at LHC, affecting average configurations in the colliding protons (nuclei), with numerous consequences for the hadronic final state. In particular, these phenomena qualitatively change the strong interaction environment for new particle production.

We have identified several directions for future theoretical research, necessary for describing the expected new phenomena at LHC. These include the resummation approaches to QCD radiation (combining logarithms of Q^2 and $1/x$), the account for energy loss in the interaction of leading partons with the small-x gluon medium, and the development of realistic models of hadron production in central pp/pA collisions with interactions close to the BDL.

An overarching goal of future theoretical research on the structure of the nucleon and small-x dynamics should be to bring together the approaches starting from "soft" (hadronic) and "hard" (partonic) physics, as envisioned in Gribov's space-time picture of high-energy interactions. We have pointed out several instances in which "soft" and "hard" dynamics match smoothly or exhibit a delicate interplay, e.g., pp elastic scattering at small impact parameters, or diffractive processes.

A natural question is what are the most promising directions for future experimental studies of small-x dynamics. This question needs to be discussed with regard to the general long-term perspectives in high-energy physics. We assume here that the decision by the DESY management to stop HERA operations in 2007 will be enacted. This would clearly be a great loss, as many insights could be obtained from further measurements at HERA in both ep and eA mode, see, e.g., the proposals put forward for the HERA III run. We shall thus focus on the possibilities offered by LHC, with comments on the possible future program at Tevatron, as well as on the electron-ion collider envisaged in the U.S. government's long-range plan (128). The small-x investigations at LHC described in the following subsections are meant to complement the studies of central inelastic pp/pA collisions (Section 7.4) and diffractive phenomena in pp/pA scattering, which are the main topic of this review.

9.2. Measurement of Parton Densities in pp and pA Collisions at LHC

Measurements at LHC could greatly expand the x-range in which the parton densities are known. This would require measurements of hard processes such as

$$
\begin{aligned}
pp \rightarrow \text{jet1} + \text{jet2} + X && \text{dijet production} \\
\text{jet} + \gamma + X, \; \gamma + \gamma + X && \text{photon production} \\
Q + \bar{Q} + X && \text{heavy quark production} \\
l^+ + l^- + X && \text{Drell-Yan pair production} \\
W^{\pm}(Z) + X && \text{weak boson production}
\end{aligned}
$$
47.

in the region where one of the colliding partons carries small momentum fraction. The cross sections of all these processes remain large down to the very edge of the LHC kinematics, corresponding to $x \approx 3 \times 10^{-7}$ for Drell-Yan pair production with $M_{\mu^+\mu^-} = 5\,\text{GeV}$ (123, 129). The main limitations come from the need to identify relatively low-$p_\perp$ jets, and from the detector acceptance. The smaller the x one wants to probe, the more forward one must look, as the momentum fractions of the colliding partons are related to the rapidities of the produced jets as

$$x_{1,2} = \frac{p_\perp}{\sqrt{s}}(e^{\pm y_1} + e^{y_2}). \qquad 48.$$

The presently planned configuration of the CMS detector would allow for the measurement of dijet production down to $x \approx 3 \times 10^{-6}$ at $p_\perp = 10\,\text{GeV}$. This would push parton distribution measurements deep into the region where unitarity effects play an important role in the dynamics of hard processes, and where evolution effect in both $\ln(1/x)$ and $\ln Q^2$ need to be taken into account. When the BDL is reached, the M^2-dependence of the cross section, $d\sigma/dx_1 dx_2$, is predicted to be much slower than $\propto 1/M^2$ as in the leading twist approximation, similarly to the case of the inclusive deep inelastic scattering (130).

If several of the reactions (47) were measured, it would allow for independent tests of the QCD factorization, which should be violated at intermediate virtualities owing to the strong interaction of the propagating system with the small-x gluon medium (Section 7). The latter will be strongly enhanced in the region of x_1 close to 1. These effects, which are of great interest in themselves, can be probed by comparing the production cross sections for fixed, large x_1 and various values of x_2, including relatively large ones where the parton densities are known.

9.3. Small-x Phenomena in Ultraperipheral Collisions at LHC

It has been long known that nuclei in high-energy collisions generate a large flux of equivalent photons, which are spread in the transverse plane over distances substantially larger than twice the nuclear radius—the maximal distance at which strong interactions are possible. Scattering processes induced by these photons are referred to as ultraperipheral collisions. They have a distinctive signature, which allows them to be separated from the more frequent events caused by strong interactions. Experimentally, one selects events in which one of the nuclei remains intact, or emits one or a few neutrons by way of dipole excitation. Such events are extremely rare in scattering at impact parameters smaller than twice the nuclear radius.

The experiments at HERA have shown that photon-induced processes provide a well-understood probe of the gluon density in the proton. At LHC, such processes could be studied up to invariant γp energies (i.e., γA energies per nucleon) exceeding the maximal HERA energy by a factor of 10. This would allow one to use dijet (charm, etc.) production to measure the gluon density in the proton/nucleus down to $x \approx 3 \times 10^{-5}$ for $p_\perp \sim$ few GeV (131, 132), as well as the diffractive gluon density. Among other things, measurements of diffractive channels would allow one to perform critical tests of the HERA observation of a large probability of gluon-induced diffraction (Section 5), and the prediction of its further enhancement for nuclear targets (see Reference (109) and references therein). Another important measurement would be the t-dependence of gluon-induced diffraction and its change with energy, using the proposed 420 m proton tagger (133). We remind the reader that the lack of direct information on the t-dependence of diffraction leads to a large uncertainty in the predictions for leading-twist nuclear shadowing (Section 5.3).

Ultraperipheral collisions would also allow one to study the coherent production of heavy quarkonia, $\gamma A \rightarrow J/\psi(\Upsilon) + A$ at $x \leq 10^{-2}$, and to investigate the propagation of small dipoles through the nuclear medium at high energies. The QCD factorization theorem predicts that the A-dependence of the amplitude for this process should change between the color transparency regime (observed at FNAL (77)), where it is $\propto A$, and the perturbative color opacity regime, where it is proportional to the leading-twist shadowed gluon density. It would be possible also to use coherent diffraction from nuclei to study the approach to the BDL in $\gamma A \rightarrow X + A$, by comparing the measured cross section to the BDL prediction (see Section 6.4). The most promising channels are J/ψ and dijet photoproduction; see Reference (134) for a review and discussion. In AA collisions, it is difficult to separate processes induced by the photons generated by the left- and right-moving nucleus. Away from zero rapidity, a low-energy contribution tends to dominate, limiting the range of x, which could be explored for production of a state with mass M to $x \geq MA/(2p_A)$, where p_A is the momentum of the colliding nuclei. However, it seems that selection of events in which the heavy nucleus undergoes a dipole excitation enhances the contribution of hard photons (135), allowing one to extend the x-range of the measurements (by a factor of up to 10 in the case of J/ψ production). The challenge is to trigger both on events with and without break-up. In pA mode, the dominant process will be the production of heavy quarkonia. Such measurements would extend the W-range of the HERA measurements by a factor of three, and make it possible to measure directly the t-dependence of the cross section in a very broad range of rapidities, using the proposed 420 m proton tagger (133), which is critical for a more accurate determination of the x-dependence of the nucleon's transverse structure (Section 4.4). Note also that in pp scattering it is possible to detect protons at very small momentum transfers, where Coulomb exchange dominates (136). This would allow one to measure exclusive photoproduction of heavy quarkonia in pp scattering with good statistics (137, 138).

9.4. Small-x Physics at RHIC and an Electron-Ion Collider

The LHC measurements described in Sections 9.2 and 9.3 will probe small-x dynamics at least down to $x \sim 10^{-5}$. However, most of these measurements are restricted to scales $Q \geq 5\,\mathrm{GeV}$, and it would be difficult to connect them with the physics at smaller scales (virtualities), $Q \sim 2$–$3\,\mathrm{GeV}$, relevant for the overall structure of central pp/pA collisions at LHC (Section 7). The gap could be filled, to some extent, by experiments at RHIC and the proposed electron-ion collider (128, 139).

Extension of the forward acceptance of the current RHIC detectors would make it possible to measure Drell-Yan pair production at $x \sim 10^{-3}$ in pp and pA/dA scattering. This would allow one to test the predictions for leading-twist nuclear shadowing and look for deviations from the leading-twist prediction in the $p_\perp$ distributions of the dileptons. Qualitatively, one expects a suppression of the low transverse momentum part of the distribution up to $p_\perp \sim p_{\perp,\mathrm{BDL}}$. As mentioned above, such measurements would also allow one to probe the role of final state interactions by varying the x-values of the leading partons in the proton. If absorption effects were significant, one would have to introduce a cut on $x_p \leq 0.3$ to suppress these effects, which would reduce somewhat the x_A-range where the parton densities can be probed.

The eRHIC design for a future electron-ion collider envisages an ep/eA collider with $\sqrt{s} \leq 100\,\mathrm{GeV}$, with significantly higher luminosity than HERA, and the ability to continuously vary the beam energies over a wide range (128,139). With such a facility one could systematically study a variety of color transparency phenomena and use them to disentangle the quark-gluon structure of hadrons and nuclei; one could also measure longitudinal cross sections, which provide stringent tests of the range of the validity of the leading-twist approximation at small x (see Sections 3, 4 and 5). In eA collisions, one could study the transition from the soft shadowing at low Q^2 to the regime of leading-twist shadowing at high Q^2, and explore whether there exists an "intermediate" regime characterized by weak coupling but large parton densities. The ability to perform such measurements with a range of nuclear beams would allow one to study these effects as a function of the nuclear thickness, reaching values 1.5 times larger than the average thickness of the heavy nuclei. No other planned facilities would be able to cover this important kinematic region. Finally, eRHIC would make it possible to measure the t-dependence of a variety of hard exclusive processes in a wide range of x, $0.1 > x \geq 0.003$. This would probe the transverse structure of the proton directly in the x-range relevant for understanding nucleon fragmentation in central pp/pA collisions at LHC.

ACKNOWLEDGMENTS

We would like to thank our colleagues, many of them collaborators of many years, for their contributions to the studies discussed in this review and many enjoyable discussions, in particular H. Abramowicz, J. Bjorken, S. Brodsky, J. Collins, J. Dainton, Yu. Dokshitzer, A. DeRoecker, H. Drescher, A. Dumitru, K. Eggert, A. Freund, V. Gribov, V. Guzey, A. Levy, L. Lipatov, M. McDermott, G. Miller, A. Mueller, A. Radyushkin, T. Rogers, W. Vogelsang, R. Vogt, H. Weigert, S. White, and M. Zhalov. This work is supported by U.S. Department of Energy Contract DE-AC05-84ER40150, under which the Southeastern Universities Research

Association (SURA) operates the Thomas Jefferson National Accelerator Facility. L.F. and M.S. acknowledge support by the Binational Scientific Foundation. The research of M.S. was supported by DOE. M.S. thanks the Frankfurt Institute for Advanced Studies at Frankfurt University for the hospitality during the time when this work was completed.

The *Annual Review of Nuclear and Particle Science* is online at
http://nucl.annualreviews.org

LITERATURE CITED

1. Gribov VN. *JETP Lett.* 41:667 (1961)
2. Gribov VN. arXiv:hep-ph/0006158.
3. Froissart M. *Phys. Rev.* 123:1053 (1961)
4. Gross DJ, Wilczek F. *Phys. Rev. Lett.* 30:1343 (1973)
5. Politzer HD, *Phys. Rev. Lett.* 30:1346 (1973)
6. Bjorken JD. *Phys. Rev.* 179:1547 (1969)
7. Feynman RP. *Phys. Rev. Lett.* 23:1415 (1969)
8. Gross DJ, Wilczek F. *Phys. Rev. D* 8:3633 (1973); *ibid. D* 9:980 (1974)
9. Dokshitzer YL. *Sov. Phys. JETP* 46:641 (1977)
10. Greisen K. *Phys. Rev. Lett.* 16:748 (1966); Zatsepin GT, Kuzmin VA. *JETP Lett.* 4:78 (1966)
11. Gribov VN, Lipatov LN. *Sov. J. Nucl. Phys.* 15:438 (1972); Lipatov LN. *Sov. J. Nucl. Phys.* 20:94 (1975)
12. Altarelli A, Parisi G. *Nucl. Phys. B* 126:298 (1977)
13. Ellis RK, Furmanski W, Petronzio R. *Nucl. Phys.* B207:1 (1982); *ibid.* B212:29 (1983)
14. Ciafaloni M. *Nucl. Phys. B* 296:49 (1988); Catani S, Fiorani F, Marchesini G. *Phys. Lett. B* 234:339 (1990)
15. Blaettel B, Baym G, Frankfurt L, Strikman M. *Phys. Rev. Lett.* 70:896 (1993).
16. Frankfurt L, Miller GA, Strikman M. *Phys. Lett. B* 304:1 (1993)
17. Frankfurt L, Radyushkin A, Strikman M. *Phys. Rev. D* 55:98 (1997)
18. Miettinen HI, Pumplin J. *Phys. Rev. D* 18:1696 (1978)
19. Frankfurt L, Miller GA, Strikman M. *Phys. Rev. D* 65:094015 (2002)
20. Gribov VN. *The Theory of complex angular momenta: Gribov lectures on theoretical physics*. Cambridge: Cambridge Univ. Press (2003)
21. Brodsky SJ, et al. *Phys. Rev. D* 50:3134 (1994)
22. Cooper-Sarkar AM, Devenish RCE, De Roeck A. *Int. J. Mod. Phys. A* 13:3385 (1998)
23. Gabareen Mokhtar A. (H1 Collab.) arXiv:hep-ex/0406036
24. Gluck M, Reya E, Vogt A. *Z. Phys. C* 67:433 (1995)
25. Cheng H, Wu TT. *Phys. Rev. D* 186:1611 (1969)
26. Frankfurt L, Koepf W, Strikman M. *Phys. Rev. D* 54:3194 (1996)
27. Low FE. *Phys. Rev. D* 12:163 (1975)
28. Nussinov S. *Phys. Rev. Lett.* 34:1286 (1975)
29. Gunion J, Soper D. *Phys. Rev. D* 15:2617 (1977)
30. McDermott M, Frankfurt L, Guzey V, Strikman M. *Eur. Phys. J. C* 16:641 (2000)
31. Diakonov D. *Prog. Part. Nucl. Phys.* 51:173 (2003)
32. Mueller AH. *Nucl. Phys. B* 335:115 (1990)
33. Frankfurt L, Guzey V, Strikman M. *J. Phys. G* 27: R23 (2001)
34. Frankfurt L, Sherman VE. *Sov. J. Nucl. Phys.* 23:581 (1976)
35. Lipatov LN. *Sov. J. Nucl. Phys.* 23:338 (1976)

36. Kuraev EA, Lipatov LN, Fadin VS. *Sov. Phys. JETP* 44:443 (1976); *ibid.* 45:199 (1977); Balitsky II, Lipatov LN. *Sov. J. Nucl. Phys.* 28:822 (1978)
37. Ciafaloni M, Camici G. *Phys. Lett. B* 430:349 (1998); Ciafaloni M, Colferai D. *Phys. Lett. B* 452:372 (1999)
38. Fadin VS, Lipatov LN. *Phys. Lett. B* 429:127 (1998)
39. Ciafaloni M, Colferai D, Salam GP, Stasto AM, *Phys. Rev. D* 68:114003 (2003)
40. Altarelli G, Ball RB, Forte S. *Nucl. Phys. B* 674:459 (2003)
41. Ciafaloni M, Colferai D, Salam GP, Stasto AM. *Phys. Lett. B* 587:87 (2004)
42. Bartels J, Loewe M. *Z. Phys. C* 12:263 (1982)
43. Muller D, et al. *Fortsch. Phys.* 42:101 (1994)
44. Abramowicz H, Frankfurt L, Strikman M. *Surveys High Energ. Phys.* 11:51 (1997); also in: SLAC Summer Inst. 1994:0539
45. Ji XD. *Phys. Rev. D* 55:7114 (1997)
46. Radyushkin A. *Phys. Rev. D* 56:5524 (1997)
47. Frankfurt L, Freund A, Strikman M. *Phys. Rev. D* 58:114001 (1998); Erratum. *Phys. Rev. D* 59:119901 (1999)
48. Collins JJ, Freund A. *Phys. Rev. D* 59:074009 (1999)
49. Collins JC, Frankfurt L, Strikman M. *Phys. Rev. D* 56:2982 (1997)
50. Diehl M. *Phys. Rept.* 388:41 (2003)
51. Belitsky AV, Radyushkin AV. arXiv:hep-ph/0504030
52. Frankfurt L, Freund A, Guzey V, Strikman M. *Phys. Lett.* B418:345 (1998); Erratum. *Phys. Lett.* B429:14 (1998)
53. Shuvaev AG, Golec-Biernat KJ, Martin AD, Ryskin MG. *Phys. Rev. D* 60:014015 (1999)
54. Musatov I, Radyushkin A, *Phys. Rev. D* 61:074027 (2000)
55. Ryskin MG. *Z. Phys. C* 57:89 (1993)
56. Frankfurt L, Koepf W, Strikman M. *Phys. Rev. D* 57:512 (1998)
57. Levy A, *Nucl. Phys. Proc. Suppl.* 146:92 (2005)
58. Martin AD, Ryskin MG, Teubner T. *Phys. Rev. D* 55:4329 (1997)
59. Frankfurt L, McDermott M, Strikman M. *JHEP* 9902:002 (1999)
60. Martin AD, Ryskin MG, Teubner T. *Phys. Lett.* B454:339 (1999)
61. Brodsky SJ, Close FE, Gunion JF. *Phys. Rev. D* 6:177 (1972)
62. McDermott M, Sandapen R, Shaw G. *Eur. Phys. J. C* 22:655 (2002)
63. Favart L. *Eur. Phys. J.* C33:S509 (2004)
64. Freund A, McDermott M. *Eur. Phys. J.* C23:651 (2002)
65. Freund A, McDermott M, Strikman M. *Phys. Rev. D* 67:036001 (2003)
66. Pobylitsa PV. *Phys. Rev. D* 66:094002 (2002)
67. Frankfurt L, Strikman M. *Phys. Rev. D* 66:031502 (2002)
68. Strikman M, Weiss C. *Proc. XII International Workshop on Deep Inelastic Scattering (DIS 2004)*, Strbske Pleso, Slovakia, Apr. 14–18 (2004) (arXiv:hep-ph/0408345)
69. Adloff C, et al. (H1 Collab.) *Phys. Lett.* B483:23 (2000)
70. Chekanov S, et al. (ZEUS Collab.) *Nucl. Phys. B* 695:3 (2004)
71. Strikman M, Weiss C. *Phys. Rev. D* 69:054012 (2004)
72. Frankfurt L, Strikman M. *Nucl. Phys. Proc. Suppl.* 79:671 (1999)
73. Frankfurt L, Strikman M, Weiss C. *Phys. Rev. D* 69:114010 (2004)
74. Bertsch G, Brodsky SJ, Goldhaber AS, Gunion J. *Phys. Rev. Lett.* 47:297 (1981)
75. Frankfurt L. Professor Habilitation thesis, Leningrad Institute for Nuclear Physics, Leningrad (1981)
76. Arneodo M. *Phys. Rept.* 240:301 (1994)
77. Sokoloff MD, et al. *Phys. Rev. Lett.* 57:3003 (1986)
78. Aitala EM, et al. (E791 Collab.) *Phys. Rev. Lett.* 86:4773 (2001); *ibid.* 86:4768 (2001)
79. Trentadue L, Veneziano G. *Phys. Lett.* B323:201 (1994)

80. Collins JC. *Phys. Rev. D* 57:3051 (1998); Erratum. *Phys. Rev. D* 61:019902 (2000)

81. Ingelman G, Schlein PE. *Phys. Lett. B* 152:256 (1985)

82. Frankfurt L, Strikman M. *Eur. Phys. J.* A5:293 (1999)

83. Buchmuller W, Gehrmann T, Hebecker A. *Nucl. Phys. B* 537:477 (1999)

84. Bartels J, Ellis JR, Kowalski H, Wusthoff M. *Eur. Phys. J.* C7:443 (1999)

85. Chekanov S, et al. (ZEUS Collab.) *Phys. Lett.* B516:273 (2001)

86. Goulianos K. *Phys. Lett.* B358:379 (1995)

87. Erhan S, Schlein PE. *Phys. Lett.* B481:177 (2000)

88. Gribov VN. *Sov. Phys. JETP* 29:483 (1969)

89. Gribov VN. *Sov. Phys. JETP* 30:709 (1970)

90. Abramovsky VA, Gribov VN, Kancheli OV. *Sov. J. Nucl. Phys.* 18:308 (1974)

91. Treleani D. *Int. J. Mod. Phys.* A11:613 (1996)

92. Alvero L, Frankfurt L, Strikman M. *Eur. Phys. J.* A5:97 (1999)

93. Frankfurt L, Guzey V, Strikman M. *Phys. Rev. D* 71:054001 (2005)

94. Piller G, Weise W. *Phys. Rept.* 330:1 (2000)

95. Kaidalov AB. *Phys. Rept.* 50:157 (1979)

96. McLerran L. arXiv:hep-ph/0402137.

97. Vogt R. *Phys. Rev. C* 71:054902 (2005)

98. Jipa A. (BRAHMS Collab.) *Acta Phys. Hung.* A22:121 (2005)

99. Guzey V, Strikman M, Vogelsang W. *Phys. Lett. B* 603:173 (2004)

100. Rogers T, Guzey V, Strikman M, Zu X. *Phys. Rev. D* 69:074011 (2004)

101. Pumplin J, et al. *JHEP* 0207:012 (2002)

102. Frankfurt L, Strikman M, Zhalov M. *Phys. Lett.* B616:59 (2005)

103. Golec-Biernat K, Wusthoff M. *Phys. Rev. D* 59:014017 (1999)

104. Bondarenko S, Kozlov M, Levin E. *Nucl. Phys.* A727:139 (2003)

105. Balitsky I. *Nucl. Phys. B* 463:99 (1996); Kovchegov YV. *Phys. Rev. D* 61:074018 (2000)

106. Frankfurt L, Guzey V, McDermott M, Strikman M. *Phys. Rev. Lett.* 87:192301 (2001)

107. Munier S, Stasto AM, Mueller AH. *Nucl. Phys.* B603:427 (2001)

108. Mueller AH, Qiu JW. *Nucl. Phys.* B268:427 (1986)

109. Frankfurt L, Guzey V, Strikman M. *Phys. Lett.* B586:41 (2004)

110. Islam MM, Luddy RJ, Prokudin AV. *Mod. Phys. Lett. A* 18:743 (2003)

111. Abe F, et al. (CDF Collab.) *Phys. Rev. Lett.* 79:584 (1997); *Phys. Rev. D* 56:3811 (1997)

112. Frankfurt L, Strikman M, Weiss C. *Ann. Phys. (Leipzig)* 13:665 (2004)

113. Berera A, et al. *Phys. Lett.* B403:1 (1997)

114. Dumitru A, Gerland L, Strikman M. *Phys. Rev. Lett.* 90:092301 (2003); Erratum. *Phys. Rev. Lett.* 91:259901 (2003)

115. Venugopalan R. arXiv:hep-ph/0412396

116. Affolder T, et al. (CDF Collab.) *Phys. Rev. D* 65:092002 (2002)

117. Donnachie A, Landshoff PV. *Phys. Lett.* B296:227 (1992)

118. Migdal AA, Polyakov AM, Ter-Martirosian KA. *Phys. Lett.* B48:239 (1974)

119. Amati D, Le Bellac M, Marchesini G, Ciafaloni M. *Nucl. Phys.* B112:107 (1976)

120. Kancheli OV. arXiv:hep-ph/0008299

121. Drescher HJ, Dumitru A, Strikman M. *Phys. Rev. Lett.* 94:231801 (2005)

122. Frankfurt L, Strikman M. *Surveys High Energ. Phys.* 14:9 (1999)

123. Lippmaa E, et al. (FELIX Collab.) SLAC-R-638; CERN-LHCC-97-45, LHCC-I10 (1997); Ageev A, et al. *J. Phys. G* 28:R117 (2002)

124. Kaidalov AB, Khoze VA, Martin AD, Ryskin MG. *Eur. Phys. J.* C31:387 (2003)

125. Khoze VA, Martin AD, Ryskin MG. *Eur. Phys. J.* C18:167 (2000)

126. Maor U. Talk presented at the Workshop

"HERA and the LHC," CERN Geneva, March 26–27 (2004)

127. Khoze VA, Martin AD, Ryskin MG. *Eur. Phys. J.* C24:581 (2002)

128. The Electron Ion Collider White Paper, BNL Rep. BNL-68933 (2002)

129. Alvero L, Collins JC, Strikman M, Whitmore JJ. *Phys. Rev. D* 57:4063 (1998)

130. Frankfurt L, Strikman M. *Phys. Rev. Lett.* 91:022301 (2003)

131. Klein SR, Nystrand J, Vogt R. *Phys. Rev. C* 66:044906 (2002)

132. Strikman M, Vogt R, White S. Preprint LBNL-57843 (2005)

133. Albrow MG, et al. CERN Rep. CERN-LHCC-2005-025 (2005)

134. Frankfurt L, Strikman M, Zhalov M. *Acta Phys. Polon.* B34:3215 (2003)

135. Baltz AG, Klein SR, Nystrand J. *Phys. Rev. Lett.* 89:012301 (2002)

136. Piotrzkowski K. *Proc. Physics at LHC*, in press

137. Khoze VA, Martin AD, Ryskin MG. *Eur. Phys. J.* C24:459 (2002)

138. Klein SR, Nystrand J. *Phys. Rev. Lett.* 92:142003 (2004)

139. Deshpande A, Milner R, Venugopalan R, Vogelsang W. *Annu. Rev. Nucl. Part. Sci.* 55:169 (2005)

Annu. Rev. Nucl. Part. Sci. 2005. 55:467–515
doi: 10.1146/annurev.nucl.55.090704.151608

Ascertaining the Core Collapse Supernova Mechanism: The State of the Art and the Road Ahead

Anthony Mezzacappa

Physics Division, Oak Ridge National Laboratory, Oak Ridge, Tennessee 37831;
email: mezzacappaa@ornl.gov

Key Words   supernovae, neutrinos, hydrodynamics, magnetohydrodynamics, equations of state

■ **Abstract**   More than four decades have elapsed since modeling of the core collapse supernova mechanism began in earnest. To date, the mechanism remains elusive, at least in detail, although significant progress has been made in understanding these multiscale, multiphysics events. One-, two-, and three-dimensional simulations of or relevant to core collapse supernovae have shown that (a) neutrino transport, (b) fluid instabilities, (c) rotation, and (d) magnetic fields, together with proper treatments of (e) the sub- and super- nuclear density stellar core equation of state, (f) the neutrino interactions, and (g) gravity are all important. The importance of these ingredients applies to both the explosion mechanism and to phenomena directly associated with the mechanism, such as neutron star kicks, supernova neutrino and gravitational wave emission, and supernova spectropolarimetry.

Not surprisingly, current two- and three-dimensional models have yet to include (a)–(d) with sufficient realism. One-dimensional spherically symmetric models have achieved a significant level of sophistication but, by definition, cannot incorporate (b)–(d), except phenomenologically. Fully general relativistic spherically symmetric simulations with Boltzmann neutrino transport do not yield explosions, demonstrating that some combination of (b), (c), and (d) is required to achieve this. Systematic layering of the dimensionality and the physics will be needed to achieve a complete understanding of the supernova mechanism and phenomenology. The past modeling efforts alluded to above have illuminated that core collapse supernovae may be neutrino driven, magnetohydrodynamically (MHD) driven, or both, but uncertainties in the current models prevent us from being able to answer even this most basic question. And it may be that more than one possibility is realized in nature. Nonetheless, if a supernova is neutrino driven, magnetic fields will likely have an impact on the dynamics of the explosion. Similarly, if a supernova is MHD driven, the neutrino transport will dictate the dynamics of stellar core collapse, bounce, and the postbounce evolution, which in turn will create the environment in which an MHD-driven explosion would occur. Thus, although reduction will allow us to sort out the roles of each of the major physics components listed above, we will not obtain a quantitative, and perhaps even

0163-8998/05/1208-0467$20.00

467

qualitative, understanding of core collapse supernovae until all components and their coupling are included in the models with sufficient realism.

CONTENTS

1. IS THERE A CORE COLLAPSE SUPERNOVA PARADIGM?

Stars more massive than ~ 10 $M_\odot$ evolve to an onion-like configuration (Figure 1, see color insert), with an iron core surrounded by successive layers of silicon, oxygen, carbon, helium, and finally hydrogen. In addition to iron group nuclei, the core is composed of electrons, positrons, photons, and a small fraction of protons and neutrons. The pressure in the core, which supports it against the inward pull of gravity, is dominated at this stage by the electrons, and the balance between the electron pressure and gravity is only marginally stable. As a result of electron capture on the free protons and nuclei in the core and as a result of nuclear dissociation under the extreme densities and temperatures, electron and thermal pressure support are reduced, and the core becomes unstable and collapses.

The velocity of infalling matter in the core increases linearly with radius, characteristic of the homologous collapse expected of a fluid whose pressure is dominated by relativistic, degenerate electron pressure. The sound speed, on the other hand, decreases with density and, therefore, radius. Thus, with increasing radius, the infall velocity eventually exceeds the local sound speed; i.e., the infall becomes supersonic. Consequently, during infall the core splits into an inner homologously and subsonically infalling core and an outer supersonically infalling outer core (stage 1 of Figure 2, see color insert).

Beginning with central densities $\sim 10^{9-10}$ g/cm^3, the collapse proceeds through nuclear matter densities ($\sim 1 - 3 \times 10^{14}$ g/cm^3). At this point the inner core undergoes a phase transition from a two-phase system of nucleons and nuclei to a one-phase system of bulk nuclear matter. One may view the inner core at this point as one giant nucleus. The pressure in the inner core increases dramatically as the result of Fermi effects and the repulsive nature of the nucleon-nucleon interaction potential at short distances. As a result of this dramatic increase in pressure, the inner core becomes incompressible and rebounds (stages 3–5 of Figure 2). Any information about the rebounding inner core would be conveyed to the outer core via pressure waves that propagate radially outward at the speed of sound. When these waves reach the point at which the infall is supersonic—i.e., the sonic point—they are swept in as fast as they attempt to propagate outward. The net result: No information about the rebounding inner core reaches the infalling outer core, which in turn sets up a density, pressure, and velocity discontinuity in the flow—i.e., a shock wave. This shock wave will ultimately be responsible for propagating outward through the star, disrupting the star in a core collapse supernova. Schematically, the shock wave is launched and energized by the rebounding inner core piston. (In Figure 2, the shock wave is represented schematically by the orange circle in stages 5–7.) If the shock were to propagate outward without stalling, we would have what has been called a prompt explosion. All of the realistic models completed to date suggest that this does not occur. Because the shock loses energy in dissociating the iron nuclei that pass through it as it propagates outward, the shock is enervated. Nuclei exist in the regions shaded with a light blue in Figure 2. The yellow regions in stages 5–7 are regions of shock compressed and heated material in which the nuclei are dissociated into nucleons. The shock loses additional energy in the form of electron neutrinos. The copious production of electron neutrinos occurs when the core electrons capture on the newly dissociation-liberated protons. These neutrinos are initially trapped but escape when the shock moves out beyond the electron neutrinosphere. This gives rise to the electron neutrino burst in a core collapse supernova, which is the first of three major phases of the three-flavor neutrino emission during these events. As a result of these two enervating mechanisms, the shock stalls in the iron core. How the shock is reenergized in a delayed shock mechanism is currently the central question in core collapse supernova theory.

At the time the shock stalls, the core configuration is composed of a central radiating object, the proto-neutron star (Figure 3), which will go on to form a neutron star or black hole. The proto-neutron star has a relatively cold inner part, composed of unshocked bulk nuclear matter, together with a hot mantle of nuclear matter that has been shocked but not expelled. The ultimate source of energy in a core collapse supernova is the $\sim 10^{53}$ erg of gravitational binding energy associated with the formation of the neutron star. This gravitational binding energy is released after core bounce over ~ 10 s in the form of a three-flavor neutrino pulse.

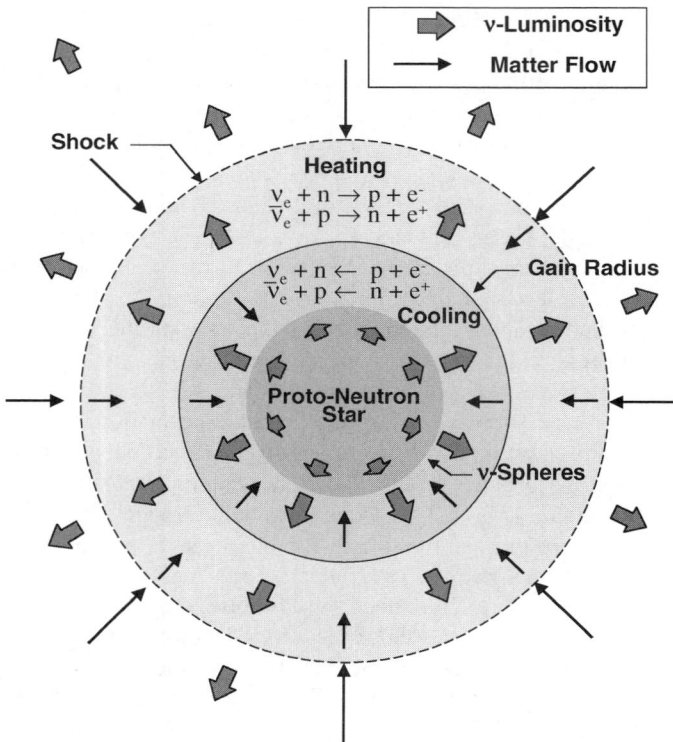

Figure 3   During the shock reheating phase, the stellar core is composed of a central radiating proto-neutron star whose surface is defined by the neutrinospheres (represented here by a single sphere) and a region above the neutrinosphere consisting of a net cooling region and a net heating region below the stalled shock, separated by the gain radius at which heating and cooling balance. Heating and cooling are mediated by electron neutrino and antineutrino absorption and emission.

This marks the second phase of the neutrino emission from a core collapse supernova. Electron neutrinos are produced during stellar core collapse by electron capture on protons and nuclei, but after bounce, in the hot proto-neutron star mantle, all three flavors of neutrinos and their antineutrinos are produced and are emitted as the mantle cools and contracts during its Kelvin-Helmholtz cooling phase. The neutrinos are emitted from their respective neutrinospheres. The neutrinospheres are defined in a way similar to the way the photosphere of the sun is defined. They are the surfaces of last scattering for each energy and flavor. Equivalently, they are at radii at which the respective neutrino depths are two thirds. In Figure 3, the neutrinospheres are represented by one surface. The neutrino luminosities during this phase are maintained at their average values $\sim 10^{52}$ erg/s by mass accretion onto the proto-neutron star (the kinetic energy of infall is converted into thermal energy when the material hits the proto-neutron star surface). After explosion is initiated, the accretion luminosity decreases dramatically, and the neutrino pulse enters its third and final stage: the exponential decay of the neutrino luminosities characteristic of neutron star formation and cooling.

The stalled supernova shock is thought to be revived, at least in part, by the charged-current absorption of electron neutrinos and antineutrinos that emerge from the proto-neutron star, a fraction of which are absorbed by protons and neutrons behind the shock. This is known as the delayed shock or neutrino-heating mechanism, originally proposed by Wilson and Bethe (1, 2). Although the total energy emitted in neutrinos is two orders of magnitude greater than what is required for the generation of an $\sim 10^{51}$ erg explosion, deciphering the precise role of this neutrino heating in the supernova mechanism is, as we will discuss, difficult.

Between the neutrinosphere and the shock, the material both heats and cools by electron neutrino and antineutrino emission and absorption. The neutrino heating and cooling have different radial profiles; consequently, this region splits into a net cooling region and a net heating region, separated by a gain radius at which heating and cooling balance. We refer to the region between the gain radius and the shock as the *gain region.*

The neutrino heating in the gain region can be written as

$$\dot{\epsilon} = \frac{X_n}{\lambda_0^a} \frac{L_{\nu_e}}{4\pi r^2} \langle E_{\nu_e}^2 \rangle \left\langle \frac{1}{F} \right\rangle + \frac{X_p}{\bar{\lambda}_0^a} \frac{L_{\bar{\nu}_e}}{4\pi r^2} \langle E_{\bar{\nu}_e}^2 \rangle \left\langle \frac{1}{\bar{F}} \right\rangle . \qquad 1.$$

The first (second) term corresponds to the absorption of electron neutrinos (antineutrinos). It depends linearly on the neutrino luminosity and inverse flux factor, which is a measure of the isotropy of the neutrino distribution, and quadratically on the neutrino spectrum.

In addition to the dependence on the three key neutrino quantities in the heating rate above, the revival of the stalled supernova shock depends on a complex interplay of neutrino heating, mass accretion through the shock, and mass accretion through the gain radius (3). It is mass accretion through the shock and gain radii that determines the amount of mass in the gain region, the former being a source of mass in the gain region and the latter being a sink. Moreover, the mass accretion through the gain radius serves to both sustain the neutrino luminosities and to undermine the pressure in the gain region, simultaneously—i.e., it serves a supporting and a detrimental role.

All three quantities in the neutrino heating rate must be computed accurately (3–8), which requires that we solve the neutrino Boltzmann neutrino transport equations (see the next section), but given the quadratic dependence on the neutrino spectrum, it is imperative that the spectrum be computed accurately. This requires the use of multineutrino energy (a.k.a. multifrequency or multigroup) transport. In the "gray" approximation, the neutrino angles and energies are integrated out, and one solves for the neutrino specific energy and flux, which are functions of the spatial coordinates only (in the case of spherical symmetry, radius). In other words, in a gray approach one does not solve for the neutrino specific energy and flux per neutrino direction cosine and energy. The fundamental shortcoming in a gray approach can easily been seen if we specialize the neutrino Boltzmann equation (5) examined in the next section to an equation that includes only absorption and then integrate over neutrino direction cosines and energies. Equation 2 gives the local rate of change of the neutrino specific energy, ϵ_R, owing to electron neutrino absorption. This can be expressed in terms of the energy mean absorption opacity, which in turn depends on the electron neutrino spectra. But in a gray approach, the neutrino spectra are not computed. Rather, they must be imposed, and therein lies the problem. If an overly hard spectrum is imposed, explosions can be induced in models artificially. Indeed,

$$\frac{\partial \epsilon_R}{\partial t} = \int d\mu dEE^3 \frac{\partial F}{\partial t} = - \int d\mu dEE^3 \tilde{\chi} F \equiv -\bar{\tilde{\chi}} \epsilon_R, \qquad 2.$$

where the energy mean absorption opacity $\bar{\tilde{\chi}}$ and the neutrino specific energy ϵ_R are given by

$$\bar{\tilde{\chi}} \equiv \frac{\int d\mu dEE^3 \tilde{\chi} F}{\int d\mu dEE^3 F} \qquad 3.$$

and

$$\epsilon_R \equiv \int d\mu dEE^3 F. \qquad 4.$$

It is evident in Equation 3 that the energy mean absorption opacity depends on the neutrino distribution function, in general, and the neutrino spectrum, in particular. In a gray approach, F is not computed and must therefore be imposed in the computation of $\bar{\tilde{\chi}}$.

The most compelling argument for the use of multifrequency neutrino transport can be made, of course, by simply taking stock of the results from all past core collapse supernova simulations. With the exception of Wilson's spherically symmetric models that invoke the doubly diffusive neutron finger instability in the proto-neutron star to boost the neutrino luminosities, no simulation to date performed with multifrequency neutrino transport has yielded an explosion. This is a sobering fact (7, 9–15).[1] And this list now includes both one- and two-dimensional simulations. Moreover, without neutron fingers, whose existence is a matter of current debate (see Section 5.1.2), Wilson does not obtain explosions either (J. Wilson, private communication). On the other hand, all of the published models in which an explosion was reported (17–22) were based on gray neutrino radiation hydrodynamics and were therefore parameterized models. In essence, the remainder of this review will delineate what must be done to fill this fundamental gap in supernova theory—that is, to obtain an unparameterized, first-principles solution to the supernova problem.

---

[1] A recent marginal exception to this trend was found by (16), where a weak explosion of a less-massive 11 $M_\odot$ progenitor was reported in the context of a two-dimensional model.

Nonetheless, owing to their dimensional reduction, simulations in two and even three spatial dimensions that implement gray neutrino transport are possible and allow us to explore the impact of physics beyond the neutrino transport, such as convection and rotation. Thus, gray and multifrequency treatments are complementary.

Generally, we anticipate that the magnetic fields in the collapsed stellar core may be amplified through a variety of mechanisms, as we will discuss, to become important in the supernova mechanism. In light of the above mentioned difficulties associated with generating core collapse supernovae via neutrino heating when realistic multifrequency neutrino transport is used, and in light of the expectation that magnetic fields may play a role in the supernova mechanism, a number of investigations have been performed and have concluded that if the magnetic fields are in fact sufficiently amplified, they may aid the neutrinos in powering the explosion or even replace them as the central vehicle whereby energy in the collapse is converted into outflow kinetic energy (23–26). In the case where explosion is powered by neutrinos, the neutrinos serve as the conduit between gravitational binding energy and internal energy behind the shock, as discussed above. In the case where explosion is driven by magnetic fields, the magnetic fields serve as the conduit between rotational energy in the collapsing core and the kinetic energy of outflow, as we will discuss later. The latter possibility would correspond to a paradigm shift. Hence, even the basic neutrino heating-mediated delayed-shock paradigm has been called into question. We might already anticipate that both the neutrinos and the magnetic fields will act in concert to produce these explosions. But an answer to these basic questions will require three-dimensional neutrino radiation magnetohydrodynamics simulations, which have not yet been performed. Even guidance on this issue from two-dimensional simulations would be welcome.

Given the central role that neutrinos play in stellar core dynamics and the core collapse supernova mechanism, and given that the simulation of three-dimensional, multiangle, multifrequency neutrino transport will by far be the most computationally intensive component in any three-dimensional supernova model, ultimately defining this problem as a peta-scale application, it is prudent to focus on some of the neutrino transport fundamentals to illustrate the essential physics and practical issues we face. To do so, we will restrict this discussion to the spherically symmetric case, which illustrates the physics and practical issues in what will already be a three-dimensional (phase space) problem.

2. A NEUTRINO TRANSPORT PRIMER

Neutrinos propagate through the proto-neutron star and interact with the protons, neutrons, and electrons in this central object via absorption, scattering, and other interaction channels. Because the cross sections for neutrino interactions are neutrino-energy dependent, with generally reduced cross sections for reduced energies, neutrinos of lower energies have longer mean free paths. The neutrinosphere is located at a point in the core at which the neutrino mean free paths become comparable to the size of the proto-neutron star. For mean free paths that are much smaller, which occur deep within the core below the neutrinosphere, the neutrinos interact with the core many times before escaping. Consequently, these neutrinos diffuse out of the core, and their transport is well described by diffusion theory. On the other hand, for mean free paths that are much larger than the size of the proto-neutron star, which occur well outside it, the neutrinos do not interact with the core material, and they stream out of the core unimpeded. Their transport at this point is well described by free streaming (radial free streaming at infinity). At the neutrinospheres, the neutrinos are not transported by diffusion, nor are they radially free streaming. Their transport is significantly more complex and is well described only by solutions of the full Boltzmann neutrino kinetic equations for the neutrino distribution functions. An additional complexity is introduced here by the fact that the neutrinosphere is neutrino energy and flavor dependent. In reality, there are multiple neutrinospheres. The neutrino interaction cross sections and, hence, the mean free paths are energy and flavor dependent. A solution to the Boltzmann equation describes the time evolution of the neutrino distribution function, which, at each instant of time and each location in space, gives the distribution of neutrinos in direction cosine and energy. Thus, the Boltzmann equation is a phase-space equation—i.e., an equation in the multidimensional space of all possible spatial locations, neutrino direction cosines, and neutrino energies. Therefore, even a one-dimensional supernova simulation, in which spherical symmetry is assumed, is in reality a three-dimensional (radius, one direction cosine, and energy) simulation. If we include emission, absorption, isoenergetic scattering of neutrinos by nucleons and nuclei, neutrino-electron scattering, and pair emission and absorption, the Boltzmann equation in spherical symmetry is

$$
\frac{1}{c}\frac{\partial F}{\partial t} + 4\pi\mu\frac{\partial(r^2\rho F)}{\partial m} + \frac{1}{r}\frac{\partial[(1-\mu^2)F]}{\partial\mu} + \frac{1}{c}\left(\frac{\partial\ln\rho}{\partial t} + \frac{3v}{r}\right)\frac{\partial[\mu(1-\mu^2)F]}{\partial\mu}
$$

$$
+ \frac{1}{c}\left[\mu^2\left(\frac{\partial\ln\rho}{\partial t} + \frac{3v}{r}\right) - \frac{v}{r}\right]\frac{1}{E^2}\frac{\partial(E^3 F)}{\partial E}
$$

$$
= \frac{j}{\rho} - \tilde{\chi}F + \frac{1}{c}\frac{1}{h^3c^3}E^2\int d\mu' R_{\text{IS}}F - \frac{1}{c}\frac{1}{h^3c^3}E^2 F\int d\mu' R_{\text{IS}}
$$

$$
+ \frac{1}{h^3c^4}\left(\frac{1}{\rho} - F\right)\!\int dE'\,E'^2 d\mu'\,\tilde{R}^{\text{in}}_{\text{NES}}F - \frac{1}{h^3c^4}F\!\int dE'\,E'^2 d\mu'\,\tilde{R}^{\text{out}}_{\text{NES}}\left(\frac{1}{\rho} - F\right)
$$

$$
+ \frac{1}{h^3c^4}\left(\frac{1}{\rho} - F\right)\!\int dE'\,E'^2 d\mu'\,\tilde{R}^{\text{em}}_{\text{PAIR}}\left(\frac{1}{\rho} - \bar{F}\right) - \frac{1}{h^3c^4}F\!\int dE'\,E'^2 d\mu'\,\tilde{R}^{\text{abs}}_{\text{PAIR}}\bar{F}.
$$

5.

In Equation 5, $F(m, \mu, E)$ is the specific neutrino distribution function, f/ρ, where f is the neutrino distribution function. $\bar{F}$ is the corresponding specific antineutrino distribution function. m is the enclosed mass, μ is the neutrino direction cosine, and E is the neutrino energy. The mass derivative term on the left-hand side of the Boltzmann equation describes the propagation of neutrinos with respect to the Lagrangian mass coordinate, m. Outwardly propagating neutrinos have $\mu > 0$, whereas inwardly propagating neutrinos have $\mu < 0$. The first μ-derivative term describes the rate of change of the neutrino propagation direction with respect to the outward radial direction as the neutrino propagates inward or outward in mass. The second μ-derivative term describes the aberration in the neutrino propagation direction measured by an observer who is instantaneously comoving with the fluid. Because the fluid is accelerating, two neighboring comoving observers will measure different direction cosines. The energy-derivative term describes the shift in the neutrino energy measured by comoving observers. This is a Doppler shift resulting from the change in the velocity, with time and/or radius, of an accelerated fluid; two neighboring comoving observers will measure different frequencies. The angular aberration and frequency shift terms in the Boltzmann equations play a critical role in the development of the neutrino distributions during stellar core collapse. This is especially true of the frequency shift term, which is responsible, in the limit in which the neutrinos are trapped in the flow and diffusive, for the increase in the neutrino Fermi energy as the neutrinos are compressed during collapse. We will refer to these terms as *observer corrections*. They are $O(v/c)$, whereas the other terms on the left-hand side of Equation 5 are $O(1)$. On the right-hand side of Equation 5, the first two terms describe the change in the neutrino distribution function resulting from the absorption and emission of neutrinos by nucleons and nuclei. The next two terms describe the isoenergetic inscattering and outscattering, respectively, of neutrinos by nucleons and nuclei. The fourth and fifth terms describe non-isoenergetic neutrino-electron scattering, and the last two terms describe pair emission and absorption.

A solution of the Boltzmann equation for the neutrino distribution function then gives an infinite hierarchy of moments of the distribution function, defined by

$$\psi^n = \int d\mu \mu^n f, \qquad 6.$$

where n = 0, 1, 2 ... n. Therefore, a solution of the Boltzmann equation is equivalent to solving the following infinite hierarchy of equations for the moments of the distribution function:

$$\frac{\partial \psi^n}{\partial t} = \int d\mu \mu^n \frac{\partial f}{\partial t} = \int d\mu \mu^n O[f], \qquad 7.$$

where n = 0, 1, ... n and where we have written the Boltzmann Equation (5) in operator form

$$\frac{\partial f}{\partial t} = O[f]. \qquad 8.$$

It is instructive to note that for each component of the stellar core—i.e., the photons, electrons, positrons, protons, neutrons, and nuclei—one can write down a kinetic equation for the component distribution function. Each kinetic equation, in turn, induces an infinite series of moment equations, as in Equation 8. Under certain conditions and assumptions, these series close and give rise to the familiar hydrodynamics equations for the component fluids. [For more on this, see Reference 27.] These conditions do not obtain for the neutrinos. Because solving the Boltzmann equation is computationally intensive even in one-dimensional simulations that assume spherical symmetry, historically a number of approximations have been implemented, ranging from very simple to rather sophisticated. A brief history will be given in the next section. For our purposes later we note that in Equation 6, ψ^0 (ψ^1) is the neutrino energy density (flux) for the neutrino energy E.

3. A BRIEF HISTORY OF NEUTRINO TRANSPORT IN STELLAR CORE COLLAPSE

Dictated mainly by available technology, the transport of neutrinos was, until recently, simulated with a variety of increasingly sophisticated approximations, from simple leakage schemes (e.g., see References 28 and 29), to two-fluid approaches (e.g., see References 30 and 31), and finally multigroup flux-limited diffusion (e.g., see References 32 and 35). Multigroup flux-limited diffusion (MGFLD) closes the neutrino radiation hydrodynamics hierarchy of Equations 6 at the level of the zeroth moment (the neutrino energy density per group) by imposing a relationship between the flux per group (the first moment) and the gradient of the zeroth moment. For example,

$$\psi^1 = -\frac{c\Lambda}{3}\frac{\partial\psi^0}{\partial r} + \dots, \qquad 9.$$

$$\Lambda = \frac{1}{1/\lambda + |\partial\psi^0/\partial r|/3\psi^0}, \qquad 10.$$

where λ is the neutrino mean free path (34). (Other forms for the flux-limiter Λ can be found in (33, 35, 36).) The term *multigroup* means that the transport is carried out for each neutrino energy separately. Because of the energy dependence of the neutrino interactions and, consequently, mean free paths, neutrinos of different energies may behave very differently. The term *flux limited* means that the flux is kept from exceeding the maximum flux, corresponding to neutrinos streaming out radially at the speed of light. If diffusion theory were used to describe the transport in this instance, the neutrino flux would be superluminal. This happens because diffusion theory assumes that the neutrinos always propagate a distance given by their mean free path, even if this distance exceeds the distance that could be traversed by the neutrinos in a time Δt, moving at the speed of light (37). Whereas the limits $\lambda \to 0$ and $\lambda \to \infty$ in Equations 7 and 8 produce the correct diffusion and free streaming fluxes, it is in the critical intermediate region around the neutrinospheres where the MGFLD approximation is a priori of unknown accuracy, necessitating comparisons with and, ultimately, the use of Boltzmann transport in supernova models. The extreme sensitivity of supernova models to

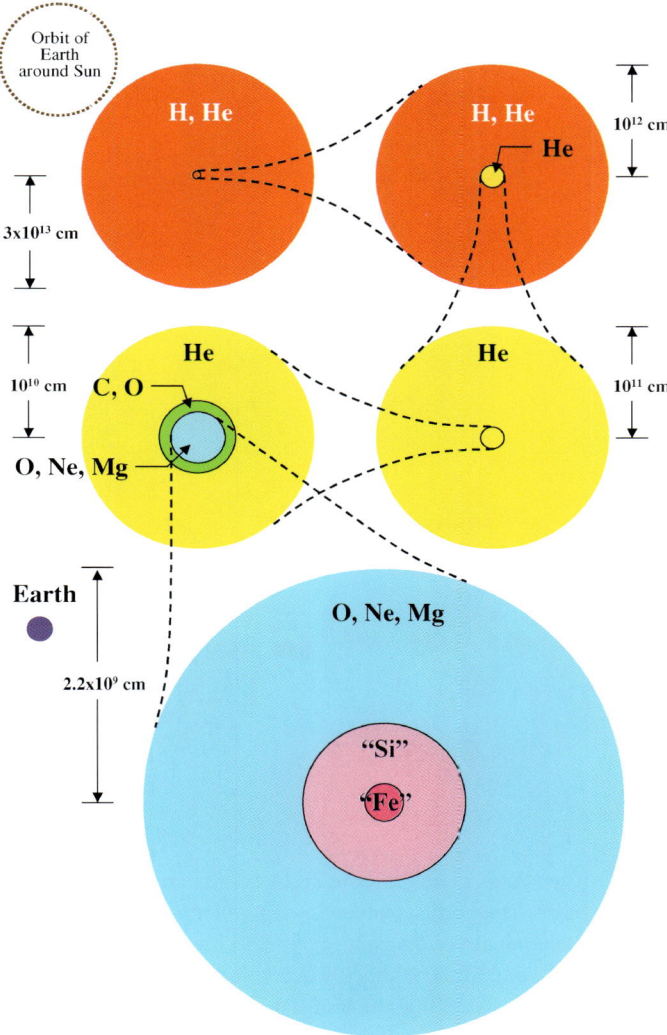

Figure 1   The structure of a core collapse supernova progenitor at the onset of stellar core collapse. The size of the iron core and star are compared with the size of the earth and its orbit around the sun, respectively.

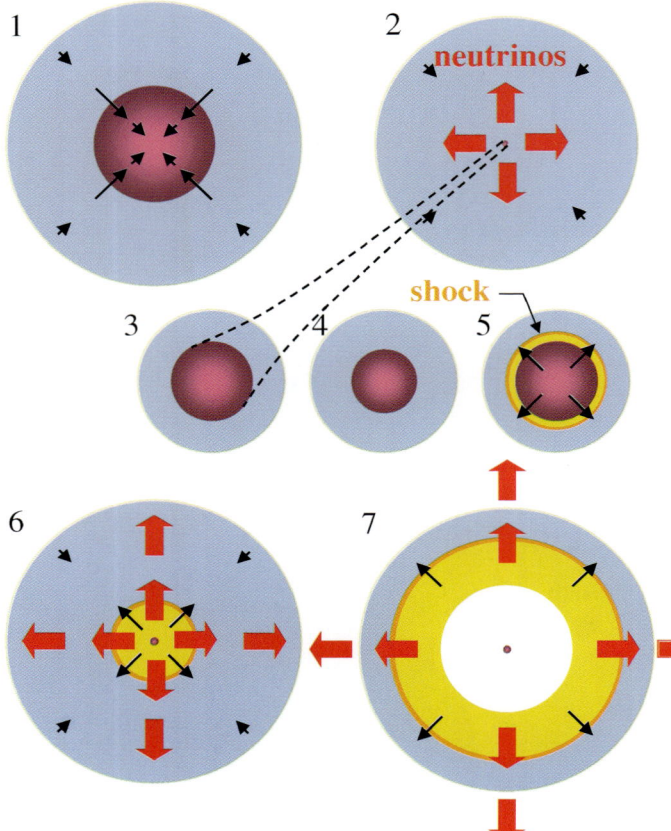

Figure 2   The stages of stellar core collapse, maximum compression, bounce, and shock formation. The core separates into an inner, subsonically collapsing core and an outer, supersonically collapsing core. The inner core bounce launches a shock wave into the outer core. The shock wave will propagate through the outer iron core and layers above it, expelling much of the layers to produce the supernova.

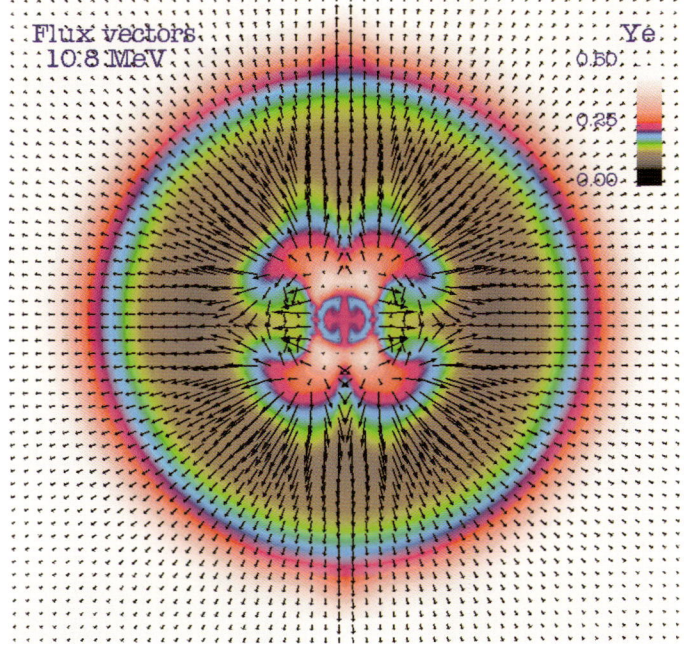

Figure 8 The correlation between the two-dimensional flow in the stellar core after core bounce and the two-dimensional neutrino radiation field are evident in this simulation performed with $O(1)$ Boltzmann neutrino transport (48). The figure shows the two-dimensional electron fraction and superimposed electron neutrino flux.

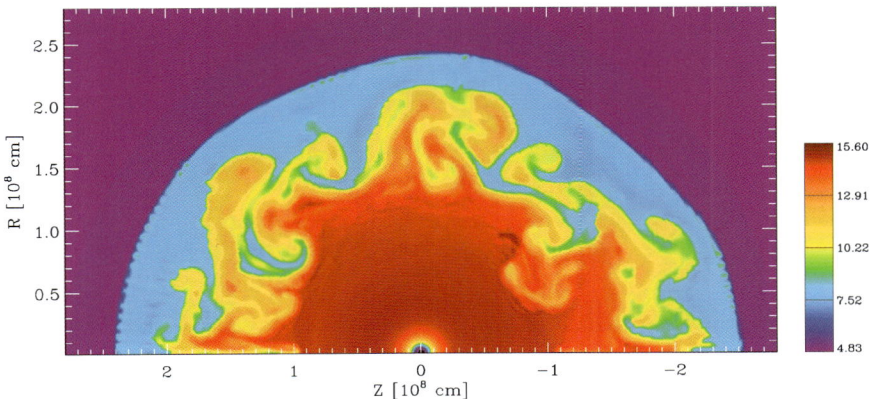

Figure 9 Snapshot of the development of neutrino-driven convection in the postshock region at 208 ms after bounce with odd-even decoupling (53). Significant perturbations in the form of striations below the shock are evident. These, in turn, have seeded neutrino-driven convection, well under development here.

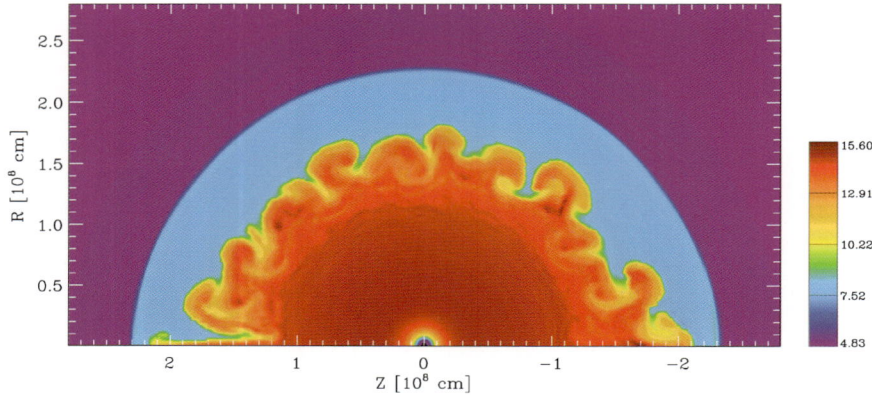

Figure 10 Snapshot of the development of neutrino-driven convection in the postshock region at 208 ms after bounce with no odd-even decoupling (53). Owing to the lack of the perturbations shown in Figure 9, the development of neutrino-driven convection is significantly less advanced in this model at this time.

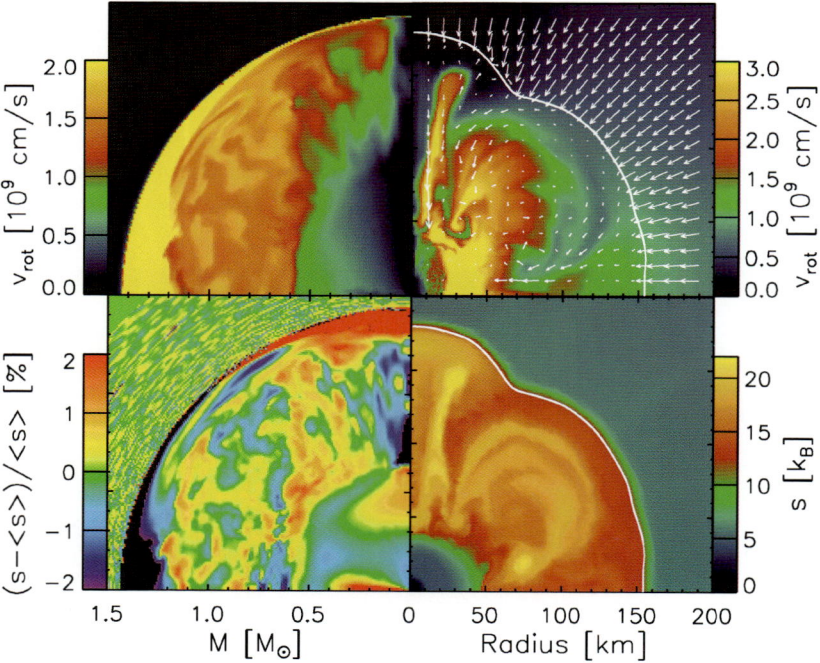

Figure 12 Four panels displaying the rotational velocity and entropy deviation, deep in the proto-neutron star (*upper* and *lower left*, respectively), and the rotational velocity and entropy, throughout the core (*upper* and *lower right*, respectively), in a two-dimensional simulation performed with sophisticated multifrequency radial-ray neutrino transport (15).

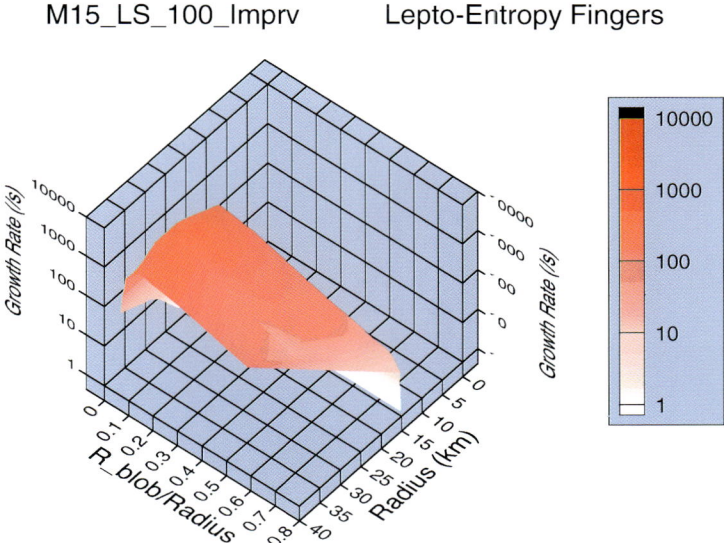

Figure 14   The growth rate for the development of lepto-entropy fingers is plotted as a function of radius and fluid element size under typical conditions found after stellar core bounce (56). The region over which LEFs develop is extensive and fairly insensitive to the size of the fluid element involved.

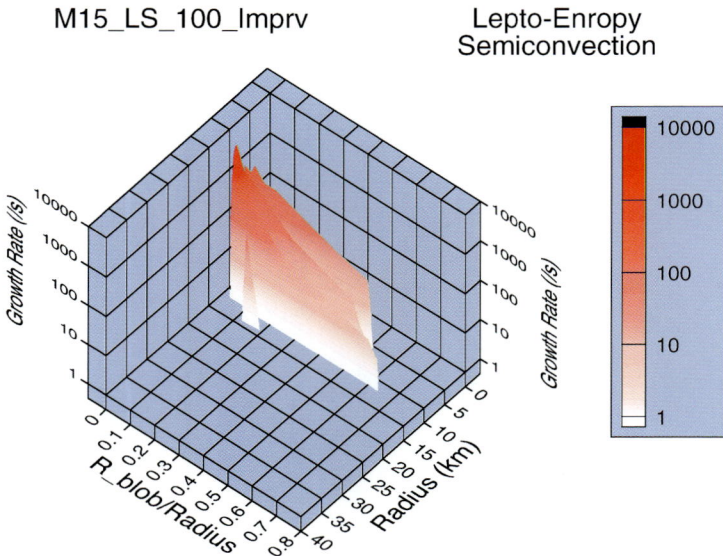

Figure 15 The growth rate for the development of lepto-entropy semiconvection is plotted as a function of radius and fluid element size under typical conditions found after stellar core bounce (56). The region over which LESC develops is confined to a narrow region deep in the stellar core after bounce. LESC also does not develop for fluid elements whose radii are a significant fraction of the radius of instability, which in this case is small.

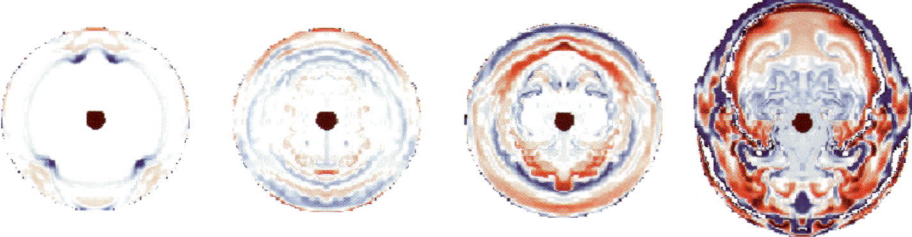

Figure 16 A series of snapshots of the two-dimensional entropy below the shock wave in a model of the accretion phase after stellar core bounce. In this model, nonspherical perturbations of the stationary shock are introduced and the consequent instability develops, further distorting the shock and leading to a bipolar explosion (60).

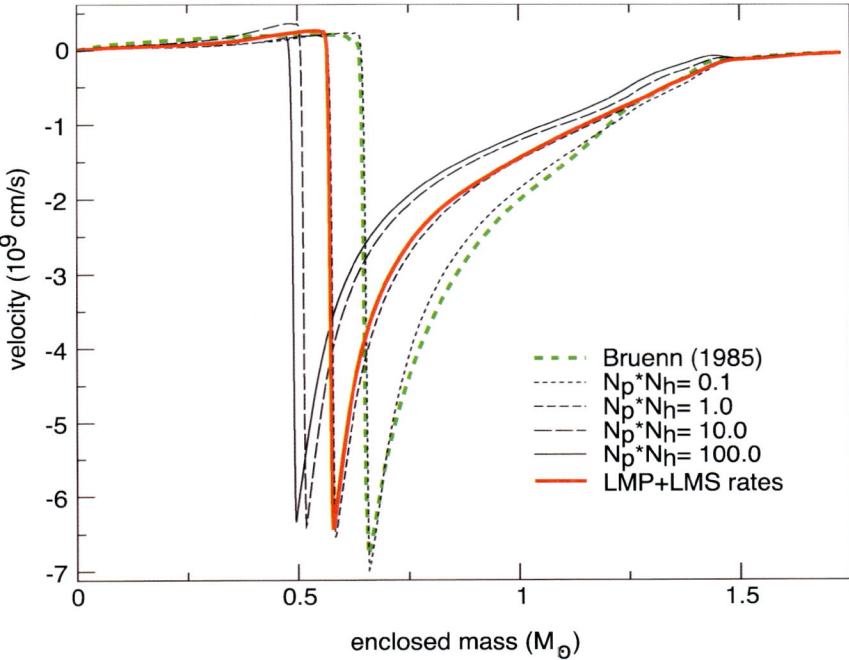

Figure 20   The postbounce velocity profile is shown for several values of the parameter used in these studies of electron capture, covering the cases in which electron capture on nuclei is negligible ($N_p \times N_H = 0.1$) or dominant ($N_p \times N_H = 1.0$ or higher) relative to electron capture on protons. Also shown are the *Bruenn (1985)* case, in which a zero-order shell model is used and, hence, for which electron capture on nuclei is supressed for $N > 40$, and the *LMP + LMS* case, in which the hybrid model of electron capture is used (81).

differences in the computed neutrino luminosities, spectra, and distributions in direction cosine (e.g., see References 3 and 5) provides an additional compelling reason to use Boltzmann neutrino transport in supernova models. In addition to Boltzmann transport, recent simulations of stellar core collapse and postbounce evolution have been performed with a multigroup variable Eddington factor (MGVEF) scheme (38, 39). In most practical implementations of this scheme, an approximation to the Boltzmann equation is solved in order to compute the Eddington factors (ratio of moments) needed to close the hierarchy of multigroup radiation moment equations, Equation 8, although a full Boltzmann solution can be performed together with a solution of the moment equations. In a MGVEF approach, the hierarchy is closed at the level of the first moment, the multigroup radiation momentum density, whereas in a MGFLD approach, only the zeroth moment is evolved. A detailed comparison of the results from Boltzmann and MGVEF simulations shows excellent agreement in the spherically symmetric case (40). Notably, in neither the Boltzmann nor the MGVEF cases was an explosion obtained. As one might imagine, the neutrino transport approximations implemented in past multidimensional simulations have, by necessity, been less sophisticated than those implemented in detailed one-dimensional simulations (e.g., see References 5, 7, 18–20, 22). One notable exception was the more recent implementation of MGVEF along radial rays in two-dimensional models (15), although this was a hybrid model coupling one-dimensional transport and two-dimensional hydrodynamics. In the next section, we will discuss ongoing efforts to develop realistic two- and three-dimensional multifrequency and multiangle neutrino transport.

4. NEUTRINO TRANSPORT: THE CURRENT STATE OF THE ART AND FUTURE CHALLENGES

After four decades of core collapse supernova research, detailed spherically symmetric simulations that now include state-of-the-art neutrino interactions, an industry standard equation of state, and multiangle, multifrequency, Boltzmann neutrino transport in full general relativity have finally been performed (41). It is important to remember that we are working in phase space. Therefore, simulations that assume spherical symmetry and require a solution of the neutrino Boltzmann equation are actually simulations in three dimensions (radius or mass, direction cosine, and energy), not one dimension (radius). When cast in this light, it is not surprising that such three-dimensional multiphysics simulations have taken four decades to complete. This is also an omen of what will be required to perform such simulations in three spatial dimensions (six phase space dimensions). Despite this seemingly overwhelming requirement, we must proceed in a systematic and steadfast manner.

A typical result for these simulations is shown in Figure 4. The shock radius reaches a maximum and then recedes with time, indicating a failure of the stellar core to explode and initiate a core collapse supernova. The results shown here are for a 13 $M_\odot$ model, an optimistic case owing to the small size of its iron core (1.2 $M_\odot$, but the outcome is the same for progenitors of 15, 20, 25, and 40 $M_\odot$, all

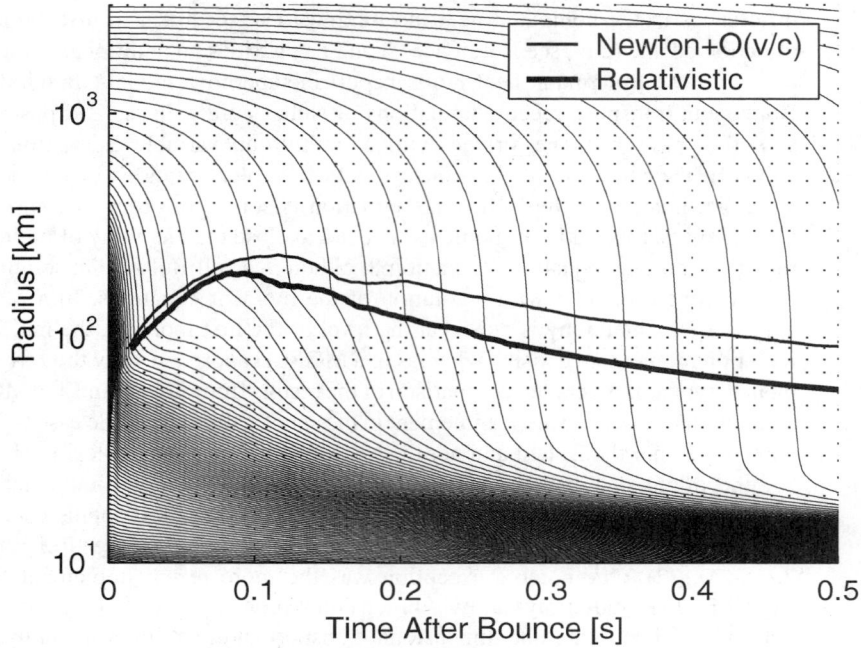

Figure 4   The shock trajectories are plotted as a function of time for both the Newtonian and fully general relativistic spherically symmetric models in the case of a 13 M$_\odot$ progenitor (41). The shock recedes with time in both cases, indicative of a failed explosion. This result is typical for all spherically symmetric models.

of which have more massive cores). The shock trajectories for both the Newtonian and general relativistic cases are shown. In the general relativistic case, owing to the increased gravitational field, the shock radius is always deeper. However, as we will discuss later, the inclusion of general relativity does not necessarily present a more pessimistic case, as might appear by considering the shock radius results out of context. In Figure 5, a more complete picture is presented for both the Newtonian and general relativistic simulations. Shown are all of the core profiles in density, entropy, electron fraction, velocity, and composition, along with the three key neutrino quantities mentioned above—luminosity, RMS energy, and mean flux factor—for both the electron neutrinos and antineutrinos.

Until these simulations were completed, failure to produce explosions in past models that used approximate treatments for the neutrino transport could have resulted from either the transport approximations or from the neglect of essential physics. We could not have known a priori which of these possibilities would in fact be realized. We now know that the transport approximations were not the cause of these failures and, as a result, we have crossed a threshold in supernova theory: New physics is needed.

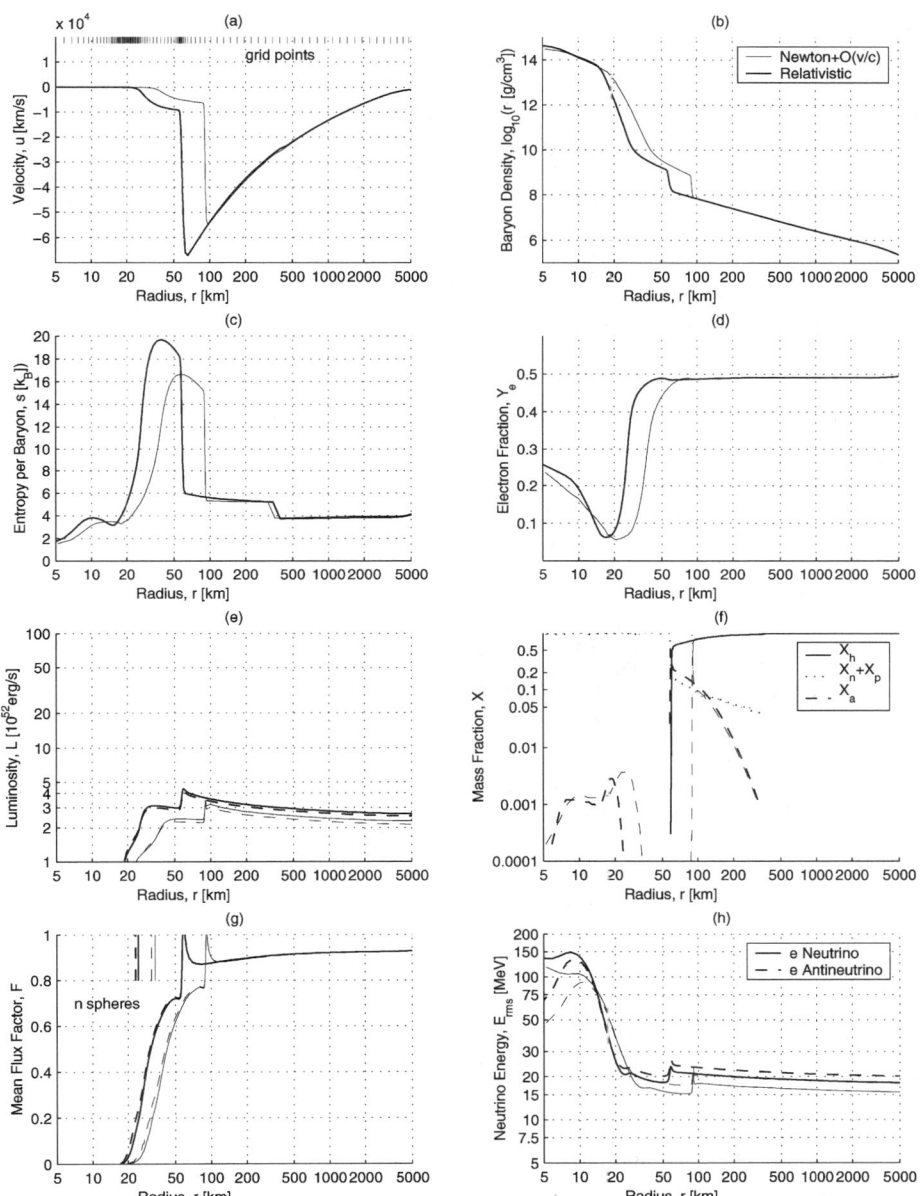

Figure 5   The velocity, density, entropy, electron fraction, electron neutrino and antineutrino luminosity, composition, electron neutrino and antineutrino flux factor and RMS energy profiles at 100 ms after bounce in Newtonian and general relativistic spherically symmetric models initiated from a 13 $M_\odot$ progenitor (41).

Nonetheless, the move from multigroup flux-limited diffusion to Boltzmann neutrino transport led to important quantitative changes in the spherically symmetric models. This alone is sufficient motivation for the development of Boltzmann neutrino transport in two- and three-dimensional models. Moreover, as we discussed in the previous section, Boltzmann transport is a first-principles, non-ad hoc treatment of the transport in core collapse supernova models. Flux-limited diffusion is an approximation, motivated by practical considerations, that exhibits fundamental uncertainties. In spherical symmetry, the Boltzmann results were used to tune the flux-limiters to better reproduce the Boltzmann results. Of course, this will be increasingly difficult as we move from spherical symmetry to axisymmetry to three-dimensional models, as the flow geometry becomes increasingly complex and the flux-limited diffusion approximation becomes increasingly difficult to implement without gross errors. Therefore, the need to develop Boltzmann neutrino transport for multidimensional models will grow as the uncertainties associated with flux-limiting increase commensurately.

In our attempt to simulate core collapse supernova explosions, we are presented with a number of underlying technical challenges that will ultimately dictate the degree to which we are confident that our simulation outcomes represent reality at all. Among these is the challenge of maintaining conservation of lepton number (for massless neutrinos) and energy in any given supernova model. Conservation of lepton number and energy are no guarantee that a model is correct. Models can be constructed, for example, to conserve total energy but may not accurately model the partition of energy among kinetic energy, internal energy, gravitational energy, etc. But any model that does not conserve lepton number and energy does not satisfy the most important quality control we can use to gauge a simulation's realism. How should we interpret the prediction of a $\sim 10^{51}$ erg explosion in a model where the total energy varies during the course of the simulation by $\sim 10^{51}$ erg or more?

To understand why ensuring conservation presents such a great technical challenge, we simply have to recall that the ultimate source of energy in a core collapse supernova is the gravitational binding energy of the remnant neutron star. This $\sim 10^{53}$ erg of energy is two orders of magnitude larger than the $\sim 10^{51}$ erg associated with the explosion. Consequently, total energy must be conserved over the course of a simulation to better than one part in 10^3. A typical simulation will be carried out over $\sim 10^{5-6}$ time steps, which requires that energy be conserved systematically to better than one part in 10^{8-9} per time step. This is a severe requirement, one that is very difficult to satisfy in a realistic supernova model.

First, we must define an energy that is conserved. For the radiation field—modulo energy losses and gains owing to the neutrino interactions with the matter—the lab frame specific radiation energy is globally conserved. In one approach, we may begin with the comoving frame specific radiation energy and flux. In the continuum limit, the lab frame specific radiation energy conservation is guaranteed by a cancellation of terms in the equations for the comoving frame specific radiation energy and flux when these equations are added together to give the evolution (conservation) equation for the lab frame specific radiation energy (the lab frame specific radiation energy, by the Lorentz transformation, is a sum of the comoving frame specific radiation energy and flux). In the $O(v/c)$ limit, for example, the cancelling terms arise from *both* the $O(1)$ and $O(v/c)$ (observer corrections: angular aberration and frequency shift) terms in the original Boltzmann equation from which the comoving frame specific energy and flux equations arise (these are the first two moments of the Boltzmann equation, defined in Section 2). We must in turn ensure that such cancellations occur in the discrete limit to ensure that energy is conserved in our simulations. This requires that we construct the discrete representation of the terms in the Boltzmann equation from which the cancelling terms arise with great care—i.e., the discrete representation of the $O(1)$ and $O(v/c)$ terms in the Boltzmann equation are not independent. This numerical feat has been achieved in the spherically symmetric case (42–44). The proliferation of $O[(v/c^2)]$ terms in the Boltzmann equation as we move from one- to three-dimensional models makes this approach increasingly difficult (44, 45). As a result, other approaches are now being considered to achieve the same end (46).

Efforts by several groups are now underway to develop Boltzmann neutrino transport for the two- and three-dimensional cases (45–48). The simulations performed thus far have been confined to the Newtonian gravity, $O(1)$ limit, restricted not only by Newtonian gravity but by the exclusion of the $O(v/c)$ terms on the left-hand side of the Boltzmann equation mentioned above (48). The observer corrections are critical in the evolution of the comoving frame neutrino distributions and in the dynamics of stellar collapse (34, 39, 42, 49, 50). They cannot be excluded from any realistic model, as Figures 6–7 show. In these spherically symmetric collapses, by the time the central density reaches a value of 1×10^{14} g/cm^3, the maximum infall velocity, inner homologous core mass, and entropy throughout the inner core are significantly different for the collapse simulation in which the $O(v/c)$ terms have been neglected (51). In addition, without the $O(v/c)$ terms, the neutrino distributions at a given energy can significantly exceed unity, which is unphysical. The $O(v/c)$ terms are responsible for properly redistributing the neutrinos in energy (to higher energy) as the collapse proceeds. If they are neglected, some neutrinos in energy groups for which $f > 1$ must be discarded so that $f = 1$ (or handled in some other ad hoc fashion), and energy and lepton number will not be conserved. Figure 7 illustrates the extent to which the conservation breaks down. At a central density of 1×10^{14} g/cm^3, we have lost nearly 3×10^{51} erg. Although discarding the neutrinos is only one ad hoc way to manage the error incurred when the $O(v/c)$ terms are neglected, these results clearly demonstrate the impact these terms have not only on the neutrino distributions, but also on the dynamics of stellar collapse and the conservation of total energy and lepton number. Not only can they not be neglected, they must be included with great care.

Nonetheless, progress has been made. Figure 8 (see color insert) shows the two-dimensional electron fraction distribution at approximately 15 ms after stellar core bounce in a simulation performed by Livne et al. (48) with an $O(1)$ two-dimensional Boltzmann solver. Most important, the anisotropies in the neutrino

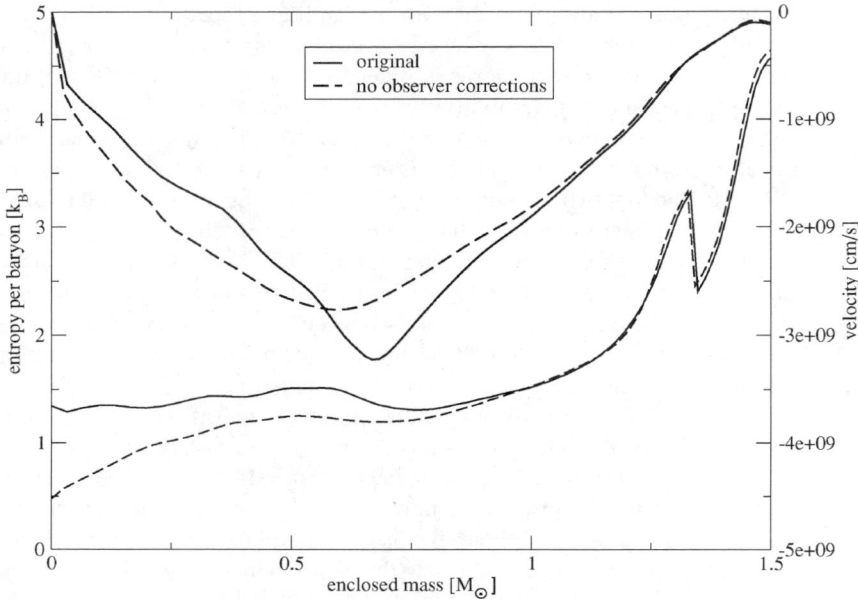

Figure 6   The entropy and velocity profiles as a function of enclosed mass at a central density of 10^{14} g/cm^3 for spherically symmetric models with and without the $O(v/c)$ observer corrections (51). The differences between the two cases are striking.

flux for 11 MeV neutrinos, owing to anisotropies in the hydrodynamics, are super-imposed in the figure. These offer a first glimpse at the two-dimensional neutrino radiation field in the stellar core after bounce.

5. TWO- AND THREE-DIMENSIONAL MODELS

As for the physics neglected in past spherically symmetric models, there are two possibilities: (a) Physics included in the models could have been treated more accurately. We have already discussed the improvements in neutrino transport that were possible and that could have resulted (although did not result) in qualitative changes in the models. There are other model components that can be improved, as we will discuss. (b) Physics essential to the explosion mechanism was not included in any approximation.

Ongoing efforts to improve the neutrino opacities and high-density equation of state fall under category (a) and are examples of efforts to develop supernova models quantitatively. We will discuss this work in more detail later, but it is unlikely that advances in this area will change the outcome of current supernova models qualitatively—that is, lead to explosion in models that currently do not explode.

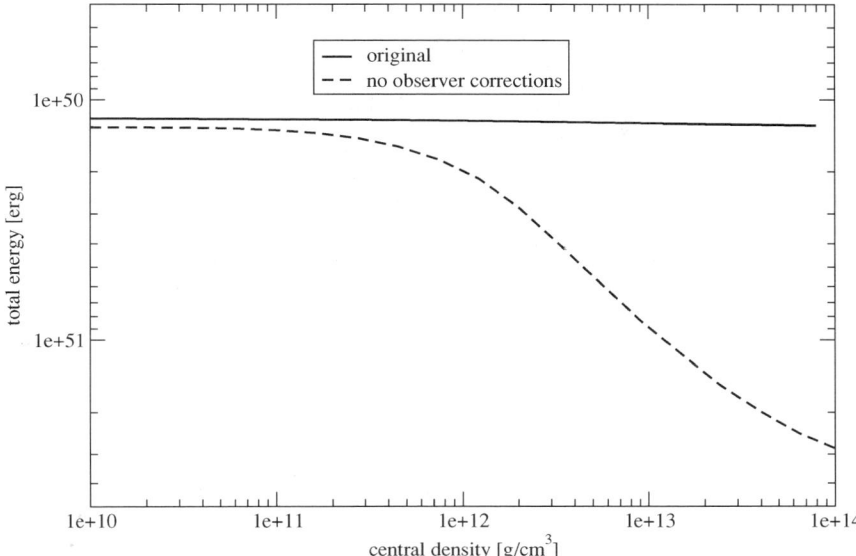

Figure 7   The total energy versus central density for the same models shown in Figure 6. The model without the observer corrections has clearly lost an energy $\sim 3 \times 10^{51}$ ergs prior to stellar core bounce.

Nonetheless, our ultimate goal is to produce explosions and to predict quantitatively all of the supernova-associated observables. Changes in the microphysics (weak interaction rates, sub- and super-nuclear density equation of state) lead to changes in the models that will affect these observables.

Ongoing efforts to include fluid instabilities, rotation, and magnetic fields in core collapse supernova models fall under category (b). Although they have fundamentally altered supernova theory, we still do not have in hand the long-sought-after mechanism by which explosions are guaranteed.

To date, supernova modelers have identified seven major components of a detailed supernova model: neutrino transport, fluid instabilities, rotation, magnetic fields, gravity, the neutrino weak interactions, and the sub- and super-nuclear equation of state. Two- and three-dimensional models developed so far have included only a subset of these known important components. Thus, multidimensional supernova modeling is in its infancy. That we do not yet have a detailed description of how massive stars explode is no great mystery. During the next five years, multidimensional supernova models will undergo a dramatic change in realism. We will be provided with far more realistic two- and three-dimensional models that will give us far better guidance on the roles of fluid instabilities, rotation, and magnetic fields, and their coupling, in generating and defining supernova explosions. Of course, quantitatively accurate models will require fully general relativistic simulations. These will not be completed in the near term. They will require a decade or more to develop. Nonetheless, efforts to include corrections for general relativity in multidimensional supernova models are underway (52).

As we discussed earlier, attempts to simulate core collapse supernova explosions are met with a number of significant technical challenges. We have already discussed one of them. A second is the challenge to understand the interplay between physics components in a tightly coupled, nonlinear, multiphysics model and to capture such interplay faithfully. For example, one cannot predict a priori how a small perturbation will affect a multiphysics, nonlinear system as it grows or decays. This knowledge is equivalent to a solution of the nonlinear partial differential equations governing the evolution of the system, which is obviously not a priori known. Such perturbations may be damped in some cases and magnified in others, with consequent feedbacks to other components of the system.

We can illustrate this point with the following example from Kifonidis et al. (53). If a strong shock is aligned or nearly aligned with the coordinate grid in a numerical simulation, certain numerical hydrodynamics schemes (e.g., the piecewise parabolic method, or PPM, schemes) will exhibit what is known as odd-even decoupling (54). In this numerical phenomenon, small perturbations—e.g., ones that may be put in by hand to seed neutrino-driven convection in a two-dimensional model initiated after bounce starting with a profile from a one-dimensional collapse simulation—will grow along the shock surface, inducing a ripple-like pattern of larger perturbations. If the latter perturbations are not damped, the subsequent development of neutrino-driven convection below the shock and, in turn, the resultant shock dynamics will be dramatically altered. Such differences are evident in Figures 9 and 10 (see color insert), which show two-dimensional entropy distributions at 208 ms after core bounce and the development of neutrino-driven convection behind the shock, with and without odd-even decoupling. In the former case, owing to the lack of numerical damping, neutrino-driven convection is seeded more strongly and develops more quickly. In turn, this leads to significant differences in the postshock flow and, subsequently, the shock radius. Clearly such differences would have an impact on neutrino shock reheating and the potential to launch an explosion and other observables. Indeed, Kifonidis et al. point out that in their models, whose focus was in part on the production of ^{56}Ni, the ^{56}Ni yield differed by 40% between the two cases. This example, among many others, reminds us of the arduous validation process we must adhere to, always, in order to ensure a faithful simulation and accurate model predictions.

5.1. Fluid Instabilities

The potential role of fluid instabilities in the post-stellar-core-bounce dynamics was first articulated decades ago and explored phenomenologically for some time in the context of spherically symmetric models. Fortunately, two- and three-dimensional models that have emerged over the past decade, albeit incomplete, have fundamentally changed our ability to model these instabilities and to assess their role in the supernova mechanism.

Before we begin discussing these, it is important to classify the regions in which the various instabilities may occur and their fundamental nature. Fluid instabilities in core collapse supernovae can be put in one of three fundamental categories: (a) convection, (b) doubly diffusive instabilities, and (c) shock wave instabilities. Convection includes both Ledoux convection, which results from both entropy and composition gradients (in this case lepton fraction gradients) in the stellar core, and Schwarzschild convection, which results from entropy gradients alone. Doubly diffusive instabilities result from a competition between the transport of entropy and leptons in the stellar core and occur in regions in which there are crossed or competing gradients in entropy and composition, with one being stabilizing and the other destabilizing. Figure 11, taken from a spherically symmetric model after stellar core bounce, shows the entropy and electron fraction gradients throughout the stellar core as a function of time. Regions of negative entropy and lepton fraction gradients, which are individually (and combined) destabilizing, and crossed gradients, which may also be destabilizing, are evident throughout the core after bounce.

Instabilities in the stellar core, whether we are discussing convection or doubly diffusive instabilities, occur throughout the region below the supernova shock wave. If they occur beneath the neutrinosphere, we refer to them as proto-neutron star (PNS) instabilities. PNS instabilities may boost the neutrino luminosities emerging from the PNS and thereby boost the neutrino energy deposition beneath the stalled shock. Essentially, these instabilities may lead to the more efficient transport of neutrinos by advection, rather than diffusion, outward in radius to the neutrinosphere.

Directly beneath the stalled shock, between the gain radius and the shock, where the material is undergoing net neutrino heating, an instability develops as the result of the heating-associated entropy gradient. This instability is referred to as neutrino-driven convection. Researchers (e.g., see References 15, 18, 19, 22) have shown that neutrino-driven convection fundamentally alters the scenario under which neutrino shock reheating occurs. Neutrino-driven convection allows for both an explosion, in which the shock wave and material behind it begin to move radially outward, and continued accretion, which maintains the neutrino luminosities sufficiently high so as to sustain the heating. In spherical symmetry, explosion and accretion are mutually exclusive. Neutrino-driven convection also acts hydrodynamically on the shock, pushing it out to larger radii and shallower points in the gravitational potential, thereby facilitating explosion, and it can boost the neutrino heating efficiency by moving heated matter upward towards the shock—in so doing, the heated matter expands, its temperature drops, and its neutrino cooling losses are reduced, thereby keeping more of the neutrino-deposited energy in the gain region.

Whereas the vigor and extent of PNS instabilities and, consequently, their impact on the supernova mechanism are currently a matter of debate, a debate that will ultimately require two- and three-dimensional radiation magnetohydrodynamics simulations to resolve, the development of neutrino-driven convection is a

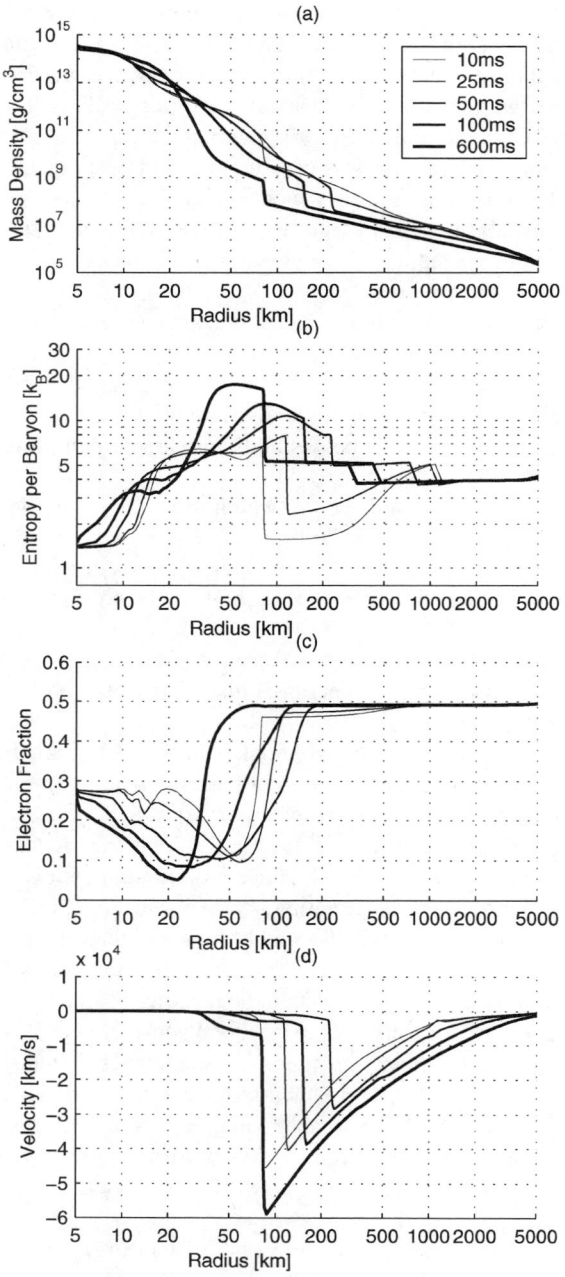

Figure 11   Core density, entropy, electron fraction, and velocity profiles at different times after bounce, taken from spherically symmetric models with Boltzmann neutrino transport (8).

characteristic feature of all two- and three-dimensional supernova models that have been performed to date, although, as we will see, it does not guarantee explosions.

5.1.1. NEUTRINO-DRIVEN CONVECTION   The results shown in Figure 12 (see color insert) are from the most advanced two-dimensional neutrino radiation hydrodynamics simulation performed to date to evolve for hundreds of milliseconds after bounce (15). In this simulation, the multifrequency moment equations for the neutrino number, energy, and momentum are solved along each radial ray on the grid. (These moment equations are derived from the Boltzmann equation, as shown earlier.) No lateral transport of neutrinos is included in the model—i.e., the model is not fully two-dimensional—although lateral advection of neutrinos between rays is included. Corrections for general relativistic gravity are also included, and the models were performed with the most complete set of known relevant neutrino interactions.

The left panels in Figure 12 show the rotational velocity and entropy variations as a function of enclosed mass, whereas the right panels show the rotational velocity and entropy as a function of radius. In the upper right panel, the velocity field is superimposed. In this panel, the existence of a large-scale flow owing to rotation-enhanced neutrino-driven convection is evident. Although neutrino-driven convection develops and is enhanced by the effects of rotation in this model, no explosion develops. The model was initiated from a 15 $M_\odot$ progenitor and the angular grid was restricted to 90 degrees. The extension to 180 degrees did not result in an explosion in this model, but led to a weak explosion in an 11 $M_\odot$ model (16). As we will discuss below, the extension to a 180 degree grid is fundamental, given that it allows for $l = 1$ hydrodynamic modes that will likely play a significant role in the explosion mechanism.

The neglected lateral transport of neutrinos in the above simulations will be particularly important in the proto-neutron star, where the neutrinos and the fluid are strongly coupled. A determination of the development of PNS instabilities and their possible impact on the supernova mechanism will require two- and three-dimensional simulations with two- and three-dimensional transport. The same is true if we want to fully explore the development and impact of neutrino-driven convection as well.

Although neutrino driven convection has been confirmed in all two- and three-dimensional simulations performed to date and was originally thought to be of importance to the supernova mechanism for the reasons indicated above, once the shock is distorted by neutrino-driven convection the subsequent postshock flow and shock dynamics are no longer dictated by it alone. Rather, the flow results from a combination of neutrino-driven convection and the effect of material inflowing through a now nonspherical (oblique) shock (55). In fact, the latter may come to dominate the definition of the flow characteristics in the postshock region. Neutrino-driven convection may act only as a seed to this latter phase of the shock's dynamics and the dynamics of the postshock flow.

5.1.2. PROTO-NEUTRON STAR INSTABILITIES   The existence and impact of fluid instabilities in the proto-neutron star is an issue of radiation magnetohydrodynamics.

As we now discuss, the coupling of the neutrino transport, the magnetic fields, or both to the stellar core flow in this region can (a) dramatically alter the development of the instabilities (e.g., convection) that have traditionally been studied and, more important, (b) lead to entirely new types of instabilities. For a detailed and comprehensive analysis of proto-neutron star instabilities, the reader is referred to (56, 57).

Nowhere is the combined role of neutrino transport and hydrodynamics in the evolution of stellar core fluid instabilities more evident than in the case of doubly diffusive instabilities. Significant progress has been made recently in better understanding the nature of these instabilities and their potential role in the core collapse supernova mechanism (56).

One doubly diffusive instability, neutron fingers, was first discussed in the context of core collapse supernovae by Smarr and collaborators (58) and, as we mentioned earlier, it was an integral component of the supernova models developed by Wilson (9, 59). Without neutron fingers, Wilson did not obtain explosions. (In spherically symmetric models, such as Wilson's models, if the assumed criterion for the existence of neutron fingers is met, the fluid in the proto-neutron star is mixed according to a phenomenological mixing-length theory.)

If a region of crossed entropy and lepton fraction gradients has higher-entropy, lower-lepton fraction material above lower-entropy, higher-lepton fraction material in the gravitational field of the proto-neutron star, and further, if entropy transport by neutrinos is more efficient than lepton transport, the core will be unstable to the development of neutron fingers, akin to salt fingers in the ocean. As shown in Figure 13, if entropy transport is indeed more efficient, the initial perturbation evolves to a low-entropy, low–lepton-fraction perturbation surrounded by a low-entropy, high–lepton-fraction medium. It will therefore be denser than the surrounding medium, and it will continue to sink. Rough arguments based on the fact that three flavors of neutrinos carry entropy whereas only one carries lepton number might suggest that entropy transport would indeed be more efficient (9, 59). However, two flavors of neutrinos—muon and tau—are inefficient in transporting entropy, and both entropy and lepton fraction are largely transported by one flavor of neutrinos, electron neutrinos. Indeed, detailed numerical experiments performed under conditions culled from core collapse supernova simulations suggest that neutron fingers are unlikely to occur in the stellar core early after bounce, during the shock reheating phase, and, based on these experiments, we should not expect them to aid in generating an explosion (56).

Although these experiments did not confirm the existence of neutron fingers in the stellar core, they led to the discovery of two new doubly diffusive instabilities [lepto-entropy fingers (LEFs) and lepto-entropy semiconvection (LESC)] that are expected to exist during this epoch and that may act, in the case of lepto-entropy fingers, to boost the neutrinosphere luminosities in lieu of neutron fingers.

To better understand these instabilities, consider the equations governing the motion of a fluid element in the proto-neutron star and the evolution of the entropy and lepton fraction contrast between the fluid element and its surroundings:

Neutron Fingers

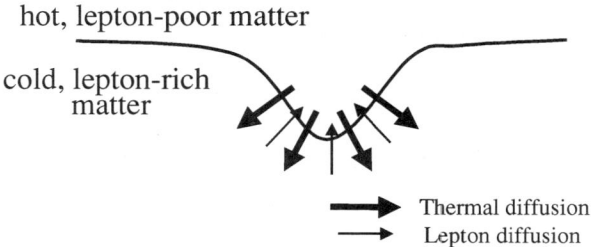

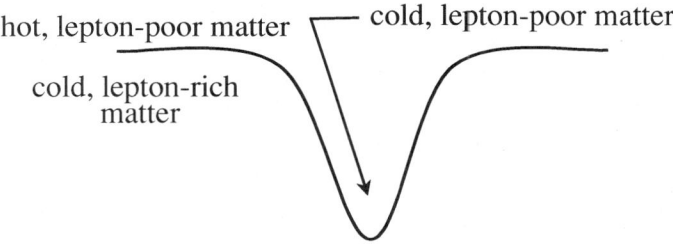

Figure 13   Schematic illustrating the development of neutron fingers under the assumption that transport of entropy between a fluid element and its surroundings dominates over lepton transport. Under these assumptions, our perturbed element becomes cold and lepton poor and is surrounded by material that is cold and lepton rich. The element will be more dense, as a result, and continue to sink.

$$\dot{\theta}_s = \Sigma_s \theta_s + \Sigma_{Y_\ell} \theta_{Y_\ell} - \frac{d\bar{s}}{dz} v, \qquad\qquad 11.$$

$$\dot{\theta}_{Y_\ell} = \Upsilon_s \theta_s + \Upsilon_{Y_\ell} \theta_{Y_\ell} - \frac{d\bar{Y}_\ell}{dz} v, \qquad\qquad 12.$$

$$\dot{v} = -\frac{g}{\rho} \left(\frac{\partial \rho}{\partial s} \right)_{p, Y_\ell} \theta_s - \frac{g}{\rho} \left(\frac{\partial \rho}{\partial Y_\ell} \right)_{p, s} \theta_{Y_\ell}. \qquad\qquad 13.$$

In Equations 9 and 10, θ_s and θ_{Y_ℓ} are the contrast in entropy and lepton fraction, respectively, between the fluid element and its surroundings, and v is the velocity of the fluid element that arises owing to acceleration through buoyancy. The coefficients in Equations 9 and 10 are the neutrino transport response functions, Σ_{s,Y_ℓ} and Υ_{s,Y_ℓ}, and the gradients of the background entropy and lepton fraction, $\bar{s}$ and $\bar{Y}_\ell$, with respect to radius (or in this case z). For example, Σ_{s,Y_ℓ} give the rate of change of θ_s and θ_{Y_ℓ}, respectively, owing to neutrino transport between the fluid element and its surroundings.

First, in the language of Equations 9 and 10, the response functions Σ_s and Υ_{Y_ℓ} are strictly zero in the case of Ledoux convection, although if they are nonzero but small, the instability may still be Ledoux rather than semiconvection or neutron fingers.

The response functions Σ_s and Υ_{Y_ℓ} are the relevant response functions for the development of neutron fingers. As Figure 13 shows in schematic form, this instability results from a competition between the gradients in entropy and lepton fraction when one is stabilizing and the other is destabilizing. In the language of Equations 9 and 10, whether or not neutron fingers develop depends on the relative magnitudes of the two response functions Σ_s and Υ_{Y_ℓ}.

The key response functions for the development of LEFs and LESC are the cross response functions, Υ_s and Σ_{Y_ℓ}. Moreover, in typical postbounce conditions in the stellar core, Υ_s is large relative to the other response functions, and positive. LEFs are caused when a stabilizing entropy gradient quickly induces (owing to a large Υ_s) a difference in lepton fraction between a fluid element and its surroundings that is subsequently destabilizing, resulting in a growing lepton finger. (An entropy contrast between the fluid element in question and its surroundings will induce neutrino transport between the two, which carries both entropy and lepton fraction.) Thus, LEFs result when lepton transport destabilizes an otherwise stable entropy gradient. LESC, on the other hand, arises when a destabilizing entropy gradient is stabilized by lepton transport, leading to a growing oscillatory mode (as opposed to growing fingers).

In the detailed analyses we have just cited and discussed, the following picture emerges: Typical conditions found in the postbounce stellar core suggest that the proto-neutron star will be stratified. There will be an innermost region unstable to LESC, an intermediate region unstable to Ledoux convection, and an outer region extending up to the neutrinospheres unstable to LEFs. This is evident in Figures 14 and 15 (see color insert), where the regions unstable to lepto-entropy fingers and lepto-entropy semiconvection are shown for conditions found in the stellar core after core bounce (56). Shown are the growth rates for both instabilities as a function of radius and the size of the fluid element involved. The growth rate for lepto-entropy fingers is significant regardless of the fluid element size (relative to the radius) and for an extended range in radius. On the other hand, lepto-entropy semiconvection is confined to a narrow region in radius and develops over a more restricted range in fluid element size (relative to the radius).

Future fully two- and three-dimensional models will be able to investigate definitively whether or not lepto-entropy fingers exist in the postbounce stellar core and whether or not they have a discernible impact on the supernova mechanism.

The above analyses apply only in the linear regime at the onset of an instability and cannot determine the subsequent nonlinear, neutrino matter–coupled evolution.

In the simulation results we showed earlier from Buras et al. (15), which were taken from two-dimensional simulations that implemented one-dimensional radial ray transport, it is evident in the lower left panel of Figure 12 that an instability is developing deep in the proto-neutron star and is surrounded by a convectively stable layer. (The authors attributed this to Ledoux convection, but in light of the above analysis, it may be a doubly diffusive instability.) Consequently, Buras et al. conclude that the instability is not important, at least in this model, in contributing to boosted neutrinosphere luminosities and shock reheating. However, the use of radial ray transport (even if lateral advection is included) and the restriction to two spatial dimensions do not allow a definitive assessment of the neutrino radiation hydrodynamics in the proto-neutron star, whether the Ledoux instability or a doubly diffusive instability is the driving instability.

The role of magnetic fields in the development of proto-neutron star instabilities was investigated in (57). Magnetic fields have a dual role: stabilization and destabilization. The essential results indicate that magnetic field strengths in excess of 10^{13} G may do both: stabilize regions unstable to convection, limiting the spatial extent of such regions, and destabilize regions to new modes of instability. However, only very strong magnetic fields ($>10^{16}$ G) are expected to significantly alter the stability properties of proto-neutron stars. These fields might be generated by dynamo action from convective activity present in the core prior to the escalation of field strengths. Once the field strength increases, convective activity is shut off. Thus, magnetic fields may affect both the spatial extent and the temporal duration of convective activity in the proto-neutron star.

5.1.3. THE STATIONARY ACCRETION SHOCK INSTABILITY (SASI)   We might not have expected surprises in the multidimensional hydrodynamics of core collapse supernova models, given that this component of the models was much more mature than others. Nonetheless, a fundamentally new hydrodynamic instability has been discovered recently (60) that may play a significant role in generating core collapse supernovae and some of their key observables (e.g., neutron star kicks, spectropolarimetry, pulsar spin up).

The postbounce stellar core flow is best characterized as an accretion flow through a quasi-stationary shock. It has been shown in two- and three-dimensional hydrodynamics studies constructed to reflect the conditions during the postbounce shock reheating epoch that nonspherical perturbations of the accretion shock lead to the development of a stationary accretion shock instability (SASI) (see Figure 16, see color insert). Recent studies (61) confirm the existence of the SASI in two-dimensional models that include radial-ray neutrino transport. The SASI is an $l = 1$ instability and results from the establishment and amplification of a standing pressure wave in the acoustic cavity defined by the shock and the neutrinosphere (62). The potential ramifications of the SASI for the supernova mechanism and phenomenology were first elaborated in (60): (a) The SASI may aid in generating the supernova explosion itself. Simulations indicate that, much like the neutrinos, the SASI may act as a conduit between gravitational binding energy and the kinetic energy of outflow. In the simulations mentioned above, explosions were generated in the absence of neutrinos, the important point being that these simulations offered a proof of principle demonstration that the SASI can serve to rechannel energy in the supernova. (In addition to aiding the explosion directly, the SASI may aid the explosion indirectly by aiding the neutrino heating (61).) (b) The SASI may also define the gross asymmetry of the explosion. Two-dimensional simulations of the SASI lead to bipolar explosions and to a self-similarity in the flow at late times with an aspect ratio consistent with the supernova spectropolarimetry data (63). Three-dimensional simulations yield a far more complex outcome, with the $l = 1$ axis rotating as the evolution proceeds (64). Moreover, these simulations offer a proof of principle that the SASI is also capable of imparting significant angular momentum to the proto-neutron star, even beginning with spherically symmetric initial conditions (64).

The discovery of the SASI must fundamentally alter the way we think about core collapse supernovae. We must certainly retire many of our past prejudices regarding the role of rotation in defining the nature of the explosion and the evolution of the angular momentum distribution in the stellar core during collapse and after bounce given the initial angular momentum distribution of the progenitor model. For example, although somewhat restricted, two-dimensional simulations indicate that core rotation may not be required to generate bipolar-like outflows (60). And in three dimensions, the SASI is capable of spinning up the proto-neutron star significantly even when we begin with spherically symmetric initial conditions. The discovery of the SASI must also force us to reexamine how we interpret core collapse supernova observations. For example, the connection one might make between supernova spectropolarimetry data and stellar core rotation and magnetic fields may be complicated by the SASI. Generally speaking, the connections between supernova observables and theory are likely far more complex than we had ever assumed in the past.

In retrospect, it should have been obvious that once the shock wave is distorted from spherical symmetry by neutrino-driven convection, the flow beneath the shock is no longer defined solely by the convection. A distorted shock will deflect radially infalling material passing through it, leading to highly nonradial flow beneath it. In fact, with time the fluid flow beneath the shock may largely be determined by the SASI, not by convection (55). Instabilities such as neutrino-driven convection are important at early times in aiding the neutrino heating, as we discussed above, and in setting the shock standoff radius while the explosion is initiated. The standoff radius will in part determine the time scale over which the SASI may develop.

Future two- and especially three-dimensional simulations will determine, ultimately, the role of the SASI in core collapse supernovae, but the simulations described above have given us a far better understanding of the turbulent flow beneath the shock, and its origins, and most important, a fundamentally new phenomenon in supernova theory.

5.2. Rotation

Neutrino transport in core collapse supernovae has been studied extensively during the last four decades, culminating recently in fully general relativistic simulations with Boltzmann neutrino transport that have essentially closed the book on spherically symmetric models [at least in the absence of neutrino mixing; (13)]. Several two-dimensional simulations have been performed during the past decade to explore the dynamics of fluid instabilities in the stellar core after bounce and their impact on the supernova mechanism, culminating in simulations that include sophisticated radial-ray neutrino transport that captures a significant amount of realism in the models (15).

In contrast, very few simulations have been performed to date that include rotation, and no contemporary, sufficiently realistic simulation has been performed that includes magnetic fields.

The work of several groups [e.g., see References 15, 22, 23, 65] has shown that rotation can significantly influence stellar core collapse and the postbounce dynamics in a variety of ways: (a) Centrifugal forces will slow collapse along the equator and, if sufficiently large, can lead to a low-density bounce. (b) Gravitational binding energy will be channeled differently during core collapse, relative to the spherically symmetric case, partitioned between the internal energy of the matter and neutrinos and the rotational energy of the core. (c) The neutrinospheres will be distorted, and the neutrino luminosities and RMS energies may be noticeably changed. (d) The preshock accretion ram pressure along the rotation axis and the equator may differ significantly. (e) Rotation will alter the development of fluid instabilities below the shock, may provide a new source of internal energy in the postshock flow that may augment the energy supplied by neutrino heating, and may have a dramatic impact on the growth of magnetic fields in the stellar core after bounce.

The simulations by Buras et al. (15) demonstrate clearly that rotation can have a dramatic effect on the postbounce shock dynamics. In this simulation, the average shock radius was increased by approximately a factor of 2 between 200 and 300 ms after bounce (see Figure 17). Moreover, owing to centrifugal forces in the equatorial plane and a decreased ram pressure along the rotation axis, a violent overturn was observed in the postshock, convective region and confined to an angular region around the rotation axis (see Figure 12). These simulations were performed on a 90 degree angular grid, which will not admit the $l = 1$ SASI instability mentioned above. Future simulations that include rotation and a 180 degree grid will allow us to more thoroughly explore how postbounce instabilities and rotation act in concert in the explosion mechanism.

Regarding point (e) above and the interplay between fluid instabilities in the postbounce stellar core and rotation, we turn to the most advanced three-dimensional core collapse supernova simulations performed to date (22). These were the first three-dimensional simulations to include neutrino transport, albeit gray transport (see the discussion on the pros and cons of gray transport in Section 1), and one of their primary goals was to discern the impact of rotation

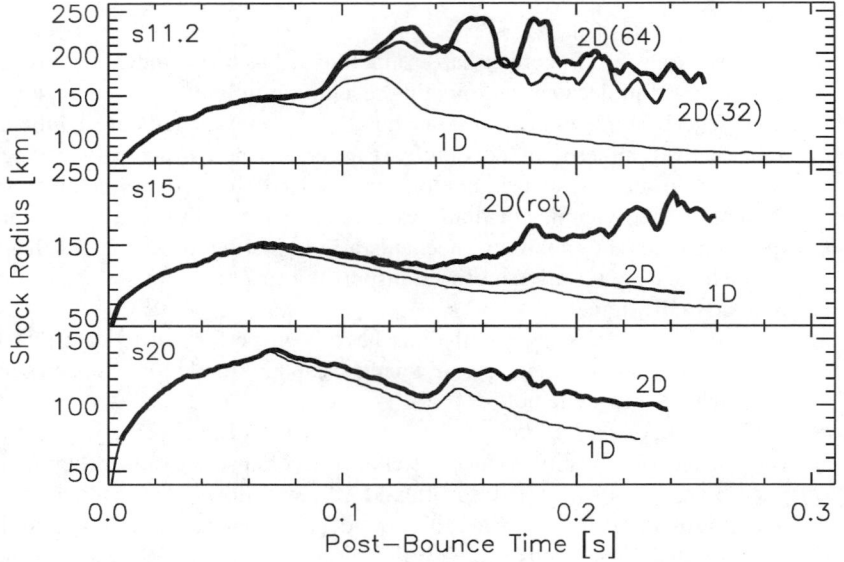

Figure 17   The shock radius versus time for a series of models initiated from different supernova progenitors, for both spherically symmetric (1D) and axisymmetric (2D) runs (15). In panel *s15*, the average shock radius for the model with rotation 2D(rot) is dramatically increased after ~150 ms after bounce relative to the model without rotation 2D. Also illustrated in these figures is the boost in shock radius owing to convection alone as we move from 1D to 2D models.

on the development of neutrino-driven convection. An accurate assessment of this interplay can only be captured by three-dimensional simulations (e.g., the rotation of the earth in conjunction with convection in the atmosphere can generate tornados, which would not be admitted in two-dimensional axisymmetric simulations of the earth's atmosphere given that such local tornadic activity would violate the imposed symmetry). Figure 18 shows how the upflows associated with neutrino-driven convection in the Fryer and Warren models are more confined to the rotation axis for the three models (SN15A-hr, SN15A, SN15B) with rapid rotation. The upflows associated with the other three models are more distributed in angle. Despite the differences that one might expect between two- and three-dimensional models, these results are in accord with those shown earlier from the two-dimensional model with rotation of Buras et al. (Figure 12).

It is also important to note, reflecting back on our discussions of the SASI, that rotation in the stellar core after bounce is not solely reliant on the differential rotation in the core prior to collapse and its subsequent evolution during collapse. For example, as we discussed earlier, the SASI may be able to generate significant angular momentum in the proto-neutron star beginning with an initially spherically symmetric state (64).

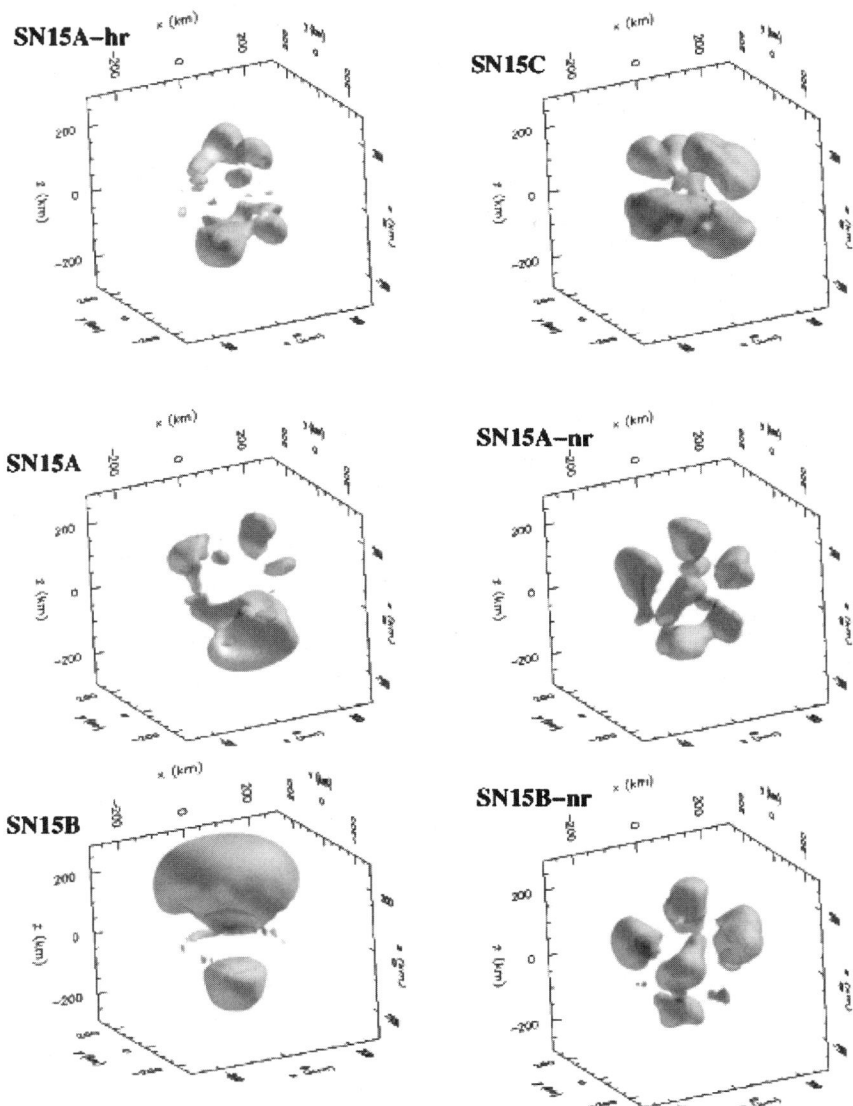

Figure 18   In this figure there are six panels of isosurfaces of upflows with velocities of 3000 km/s (22). Three models (SN15A-hr, SN15A, SN15B) have rapid rotation, and three models (SN15C, SN15A-nr, and SN15B-nr) have slow or no rotation. In the case of rapid rotation, the upflows associated with neutrino-driven convection are confined to angular regions around the rotation axis. For the slow or nonrotating cases, the upflows are distributed in angle.

Finally, much as the fluid instabilities in the stellar core and core rotation are inextricably linked, the development of the magnetic fields in the stellar core is dependent on both, as will be shown in the next section.

5.3. Magnetic Fields

Several groups [e.g., see References 23, 24, 57, 65, 66] have shown that, much like rotation, magnetic fields may influence stellar core collapse and the supernova mechanism in a number of ways: (a) The development of significant magnetic pressure may have an impact on stellar core collapse and the postbounce flow. Magnetic fields may (b) alter the development of fluid instabilities in the stellar core after bounce, (c) provide additional channels for the generation of internal energy through viscous dissipation, (d) alter the weak interactions in the stellar core, and (e) provide a conduit through which rotational energy in the collapsed core may be channeled into the outflows of explosion.

Arguably, the fundamental question is whether the magnetic fields will organize into large-scale configurations that will help drive and collimate outflows from the stellar core [point (e) above]. The pioneering simulations of (25) and (24) were the first to explore the evolution of stellar core magnetic fields during core collapse and their impact on the explosion mechanism. These simulations exhibited the development of a magnetic bubble deep in the core owing to the dramatic increase in magnetic pressure close to the rotation axis as core field lines are dragged inward and compressed. This magnetic bubble led to buoyant, bipolar outflows that culminated in bipolar explosions [the LeBlanc-Wilson (LW) jet shown in Figure 19].

Owing to stellar core differential rotation, an initially poloidal field threading the stellar core could be wound up into a potentially significant toroidal field. Moreover, the field may be wound up quickly before it has a chance to expand vertically, later expanding in a spring-like fashion along the rotation axis. In so doing, the field would evolve into an open helix, as shown in Figure 19 (26). In this way, material in the core could be driven outward along the rotation axis in a spring and fling manner, the latter arising because the material is accelerated along the rotating field lines by centrifugal forces. The so-called hoop stresses would serve to collimate the flow.

The most advanced simulations of stellar core collapse to include magnetic fields were performed more than twenty years ago. Symbalisty (24) concluded that inordinately large rotation and magnetic fields strengths were required for an explosion to develop through the original Leblanc-Wilson mechanism. The magnetic fields in the stellar core can be amplified in the way suggested by Leblanc and Wilson or through wrapping, as described above, but the growth of field strength through other mechanisms must be considered.

Magnetic fields during stellar core collapse may be amplified in one of four basic ways: (a) through collapse, as in the Leblanc-Wilson scenario; (b) by wrapping; (c) through a dynamo effect combining the action of fluid instabilities, such as

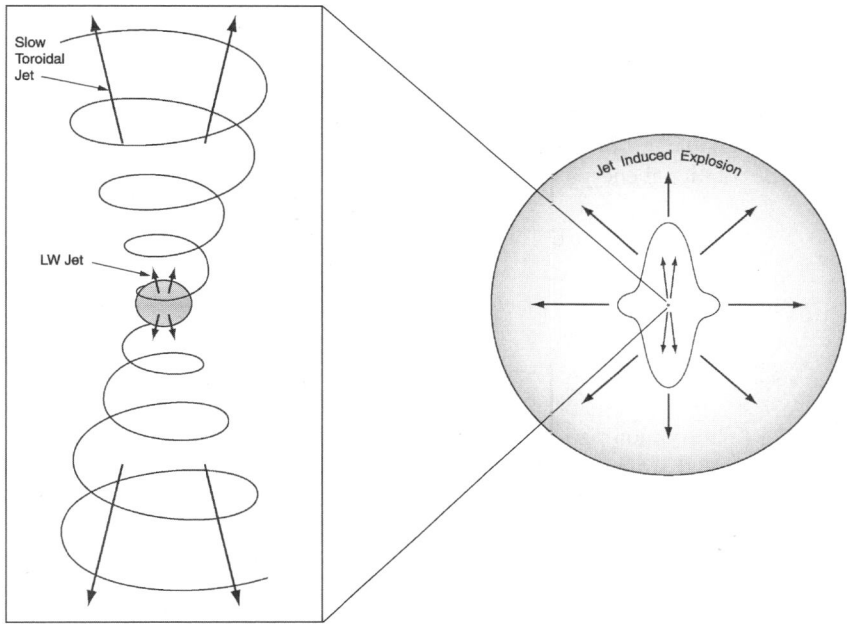

Figure 19   This schematic illustrates one possible organized configuration of the magnetic field after stellar core bounce. The field may be wound up by rotation during and after bounce, initially confined like a compressed spring near the equatorial plane. Subsequent expansion of the magnetic field lines may then lead to this helical field. The fields may drive and collimate bipolar outflows along the rotation axis in a so-called spring and fling mechanism (26).

convection, and rotation; and (d) through shear (the magnetorotational instability discussed below).

At the time of the early simulations by Leblanc, Wilson, and Symbalisty, the magnetorotational instability (MRI) was not yet discovered. The MRI was first discovered in the context of the differential rotation in accretion disks (67) and is believed to be the mechanism whereby angular momentum is transported in such disks. But it may also operate in a differentially rotating core after stellar core bounce. Field amplification through the MRI is fundamentally different than through compression and wrapping. In the case of the MRI, the field amplification occurs through the stretching of the field lines in a strongly differentially rotating core.

Akiyama et al. (23) were the first to propose that the MRI could be important in the core collapse supernova context. They argued that with sufficient differential rotation, the MRI could amplify the magnetic field strengths in the stellar core exponentially quickly, rather than linearly, and lead to magnetohydrodynamic luminosities $\sim 10^{52}$ erg/s, rivaling the neutrino luminosities, although as we now discuss, the magnetic fields are not expected to power the explosion directly.

In the scenarios delineated above, the magnetic fields act primarily as a conduit, channeling rotational energy into outflows (26). Substantial rotational energy may be available, as the following simple estimates illustrate, to power such outflows. The rotational energy in the proto-neutron star after bounce is (26)

$$E_{\text{Rotation}} \approx 9 \times 10^{50} \text{ erg} \left(\frac{M_{\text{NS}}}{1.5 \text{ M}_\odot} \right) \left(\frac{P_{\text{PNS}}}{25 \text{ ms}} \right)^{-2} \left(\frac{R_{\text{PNS}}}{50 \text{ km}} \right)^2. \qquad 14.$$

After deleptonization and contraction the rotational energy is

$$E_{\text{Rotation}} \approx 2 \times 10^{52} \text{ erg} \left(\frac{M_{\text{NS}}}{1.5 \text{ M}_\odot} \right) \left(\frac{P_{\text{PNS}}}{1 \text{ ms}} \right)^{-2} \left(\frac{R_{\text{NS}}}{10 \text{ km}} \right)^2. \qquad 15.$$

For final rotation periods less than $\sim$5 ms, substantial rotational energy becomes available that might be tapped (via the magnetic fields) to launch the supernova. Of course, it may be difficult to achieve this.

Improved progenitor models that include both rotation and magnetic braking yield iron cores that rotate at only $\sim$0.1 rad/s (68). Without magnetic breaking, the core rotation rates are 1–2 orders of magnitude faster. Therefore, magnetic fields lead to competing outcomes: For the MRI to increase field strengths to magnitudes that would be dynamically significant, perhaps dominant, would require significant differential rotation in the core. On the other hand, magnetic braking in the core tends to slow the stellar core rotation prior to collapse. Of course, two- and three-dimensional progenitor models will be required to better determine the differential rotation of the core at the start of collapse.

In short, magnetic fields will likely play an important role in supernova dynamics. Even initially small magnetic fields in the stellar core may be amplified quickly after core bounce, through a variety of mechanisms, to participate in the supernova dynamics. However, determining their precise role will require realistic two- and three-dimensional simulations, which have not yet been performed.

6. NEUTRINO WEAK INTERACTIONS

Improvements in modeling the macrophysics of stellar core radiation magnetohydrodynamics must be matched by improvements in modeling the neutrino weak interactions in the stellar core and the stellar core equation of state. In the end, a simulation is only as good as its weakest link.

6.1. Electron Capture on Nuclei

First among the weak interactions of concern is electron capture on nuclei. [For an extensive discussion of the physics of electron capture on nuclei in this context, the reader is referred to (69–74).] Electron capture during stellar core collapse determines the extent to which the core is deleptonized during collapse, which in turn sets the mean electron fraction in the core and the size of the inner homologous core at bounce (75). The size of the inner core determines the mass at which the shock forms and the energy imparted to it—in short, the initial conditions for the postbounce core evolution, shock revival, and consequent supernova explosion.

Electron capture on nuclei is dominated by Gamow-Teller transitions. Until recently, the standard weak interactions used in supernova simulations included electron capture on nuclei as described by a zero-order shell model (70, 71, 76). In this model, nucleons in the nucleus are assumed to be independent, and electron capture is Fermi blocked for nuclei with $N > 40$ (71). The net result: In this approximation the electron capture during core collapse is dominated by capture on free protons, not nuclei. Actually, the opposite is the case.

The nucleus is an interacting many-body system. Correlations owing to the residual interaction between nucleons excite nucleons in the nucleus and unblock these transitions (this is known as configuration mixing) (77). Nucleons are assumed to move in a mean field. The residual interaction corrects the mean field to better represent the complete interaction experienced by each nucleon. Moreover, as the core becomes neutronized during stellar core collapse via electron capture on protons and nuclei, the nuclei increase in mass. The nuclear size results from a competition between Coulomb and surface effects in the nucleus, the latter of which favors larger nuclei (78). In fact, electron capture rates will be needed for nuclei well above mass 100 (79). This will present a significant practical challenge. Nuclear structure theorists will have to perform these rate calculations, taking into account configuration mixing, in large Hilbert spaces.

Current nuclear structure calculations are typically performed at zero temperature, focused on ground state properties, not at the finite temperatures found in the stellar core during collapse, which are sufficient to excite nuclei and unblock the Gamow-Teller transitions (71, 80). As we will describe below, hybrid models have been constructed to contend with both finite temperatures and configuration mixing.

Once the neutrinos become trapped in the core during stellar core collapse, the inverse reaction of electron-neutrino capture on nuclei becomes significant, and, eventually, electron capture and its inverse reaction reach an equilibrium, at which point the electron fraction in the core is set by the nuclear partition functions, which must therefore be determined self-consistently (e.g., see References 70 and 74).

To advance the state of the art in nuclear structure theory to satisfy all of the above requirements will take considerable effort, and advances must be made through the implementation of models of intermediate sophistication, as we will describe. But before we delve into the specifics of recent advances in the study of core collapse when increasingly sophisticated nuclear theory has been used, it is instructive to consider what parameterized studies tell us.

Simulations that were designed to explore the sensitivity of stellar core dynamics to the uncertainties of the nuclear electron capture rates have demonstrated that a factor of 10 change in the rates has a dramatic impact on the dynamics of core collapse (79). This is clearly illustrated in Figure 20 (see color insert). Shown are the velocity profiles after bounce for a number of runs in which the electron capture rate on nuclei was varied by factors of 10. The shock is the discontinuity in each of the profiles. The simulation with $N_p \times N_h = 0.1$ effectively sets the capture rate on nuclei to zero, which is equivalent to using the zero-order shell model. When the capture rate on nuclei is on average increased by a factor of 10 ($N_p \times N_h = 1.0$), the mass at which the shock forms is reduced, as expected with increasing deleptonization of the core, by 0.1 $M_\odot$. This is a significant result. The shock loses $\sim 10^{51}$ erg of energy—i.e., an explosion energy—for every 0.1 $M_\odot$ of nuclei it dissociates as it propagates through the core. Increasing the capture rate on nuclei further ($N_p \times N_h = 10.0, 100.0$) induces additional changes in the shock formation mass, with the increased opacity of the reverse channel (neutrino capture on nuclei) as the parameter values are increased tamping the effect.

Recent simulations (81) of stellar core collapse were performed with a hybrid model of electron capture on nuclei (82). In this model, shell model Monte Carlo (SMMC) methods are used to determine the thermal population of single-particle states in an assumed mean field. The random phase approximation (RPA) is then used to determine the effect of a subset of correlations among the nucleons in this configuration owing to their residual interaction. In RPA, single-particle–single-hole excitations are included, all other correlations are excluded. Using the hybrid model rather than a zero-order shell model demonstrates that electron capture during stellar core collapse is dominated by capture on nuclei, not protons, and that a more sophisticated treatment of capture leads to (a) dynamically significant changes in the size of the inner homologous core (and therefore, where the shock is launched) and (b) the preshock stellar core profiles through which the shock will move, which will affect the subsequent shock dynamics, development of core fluid instabilities, and nucleosynthesis. This is clear in Figure 20. The results of stellar core collapse with the hybrid model electron capture rates are shown by the profile labeled *LMP+LMS rates*. Moreover, the curve *Bruenn (1985)* shows the results from core collapse performed with the zero-order shell model. The increased sophistication in the description of electron capture on nuclei from the zero-order shell model to the hybrid model gives rise to a significant change in the shock formation mass, equivalent to the change associated with a factor of 10 change in the parameterized rates.

In conclusion, it is no longer a question of whether or not more accurate rates for electron capture on nuclei are necessary. The question is: What is the path forward in light of the fact we must contend with heavy nuclei and finite temperatures? The rates clearly must be computed with greater accuracy than in the past.

6.2. Other Important Weak Interactions

Another compelling recent development in the theory of weak interactions in core collapse supernovae was the discovery that nucleon–nucleon bremsstrahlung and neutral-current neutrino–antineutrino annihilation are in fact the most important source of muon and tau neutrinos and antineutrinos in the proto-neutron star mantle after core bounce (83, 84). Prior to this discovery, the sole mechanism whereby neutrinos and antineutrinos of these two flavors were produced in the models was through the annihilation of electrons and positrons (34). The latter has now been shown to be insignificant in the production of these neutrinos relative to the contributions from bremsstrahlung and neutral-current neutrino–antineutrino annihilation. Figure 21 clearly demonstrates this. The differential production rate of muon and tau neutrino pairs is more than doubled as we move from production via electron–positron to electron neutrino–antineutrino pair annihilation. Despite the

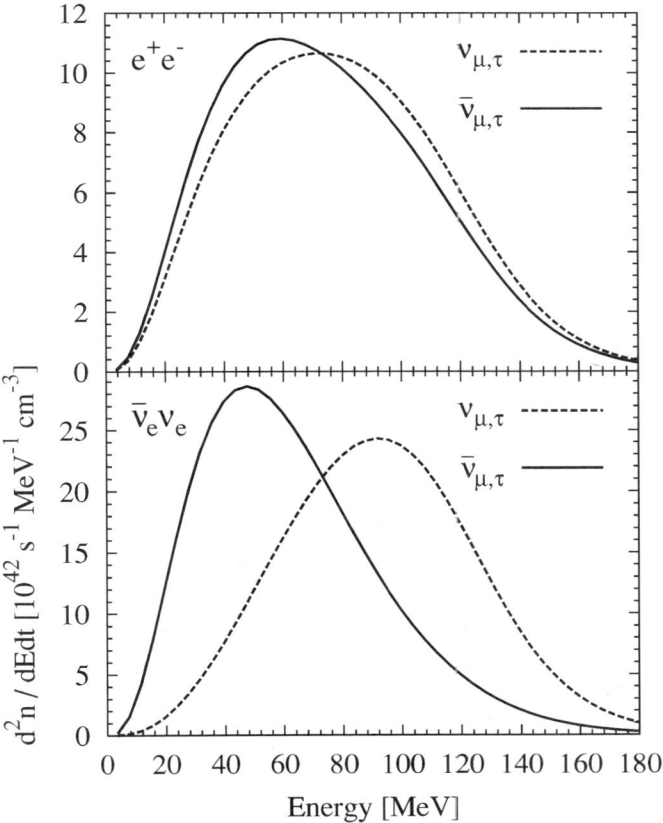

Figure 21   The differential production rate for muon and tau neutrino pairs from both electron–positron annihilation and electron neutrino–antineutrino annihilation are shown as a function of neutrino energy (84). Clearly, the production for conditions found in the shocked proto-neutron star mantle is dominated by electron neutrino–antineutrino annihilation, contrary to what was assumed and implemented in past models.

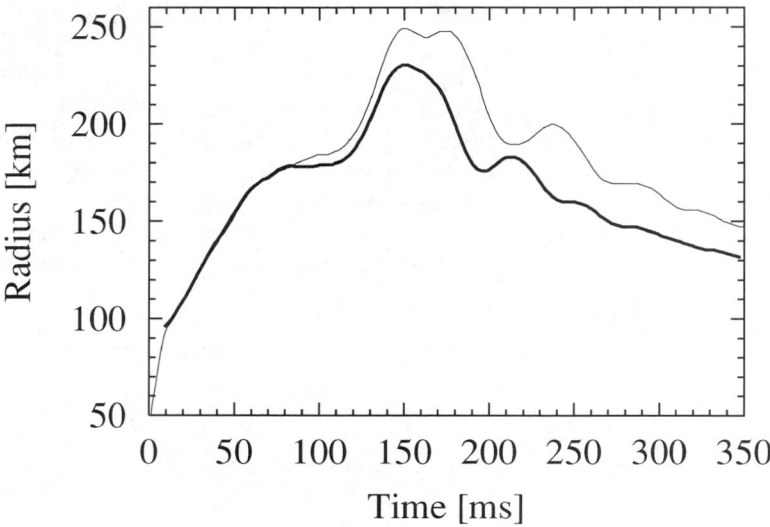

Figure 22   The shock radii versus time are plotted for two simulations in which electron neutrino–antineutrino annihilation is (*heavy line*) and is not (*light line*) included. The difference in shock radius at ∼175 ms after bounce (∼50–60 km) cannot be ignored, despite the fact that in both cases the shock recedes with time and no explosion is launched (84).

dramatic change in the production rate, Figure 22 demonstrates that the inclusion of electron neutrino–antineutrino annihilation as a source of muon and tau neutrinos and antineutrinos in the models does not qualitatively alter the outcomes (i.e., lead to an explosion). Nonetheless, even with all of the feedbacks present in a supernova model, the new emissivity channel introduces significant quantitative changes in the model. For example, in Figure 22, between 150 and 200 ms after bounce, the shock radius is nearly 50 km less in the case where electron neutrino pair annihilation is included.

In addition to the realization that new neutrino interaction channels may be important in core collapse supernovae, such as in the cases mentioned above, significant improvements have been made recently in the computation of the weak interaction rates traditionally used in supernova models. The inclusion of nucleon recoil, Fermi blocking (a form of correlation), and degeneracy in the charged- and neutral-current interactions of neutrinos on free nucleons is one example (85–88). These improvements led to significant changes in the neutrino emissivities and opacities (factors of 2–3) and the realization that energy exchange in neutrino-nucleon scattering in the pre- and post-bounce core cannot be ignored (89).

Generally, the naive picture of neutrinos interacting on free nucleons and nuclei is now being replaced by the more accurate picture of neutrinos interacting with correlated nuclei and nucleons at sub- and super-nuclear density. Calculations of neutrino interactions on nuclei during stellar core collapse that include correlations among nuclei (90), on the extended correlated structures present in the nuclear "pasta" phase during the transition from nuclei to nuclear matter (91), and on the strongly interacting correlated nucleons in the proto-neutron star (85–88) have either been completed or are well underway [for a review, the reader is referred to (92, 93)]. Nonetheless, more remains to be done, and the calculations will be increasingly challenging.

However, although the overall neutrino emissivities and opacities can be dramatically altered when either new channels are included or improved rates are calculated, the impact of new and/or improved neutrino emissivities and opacities on the proto-neutron star evolution and supernova mechanism can only be determined when they are included in the postbounce neutrino radiation hydrodynamics models with all of the concommitant feedbacks. We cannot know a priori whether a large change in the weak interaction rates will lead to a large or small change in the models. Both scenarios are possible. Improved electron capture rates on nuclei led to significant changes in the models, and the inclusion of muon and tau neutrino production via electron neutrino–antineutrino pair annihilation led to moderate changes in the models. On the other hand, when ion-ion correlations during stellar core collapse (94) or correlations among nucleons in the proto-neutron star during the first $\sim$1 sec after bounce (86) are included, the changes in the models are negligible.

Finally, any of the above weak interaction rates that involve a nuclear force model—e.g., neutrino scattering on interacting nucleons at high densities in the stellar core before and after bounce—must be computed self-consistently with the nuclear equation of state (86), in which a nuclear force model is assumed.

7. EQUATION OF STATE

7.1. Energy Per Particle

Foundational to any determination of the equation of state of matter in the stellar core during core collapse and after core bounce, at both sub- and super-nuclear densities, is the determination of the energy per particle [for an overview of what we are about to discuss, the reader is referred to (95, 96)]. Given this quantity, the equation of state is readily derived from standard thermodynamics. However, our ability to compute this quantity in what is generally a strongly interacting environment is, of course, limited. Nonetheless, at present there are several fundamentally different approaches that have been adopted: (a) semi-empirical compressible liquid drop models, (b) phenomenological models, and (c) realistic models. Moreover, there are both nonrelativistic and relativistic versions for models in categories (b) and (c), in which a nuclear force model is assumed in the determination of V_{NN}, the nucleon-nucleon interaction potential.

In the compressible liquid drop model, the energy per particle is assumed to be expressible as a polynomial in the symmetry (neutron to proton ratio) and compressibility (density relative to the saturation density of symmetric nuclear matter) of the matter. The starting points are the empirically determined values of the symmetry energy and compressibility for symmetric nuclear matter at saturation. The compressible liquid drop model is, of course, appropriate only for sub-nuclear densities.

In the phenomenological models, a parameterized (e.g., Skyrme) nucleon-nucleon interaction potential is assumed, where the parameters are fit to reproduce observables such as ground state properties of nuclei, symmetric nuclear matter at saturation, neutron star masses and radii, etc. There are a total of 10–15 adjustable parameters in such a parameterization, fit to hundreds of data points. However, there are, for example, 90 different Skyrme parameterizations of the nucleon-nucleon interaction potential, although only about one third of them are able to reproduce neutron star properties. Phenomenological models are capable of describing both sub- and super-nuclear density matter.

In the realistic models, a nucleon-nucleon interaction potential with 40–60 adjustable parameters fit to several thousand data points from free nucleon-nucleon scattering and properties of the deuteron is assumed. However, when applied to dense, interacting matter, the potentials must obviously be renormalized using Brueckner-Hartree-Fock and Dirac-Brueckner-Hartree-Fock techniques in the nonrelativistic and relativistic cases, respectively. Moreover, realistic potentials are computed only for symmetric nuclear matter and pure neutron matter. Interpolation is then required to describe the matter present in the cores of supernovae and in neutron stars, which is neither symmetric nor purely neutronic. Although realistic models can be used to describe matter at super-nuclear densities, they cannot yet be used at sub-nuclear densities and, therefore, in a self-consistent description of matter in the stellar core at both sub- and super-nuclear densities during stellar core collapse.

Despite the uncertainties in our knowledge of the equation of state (EOS) of matter in stellar cores at both sub- and super-nuclear densities, the availability of such a diverse set of models allows us to explore the sensitivity of the supernova mechanism and quantitative predictions of supernova models to variations in this physics. Indeed, core collapse and postbounce simulations using the Lattimer-Swesty compressible liquid drop EOS (97), the Shen et al. relativistic (TM1) mean field EOS (98), and the Hillebrandt-Wolff nonrelativistic (Skyrme) mean field EOS (99) [the latter two equations of state fall under category (b) above] have been performed recently and yield, for example, three-flavor neutrino luminosities that differ by tens of percent between 50 and 150 ms after bounce (16), as shown in Figure 23. As we have discussed, variations in the neutrino luminosities of this magnitude are significant. In addition, in the same comparisons, the shock radii exhibited noticeable differences as a function of time in the models, as shown in Figure 24.

While qualitative changes in the models—e.g., the generation of an explosion versus a dud—did not occur when a diverse set of equations of state were used, significant quantitative changes in the model predictions did occur. Therefore,

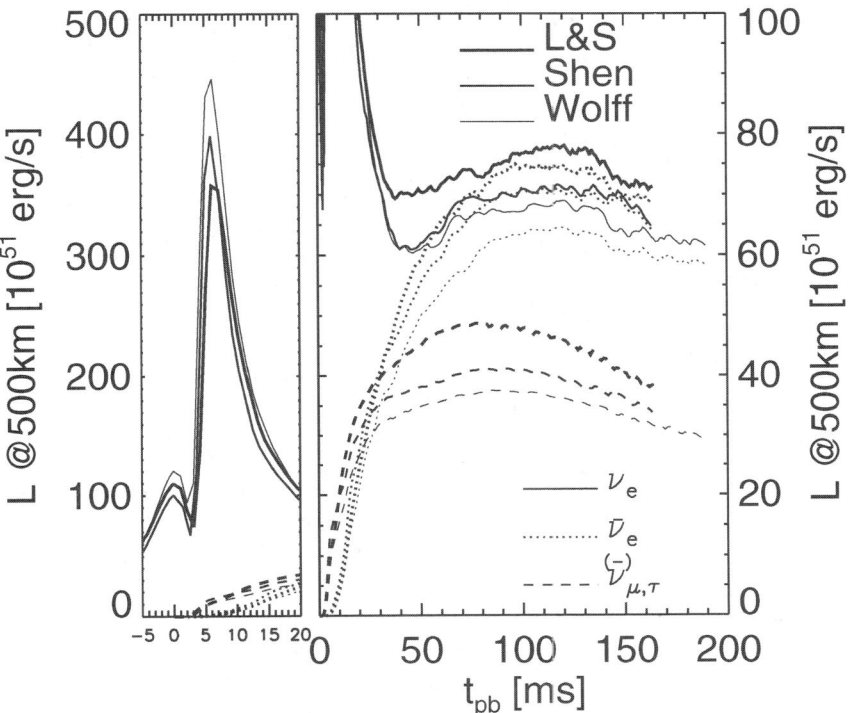

Figure 23   The electron, muon, and tau neutrino and antineutrino luminosities are plotted as a function of time for three simulations that each implement a different equation of state from the three (disparate) equations of state discussed in the text (16). Maximum differences of 15–30% in the neutrino luminosities are evident, depending on the flavor, as we change from one equation of state to another. As we discussed, such differences are dynamically significant for the shock neutrino reheating.

further development of the equation of state in core collapse supernovae, and such comparisons, must continue.

Two final points should be made before concluding this subsection. First, at 2–3 times the saturation density of symmetric nuclear matter, heavy baryons (e.g., hyperons), meson condensates (e.g., pion and kaon condensates), and possibly deconfined quarks may be present. Our discussions above, on the other hand, focused solely on different approaches at calculating the nucleon-nucleon interaction potential. The interactions between nucleons and the heavy baryons, and between the heavy baryons themselves, are not as well known (100), an uncertainty which also casts uncertainty on the densities at which we might expect these more exotic constituents to be present in our stellar cores. For a path forward, the reader is referred to (101). Second, during stellar core collapse at both sub- and super-nuclear

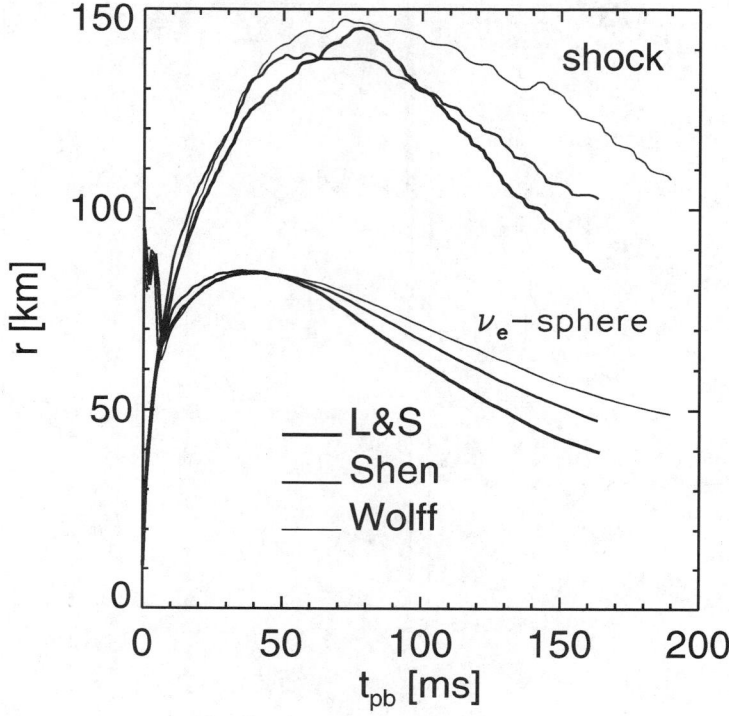

Figure 24   The shock and neutrinosphere radii as a function of time are plotted for three simulations that each implement a different equation of state from the three (disparate) equations of state discussed in the text (16). In particular, the spread in the shock radii at $t \sim 150$ ms after bounce is striking.

densities, the equation of state must be developed for temperatures that reach tens of MeV. To date, only a few equations of state have been developed for finite temperature (97–99, 102).

7.2. Inhomogeneous Matter

In all simulations of stellar core collapse thus far performed, a mean-nucleus approximation has been used in the determination of the sub-nuclear density equation of state. In this approximation, the thermodynamic state of the core is computed assuming the core is composed of a single representative nucleus, as opposed to the ensemble of nuclei actually present. Although the thermodynamic state at sub-nuclear densities is well approximated in such a mean nucleus approach, weak interactions on the nuclei actually present in the core will vary significantly from nucleus to nucleus and will, consequently, not be well approximated by computing the weak interactions on a mean nucleus. Therefore, hybrid computations using an

NSE mean-nucleus equation of state in combination with an NSE network must be performed. The progress detailed above in the implementation of improved electron capture rates on nuclei during stellar core collapse involved such a hybrid implementation (82, 81).

In addition to the use of a mean nucleus rather than an ensemble of nuclei, we also have to consider what restrictions we have placed in the past on the nature of these nuclei. In particular, in all of the industry standard equations of state used in supernova models performed to date (and discussed above), the nuclei are assumed to be spherical in shape and organized in a body-centered-cubic lattice during stellar core collapse. In the critical transition region between inhomogeneous matter composed of nuclei and nucleons and homogeneous nuclear matter—the so-called pasta phase—an ensemble of complex shapes are anticipated (spheres, tubes, sheets, etc.) (103). Clearly, the assumption that nuclei are spherical is not compatible with the fact that we expect such complex shapes. Three-dimensional models that abandon the restriction to spherical nuclei have been developed for $T = 0$ (104). Efforts are now underway to develop a three-dimensional equation of state at finite temperature (105), for both homogeneous (super-nuclear density) and inhomogeneous (sub-nuclear density) matter. Of course, this development must be accompanied by efforts to calculate the interactions of neutrinos on the complex structures that will be admitted by such an equation of state, such as was done in (91).

8. GRAVITY

One should expect the gravitational fields around proto-neutron stars to deviate significantly from their Newtonian values and lead to significant changes in the collapse and postbounce hydrodynamics and neutrino transport. Detailed comparisons of Newtonian and fully general relativistic collapse have been performed in spherical symmetry (12, 13), and they confirm this expectation.

General relativistic effects can be expected to substantially modify the hydrodynamics of the core at high densities. An extreme example of this, of course, is the possibility that in the general relativistic limit one can have continued collapse and the formation of an event horizon. The neutrino transport will also be modified by general relativity, directly through gravitational redshift and aberration, and indirectly through its strong coupling to the general relativistic modified hydrodynamics.

The results from models beginning with 15 and 25 $M_\odot$ progenitors show that (a) the neutrinosphere, gain, and shock radii and (b) the infall velocities can differ by as much as a factor of 2 between the Newtonian and general relativistic models (12). This is evident in Figures 25 and 26.

Considering now the neutrino quantities: Switching from Newtonian to general relativistic hydrodynamics increases the luminosity and rms energy of all neutrino flavors during the shock reheating epoch (12, 13). This arises because of the

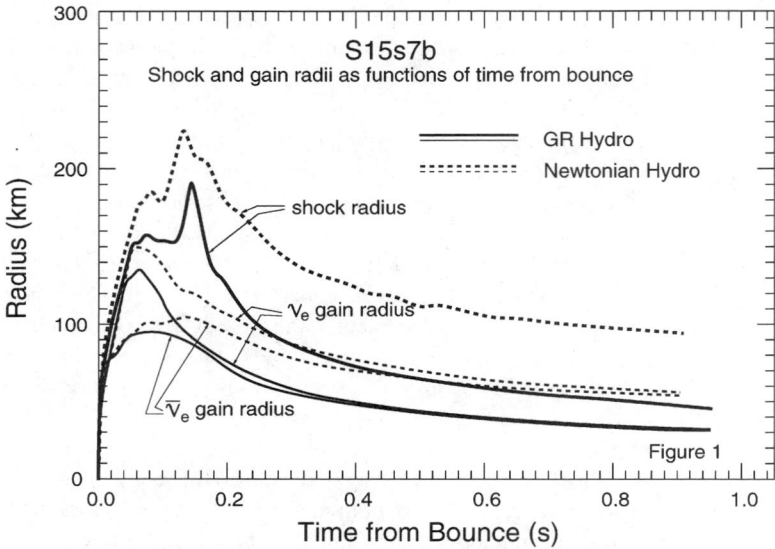

Figure 25 The shock, electron neutrinosphere, and electron antineutrinosphere radii as a function of time for two simulations, one Newtonian and one fully general relativistic, initiated from a 15 $M_\odot$ progenitor. Owing to the increased gravitational field, the postbounce stellar core configuration is far more compact in the general relativistic case for postbounce times greater than ~200 ms (12).

more compact core structures that develop with general relativistic hydrodynamics, which yield neutrinospheres at smaller radii and higher temperatures. Switching from Newtonian to general relativistic transport reduces the luminosities and rms energies of all three neutrino flavors during the shock reheating epoch because of the gravitational redshift of the neutrinos as they propagate out to large radii (12). Generally, the reduction in neutrino luminosities and rms energies when switching from Newtonian to general relativistic transport does not fully compensate for the increase in these quantities when switching from Newtonian to general relativistic hydrodynamics. Therefore, the net effect in switching from a fully Newtonian to a fully general relativistic simulation in most cases is an increase in both the luminosities and rms energies of neutrinos of all flavors during the shock reheating epoch (12, 13). This is evident in Figure 5.

It is important to note the complex feedbacks active during stellar core collapse and the postbounce evolution that are made evident through these detailed comparisons. It would be simplistic to assume that general relativity, in increasing the gravitational potential and, hence, the compactness of the postbounce stellar core configuration and in leading to the gravitational redshift of neutrinos, would lead to a postbounce state far less conducive to explosion. While the compactness is increased by including general relativistic gravity in the models, so too are the

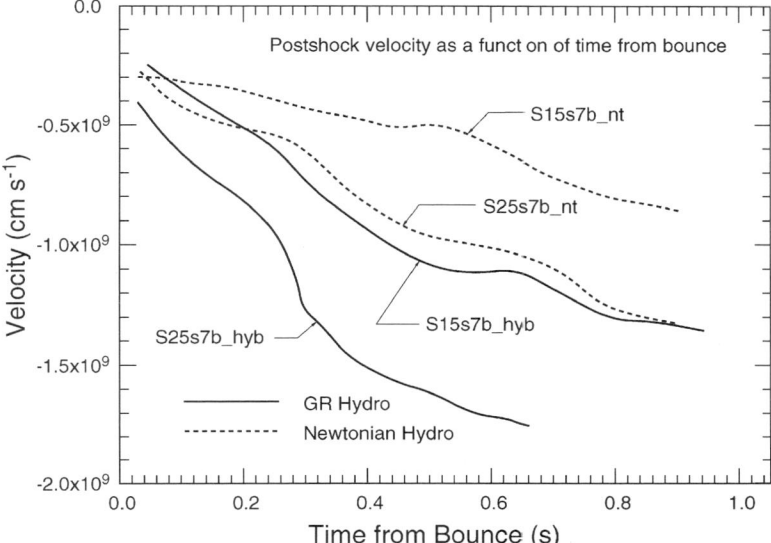

Figure 26 The postshock velocity is plotted as a function of time for Newtonian and general relativistic models initiated from 15 and 25 $M_\odot$ progenitors. The postshock velocity in the general relativistic case can be as much as a factor 2 larger (12).

neutrino luminosities and rms energies. It remains to be seen, in the case of a complete model with general relativity and explosion, what final role general relativity will play (13).

In either event, the results from these complete spherically symmetric models strongly suggest that Newtonian multidimensional models can be viewed only as a stepping stone to fully general relativistic multidimensional models.

9. NEUTRINO FLAVOR TRANSFORMATION

Neutrinos have mass and, therefore, mix in flavor. Observations of solar and atmospheric neutrinos and experiments at LSND indicate there may be as many as three independent values of the difference in the square of the neutrino masses (δm^2) and four mixing angles, which would require three active and at least one sterile neutrino. Although we await confirmation (or not) of the LSND findings, the data already strongly suggest that neutrino mixing should be included in core collapse supernova models. Neutrino mixing may significantly affect one or more of the following: the supernova mechanism, supernova nucleosynthesis, and terrestrial supernova neutrino detection.

As we discussed earlier, all three active neutrino flavors are involved in core collapse supernova dynamics. Electron, muon, and tau neutrinos and their antineutrinos are produced primarily through thermal emission, nucleon-nucleon bremsstrahlung, and neutral-current neutrino-neutrino annihilation in the hot mantle of the proto-neutron star after core bounce. Owing to the lack of charged-current interactions among the muon and tau neutrinos and antineutrinos, electron neutrinos and antineutrinos decouple at lower densities given their larger total interaction cross sections. Decoupling at lower densities and, consequently, lower temperatures result in softer relative spectra for the electron flavor neutrinos (106). Herein lies the essential relevance of neutrino mixing to the supernova mechanism: If flavor conversion between electron flavor and muon/tau flavor neutrinos were to occur below the supernova shock wave in the neutrino heating epoch after stellar core bounce, the neutrino heating behind the shock, which is mediated predominantly by the charged-current absorption of electron neutrinos and antineutrinos, could be significantly increased (107, 108). The softer electron neutrino flavor spectra would be replaced by the harder muon/tau neutrino flavor spectra in this region.

Even if neutrino flavor conversion did not occur deep in the stellar core—in or above the proto-neutron star after core bounce—and, therefore, was not a major factor in initiating the explosion, our ability to use the next Galactic supernova neutrino detection to improve supernova models, to better understand the complex dynamics of core collapse supernovae, and to cull fundamental nuclear and particle physics from it would be severely compromised if we did not have a way to predict the flavor conversion that could occur between the source supernova and terrestrial detectors.

The argument to consider neutrino mixing in the core collapse supernova context, particularly with an eye toward the explosion mechanism, has been made even more compelling recently with the discovery that neutrino mixing may occur deep in the stellar core after bounce at small values of δm^2 (109, 110). This mixing may arise as a result of the neutrino background in and above the proto-neutron star, which would be fundamentally different than MSW (111) mixing induced by the matter background or vacuum mixing. Neutral-current neutrino-neutrino forward scattering increases the neutrino effective mass, much as charged-current electron-neutrino scattering increases the electron-neutrino effective mass in the MSW case. The net result may be near maximal mixing of neutrino flavors in the environment of the proto-neutron star after bounce (109, 110), with obvious ramifications for the supernova mechanism, nucleosynthesis, and neutrino signatures. Pantaleone (112, 113) was the first to recognize that a background neutrino, in a superposition of flavor states, would lead to an off-diagonal refractive index in flavor space if neutrino-neutrino interactions are taken into account. This idea was then further developed by Sigl and Raffelt (114).

Matter-enhanced neutrino mixing has been included in an approximate way in past (spherically symmetric) models of core collapse supernovae using a simple Landau-Zener prescription (107, 108, 115). In such a prescription, the probability of mixing in the resonance region as a function of neutrino energy is computed, and the commensurate fraction of neutrinos is simply converted from one flavor to the other. This prescription is easily implemented in spherically symmetric models, where a resonance region is a spherical shell.

Although the experimental evidence for neutrino mixing is now clear, and although a number of past exploratory studies have elucidated some of the possible ramifications neutrino mixing may have for core collapse supernova dynamics, the precise impact of such flavor transformation remains to be determined. Neutrino mixing is a coherent, quantum mechanical phenomenon, unlike the incoherent collisional phenomena included in the Boltzmann kinetic equations discussed earlier. A more complete (quantum kinetic) treatment of neutrino transport in stellar cores beyond (classical) Boltzmann transport will be needed if we are to accurately and fully explore the impact of neutrino mixing on core collapse supernova dynamics.

10. CONCLUSION

Three-dimensional, general relativistic, radiation magnetohydrodynamics models of core collapse supernovae must be developed before we can be confident that we truly understand the supernova mechanism and can accurately predict all of the associated supernova observables. This will take a decade or more of systematic research. We currently do not have models of this level of sophistication even in two spatial dimensions, let alone three, and three-dimensional models with multiangle, multifrequency neutrino transport will likely require peta-scale supercomputers, which will not be available until 2009 or later.

Nonetheless, there is much work to be done on the road forward. We have no two-dimensional models that implement two-dimensional multifrequency neutrino transport and all of the relevant terms in the transport equations—i.e., realistic two-dimensional transport. There are no simulations, two- or three-dimensional, that include magnetic fields and realistic neutrino transport. And all of the two- and three-dimensional models performed to date are largely confined to the Newtonian limit. Moreover, the precise implications of neutrino mixing are essentially unexplored in any supernova model.

Turning to the input nuclear and weak interaction physics, nuclear structure theory and the theory of nuclear matter in supernovae have certainly been challenged by the requirements of realistic supernova models, as we have discussed. Responding to these challenges will require a concomitant long-term effort on the part of the nuclear physics community to advance nuclear structure models for heavy nuclei and to pin down the high-density equation of state with sufficient accuracy, where what constitutes "sufficient" is yet to be fully determined.

But despite the long road ahead, this is a time for optimism, not pessimism— much progress has been made. General relativistic spherically symmetric models with multiangle, multifrequency neutrino transport have finally been developed, and far more realistic two-dimensional Newtonian models (with two-dimensional multifrequency neutrino transport) are within reach. Moreover, two-dimensional simulations that will soon include magnetic fields will begin to fill voids that currently exist in supernova theory. Myriad simulations that have been performed to date demonstrate the need for continued study and continued improvements in our treatment of the physics we have included thus far in the models, to improve the models quantitatively. Once explosions become routine in core collapse supernova models, the models will be gauged by their ability to make accurate predictions of: (a) the synthesis of the elements; (b) the emission of neutrinos, gravitational waves, and gamma rays; (c) neutron star kicks; and (d) spectropolarimetry, etc. Our ability to use core collapse supernovae as laboratories will depend on our ability to predict supernova observables with precision.

Looking back at all of the past results we have in the supernova literature, we must be guided at the moment by two rules of thumb: (a) How we model each physics component in a supernova model does matter. The lack of robust explosions in any model that implements multifrequency, and hence more realistic, neutrino transport and the significant quantitative changes we have witnessed in past models as other physics components have been improved are a testament to this. (b) Even if we obtain explosions in, for example, two-dimensional Newtonian models with some set of the relevant physics and with this physics modeled realistically, history tells us that the addition of new physics or the switch to three dimensions may change the outcomes in these models qualitatively. Therefore, we cannot know the supernova mechanism is in hand until all components of the model are included with sufficient realism and the models are performed in three dimensions.

The stakes are high, and so we must forge ahead, patiently and systematically.

ACKNOWLEDGMENT

A.M. is supported at the Oak Ridge National Laboratory, which is managed by UT-Battelle, LLC for the DOE under contract DE-AC05-00OR22725. He is also supported in part by a DOE Scientific Discovery through Advanced Computing grant. A.M. acknowledges many invaluable discussions with colleagues, especially John Blondin, Steve Bruenn, Christian Cardall, George Fuller, Raph Hix, Matthias Liebendörfer, Bronson Messer, Jirina Stone, and Michael Strayer. He also grate-fully acknowledges permission from Chris Fryer, Thomas Janka, Eli Livne, Ewald Müller, and Craig Wheeler to use figures from their published work.

**The *Annual Review of Nuclear and Particle Science* is online at
http://nucl.annualreviews.org**

LITERATURE CITED

1. Wilson JR. In *Numerical Astrophysics*, ed. JM Centrella, JM LeBlanc, RL Bowers, p. 422. Boston: Jones & Bartlett (1985)
2. Bethe HA, Wilson JR. *Astrophys. J.* 295:14 (1985)
3. Janka HT. *Astron. Astrophys.* 368:527 (2001)
4. Burrows A, Goshy J. *Astrophys. J. Lett.* 416:L75 (1993)
5. Janka HT, Müller E. *Astron. Astrophys.* 306:167 (1996)
6. Messer OEB, Mezzacappa A, Bruenn SW, Guidry MW. *Astrophys. J.* 507:353 (1998)

7. Mezzacappa A, et al. *Astrophys. J.* 495:911 (1998)
8. Mezzacappa A, et al. *Phys. Rev. Lett.* 86:1935 (2001)
9. Wilson JR, Mayle RW. *Phys. Rep.* 227:97 (1993)
10. Swesty F, Lattimer J. *Astrophys. J.* 425:195 (1994)
11. Rampp M, Janka HT. *Astrophys. J.* 539: L33 (2000)
12. Bruenn SW, DeNisco KR, Mezzacappa A. *Astrophys. J.* 560:326 (2001)
13. Liebendörfer M, et al. *Phys. Rev. D* 63: 103004 (2001)
14. Thompson TA, Burrows A, Pinto PA. *Astrophys. J.* 592:434 (2003)
15. Buras R, Rampp M, Janka HT, Kifonidis K. *Phys. Rev. Lett.* 90:241101 (2003)
16. Janka HT, et al. In *Supernovae*, ed. J Marcaide, K Weiler, p. 253. Berlin: Springer-Verlag (2004)
17. Herant M, Benz W, Colgate SA. *Astrophys. J.* 395:642 (1992)
18. Herant M, et al. *Astrophys. J.* 435:339 (1994)
19. Burrows A, Hayes J, Fryxell BA. *Astrophys. J.* 450:830 (1995)
20. Fryer CL, Heger A. *Astrophys. J.* 541:1033 (2000)
21. Fryer CL, Warren MS. *Astrophys. J.* 574: L65 (2002)
22. Fryer CL, Warren MS. *Astrophys. J.* 601: 391 (2004)
23. Akiyama S, Wheeler JC, Meier DL, Lichtenstadt I. *Astrophys. J.* 584:954 (2003)
24. Symbalisty EMD. *Astrophys. J.* 285:729 (1984)
25. LeBlanc JM, Wilson JR. *Astrophys. J.* 161: 541 (1970)
26. Wheeler JC, Meier DL, Wilson JR. *Astrophys. J.* 568:807 (2002)
27. Balescu R. In *Equilibrium and Nonequilibrium Statistical Mechanics*. New York: Wiley-Intersci. (1975)
28. Van Riper KA, Lattimer JM. *Astrophys. J.* 249:270 (1981)
29. Baron EA, Cooperstein J, Kahana S. *Nucl. Phys.* A440:744 (1985)
30. Hillebrandt W, Nomoto K, Wolff R. *Astron. Astrophys.* 133:175 (1984)
31. Cooperstein J, Van denHorn LJ, Baron EA. *Astrophys. J.* 309:653 (1986)
32. Arnett WD. *Astrophys. J.* 218:815 (1977)
33. Bowers RL, Wilson JR. *Astrophys. J. Suppl.* 50:115 (1982)
34. Bruenn SW. *Astrophys. J. Suppl.* 58:771 (1985)
35. Myra ES, et al. *Astrophys. J.* 318:744 (1987)
36. Levermore CD, Pomraning GC. *Astrophys. J.* 248:321 (1981)
37. Mihalas D, Mihalas B. In *Foundations of Radiation Hydrodynamics*. New York: Oxford Univ. Press (1975)
38. Burrows A, et al. *Astrophys. J.* 539:865 (2000)
39. Rampp M, Janka HT. *Astron. Astrophys.* 396:361 (2002)
40. Liebendörfer M, Rampp M, Janka HT, Mezzacappa A. *Astrophys. J.* 620:840 (2005)
41. Liebendörfer M, et al. *Phys. Rev. D* 63: 103004 (2001)
42. Mezzacappa A, Bruenn SW. *Astrophys. J.* 405:669 (1993)
43. Liebendörfer M, et al. *Astrophys. J. Suppl.* 150:263 (2004)
44. Mezzacappa A, et al. In *Stellar Collapse*, ed. C Fryer, p. 99. Dordrecht: Kluwer Acad. (2004)
45. Cardall CY, Mezzacappa A. *Phys. Rev. D* 68:023006 (2003)
46. Cardall CY, Lentz E, Mezzacappa A. *Phys. Rev. D* (2005) In press
47. Cardall CY. In *Numerical Methods for Multidimensional Radiative Transfer Problems.* ed. R Rannacher, R Wehrse. Berlin: Springer-Verlag (2004)
48. Livne E, et al. *Astrophys. J.* 609:277 (2004)
49. Mezzacappa A, Bruenn SW. *Astrophys. J.* 405:637 (1993)
50. Mezzacappa A, Bruenn SW. *Astrophys. J.* 410:740 (1993)
51. Mezzacappa A, Messer O. *Astrophys. J.* (2005) In press
52. Marek A, et al. astro-ph/0502161 (2005)

53. Kifonidis K, Plewa T, Janka HT, Müller E. *Astron. Astrophys.* 408:621 (2003)
54. Quirk J. *Int. J. Num. Methods Fluids* 18:555 (1994)
55. Blondin J. In *Open Issues in Core Collapse Supernova Theory*, ed. A Mezzacappa, G Fuller. Singapore: World Sci. (2005) In press
56. Bruenn S, Raley E, Mezzacappa A. *Astrophys. J.* (2005) In press
57. Miralles JA, Pons JA, Urpin VA. *Astrophys. J.* 574:356 (2002)
58. Smarr L, Wilson JR, Barton RT, Bowers RL. *Astrophys. J.* 246:515 (1981)
59. Wilson JR, Mayle RW. *Phys. Rep.* 163:63 (1988)
60. Blondin JM, Mezzacappa A, DeMarino C. *Astrophys. J.* 584:971 (2003)
61. Janka HT, et al. *Nucl. Phys. A.* (2005) In press
62. Blondin J, Mezzacappa A. *Astrophys. J.* (2005) Submitted
63. Wang L, Howell DA, Höflich P, Wheeler JC. *Astrophys. J.* 550:1030 (2001)
64. Blondin J, Mezzacappa A. *Nature* (2005) Submitted
65. Thompson TA, Quataert E, Burrows A. *Astrophys. J.* 620:861 (2005)
66. Duan H, Qian Y. *Phys. Rev. D* 69:123004 (2004)
67. Balbus S, Hawley J. *Astrophys. J.* 376:214 (1991)
68. Heger A, Woosley S, Spruit H. *Astrophys. J.* 626:350 (2005) Submitted
69. Fuller GM, Fowler WA, Newman MJ. *Astrophys. J. Suppl.* 42:447 (1980)
70. Fuller GM, Fowler WA, Newman MJ. *Astrophys. J.* 252:715 (1982)
71. Fuller GM. *Astrophys. J.* 252:741 (1982)
72. Fuller GM, Fowler WA, Newman MJ. *Astrophys. J. Suppl.* 48:279 (1982)
73. Fuller GM, Fowler WA, Newman MJ. *Astrophys. J.* 293:1 (1985)
74. Pruet J, Fuller GM. *Astrophys. J. Suppl.* 149:189 (2003)
75. Yahil A, Lattimer JM. In *NATO ASIC Proc. 90: Supernovae: A Survey of Current Research,* ed. MJ Rees, RS Stoneham, p. 53. Dordrect: Reidel. (1982)
76. Bethe H, Brown G, Applegate J, Lattimer J. *Nucl. Phys.* A324:487 (1979)
77. Langanke K, Kolbe E, Dean DJ. *Phys. Rev. C* 63:032801 (2001)
78. Cooperstein J, Baron E. In *Supernovae*, ed. A Petschek, p. 213. New York: Springer-Verlag (1990)
79. Messer O, Hix W, Liebendörfer M, Mezzacappa A. *Astrophys. J.* (2005) Submitted
80. Cooperstein J, Wambach J. *Nucl. Phys.* A420:591 (1984)
81. Hix WR, et al. *Phys. Rev. Lett.* 91:201102 (2003)
82. Langanke K, et al. *Phys. Rev. Lett.* 90: 241102 (2003)
83. Hannestad S, Raffelt G. *Astrophys. J.* 507: 339 (1998)
84. Buras R, et al. *Astrophys. J.* 587:320 (2003)
85. Reddy S, Prakash M, Lattimer JM. *Phys. Rev. D* 58:013009 (1998)
86. Reddy S, Prakash M, Lattimer JM, Pons JA. *Phys. Rev. C* 59:2888 (1999)
87. Burrows A, Sawyer RF. *Phys. Rev. C* 58: 554 (1998)
88. Burrows A, Sawyer RF. *Phys. Rev. C* 59: 510 (1999)
89. Thompson TA, Burrows A, Horvath JE. *Phys. Rev. C* 62:035802 (2000)
90. Horowitz CJ. *Phys. Rev. D* 55:4577 (1997)
91. Horowitz CJ, Pérez-García MA, Piekarewicz J. *Phys. Rev. C* 69:045804 (2004)
92. Burrows A, Thompson T. In *Stellar Collapse*, ed. C Fryer, p. 133. Dordrecht: Kluwer Acad. (2004)
93. Burrows A, Reddy S, Thompson T. *Nucl. Phys. A.* (2005) In press
94. Bruenn SW, Mezzacappa A. *Phys. Rev. D* 56:7529 (1997)
95. Rikovska Stone, et al. *Phys. Rev. C* 68: 034324 (2003)
96. Stone J. In *Open Issues in Core Collapse Supernova Theory*, ed. A Mezzacappa, G Fuller. Singapore: World Sci. (2005) In press
97. Lattimer J, Swesty FD. *Nucl. Phys.* A535:331 (1991)

98. Shen H, Toki H, Oyamatsu K, Sumiyoshi K. *Nucl. Phys.* A637:435 (1998)
99. Hillebrandt W, Wolff RG. In *Nucleosynthesis: Challenges and New Developments*, ed. D Arnett, J Truran, p. 131. Chicago: Univ. Chicago Press (1985)
100. Heiselberg H, Pandharipande V. *Annu. Rev. Nucl. Part. Sci.* 50:481 (2000)
101. Barnes F. In *Open Issues in Core Collapse Supernova Theory*, ed. A Mezzacappa, G Fuller. Singapore: World Sci. (2005) In press
102. Onsi M, Przysiezniak H, Pearson JM. *Phys. Rev. C* 55:3139 (1997)
103. Ravenhall DG, Pethick CJ, Wilson JR. *Phys. Rev. Lett.* 50:2066 (1983)
104. Magierski P, Heenen PH. *Phys. Rev. C* 65:045804 (2002)
105. Newton W. PhD thesis. Oxford Univ. Press (2005) In press
106. Burrows A, Thompson T. In *Stellar Collapse*, ed. C Fryer. Dordrecht: Kluwer Acad. (2004)
107. Fuller GM, Mayle R, Wilson JR, Schramm D. *Astrophys. J.* 322:795 (1987)
108. Fuller GM, Mayle R, Meyer BS, Wilson JR. *Astrophys. J.* 389:517 (1992)
109. Qian YZ, Fuller G. *Phys. Rev. D* 52:656 (1995)
110. Fuller G, Qian YZ. *Phys. Rev.* (2005) Submitted
111. Mikheyev SP, Yu A. *Sov. J. Nucl. Phys.* 42:913 (1985)
112. Pantaleone J. *Phys. Lett.* B287:128 (1992)
113. Pantaleone J. *Phys. Lett.* B342:250 (1995)
114. Sigl G, Raffelt G. *Nucl. Phys.* B406:423 (1993)
115. Mezzacappa A, Bruenn SW. In *The Identification of Dark Matter*, ed. N Spooner, V Kudryavtsev, p. 655. Singapore: World Sci. (1999)

Annu. Rev. Nucl. Part. Sci. 2005. 55:517–54
doi: 10.1146/annurev.nucl.53.041002.110533
Copyright © 2005 by Annual Reviews. All rights reserved

Direct Photon Production in Relativistic Heavy-Ion Collisions

Paul Stankus

Oak Ridge National Laboratory, Oak Ridge, Tennessee 37831;
email: stankus@mail.phy.ornl.gov

Key Words   thermal, QCD, QGP, radiation, plasma

■ **Abstract**   We examine the uses of direct photons in diagnosing the highly excited state of nuclear matter created in high-energy nuclear collisions. The traditional focus has been on direct photons as thermal radiation from the excited state, but we also explore the many other roles direct photons can play. We review experimental and theoretical techniques as well as the history of direct photon measurements in heavy-ion collisions and their interpretation.

CONTENTS

1. INTRODUCTION

Since its beginnings in the mid-1980s, a prime goal of the experimental program of colliding nuclei at high energies has been to create—and then study—highly excited, strongly interacting (i.e., hadronic or nuclear) matter over an extended region of space and time. Interest had been raised since the mid-1970s that such a region, if it could be created in the laboratory, might show interesting properties, and these in turn would elucidate the nature of strong interactions in a nonperturbative regime. With a description in terms of QCD, widely taken to be the fundamental theory of strong interactions, the highly excited state of nuclear/hadronic matter is now generally referred to as a quark-gluon plasma, or QGP.

Well in advance of experiments, it was proposed that electromagnetic radiation would be a useful diagnostic of the QGP state. In 1976, for example, Feinberg (1) suggested that copious photon production would be a distinct feature of highly excited hadronic matter, and speculated that the number of photons could even exceed the number of hadrons in the final state after a QGP had cooled and decayed. The basic argument was straightforward: because electromagnetic interactions are much weaker than strong interactions, an hadronic system should be essentially transparent to photons, and so a large and long-lived QGP can radiate photons into the final state from throughout its entire volume and over its entire lifetime. Models of QGP radiation have been greatly refined over the last several decades, but the essential point has held true, that final-state photons are interesting because they can carry information from the entire history of a QGP's evolution (2, 3).

1.1. What Is Meant by "Direct"?

Observing direct radiation from a QGP state has been a central part of the motivation to measure photon production in relativistic heavy-ion collisions. However, direct photons of other kinds can also play important and interesting roles in diagnosing a QGP state. We explore the full palette of opportunities in this review. A sensible place to start is by defining the term direct photon, along with the related, but not synonymous, terms prompt photon, isolated photon, and thermal photon.

Sensibly enough, the term direct photon is used to indicate photons that emerge directly from a particle collision; they are distinguished from decay photons, which emerge as the daughters of long-lived secondaries which decay electromagnetically, such as $\pi^0 \to \gamma\gamma$ or $\Sigma^0 \to \Lambda\gamma$. For photons of interest produced in high-energy hadron collisions, i.e., hard gamma rays with energies $E > \sim 100$ MeV, the large majority will typically be decay photons, and so sifting out the direct photons is always an experimental challenge (see Section 3 for details). In Figure 1 we illustrate the basic suite of processes which can create direct photons in hadron collisions.

Direct photons can be created in hard scatterings of incoming partons. Panel A of Figure 1 shows the two perturbative QCD (pQCD) diagrams that produce final-state photons at lowest order: the gluon-photon Compton process (upper) and quark-antiquark annihilation (lower). A quark can also radiate photons after its initial hard scattering as part of the jet fragmentation process; this is depicted in the upper diagram of Panel B of Figure 1. To lowest order these are all the relevant processes for direct photon production in elementary hadron-hadron collisions (e.g., p + p or π + p) taking place in vacuum. Their rates can in principle be calculated in pQCD and so they are sometimes called pQCD photons (also prompt photons; see below).

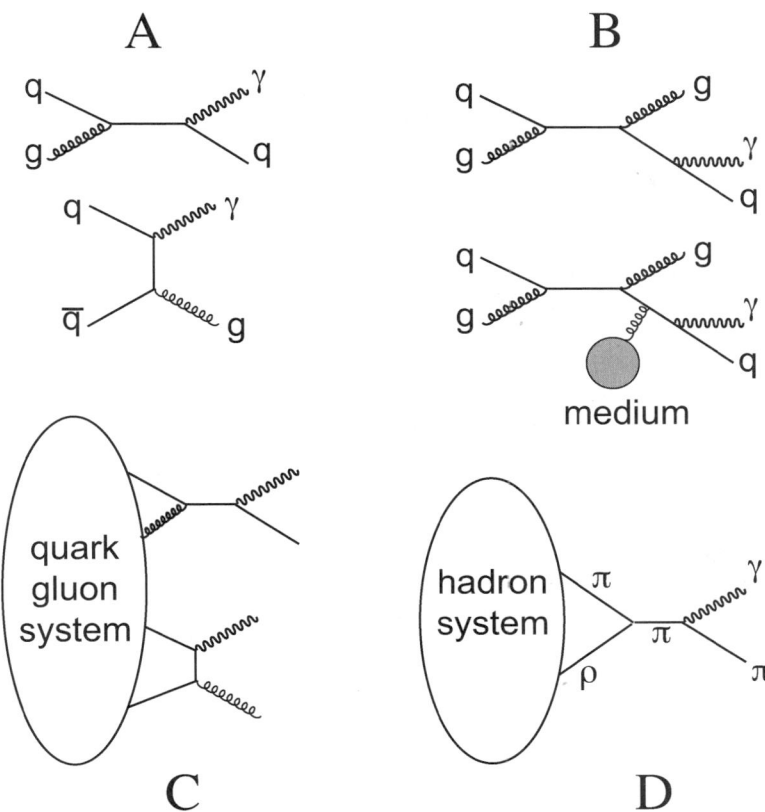

Figure 1 Illustration of the variety of processes which produce direct photons in hadron collisions. (A) Scattering between incoming partons. (B) Photons radiated by outgoing scattered partons, as part of the jet fragmentation process. (C) Scattering between quarks and gluons from a multi-collisional quark/gluon system. (D) Scattering between hadrons from a hadron system.

When a dense strongly interacting medium is present, however, more possibilities are opened up. If a scattered quark is traveling and fragmenting in a space-time co-occupied by such a medium, then photon radiation can be induced by parton-medium interactions, as in the lower diagram of Panel B of Figure 1. When a QCD medium is sufficiently hot, dense, and extended, then its constituents can scatter off each other and produce photons in the process. If the constituents are quarks and gluons, as shown in Panel C, then the lowest-level diagrams are the same as in Panel A. If the appropriate degrees of freedom in the medium are hadrons, then direct photons can be produced through interactions such as $\pi + \rho \rightarrow \pi + \gamma$, shown in Panel D. For a medium in thermal equilibrium, whether quark-gluonic, or hadronic (or other), its radiation is termed thermal photons. For a medium which is self-interacting but has not yet reached equilibrium, its radiation could be termed pre-equilibrium photons, though the term is not yet in common use.

Now, a few more words on terminology. The final state of diagrams like those in Panel A of Figure 1 is a single direct photon recoiling against a jet. Photons in such a final state are termed isolated photons, or sometimes single photons, because the scattering produces no hadronic energy in a direction near the photon's. By contrast, photons from the processes in Panel B would always have hadrons accompanying the photon and are termed fragmentation photons or bremsstrahlung photons (and, in some older literature, anamolous direct photons). In high-energy physics it has been traditional to measure, and to treat theoretically, isolated photons as objects of primary interest, to the extent that isolated and direct are used almost interchangeably in some high-energy literature. In relativistic heavy-ion collisions, however, the final-state phase-space density of all hadrons is so high that measuring isolated photons is a practical impossibility, and so they are rarely discussed.

The term prompt photon is commonly heard, but its usage is not perfectly uniform. In the simplest definition, prompt refers to photons created on short time scales and so is synonymous with direct. In the more common usage, which we adopt in this review, prompt is used to distinguish photons from the processes in Panels A and B, which can be traced directly to incoming partons, from the processes in Panels C and D, which happen later in the evolution of a collision. This makes prompt photons equivalent to what we termed pQCD photons above. We can then state neatly that direct photons are the sum of prompt, pre-equilibrium, and thermal components.

1.2. Structure of This Review

All the varieties of direct photons described here have relevance to the diagnosis of a highly excited hadronic medium, or QGP, being created in relativistic heavy-ion collisions. In Section 2 we elaborate on each of these sources and its utility. In Section 3 we detail experimental techniques and review results of direct photon measurements in heavy-ion collisions, including both fixed-target and collider experiments. We also consider what lies ahead for the current programs.

Direct photon production in heavy-ion collisions is a broad subject with a substantial history, and we cannot hope to cover everything of relevance in a single article. The intent of this review is to be instructive, rather than encyclopedic, and we apologize to those whose work we could not find space to include or have overlooked in error. For more technically oriented reviews, see References 4–6. For useful, recent short summaries see References 7–9.

2. SOURCES AND MOTIVATIONS

2.1. Prompt pQCD Processes

To understand the roles that prompt pQCD photons can play in relativistic heavy-ion collisions, it is useful first to picture a collision's geometry. In Figure 2, we

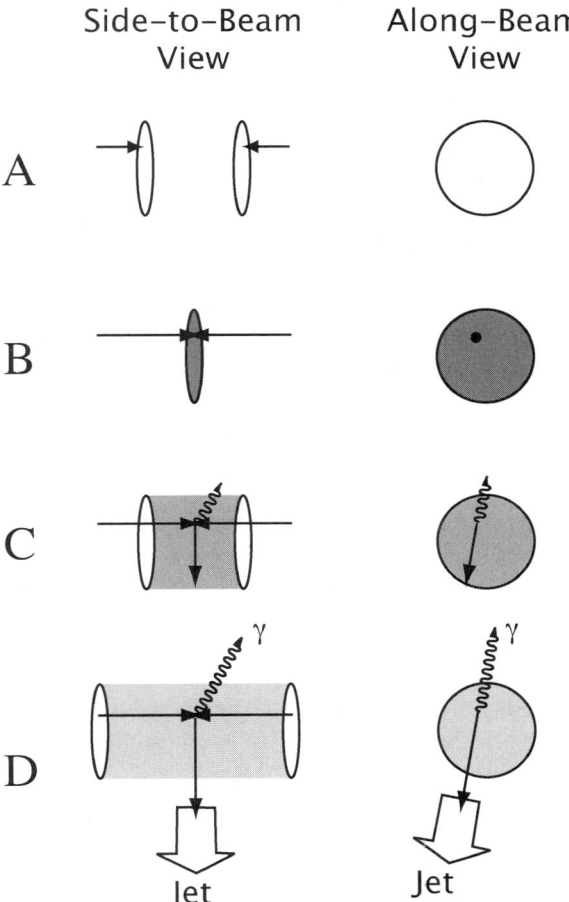

Figure 2   The geometry of a hard parton scattering to a jet+photon final state, taking place within a heavy-ion collision. (A) The incoming nuclei, before the collision, shown with Lorentz contraction along the beam direction. The arrows depict the paths of two partons within the nuclei. (B) At the point of collision the incoming nuclei overlap, and the partons undergo a hard scattering. The dark shade of the overlap region indicates a high energy density in an excited region. (C) As the remnants of the nuclei recede, the excited region expands and its energy density decreases, as indicated by the lighter shade. During the same interval the photon and scattered parton travel through the medium. (D) The medium continues to expand and cool. By this time the photon and the parton have exited the medium; in the final state the parton will be resolved as a jet of hadrons.

show a schematic picture of the time evolution of a heavy-ion collision in which the nuclei collide head-on and a hard parton-parton scattering takes place, which generates a jet and a photon. The scattering shown here is of the type in Panel A of Figure 1, where the jet (either quark or gluon) is balanced against a single photon. We can see that both the photon and the scattered parton, which later becomes a jet, must propagate through the QCD medium created by the collision.

In this example the photon is likely to emerge unscathed by the medium, while the parton/jet could have a significant interaction if the medium is sufficiently dense. Indeed, there is strong evidence that such dense media are created in heavy ion collisions at RHIC[1] energies. One of the most intriguing early results reported by the RHIC experiments was that of jet quenching: the yield of high-p_T hadrons in Au+Au collisions was greatly depleted, by a factor of up to 5, compared to a nominal scaling expectation (see below). Because jet fragmentation is thought to be the primary source of high-p_T hadrons, this depletion has been interpreted as strong parton/jet–medium interactions.[2]

The phenomenon of jet quenching has great potential as a diagnostic of the dense hadronic medium created in these collisions. Such interpretation is not without ambiguities, but several of these can be resolved with the help of direct photons.

2.1.1. PARTON FLUX, NUCLEAR STRUCTURE FUNCTIONS   If we know the yield for some parton-parton hard scattering process in a p+p (or a nucleon+nucleon, hereafter N+N) collision, then how do we predict the yield for that process in a nuclear collision? The simplest approach, sometimes called point-like scaling or binary-collision scaling, follows from two assumptions: (*a*) Each nucleus is treated as a cloud of independent nucleons, each of which collides with nucleons in the opposing nucleus, possibly more than once. (*b*) Each of those N+N collisions is taken to be equivalent to an elementary N+N collision, even if one or both of the nucleons collides multiple times.[3] Point-like scaling is the baseline assumption used in discussing yields of hard pQCD processes in nuclear collisions ("the nominal expectation" mentioned above).

---

[1]The acceletor venues for heavy-ion collisions discussed in this review are RHIC (Relativistic Heavy-Ion Collider) and AGS (Alternating Gradient Synchrotron) at BNL, and SPS (Super Proton Synchrotron) and LHC (Large Hadron Collider) at CERN. See Section 3.2 for parameters and details.

[2]Although the effect at RHIC is dramatic, opinions differ as to whether jet quenching has been observed in hadron spectra from Pb+Pb collisions at CERN-SPS fixed-target energies, $\sqrt{s_{NN}} = 17$ GeV; see Reference 10.

[3]The justification for the latter assumption is twofold. First, hard parton scatterings are high-Q^2 processes that materialize quickly, before the soft processes that make up most of the N+N cross section, and so are unaffected by how many times a nucleon interacts. Second, because hard scatterings are very rare within each N+N interaction (hence point-like, meaning small cross section) it is reasonable to add their probabilities from different N+N collisions.

The dramatic departure from point-like scaling for high-p_T hadrons is the signature of jet quenching, and the most exciting interpretation for the violation is as an effect of a dense created medium on a scattered parton/jet. But point-like scaling could, in principle, be violated for other reasons as well. One frequently discussed possibility (11), for example, is that the combination of high $\sqrt{s_{NN}}$ and large nuclear size in RHIC Au+Au collision would cause the initial state of the incoming nuclei to be manifested as a color-glass condensate (CGC), a largely coherent state completely at odds with the assumption of independent nucleons described above. One effect of the CGC would be a substantial decrease in the flux of partons (compared with point-like scaling), which could lead to a decrease in the yield of high-p_T hadrons.

High-energy direct photons, which are produced primarily through processes like those shown in Figure 1A, allow us to separate the influence of initial-state effects (effects on partons before a hard scattering) from those of final-state effects (after the hard scattering, such as a created medium), because they are sensitive to the former but not the latter. As suggested in Figure 2, they can play the role of a control sample. A measurement such as the ratio between the yield of hadrons and the yield of pQCD photons, therefore, can indicate the presence of jet quenching directly without the need to invoke point-like scaling as a baseline (see Section 3.5.1).

At the same time, an absolute measurement of the yield of pQCD photons in relativistic heavy-ion collisions can serve to test the point-like scaling assumption for hard processes, effectively measuring the incoming parton flux in the incoming nuclei. A more exact statement is that they would allow the structure functions (equivalently, parton distribution functions) for nuclei to be compared with those for nucleons. This is one way that the presence of a CGC could be directly revealed, for example.[4]

These kinds of measurements are in principle very attractive for investigating the effects of nuclei on partons, but the interpretation of single inclusive direct photons in pQCD is not at present a settled matter theoretically (12), even for elementary collisions. It may be some time, then, before a reliable connection can be drawn between direct photons and parton distributions.

2.1.2. GAMMA-TAGGED JETS   The fragmentation process can be described through jet fragmentation functions, which relate the energy of a scattered parton to the yields and spectra of the hadrons (and photons) which constitute its jet. Jet quenching, then, can be described (13) as a (severe) modification to these fragmentation functions brought on by the dense created medium.

In elementary hadron-hadron collisions, jets are identified as concentrations of hadronic energy in a narrow angular cone. The total energy of a jet can be

---

[4]Note that measurement of the Drell-Yan process $q\bar{q} \rightarrow l^+l^-$, which occurs through the production of a virtual photon, has the same advantages and can be seen as complementary to the measurement of single direct photons.

determined through calorimetery, and, in principle, fragmentation functions could be studied by comparing the spectra of particles in the cone to the total jet energy.[5]

This approach would be problematic in nuclear collisions, however, for two reasons. First, the calorimetric measurement of jet energies, and even the identification of jets as energy in a cone, will be greatly affected by the high general level of hadronic energy created in A+A collisions (and by the fluctuations of same). Second, without a good model for the jet quenching process, we cannot be sure that all of a parton's energy will even survive in hadrons correlated with the parton's direction. It is entirely possible that a significant portion of the quenched energy may be transferred to the dense medium and then redistributed over a wide range of angles.

If, however, a hard-scattered parton is balanced against a direct photon, as depicted in Figure 2, then the parton's energy can be reconstructed by measuring the photon. If we assume that the incoming partons have zero total momentum transverse to the beam axis, then the transverse momenta of the outgoing parton and photon must be equal and opposite. If the p_T of the photon is measured and the polar angle of the parton is known[6] then the parton's total energy can be uniquely fixed, having been tagged by the photon.

This program for measuring medium effects on partons/jets opposite to a direct photon has been discussed widely [see for instance Wang and Huang (13), and also (14, 15)], and although it is in principle extremely promising, it does suffer from both experimental and theoretical difficulties. Experimentally, jet+photon final states are very rare among hard scatterings—the vast majority produce jet+jet final states—so this measurement will require much higher statistics than did the original quenching measurements. Also, the measurement requires the identification of direct photons on an individual, photon-by-photon basis, which can be challenging in the environment of a nuclear collision (see Section 3.4.2). Theoretically, there is strong evidence (16) that the hard scatterings which produce direct photons have nonzero total transverse momentum (intrinsic k_T) among their high-p_T outgoing particles, which complicates the energy tagging. However, these difficulties can all be reasonably controlled, and jet–direct photon correlations are actively being pursued, and their results eagerly awaited, at the time of this writing.

2.2. Thermal Radiation

2.2.1. THERMODYNAMICS OF QCD   Early speculations that highly excited, strongly interacting matter might show interesting properties was put on a more quantitative footing with the advent of the lattice QCD computational technique (17). Roughly

---

[5]In practice fragmentation functions are usually studied in $e^+ + e^- \rightarrow$ jet(s) events, or deeply inelastic l+p collisions, where the energies of scattered partons are known without measuring their daughter hadrons.

[6]Even though the full jet cannot be reconstructed, the direction of any (surviving) high-energy hadron is a reasonable approximation to the direction of its jet's parent parton.

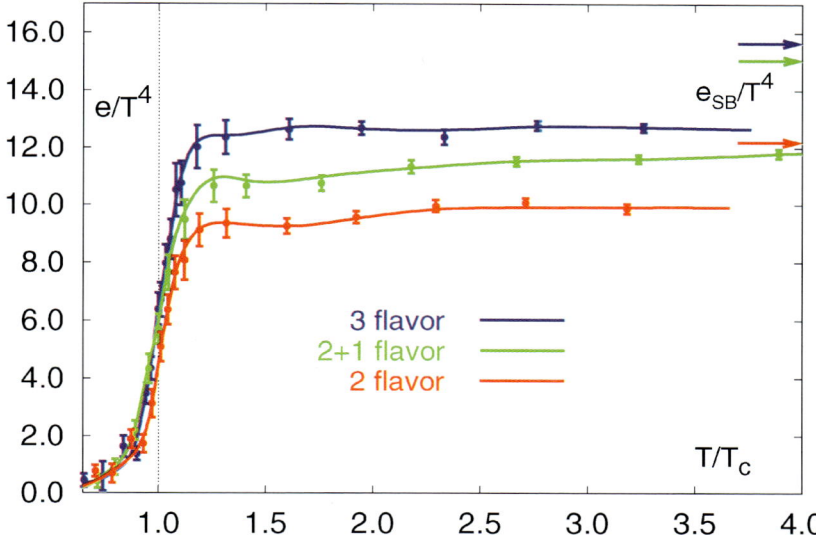

Figure 3 Lattice QCD results for normalized energy density ε/T^4 versus temperature, for the assumption of three colors, zero net baryon number, and different numbers of quark flavors (from Karsch (18)). The steep rise suggests a phase transition. The arrows (*right*) mark the Stefan-Boltzmann value for each assumption, i.e., the energy density ε_{SB} that an ideal relativistic gas with the corresponding number of particle types would have.

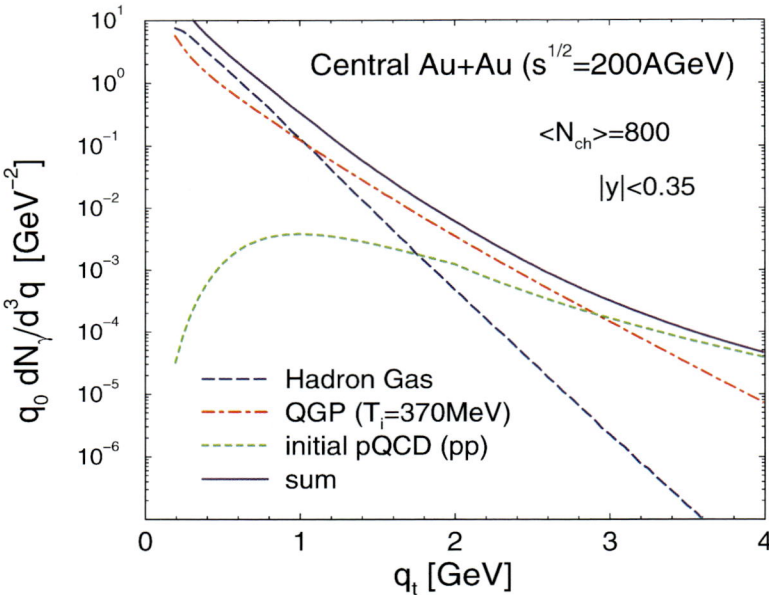

Figure 9 Predicted direct photon spectrum from central RHIC Au+Au collisions, showing the contributions from initial prompt pQCD scatterings and thermal radiation from QGP and hadron gas phases [from Reference (37)].

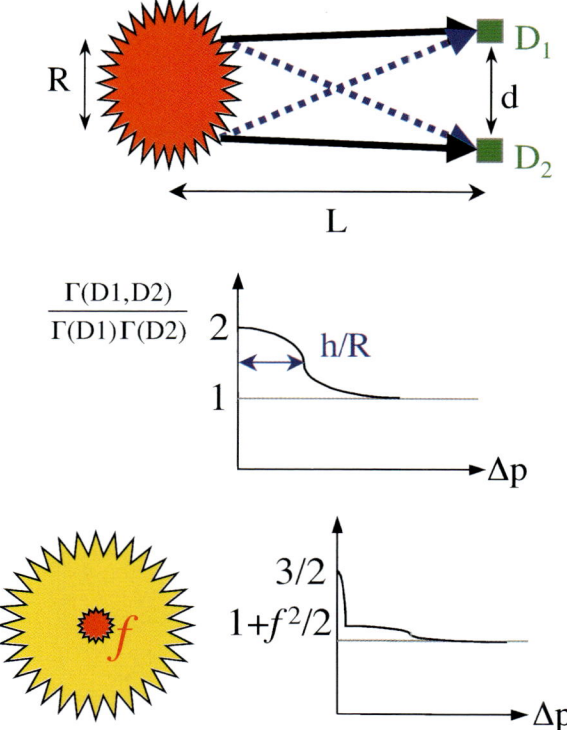

Figure 10 (*Upper row*) Two possible combinations cf paths (solid vs. dotted lines) for two photons from points on a radiating source to detectors D_1 and D_2. (*Middle row*) An enhancement in the pair coincidence rate is observed when $R \, \Delta p < \sim \hbar$, where R is the size of the source and Δp is the relative momentum between the two photons. (*Bottom row*) When the source has pieces at two different lengths scales, the enhancement pattern exhibits features at two different momentum scales.

speaking, lattice QCD simulates numerically a thermal ensemble of QCD field configurations. Ensemble averages can then reveal the thermodynamic properties—pressure, energy density, etc.—which a thermal system governed by the QCD Lagrangian (including, presumably, any locally equilibrated matter created in a heavy-ion collision) would have at any particular temperature.

This simple description belies the fact that lattice QCD calculations are highly technical and involve a number of approximations, and their results should not be mistaken for a closed-form description of QCD reality. There is no doubt, though, that the practice and technology of lattice QCD have made great advances in the last two decades, and they remain very active areas today. See (18–20) for recent general discussions of these techniques.

On very general grounds we can picture two limits for thermalized QCD: (*a*) At low temperatures, for example, well below the mass of a pion, we expect a QCD medium to look like a gas of (color-singlet) hadrons. (*b*) At very high temperatures, for example, tens of GeV or above, QCD color interactions weaken owing to asymptotic freedom, and a QCD medium should look like a nearly ideal gas of quarks and gluons. One might imagine that the change-over from one regime to the other could be gradual as a function of temperature. What lattice QCD reveals, however, is a sharp transition between two such regions, strongly suggestive of a phase transition at a critical temperature of about $T_C \simeq 170$ MeV. Results of this kind detailing a QCD phase transition have been a major source of motivation for the high-energy heavy-ion experimental program since its beginnings.

Figure 3 (see color insert) shows a modern calculation of such a transition, tracking the scaled energy density ε/T^4 versus temperature. For a relativistic Bose or Fermi gas, this normalized energy density is proportional to the number of different particle species, including the multiplicity of spin and color states; such a count is also (loosely) called the number of degrees of freedom in the system. The steep increase at $T = T_C$, followed by the plateau, can be interpreted[7] as the opening from hadron to quark and gluon degrees of freedom. Whether or not this interpretation is completely justified (see Section 2.2.2 below), it has become standard to refer to these two phases of thermalized QCD as the QGP phase (or sometimes quark phase) and the hadron-gas or HG phase (sometimes also hadron-resonance gas, HRG).

Calculation of the rate of thermal direct photon emission from an equilibrated QGP and an equilibrated HG, which we discuss in Section 2.2.2 and Section 2.2.3, are the basic ingredients required to predict the spectra we would expect to see from a heavy ion collision. A full prediction also requires a detailed model of the collision's evolution; we return to this subject in Section 3.3.

---

[7]This interpretation is somewhat informal; more precisely, lattice QCD results describe two transitions in this region: deconfinement, which allows open color degrees of freedom, not just color singlets; and restoration of chiral symmetry, in which the quarks have no QCD-generated mass and their masses revert to their bare current (Higgs-generated) masses.

2.2.2. THERMAL QUARK-GLUON PLASMA RADIATION   Calculating the emission of direct photons from a thermalized QGP, or HG, is an exercise in thermal field theory, whose techniques date back to the 1950s (21). The case of thermal QCD in particular has seen increased work and attention since the 1980s, in coincidence with the experimental program of high-energy nuclear collisions and the expectation of studying the QGP. Here we review briefly the basic theoretical ingredients relevant for real photon production and describe some representative calculations.

It should be noted that all calculations for the production of real photons are also immediately relevant to the production of virtual photons. Virtual photons are observable as continuum dilepton pairs, and the theory and observation of dileptons produced in heavy-ion collisions is a rich and important topic. Real and virtual photons should be thought of as two facets of the single subject of thermal electromagnetic radiation; and any calculation that compares with data for one of these observables—thermal direct photons or thermal dileptons—is, in principle, not complete until it is extended to compare with the other. We cannot even touch on the subject of dileptons in this review, but interested readers can start with (22) and references therein.

2.2.2.1. Kinetic theory   The most immediately intuitive approach to calculating the rate of photon production in a thermal system is through kinetic theory. We identify a specific process by which the thermalized particles can produce a photon, and then fold the amplitude for that process in with the particles' thermal phase-space distributions.

For a process of the type $1 + 2 \rightarrow 3 + \gamma$, whose lowest-order diagrams are shown in Figure 1C, we denote $p^{\mu} = (E, \vec{p})$ as the four-momentum of the photon and $p_i^{\mu} = (E_i, \vec{p}_i)$ as the four-momenta of particles $i = 1, 2, 3$. If we define R_{γ} as the rate of photons produced per volume per time, then the contribution to the rate from this process can then be written as:

$$E \frac{dR_{\gamma}}{d^3 p} = \frac{1}{2(2\pi)^3} \int \frac{d^3 p_1}{2(2\pi)^3 E_1} \frac{d^3 p_2}{2(2\pi)^3 E_2} f_1(E_1) f_2(E_2)$$

$$(2\pi)^4 \delta^{(4)} \left(p_1^{\mu} + p_2^{\mu} - p_3^{\mu} - q_1^{\mu} \right) |\mathcal{M}|^2$$

$$\frac{d^3 p_3}{2(2\pi)^3 E_3} [1 \pm f_3(E_3)]. \qquad 1.$$

Here $\mathcal{M}$ is the amplitude for the process; the $f_i()$ are the appropriate (Bose-Einstein or Fermi-Dirac) thermal phase space distributions of the strongly interacting particles; and the $\pm$ implies either a Bose enhancement or a Pauli blocking (respectively) for the production of final-state particle 3.

The full rate is then obtained by summing over the rates for all possible photon-producing processes, and to be fully correct the amplitude for each process must be calculated over all orders of the strong coupling. (Because we only consider single photon emission, these are only valid to order e^2 in electromagnetic coupling.) An advantage of the kinetic theory formulation, though, is that it is possible to make an approximate calculation perturbatively, starting with the simplest processes calculated at finite order. For this reason the kinetic approach has been widely used to predict photon rates from a QGP.

2.2.2.2. Photon self-energy A more formal, nonperturbative formulation is to recognize that the production rate of real, on-shell photons that escape a thermal system can be related (23) to the imaginary part of the in-medium, on-shell photon self-energy Π. For temperature T and photon energy E we have (24)

$$E\frac{dR_\gamma}{d^3p} = -\frac{2}{(2\pi)^3}\mathrm{Im}\Pi_\mu^{R,\mu}\frac{1}{e^{E/T}-1}.$$ 2.

This relation is valid to order e^2 in the electromagnetic coupling, because it only considers the production of one photon, but in principle should be correct to all orders in the strong coupling.

The relation can be motivated, at a simple level, as follows (23, 25). The imaginary part of a particle's self-energy, or equivalently the imaginary part of its propagator, is associated with the disappearance of the particle's amplitude with time; in a zero-temperature vacuum it is directly related to the particle's decay rate. On-shell photons do not decay in vacuum, but in a medium they can disappear through processes which are the reverse of those shown in Figure 1C. But the rate at which a thermal system absorbs a particular particle at a specific energy and momentum must be balanced by the rate at which the system produces that same particle, so the connection between R_γ and $\mathrm{Im}\Pi$ in Equation 2 emerges naturally.

The photon self-energy can be represented as an expansion in diagrams, and so evaluated perturbatively. One example at the two-loop level is shown in Figure 4; if the photon is in a thermal medium, the effects of the medium enter through the use of thermal propagators for the internal lines. The diagram expansion allows us to see a direct connection between the self-energy and kinetic theory representations of the photon rate: in Figure 4, the diagrams on either side of the cut line have exactly the same topology as the low-order photon production diagrams shown in Figure 1A and Figure 1C. A similar correspondence will hold for the higher-order photon production processes, and so perturbative evaluations in the two frameworks can be seen as essentially the same calculation.

2.2.2.3. Prototypical calculations Evaluating the rate R_γ using only the lowest-order photon production diagrams gives a result (4, 27, 28) whose essential features are shown here:

$$E\frac{dR_\gamma}{d^3p} \propto \alpha\alpha_s T^2 e^{-E/T}\log\frac{ET}{k_c^2},$$ 3.

where $\alpha = e^2/4\pi$ and $\alpha_s = g^2/4\pi$ are the electromagnetic (EM) and strong QCD couplings, and the proportionality factor will involve the number of quarks

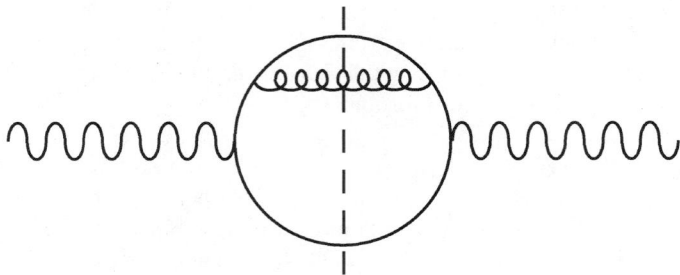

Figure 4   One QCD diagram that contributes to the photon self-energy. Each half of the diagram on either side of the cut (*vertical dashed line*) has the same topology as the lowest-order photon production diagrams in Figure 1A, C.

and the (squares of) their electrical charges. The result is infrared divergent in the mass of the intermediate quark line, and here is regulated with some infrared cutoff k_c.

The simplest choice for k_c would be the quark current mass. But this is unsatisfactory, in that the rate would be controlled and made finite by physics outside QCD, and no thermal medium information would enter into the quark propagator. Kajantie and Ruuskanen (26) had argued that the intermediate quarks should acquire an effective thermal mass in the medium; and the hard-thermal-loop (HTL) resummation of Braaten and Pisarski (29) showed that very soft modes of intermediate quarks and gluons would be suppressed in-medium, leading to a natural[8] cutoff $k_c \sim gT$. The effective thermal mass was incorporated into the lowest-order calculation by Baier et al. (28), and the HTL cutoff was incorporated by Kapusta et al., who arrived at the result (27)

$$E\frac{dR_\gamma}{d^3p} = \frac{5}{9}\frac{\alpha\alpha_s}{2\pi^2}T^2 e^{-E/T}\log\left(\frac{2.912}{g^2}\frac{E}{T}\right) \qquad 4.$$

for two light quark flavors in the limit $E \gg T$; g and α_s are evaluated at the scale of the temperature.

The result of Equation 4 was used in many models of heavy-ion collisions to predict photon rates from the QGP phase. But it was subsequently pointed out by Aurenche et al. (30) that higher-order diagrams can contribute to the hard thermal photon rate at order $\alpha\alpha_s$ and so add significantly to the yield. Recently Arnold, Moore, and Yaffee have extended (31) the rate calculations to include the the Landau-Pomeranchuk-Migdal coherence effect and presented (32) complete leading-order results for QGP photon radiation.

---

[8]The HTL result is sometimes described loosely as in-medium quarks (and gluons) acquiring an effective mass. But these quarks do not acquire all the properties we would associate with mass—for example, they do not violate chiral symmetry—and so it is more accurate to describe the HTL result as a cutoff.

2.2.2.4. Photon rates in lattice QCD   In Section 2.2.1 we described the very high temperature limit of QCD as approaching an ideal, weakly interacting gas of quarks and gluons. The graph in Figure 3 (see color insert) shows, however, that this limit is not reached even for $T \sim 4T_C$—i.e., at temperatures approaching a GeV higher than are commonly expected to occur in heavy-ion collisions—since the energy density remains well below its ideal Stefan-Boltzmann value. This brings up the question of whether quarks and gluons are actually the appropriate degrees of freedom in a QGP phase near T_C for calculations in finite order. It has even been suggested (see 34 and references therein) that a variety of exotic massive states may appear in such a phase.

A complete calculation of thermal EM radiation with quarks and gluons to all orders would presumably account for the behavior of a QGP (or any other strongly interacting system) correctly. But because this cannot be accomplished in a perturbative expansion, an alternative approach is to calculate EM radiation within the framework of lattice QCD. The first lattice calculations of thermal dilepton rates have been carried out quite recently (33), and it is interesting to note that their results differ (35) from the pertubative calculation results at low photon energies. At the time of this writing, study of the subject is just beginning, and the validity of some of the techniques is being explored, but we are not aware of any fundamental reason why EM radiation should not eventually be on a par with other lattice QCD predictions.

2.2.3. THERMAL HADRON-GAS RADIATION   In the early days of the relativistic heavy-ion collision program, there was an informal presumption that a QGP would produce a great deal more photon radiation than would a hadron gas at the same temperature. The (nonquantitative) reasoning ran along these lines: Because the quarks in a QGP are massless (due to the chiral symmetry restoration), they and their electrical charges will be present in large numbers, and because they are able to interact strongly (owing to deconfinement), we would expect to have a lot of charges being scattered and so a large amount of EM radiation—large, that is, relative nonquantitatively to a collection of massive hadrons bound up into color singlets. It came as something of a surprise, then, when Kapusta, Lichard, and Seibert (KLS) made the first serious calculation (27) of photon radiation from a thermal hadron gas and announced that at relevant temperatures (~ 200 MeV), the HG and QGP would produce very similar spectra of radiated photons.

The technology for calculating thermal radiation from a hot HG is the same thermal field theory as discussed above in Section 2.2.2. The same two formulations, kinetic theory and photon self-energy, will apply, but now with hadron processes and diagrams like those in Figure 1D. The programs are in principle very similar, but there are two factors that further complicate the thermal hadron gas case.

First, hadrons are more complicated objects than quarks or gluons. We know that the natural lineshape of the ρ resonance is broad, with a mass of 770 MeV and a width of 150 MeV. Further, the properties of the ρ are expected to change significantly in a dense hadronic medium in ways that are not straightforward to calculate. These complications—which are not limited just to the ρ—must be taken into account in order to make any realistic estimate of electromagnetic radiation from a hadron gas.

Second, while there is only one gluon and two (or three) flavors of light quark, there are many different hadrons, and it is not always immediately apparent which might be important in calculating photon production. For example, the KLS calculation included processes involving π, ρ, and η mesons, and the decay $\omega \rightarrow \pi^0\gamma$. But shortly thereafter Xiong, Shuryak, and Brown pointed out (36) that processes involving the $A_1(1260)$ resonance provided new channels, which would significantly raise the rate for photon production. Later refinements included: processes with strange mesons and resonances; processes with ω's in the t-channel; and the effects of nonzero net baryon density. So the fidelity of hadron gas calculations depends on the level of sophistication one is able to bring to them, even in something as basic as the choice of which degrees of freedom to consider.

Recently, reasonably state-of-the-art calculations for radiation from a thermal hadron gas were presented (37, 38) by Turbide, Rapp, and Gale, considering all the effects listed above and including a massive Yang-Mills model of hadron interactions. Their conclusion, interestingly, is essentially the same as that of KLS, that the thermal photon spectra from a hadron gas are very similar to that from a QGP at the same temperature; Steffen and Thoma (39) reached a similar conclusion. This outcome is somewhat ironic, perhaps, after a decade's worth of increased sophistication in both HG and QGP calculations.

2.3. Pre-Equilibrium Radiation

As mentioned in Section 1, between the prompt source and the thermal source lies the pre-equilibrium source of photons, referring to processes that are not associated with one particular parton scattering but that take place before local thermal equilibrium obtains. Pre-equilibrium photons are one of the few observables that could provide direct information on this not-well-understood—but potentially novel and interesting—stage of a heavy-ion collision.

The simplest version of a pre-equilibrium state is one that is close to, but not quite in, local equilibrium; an example would be a sytem which is equilibrated kinetically but not chemically. In the "hot-glue" scenario (40), for instance, gluons equilibrate quickly in a nuclear collision but the quarks lag in coming up to their equlibrium density; as a result (40, 41) EM radiation is less intense than the radiation from fully equilibrated quarks and gluons at the same (kinetic) temperature. However, at fixed energy density, an underpopulation of quarks is compensated by an increase in the kinetic temperature, making the photon rates similar to those from true equilibrium (42).

Descriptions of such near-equilibrium states offer no insight on what processes bring about kinetic equilibrium initially. The subject of pre-equilibrium photon radiation has been treated in a number of approaches, including parton cascades (43), microscopic transport (44), and nonequilibrium field theory (45, 46), and found to be a potentially significant contribution to the direct photon rate. However, the overall subject of the pre-equilibrium stage of a heavy-ion collision, and how it evolves to local equilibrium, is decidedly underexplored at present.

2.4. Parton-Medium Interactions

The phenomenon of jet quenching in heavy-ion collisions (see Section 2.1) provides strong evidence for, and in principle a diagnostic of, a dense medium created in those collisions. Direct photons associated with jets (fragmentation or bremsstrahlung photons) offer a new window into parton-medium interactions.

Photons appear, at some level, alongside hadrons when a scattered parton fragments into a jet, as in the vacuum diagram of Figure 1B. If partons lose energy in the medium before they fragment, then the population of high-p_T fragmentation photons will be suppressed along with high-p_T hadrons. Jeon et al. (47) estimated that this could lead to as much as a 20–40% reduction in the prompt photon spectrum from a RHIC Au+Au collision for $p_T > 3$ GeV/c. Zakharov argued (49), by contrast, that a dense medium acting on an energetic scattered quark would induce additional photon bremsstrahlung as in the in-medium diagram of Figure 1B. The net effect would be a substantial increase in prompt photons in RHIC and LHC collisions in a range $5 < p_T < 15$ GeV/c.

Fries, Müller & Srivastava (48) noted that scattered fast quarks could create photons by annihilating on antiquarks in the created medium, or through a Compton scattering with a gluon in the medium. These interactions would provide a direct measure of the density of quarks and gluons in the medium. The same mechanisms could also lead to the production of high-p_T, high-mass dilepton pairs (50). Fries et al. predicted that this process would be the dominant source of direct photons for $p_T < 6$ GeV/c in RHIC and LHC collisions.

These pictures of enhanced photon production from parton-medium interactions imply an interesting experimental signature: As the parton travels through a longer or denser medium, the number of prompt photons will increase while the number of high-p_T hadrons decreases. This anti-correlation could be a unique way to observe these medium-induced photons apart from other kinds of direct photons. Angular correlations among photon-hadron pairs are another natural way to observe jet fragmentation photons, and results along these lines at RHIC are eagerly awaited.

2.5. Other Electro-Weak Processes

As mentioned above, the production of real photons in a QCD system is intimately related to the production of virtual photons; and with only a little latitude, we can consider the entire electro-weak sector as part of the same subject. Taking that latitude, we here make some brief mention of how the production of heavy electroweak $W^{\pm}$ and Z bosons may be of interest in relativistic heavy-ion collisions at collider energies.

2.5.1. PARTON DISTRIBUTIONS AND JET TAGGING   Beam energies at RHIC and LHC will be high enough to create, for the first time ever, real (i.e., s-channel) Ws and Zs in collisions involving nuclei in the laboratory. For a $2 \rightarrow 1$ parton process, the mass of the created particle is related to the Bjorken x of the partons by $M^2 = x_1 x_2 s$. Heavy electroweak boson production at RHIC, then, is a fairly high-x process with average $\sqrt{x_1 x_2} \sim 0.4$, and at LHC a more modest $\sqrt{x_1 x_2} \sim 0.015$.

Partonic cross sections for W and Z production are straightforward to calculate and so measuring their production in collisions with nuclei provides a measure (51) of nuclear effects—shadowing at LHC, Fermi motion at RHIC—on quark and antiquark distributions at very high $Q^2 \sim 10^4$ GeV^2 scales, much higher than have ever been explored in nuclei before. These nuclear effects can be studied in either $p+A$ or $A+A$ systems; but $A+A$ has the advantage that their geometric dependence, i.e., variation with nuclear impact parameter, can be more readily mapped out, which will provide important information as to their origin.

Heavy electroweak bosons can also play roles in the study of jet production and jet quenching in nuclear collisions. It has been been suggested (52) that because Zs can be reconstructed with great accuracy through their dilepton channel, very high-p_T Zs could serve to tag high-energy jets in a manner similar to the photon tagging of jets discussed in Section 2.1.2. Also, Ws and Zs will provide a steady, calculable[9] supply of pairs of high-energy quark/antiquark jets through their $q\bar{q}$ decay. Departures from the expected appearance of these jet pairs will provide a measure of the dense medium's effect on (anti)quark jets separately from the more common gluon jets.

2.5.2. DENSE-MEDIUM Z MODIFICATION   The heavy electro-weak bosons have the potential to serve as dedicated early-time probes in a heavy-ion collision. A Z will be formed quickly in an initial hard scattering ($\hbar/M_Z \sim 10^{-3}$ fm/c) and live for only a short time ($\hbar/\Gamma_Z < 0.1$ fm/c), and so will experience only the earliest equilibrated and pre-equilibrium stage of the collision. Given this, it is interesting to ask whether the properties of the Z should be measurably[10] modified in a hot/dense QCD medium.

Changes in the mass and/or decay width of the Z follow from how its propagator, or self-energy, is modified in the thermal medium. This is essentially the same calculation described in Section 2.2.2 for photons, and it can be calculated perturbatively with diagrams like the diagram in Figure 4, with a Z line in place of the photon line. (Semi-classically we can think of the Z as undergoing collisional broadening by the medium; the connection between thermal self-energies and multiple scattering in a medium has been discussed (54).)

Kapusta & Wong (53) calculated these modifications at the two-loop level and concluded that the effects would be small. For temperature $T = 1$ GeV, the

---

[9]In LHC collisions we would expect pairs of jets with invariant mass equal to either M_W or M_Z, whereas at RHIC the antiquark distribution is falling so steeply at high x that the W and Z lineshapes are greatly distorted on the low-mass side. But due simply to the rates that can be expected this measurement is probably practical only at the LHC anyway.

[10]Although the medium effects on $W^\pm$ and Z will be very similar, we focus here on the Z since its mass and decay width can be measured much more accurately through its dilepton decay channel.

expected increase in the Z width would be on the order of 1 MeV, and the change in mass even smaller, compared with the natural (vacuum) width of 2.5 GeV. Technical objections to this calculation were raised by Aurenche et al. over the handling of divergences (see Reference 55 for the comment and Reference 56 for the reply). The subject of divergences in thermal boson self-energies was examined in detail by Majumder and Gale (57), who presented results that differed from those in (53) by logarithmic factors. However, there is no practical reason at present to doubt Kapusta and Wong's basic conclusion that effects on the Z from a thermalized QGP that can be produced in heavy-ion collisions will be unobservable.

It remains an open question, however, whether the pre-equilibrium medium might have a larger, measurable effect on the electro-weak bosons—although this is primarily because there is currently no good theory of what physics bridges the initial scatterings and the initial locally thermalized state (even the use of the word "medium" for the pre-equilibrium stage may be an overstatement; see also Section 2.3). We are not aware of any calculations along these lines[11], but it remains true on general grounds that Z would constitute a focused probe of the very earliest stages of a heavy-ion collision.

3. EXPERIMENTAL TECHNIQUES AND RESULTS

3.1. The Statistical Subtraction Technique

The statistical subtraction technique embodies the very definition of direct photons: (a) measure the spectrum of all photons $\gamma^{Incl}(p_T)$ inclusively; (b) calculate the spectrum of photons $\gamma^{Decay}(p_T)$ produced in hadronic decays; and (c) subtract the decay contribution from the inclusive spectrum, and the difference must be the direct spectrum as

$$\gamma^{Dir}(p_T) = \gamma^{Incl}(p_T) - \gamma^{Decay}(p_T). \qquad 5.$$

Although this program is in principle well defined, the experimental practice has challenges at each step. (Here we can only scratch the barest surface of the experimental details; see Reference 58 for an exhaustive discussion.)

3.1.1. PHOTON DETECTION   There are two general approaches to detecting and reconstructing final-state photons. First, their energy and position can be measured in a transversely segmented electromagnetic calorimeter. Second, they can undergo conversion in a thin layer of material into an electron-positron pair, which are then tracked as charged particles. The first method works well for photons with high lab energy, because the fractional resolution of calorimeters improves with energy; but it will become difficult, typically, for photons with lab energies below 0.5–1.0 GeV.

---

[11]This does not mean such calculations could not be attempted today. The possibility that pre-equilibrium gluons could affect the Z propagator could be addressed in a parton cascade framework, for example.

The conversion method will provide better resolution at these low photon energies, but will lose resolution with increasing lab energy. Additionally, detection through conversion will suffer a loss of efficiency because the converter material layer must be kept thin, on the order of a few percent of a radiation length, so as not to degrade the charged particle tracking. Overall the segmented-calorimeter approach is more common in photon detection experiments.

3.1.2. MESON RECONSTRUCTION   In all hadron collisions the primary source of decay photons is from the decay of π^0's, which have a 99% branching ratio to two photons and typically produce 85–90% of all decay photons. Other important sources include the decays of the η (5–10% of decay photons) and $\eta\prime$ and ω (on the order of 1% each). For the event sample of interest, the spectra of all these mesons must be measured, or very well estimated, in order to calculate the spectrum of decay photons.

The most common method of reconstructing π^0s is by detecting the photons of their primary decay $\pi^0 \rightarrow \gamma\gamma$. Photon pairs with a (combined) p_T in some specified range are binned in pair invariant mass $m_{\gamma\gamma}$, and π^0s appear as a peak at $m_{\gamma\gamma} = m_{\pi^0}$ on top of a combinatorial background. The signal-to-background ratio (S/B) in the peak region is reduced at low p_T and in more central nuclear collisions with high photon multiplicities; yields of π^0 peaks have been reliably extracted for S/B below 1%. Systematic uncertainties on π^0 peak extraction are usually[12] the dominant source of error in direct photon measurements using the subtraction method.

To keep relative normalization uncertainties at an acceptable level, the spectrum of π^0s should be measured in the same experiment and over the same event sample as the inclusive photons. The heavier mesons can in principle be reconstructed in a similar way, but their statistical significance will be reduced by their lower yields and low branching ratios to all-photon final states. A more precise estimate of their spectra can be made by assuming some scaling behavior[13] relative to the π^0 spectrum and then normalizing to meson/π^0 ratios measured at a few points in p_T or taken from elementary hadron collisions.

3.1.3. DECAY SUBTRACTION   Once the spectra of the contributing mesons are known, the spectrum of their decay photons $\gamma^{Decay}(p_T)$ can be calculated straightforwardly. With the inclusive photon spectrum $\gamma^{Incl}(p_T)$, one can then in principle calculate $\gamma^{Dir}(p_T)$ as per Equation 5.

However, it is advantageous for a number of reasons to work instead with the ratio of spectra $\gamma(p_T)/\pi^0(p_T)$ (or γ/π^0 for short). A number of systematic

[12]There are approaches to measuring the spectra of π^0s by inferring them from the spectra of charged pions; see Section 3.2 for examples. But these methods generally do not yield the same sensitivity in direct photon measurement as does the reconstruction of π^0s.
[13]A standard assumption is that the meson spectra have the same shape when plotted against the transverse mass $m_T = \sqrt{p_T^2 + m^2}$.

uncertainties can be reduced or eliminated this way; for example, the effects of uncertainty in the overall, linear scale of the reconstructed energy will cancel directly in this ratio, assuming that the photons and π^0s are measured in the same detectors (note that the effects from any potential nonlinearity in the detectors' energy response can still be important). Also, γ/π^0 is more convenient to examine, because although the individual spectra may vary by several orders of magnitude over a few GeV/c range in p_T, the γ/π^0 ratio will typically vary by substantially less than one order of magnitude. In fact, if the π^0 spectrum follows a power law, i.e., $\propto (p_T)^{-n}$ for some power n, then at high $p_T \gg m_\pi$ the γ/π^0 ratio will simply go to a constant.

The earlier generation of direct photon analyses (see Reference 60, for example) accordingly focused on the γ/π^0 ratio as the relevant observable and reconstructed the direct photon spectrum as

$$\gamma^{Dir}(p_T) = \left[\frac{\gamma^{Incl\ Meas}(p_T)}{\pi^{0\ Meas}(p_T)} - \frac{\gamma^{Decay\ Sim}(p_T)}{\pi^{0\ Sim}(p_T)}\right] \pi^{0\ Meas}(p_T). \qquad 6.$$

Here we have explicitly distinguished between several spectra: $\gamma^{Incl\ Meas}$ and $\pi^{0\ Meas}$ are the measured inclusive spectra, with their statistical fluctuations. The measured $\pi^{0\ Meas}$ is then represented by $\pi^{0\ Sim}$, typically a smoothed functional form from which the spectra of the heavier mesons and the decay photons $\gamma^{Decay\ Sim}$ can then be calculated.

A later refinement (63) was to focus on the so-called double ratio observable, here termed R:

$$R(p_T) \equiv \frac{\gamma^{Incl\ Meas}(p_T)/\pi^{0\ Meas}(p_T)}{\gamma^{Decay\ Sim}(p_T)/\pi^{0\ Sim}(p_T)}. \qquad 7.$$

The double ratio $R(p_T)$ will now, in principle, be a measure of $\gamma^{Incl}(p_T)/\gamma^{Decay}(p_T) = 1 + \gamma^{Direct}/\gamma^{Decay}$. The presence of direct photons will be revealed as a departure from $R = 1$, whereas the null result of no detectable direct photons will present itself as R being consistent with 1 over all p_T.

An example of the latter case is shown in Figure 5. Direct photons are not detectable from these data, meaning that one cannot set a nonzero lower limit on their yield at any p_T. However, it should be noted that it is always possible to set an upper limit on the yield at any p_T. (See Figure 7 for the limits corresponding to Figure 5, and the discussion in Section 3.2.) An example of double ratios which rise significantly above 1 are shown in Figure 6, and the corresponding direct photon spectra is shown in Figure 12.

3.2. Fixed-Target Results

The experimental program of high-energy heavy-ion collisions can be divided into three successive stages:

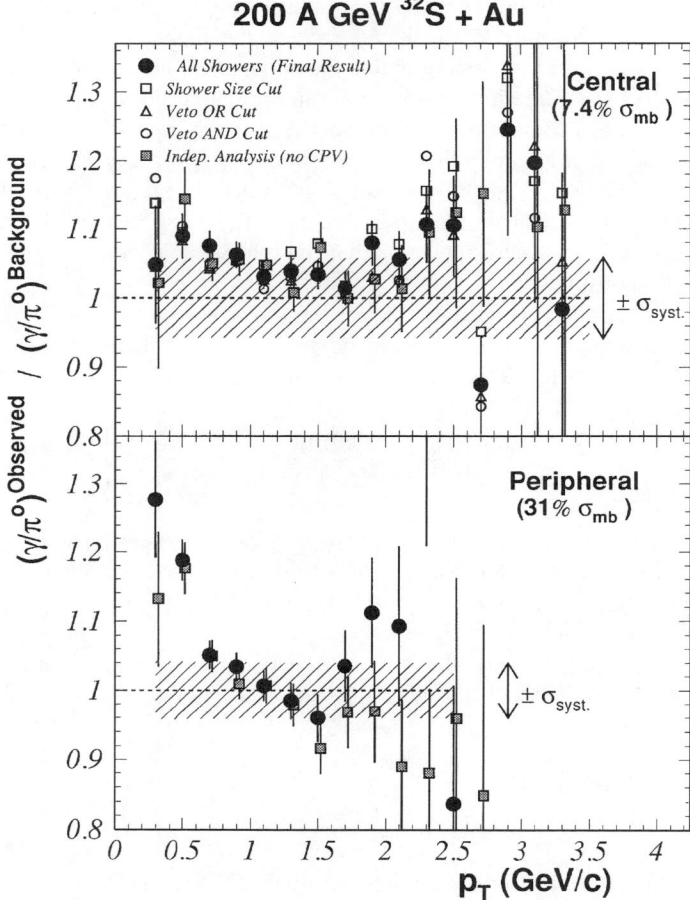

Figure 5 The double ratio (see text) versus p_T for central and peripheral S+Au collisions at $\sqrt{s_{NN}} = 19$ GeV at the CERN-SPS (63). In both event samples the double ratio is consistent, to within errors, with 1 at all p_T. This means that no lower limit can be set on the yield of direct photons from these data; the upper limits are shown in Figure 7.

1. Fixed-target experiments with beams of lighter nuclei,[14] starting in the mid-1980s: ^{16}O and ^{28}Si beams were accelerated at the BNL-AGS to 14.6 AGeV/c ($\sqrt{s_{NN}} = 5.4$ GeV), and ^{16}O at 60 AGeV/c ($\sqrt{s_{NN}} = 11$ GeV) at the CERN-SPS followed by ^{16}O and ^{32}S beams at 200 AGeV/c ($\sqrt{s_{NN}} = 19$ GeV).

---

[14]Linguistic purists may object to hear a light nucleus being referred to as a heavy ion; but the term "heavy ion" is an archaic one, used to denote any largely or fully ionized nucleus.

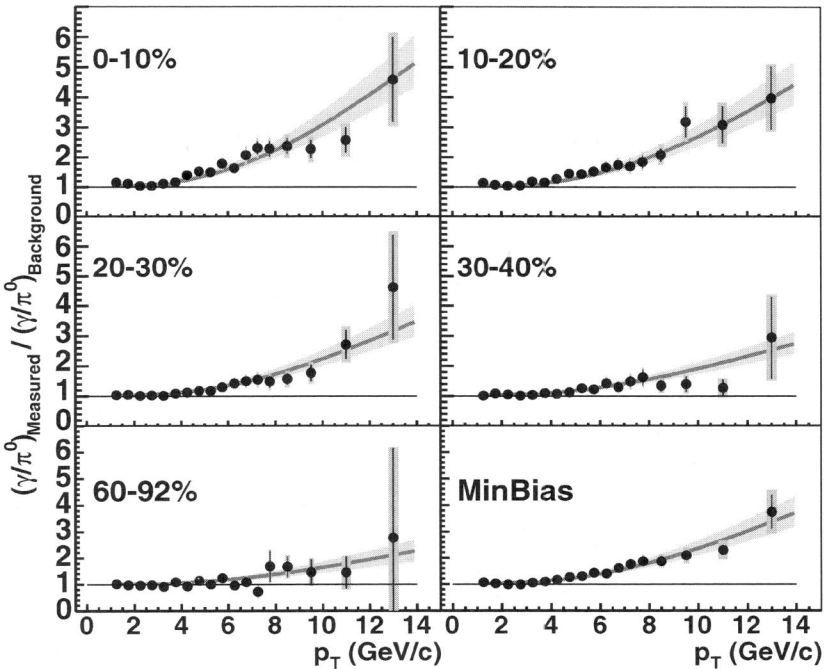

Figure 6   The double ratio versus p_T for various centralities in Au+Au collisions at RHIC, $\sqrt{s_{NN}} = 200$ GeV, as measured by the PHENIX experiment (93) and compared to NLO pQCD calculations. The resulting spectra are shown in Figure 12.

2. Fixed-target experiments with beams of heavy nuclei, starting in the early 1990s: ^{192}Au at 11.7 AGeV/c ($\sqrt{s_{NN}} = 4.8$ GeV) at the AGS, and ^{208}Pb at 160 AGeV/c ($\sqrt{s_{NN}} = 17$ GeV) at the SPS.

3. Collider experiments, which started in the year 2000 with Au+Au at BNL-RHIC at energies of $\sqrt{s_{NN}} = 130$ GeV, followed by Au+Au at $\sqrt{s_{NN}} = 200$ GeV and then a number of other combinations of nuclei and energies. Pb+Pb collisions at the CERN-LHC at $\sqrt{s_{NN}} = 5.5$ TeV should be available before the end of this decade.

In this section we summarize the results of direct photon measurements in SPS fixed-target experiments (none were reported in AGS experiments). Recent RHIC results are reviewed in Section 3.5.1, and future LHC prospects are discussed in Section 3.5.2.

3.2.1. LIGHT-NUCLEUS BEAMS   All the direct photon results from SPS experiments described in this section used some form of the statistical subtraction method (see Section 3.1); other more recent SPS results obtained at very low p_T are discussed

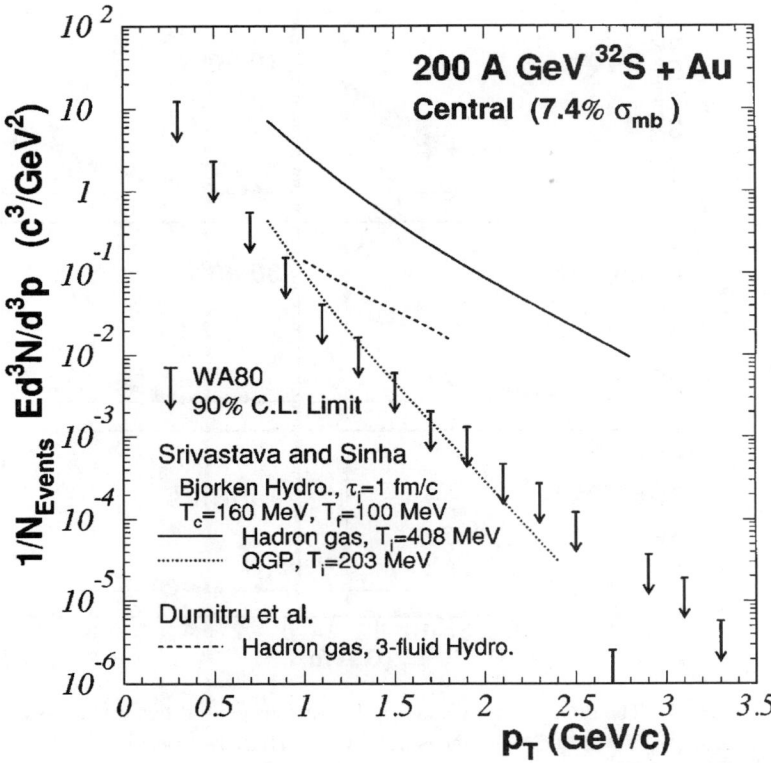

Figure 7 Upper limits on direct photon production in central S+Au collisions at $\sqrt{s_{NN}} = 19$ GeV as measured by the WA80 experiment (63). The bars at the foot of each arrow mark the 90% C.L. upper limit on the differential yield per event (the lengths of the arrows have no significance).

in Section 3.4.1. None of the lighter beam experiments observed a positive signal for direct photons, and so were able to report only upper limits.

3.2.1.1. HELIOS/NA34 The HELIOS/NA34 experiment investigated (59) direct photon production from p, O and S beams on Pt and W targets. Photons were measured through the conversion method using a thin (5.7% of a radiation length) iron plate downstream from the target followed by a charged particle tracking system, covering the rapidity range $1.0 < y < 1.9$ in the lab. The π^0 spectrum was assumed to be equal to the π^- spectrum, which was measured using negative tracks in the same experiment. The ratios of heavier mesons to π^0s were taken from ISR results, with conservative errors; the dominant contribution to the uncertainty on decay photons was stated to be uncertainty on η production.

Inclusive photon spectra, normalized to the total number of π^0s, were seen to be consistent with the predicted spectrum of decay photons over a range $0.1 < p_T < 1.5$ GeV/c for all beam-target combinations. The ratio $r_\gamma \equiv \gamma_{All}/\pi^0$, defined as integral of all photons with $p_T > 0.1$ GeV/c (and $p_T > 0.6$ GeV/c in one case) normalized to the number of pions, was reported for different collision centralities and was always consistent with the result calculated for decay photons. Though they were not reported, it would be possible from knowledge of the errors (4–11% statistical and 9% systematic on the photon data, and 9% on the decay estimate) to calculate upper limits on the integral of the direct photon spectrum from this comparison. However, it should be realized that any estimate of (or constraint on) the integral of a steeply falling p_T spectrum is, in effect, information about only the lowest p_T end of that spectrum.

3.2.1.2. WA80 O beam The WA80 experiment measured photons, π^0s, and ηs simultaneously using a highly segmented lead-glass EM calorimeter. High-energy photons deposit their energy in the form of an EM shower spread in a regular pattern across several calorimeter modules, from which the photon's energy and position can be reconstructed. This reconstruction is complicated in heavy-ion collisions because the high phase-space density of particles leads to significant overlapping of showers, which must then be separated through a clustering algorithm. Stable hadrons will also create hadronic showers in the calorimeter, which must be distinguished from photon-induced EM showers. Discrimination against hadrons is possible based on the shower shapes and with a charged particle veto hodoscope used in WA80; but some remaining hadron contamination, particularly from antineutrons, has to be subtracted away along with the decay photons.

The WA80 experiment (60) reported on a search for direct photons from collisions of p and O beams on C and Au targets, using the in-situ measurements of photons and π^0s at all p_T, and the η/π^0 ratio in a single p_T bin. The γ/π^0 ratio was measured as a function of p_T in $0.4 < p_T < 2.4$ GeV/c, for all four beam-target combinations and in separate peripheral and central samples for O+Au. The ratio was systematically higher than γ/π^0 predicted for decay photons in O+C, O+Au and central O+Au, but the excesses were not significant compared to the overall uncertainties. For O+Au collisions, upper limits were set on the direct photon spectrum and not just on a single integral: at 90% C.L. direct photons constituted no more than 15% of the inclusive photon yield at each p_T in the range $0.4 < p_T < 2.4$ GeV/c.

3.2.1.3. CERES/NA45 The CERES/NA45 experiment carried out an ambitious program of measuring e^+e^- pairs produced in primary p+A and A+A interactions, by identifying and tracking the $e^\pm$ using two ring-imaging Cherenkov (RICH) detectors with an axisymmetric magnetic field in between. Photons which convert inside the primary target are detectable as e^+e^- pairs with vanishing initial opening angle, and a search for direct photon production in S+Au collisions was carried out (61).

The p_T spectrum of photons over the range $0.4 < p_T < 2.0$ GeV/c, and within $2.1 < y < 2.65$, was examined, normalized to the pseudorapidity density of charged particles $dN_{ch}/d\eta$. The π^0 spectrum was inferred in two stages: the shape of the spectrum was constrained to describe the shapes of pion spectra (charged and neutral) from several other S+Au experiments, including the effects of decays of heavier particles; and the magnitude of the π^0 spectrum, relative to $dN_{ch}/d\eta$, was extrapolated from the same ratio in $p+p$ collisions, including a correction for additional baryon stopping in A+A. No evidence was seen for an excess of photons above the calculated decay spectrum. An upper limit was reported on the spectrum integrated over $0.4 < p_T < 2.0$ GeV/c, that at a 90% C.L. direct photons constituted no more than 14% of the integral of inclusive photons in central S+Au collisions. Note that, as mentioned above regarding the HELIOS results, a measurement of such an integral is equivalent to a measurement of the spectrum at the low end of the p_T range.

Additionally, another limit on thermal direct photon production in CERES was derived based on a particular model of thermal production. Since thermal photons are created in the re-scatterings of produced particles, be they quarks and gluons or hadrons, one might expect that the rate of photon production in a volume at some time would go as the square of the density of produced particles in that volume at that time. One might then expect—very naively—that the final-state number of thermal photons would vary as the square of the final-state number of charged particles. If this quadratic relation is assumed to hold perfectly, then the observed relation between the number of photons and N_{ch} as a function of centrality in CERES can be used to set a limit on thermal photons as a fraction of all photons. Such a limit is somewhat lower than the limit from the subtraction method, but it is difficult to know exactly how to use this information in more realistic models.

3.2.1.4. WA80 S beam   The WA80 experiment continued its investigations with an expanded calorimeter array to measure photon π^0 and η production in S beam collisions. A preliminary result (62) indicating a significant excess in the γ/π^0 ratio above the decay expectation in central S+Au collisions generated a considerable amount of excitement. However, the final result, measured with the double ratio shown in Figure 5, showed no significant excess of inclusive over decay photons in peripheral or central collisions. Upper limits on the direct photon yield in central S+Au collisions were reported as a function of p_T over $0.5 < p_T < 2.5$ GeV/c; see Figure 7.

In a historical aside, it is interesting to note that the information of upper limits on a direct photon spectrum was not always received in a completely rigorous fashion. At first, some researchers derided limits as indicating only that direct photon results were "missing," but that view was belied when the WA80 upper limits were shown to provide a significant constraint on theoretical models (see Section 3.3). Others were unused to dealing with having information presented in the form of single-sided limits. For example, in a thorough and well-regarded 1998 review (64) of the field, the state of direct photon measurements in S+Au were described by the sentence, "Within the reported systematic errors the results of the WA80 and the CERES/NA45 collaborations are compatible with each other." While true, the statement is tautological: two upper limits (or two lower limits) on the same quantity, or on overlapping quantities, are always compatible with each other.

3.2.2. HEAVY-NUCLEUS BEAMS   The CERN-SPS fixed-target heavy ion program continued with ^{208}Pb beams, which were used by upgraded versions of most of the lighter beam experiments. The WA98 experiment, descended from WA80, carried out the same program of measuring photons π^0's and η's in Pb+Pb collisions using a highly-segmented lead-glass calorimeter together with a charged particle veto shield. Spectra were examined over $0.5 < p_T < 3.5$ GeV/c, and the inclusive photons were seen to exceed decay photons at the level of about 2σ in the range $2.0 < p_T < 3.5$ GeV/c. Thus, for the first time ever in heavy-ion collisions, a direct photon p_T spectrum was reported (58, 65), and not just single-sided limits. These results are shown in Figure 8, and their significance is discussed in Section 3.3.

3.3. Theoretical Interpretation

In the early days of high-energy heavy-ion collisions the first diagnostic goal was simply to show that a multi-collisional, plausibly thermalized state had actually been formed in a collision, and it was hoped that any observation of thermal direct photons would be strong evidence for such a state. So the negative, limits-only results from the early lighter beam experiments were met with some disappointment. But they did spur the development of much more rigorous treatments for diagnosing the states in a collision through direct photons.

To connect a model of a collision to the direct photon spectrum one must, in principle, include all the sources listed in Section 2. A typical calculation would come in two parts: (*a*) predicting the pQCD prompt photons through standard techniques and (*b*) predicting the thermal radiation from the QGP and HG phases of matter in the collision as it evolves. Figure 9 (see color insert) shows the result of such a calculation (37), with these three components[15] and their sum. The features shown here are typical: The HG radiation dominates the total number of photons due to that phase's larger volume and longer lifetime; but the QGP radiation dominates at higher photon energy due to its higher temperature, reflecting the $e^{-E/T}$ factor in rate expressions like Equations 3 and 4. The prompt component has even fewer photons, but owing to its flatter power-law shape it will come to dominate at the highest photon energies.

The framework for calculating the thermal components is typically a hydrodynamical description of the collision and its evolution. These models are fast evolving; we reprise only the barest details here. Interested readers can consult (66)

---

[15] A full calculation would also include also pre-equilibrium and parton-medium-interaction photons; but these sources are typically not included other than by authors highlighting specific models for them.

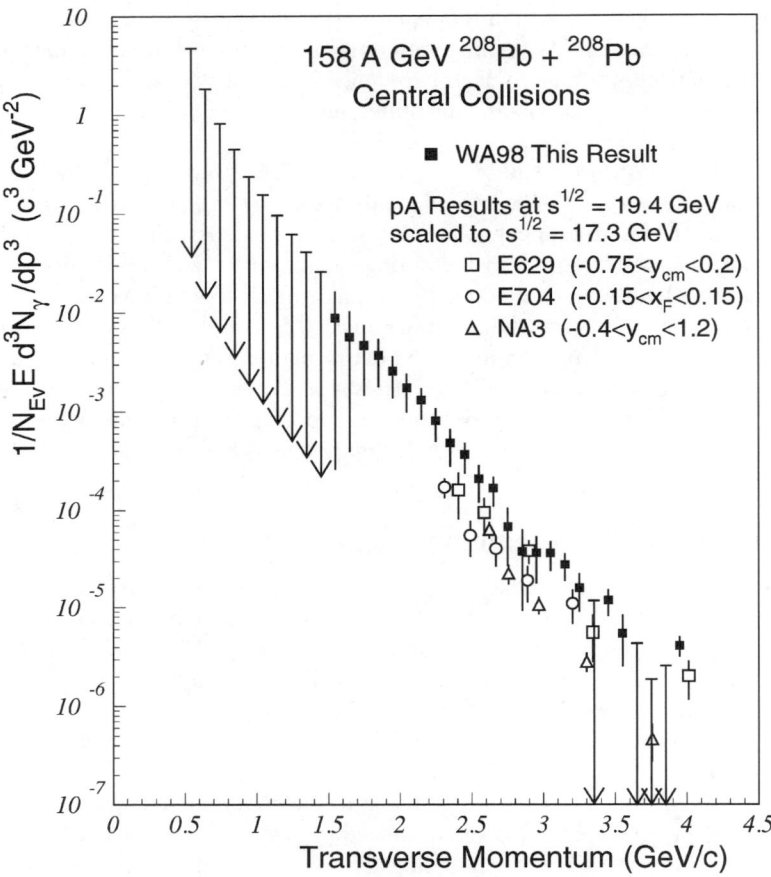

Figure 8   Spectrum of, and upper limits on, direct photon production in central Pb+Pb collisions at $\sqrt{s_{NN}}$ =17 GeV as measured by the WA98 experiment (65). The points mark the differential yield per event, with standard errors. The bars at the foot of each arrow mark the 90% C.L. upper limit on the yield (the lengths of the arrows have no significance). The hollow points are direct photon yields from elementary collisions at similar energies, which have been re-scaled to the Pb+Pb beam energy and also adjusted via binary scaling to the equivalent yield in central Pb+Pb collisions.

for more detail. The basic ingredients are an initial condition of energy density as a function of position at the time of initial local thermalization, and an equation of state (EOS) relating pressure and temperature to energy density, including any phase transitions (these are typically taken from lattice QCD, for example). The density is then evolved according to energy-momentum conservation, giving each point in space-time a local fluid velocity and a temperature in the local fluid's rest frame. The radiation from each fluid element can then be computed in its rest frame using a form like Equation 4 and boosted to a common observer's frame to be tallied. The initial conditions are in principle unknown, but can be constrained by comparison to hadron data such as the distribution of final-state transverse energy.

The simple goal of these studies, when compared to data, is to constrain some property of the thermalized state, for example, to show that its density and/or temperature were above the threshold for QGP production. The more ambitious goal is to constrain the equation of state and so learn the properties of hot, thermal QCD matter. A typical exercise is to run a calculation with a few different sets of EOS and initial conditions, and see if any can be ruled out. Less common, but more valuable, are parametric studies exploring which regions of parameter space can, and cannot, be ruled out by comparison to the data.

3.3.1. S BEAM RESULTS   A common choice for two EOSs is one with a QGP state as described by lattice QCD and a phase transition at $T = T_C$ to a hadron gas, and one with no phase transition and a pure hadron gas at all temperatures. The hope is, that a no-transition scenario being ruled out can provide evidence for a QCD phase transition taking place in the collision. An example is visible in Figure 7, where photons from these two scenarios calculated by Srivastava & Sinha (67) are compared to the WA80 upper limits on direct photons in S+Au. The pure HG scenario used here incorporated a limited number of hadron species, and this low number of degrees of freedom combined with the assumed initial energy densities resulted in a very high temperature, whose photon rates can easily be ruled out by the data. The scenario with the initial QGP, having a large number of degrees of freedom, is initially cooler and its photons are in much better agreement with the data. A similar conclusion was drawn from the calculations of Dumitru et al. (68).

This comparison does not rule out a pure HG scenario so much as it rules out an HG with artificially few degrees of freedom. Srivastava and Sinha revisited (69) the comparison with S+Au data using an HG incorporating all hadrons with masses below 2.5 GeV/c^2 and concluded that the photons from a no-transition scenario with this richer HG[16] were similar to those from the QGP scenario and were consistent with the data. The implication, noted also by Sarkar et al. (70), was that a blunt transition/no transition distinction could not be drawn from the S+Au data.

A more supple analysis by Sollfrank et al. (72) concluded that WA80 limits ruled out a temperature higher than 250 MeV for the initial phase of a S+Au collision, regardless of its composition. This "cool" initial state, in turn, implies a minimum on the number of degrees of freedom opened in the initial state and so allows us to make contact with the thermodynamics of Figure 3. Inasmuch as a high number of initial degrees of freedom can be taken as evidence for a phase

---

[16]It was noted, however, that these "full" hadron gases have number densities exceeding several/fm^3 at temperatures of 200 MeV and above, which may be unphysical; see also Bebie (71).

transition, this evidence gets stronger when fewer photons are observed, in an interesting inversion of the original presumption for direct photons.

3.3.2. PB BEAM RESULTS The positive observation (that is, both upper and lower limits) of direct photons in Pb+Pb reported by WA98 triggered a new, expanded round of theoretical interpretation. No single conclusion has emerged at present, and the sophistication in the treatment of various elements varies widely in different calculations. We will step through some of the main conclusions from a sample of authors.

Dumitru et al. (76) examined prompt pQCD production as the source of photons in Pb+Pb at high $p_T > 3$ GeV/c, particularly the evidence for k_T broadening. As measured by the parton intrinsic transverse momentum parameter $\langle k_T^2 \rangle$, they found the WA98 data best fit by values in the range 1.8–2.4 GeV^2/c^2, which is up to 1 GeV^2/c^2 larger than a standard value of 1.3 GeV^2/c^2 called for in $p+p$ photon data. They also concluded that prompt photons could not account for the spectrum at $p_T < 2.5$ GeV/c, and so a thermal component is required. Steffen and Thoma (39) examined thermal production with a small variety of initial state assumptions, including maximum temperatures up to 235 MeV, and drew the complementary conclusion that a prompt component was necessary to explain the data above $p_T > 2$ GeV/c.

Srivastava and Sinha (77) varied the initial thermalization time τ_0 and found a good fit to the data with a short $\tau_0 = 0.2$ fm/c for an initial QGP phase with a correspondingly high temperature of 335 MeV. Huovinen et al. (73) explored different thermalization times and different longitudinal hydrodynamical models, and concluded that the data could be fit with either a scenario with a phase transition or with a hadron-gas-only scenario, but noted that in the latter case the required maximum initial temperature is well above T_C.

Gallmeister et al., motivated by dilepton data, exhibited (74) a fit to the WA98 data with a prompt source plus a simplified, single-temperature, spherically expanding thermal source. Alam et al. (75) incorporated a nonzero initial radial expansion profile and were able to fit the WA98 data with both a QGP initial state and a chirally-symmetric HG initial state, both at modest initial temperatures very close to 200 MeV.

Turbide et al. (37) examined the WA98 photon data as part of their study of hadron gas radiation. Their baseline QGP/HG scenario with a thermalization time of $\tau_0 = 1.0$ fm/c, initial temperature of 205 MeV, and no additional nuclear k_T broadening underpredicted the data in the range $p_T > 1.5$ GeV/c. They concluded that the gap could be filled either by shortening the thermalization time and increasing the initial temperature to 270 MeV, or by adding additional broadening of $\langle \Delta k_T^2 \rangle = 0.3$ GeV^2/c^2 to the prompt production.

What emerges is a kind of rough parameter space in which one might vary (*a*) additional parton nuclear broadening; (*b*) initial thermalization times, and hence initial temperatures; (*c*) a QGP/HG versus pure HG EOS; and possibly (*d*) an initial transverse flow profile. Different authors have shown that various points in this parameter space can reproduce the WA98 data, but what's missing is a thorough parametric study detailing which segments of parameter space can be ruled out by the data. Such a study has the potential to constrain the thermodynamics of QCD in the SPS Pb+Pb system, and so may hold great promise. Despite the volume of work done so far, there is still more to be learned from the WA98 direct photon data.

3.4. Other Techniques

3.4.1. GAMMA-GAMMA HBT   A different approach to fixing the relative rates of direct vs. decay photons takes advantage of the fact that the spatial distribution of the sources of the two populations differ widely. Direct photons from a hadron collision are all generated from a volume whose size is set by the scale of strong interactions and the size of the nucleus, on the (rough) order of femtometers (fm). The decay photons are born in a volume with typical dimension on the order of the decay length for an EM decay—25 nanometers (nm) for π^0—which is many orders of magnitude larger. The distributions of single inclusive photons are not sensitive to these shapes, but the distributions of pairs of photons can be. Specifically, the Hanbury-Brown-Twiss (HBT) technique of identical boson interferometry can relate the size and shape of a source distribution to the intensity and shape of a photon pairs distribution as a function of their relative momentum.

The HBT technique was first employed (78) to measure the sizes of distant stars. But over the last several decades it has become very highly developed (79, 80) for reconstructing the shapes of secondary source distributions from high-energy particle collisions, which are smaller than stars by more than 20 orders of magnitude. We will give only the briefest outline of the technique here, highlighting the relevance to measuring direct photons (see References 81–84 for more details on this application).

Figure 10 (see color insert) illustrates the basic idea. A source is incoherently radiating some kind of particle from some finite region of space. We consider the case that two identical particles with very similar energies are emitted very close together in time from two separate points on the source and are then observed in coincidence in two detectors D_1 and D_2. Classically we would expect that the rate of pair coincidences would follow from the rates of singles in each detector, $\Gamma(D_1, D_2) \propto \Gamma(D_1)\Gamma(D_2)$.

Quantum mechanics alters this picture, however. For a coincidence to occur, there are two possible combinations of paths that the particles can take from the source points to the detectors, shown as solid and dashed lines in Figure 10. Because the two particles of the pair are identical, these two possibilities are indistinguishable—they have identical initial conditions and identical final states—and so we must add their quantum mechanical amplitudes in order to predict the rate of coincidence events. We do not recap the mathematics here, but the basic result is that for identical boson pairs the coincidence rate will be enhanced for pairs with small relative momentum Δp, specifically when Δp is on the order of (or smaller than)$\hbar/R$ where R is the size of the source. This is depicted schematically in the middle row of Figure 10. First we scale the coincidence rate by the product of the singles rates to form a correlation function $\Gamma(D_1, D_2)/\Gamma(D_1)\Gamma(D_2)$ as a function of pair Δp and normalize it to 1 at high $\Delta p \gg \hbar/R$. The correlation function will then show a peak for $\Delta p <\sim \hbar/R$ and approach a value of 2 as $\Delta p \to 0$ for pairs of spinless bosons.[17] For photons, the correlation function peaks at a value of 3/2—the strength of the enhancement is reduced by half because two photons act as identical particles only if they have the same polarization, and this will be true of only half the pairs.

The spatial source distribution of photons emerging from a high-energy hadron collision, however, has pieces at two length scales: a core on the fm scale, from which the direct photons are produced, and a halo on the nm scale, from which the decay photons are produced. The photon-photon correlation function for this source will also have features on two scales. Photon pairs where both are from the core, i.e., direct-direct pairs, will show an enhanced coincidence rate for momentum differences on the order of $\Delta p \sim \hbar/\text{fm} \sim 100 \text{ MeV}/c$; whereas the other pairs, the direct-decay and decay-decay, will show enhancement on the much smaller scale of $\Delta p \sim \hbar/\text{nm} \sim 100 \text{ eV}/c$. The resulting correlation function is depicted schematically in the bottom row of Figure 10, rising to its ideal height of 3/2 at $\Delta p = 0$.

In practice the narrower feature is never seen in high-energy collisions, for two reasons. First, pairs of hard gamma rays with these tiny values of Δp would have an opening angle so small that it would be extremely difficult to detect them separately. Second, the phase space for pairs with tiny Δp is so small that they would essentially never be observed in a typical experiment. So in practice we would expect the measured photon-photon correlation function to have a single-peak shape over Δp, but rising to an apparent peak value (well) below 3/2.

If f is the fraction all photons which are direct, then the direct-direct pairs are a fraction f^2 of all pairs. The low-Δp enhancement will be reduced by exactly this fraction, and the wide feature in the correlation function will (in the ideal case) have a peak value of $1 + f^2/2$ as indicated in Figure 10. So the peak value of the correlation function, combined with a measure of the inclusive photon rate, can in principle yield a measurement of the direct photon rate at that energy.

Extracting direct photon yields via photon-photon HBT has great appeal because it does not require any measurement of π^0's or other hadrons; but it is experimentally very challenging. First results for direct photon yields extracted this way at low p_T in Pb+Pb were recently reported by the WA98 experiment [86, see also Reference 85 (Peressounko)]. The data are shown in Figure 11 and lie

---

[17]This is true in an idealized situation, while the actual peak values are affected by whether the source is truly incoherent and by residual strong and EM interactions between the particles. Because photons do not suffer the latter, the case of photon-photon HBT will be closer to ideal than hadron-hadron HBT.

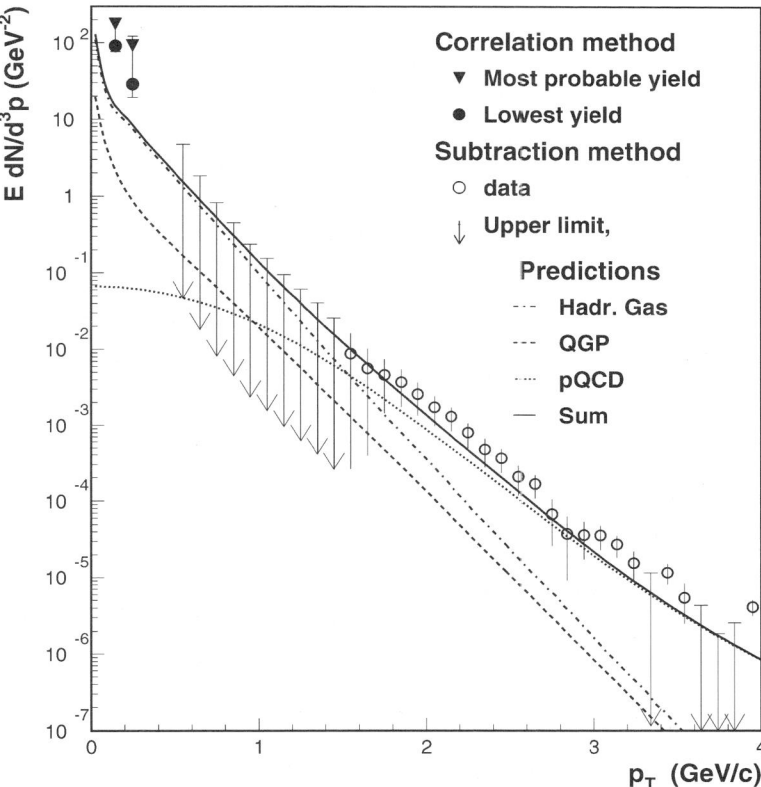

Figure 11 Direct photon spectrum from Pb+Pb collisions at CERN-SPS ($\sqrt{s_{NN}}$ =17 GeV), now extended to lower p_T with the photon-photon HBT technique (86). The curves are direct photon rates as calculated (37) for different sources. The solid line is the sum, which is in reasonable agreement with the higher-p_T results but badly underpredicts the rate at lower p_T.

well above a typical theoretical expectation. We would expect that direct photons at these low p_T would be primarily hadron-gas radiation, and the shapes of the photon-photon correlation functions indicate a source size that would be consistent with the later, expanded stages of a heavy-ion collision. The implications of the low-p_T yield and the correlation shapes for HG dynamics are presently under very active investigation (9, 85, 88).

3.4.2. PHOTON-BY-PHOTON IDENTIFICATION   The techniques described above, statistical subtraction and photon-photon HBT, can give information only on the average yield of direct photons over a large event sample. In order to make a hadron-direct photon pair measurement, for example to measure photon-tagged jets

(Section 2.1) or to search for medium-induced photon jet fragments (Section 2.4), it is necessary to identify direct photons individually, on a photon-by-photon basis, with reasonable efficiency.

3.4.2.1. Isolation   As mentioned in Section 1, a standard approach in elementary particle collisions is to measure photons that have no accompanying energy, either hadronic or EM, within some angular range. Such an object, defined as an isolated photon, is reasonably assured to be a direct photon, at least at very high momentum, because the parent hadron of a decay photon would be part of a jet and other hadrons from the jet would be nearby at close angles. Isolated photons can be measured in calorimeters with relatively coarse segmentation and lower energy resolution, since fine precision on their position and energy is not required. Treating isolated photons theoretically is complicated, because the definition of the isolation cut and the properties of jet fragmentation enter into calculations of their rates, and the definition will exclude fragmentation/bremsstrahlung photons.

Although isolated photons are quite standard as observable objects in elementary collisions, it is hopeless to search for them from high-energy nuclear collisions because the high densities of produced particles will essentially always deposit some energy within any standard isolation cone around a photon.

3.4.2.2. Sister-tagging   An alternative approach to individual direct photon identification that might be tractable in heavy-ion collisions is to veto a candidate photon if an accompanying "sister photon" is found, which indicates that both could have been produced in an hadronic decay. Operationally one would, for each candidate photon, run through all pairs with other photons in the event and check to see if the invariant mass $m_{\gamma\gamma}$ of any pair was consistent with the mass of a hypothetical parent π^0 or η. Though the term is not in common use, this method could reasonably be termed sister-matching or sister-tagging. The approach requires an EM calorimeter with sufficient energy and position resolution to provide good accuracy in $m_{\gamma\gamma}$ in order to minimize efficiency losses from accidental rejections, though these will always be present at some level. Also, some false positive identification of decay photons as direct will always occur at some level, for two reasons: Not all potential parent decays can be identified in a two-photon channel; and the opening angle between two photon daughters can, regardless of parent energy, always[18] be as large as $180°$, i.e., perfectly back-to-back, so without true 4π coverage the sister always has some chance to be lost.

Sister-tagging is also used to measure inclusive direct photons in elementary high-energy collisions (16, 89), and direct photons identified this way are easier to treat theoretically than are isolated photons. Its utility for identification of individual direct photons in heavy-ion collisions has not yet been exhibited, but work continues actively.

---

[18]This interesting property is unique to two-body decays with massless daughters.

3.5. Collider Experiments: First Results and Future Prospects

3.5.1. FIRST RESULTS AT RHIC   Heavy-ion collisions at RHIC have been examined by four experiments starting in the year 2000, and the wealth of data accumulated so far has provided strong evidence (90) that a dense, thermalized state is formed early in Au+Au collisions at $\sqrt{s_{NN}} = 200$ GeV. Many questions are not yet answered, such as whether deconfinement can be shown to have occured and what might be the mechanism through which fast initial thermalization is achieved. The measurement of direct photons can offer unique information in furthering this investigation.

RHIC A+A collisions are expected to provide a rich field for direct photon physics, because all of the components we have discussed—prompt, pre-equilibrium, thermal, and parton-medium-interaction induced—have been predicted to appear at significant, observable levels. There are also several advantages to measuring and interpreting direct photons at RHIC compared with the SPS. On the theoretical side, the spectrum of π^0s and prompt direct photons measured in $p+p$ collisions at RHIC by the PHENIX experiment (91, 92) have been well-reproduced by next-to-leading order (NLO) pQCD calculations. The agreement is good even down to the low p_T ranges, where disagreement was seen between data and theory (12, 16) for lower beam energies, suggesting that the prompt component is under reasonable control at RHIC. On the experimental side, the depletion of high-p_T π^0's and other hadrons in A+A collisions greatly reduces the yield of decay photons at RHIC, which greatly eases the task of extracting direct photons (see Figure 6).

The PHENIX experiment has recently reported (93, 94) spectra of direct photons in Au+Au collisions, for a variety of centralities, obtained with the statistical subtraction method. The direct photon spectra are shown in Figure 12, and compared to the NLO calculation for $p + p$ collisions at the same $\sqrt{s_{NN}}$, adjusted according to binary scaling for that centrality. Interestingly, the binary-scaled NLO calculation is consistent with the direct photon data at all centralities, suggesting, at least at first glance, that there are no large enhancements from other sources compared to the prompt source. The fact that direct photons obey point-like scaling whereas high-p_T hadrons do not is reflected in the great increase in the γ/π^0 ratio with centrality seen in Figure 6. This provides a dramatic confirmation that the depletion of high-p_T hadrons, i.e., jet quenching, is due to a final-state effect on scattered partons, as discussed in Section 2.1.

What the implications might be of small yields from the other sources, exactly how much room remains between the PHENIX data and the calculated prompt yield, and how reliable a prompt calculation can be are all questions that we expect to be investigated with great interest in the near future. In a very recent paper (95) d'Enterria and Peressounko employed a hydrodynamical model of RHIC Au+Au collisions, as constrained by RHIC hadron data, to calculate a thermal direct photon

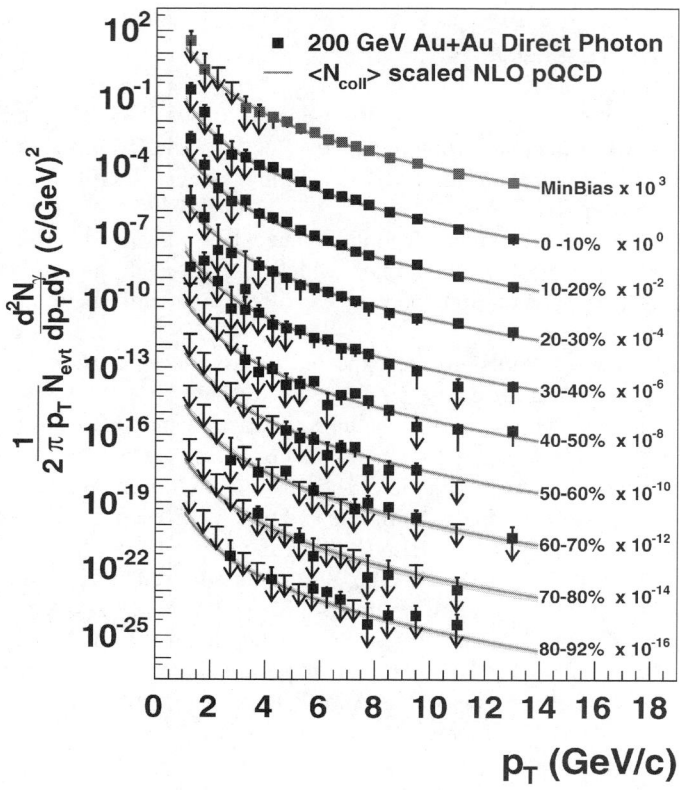

Figure 12 Spectrum of, and upper limits on, direct photon production in Au+Au collisions at RHIC, $\sqrt{s_{NN}} = 200$ GeV, as measured by the PHENIX experiment (93). The points mark the differential yield per event, with standard errors. The bars at the foot of each arrow mark the 90% C.L. upper limit on the yield (the lengths of the arrows have no significance).

spectrum. They reported that the thermal photons should dominate the prompt for $p_T < 3$ GeV/c, and that the sum spectrum is consistent with the PHENIX data for central and peripheral collisions over the whole p_T range down to 1 GeV/c. They also exhibit a relation between the local slope of the thermal photon spectrum and the initial temperature of the system, calculated through the hydrodynamical model, and propose a method by which temperature information can be combined with entropy density information (from the measured charged particle multiplicity) to constrain the number of degrees of freedom in the initial state. This approach is similar in spirit to that of Sollfrank et al. (72), who combined temperature and energy density to investigate degrees of freedom.

Many experimental advances are expected soon in direct photons at RHIC. The direct photon spectra discussed here were derived from only a small fraction of the now-available A+A and $p+p$ data, so great improvements in statistical accuracy should be possible. Also, the very large $\gamma^{Direct}/\gamma^{Decay}$ ratio, as a result of jet quenching of mesons, means that identifying individual direct photons is much easier, which greatly facilitates measurements like γ^{Direct}-tagged jets and γ^{Direct}-hadron correlations. All in all, there is every reason to expect that direct photons will soon be a unique and important part of the investigation of RHIC collisions.

3.5.2. FUTURE PROSPECTS AT LHC   Beyond RHIC, the next stage on the beam-energy frontier in heavy-ion collisions is the CERN-LHC. The LHC will start providing Pb+Pb collisions at $\sqrt{s_{NN}} = 5.5$ TeV before the end of this decade, an increase by a larger factor over RHIC ($\times 27$) than RHIC was over SPS ($\times 10$). Unlike RHIC, the LHC will be primarily dedicated to high-energy particle physics. As currently planned, the LHC heavy-ion program will run for one month out of the year, with the balance going for $p+p$ collisions at top energy of $\sqrt{s} = 18$ TeV.

The role of direct photons in nuclear collisions at the LHC is discussed in great detail in the "Photon Physics" chapter of the *CERN Yellow Report on Hard Probes in Heavy Ion Collisions at the LHC* (96). We touch here on some general considerations to keep in mind.

Higher energy scales in LHC collisions will change the balance between different sources of direct photons, but it is not yet clear how. The prompt component will have a much greater yield and a flatter spectrum from LHC collisions than at RHIC. But initial temperatures are also expected to be higher—numbers between 300 MeV and 800 MeV have been mentioned—which could still lead to a QGP thermal component being dominant in some p_T range. Various authors have reached different conclusions depending on the particulars of their assumptions.

Prompt photons with p_Ts of few–10 GeV/c will probe much lower Bjorken x's in the incoming nuclei, and so deepen the investigation of possible coherent/saturated gluon initial states such as the CGC. At the same time, high energy jets with many 10s of GeV will be produced at significant rates at LHC, and so photon-jet, or possibly Z-jet, tagging will be of great importance in studying possible modifications to their fragmentation.

Finally, pre-equilibrium processes are likely to be of greater significance at LHC. Parton scatterings at LHC are overwhelmingly dominated by gluons—the machine is sometimes described as a "gluon collider"—and so the possibility that early thermalized states may be out of chemical equilibrium, gluon rich but quark poor, must be kept in mind. In our opinion, the as-yet-unknown physics of the pre-equilibrium stage may be the next great subject of study in heavy-ion collisions, yielding greater surprises than the study of the thermal QGP. And whatever happens in those very early stages, the same general principle applies, that direct photons may carry and preserve the best diagnostic information about them.

The *Annual Review of Nuclear and Particle Science* is online at
http://nucl.annualreviews.org

LITERATURE CITED

1. Feinberg EL. *Nuovo Cim.* A34:391 (1976)
2. Shuryak EV. *Phys. Lett.* B78:150 (1978)
3. McLerran LD, Toimela T. *Phys. Rev. D* 31: 545 (1985)
4. Peitzmann T, Thoma MH. *Phys. Rep.* 364: 175 (2002)
5. Gale C, Haglin KL. arXiv:hep-ph/0306098
6. Alam J, Sinha B, Raha S. *Phys. Rep.* 273: 243 (1996)
7. Aurenche P. arXiv:hep-ph/0201011
8. Gale C. *Nucl. Phys.* A698:143 (2002)
9. Rapp R. *Mod. Phys. Lett.* A19:1717 (2004)
10. d'Enterria D. *Phys. Lett.* B596:32 (2004)
11. McLerran LD. *Lect. Notes Phys.* 583:291 (2002)
12. Aurenche P, et al. *Eur. Phys. J.* C9:107 (1999); Aurenche P, et al. *Phys. Rev. D* 55: 1124 (1997)
13. Wang XN, Huang Z. *Phys. Rev. C* 55:3047 (1997)
14. Srivastava DK, Gale C, Awes TC. *Phys. Rev. C* 67:054904 (2003)
15. Arleo F, Aurenche P, Belghobsi Z, Guillet JP. *J. High Energy Phys.* 0411:009 (2004)
16. Apanasevich L, et al. (Fermilab E706 Collab.) *Phys. Rev. D* 70:092009 (2004)
17. Creutz M. *Phys. Rev. D* 15:1128 (1977)
18. Karsch F. *Lect. Notes Phys.* 583:209 (2002)
19. Laermann E, Philipsen O. *Annu. Rev. Nucl. Part. Sci.* 53:163 (2003)
20. Petreczky P. *Nucl. Phys. Proc. Suppl.* 140: 78 (2005)
21. Matsubara T. *Prog. Theor. Phys.* 14:351 (1955)
22. Rapp R, Wambach J. *Adv. Nucl. Phys.* 25:1 (2000); Rapp R. *J. Phys.* G31:S217 (2005)
23. Weldon HA. *Phys. Rev. D* 28:2007 (1983)
24. Gale C, Kapusta JI. *Nucl. Phys.* B357:65 (1991)
25. Thoma MH. *Phys. Rev. D* 51:862 (1995)
26. Kajantie K, Ruuskanen PV. *Phys. Lett.* B121:352 (1983)
27. Kapusta JI, Lichard P, Seibert D. *Phys. Rev. D* 44:2774 (1991). Erratum. *Phys. Rev. D* 47:4171 (1993)
28. Baier R, Nakkagawa H, Niegawa A, Redlich K. *Z. Phys. C* 53:433 (1992)
29. Braaten E, Pisarski RD. *Nucl. Phys.* B337:569 (1990); Braaten E, Pisarski RD. *Phys. Rev. D* 42:2156 (1990); Braaten E, Pisarski RD. *Phys. Rev. D* 45:1827 (1992)
30. Aurenche P, Gelis F, Kobes R, Zaraket H. *Phys. Rev. D* 58:085003 (1998); Aurenche P, Gelis F, Kobes R, Zaraket H. *Phys. Rev. D* 60:076002 (1999)
31. Arnold P, Moore GD, Yaffe LG. *J. High Energy Phys.* 0111:057 (2001)
32. Arnold P, Moore GD, Yaffe LG. *J. High Energy Phys.* 0112:009 (2001)
33. Karsch F, et al. *Phys. Lett. B* 530:147 (2002)
34. Brown GE, Lee CH, Rho M, Shuryak E. *Nucl. Phys.* A740:171 (2004)
35. Karsch F, et al. *Nucl. Phys.* A715:701 (2003)
36. Xiong L, Shuryak EV, Brown GE. *Phys. Rev. D* 46:3798 (1992)
37. Turbide S, Rapp R, Gale C. *Phys. Rev. C* 69: 014903 (2004)
38. Turbide S, Rapp R, Gale C. *Int. J. Mod. Phys.* A19:5351 (2004)
39. Steffen FD, Thoma MH. *Phys. Lett.* B510: 98 (2001)
40. Shuryak EV, Xiong L. *Phys. Rev. Lett.* 70: 2241 (1993)
41. Traxler CT, Vija H, Thoma MH. *Phys. Lett.* B346:329 (1995); Traxler CT, Thoma MH. *Phys. Rev. C* 53:1348 (1996)
42. Gelis F, Niemi H, Ruuskanen PV, Rasanen SS. *J. Phys.* G30:S1031 (2004)
43. Srivastava DK, Geiger K. *Phys. Rev. C* 58: 1734 (1998)
44. Dumitru A, et al. *Phys. Rev. C* 57:3271 (1998)

45. Boyanovsky D, de Vega HJ. *Phys. Rev. D* 68:065018 (2003); Boyanovsky D, de Vega HJ. *Nucl. Phys.* A747:564 (2005)

46. Serreau J. *J. High Energy Phys.* 0405:078 (2004)

47. Jeon SY, Jalilian-Marian J, Sarcevic I. *Phys. Lett.* B562:45 (2003)

48. Fries RJ, Müller B, Srivastava DK. *Phys. Rev. Lett.* 90:132301 (2003)

49. Zakharov BG. *JETP Lett.* 80:1 (2004) [*Pisma Zh. Eksp. Teor. Fiz.* 80:3 (2004)]

50. Srivastava DK, Gale C, Fries RJ. *Phys. Rev. C* 67:034903 (2003); Gale C, Awes TC, Fries RJ, Srivastava DK. *J. Phys. G30:* S1013 (2004)

51. Vogt R. *Phys. Rev. C* 64:044901 (2001)

52. Kartvelishvili V, Kvatadze R, Shanidze R. *Phys. Lett.* B356:589 (1995)

53. Kapusta JI, Wong SMH. *Phys. Rev. D* 62: 037301 (2000)

54. Kapusta JI, Wong SMH. *Phys. Rev. D* 64: 045008 (2001)

55. Aurenche P, et al. *Phys. Rev. D* 65:038501 (2002)

56. Kapusta JI, Wong SMH. *Phys. Rev. D* 65: 038502 (2002)

57. Majumder A, Gale C. *Phys. Rev. C* 65: 055203 (2002)

58. Aggarwal MM, et al. (WA98 Collab.) arXiv:nucl-ex/0006007 *Phys. Rev. C.* Submitted

59. Akesson T, et al. *Z. Phys. C* 46:369 (1990)

60. Albrecht R, et al. (WA80 Collab.) *Z. Phys. C* 51:1 (1991)

61. Baur R, et al. (CERES Collab.) *Z. Phys. C* 71:571 (1996); Kampert KH, et al. *Z. Phys. C* 74:587 (1997); Baur R, et al. *Z. Phys. C* 74:593 (1997)

62. Santo R, et al. (WA80 Collab.) *Nucl. Phys.* A566:61C (1994)

63. Albrecht R, et al. (WA80 Collab.) *Phys. Rev. Lett.* 76:3506 (1996)

64. Bass SA, Gyulassy M, Stocker H, Greiner W. *J. Phys.* G25:R1 (1999)

65. Aggarwal MM, et al. (WA98 Collab.) *Phys. Rev. Lett.* 85:3595 (2000)

66. Huovinen P, Ruuskanen V. *Annu. Rev. Nucl. Part. Sci.* 56:In press (2006)

67. Srivastava DK, Sinha B. *Phys. Rev. Lett.* 73:2421 (1994)

68. Dumitru A, et al. *Phys. Rev. C* 51:2166 (1995)

69. Srivastava DK, Sinha BC. *Eur. Phys. J.* C12:109 (2000). Erratum. *Eur. Phys. J.* C20:397 (2001)

70. Sarkar S, Roy P, Alam J, Sinha B. *Phys. Rev. C* 60:054907 (1999)

71. Bebie H, Gerber P, Goity JL, Leutwyler H. *Nucl. Phys.* B378:95 (1992)

72. Sollfrank J, et al. *Phys. Rev. C* 55:392 (1997)

73. Huovinen P, Ruuskanen PV, Rasanen SS. *Phys. Lett.* B535:109 (2002)

74. Gallmeister K, Kampfer B, Pavlenko OP. *Phys. Rev. C* 62:057901 (2000)

75. Alam J, Sarkar S, Hatsuda T, Nayak TK, Sinha B. *Phys. Rev. C* 63:021901 (2001)

76. Dumitru A, et al. *Phys. Rev. C* 64:054909 (2001)

77. Srivastava DK, Sinha B. *Phys. Rev. C* 64: 034902 (2001)

78. Hanbury Brown R, Twiss RQ. *Philos. Mag.* 45:663 (1954)

79. Lisa M, Pratt S, Soltz R, Wiedemann U. *Annu. Rev. Nucl. Part. Sci.* 55:In press (2005)

80. Baym G. *Acta Phys. Polon.* B29:1839 (1998). arXiv:nucl-th/9804026

81. Srivastava DK, Kapusta JI. *Phys. Rev. C* 48:1335 (1993); Srivastava DK, Kapusta JI. *Phys. Rev. C* 50:505 (1994)

82. Marques FM, et al. *Phys. Rept.* 284:91 (1997)

83. Peressounko D. *Phys. Rev. C* 67:014905 (2003)

84. Bass SA, Müller B, Srivastava DK. *Phys. Rev. Lett.* 93:162301 (2004)

85. Alam J, et al. *Phys. Rev. C* 67:054902 (2003)

86. Aggarwal MM, et al. (WA98 Collab.) *Phys. Rev. Lett.* 93:022301 (2004)

87. Peressounko D, (WA98 Collab.) *J. Phys.* G30:S1065 (2004)

88. Srivastava DK. *Phys. Rev. C* 71:034905 (2005)

89. Bonesini M, et al. (WA70 Collab.) *Z. Phys. C* 38:371 (1988)

90. Arsene I. *Nucl. Phys. A* 757:1 (2005). Adcox K. *Nucl. Phys. A* 757:184 (2005). Back BB. *Nucl. Phys. A* 757:28 (2005). Adams J. *Nucl. Phys. A* 757:102 (2005)

91. Adler SS, et al. (PHENIX Collab.) *Phys. Rev. Lett.* 94:232301 (2005) arXiv:nucl-ex/0503003

92. Adler SS, et al. (PHENIX Collab.) *Phys. Rev. Lett.* 91:241803 (2003)

93. Adler SS, et al. (PHENIX Collab.) arXiv:nucl-ex/0503003

94. Frantz J, (PHENIX Collab.) *J. Phys. G* 30:S1003 (2004)

95. d'Enterria D, Peressounko D. arXiv:nucl-th/0503054

96. Arleo F, et al. arXiv:hep-ph/0311131

Annu. Rev. Nucl. Part. Sci. 2005. 55:555–88
doi: 10.1146/annurev.nucl.55.090704.151505

TOOLS FOR THE SIMULATION OF HARD HADRONIC COLLISIONS

Michelangelo L. Mangano
TH Unit, PH Department, CERN, 1211 Geneva 23, Switzerland;
email: Michelangelo.Mangano@cern.ch

Timothy J. Stelzer
Department of Physics, High Energy Physics, University of Illinois, Urbana,
Illinois 61801; email: tstelzer@uiuc.edu

Key Words QCD, Monte Carlo, Tevatron, LHC

■ **Abstract** This review gives a pedagogical introduction to the current status and ongoing progress in the development of QCD-based Monte Carlo tools for the calculation and simulation of high-Q^2 processes in hadronic collisions.

CONTENTS

1. INTRODUCTION

The interpretation of data from high-energy particle colliders and their use to extract the measurement of fundamental physical parameters, or to infer the possible existence of new physics and study its properties, rely heavily on the theoretical modeling of the outcome of these collisions. In recent years, a great effort has therefore been put into the development of tools enabling the description of the final states resulting from high-energy collisions. These tools are called Monte Carlo (MC) codes because the state-of-the-art knowledge about Quantum Chromo Dynamics (QCD) is implemented in them using numerical MC techniques. The high accuracy of the measurements carried out at the LEP and SLC e^+e^- colliders has prompted great improvements in the theoretical understanding of QCD in e^+e^- annihilations (1). The Collider Detectors at Fermilab (CDF) (2) and D0 (3) experiments at the Fermilab proton-antiproton collider (4), and the prospects for very high statistics and precise measurements at the forthcoming CERN proton-proton Large Hadron Collider (LHC) (5–8), have recently shifted the efforts of theorists to the much more complex case of hadronic collisions. This is where the most remarkable progress has been seen in the past few years. MC development is a very technical topic. This review is intended to provide a pedagogic and qualitative introduction to the main features of event generation, stressing the underlying physical ideas and presenting the main features, as well as limitations, of the different approaches. It is not aimed at the developers of these tools, but rather at the young students and researchers using them for their phenomenological or experimental analyses. More complete and technical overviews can be found elsewhere (9–14). After a general introduction, we cover the topics of parton-level event generation, shower evolution and hadronization, the underlying event, higher-order corrections, and the simulation of physics beyond the standard model (SM). Topics that we have no space to cover in detail are briefly mentioned in the final section, together with our conclusions.

The scope of this review is limited to the study of hard hadron-hadron interactions, processes where a proton-(anti)proton pair collides at high center-of-mass energy ($\sqrt{S}$, typically larger than several hundred GeV) and undergoes a strongly inelastic interaction, with momentum transfers between the participants in excess of several GeV. The outcome of this hard interaction could be the simple scattering at large angle of some of the hadron's elementary constituents, their annihilation into new massive resonances, or a combination of the two. In all cases the final state consists of a large multiplicity of particles, associated to the evolution of the fragments of the initial hadrons, as well as of the new states produced. The main goal of a MC event generator (15–19) is to provide a complete picture of these final states: the description of the particle types and momenta on an event-by-event basis, as well as the absolute production rates for the different possible processes. An event generator should therefore "produce" events imitating Nature's behavior in a real experiment. As discussed below, the fundamental physical concept that makes this program possible is "factorization," the ability to isolate separate independent phases of the overall collision. These phases are dominated by different dynamics, and the most appropriate techniques can be applied to describe each of them separately. In particular, factorization allows one to decouple the complexity of the proton structure and of the final-state hadron formation from the elementary nature of the perturbative hard interaction among parton constituents.

Figure 1 (see color insert) illustrates how this works. As the left proton travels freely before coming into contact with the hadron coming in from the right, its constituent quarks are held together by the constant exchange of virtual gluons (e.g., gluons a and b in the picture). These gluons are mostly soft, because any hard exchange would cause the constituent quarks to fly apart, and a second hard exchange would be necessary to reestablish the balance of momentum and keep the proton together. This can be seen with a simple calculation, which shows that the exchange of gluons with virtuality Q between quarks inside a proton has a probability suppressed by some power of m_p/Q, m_p being the proton mass. Gluons of high virtuality (gluon c in the picture) prefer therefore to be reabsorbed by the same quark, within a time inversely proportional to their virtuality, as prescribed by the uncertainty principle. The state of the quark is, however, left unchanged by this process. Altogether this suggests that the global state of the proton, although defined by a complex set of gluon exchanges between quarks, is nevertheless determined by interactions which have a time scale of the order of $1/m_p$. When seen in the laboratory frame where the proton is moving with energy $\sqrt{S}/2$, this time is furthermore Lorentz dilated by a factor $\gamma = \sqrt{S}/2m_p$. If we disturb a quark with a probe of virtuality $Q \gg m_p$, the time frame for this interaction is so short $(1/Q)$ that the interactions of the quark with the rest of the proton can be neglected. The struck quark cannot negotiate with its partners a coherent response to the external perturbation: It simply does not have the time to communicate to them that it is being kicked away. On this time scale, only gluons with energy of the order of Q can be emitted, something which, to happen coherently over the whole proton, is suppressed by powers of m_p/Q (this suppression characterizes the so-called elastic form factor of the proton). In this figure, the hard process is represented by the rectangle labeled HP. In this example a head-on collision with a gluon from the opposite hadron, leads to a $qg \to qg$ scattering with a momentum exchange of the order of Q. This and other possible processes can be calculated from first principles in perturbative QCD.

When the constituent is suddenly deflected, the partons that it had recently radiated cannot be reabsorbed (as happened to gluon c earlier) because the constituent is no longer there waiting for the partons to come back. This is the case, for example, of the gluon d emitted by the quark, and of the quark e from the opposite hadron; the emitted gluon got engaged in the hard interaction. The number of "liberated" partons will depend on the hard scale Q: the larger Q, the more sudden the deflection of the struck parton, and the fewer the partons that can reconnect before its departure (typically only partons with virtuality larger than Q).

Summarizing the situation up to this point: a hard scattering takes place over such a short time scale that the details of how the proton is held together, and of what will happen to it after the hard collision, are irrelevant. The internal proton dynamics is by and large frozen during the hard process, which can therefore be described as an interaction among fundamental, freely moving constituents. The precise state of the struck parton, however, depends on the scale Q of the process, because at large Q it will have given out a larger fraction of its momentum to radiated gluons. The change as a function of Q is driven by the behavior of emission of gluons of virtuality Q: Once again, a process which decouples from the complex dynamics responsible for the overall structure of the proton, and which can be calculated from first principles in perturbative QCD (20–22). Because the state of the struck quark depends only on the scale Q at which it is being probed, and not on the nature of the probe, it can be measured in one experiment and the information can then be extrapolated for use in other experiments.

After the hard process, the partons liberated during the evolution prior to the collision and the partons created by the hard collision will themselves emit radiation, as discussed in Section 3. The radiation process, governed by perturbative QCD, continues until a low virtuality scale is reached (the boundary region labeled with a dotted line, H, in Figure 1). To describe this perturbative evolution phase, proper care has to be taken to incorporate quantum coherence effects, which in principle connect the probabilities of radiation off different partons in the event. As discussed in Section 3, once the low virtuality scale is reached the memory of the hard-process phase has been lost, once again as a result of different time scales in the problem, and the final phase of hadronization takes over. Because of the decoupling from the hard-process phase, the hadronization is assumed to be independent of the initial hard process, and its parameterization, tuned to the observables of some reference process, can then be used in other hard interactions (universality of hadronization). Nearby partons merge into color-singlet clusters (the grey blobs in Figure 1), which are then decayed phenomenologically into physical hadrons. To complete the picture, we need to understand the evolution of the fragments of the initial hadrons. As shown in the figure, this evolution cannot be entirely independent of what happens in the hard event, because at least color quantum numbers must be exchanged to guarantee the overall neutrality and conservation of baryon number. In our example, the gluons f and g, emitted early on in the perturbative evolution of the initial-state, split into $q\bar{q}$ pairs which are shared between the hadron fragments [whose overall interaction is represented by the oval labeled UE (underlying event)] and the clusters resulting from the evolution of the initial state.

The hard-process phase in this description of the collision is where the selection of the phenomenon of interest is performed, and the point where the event generation starts. The creation of Z bosons or of new particles is simulated by selecting the respective production channels and calculating, to the desired level of accuracy in perturbation theory, the corresponding matrix elements (ME). The reconstruction of the partons emitted during the evolution of the constituent quarks toward the hard scattering, and of those emitted after, is performed using the hard event as a boundary condition for the matching of the three phases described above. In the particular case of events with several hard jets in the final state, the additional problem of sorting out a possible double counting between configurations where the jets are already present in the calculation of the hard process and configurations where they emerge from the perturbative evolution, needs to be addressed. In the following sections, we cover each of these phases in detail, beginning with the generation of the hard process.

2. PARTON-LEVEL EVENT GENERATORS

Current perturbative calculations in the SM of high-energy collisions are limited to fundamental constituents of the theory, such as quarks, leptons, and gauge bosons. Hence the simulation of a hard hadronic collision starts from the generation of a parton-level (PL) event, i.e., an interaction between the elementary constituents of the proton, quarks, and gluons. Under the simplifying assumption that a hard parton evolves into an individual observable jet, the PL events can be used to evaluate the production rates for complex multijet final states implementing realistic detector acceptances. Furthermore, the PL event is used as a seed for the development of a complete description of the final state via the showering and hadronization algorithms. PL event generators are therefore the core of the comparison of experiment with theory, and most analyses begin with a PL study.

The fully differential cross section for a general hadron-hadron collision is given by:

$$\frac{d\sigma}{d\Omega dx_1 dx_2} = \frac{1}{Flux} \sum_{i,j,k,\ldots n} f_i(x_1, \mu_F) f_j(x_2, \mu_F) \times \overline{\sum} |M(i, j \to k, l, \ldots n)|^2.$$

1.

The function f_i (known as parton distribution function, PDF) represents the number density of parton i with momentum fraction x in a proton probed at a scale μ_F. The outer summation is over all partons within the proton, and the inner summation represents the average and sum over helicity and colors. $|M^2|$ represents the square of the matrix element for the process of interest.

A PL event generator for hadron colliders must correctly implement all of the components shown in Equation 1. It must identify and sum over all of the contributing partons in the initial state and final state, as well as summing over their possible helicity and color. The scattering amplitude $|M(i, j \to k, l, \ldots n)|^2$ must be calculated and convoluted with the appropriate PDFs $f_i(x, \mu_F)$. To be an event generator, the program must not only integrate this function over phase space, but provide unweighted events according to the fully differential distribution including the partons flavor, helicity, momentum, and color.

Even though partons are fundamental constituents of QCD, matrix elements for the hard scattering process can only be calculated in a perturbative expansion in powers of the coupling constant $\alpha_s(\mu_R)$, which is small at high-energy scales. This expansion is organized in terms of Feynman diagrams, providing a consistent, gauge invariant, result. The truncation of the perturbative expansion for the cross section to a finite order in $\alpha_s(\mu_R)$ introduces an intrinsic uncertainty in the predictions. This is manifest in an explicit dependence on the choice of the renormalization (μ_R) scale of the coupling constant α_s and the factorization scale (μ_F) of the parton densities. For a process that at tree-level (i.e., leading-order, LO) is proportional to α_s^N (i.e., production of N jets), the dependence of the scattering amplitude on μ_R is given by

$$\Delta|M^2| = \frac{\Delta\mu_R^2}{\mu_R^2} N b_0 \alpha_s \alpha_s |M^2|, \qquad\qquad 2.$$

where $b_0 = (33 - 2n_f)/12\pi$ is the coefficient of the QCD 1-loop β function (n_f is the number of quark flavors with mass smaller than μ_R). For $m_b < \mu_R < m_t$, $b_0 \sim 0.6$. It is customary to consider variations of μ_R in the range $\mu_0/2 < \mu_R < 2\mu_0$, where μ_0 is the "typical" hard scale of the process. For example, $\mu_R^2 \sim M_{\ell\ell}^2$ (the invariant mass of the dilepton pair) for Drell-Yan (DY) production and $\mu_R \sim \langle E_T \rangle$ (average transverse energy of the jet) for jet production. As a result, the uncertainty on $\Delta|M^2|/|M^2|$ is of the order $4N\alpha_s$, a value which for large N, as in the case of multijet production, can be much larger than 100%. In this case absolute rate predictions are rather inaccurate, although experience shows that the shapes of the distributions are much more precise and reliable. Inclusion of higher-order corrections reduces the scale dependence by an extra factor of α_s for each additional loop. However, loop calculations are technically much more difficult as they involve cancellation of divergences between different diagrams. Generating unweighted events from loop calculations provides yet another layer of complexity, as discussed in Section 5. Therefore, most of the available event generators for hadron collisions, particularly in the case of multijet processes, are based on lowest-order (LO) matrix-element calculations.

Although in principle the method for calculating tree-level PL matrix elements has been known for a long time, the task is complicated by the need to sum large numbers of Feynman diagrams (23) and by the large variety of different processes, including processes beyond the SM. Several tools have been introduced to deal with these complexities (24–34). The large number of tools recently developed for simulating high-energy QCD collisions is encouraging, as it is likely that a tool exists for the analysis the user wishes to perform. However, the choice may also be intimidating, as the user needs to select the appropriate tool for the study of interest. A current and detailed list of the available tools is documented in the HEPCODE listing (35). It is reassuring that the results of the workshop "MC tools for the LHC" demonstrated that the PL generators agree in their region of overlap (36). Table 1 shows a comparison between several PL event generators over a variety of interesting processes. Any attempt to enumerate which tool is best for which analysis would soon become obsolete. Hence this section focuses on the general questions one should consider when choosing a matrix-element simulation package.

PL event generators fall roughly into two categories. On one hand, there are dedicated codes that contain a library of processes that have been optimized for analysis by the authors. On the other hand, there are general purpose programs that write code to produce PL events for an arbitrary process. Independent of the choice of dedicated code or computer-generated code, several important physics issues must be considered to ensure that one is using the proper tool.

The first important consideration is whether the event generator is designed to be interfaced with a showering and hadronization package to produce a full

TABLE 1 Comparison of parton level cross sections calculated by a variety of tools studied at the MC Tools for the LHC Workshop

σ (pb)	Number of jets				
$e^- \bar{\nu}_e + n$ QCD jets	0	1	2	3	4
ALPGEN (33)	3904(6)	1013(2)	364(2)	136(1)	53.6(6)
AMEGIC++ (31)	3908(3)	1011(2)	362.3(9)	137.5(5)	54(1)
CompHEP (29)	3947.4(3)	1022.4(5)	364.4(4)		
GR@PPA (32)	3905(5)	1013(1)	361.0(7)	133.8(3)	53.8(1)
HELAC/PHEGAS/JetI (30)	3786(81)	1021(8)	361(4)	157(1)	46(1)
MadEvent (34)	3902(5)	1012(2)	361(1)	135.5(3)	53.6(2)
$t\bar{t} + n$ QCD jets	0	1	2	3	4
ALPGEN	755.4(8)	748(2)	518(2)	310.9(8)	170.9(5)
AMEGIC++	754.0(8)	748.7(7)	519(1)		
CompHEP	757.8(8)	752(1)	519(1)		
HELAC/PHEGAS/JetI	745(5)	711(7)	515(5)		
MadEvent	754(2)	749(2)	516(1)	306(1)	
$e^- \bar{\nu}_e + b\bar{b} + n$QCD jets	0	1	2	3	4
ALPGEN	9.34(4)	9.85(6)	6.82(6)	4.18(7)	2.39(5)
AMEGIC++	9.37(1)	9.86(2)	6.87(5)		
CompHEP	9.415(5)	9.91(2)			
GR@PPA	9.31(1)	9.80(2)			
HELAC/PHEGAS/JetI	9.88(11)				
MadEvent	9.32(3)	9.74(1)	6.80(2)		

The cross sections are for pp collisions at 14 TeV collider energy. Details of the cuts and scale choices can be found in Reference (36). The quoted errors are from the integration statistics. The programs ALPGEN and GR@PPA are examples of dedicated codes, whereas the others are multipurpose codes.

event simulation. Although PL calculations provide excellent insight into the basic features of the process, before an experimental analysis can be performed it will be necessary to generate a complete event simulation. As we shall discuss later, an essential ingredient to allow the shower evolution of a PL event is knowledge of the color and flavor structure of the event, which are therefore mandatory information that the PL code should make available. Many available codes write out event information in the Les Houches format (37), which was defined to provide easy integration of the PL calculations into a full event simulations.

A second important consideration is how the tool handles unstable particles. It is usually safest to use a package that includes the decay of unstable particles incorporating any interference effects and spin correlations. Single particle distributions such as p_T distributions are typically insensitive to these effects, and it is acceptable to treat the intermediate particles as "stable" and have the decay handled when the PL event is evolved into a full event. However, some multiparticle distributions are sensitive to these effects; this is demonstrated by the spin correlation analysis of the final state leptons in dilepton $t\bar{t}$ events (38).

The treatment of the bottom quark mass is another area that should be considered. In many analyses (and many PL generators), the mass of the bottom quark may be safely ignored. However, if the analysis allows for very soft b jets, or b-pairs that are very collinear, the mass effects will play an important role in the event samples.

In some analyses, one has to consider the effect of interference between diagrams with different intermediate states, as well as between electroweak and QCD diagrams. An example of this is analyses with e^+e^- final state, where it is crucial to add the contribution of off-shell photons to the resonant Z decays. Another example is single-top production in association with a $\bar{b}$ quark via an off-shell W ($\mathcal{O}(\alpha_W^2)$), where the interference of the $Wb\bar{b}$ final state with the nonresonant direct production of $Wb\bar{b}$ ($\mathcal{O}(\alpha_s^2\alpha_W)$) could be non-negligible. It is usually safest to specify the complete partonic final state such that different intermediate states are allowed to contribute.

The above illustrates the types of issues to consider in choosing an event simulation tool. The specific relevant issues will depend on details of the analysis being performed. The following sections discuss some of the advantages of dedicated codes and computer-generated codes one may want to consider when determining which tool is best for one's analysis.

2.1. Dedicated Codes

Dedicated codes are created by experts interested in studying a particular process or group of processes. One advantage of such a tool is that if one is using it to generate events for an analysis similar to the author's intention, it will already be optimized. The authors will likely have incorporated all of the relevant processes required for the analysis. They will also have optimized scale and other input parameter choices such that in many cases, one will have the event sample required "straight out of the box." Furthermore, the documentation for the package will often contain examples and results for event generation that are directly applicable to one's own study.

An example shows the approach one might follow in choosing a dedicated package for studying top production at the Large Hadron Collider (LHC). Checking the available programs from Reference (35), the TopRex package (39) would be of immediate interest, having a full set of top-related processes, including single top production, production via Higgs bosons, and anomalous top couplings, as well as an interface to PYTHIA for a complete event simulation.

The analysis will most likely require simulation of not only the signal, but also the backgrounds. Again referring to the HEPCODE website, one finds the ALPGEN (33) package that has both the signal for $t\bar{t}$ production, which one could compare with TopRex, as well as the dominant backgrounds from W+jet production. The program documentation shows that in addition to generating the parton momentum, ALPGEN also provides events in the Les Houches format necessary to shower the events for a full event simulation. Looking at the unstable particles, we see ALPGEN decays both the top quarks and the W boson including the spin correlations. The b-quark mass is also included in the background processes, which is reassuring because b-tagging is an important component to many top quark analyses.

Top quark physics is one of the clear priorities of collider physics, and hence considerable effort has gone into creating dedicated packages that contain the relevant physics for generating signal as well as background events. These dedicated packages allow a user to quickly reproduce the results illustrated in the manual, and then proceed to develop a data set suitable for study in their own analysis.

2.2. General Purpose Programs

Although dedicated tools can often provide the PL events necessary for an analysis, it is often useful to use one of the general purpose programs to create code for event generation. Unlike dedicated codes that have subprocesses specifically coded into them, general purpose programs contain information about the particles and interactions of the model. When a user requests a process (e.g., $pp \rightarrow jjb\bar{b}$, where j represents jet), the generator determines which subprocesses contribute (e.g., $u\bar{u} \rightarrow d\bar{d}b\bar{b}$, $ud \rightarrow udb\bar{b}$, . . .), and then for each subprocess generates code to calculate the scattering amplitude on the basis of the complete set of Feynman diagrams. The program must also write code to sum over the subprocesses, helicity, and color and integrate over phase space to provide the cross section and produce fully differential unweighted events.

The most common motivation for using a general purpose code to create your own event simulator is that a dedicated package for the relevant process is not available. This is especially common for processes that have not yet generated the interest necessary for the creation of a dedicated code. Even when dedicated codes exist for the processes of interest, it is often desirable to create your own with the general purpose program, as it provides the flexibility to create all of the signals and backgrounds within the same framework. With this increased flexibility comes an increase in responsibility to identify the relevant processes for signals and backgrounds, and to make appropriate choices for input parameters. This too can be an advantage: Being faced with decisions about the subprocesses and Feynman diagrams often leads to greater insight that can benefit the analysis.

Consider for example backgrounds to the dilepton channel of $t\bar{t}$ production. The MadEvent website (40) allows users to specify the process they wish to study, in this case $pp \rightarrow e^{+}\nu_{e}\mu^{-}\bar{\nu}_{\mu}b\bar{b}$ where there are a maximum of four electroweak vertices. The site automatically creates an event generator code and returns a list of subprocesses and associated Feynman diagrams. The generated Feynman diagrams reveal that in addition to the anticipated background diagrams from weak boson production and decay, diagrams from top quark production were also included, as they satisfied the user request. For a background study these diagrams are probably not wanted, so the user could make a second request explicitly forbidding top quarks from the diagrams. A second Web form allows the user to specify the parameters used for the simulation. Once submitted, the site will generate events for analysis and provide plots of the basic distributions.

Many of the general purpose programs include extensions of the SM such as the Minimal Supersymmetric Standard Model (MSSM). Once these have been specified, the program generates cross sections, events, and distributions just as it does for the SM. As always, this additional flexibility comes at the price of additional responsibility. In addition to checking the self-consistency of the new interactions, authors entering new models must also be consistent with the interactions already programmed into the system, and careful checks are essential.

As mentioned in the Introduction, the PL prediction is a good starting point for an analysis; however, a full event simulation requires that the PL hard scattering be interfaced to a showering and hadronization routine. Therefore, most of the automated generators provide events that include the color information necessary to incorporate the events into a shower Monte Carlo such as PYTHIA (16) or HERWIG (11, 17). There are several important considerations when combining PL-generated events with showering algorithms. In addition to assigning the correct particles, momentum, and color flow, one has to be careful to avoid double counting. This double counting can arise when the same final state can be generated both by the parton level calculation, and by the showering algorithm. For example a W boson produced in association with two jets could be generated at the parton level by producing a W and two high-energy gluons. It could also be generated by combining a parton level event with a W boson and one hard gluon, and then the second jet could be produced by the showering algorithm. The removal of double counting can be accomplished with a procedure called matching (discussed in more detail in Section 5.3). It is worth noting that the automated PL event generator AMEGIC++ was designed to automatically incorporate matching under the SHERPA framework (19). Hence, PL events generated with AMEGIC++ can easily be evolved into fully exclusive hadronic final states.

3. PARTON-SHOWER GENERATORS

The PL hard scattering matrix element describes the hadronic collision at the smallest scales in time and distance. On these scales, the colliding partons can be considered free and the perturbative expansion provides a reliable prediction of the scattering. The hard scattering of the partons results in the acceleration of "color charge." Just as accelerating electrically charged particles produce photon radiation, accelerating colored partons produce QCD radiation. For high-energy scattering, the effect of this radiation is significant, resulting in dozens of additional partons being associated with the event. The distribution of this radiation is predicted by QCD and approximately implemented in parton shower codes. As the colored particles continue to radiate and move further apart, the QCD force grows and confinement effects result in the hadronization of the colored partons into color-singlet hadrons. Hadronization occurs at larger nonperturbative scales and is therefore typically implemented in event generators using phenomenological models. We begin this section with a review of the physics of shower evolution algorithms, with an emphasis on the need for the complete history of the color charge, and conclude with a discussion of the models for hadronization of the final state partons.

3.1. The Shower-Evolution Algorithms

Figure 2a shows the Feynman diagram for $q\bar{q}$ production from an off-shell photon. Production of the two colored quarks traveling away from each other results in the acceleration of color charge, and hence the radiation of gluons. The distribution of this gluon radiation can be calculated from the Feynman diagrams shown in Figure 2b. Notice that these diagrams are exactly the diagrams corresponding to the splitting of a quark into a quark+gluon. Radiation due to the acceleration is not limited to a single emission, and similar to electromagnetic radiation many gluons are radiated. However, unlike the photon which is neutral, the gluon carries color charge. Hence the very act of radiating a gluon results in an accelerated color change and therefore influences future emissions. Below we review the essential features based on the discussion in Reference (12).

The radiation pattern due to the accelerated color charge is dominated by gluon contributions when the gluon is collinear with the momentum of one of the quarks. The evolution equations (20–22) provide the formalism for calculating these leading contributions and can be written in a form to express the probability for an emission:

$$dP_a(z, \mu^2) = \frac{d\mu^2}{\mu^2} \frac{\alpha_s}{2\pi} P_{a \to bc}(z)\, dz \qquad\qquad 3.$$

Equation 3 shows the probability that parton a will split into partons b and c at a virtuality scale μ^2 and with parton b carrying a fraction z of the parton a's momentum. Parton a is commonly referred to as the parent, and partons b and c are referred to as the daughters. The total emission probability expressed in Equation 3 diverges logarithmically in the soft ($z = 1, 0$) and collinear ($\mu = 0$) regions. This is not a problem, because detectors will not be able to resolve two partons very close together, thus introducing an effective cutoff. A cutoff μ_0, of the order of several hundred MeV, is therefore introduced to screen the $\mu \to 0$ region. This cutoff automatically constrains the z range in the domain

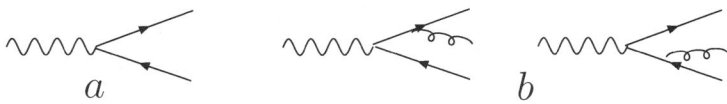

Figure 2   Production of $q\bar{q}$ pair from an off-shell photon (a) and leading diagrams for subsequent gluon radiation (b).

$\mu_0/\mu < z < 1 - \mu_0/\mu$. Even with this cutoff, the total emission probability is typically greater than one, because there will typically be several emissions above the cutoff scale. To generate an exclusive shower state, the emissions are typically ordered in Q^2 using the Sudakov form factor (41):

$$P_a^{no}(Q_{max}^2, Q^2) = \exp\left(-\int_{Q^2}^{Q_{max}^2}\int_{z_{min}}^{z_{max}} dP_a(z', Q'^2)\right), \qquad 4.$$

where $P_a^{no}(Q_{max}^2, Q^2)$ is the probability that no emissions occur from partons between Q_{max}^2 and Q^2. So one can express the probability that the highest Q^2 emission will occur, as the product of P and P_a^{no}.

It is natural to associate the evolution variable Q^2 with the virtuality μ^2 of the parent parton. Because the formalism of evolution equations makes it possible to determine only the leading behavior in the collinear limit, the precise relation between Q^2 and μ^2 is determined up to terms that affect subleading effects. It is therefore customary to select a relation of the form $Q^2 = f(z)\mu^2$. Different choices for the function $f(z)$ result in the different showering algorithms commonly employed. PYTHIA chooses $Q^2 = \mu^2$ and orders emissions according to the invariant mass of the parent. HERWIG chooses $Q^2 = \mu^2/(z(1-z))$. Expanding the expression for μ^2 one obtains $Q^2 = (P_b + P_c)^2/(z(1-z))$, which can be written as $2E_a^2(1 - \cos\theta)$, where E_a is the energy of the parent parton and θ is the angular separation between the quarks. Thus HERWIG orders emissions according to the parent energy and angle of emission. Similarly, expanding ARIADNE's (18) choice of $Q^2 = \mu^2 z(1-z)$ shows that it orders emissions according to the transverse momentum with respect to the radiating color dipole. As mentioned above, all these choices are equivalent at the level of leading double logarithms: One power of logarithm comes from the integration over the region of soft emission ($\log(1 - z_{max})$); another power comes from the collinear region ($\log(\mu/\mu_0)$). However, they lead to different estimates of subleading effects. Identifying the best set of evolution variables is therefore crucial to improve the accuracy of the approximations, by capturing as much as possible of the effects beyond the leading double logarithmic level (11, 12, 42).

The preceding radiation algorithm contains the dominant physics for radiation initiated from a single Feynman diagram. An important feature of the QCD radiation that is not explicitly included in the above equations is the idea of coherence or interference between two diagrams as shown in Figure 2b. This effect can be approximated by implementing explicit angular ordering to the radiation. Conceptually this can be understood by examining Figure 3a (see color insert), which shows a Feynman diagram for the emission of a gluon from a quark line. The quark momentum is denoted by l and the gluon momentum by k, θ is the opening angle between the quark and antiquark, and α is the angle between the nearest quark and the emitted gluon. We work in the double-log-enhanced soft $k^0 \ll l^0$ and collinear $\alpha \ll 1$ region. The internal quark propagator $p = (l + k)$ is off-shell, setting the time scale for the gluon emission:

$$\Delta t \simeq \frac{1}{\Delta E} = \frac{l^0}{(k+l)^2} \rightarrow \Delta t \simeq \frac{1}{k^0 \alpha^2}.$$ 5.

To resolve the quarks, the transverse wavelength of the gluon $\lambda_\perp = 1/E_\perp$ must be smaller than the separation between the quarks $b(t) \simeq \theta \Delta t$, giving the constraint $1/(\alpha k^0) < \theta \Delta t$. Using the results of Equation 5 for Δt, we arrive at the angular ordering constraint $\alpha < \theta$. Gluon emissions at an angle smaller than θ can resolve the two individual color quarks and are allowed; emissions at greater angles do not see the color charge and are therefore suppressed. In processes involving more partons, the angle θ is defined not by the nearest parton, but by the color-connected parton (e.g., the parton that forms a color singlet with the emitting parton). Because the color connections are so important to the event evolution, the color connections are sometimes drawn in place of the Feynman diagram. Figure 3 shows the color connections for the $q\bar{q}$ event after the gluon is emitted. Color lines begin on quarks and end on antiquarks. Because gluons are color octets, they contain the beginning of one line and the end of another.

The above derivation is valid for the emission of either photons or gluons. However, because the gluon itself carries color charge, there are other important implications of the angular constraint. The color connections shown in Figure 3b demonstrate how each time a gluon is radiated, it has the effect of narrowing the gap between the color-connected partons, hence reducing the invariant mass of the pair. This leads to the so-called color preconfinement, which is the key to the validity of the hadronization models which we discuss in the next section.

Because HERWIG explicitly uses the angular separation as its radiation ordering parameter, it implements the effects of coherence through angular ordering. PYTHIA (16) currently implements the color coherence effect with an explicit veto. ARIADNE's (18) radiation is based on a dipole formalism in which color-connected pairs radiate coherently to start.

Researchers are working to identify the optimal evolution variable. The goal is to improve the treatment of large-angle radiation, as well as the radiation off heavy quarks. New ideas have been explored recently and are incorporated in the C++ incarnation of the HERWIG code (HERWIG++) (42) and in the latest version of PYTHIA (12).

The preceding discussion is valid for final state showers. There is an analogous machinery for the development of the initial state showers known as backward evolution (43, 44). As the name implies, initial state showers are built up backward in time, starting with the hard interaction and then building up the shower by radiating with increasing virtuality.

To summarize, accelerating color charges radiate, and this radiation pattern can be predicted by perturbative QCD. Because gluon radiation carries color, the separation between color connected objects is continually decreasing, laying the foundation for the partons from the shower to cluster into color singlet pairs which hadronize. To set the proper initial condition for the development of a color-coherent shower evolution, events generated by PL calculations of the hard scattering process must provide not only the partons' flavor and momentum, but also their color connections.

3.2. Hadronization Models

As the partons radiate and travel further apart from the central interaction, the confining effects of QCD become important. The event generation enters the non-perturbative regime, where calculations based on first-principles are absent and we need to resort to parameterizations of phenomenological models based on reasonable assumptions. Fortunately, the factorization assumption and color pre-confinement tell us that the hadronization of the partons should be independent of the hard scattering process. The validity of the factorization assumption as a good approximation, and the effectiveness of preconfinement, are supported by Figure 4 (see color insert) (taken from Reference 42). The figure shows the results of the simulation of the shape of the invariant-mass distribution of pairs of color-connected partons at the end of the perturbative evolution (i.e., the clusters), performed for e^+e^- collisions at different center of mass energies. The cluster-mass spectrum is peaked at very low values $\simeq 1$ GeV (proving the preconfinement induced by the perturbative evolution) and is independent of the energy of the hard interaction (proving the independence of the spectrum of color-singlet clusters, i.e., the seeds of hadronization, from the hard process). Therefore, the parameters used to describe a hadronization model can be fit (45) to the results of one experiment (or one specific center of mass energy) and then applied to another. An example of this is shown in Figure 5, which shows an excellent agreement between the multiplicity of charged particles measured by the experiment OPAL at LEP (46)

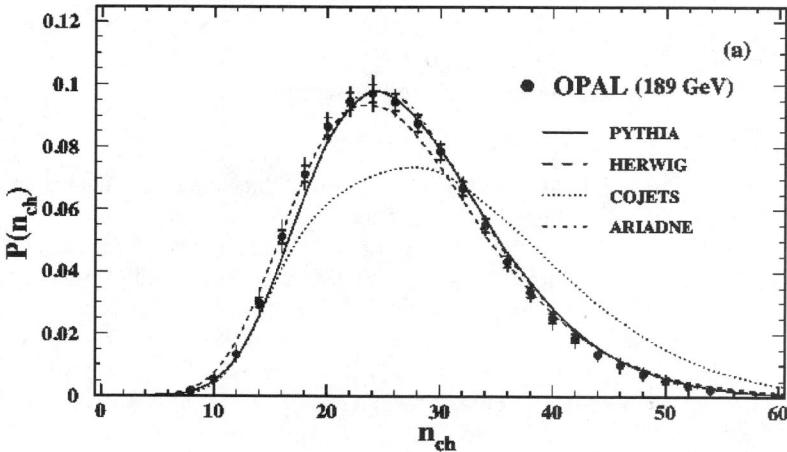

Figure 5   Distribution of the charged-particle multiplicity in e^+e^- collisions at $\sqrt{S} = 189$ GeV, compared with the predictions of various MC codes.

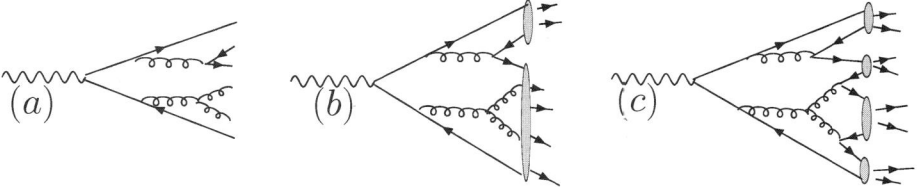

Figure 6   Possible radiation pattern from a $q\bar{q}$ pair (a) and illustration of string fragmentation (b) and cluster hadronization (c).

at the center of mass energy $\sqrt{s} = 189$ GeV with the prediction of the PYTHIA MC tuned at $\sqrt{s} = 91.2$ GeV. We provide a summary of some essential elements of various hadronization models below. A more detailed description may be found in Reference (9).

The simplest model for fragmentation is based on the hypothesis of local parton-hadron duality (47). It is based on the intuitive assumption that the basic flow of momentum and quantum numbers found at the PL is maintained at the hadron level. As we saw in the previous section, and proven in Figure 4, QCD predicts that development of the parton shower will result in color preconfinement and a decrease in the invariant mass of the color-connected partons. These color-connected pairs form the seed for the hadronization, so it is natural to expect the quark content of the hadrons in a region of the detector to be similar to that of the partons before hadronization. This simple idea is implemented quantitatively in the two most common hadronization models: string fragmentation (48–51), used in PYTHIA, and the cluster model (52–54), used in HERWIG. These models are conceptually illustrated in Figure 6. The first diagram represents the parton configuration after showering is completed, arranged such that adjacent partons are color connected. Figure 6b illustrates the organization of partons for string fragmentation, which is based on the idea that the confinement comes from a linearly increasing color field represented by a string. Notice that the endpoints of each string are quarks, but gluons can form *kinks* in the string. As the partons travel further apart, the potential energy of the string increases until it is large enough to produce a $q\bar{q}$ pair. The appearance of the $q\bar{q}$ pair breaks the string in two, according to a parameterized fragmentation function. The process is iterated until individual string segments can be associated to hadrons. There are several adjustable parameters in the fragmentation function that can be tuned on one set of data and are then used to predict another.

Cluster hadronization is based on the observation that, after showering, the distribution of invariant masses for color-connected partons is independent of the scale of the original hard interaction. This is illustrated in Figure 4, which shows the distribution of cluster masses over a wide range of collision energies. Because the number of clusters depends on the collision energy, each curve has been normalized to emphasize the similarity in shapes. The universality of this distribution and its peak at low invariant mass motivates the idea of clustering the $q\bar{q}$

pairs and then decaying each cluster into a pair of hadrons. Figure 6c illustrates the cluster hadronization model for the same parton configuration shown in Figure 6a. Notice the gluons that existed after the parton shower are first split into $q\bar{q}$ pairs, and then these quarks form clusters. This procedure reproduces the expected steeply falling distribution of mass clusters, with the primary dependence being only on the cut-off scale of the parton shower.

The comparison of accurate data from LEP with the above approaches results in excellent agreement, as reviewed, for example, in Reference (1). Recent modifications to the clustering algorithm consider also the possibility of color recombination (55): since there are only three possible colors, a quark-antiquark pair which is not color connected could nevertheless be in a color-singlet state. This will happen with a probability proportional to 1/9, as only one out of the nine possible color configurations of a $q\bar{q}$ pair corresponds to a color singlet state. These modified algorithms allow therefore the formation of color-singlet clusters among low-invariant-mass $q\bar{q}$ pairs which are not color connected. Currently no evidence that this added realism improves the agreement with data exists, but it is important to have such more flexible schemes available, should future data require more sophisticated models.

4. THE STRUCTURE OF THE UNDERLYING EVENT

The least understood aspect of hadronic collisions is the description of the underlying event (UE), namely the fate of the fragments of the hadrons left over from the primary hard interaction. One could naively expect that these fragments will continue traveling at a very small angle with respect to the beam direction and end up undetected down the beam pipe. In general, however, they are not color neutral, and the neutralization of their color forces long-range interactions with the partons involved in the hard process itself, therefore leading to particles evenly distributed throughout the full rapidity range. The presence of these particles not directly related to the hard collision can affect the interpretation of the data, for example, by contributing to the energy associated to the jets and biasing the measurement of the jet energies.

Recent data from the Tevatron provide inputs to develop and test quantitative models for the UE. The most significant progress comes from the conclusive evidence for multiple interactions during hadronic collisions. Early evidence of double parton scatterings in 4-jet final states was obtained by the AFS experiment at the ISR (56) and then by CDF (57). Independent and unambiguous confirmation followed with a CDF analysis of γ+3jet final states (58), which identified a set of events due to the superposition of a γj and a jj (jet jet) subprocess (as shown in Figure 7).

Taking the cross section to be equal to $\sigma(jj) \times \sigma(\gamma j)/\sigma_{eff}$, CDF extracted a value of σ_{eff} of the order of 14.5 ± 2.5 mb (58). This is consistent with the expectation that the phenomenological parameter σ_{eff} be of the order of the inelastic cross

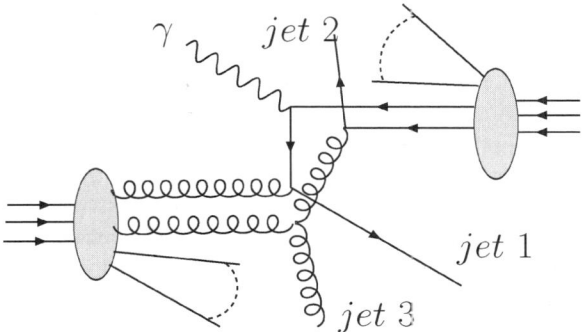

Figure 7 Example of double parton scattering event, leading to a $\gamma+3$-jet final state.

section. These results imply that as two partons from the two hadrons are engaged in the hardest process, the spectator quarks and gluons will themselves undergo, with non-negligible probability, a hard collision. As the hardness of this secondary collision, namely the p_T of the $2 \to 2$ process, decreases, its probability increases and, in LO perturbation theory, diverges faster than $1/p_T^2$ as p_T becomes of the order of Λ_{QCD}. This divergence is unphysical, because in the complete theory it is screened by the neutrality of the overall color field of the proton. Nevertheless there is an intermediate range at the borderline between the applicability of perturbation theory and the onset of nonperturbative phenomena where the probability of these secondary "semihard" collisions is of order 1. It is therefore reasonable to expect that these semihard multiple scatterings play an important role in the development of the UE. Following these ideas, a model of the UE based on the superposition of several low-p_T $2 \to 2$ scatterings had been proposed and implemented in PYTHIA many years ago (59). Here a lower threshold for p_T is introduced, p_T^{min}, and the average number of $2 \to 2$ partonic scatterings with $p_T > p_T^{min}$ during a pp collision is given by $\langle n \rangle = \sigma(p_T > p_T^{min})/\sigma_0$. In the simplest version of this model, the individual scatterings are independent and their multiplicity follows a Poisson distribution: $P(n) = \langle n \rangle^n \exp(-\langle n \rangle)/n!$. As in the case of the parameter σ_{eff} introduced earlier, σ_0 should be related to the total inelastic, nondiffractive cross section σ_{ND}. At the Tevatron, σ_{ND} is measured to be 50.9 ± 1.5 mb (60), a factor of 3 larger than the value of σ_{eff} which CDF has measured in the $\gamma + 3$jet study of hard multiple collisions. There is no complete understanding of this discrepancy. It has been argued (61), for example, that this mismatch is due to a large component of color-singlet Pomeron-like exchanges, which do not resolve the quark structure of the proton. It also cannot be excluded that the value of σ_{eff} depends on the scale of the primary hard process, because harder collisions are expected to have smaller impact parameters than do softer collisions. In other words, σ_{eff} or σ_0 should represent only the inelastic cross-section for $p\bar{p}$ collisions occurring at very small impact parameter, in view of the presence of the short-distance hard process which forces the two hadrons to be close in the transverse plane.

Explicit calculations of the $gg \to gg$ rate indicate that, at the Tevatron (LHC), $\sigma(p_T > p_T^{min})$ attains a value of several tens of mb for p_T^{min} as large as 3 GeV (5 GeV), supporting the assumption built into these models that for σ_0 values in the range 10–50 mb, $\langle n \rangle$ is of order 1 in the semihard, perturbative domain.

Similar models have been developed in the context of photoproduction reactions at HERA, where the real photon emitted by the electron beam contains a hadronic component. Both JIMMY (62), a model built in the framework of HERWIG, and a two-component dual parton model (63) show better agreement with data than the simulations without multiple parton scattering. More detailed models are also being explored. Reference (64) added a soft component (particles with $p_T < p_{Tmin}$) to the semihard component (2 → 2 scatterings with $p_T > p_{Tmin} \sim 2 - 3$ GeV) described by JIMMY. The original PYTHIA model has been improved recently (65) with the inclusion of parton showers for all secondary interactions, as well as flavor, momentum, and color correlations among beam remnants and the hard-interaction participants. Odagiri (61) studied a model based on HERWIG and multiple-parton scatterings, with the number of scatters given by a Poisson distribution, $p_{Tmin} = 3$ GeV and $\sigma_{eff} = 14.5$ mb, as measured by CDF (58). Fitting the CDF UE data (66) requires an increase in the value of the maximum cluster mass in HERWIG's cluster model for hadronization, leading to the prediction of an enhanced baryon yield. As suggested in Reference (61), a direct measurement of the proton-to-pion ratio as a function of the various observables probing the UE structure could give precious information on the hadronization of the final state of semihard collisions. This is particularly interesting because the scale of semihard interactions is so close to the confinement region that large corrections to the factorization theorem can be present, and might require modifications to the naive cluster and string models, or to the value of their universal phenomenological parameters.

A description of the UE based on multiple semi-hard collisions leads to features of the final state that differ significantly from those expected in models based on purely soft interactions among the hadron fragments (as employed, e.g., in the standard HERWIG and ISAJET). For example, an UE model based on multiple interactions predicts a power-like fall-off of the p_T spectrum of charged tracks. Furthermore, one expects that on an event-by-event basis, the UE will show the correlations associated to the production of back-to-back 2 → 2 semihard scatterings (67), rather than flat and uniform distributions in rapidity and azimuth. The comparison with experimental data is therefore crucial to assess the validity of these models and to determine the value of their input parameters, p_T^{min} and σ_0. Their values affect both the number of additional events and the spectrum of the UE particles, and it is therefore not obvious that a global fit of all properties can be achieved. Furthermore, contrary to the parameters controlling the hadronization in the cluster or string models, there is no reason to expect that p_T^{min} and σ_0 should be independent of $\sqrt{s}$.

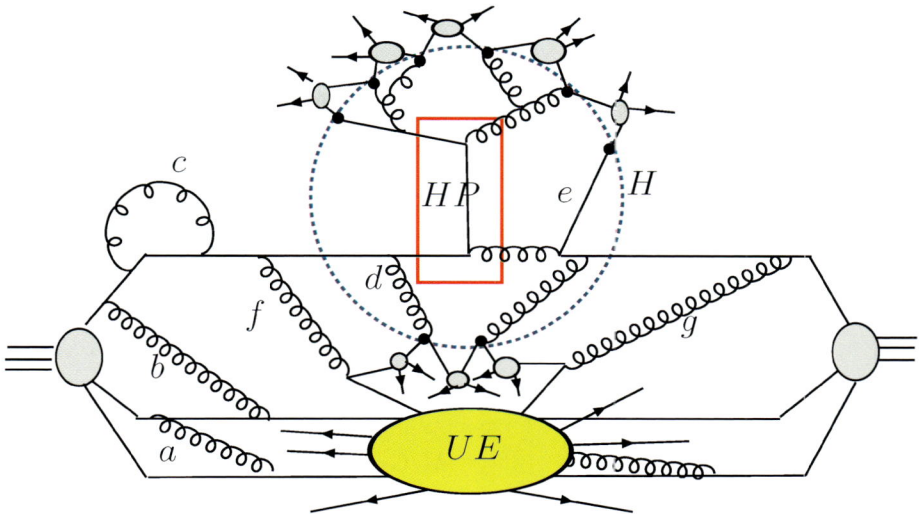

Figure 1 General structure of a hard proton-proton collision. HP, hard process; UE, underlying event. See text for details.

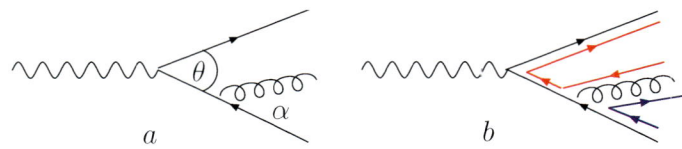

Figure 3 Radiation of $q\bar{q}$ pair produced by an off-shell photon.

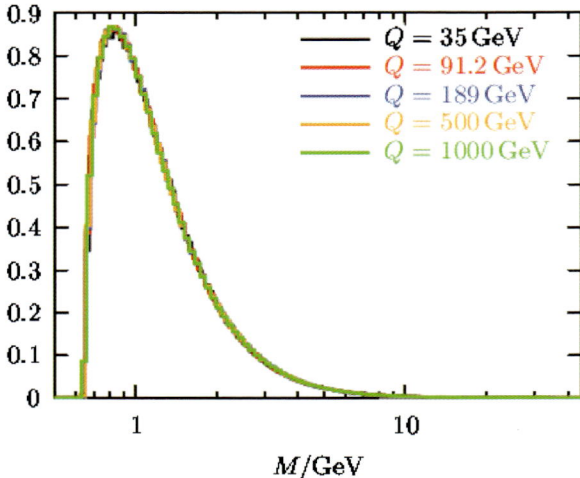

Figure 4 Distribution of cluster masses for $e^+e^- \to$ *hadrons* simulated at different values of $\sqrt{S} = Q$.

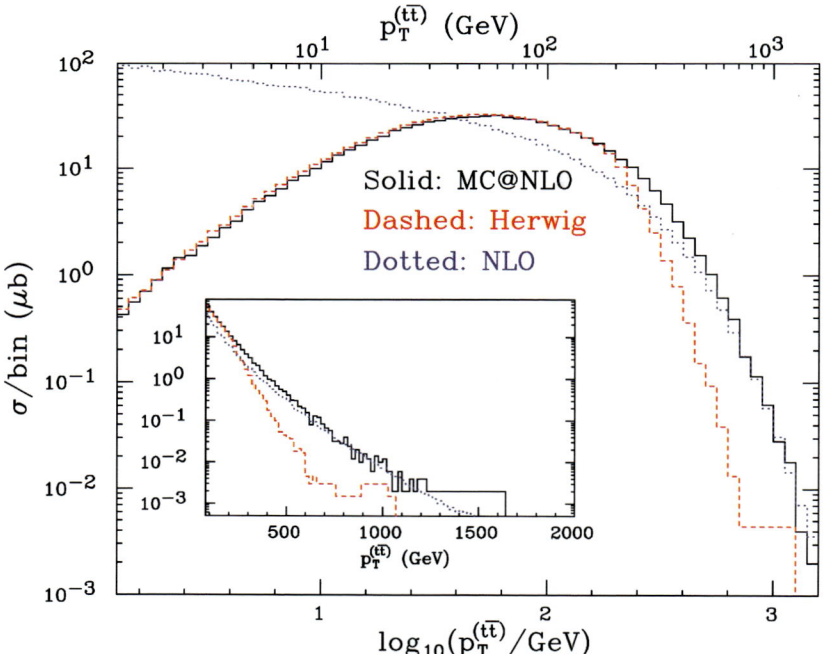

Figure 9 Transverse momentum distribution of top quark pairs using three different approaches: the LO HERWIG MC, the PL NLO MC, and the merging of the two into MC@NLO. Figure from Reference 77.

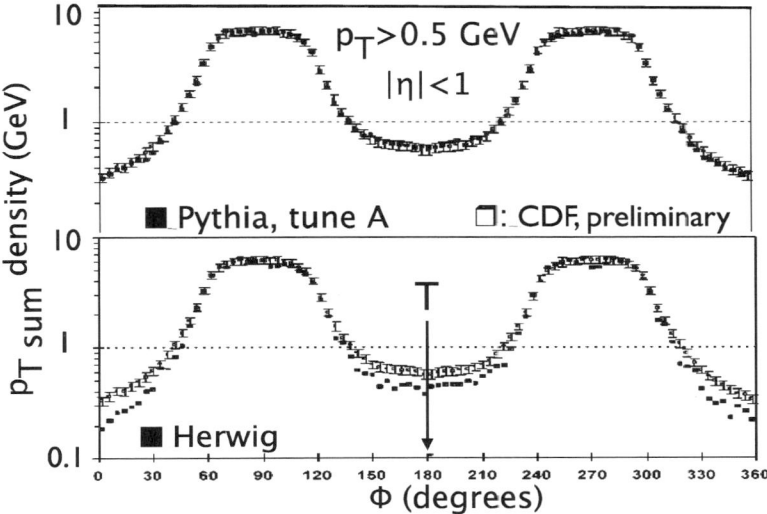

Figure 8   Distribution of the transverse momentum density relative to the direction of the largest-p_T track T in the region transverse to the event leading jet at CDF (67), $\sqrt{S} = 1.96$ TeV.

CDF has performed several detailed studies of the properties of the UE in inclusive jet samples (66). They define regions transverse to the direction of the leading jet in an event, and they study the charged track multiplicity and transverse momentum deposited in these regions. This is done as a function of the energy of the leading jet, and as a function of the relation between the leading and the recoiling jets. Additional information can be obtained by selecting events with or without a third jet, and by comparing the UE activity in different regions away from the primary jets. The results of these studies (67) add support to the belief that multiple interactions play a major role in shaping the structure of the UE. This is shown, for example, in Figure 8 (taken from Reference (67)). Here one considers events with a leading jet of transverse energy $30 < E_T/\text{GeV} < 70$ and searches for the largest-p_T track (T) in the region $30° < |\phi(T) - \phi_{jet}| < 120°$, namely the region transverse to the leading jet and its recoil. Setting the origin of the azimuthal coordinate such that $\phi(T) = 180°$, one plots the distribution of the total transverse momentum density of tracks with $p_T > 0.5$ GeV and pesudorapidity $|n| < 1$, excluding T itself. The preliminary data from CDF are compared to the predictions of the PYTHIA which has been tuned (66) (upper plot) and of HERWIG. The peaks at 90° and 270° correspond to the two leading jets in the event. The behavior in the regions $\phi \sim 0°$ and 180° shows a larger activity than predicted by HERWIG. The better agreement of PYTHIA, which contains multiple interactions, is interpreted as evidence for their presence.

Tuning of the MC parameters to optimize the agreement with the data is possible (66–68) for models like PYTHIA which incorporate a description of multiple scatterings. However there is a lack of general agreements for all features of the final states, with tunings appropriate for a given observable failing to provide good fits of others (66). It is also necessary to check whether these tunings are sufficient to describe correctly final states for other hard processes, e.g., multijets, Drell-Yan, W+jets and $t\bar{t}$ events. To what extent these tunings allow a correct extrapolation to the LHC energies remains to be seen and will have to be assessed when the first, low-luminosity (i.e., without overlapping events) data becomes available. The extrapolation from 1800 to 630 GeV and vice versa shows weaknesses of these models (66), and the jump to 14 TeV remains one of the most problematic aspects of predicting the structure of final states at the LHC (67, 69).

5. HIGHER-ORDER PERTURBATIVE CORRECTIONS

Although the description of the hadronization and the UE can only be improved by introducing new models and enough parameters to be tuned on the data, the accuracy of the perturbative part of the event generation can be increased by adding higher-order (HO) terms that are calculable from first principles in perturbative QCD.

In the case of the matrix elements for the hard scattering which starts the shower, the inclusion of HO terms is required to reduce the dependence on the renormalization and factorization scale dependence of the inclusive cross sections. Typically, HO corrections introduce also subprocesses which were absent at the lower order (LO). For example, LO production of Z bosons only involves $q\bar{q}$ initial states, whereas at HO qg and then gg, initial states are present. The inclusion of real-emission HO corrections is also required to improve the collinear and soft approximation used by the shower MC to describe the emission of extra jets. This is because the angular-ordering prescription, introduced to implement quantum coherence effects for large-angle gluon radiation, will incorrectly suppress hard large-angle emissions, underestimating the rate of multijet final states.

In the shower evolution, virtual diagrams are needed to cancel the infrared (IR) singularities due to the real emission of soft gluons. As in the case of real emission, they are evaluated in the soft approximation, and are resummed to all orders into the Sudakov form factors, which enforce the unitarity of the shower evolution. In a full next-to-leading order (NLO) calculation of the matrix element, virtual effects are instead included exactly, although only at a fixed order.

Because HO corrections to the LO matrix elements are already partly accounted for by the shower algorithm, their consistent implementation in an improved event generator requires that no single piece of the calculation is included twice. Many recent efforts have focused on this problem and are briefly summarized here. We shall cover separately the proposed approaches to two aspects of this problem: the merging with a shower MC of complete NLO matrix-element calculations, and of matrix elements for multi-parton final states.

5.1. Inclusion of NLO Corrections in Matrix-Element Generators

At NLO, one needs to account for both real-emission diagrams and virtual corrections. Virtual corrections to an N-body final state give still an N-body final state, while real emissions lead to an (N+1)-body final state. An NLO event generator must therefore generate both N-body and (N+1)-body final states. The two cannot be treated separately, however, because the contribution of the virtual part to the cross section is equal to minus infinity, whereas the real part has divergences for collinear and soft configurations. The structure of the cross section in these regions can be parameterized in an idealized form as follows:

$$\frac{d\sigma}{dx} = \frac{f(x)}{(1-x)_+},$$ 6.

where by definition of $(\ )_+$:

$$\int_0^1 dx \frac{d\sigma}{dx} = \int_0^1 dx \frac{f(x) - f(1)}{(1-x)}.$$ 7.

The limit for the generic phase-space variable $x \to 1$ corresponds in this context to kinematical configurations where the (N+1)-body final state degenerates (via soft or collinear emission) to an N-body final state. In the case of collinear emission, for example, $x = \cos\theta$. For heavy quark (Q) pair production, $x = m_{Q+\bar{Q}}^2/\hat{s} \to 1$ (where $\sqrt{\hat{s}}$ is the invariant center-of-mass energy of partonic initial state) corresponds to the soft-emission limit. $f(x)$ is continuous at $x = 1$; therefore any integral over an arbitrary range of x, including possibly $x = 1$, is finite. This is the essence of the virtual/real cancellation of infinities. If we formally remove the integration from Equation 7, we "define" a nonsingular differential cross section as

$$\frac{d\sigma}{dx} = \frac{f(x) - f(1)}{(1-x)}.$$ 8.

One can interpret this relation as follows: for any given (N+1)-body kinematical configuration $C(x)$ ($x \neq 1$), whose weight is given by $f(x)/(1-x)$, we can associate an N-body virtual configuration $C(1)$ with weight $-f(1)/(1-x)$, whose kinematics is obtained by the $x \to 1$ limit of $C(x)$. These virtual configurations are typically called "counter-events." One can construct an event generator by adding to each (N+1)-body final state its corresponding N-body counter-event. Both events will have a finite weight, although the counter-event's is negative. Because their kinematics is different, event and counter-event will typically populate different bins of our histograms. For bin sizes sufficiently large (namely for observables sufficiently inclusive), summing events over the full phase-space will nevertheless lead to positive rates in each bin. This is not guaranteed to happen when the bins are too small. For example, let us consider the integral of the cross section in a bin covering the range $1 - \epsilon < x < 1$, with ϵ small:

$$\int_{1-\epsilon}^{1} d\sigma = \int_{1-\epsilon}^{1} dx \frac{f(x)}{(1-x)} - \int_{0}^{1} dx \frac{f(1)}{(1-x)}. \qquad 9.$$

The first contribution is from the $N+1$-body final states, the second from the counter-events which all accumulate at $x = 1$. Simple algebra leads to the following result:

$$\int_{1-\epsilon}^{1} d\sigma = C + f(1) \log \epsilon, \qquad 10.$$

where C is a finite constant when $\epsilon \to 0$. Because $f(1)$ is positive, when the bin-size ϵ is small enough, the integral becomes negative. This is an indication that radiative corrections in this small corner of phase-space are large. In other words, if we try to push the calculation in a region of low inclusivity, probing the final state structure with very fine resolution, the fixed-order perturbative approximation breaks down. HO corrections become very large and have to be included to restore the positivity of the cross section in that bin. The smaller the bin, the larger the number of orders required.

In concrete applications, life is complicated by the identification of the functions $f(x)$, the interplay of soft and collinear singularities, and the description of the phase-space in terms of suitable variables; however, the above simplified description captures the main features of the technique. It has been used, in various implementations, for the development of NLO matrix-element (ME) event generators covering several of the interesting LHC processes. Examples include 2-jets (70) and 3-jets (71), heavy quarks (72), and vector boson (73) production. For a detailed list of available tools, see Reference (13). The extension of these techniques to next-to-next-to-leading-order (NNLO) calculations is under study and has led to the first encouraging results (74, 75).

5.2. NLO Corrections in Shower MCs

The necessity to include NLO corrections in ME generators is twofold. On one hand, only shower MCs provide a representation of the final state complete enough to allow realistic detector simulations. Inclusion of the NLO matrix elements for the hard process, which will provide cross sections with full NLO accuracy, is a natural improvement of these essential tools. On the other hand, as mentioned in the previous subsection, the inclusion of NLO effects in fixed-order ME MCs leads to distributions which are not positive definite, thus calling for a tool where these large (and possibly negative) logarithmic effects that arise at any fixed order in some corners of phase-space can be properly resummed. This goal can be achieved via the inclusion of the NLO MEs in the shower MC. A priori, one may expect this task to be ill defined, as shower MCs already incorporate part of the NLO effects: they have real emissions, as well as virtual effects included in the Sudakov form factors. The naive introduction of NLO MEs would then lead to double count-ing. Recent work by Frixione & Webber (76) showed how this merging can be done very effectively in what they called a MC@NLO. One starts by identifying the analytic form of the approximation used by the shower MC to describe real emission and the LO virtual correction contained in the Sudakov form factor. One can then subtract these expressions from the NLO matrix elements. Because the shower approximation has the correct residue for all singular contributions, the subtracted NLO matrix elements are finite. In the simple formalism used in the previous section, one can represent the subtraction from the real emission term as follows:

$$\frac{d\sigma}{dx} = \frac{f(x) - f_{MC}(x)}{(1-x)}, \qquad 11.$$

where $f_{MC}(x)$ is the approximate MC expression for the real emission matrix element, with the condition that $f_{MC}(1) = f(1)$. In this way the $x \to 1$ singularity is not removed by merging with the virtual correction, but by letting the shower algorithm handle it and absorb it into the Sudakov form factor. As for the virtual part, the singular contribution is all contained in the shower approximation, and what is left for the NLO correction to describe is just a finite term with the N-body, Born-like, kinematics. One is still left with positive and negative weight events, because the difference between the exact nonsingular terms from the full NLO calculation and those used in the shower can have either sign. However, because the residual positive and negative weights are bounded, one can define an unweighting procedure whereby positive-weight events are unweighted against the maximum positive weight, and negative-weight events are unweighted against the minimum negative weight. This procedure has been implemented in MC@NLO codes (76, 77) describing heavy-quark pair, Higgs, DY, and gauge boson pair production. There is no obstacle to extending it to all remaining processes known at NLO. Other approaches have also been developed into alternative codes for DY (78) and vector boson pair (79) production.

The inclusion of NLO corrections in the shower MC guarantees that total cross sections generated by the MC reproduce those of the NLO ME calcula-tion, thereby properly including the K factors and reducing the systematic un-certainties induced by renormalization and factorization scale variations. At the same time, however, the presence of the HO corrections generated by the shower will improve the description of the NLO distributions, leading to departures from the PL NLO result. Figure 9 (see color insert) shows the p_T spectrum of a $t\bar{t}$ pair resulting from the pure NLO calculation, from the LO shower, and from the MC@NLO improvement. At large p_T, a region dominated by the NLO effects, MC@NLO faithfully reproduces the hard, large-angle emission distribution given by the NLO matrix elements. At small p_T, a region dominated by multiple radi-ation and HO effects, the MC@NLO departs significantly from the NLO result, while properly incorporating the Sudakov resummation effects only available via the initial state shower evolution.

5.3. Merging Multijet ME Generators and Shower Evolution

The inclusion of NLO MEs in the shower MCs guarantees the correct description of the emission of one extra hard parton (ultimately a jet) from the Born, LO process. In the case of NLO corrections to the dijet final states, for example, this means that all topologies with up to three jets will be accurately described. To go beyond this in a NLO framework, however, requires the knowledge of NLO MEs which may not be available in the case of the highest multijet final states relevant to several LHC studies. As mentioned earlier, the description of multijets obtained from the shower evolution is inaccurate, because hard radiation at large angle is suppressed by the angular ordering prescription. One therefore needs an approach in which multi-parton events generated using an exact LO ME calculation can be consistently evolved into multijet final states via a shower MC. As in the case of the inclusion of NLO corrections, the main problem is double counting. We describe this problem by discussing a specific example (80): Consider inclusive production of 3-jet events, with jets defined by a cone of size R_j in $\eta - \phi$ space and transverse energy larger than E_T^{min}. We can generate these events by generating PL configurations with partons separated by $\Delta R_{jj} > R_j$ and letting the shower evolve them into jets. By and large, there is a one-to-one correspondence between the generated hard partons and the jets, and the angular distributions of the three jets correctly include all interference effects among the various diagrams. There are configurations, however, where the correspondence is not guaranteed. Take for example events where two jets have E_T much larger than that of the third: $E_{T1} \sim E_{T2} \gg E_{T3} \simeq E_T^{min}$. These events can be generated in two independent ways. On the one hand, we can start from three partons with the kinematics of the three jets (see the left plot in Figure 10). After evolution, the partons will generate the desired jets. On the other hand (right plot in Figure 10), we can start from configurations where the two leading partons 1 and 2 create jets 1 and 2, but parton 3 is too soft to generate its own jet. Hard radiation by one of the two leading jets may lead, with probability α_s, to the generation of jet 3 with the desired energy. In other words, the shower evolution in these cases could produce jets with transverse energy larger than that of partons already present in the hard event. Although the probability of this happening is parametrically of order α_s relative to the LO process, configurations with two hard partons and a much softer one are enhanced by large logarithms. This is the result of the large phase-space available for the emission of a soft parton in a

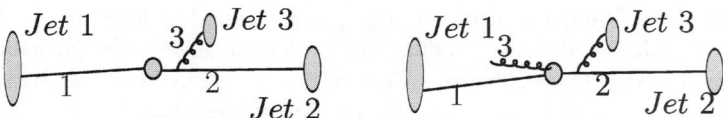

Figure 10   Example of double counting for 3-jet final states.

hard event. One can estimate that this enhancement is of order $\log(E_{T3}/E_T^{min})$. As a result, the overall probability that the third jet is emitted from the shower becomes a number of order $\alpha_s \log(E_{T3}/E_T^{min})$. This dependence on E_T^{min} of the cross section for 3-jet events well above the E_T^{min} threshold is an unphysical result caused by the double counting of equivalent configurations that can be produced both via the parton shower and the matrix element. To solve this problem, and remove the E_T^{min} dependence, one has to carefully check that a given phase-space configuration is generated only once: either directly by the PL event or by the shower evolution.

Two general approaches have recently been developed to address the problem of double counting. There is a trade-off between the accuracy of the approach and the ease of implementation. The first (18, 81) and most accurate is generically known as "matrix-element correction" technique (MEC), as it corrects the approximate ME for the emission of the hardest gluon in a given process by using the exact LO ME. The second is known as CKKW (82); its goal is to implement multijet ME corrections at the leading (LL), or next-to-leading (NLL), logarithmic level. In the MEC, one starts by identifying analytically the phase-space region Ω covered by the shower algorithm. In the case of $e^+e^- \to q\bar{q}g$, for example, this is given by a subset of the full phase-space domain Δ defined by $\Delta = [1 \leq x_1 + x_2 \leq 2] \cap [x_i \leq 1]$, where $x_i = 2E_i/\sqrt{S}$. Ω contains the singular regions corresponding to soft ($x_i = 0$) and collinear ($x_i = 1$) gluon emission. The integral σ_{Ω^0} of the 3-body cross section over the complement of Ω, $\Omega^0 = \Delta - \Omega$ is therefore free of singularities and finite. The integral over Ω, σ_Ω, is also finite, once the virtual corrections at the edge of phase-space are included. The MEC MC generation then works by deciding on an event-by-event basis whether to generate the event in Ω^0 or in Ω, based on the relative value of the respective cross sections. If the event falls in Ω, one generates a LO $e^+e^- \to q\bar{q}$ event. If it falls in Ω^0, one generates a $e^+e^- \to q\bar{q}g$ event. In both cases, the events are then evolved through the shower. Small adjustments should then be made in the first case to ensure a proper continuity across the boundary between the two domains. This technique has been applied also to DY production (83), and to top decays (84). Its extension to more complicated processes, however, is made particularly difficult by the need to provide an analytic description of Ω. When there are more than 3 colored partons in the process (as e.g., in dijet or heavy quark pair production at the LHC), this becomes very hard and impractical.

The CKKW approach circumvents this problem by limiting its precision goal to NLL accuracy. In this approach the double counting is removed not by exactly separating a priori the domains Ω and Ω^0, but by a probabilistic rejection procedure applied to events falling in the overlap, to ensure that a given phase-space configuration is only counted once. The generation is carried out in a phase-space domain defined by a Durham-like jet algorithm suitable for hadronic collisions (85); a resolution scale variable for two partons is defined by:

$$k_{ij} = min(E_{T,i}, E_{T,j})R_{ij},$$ 12.

if both i and j are in the final state, or by $k_{iJ} = E_{T,i}$ if J is in the initial state. R_{ij} is a measure of separation in the transverse plane, for example the standard $\sqrt{\Delta\phi^2 + \Delta\eta^2}$ measure. An N-parton final state is then classified as an N-jet event if $k_{ab} > k_0$ for all possible parton pairings. k_0 is a resolution threshold, introduced as a parameter necessary to group events in samples of different jet multiplicity. The physical results obtained at the end of the generation, however, should not depend on the choice of k_0.

One starts by generating samples of multi-parton events of different multiplicities N, using the exact LO ME and renormalization/factorization scale fixed to the resolution scale k_0. Events are extracted from the different N-jet samples with probability proportional to the sample cross section. The jet algorithm can then be used to define a tree structure for the event. The two partons i, j with the smallest k_{ij} are clustered into a single virtual parton ℓ, provided the parton types and flavors of i and j can be merged. The procedure is repeated after removing i, j and adding ℓ to the list of partons. The clustering continues until one gets a $2 \to 1$ or $2 \to 2$ process. The sequence of $k_{ab} = q_i$ values at the nodes i of the resulting tree is used to construct the reweighting factor:

$$w = \prod_{i=1}^{n} \frac{\alpha_s(q_i)}{\alpha_s(k_0)} \times \prod_{(k,l)} \frac{\Delta(q_k, k_0)}{\Delta(q_l, k_0)}, \qquad 13.$$

where the product of Sudakov form-factor Δ ratios goes over the pairs of nodes connected by a line, with $q_k > q_l$. By construction, $w \leq 1$. The event is then kept with probability w, or rejected. This Sudakov rejection is meant to eliminate from the N-jet rate events with radiation at scales larger than k_0, i.e., events which should be described with a higher-order matrix element. The events which survive this rejection are showered, using the tree structure determined above as a color-flow pattern. Shower emissions at a scale larger than k_0 are vetoed, because these emissions have been removed by the Sudakov weight. In the case of e^+e^- collisions (82, 86) one can prove that this algorithm correctly reproduces the weight of an event to NLL accuracy, and rates are independent of k_0. In hadron collisions (87) such a proof is still missing, but this framework provides a very good starting point for further developments. The results of the first studies in the case of W+jets production can be found in (88, 89).

An alternative proposal, accurate to LL, was proposed in (80, 90). Here one matches the partons from the ME calculation to the jets reconstructed after the perturbative shower. PL events are defined by a minimum E_T threshold E_T^{min} for the partons, and a minimum separation among them, $\Delta R_{jj} > R_{min}$. A tree structure is defined in analogy with the CKKW algorithm, thus defining the scales at which the various powers of α_s are calculated. However, no Sudakov reweighting is applied. Rather, events are showered, without any hard-emission veto during the shower. After evolution, a jet cone algorithm with cone size R_{min} and minimum transverse energy E_T^{min} is applied to the final state. Starting from the hardest parton, the jet which is closest to it in (η, ϕ) is selected. If the distance between the parton and the jet centroid is smaller than R_{min}, the parton and the jet match. The matched jet

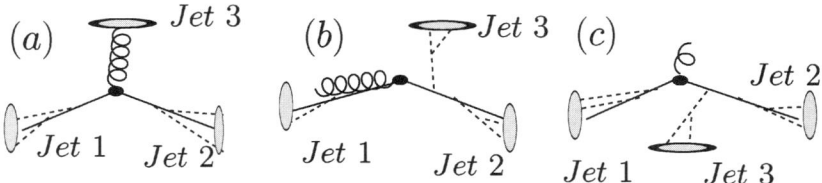

Figure 11   Examples of parton-jet matching. The three partons in (a) match the final state jets; those in (b) and (c) do not.

is removed from the list of jets, and matching for subsequent partons is performed. The event is fully matched if each parton has a matched jet. The algorithm is illustrated in Figure 11. The two solid lines and the gluon line, emerging from the filled vertex, represent partons present at the matrix-element level. The dotted lines represent partons emitted by the shower. The shaded blobs represent reconstructed jets. In the first case (a), the three original partons match three independent jets, and the event is kept. In case (b), the hard gluon cannot find a matching jet because the closest one (jet 1) has already been matched to the hard quark, while jet 3, associated to initial- or final-state, is too far. This event does not match, and is rejected. Rejection of this class of events removes double counting of the leading double logarithms, associated to the cases where the gluon is both collinear to the quark and soft. In case (c) the hard gluon is too soft to generate its own jet, and again it is too far from jet 3. This removes double counting of some single (NLL) logarithms. For events which satisfy matching, it is also required that no extra jet, in addition to those matching the partons, be present. Events with extra jets are rejected, a suppression replacing the Sudakov reweighting used in the CKKW approach. Events obtained by applying this procedure to PL sets with increasing multiplicity can then be combined to obtain fully inclusive samples spanning a large multiplicity range. The distributions of observables measured on this inclusive data set should not depend on the value of the parameters E_T^{min} and R_{min}, similar to the k_0 independence obtained in the CKKW approach.

Additional improvements in the description of multijet final states may soon come from the development of new shower algorithms, being developed in the context of the new generation of C++ codes, such as HERWIG++ (42) and SHERPA (19). Preliminary results for e^+e^- collisions, comparisons with LEP/SLC data and examples of the improved description of multijet final states, can be found in Reference (42).

6. MODELING PHYSICS BEYOND THE STANDARD MODEL

The discovery of physics beyond the SM (BSM) at the Tevatron or at the LHC will rely on the accurate understanding of the SM backgrounds. This understanding depends on the study of control samples which requires a rigorous validation of the

SM MC tools described in the previous sections. The identification of the origin of any deviations from the SM, and the extraction of the properties and parameters controlling these phenomena, will rely on the ability to properly simulate the features of the final states produced in the candidate new theories. The spectrum of possible new phenomena is very large, but their description is usually simpler than the modeling of their SM backgrounds. A new particle N is typically produced in simple Drell-Yan-like processes ($q\bar{q} \to N$ or $gg \to N$). This covers, for example, the cases of new gauge bosons (massive graviton resonances in extra-dimension models or technicolor states), in association with a single parton or gauge boson (e.g., $gg \to gG$, where G represents the tower of Kaluza-Klein gravitons in the extra dimensional models discussed in Reference 91), or in pairs ($gg/q\bar{q} \to N\bar{N}$, as in the case of leptoquarks, supersymmetric particles, etc.). The decay of the new particles is usually very fast, and if they are strongly interacting as in the case of squarks and gluinos, their large mass suppresses QCD radiation before their decay. The hard process is therefore simple to describe, and the QCD radiation emitted by N or from its decay products can be described using the same formalism developed for the SM QCD processes. NLO total production rates can be calculated (and for most cases are already known) using cross section evaluators, but the kinematical properties of the final states and their QCD evolution are usually not affected by these effects. As a result, there are no new subtleties associated to HO corrections, extra jet emissions, double counting, etc. Some new problems posed by particular classes of BSM models are mentioned below.

The main issue in the modeling of BSM phenomena is typically the accurate implementation in the MC code of the various parameter options and the complete classification of all possible production and decay mechanisms. This is a rather trivial task in the case of the gauge bosons carrying new gauge interactions. They are treated like standard DY processes and are defined by the list of the axial and vector couplings to quarks and leptons. The modeling of Z' or W' production and decay is therefore a standard feature in most MC codes.

Things can become extremely complicated in the case of supersymmetry (SUSY), owing to the complexity of the spectrum and to the multitude of the possible parameters. Other more exotic phenomena, such as the decay of black holes, require speculations which cannot be fully controlled today by first principles. Because these problems are usually addressed by the community of BSM model builders and phenomenologists, tools have been developed to facilitate the translation of their results into building blocks for the event generation. In this section we shall briefly review these tools and provide an overview of what is available in the form of full MC event generators for new physics. More details, as well as results obtained from these tools in the simulation of BSM phenomena at the Tevatron and LHC, can be found in Reference (92).

SUSY models (for a review see (93)) limited to the minimal spectrum extension of the SM and to R-parity conserving interactions (a.k.a. MSSM) contain 105 new parameters, all associated to SUSY breaking interactions and masses. In practical studies the number of parameters is reduced by assuming relations motivated by theoretical models about the nature of the SUSY breaking mechanism. Common schemes, for example, assume the unification of gauge interactions (GUT) at a large scale and impose these GUT relations to SUSY breaking masses and couplings. These relations are evolved according to the renormalization group to their values at the weak scale, providing the numerical input for the parameters to be used in the simulation of SUSY production and decay at the Tevatron and LHC. Several numerical codes exist (94) which evaluate the model parameters as a function of generic GUT inputs and use them to calculate couplings, spectra, and decay widths. The validation of these tools is crucial to ensure that different codes using different sources for their parameters describe the same models when using the same inputs. This work has started recently, leading to a Web-based utility that automatically performs these comparisons (95). To ensure that generic MC codes can access and use unambiguously the results of these programs, a common standard for the labeling of SUSY particles and parameters has been recently introduced (known as the SUSY Les Houches accord, SLHA (96)). This standard specifies a unique set of conventions for SUSY extensions of the SM, together with generic file structures for (*a*) SUSY model specifications and input parameters, (*b*) EW-scale SUSY mass and coupling spectra, and (*c*) decay tables, providing a universal interface between spectrum calculation programs, decay packages, and event generators.

When (and if) SUSY is discovered, establishing the nature and proving the correct spin assignment of the new particles will require an accurate description of the SUSY particle decays, including all the relevant spin correlations among decay products. Techniques have been developed to incorporate in a systematic way these spin correlations in the MCs (97). They are now part of the HERWIG MC. Another subtlety emerges when dealing with baryon-number violating vertices present in some SUSY R-parity violating interactions (93). In the presence of these vertices, the standard hadronization models which assume color-singlet formation via connections between color-anticolor lines are not valid. New techniques have been formulated (98), where the baryon-number violating vertex is associated with the appearance of a junction in the color confinement field. This approach is now implemented in PYTHIA.

The simulation of SUSY particle production and decays in hadronic collisions is currently available, with different levels of sophistication, in most of the codes discussed earlier in this paper: ISAJET, PYTHIA, HERWIG, SUSYGEN (99), GRACE (100), CompHEP, and SHERPA. Other BSM processes can be found in one generator or another, but are less often a standard feature. The most complete code in this respect is probably PYTHIA, which includes, among other things, extended Higgs sectors, 4th generation fermions, new gauge bosons, left-right symmetric models and doubly charged Higgs bosons, leptoquarks, compositeness and anomalous couplings, excited fermions, technicolor, and some extra-dimension phenomena (graviton resonances and associated graviton plus jet production). Black hole production and decay can be simulated by HERWIG via the ancillary code Charybdis (101). Processes involving massive spin-2 particles, such as those appearing in extra dimensional models, can be simulated by SHERPA, where the helicity formalism for spin-2 particles has been incorporated (102) in its matrix-element generator AMEGIC++.

Aside from minimal SUSY, where efforts have been made to introduce uniform standards and conventions to be applied to all generators, the description of BSM processes is otherwise still fragmented. This is particularly true of the models emerged most recently, such as alternative formulations of SUSY breaking or theories with extra dimensions. Some efforts will be required in the future to formulate frameworks such as the SLHA that could cover the other most interesting cases of BSM phenomena. These frameworks will make it much easier for the model builders to enable the MC developers to include their future predictions into generators and to facilitate the complete study of these phenomena.

Among all the difficulties appearing in the simulation of hadronic collisions, those associated to the proper treatment of BSM processes appear, however, to be among the simplest. Once some new phenomenon is identified, suitable MC generators will be readily available and up to the task.

7. CONCLUDING REMARKS

As discussed in the previous sections, the prediction of physical observables for hadronic collisions requires several independent inputs. Several of them we could not cover in any detail, although they are critical to the accuracy of the predictions. First and foremost are the PDFs that parameterize the partonic densities inside the proton. Their extraction from large sets of data in both lepton-hadron and hadron-hadron collisions has recently led to extremely accurate fits, thanks to both the quality of the data and the availability of NNLO theoretical inputs (105, 106). In particular, PDF sets obtained recently (107) include an estimate of the systematic uncertainties related to both experimental inputs and theoretical approximations.

A full description of the event final state requires the modeling of the hadron decays. The event generators described in this review contain decay tables for a large fraction of the known hadronic resonances and make use of ad-hoc packages specifically tuned to particularly important classes of decay modes. Examples of these are EvtGen (108), which emphasizes decays of B mesons, and TAUOLA and PHOTOS (109), which deal with decays of τ leptons and with photon radiation in particle decays. Tools for comparing the kinematical distributions of different decay packages, such as MCtester (110), have also been developed.

Ancillary tools help the development and testing of the event generators. In view of the complexity of physics at the LHC, their role will become more and more relevant in the future. We mention here JETWEB (111), a Web-based interface and database for MC tuning and validation. The aim of the package is to allow rapid and reproducible comparisons to be made between detailed measurements at high-energy physics colliders and general physics simulation packages. The package includes a relational database, a Java servlet query and display facility, and clean interfaces to simulation packages and their parameters.

As the beginning of LHC data taking approaches, it is reassuring to note the rapid and substantial progress in the field of event generators. Most of the topics reviewed here cover progress that has taken place in the past three or four years, and work is under way to address yet more issues. Some of these will be settled by investing more manpower in the application of known techniques (for example the inclusion of increasingly more NLO processes into the MC@NLO codes). Other processes require firmer theoretical control before they will lead to practical progress in the MCs (for example the development of a MC@NNLO). In other cases, the recent ideas already incorporated in MCs require a consolidation of both theoretical ideas and a validation or tuning of their MC implementation. This is the case, for example, of the various approaches to the problem of merging higher-order tree-level matrix elements with shower MCs, or of the modeling of the UE. Systematic comparisons with the Tevatron data should be performed, a task which, in principle, is possible and could be completed on time for the LHC startup.

The complexity of hadronic collisions is such that no guarantee exists that the framework presented in this review can be used to achieve arbitrary accuracy. The violation of factorization is known to occur at the level of $1/Q$ corrections, as we discussed in the introduction. The size of these effects depends on the selected observable. Open issues include how much of these $1/Q$ effects is captured by the modeling of hadronization implemented in the current MCs, how much needs to be added separately, and whether and how these additional contributions can be incorporated in the standard MC frameworks discussed in this review. Another crucial limitation of the shower MCs is the implementation of quantum coherence, which, together with the algorithm used to develop the parton shower, is strictly valid only in the limit of an infinite number of colors. In addition, phenomena are expected at the LHC that may call for more sophisticated descriptions of the event evolution. For example, the factorization formalism we discuss will likely be insufficient to describe the effects of potentially large logarithms associated to small-x (where $x \sim Q^2/S$) effects (112) (for example production of jets or b quarks at small transverse momentum). Fortunately, tools in this direction are being developed (113, 114), although they have mostly been developed for ep collisions and are not yet a standard component of the LHC tools. Likewise there is a full range of phenomena (soft and hard diffraction, elastic scattering) that lack the solid and complete modeling available for the hard processes. Should the LHC data call for improved theoretical understanding or accuracy, it is likely, as occurred at LEP, that the data will provide the necessary direction and incentive for this progress.

**The *Annual Review of Nuclear and Particle Science* is online at
http://nucl.annualreviews.org**

LITERATURE CITED

1. Bethke S. *Phys. Rep.* 403-404:203 (2004)
2. Abe F, et al. [CDF] *Nucl. Instrum. Methods A* 271:387 (1988)
3. Abachi S, et al. [DO] *Nucl. Instrum. Methods A* 338:185 (1994)
4. hep-ph/0010338; hep-ph/0003154; hep-ph/0201071

5. CERN/LHCC/99-14 and 99-15. CERN/ LHCC/94-38. CERN-LHCC-98-4

6. Mangano ML. *Comments Mod. Nucl. Part. Phys.* 2:A153 (2002)

7. Altarelli G, Mangano ML, eds. *Proc. CERN Workshop SM Phys.*; hep-ph/ 0005025; hep-ph/0003275; hep-ph/000 3033; hep-ph/0003238; hep-ph/0003142

8. Gianotti F, et al. hep-ph/0204087 (2002)

9. Ellis RK, Stirling WJ, Webber BR. *Camb. Monogr. Part. Phys. Nucl. Phys. Cosmol.* (1996)

10. Webber BR. *Annu. Rev. Nucl. Part. Sci.* 36:253 (1986)

11. Marchesini G, Webber BR. *Nucl. Phys. B* 238:1 (1984); Marchesini G, Webber BR. *Nucl. Phys. B* 310:461 (1988)

12. Sjostrand T, Skands PZ. hep-ph/0408302

13. Dobbs MA, et al. hep-ph/0403045

14. Seymour M. *Lectures on "MC event generators for LHC physics."* http://hepwww. rl.ac.uk/theory/seymour/slides/

15. Paige FE, Protopopescu SD, Baer H, Tata X. hep-ph/9810440; hep-ph/0001086

16. Sjostrand T. *Comput. Phys. Commun.* 82:74 (1994); Sjostrand T, et al. *Comput. Phys. Commun.* 135:238 (2001); Sjostrand T, Lonnblad L, Mrenna S, Skands P. hep-ph/0308153

17. Marchesini G, et al. *Comput. Phys. Commun.* 67:465 (1992); Corcella G, et al. *JHEP* 0101:010 (2001)

18. Lonnblad L. *Comput. Phys. Commun.* 71: 15 (1992)

19. Gleisberg T, et al. hep-ph/0311263; Kuhn R, Krauss F, Ivanyi B, Soff G. *Comput. Phys. Commun.* 134:223 (2001)

20. Gribov VN, Lipatov LN. *Yad. Fiz.* 15:781 (1972); Gribov VN, Lipatov LN. *Sov. J. Nucl. Phys.* 15:438 (1972)

21. Altarelli G, Parisi G. *Nucl. Phys. B* 126:298 (1977)

22. Dokshitzer YL. *Sov. Phys. JETP* 46:641 (1977); Dokshitzer YL. *Zh. Eksp. Teor. Fiz.* 73:1216 (1977)

23. Mangano ML, Parke SJ. *Phys. Rep.* 200:301 (1991)

24. Berends FA, Giele WT, Kuijf H. *Phys. Lett. B* 232:266 (1989); Berends FA, Kuijf H, Tausk B, Giele WT. *Nucl. Phys. B* 357:32 (1991)

25. Ishikawa T, et al. KEK-92-19

26. Stelzer T, Long WF. *Comput. Phys. Commun.* 81:357 (1994)

27. Caravaglios F, Moretti M. *Phys. Lett. B* 358:332 (1995); Caravaglios F, Mangano ML, Moretti M, Pittau R. *Nucl. Phys. B* 539:215 (1999); Mangano ML, Moretti M, Pittau R. *Nucl. Phys. B* 632:343 (2002)

28. Draggiotis P, Kleiss RH, Papadopoulos CG. *Phys. Lett. B* 439:157 (1998); hep-ph/0202201; Draggiotis PD, Kleiss R. *Eur. Phys. J.* C17:437 (2000)

29. Pukhov A, et al. hep-ph/9908288

30. Kanaki A, Papadopoulos CG. *Comput. Phys. Commun.* 132:306 (2000); Papadopoulos CG. *Comput. Phys. Commun.* 137:247 (2001)

31. Krauss F, Kuhn R, Soff G. *JHEP* 0202: 044 (2002)

32. Tsuno S, et al. hep-ph/0204222

33. Mangano ML, et al. *JHEP* 0307:001 (2003)

34. Maltoni F, Stelzer T. *JHEP* 0302:027 (2003)

35. CEDAR. http://www.cedar.ac.uk/hep code/

36. CERN. http://agenda.cern.ch/fullAgenda. php?ida=a031457

37. Boos E. et al. hep-ph/0109068

38. Mahlon G, Parke SJ. *Phys. Rev. D* 55:7249 (1997)

39. Slabospitsky SR, Sonnenschein L. *Comput. Phys. Commun.* 148:87 (2002)

40. Malton F, Stelzer T. http://madgraph. hep.uiuc.edu/

41. Sudakov VV. *Sov. Phys. JETP* 3:65 (1956); Sudakov VV. *Zh. Eksp. Teor. Fiz.* 30:87 (1956)

42. Gieseke S, et al. hep-ph/0311208; Gieseke S, Stephens P, Webber B. hep-ph/0310083

43. Sjostrand T. *Phys. Lett. B* 157:321 (1985); Bengtsson M, Sjostrand T, van Zijl M. *Z. Phys. C* 32:67 (1986)

44. Gottschalk TD. *Nucl. Phys. B* 277:700 (1986)

45. Hemingway RJ. *OPAL Tech. Note* 652
46. Abbiendi G, et al. *Eur. Phys. J.* C16:185 (2000)
47. Azimov YI, Dokshitzer YL, Khoze VA, Troian SI. *Yad. Fiz.* 40:1284 (1984)
48. Artru X, Mennessier G. *Nucl. Phys. B* 70:93 (1974)
49. Andersson B, Gustafson G, Ingelman G, Sjostrand T. *Phys. Rep.* 97:31 (1983)
50. Andersson B, Gustafson G, Soderberg B. *Z. Phys. C* 20:317 (1983)
51. Sjostrand T. *Nucl. Phys. B* 248:469 (1984)
52. Field RD, Wolfram S. *Nucl. Phys. B* 213:65 (1983)
53. Webber BR. *Nucl. Phys. B* 238:492 (1984)
54. Gottschalk TD. *Nucl. Phys. B* 214:201 (1983)
55. Winter JC, Krauss F, Soff G. *Eur. Phys. J.* C36:381 (2004)
56. Akesson T, et al. *Z. Phys. C* 34:163 (1987)
57. Abe F, et al. *Phys. Rev. D* 47:4857 (1993)
58. Abe F, et al. *Phys. Rev. D* 56:3811 (1997)
59. Sjostrand T, van Zijl M. *Phys. Rev. D* 36:2019 (1987)
60. Abe F, et al. *Phys. Rev. D* 50:5550 (1994)
61. Odagiri K. *JHEP* 0408:019 (2004)
62. Butterworth JM, Forshaw JR, Seymour MH. *Z. Phys. C* 72:637 (1996)
63. Engel R. *Z. Phys. C* 66:203 (1995)
64. Borozan I, Seymour MH. *JHEP* 0209:015 (2002)
65. Skands P, Sjostrand T. *Eur. Phys. J.* C33:S548 (2004); Sjostrand T, Skands PZ. *JHEP* 0403:053 (2004)
66. Field RD. *Proc. APS/DPF/DPB Summer Study Future Part. Phys.*, ed. N Graf, eConf C010630. 501 (2001); Affolder T, et al. *Phys. Rev. D* 65:092002 (2002); Acosta D, et al. *Phys. Rev. D* 70:072002 (2004)
67. Field R. Presented at TeV4LHC Workshop, FNAL Sept. 2004 http://www.phys.ufl.edu/rfield/cdf/Tev4LHC9-16-04.pdf
68. Dobbs M, et al. hep-ph/0403100
69. Butterworth JM, Carli T. hep-ph/0408061
70. Giele WT, Glover EWN, Kosower DA. *Nucl. Phys. B* 403:633 (1993); Frixione S. *Nucl. Phys. B* 507:295 (1997)
71. Nagy Z. *Phys. Rev. D* 68:094002 (2003); Kilgore WB, Giele WT. hep-ph/0009193
72. Mangano ML, Nason P, Ridolfi G. *Nucl. Phys. B* 373:295 (1992)
73. Giele WT, Glover EWN, Kosower DA. *Phys. Lett. B* 309:205 (1993); Campbell J, Ellis RK. *Phys. Rev. D* 65:113007 (2002)
74. Glover EWN. *Nucl. Phys. Proc. Suppl.* 116:3 (2003)
75. Weinzierl S. *JHEP* 0303:062 (2003); Frixione S, Grazzini M. hep-ph/0411399
76. Frixione S, Webber BR. hep-ph/0204244; Frixione S Webber BR. hep-ph/0309186
77. Frixione S, Nason P, Webber BR. *JHEP* 0308:007 (2003)
78. Kurihara Y, et al. *Nucl. Phys. B* 654:301 (2003)
79. Dobbs M. *Phys. Rev. D* 64:034016 (2001)
80. Mangano M. Presented at FNAL Workshop MC Tuning, Nov. 2002. http://cepa.fnal.gov/CPD/MCTuning/15nov2002.html
81. Seymour MH. *Comput. Phys. Commun.* 90:95 (1995)
82. Catani S, Krauss F, Kuhn R, Webber BR. *JHEP* 0111:063 (2001)
83. Corcella G, Seymour MH. *Nucl. Phys. B* 565:227 (2000); Miu G, Sjostrand T. *Phys. Lett. B* 449:313 (1999); Lonnblad L. *Nucl. Phys. B* 458:215 (1996)
84. Corcella G, Seymour MH. *Phys. Lett. B* 442:417 (1998); Corcella G, Mangano ML, Seymour MH. *JHEP* 0007:004 (2000)
85. Catani S, Dokshitzer YL, Seymour MH, Webber BR. *Nucl. Phys. B* 406:187 (1993)
86. Lonnblad L. *JHEP* 0205:046 (2002)
87. Krauss F. *JHEP* 0208:015 (2002)
88. Mrenna S, Richardson P. *JHEP* 0405:040 (2004)
89. Krauss F, Schalicke A, Schumann S, Soff G. *Phys. Rev. D* 70:114009 (2004)
90. Mangano ML. Presented at FNAL MC4Run2 Workshop June 2004. http://cepa.fnal.gov/patriot/mc4run2/

91. Arkani-Hamed N, Dimopoulos S, Dvali GR. *Phys. Lett. B* 429:263 (1998)
92. Allanach BC, et al. hep-ph/0402295
93. Haber HE. *Phys. Lett. B* 592:1003 (2004)
94. Baer H, Paige FE, Protopopescu SD, Tata X. hep-ph/9305342; Allanach BC. *Comput. Phys. Commun.* 143:305 (2002); Djouadi A, Kneur JL, Moultaka G. hep-ph/0211331; Porod W. *Comput. Phys. Commun.* 153:275 (2003); Muhlleitner M, Djouadi A, Mambrini Y. hep-ph/0311167
95. Allanach BC, Kraml S, Porod W. *JHEP* 0303:016 (2003); http://cern.ch/kraml/comparison
96. Skands P, et al. *JHEP* 0407:036 (2004)
97. Richardson P. *JHEP* 0111:029 (2001)
98. Sjostrand T, Skands PZ. *Nucl. Phys. B* 659:243 (2003)
99. Katsanevas S, Morawitz P. *Comput. Phys. Commun.* 112:227 (1998)
100. Tanaka H, et al. *Nucl. Instrum. Methods A* 389:295 (1997)
101. Harris CM, Richardson P, Webber BR. *JHEP* 0308:033 (2003)
102. Gleisberg T, et al. *JHEP* 0309:001 (2003)
103. Giele W, et al. hep-ph/0204316
104. Deleted in proof
105. Moch S, Vermaseren JAM, Vogt A. *Nucl. Phys. B* 688:101 (2004); Moch S, Vermaseren JAM, Vogt A. *Nucl. Phys. B* 691:129 (2004)
106. Stirling WJ. arXiv:hep-ph/0411372
107. Giele WT, Keller SA, Kosower DA. hep-ph/0104052; Giele WT, Keller S. *Phys. Rev. D* 58:094023 (1998); Pumplin J, et al. hep-ph/0201195; Botje M. *Eur. Phys. J.* C14:285 (2000); Alekhin SI. hep-ex/0005042; Alekhin S. hep-ph/0211096; Martin AD, Roberts RG, Stirling WJ, Thorne RS. *Eur. Phys. J.* C23:73 (2002)
108. Lange DJ. *Nucl. Instrum. Methods A* 462:152 (2001)
109. Golonka P, et al. hep-ph/0312240
110. Golonka P, Pierzchala T, Was Z. *Comput. Phys. Commun.* 157:39 (2004)
111. Butterworth JM, Butterworth S. *Comput. Phys. Commun.* 153:164 (2003)
112. Andersson B, et al. *Eur. Phys. J.* C25:77 (2002)
113. Jung H. *Comput. Phys. Commun.* 143:100 (2002); Jung H, Salam GP. *Eur. Phys. J.* C19:351 (2001)
114. Kharraziha H, Lonnblad L. *JHEP* 9803:006 (1998)

CUMULATIVE INDEXES

CONTRIBUTING AUTHORS, VOLUMES 46–55

589

CHAPTER TITLES, VOLUMES 46–55

592

Instrumentation and Techniques

Nuclear Reaction Mechanisms–Light Particles

Nuclear Structure

Particle Spectroscopy

Weak and Electromagnetic Interactions

Annual Reviews – Your Starting Point for Research Online
http://arjournals.annualreviews.org

- Over 900 Annual Reviews volumes—more than 25,000 critical, authoritative review articles in 32 disciplines spanning the Biomedical, Physical, and Social sciences— available online, including all Annual Reviews back volumes, dating to 1932

- Current individual subscriptions include seamless online access to full-text articles, PDFs, Reviews in Advance (as much as 6 months ahead of print publication), bibliographies, and other supplementary material in the current volume and the prior 4 years' volumes

- All articles are fully supplemented, searchable, and downloadable— see http://nucl.annualreviews.org

- Access links to the reviewed references (when available online)

- Site features include customized alerting services, citation tracking, and saved searches

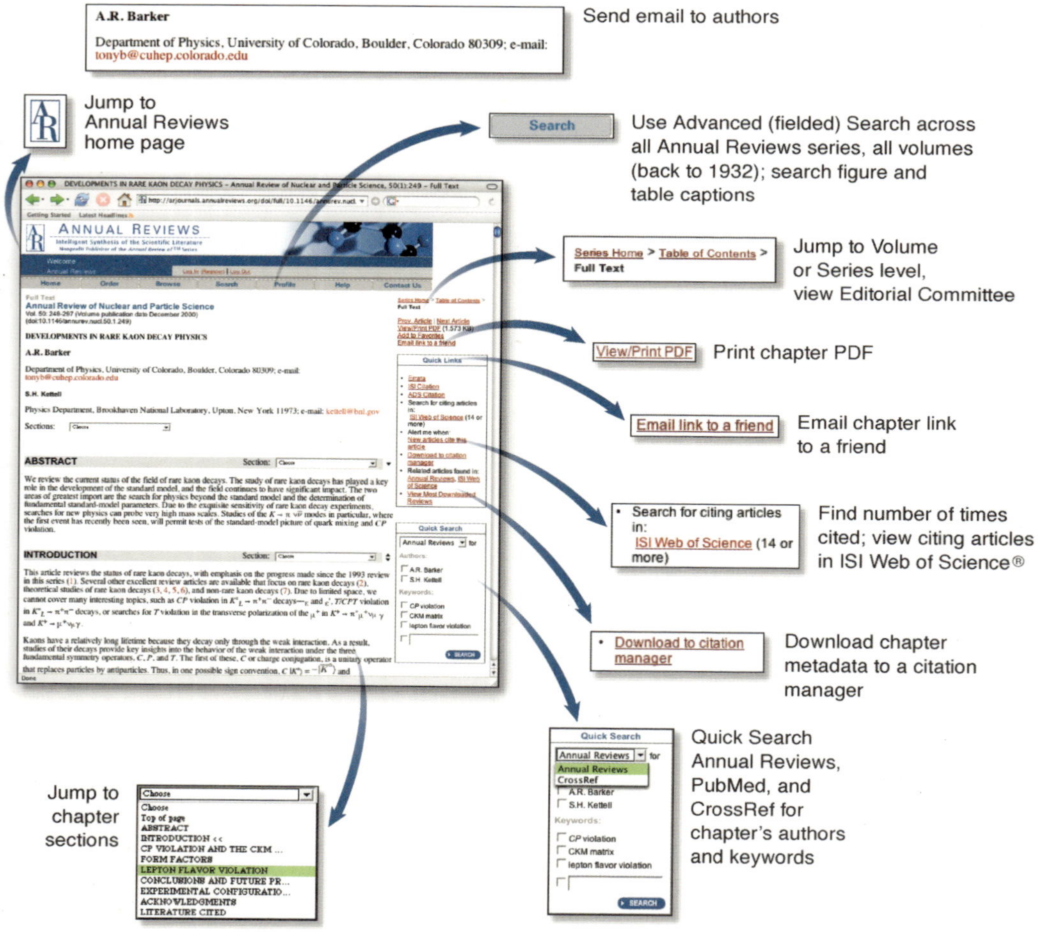

Send email to authors

Jump to Annual Reviews home page

Search — Use Advanced (fielded) Search across all Annual Reviews series, all volumes (back to 1932); search figure and table captions

Jump to Volume or Series level, view Editorial Committee

View/Print PDF — Print chapter PDF

Email link to a friend — Email chapter link to a friend

Find number of times cited; view citing articles in ISI Web of Science®

Download to citation manager — Download chapter metadata to a citation manager

Quick Search Annual Reviews, PubMed, and CrossRef for chapter's authors and keywords

Jump to chapter sections